# Neurotrophic Factors

# Neurotrophic Factors

Edited by

**Sandra E. Loughlin**
**James H. Fallon**

*Department of Anatomy and Neurobiology*
*College of Medicine*
*University of California, Irvine*
*Irvine, California*

**Academic Press, Inc.**
Harcourt Brace Jovanovich, Publishers

San Diego  New York  Boston  London  Sydney  Tokyo  Toronto

This book is printed on acid-free paper. ∞

Copyright © 1993 by ACADEMIC PRESS, INC.
All Rights Reserved.
No part of this publication may be reproduced or transmitted in any form or by any means, electronic or mechanical, including photocopy, recording, or any information storage and retrieval system, without permission in writing from the publisher.

Academic Press, Inc.
1250 Sixth Avenue, San Diego, California 92101-4311

*United Kingdom Edition published by*
Academic Press Limited
24–28 Oval Road, London NW1 7DX

Library of Congress Cataloging-in-Publication Data

Neurotrophic factors / edited by Sandra E. Loughlin and James H. Fallon.
  p. cm.
  Includes bibliographical references and index.
  ISBN 0-12-455830-5
  1. Neurotrophic functions.  2. Growth factors.  I. Loughlin, Sandra E.  II. Fallon, James H.
QP363.N52  1992
591.1'88–dc20                                           92-11683
                                                              CIP

PRINTED IN THE UNITED STATES OF AMERICA
92  93  94  95  96  97    MV    9  8  7  6  5  4  3  2  1

# Contents

Contributors xvii

Preface xxi

## 1 Functional Implications of the Anatomical Localization of Neurotrophic Factors   1
James H. Fallon and Sandra E. Loughlin

    I. Introduction   1
   II. Neurotrophic Factors Included in This Volume   2
  III. Review of Cellular Events of Neurotrophic Factor Systems   4
  IV. Modes of Secretion and Transport of Neurotrophic Factors   9
   V. Neurotrophic Factor Distribution in Central Neural Systems   11
      A. Approaches in Neuroanatomy   11
      B. Regional Analysis   11
      C. Systems Analysis   14
  VI. Summary   21
      References   22

## 2 Neurotrophic Factors: What Are They and What Are They Doing?   25
Franz Hefti, Timothy L. Denton, Beat Knusel, and Paul A. Lapchak

    I. History   25
   II. Definition   26

v

III. Physiological Functions of Neurotrophic Factors  28
   A. Synthesis and Release  28
   B. Receptors and Transducing Mechanisms  30
   C. Cellular Actions of Neurotrophic Factors  32
   D. Are Neurotrophic Factors Identical to Target-Derived Retrograde Messengers?  33
   E. Are Neurotrophic Factors Regulators of Developmental Neuronal Survival?  35
   F. Are Neurotrophic Factors Developmental Differentiation Factors?  36
   G. Selectivity of Neurotrophic Factors and Temporal Windows of Responsiveness during Development  38
   H. Function of Neurotrophic Factors in the Adult Nervous System  39
   I. Neurotrophic Factors and Aging  40
IV. Pharmacological Exploitation of Neurotrophic Factors  40
V. Conclusions  41
   References  42

## 3  Synergy, Retrograde Transport, and Cell Death  51
Ian A. Hendry and Michael F. Crouch

I. Introduction  51
II. Cell Death  51
   A. Target-Controlled Neuronal Survival  51
   B. In Vivo Survival Factors  52
   C. Trans-Synaptic Influences  54
   D. Antibodies  55
III. Axoplasmic Transport  55
   A. Specific Transport  56
   B. Nonspecific Transport  57
   C. Blockade of Axoplasmic Transport  57
IV. Interactions between Neuronally Active Molecules  58
   A. Synergy  58
V. Mechanisms of Interaction  65
   A. Synergy between Second Messengers  66
   B. Induction of Receptors  68
   C. Multifactorial Response  68
   D. Gene Induction  69
VI. Other Mechanisms to Achieve a Retrograde Message  70
   A. G-Protein Translocation  70
   B. Phosphorylated Proteins  72

C. Myristoylation-Demyristoylation   72
VII. Conclusions   72
References   73

## 4  Primary Response Gene Expression in the Nervous System                89

Alaric T. Arenander and Harvey R. Herschman

I. Introduction   89
II. *fos* and *jun* Genes as Prototypical Primary Response Genes   90
   A. c-*fos* and c-*jun*   90
   B. *fos* and *jun* Gene Families   91
   C. c-*fos* Induction Pathways   92
   D. Inhibitory Control of *fos*   93
   E. Cross-Talk   94
III. Other Primary Responses Genes   95
   A. *egr1*   95
   B. *krox20*   95
   C. *nur77*   95
   D. *scip*   96
IV. In Vitro Expression of Primary Response Genes in Neural Systems   96
   A. Neuronal Cells   96
   B. Glial Cells   98
V. In Vivo Expression of Primary Response Genes   99
   A. Seizures   101
   B. Kindling and Primary Response Gene Expression   103
   C. Long-Term Potentiation   103
   D. Mapping of Neuronal Networks   104
   E. Sensory Stimulation   105
   F. Pain Systems   106
   G. Lesions   108
   H. Neuroactive Agents   109
   I. Neuroendocrine Systems   110
   J. Biological Rhythms   111
VI. Secondary Response Genes as Targets of Primary Response Genes   112
   A. Proenkephalin   113
   B. Nerve Growth Factor   113
   C. Tyrosine Hydroxylase   114
   D. Transin   115
VII. Specificity of Primary Response Gene Expression   115
   A. Quantitative Parameters   116

B. Cell-Type Restriction of Primary Response Gene Expression   117
References   118

## 5  Nerve Growth Factor and Related Substances: Structure and Mechanism of Action    129
Joseph G. Altin and Ralph A. Bradshaw

I. Introduction   129
II. NGF Structure   130
   A. Sequences of NGF and Associated Subunits   130
   B. Sequences of Other Neurotrophins   137
   C. Three-Dimensional Properties   137
III. Receptors   138
   A. Properties   138
   B. *trk* as an NGF Receptor   140
IV. Mechanism of Action of NGF   141
   A. General Responses   141
   B. Initial Signal Transduction   144
   C. Signal Transmission   146
   D. Induction of Primary Response Genes   157
V. Summary and Conclusions   161
References   163

## 6  Regulation of Nerve Growth Factor Expression    181
Robert H. Edwards

I. Expression in the Target   182
   A. Biologic Regulation   182
   B. Mechanisms of Regulation   188
   C. Biologic Significance of Regulated NGF Expression   196
II. Expression in Neural Supporting Cells after Injury   196
   A. Observations in Peripheral Nerve   196
   B. Mechanisms of NGF Induction   197
   C. Biologic Significance   199
III. Expression in Early Development and Nonneural Settings   199
IV. Biologic Significance of NGF Regulation: Strategies for Manipulation in Vivo   200
References   202

**7 Nerve Growth Factor: Actions in the Peripheral and    209
Central Nervous Systems**

Frank M. Longo, David M. Holtzman, Mark L. Grimes, and William C. Mobley

  I. Introduction   209
 II. NGF Actions in the Peripheral Nervous System   210
     A. Discovery of NGF   210
     B. NGF and NGF Receptors in the Peripheral Nervous System   211
     C. Regulation of Developmental Cell Death by NGF   217
     D. NGF and Differentiation of Peripheral Nervous System Neurons   223
     E. NGF in the Adult Normal and Adult Lesioned Peripheral Nervous System   230
     F. Summary of NGF Actions in the Peripheral Nervous System   232
III. NGF Role in the Central Nervous System   232
     A. Distribution of NGF and the NGF Receptor in the Central Nervous System   232
     B. NGF as a Trophic Factor for Cholinergic Neurons of the Basal Forebrain and Caudate Putamen   235
     C. Other Populations in the Central Nervous System   237
     D. Role of NGF in the Lesioned Central Nervous System   238
     E. Role of NGF in Central Nervous System Diseases   239
        References   241

**8 Brain-Derived Neurotrophic Factor: An    257
NGF-Related Neurotrophin**

Ronald M. Lindsay

    I. Introduction   258
   II. NGF: Prototypical Neurotrophic Factor   258
  III. Search for Other Neurotrophic Factors   260
   IV. Discovery of Neurotrophic "Activities" Distinct from NGF   260
    V. Purification of BDNF from Porcine Brain   261
   VI. Neuronal Specificity of BDNF   262
       A. Peripheral Nervous System Neurons: Distinct and Overlapping Specificity of BDNF and NGF   262
       B. CNS Neurons: Distinct and Overlapping Specificity of BDNF and NGF   265
  VII. BDNF Prevents Neuronal Death in Vivo   267
 VIII. Molecular Cloning of Porcine, Mouse, Rat, and Human BDNF   269

IX. Consequences of the Molecular Cloning of BDNF   270
  A. Distribution of BDNF mRNA in Brain Regions and Peripheral Tissues   270
X. Binding of BDNF to Specific Receptors   272
XI. Neurotrophin Family of NGF-Related Neurotrophic Factors—NGF, BDNF, NT-3, and Others   274
XII. Clinical Perspective   275
XIII. Conclusions   276
  References   277

## 9  Biochemistry and Molecular Biology of Fibroblast Growth Factors    285
Kenneth A. Thomas

I. Introduction   285
II. Structure and Expression of FGF Genes   285
III. Protein Structures   287
IV. Receptors   292
V. Inhibitors   295
VI. Signal Transduction   296
VII. Biologic Functions   299
  References   303

## 10  Fibroblast Growth Factors: Their Roles in the Central and Peripheral Nervous System    313
Klaus Unsicker, Glaudia Grothe, Gerson Lüdecke, Dörte Otto, and Reiner Westermann

I. Introduction: The FGF Gene Family   313
II. Localization of the FGF Gene Family in Neural Cells and Tissues   314
III. FGF Receptors on Neural Cells   317
IV. FGFs and Their Effects on Neurons   319
  A. In Vitro Effects   319
  B. In Vivo Effects   320
V. FGFs and Their Effects on Glial Cells   321
VI. FGFs in the Lesioned Nervous System: Regulation and Effects   322
VII. FGFs in Neural Tumors   325
VIII. FGFs in Cell Lineage Decisions   326
IX. FGFs and Their Interactions with Other Growth Factors   327
X. Conclusions   327
  References   329

## 11 Epidermal Growth Factor: Structure, Expression, and Functions in the Central Nervous System    339
Richard Morrison

    I. Identification   339
    II. Molecular Structure and Homologies   339
    III. EGF Localization in Brain   345
    IV. EGF Receptor   346
    V. EGF Receptor Expression in Brain   347
    VI. EGF and Neuroglia   349
    VIII. EGF and Neuronal Function   350
        References   353

## 12 Transforming Growth Factors Alpha and Beta    359
Pauli Puolakkainen and Daniel R. Twardzik

    I. Introduction   359
    II. Historical Background   359
    III. Assays for TGF$\alpha$   361
        A. Mitogenic Assay of TGF$\alpha$   361
        B. Radioreceptor Assay   361
        C. Other Assays for TGF$\alpha$   361
    IV. Structure and Properties of TGF$\alpha$   362
        A. Structure of TGF$\alpha$ Precursor   362
        B. Structure of TGF$\alpha$   363
        C. Synthesis, Release, and Activation of TGF$\alpha$   364
        D. Mode and Mechanism of Action of TGF$\alpha$ and Its Regulation   364
        E. Receptor Binding of TGF$\alpha$   365
    V. Molecular Biology and Genetics of TGF$\alpha$   365
        A. TGF$\alpha$ Expression, Tissue Distribution, and Role in Transformation   366
    VI. Effects of TGF$\alpha$ in Vivo and in Vitro   367
    VII. Clinical Implications   367
    VIII. TGF$\alpha$ and Neurobiology   369
    IX. Assays for TGF$\beta$   369
        A. Assay for the Stimulation of Anchorage-Independent Growth by TGF$\beta$   369
        B. Growth Inhibition and Differentiation Assays   369
        C. Immunological Assays for TGF$\beta$   370
    X. Structure and Properties of TGF$\beta$   370
        A. TGF$\beta$ Gene Family   370
        B. TGF$\beta$ Precursor   370

C. Mature TGFβ   371
D. Synthesis, Activation, and Release of TGFβ and Its Regulation   372
E. Mode and Mechanism of Action of TGFβ   373
F. Receptor Binding of TGFβ   374
XI. Molecular Biology and Genetics of TGFβ   374
A. TGFβ Expression, Tissue Distribution, and Role in Transformation   375
XII. Effects of TGFβ in Vitro and in Vivo   375
A. Growth Stimulation   375
B. Growth Inhibition   376
C. Cell Behavior   376
D. Cell Differentiation   376
E. Effects on Extracellular Matrix   377
F. In Vivo Effects of TGFβ   377
XIII. TGFβ and Other Growth Factors   378
XIV. Clinical Implications of TGFβ   378
A. Bone Remodeling   378
B. Wound Healing   378
XV. TGFβ and Neurobiology   379
XVI. Conclusions and Future Aspects   379
References   380

## 13   Insulin-Like Growth Factors in the Brain   391

Derek LeRoith, Charles T. Roberts Jr., Haim Werner, Carolyn Bondy, Mohan Raizada, and Martin L. Adamo

I. Introduction   391
II. Insulin-Like Growth Factors in the Brain   392
A. Insulin   392
B. Insulin-Like Growth Factor I   393
C. Insulin-Like Growth Factor II   395
D. Physiologic Regulation of the Expression of Insulin-Like Peptides in Brain   396
III. IGF-Binding Proteins in the Brain   397
IV. Insulin and IGF Receptors   399
A. Insulin and IGF Receptors in Brain   400
B. Developmental Expression of Brain Receptors   401
C. Regional Localization of Brain Receptors   401
V. Postreceptor Mechanisms   403
VI. Biologic Action of Insulin and IGFs   403
A. Brain Glucose Metabolism   404
VII. Hypothesis   405
References   407

## 14 Neurobiology of Insulin and Insulin-Like Growth Factors — 415
Douglas N. Ishii

  I. Introduction   415
 II. Neural Circuitry and Insulin-Like Growth Factors   416
     A. Neurite Outgrowth In Vitro   416
     B. Development of Neuromuscular Synapses   417
     C. Regeneration of Motor Synapses   419
     D. Sprouting of Motor Nerve Terminals   420
     E. Relationship between Synapse Development, Regeneration, and Sprouting   420
     F. Regeneration of Sensory Axons   421
III. IGFs and Insulin Viewed within the Context of Neurotrophic Theory   421
     A. NGF   421
     B. IGFs and Insulin   421
 IV. Shared Biochemical Pathway for Neurite Outgrowth   423
     A. Receptors   423
     B. Phosphorylation   424
     C. Is Retrograde Axonal Transport of Ligand Essential?   424
     D. Shared Pathway Regulating Neurite Outgrowth   425
  V. Other Neurobiological Roles   430
     A. Mitogenic   430
     B. Survival   430
     C. Neurotransmitters   431
     D. Other Actions   431
 VI. Pathophysiology   431
     A. Disturbances during Early Development   431
     B. Disturbances in Later Life   432
     References   433

## 15 Ciliary Neuronotrophic Factor — 443
Marston Manthorpe, Jean-Claude Louis, Theo Hagg, and Silvio Varon

  I. Historical Background: 1940–1975   443
     A. Target-Derived Protein Factor Concept   443
 II. From Ciliary Neuron Factor Concept to First Identification: 1976–1980   444
     A. Need for Purified Cholinergic Neuronal Cultures   444
     B. Heart Muscle-Derived "Parasympathetic Factor" for Ciliary Cholinergic Neurons   445
     C. "Skeletal Muscle-Derived Factor" for Cholinergic Ciliary Neurons   446

D. Ciliary Neuron Target Tissue (Eye) Is a Better Source for Purification of the Survival Promoting Substance   446
  E. Identification of Chick CNTF, a Trophic Factor Protein for Ciliary Ganglionic Neurons   447
III. Molecular Studies Leading to Mammalian CNTF Purification: 1981–1990, and Gene Cloning: 1989–1991   448
  A. Mammalian CNTF Purification   448
  B. CNTF Sequencing, Cloning, and Expression   449
IV. Biologic Properties   452
  A. CNTF-Responsive Cells   453
  B. CNTF Sources   456
  C. Physiologic Functions of CNTF   459
V. Summary and Perspectives   464
  References   465

# 16 Skeletal Muscle-Derived Neurotrophic Factors and Motoneuron Development   475

James L. McManaman and Ronald W. Oppenheim

I. In Vitro Evidence for Actions of Skeletal Muscle Neurotrophic Factors   476
II. In Vivo Evidence for Actions of Skeletal Muscle Neurotrophic Factors   478
III. Isolation and Identification of Skeletal Muscle-Derived Trophic Factors   479
IV. Effects of Purified Factors on Motoneuron Survival in Vivo   482
  References   483

# 17 Growth Factors For Myelinating Glial Cells in the Central and Peripheral Nervous Systems   489

Ellen J. Collarini and William D. Richardson

I. Introduction   489
II. Central Nervous System Glia   490
  A. Oligodendrocyte Type-2 Astrocyte Lineage   490
  B. Role of PDGF in Oligodendrocyte Development   491
  C. Regulation of SCIP Transcription Factor by Growth Factors   494
  D. Sources of PDGF in the CNS: Neurons versus Glia   494
  E. Role of FGF in 0–2A Lineage Development   496
  F. Insulin, Insulin-Like Growth Factors, and the 0–2A Lineage   497

III. Peripheral Nervous System Glia   497
   A. Interactions between Schwann Cells and Neurons   498
   B. cAMP-Dependent Regulation of Schwann Cell Development   498
   C. Synergy between Polypeptide Growth Factors and cAMP   499
   D. Autocrine Regulation of Schwann Cell Growth   500
IV. Conclusions and Outstanding Questions   501
References   503

## 18  Adhesion Factors   509
Hans W. Müller

I. Introduction   509
II. Laminin and Fibronectin   510
III. Potentiation of Neurotrophic Activity of Peptide Growth Factors by Laminin and Fibronectin   511
IV. Proteoglycans   512
V. Active Complexes of Laminin and Fibronectin with Proteoglycans   515
VI. Cell–Cell Contact-Mediated Neuronal Survival   516
VII. Purpurin and Apolipoproteins   517
VIII. Gangliosides   519
IX. Conclusions   520
References   521

## 19  Instructive Neuronal Differentiation Factors   527
Paul H. Patterson

I. Phenotypic Plasticity   527
II. Cholinergic Differentiation Factor/Leukemia Inhibitory Factor   528
III. Ciliary Neurotrophic Factor   533
IV. Membrane-Associated Neurotransmitter-Stimulating Factor   535
V. Sweat Gland Factors   535
VI. Factors Acting on Motor Neurons   537
VII. Noradrenergic Factors   539
VIII. Peptidergic Factors   541
IX. Corticosteroid   541
X. Gonadal Steroids   542
XI. Morphological Factors   543
XII. Melanization Factors   544

XIII. Neuronal Activity   545
XIV. Other Factors   546
XV. Mutants in Invertebrates   547
XVI. Hematopoietic Analogy   548
References   550

## 20  Neurotransmitters as Neurotrophic Factors           565
Frances M. Leslie

I. Introduction   565
II. Amino Acid Transmitters   566
   A. Glutamate   566
   B. γ-Amino Butyric Acid   570
III. Monoamines   573
   A. Serotonin   573
   B. Catecholamines   575
   C. Acetylcholine   579
IV. Neuropeptides   582
   A. Opioid Peptides   582
   B. POMC-Derived Peptides   584
   C. Other Peptides   584
V. Conclusions   586
References   587

Index                                                    599

# Contributors

*Numbers in parentheses indicate the pages on which the authors' contributions begin.*

**Martin L. Adamo** (391), Diabetes Branch, National Institute of Diabetes and Digestive and Kidney Diseases, National Institutes of Health, Bethesda, Maryland 20892

**Joseph G. Altin** (129), Division of Cell Biology, John Curtin School of Medical Research, Australian National University, Canberra, ACT, 2601 Australia

**Alaric T. Arenander** (89), Department of Anatomy and Cell Biology, The Mental Retardation Research Center, and The Laboratory of Biomedical and Environmental Sciences, UCLA School of Medicine, Los Angeles, California 90024

**Carolyn Bondy** (391), Developmental Endocrinology Branch, National Institute of Child Health and Human Development, National Institutes of Health, Bethesda, Maryland 20892

**Ralph A. Bradshaw** (129), Department of Biological Chemistry, College of Medicine, University of California, Irvine, Irvine, California 92717

**Ellen J. Collarini** (489), Department of Biology, University College of London, London WC1E 6BT, United Kingdom

**Michael F. Crouch** (51), Division of Neuroscience, John Curtin School of Medical Research, Australian National University, Canberra ACT 2601, Australia

**Timothy L. Denton** (25), Division of Neurogerontology, Andrus Gerontology Center, University of Southern California, Los Angeles, California 90089

**Robert H. Edwards** (181), Department of Neurology, UCLA School of Medicine, Los Angeles, California 90024

**James H. Fallon** (1), Department of Anatomy and Neurobiology, College of Medicine, University of California, Irvine, Irvine, California 92717

**Mark L. Grimes** (209), Department of Neurology, University of California, San Francisco, San Francisco, California 94121

**Glaudia Grothe** (313), Department of Anatomy and Cell Biology, University of Marburg, Robert-Koch-Straβe 6, D-3550 Marburg, Germany

**Theo Hagg** (443), Department of Biology, University of California, San Diego, La Jolla, California 92093

**Franz Hefti** (25), Andrus Gerontology Center, University of Southern California, Los Angeles, California 90089

**Ian A. Hendry** (51), Division of Neuroscience, John Curtin School of Medical Research, Australian National University, Canberra ACT 2601, Australia

**Harvey R. Herschman** (89), Department of Anatomy and Cell Biology, Department of Biological Chemistry, The Mental Retardation Research Center, and The Laboratory of Biomedical and Environmental Sciences, UCLA School of Medicine, Los Angeles, California 90024

**David M. Holtzman** (209), Department of Neurology, University of California, San Francisco, San Francisco, California 94121

**Douglas N. Ishii** (415), Departments of Physiology and Biochemistry, Colorado State University, Fort Collins, Colorado 80523

**Beat Knusel** (25), Division of Neurogerontology, Andrus Gerontology Center, University of Southern California, Los Angeles, California 90089

**Paul A. Lapchak** (25), Division of Neurogerontology, Andrus Gerontology Center, University of Southern California, Los Angeles, California 90089

**Derek LeRoith** (391), Diabetes Branch, National Institute of Diabetes and Digestive and Kidney Diseases, National Institutes of Health, Bethesda, Maryland 20892

**Frances M. Leslie** (565), Department of Pharmacology, College of Medicine, University of California, Irvine, Irvine, California 92717

**Ronald M. Lindsay** (257), Regeneron Pharmaceuticals Inc., Tarrytown, New York 10591

**Frank M. Longo** (209), Department of Neurology, University of California, San Francisco, San Francisco, California 94121

**Sandra E. Loughlin** (1), Department of Anatomy and Neurobiology, College of Medicine, University of California, Irvine, Irvine, California 92717

**Jean-Claude Louis** (443), Department of Biology, University of California, San Diego, La Jolla, California 92093

**Gerson Lüdecke** (313), Department of Anatomy and Cell Biology, University of Marburg, Robert-Koch-Straβe 6, D-3550 Marburg, Germany

**Marston Manthorpe** (443), Department of Biology, University of California, San Diego, La Jolla, California 92093

**James L. McManaman**[1] (475), Department of Neurology, Division of Neuroscience and Program in Cellular and Molecular Biology, Baylor College of Medicine, Houston, Texas 77030

**William C. Mobley** (209), Departments of Neurology, Pediatrics, and the Neuro-

---

[1]Present affiliation: Department of Neuroscience, Synergen, Inc., Boulder, Colorado 80301.

science Program, University of California, San Francisco, San Francisco, California 94121

**Richard Morrison** (339), R. S. Dow Neurological Sciences Institute and Comprehensive Cancer Center, Portland, Oregon 97209

**Hans W. Müller** (509), Molecular Neurobiology Laboratory, Department of Neurology, University of Düsseldorf, Moorenstr. 5, D-4000 Düsseldorf, Germany

**Ronald W. Oppenheim** (475), Department of Neurobiology and Anatomy and Neuroscience Program, The Bowman Gray School of Medicine, Wake Forest University, Winston-Salem, North Carolina

**Dörte Otto** (313), Department of Anatomy and Cell Biology, University of Marburg, Robert-Koch-Straße 6, D-3550 Marburg, Germany

**Paul H. Patterson** (527), Division of Biology, California Institute of Technology, Pasadena, California 91125

**Pauli Puolakkainen** (359), Bristol-Myers Squibb Pharmaceutical Research Institute—Seattle, Seattle, Washington 98121

**Mohan Raizada** (391), Department of Physiology, University of Florida, Gainesville, Florida 32610

**William D. Richardson** (489), Department of Biology, University College of London, London WC1E 6BT, United Kingdom

**Charles T. Roberts Jr.** (391), Diabetes Branch, National Institute of Diabetes and Digestive and Kidney Diseases, National Institutes of Health, Bethesda, Maryland 20892

**Kenneth A. Thomas** (285), Department of Biochemistry, Merck Research Laboratories, Rahway, New Jersey 07065

**Daniel R. Twardzik** (359), Bristol-Myers Squibb Pharmaceutical Research Institute—Seattle, Seattle, Washington 98121

**Klaus Unsicker** (313), Department of Anatomy and Cell Biology, University of Marburg, Robert-Koch-Straße 6, D-3550 Marburg, Germany

**Silvio Varon** (443), Department of Biology, University of California, San Diego, La Jolla, California 92093

**Haim Werner** (391), Diabetes Branch, National Institute of Diabetes and Digestive and Kidney Diseases, National Institutes of Health, Bethesda, Maryland 20892

**Reiner Westermann** (313), Department of Anatomy and Cell Biology, University of Marburg, Robert-Koch-Straße 6, D-3550 Marburg, Germany

# Preface

Although it has been known since the early 1950s that growth factors affect neural tissues, it is only in the past decade that some characterized growth factors have been shown to be localized in the central nervous system. In the past five years, we have seen an explosion in research into growth factors, their receptors and their functional roles in development, reorganization, and pathology. This is thus an appropriate time to comprehensively review the information on neurotrophic factors in mammalian nervous systems.

This volume provides not only current reviews, but also attempts to provide a conceptual framework for understanding the rich spectrum of actions of neurotrophic factors. The members of the major growth-factor families, including the neurotrophins, the epidermal growth factor family, and the fibroblast growth factors, are comprehensively reviewed, along with a number of more recently discovered factors. Multifaceted approaches to the study of growth-factor effects on specific and defined neuronal groups are discussed. The actions of neurotrophic factors on first, second, and third messenger systems, as well as their convergence onto primary response genes are examined. Synergistic effects and interactions between neurotrophic factors are described, especially as they affect the complex glial–neuronal mixture characteristic of the nervous system. In addition to the traditionally defined neurotrophic factors, the growth-promoting effects of neurotransmitters, neuronal activity, and interactions with the extracellular matrix are considered. Thus, this volume brings together specific reviews of known growth factors and offers a conceptual discussion of their anatomical distributions, modes of actions, and interactions.

The editors gratefully acknowledge the contributions of Nanette Canepa in preparation of the index. Kathleen Cooper of the American Parkinson Disease Association SCC continues to serve as an inspiration to us all.

# 1 Functional Implications of the Anatomical Localization of Neurotrophic Factors

James H. Fallon and Sandra E. Loughlin

## I. INTRODUCTION

This volume is devoted to the analysis and review of a class of growth factors that acts on neural tissue, namely, neurotrophic factors. The history of growth factor and neurotrophic factor research can be presented differently, depending on the definitions used and the scientific discipline of the writer. As neuroanatomists, we trace the first conceptualization of neuronal trophic factors to the visionary neuroanatomist, Ramón y Cajal (1882, 1891; De Felipe and Jones, 1991). This concept, together with studies by Hamburger (1934) on neuronal development, had its foundation in a developmental zeitgeist of 19th century European embryologists (see Chapter 2). Growth factor research per se is believed to have had its experimental genesis in 1916 when Robertson coined the term "tethelin" for the growth-stimulating activity of pituitary extracts in vitro (reviewed by Burgess, 1989). A decade later, insulin was recognized for its broad growth-promoting effects (Banting and Best, 1922). Growth factors were shown to be present in the brain over 50 years ago, when it was demonstrated that brain homogenates contained factors that stimulated fibroblasts to divide in vitro (Trowell et al., 1939; Hoffman, 1940). This fibroblast growth factor activity later was purified partially and was shown to have effects on many other tissues (Gospodarowicz, 1974). During the 1950s, the landmark studies by Levi-Montalcini and Hamburger (1953) established a conceptual and experimental blueprint for subsequent studies in the field of growth factors in the nervous system.

Despite a wealth of in vitro and in vivo molecular, biochemical, physiological, and pharmacological data generated on growth factors over the past 50 years, relatively little anatomical information on the localization of growth factors has been available until the past decade. This has been especially true for neurotrophic factors in the central nervous system. For example, epidermal growth factor was isolated 30 years ago by Cohen (1962) but the anatomical localization of epidermal growth factor-like immunoreactivity in the brain was not known until the 1980s (Fallon et al., 1984). This lag in morphological information is, in part, due to the low levels of neurotrophic factors present in the brain. However, new sensitive immunocytochemical, in situ hybridization, and receptor autoradiographic techniques have afforded the neuroanatomist powerful tools for locating neurotrophic factor systems in the brain. With these techniques, an abundance of neuroanatomical information has been gathered at the light microscopic level in the past few years that has confirmed predictions by Ramón y Cajal on the presence and importance of trophic factors in the brain.

In this introductory chapter, we will present some neuroanatomical perspectives on the roles of neurotrophic factors in the brain. Because of the sheer volume of the newly available anatomical information on the distribution of neurotrophic factors in the nervous system, it is impractical to present atlas mappings of all known neurotrophic factors. This chapter will stress some key morphological issues concerning neurotrophic factors in the central nervous system and how knowledge of the distribution of these growth factor systems is critical to our understanding of how these substances may be involved in developing, adult, aging, and damaged nervous systems.

## II. NEUROTROPHIC FACTORS INCLUDED IN THIS VOLUME

Growth factors include substances that stimulate cells to divide (hyperplasia) or increase in size (hypertrophy). Many growth factors are now known to exist (Table 1). Trophic factors include those substances that have effects on cell differentiation, survival, phenotypic expression, and plasticity, as well as on cell hypertrophy, for example, neurite extension (also considered a "trophic" action). Neurotrophic factors, a subset of growth factors acting on neural tissue, have been defined in a myriad of ways ranging from very restrictive to very general. Restrictive definitions are based on the nerve growth factor model, which recognizes only the specific aspects of developmental events, such as cell survival and neurite extension, that are supported by proteins and peptides. This restrictive definition of a neurotrophic substance parallels the restrictive definition of and requirements for a neurotransmitter, which were based on experimental paradigms developed for acetylcholine. Coincidentally, nerve growth factor and acetylcholine not only have forged our fundamental definitions of a neural growth factor and a neurotransmitter, respectively, but also are associated closely in brain function. As more putative "neurotransmitters" and "neurotrophic factors" have been discovered and analyzed, it has become clear that broader definitions of each group are necessary to capture the range of subtle and profound actions of these neuromessengers.

A focused definition that integrates both traditional and modern components

TABLE 1

**Examples of growth factors**

| Family | Specific examples |
|---|---|
| Neurotrophin | Nerve growth factor (NGF)<br>Brain-derived neurotrophic factor (BDNF)<br>Neurotrophin 3 (NT-3) |
| EGF | Epidermal growth factor (EGF)<br>Transforming growth factor alpha (TGFα)<br>Vaccinia virus growth factor<br>Amphiregulin (AR)<br>Schwannoma-derived growth factor (SDGF) |
| FGF | Acidic fibroblast growth factor (aFGF)<br>Basic fibroblast growth factor (bFGF)<br>INT-2, FGF-5, FGF-6, KGF, HST/KGF |
| Insulin-like | Insulin<br>Insulin-like growth factors (somatomedins)<br>Relaxin |
| Others | Growth hormone (GH)<br>Platelet-derived growth factor (PDGF)<br>Mast cell growth factor (MGS)<br>Colony stimulating factors<br>Ciliary neurotrophic factor (CNTF)<br>Glial maturation factor<br>Protease nexin I, II<br>Sweat gland factor<br>Cholinergic neuronal differentiation factor (CDF)<br>Muscle-derived growth factors (MDGF)<br>Striatal-derived neuronotrophic factor<br>Transforming growth factor beta (TGFβ)/inhibin/activin family<br>Membrane-associated neurotransmitter stimulating factor (MANS)<br>Thrombin<br>Entactin<br>Erythropoietin<br>Neurite inducing factor<br>Stem cell factor (SCF)<br>Interleukin 1,3,6<br>Glial-derived nexin<br>Heparin-binding NF |
| Extracellular matrix/adhesion factors | Laminin<br>Fibronectin<br>Purpurin<br>Apolipoproteins<br>Gangliosides |
| Transmitters | Neurotransmitters<br>Neuropeptides |
| Nonpeptide hormones | Steroid<br>T3/T4<br>Ion fluxes<br>Neuronal activity |

of neurotrophic activity is presented by Hefti and co-workers in this volume (Chapter 2). This definition states that "neurotrophic factors are endogenous, soluble proteins regulating survival, growth, morphological plasticity, or synthesis of proteins for differential functions of neurons." This volume includes chapters on soluble proteins that fulfill these formal criteria, as well as on other protein and nonprotein factors that could be considered neurotrophic factors based on their trophic and trophic-related functions.

The neurotrophic factors (as just defined) discussed in this volume include nerve growth factor (NGF; Chapters 5, 6, and 7), brain-derived neurotrophic factor (BDN; Chapter 8), ciliary neurotrophic factor (CNF; Chapter 15), fibroblast growth factors (FGF; Chapters 9 and 10), insulin and insulin-like growth factors (IGF; Chapters 13 and 14), epidermal growth factor (EGF; Chapter 11), transforming growth factors $\alpha$ and $\beta$ (TGF; Chapter 12), and skeletal muscle-derived neurotrophic factors (Chapter 16). The effects of a number of growth factors, especially platelet-derived growth factor, on glial cells are also described (Chapter 17). In addition to the roles of these factors, the roles of adhesion factors (Chapter 18), neurotransmitters, and neuropeptides (Chapter 20) as neurotrophic factors are included by virtue of their important neurotrophic activity. These chapters focus on our present knowledge of the neurotrophic factors (first messengers), their receptor systems, and associated transducing mechanisms (second messengers). Many neurotrophic factors effect phenotypic responses in cells through rapid ligand-induced primary response genes (third messengers). These cellular mechanisms are discussed in Chapter 4. Synergistic interactions between neurotrophic factors and intrinsic cellular systems, especially with respect to cell death and survival, are examined in Chapter 3. The role of neurotrophic factors and other cellular mediators as phenotype-specifying factors is developed in Chapter 19. These factors include neurotransmitters, neuropeptides, steroids, and membrane-bound substances, as well as neuronal activity. Thus, links are formed between neurotrophic activity and related cellular mechanisms.

## III. REVIEW OF CELLULAR EVENTS OF NEUROTROPHIC FACTOR SYSTEMS

Each of the chapters in this volume details specific molecular and cellular aspects of neurotrophic factors. In this section, a brief overview of these events is summarized (see Fig. 1). Although the major cellular events of trophic factor functions are similar for all tissue types, the highly specialized structure and function of central neural tissue presents unique problems for neurotrophic factor transport, availability, synergy, and regulation. The two major cell types in central neural tissue are neurons and glia. In Fig. 1, neuron A is shown receiving an axonal terminal from neuron B. Neuron B, in turn, projects to neuron C. A glial cell is shown sending out processes that communicate with neuron A and with a blood vessel. The nucleus of neuron A contains DNA and the cytoplasm of neuron A is shown to contain rough endoplasmic reticulum (8), Golgi apparatus (10), and polyribosomes (9). Retrograde and anterograde transport are highlighted by arrows outside the axon on neuron A.

In Fig. 1, neuron A is depicted containing membrane receptors for neurotrophic factors (1). After binding of the factor to its receptor, signal transduction proceeds through second messenger systems that may alter a number of events through control of gene expression (see, for example, Chapter 5). These events include alterations in synthesis of neuromodulators and growth factors, regulation of differentiation and survival, inhibition or facilitation of inherent death programs

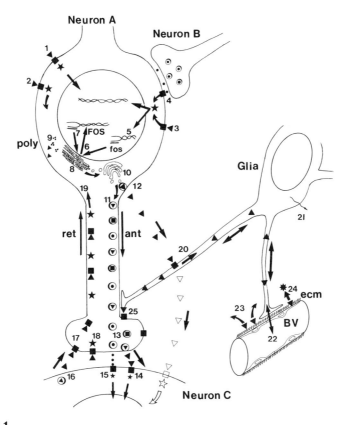

FIGURE 1

Cellular events involved in neurotrophic factor activity. Neurotrophic factors, designated generally as growth factors, are indicated by triangles (▲). Receptors are indicated by squares (■). Neurotransmitters and neuropeptide modulators are indicated by dots (•). Second messenger-associated substances such as G proteins, cAMP and protein kinase, are indicated by stars (★). Vesicles are indicated by circles (○). Abbreviations: BV, blood vessel; ret, retrograde transport; ant, anterograde transport; ecm, extracellular matrix; poly, polyribosomes. See text for discussion.

(Chapter 3), modulation of morphological plasticity, and regulation of retrograde transport through stabilization of tubulin mRNA (Chapter 14). Neurotrophic factor signal transduction may also regulate membrane (ion fluxes), cytosolic, and cytoskeletal (2) function. The effects of neurotrophic factors on cytoskeletal elements actually may be detrimental to the recovery of damaged neural systems. For example, Butcher and Woolf (1989) have proposed that, in some cases (e.g., Alzheimer's disease), neurotrophic factors may precipitate a pathologic cytoskeletal cascade that accelerates neuronal degeneration. Regulation and buffering of ion concentration may be an important function of neurotrophic factors. The control of $Ca^{2+}$ levels through modulation of calcium-binding proteins (calmodulin, 28k calbindin, parvalbumin), enzymes (protein kinase C, mitochondrial systems, $Ca^{2+}$ ATPase), and membrane $Ca^{2+}$ channels may be how neurotrophic factors afford protection against excitotoxins.

Interactions between neurotrophic factors and other neurotrophic agents (3,4) are known to occur (Chapter 3). Some neurotransmitters and neuropeptides that mediate neurotransmission at synapses possess neurotrophic activity (4) (Chapter 20). Synergy between neurotrophic factors, and between neurotrophic factors and neurotransmitters with neurotrophic activity, may be mediated through convergence on second and third messenger systems, where interactions can occur at the level of the second messengers or the phosphorylation of substrates (Chapter 4). Convergence of neurotrophic factor cascades also occurs at the level of the third messenger systems (primary response gene products), as indicated in Fig. 1 (5,6,7). Thus, a number of neurotrophic factors are able to stimulate primary response genes such as c-*fos* and c-*jun* (5). Primary response genes are transcribed to mRNAs that are translated to proteins, for example, FOS and JUN (6,8), which then activate target genes (7), leading to further cell-specific synthesis of new proteins on polyribosomes (9) or rough endoplasmic reticulum (8).

Neurotrophic factors that contain consensus secretory signal sequences for insertion through membranes (e.g., EGF, TGFα, TGFβ, NGF) are synthesized in the rough endoplasmic reticulum (8); further posttranslational modifications are carried out in the vesicular system of the Golgi apparatus (10), from which the factors may be transported and released from vesicles (11,12,13). Neurotrophic factors lacking a consensus secretory signal sequence (e.g., aFGF, bFGF, CNTF, IL-ls) are thought to be synthesized in polyribosomes and released into the cytosol (Janet et al., 1987; Chapters 9 and 10). The absence of secretory leader sequences and glycosylation in these neurotrophic factors poses an obvious problem. How are they released from cells? Holocrine secretion after cell death and lysis (see Fig. 2) is one probable mode of "secretion." Other physiological modes of release are possible, but as yet undefined. There are, interestingly, many other examples of proteins of this type. For example, annexin 1, which also contains no signal secretory sequence, is nevertheless secreted by the merocrine portion of the prostate gland (Christmas et al., 1991).

Neurotrophic factors synthesized in neurons may be secreted at a number of

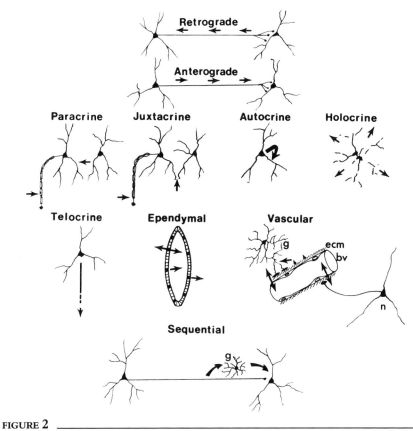

FIGURE 2

Modes of secretion and intercellular communication mechanisms of neurotrophic activity. Abbreviations: g, glial; n, neuronal; bv, blood vessel; ecm, extracellular matrix.

sites, for example, cell somata (12), dendrites (not illustrated), axons, and axon terminals (13). The problem of long distance availability of trophic factor action is obviated in neural tissue by the elaboration of the axon and its numerous terminal branches. For example, dopamine neurons of the substantia nigra of the rat project axons 10 mm or more in length to the caudate putamen (nigrostriatal pathway), where each axon may give rise to 500,000 terminals (Anden et al., 1966). In addition, a single dopamine neuron may give rise to axons that collateralize to widely divergent forebrain targets such as caudate putamen, septum, and cortex (Fallon and Loughlin, 1982). Thus, a single neuron could provide neurotrophic factors to many brain regions, as well as retrogradely transport (Fig. 1) neurotrophic factor–receptor complexes (17) or related messengers (18) (Chapters 3 and 14) from many brain regions. Therefore, the elaborate axonal (and dendritic) processes in neural tissue provide a unique morphological substrate for neurotrophic factor access and availability.

Highly branched axonal systems also provide one clue for apparent anatomical mismatches between neurotrophic factors and their receptors. For example, neuron A may project to neuron C. Neuron C releases a growth factor (16) that binds to a receptor on the axon terminal of neuron A (17,18). This case shows no anatomical mismatch between the neurotrophic factor and its receptor. However, if neuron A also projects an axon collateral to another brain site that lacks the neurotrophic factor synthesized by neuron C, the axons and axon terminals of neuron A in this other brain site may still express receptors for the neurotrophic factor secreted by neuron C. This would result in an apparent mismatch between the neurotrophic factor and its receptor.

Glia also synthesize and respond to neurotrophic factors (20,21,22). Like neurons, glia display regional heterogeneity (Chapter 19); a subset of astrocytes contain TGFα-precursor immunoreactivity (Fallon et al., 1990). Glia not only synthesize and release neurotrophic factors that may interact with neurons (25), the vasculature (22), and other glia, they also may be key intermediaries of neurotrophic activity between neurons. This "sequential" signaling may proceed from neuron A to neuron C, through an intermediary glial step (20). Vascular–neuronal trophic interactions may also be mediated by intervening glia. Glucose uptake and extracellular ion concentrations (e.g., $Ca^{2+}$) are also regulated by glia (21) and may be under neurotrophic factor control.

Neurotrophic factor availability may be restricted or enhanced by adhesion factors, for example, in the extracellular matrix of neural cells and blood vessels (Fig. 1). The extracellular matrix may provide binding or storage sites for active neurotrophic factors (23) or may mask neurotrophic factors in a way that renders them inactive unless there is injury at this site (24). FGFs are sequestered by extracellular matrix heparan proteoglycans that may act as temporary storage sites of FGF (Chapter 4) and participate in the binding of FGF to its receptor (Rapraeger et al., 1991). FGF could be released after anatomical disruption of these sites, leading to neurotrophic and angiogenic effects.

Another potential intercellular communication system for neurotrophic factors may be mediated by "juxtacrine" interactions (Ankelesaria et al., 1990; Chapter 12) (25 in Fig. 1; see also Fig. 2). In this model of interaction, the neurotrophic factor remains inserted in the membrane of the cell that synthesizes the factor (in Fig. 1, the glial cell is the cell synthesizing the neurotrophic factor) but is still able to bind the receptor located in the responsive cell (neuron A, Fig. 1). This hypothesis has been put forward by Wong et al. (1989) and Ankelesaria et al. (1990) for the binding of TGFα to the EGF receptor in adjacent cells in vitro (Chapter 12). Thus, it is possible that neurotrophic factor interactions in the central nervous system are mediated by similar intercellular mechanisms.

Many of the molecular and cellular events relating to trophic factor function have been studied in peripheral tissues and in vitro. Few of these events, however, have been examined for neurotrophic factors in the central nervous system and much of our specific knowledge remains fragmentary. Several questions regarding the functional morphology of neurotrophic factors have been presented in this section, but we are hampered by a lack of ultrastructural data on the localization of neurotrophic factor and their related molecular systems. At present, much of our morphological information is emerging from light microscopic, immunocytochemical, in situ hybridization, and receptor autoradiographic studies. These studies offer insight into some basic questions. Are the neurotrophic factors, their receptors, and the synthetic cellular machinery present in neurons or glia? Even this question is often difficult to answer directly with the techniques used in in situ hybridization and receptor autoradiographic techniques. Electron microscopic studies will be necessary to answer this most basic question. Further, the highly differentiated and compartmentalized nature of central neural tissue poses more difficult morphological questions. The dendritic and axonal arborizations of neurons are elaborate and heterogeneous. Neurotrophic factor molecular systems may be active only in restricted segments of the dendrites of a neuron. In order to determine the precise site of action and interaction of a particular neurotrophic factor, it is important to determine the subcellular localization of the various neurotrophic factors, their binding proteins and relevant adhesion factors, and their receptor, second messenger, and third messenger systems. A single neuron, responsive to several neurotrophic factors, may have multiple domains of functional activity depending on the local neural, glial, vascular, and extracellular matrix relationships of portions of its dendritic and axonal processes. Further, the neurotrophic factor effects on morphological plasticity must be studied in vivo by quantitative electron microscopic techniques.

## IV. MODES OF SECRETION AND TRANSPORT OF NEUROTROPHIC FACTORS

The specialized nature of neural tissue affords a variety of modes of secretion, transport, and availability of neurotrophic factors. As discussed in the previous section, features of neural cells such as extensive axonal arborization of neurons and neuronal–glial–vascular arrangements provide numerous means of long- and short-distance interactions between neurotrophic factor-releasing and -responsive cells. These modes of secretion and transport are summarized in this section (Fig. 2).

The most widely cited mode of neurotrophic factor transport is retrograde transport of the ligand–receptor complex or its second messenger (Retrograde, Fig. 2) (Chapter 3). The classic example of a neurotrophic factor–receptor complex undergoing this mode of transport is NGF. The second mode of transport of neurotrophic factors is anterograde axonal transport (Anterograde, Fig. 2). This mode has been observed for bFGF in the visual pathway from retinal ganglion cells to the lateral geniculate nucleus and superior colliculus (Ferguson et al., 1990).

Neurotrophic factors also are thought to act locally in the central nervous system through several mechanisms, including local actions of soluble factors on adjacent neurons and glia (paracrine, Fig. 2) and the binding of membrane-bound neurotrophic factors and their receptors on adjacent cells (juxtacrine). Local actions also might include the self-regulation of production by the binding of a neurotrophic factor to its receptors in the same cell or by nuclear targeting of a cytosolic neurotrophic factor (autocrine). One hypothetical subset of autocrine regulation is termed "intracrine" (Chapter 8) and refers to purely intracellular regulatory mechanisms. Recently, Clevenger et al. (1991) have provided evidence that peptide hormones may function in the nucleus without first binding to a cell-surface receptor, lending credence to the concept that some growth factors may function in an intracrine manner. After the death or anatomical disruption of a cell, a neurotrophic factor such as aFGF, which lacks a secretory signal sequence, can be released from the dying cell (holocrine). Factors secreted by a cell may gain access to the extracellular space, through which they diffuse to act at great distances from the secretory cell (telocrine). This mode of action has not been demonstrated convincingly in the central nervous system.

Some modes of transport unique to the central nervous system include movement of neurotrophic factors between the cerebrospinal fluid and the parenchyma of the brain and spinal cord. These one-way and two-way movements of factors occur through ependymal cells and modified ependymal cells called tanacytes (ependymal). Communication between the blood and the central nervous system occurs selectively across blood vessels participating in the blood–brain barrier as well as across circumventricular organs lacking this barrier. Release of neurotrophic factors into the blood in these regions would constitute a neuroendocrine mode of secretion.

Selective transport between the blood and the central nervous system is regulated by endothelial cells of blood vessels and transport mechanisms of astrocytes (vascular). Both glial and neuronal interactions with blood vessels are restricted, in part, by the basal laminae and associated extracellular matrix elements. The movement of insulin and insulin-like growth factors across the blood–brain barrier and through circumventricular organs lacking a blood–brain barrier is discussed by Le Roith and colleagues (Chapter 13).

Finally, an important mode of neurotrophic factor transport is through a combination of neuronal–glial interactions (sequential). For example, neurotrophic factors derived from one neuronal population first may act locally on glia that, in turn, release a second neurotrophic factor that regulates functions in a separate population of neurons. This possibility is raised by a number of experimental findings. For example, EGF and bFGF increase dopamine uptake in substantia nigra dopamine neurons in vitro only when glia are present (Knusel et al., 1990). This issue is discussed further in a subsequent section of this chapter.

## V. NEUROTROPHIC FACTOR DISTRIBUTION IN CENTRAL NEURAL SYSTEMS

### A. Approaches in Neuroanatomy

The molecular and cellular distribution of neurotrophic factor systems was discussed in Section III. In this section, issues relating to the broader distribution of neurotrophic factors in the central nervous system will be discussed. One approach used in neuroanatomical research involves a *regional analysis* of the central nervous system, which considers areas of the brain and spinal cord with respect to common location, structure, development, and patterns of connectivity (e.g., thalamus, cortex). A second approach is the study of *neural systems,* which examines the neuraxis with respect to a common functional system (e.g., visual system, extrapyramidal motor system, limbic system). A third parallel approach, *chemical neuroanatomy,* examines the central nervous system with respect to a particular neurochemical (e.g., acetylcholine, dopamine, enkephalin, neurotrophic factors). All three approaches have provided important perspectives on the functions of neurotrophic factors in the central nervous system.

### B. Regional Analysis

The hippocampus is a brain region that has received a great amount of attention in neuroanatomical studies of neurotrophic factors. The hippocampus is a phylogenetically ancient form of cortex (archicortex) with a relatively simple laminated structure. Its cell types and their connections have been studied in great detail for over 50 years (Lorente de Nó, 1934). The hippocampus has received particular attention in neurotrophic factor research because it has been known to be a major source of nerve growth factor for cholinergic neurons of the nucleus basalis of Meynert, which degenerate in patients with Alzheimer's disease (Wilcock et al., 1983; Whitehouse et al., 1987).

The most popular working hypothesis for the past decade has been that nerve growth factor is synthesized and released by cells in the hippocampus, binds to a nerve growth factor receptor on terminals of cholinergic neurons, and is retrogradely transported as a ligand–receptor complex to cell bodies in the nucleus basalis, where a second messenger signal mediates further trophic activity. Several findings from various disciplines have challenged, or at least broadened, this model of the central neural growth factor systems. These new findings are discussed in the chapters in this volume, but some of the recent neuroanatomical information is discussed here.

First, nerve growth factor is present in other cholinoceptive regions of the brain (Shelton and Reichardt, 1986) and some cholinergic brain areas do not respond to NGF (Knusel and Hefti, 1988). Second, cholinergic neurons are positioned to be under the influence of nerve growth factor that is synthesized in the local environs of their cell bodies (Lauterborn et al., 1991) as well as at distant sites such as the hippocampus. Third, some cholinoceptive brain regions do not contain nerve growth factor but may contain a separate neurotrophic factor instead. Fourth, a cholinoceptive region such as the hippocampus contains not only nerve growth factor, but also other members of the neurotrophin family (BDNF, NT-3) as well as members of other families of neurotrophic factors. This point is illustrated in Fig. 3. These illustrations are derived from in situ hybridization and/or immunocyto-

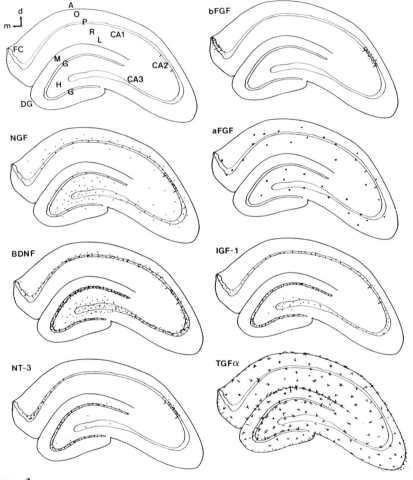

FIGURE 3

Coronal sections of hippocampus of the right side of the rat brain illustrating the distribution of neurotrophic factors in this structure. All distributions are in neurons, with the exception of TGFα-precursor, which is localized in glia. The top left panel indicates regions of the hippocampus, including the dentate gyrus (DG) with its molecular (M), granule cell (G), and hilar (H or CA4) layers. Ammon's horn includes the CA1, CA2, and CA3 regions, each with an alveus (A), stratum oriens (O), pyramidal layer (P), stratum radiatum (R), and stratum lacunosum (L). The fasciola cinerium (FC) is also illustrated.

chemical studies on the localization of several neurotrophic factor systems, including NGF (Ernfors et al., 1990; Maisonpierre et al., 1990; Senut et al., 1990); BDNF (Ernfors et al., 1990; Hofer et al., 1990; Maisonpierre et al., 1990; Phillips et al., 1990; Wetmore et al., 1990), neurotrophin-3 (NT-3) (Ernfors et al., 1990; Maisonpierre et al., 1990; Phillips et al., 1990), acidic FGF (aFGF) (Fallon et al., 1991), basic FGF (bFGF) (Pettman et al., 1986; Wilcox and Unnerstall, 1991), IGF-I (Noguchi et al., 1987), and TGFα precursor (Fallon et al., 1990; J. H. Fallon and S. E. Loughlin, unpublished observations).

As illustrated in Fig. 1, NGF (mRNA and peptide) is localized in scattered pyramidal and nonpyramidal neurons throughout the dentate gyrus, hilus, and Ammon's horn (fields CA1, CA2, CA3). The second member of the neurotrophic family, BDNF, is localized in the same areas as NGF, but a higher proportion of dentate granule cells and Ammon's horn pyramidal neurons exhibits BDNF mRNA expression. The third member of the neurotrophic family, NT-3, has a more restricted distribution, primarily in dentate granule cells and pyramidal neurons of CA2 and medial CA1. Other levels of complexity of these distributions are present. For example, NT-3 mRNA is highly expressed early in development, BDNF mRNA is expressed later in development after neurogenesis is complete, and NGF mRNA is expressed more consistently throughout development (Maisonpierre et al., 1990). The heterogeneity of receptor subsystems for the neurotrophins adds to the complexities of this neurotrophic system (Chapters 5 and 8).

The hippocampus also contains other neurotrophic activity. For example, bFGF is localized in pyramidal neurons of CA2 and FC (fasciola cinerium). aFGF is sparse, scattered, and found mostly in nonpyramidal neurons. IGF-I is localized in a moderate number of dentate granule cells and in Ammon's horn pyramidal neurons in CA1, CA2, and CA3. In contrast to these neurotrophic factors, TGFα-precursor is present in a subpopulation of astrocytes throughout the hippocampus.

The plots in Fig. 3 illustrate the rich and heterogeneous distribution of neurotrophic factors in neurons and glia of the hippocampus. Such a regional analysis of neurotrophic factor distribution in the central nervous system highlights the rich diversity of potential growth factor interactions in one brain region and helps isolate specific subregions that may be particularly susceptible to, or resistant to, certain pathological processes. For example, the CA2 region is endowed with a variety of neurotrophic factors (Fig. 3). The CA2 region is particularly resistant to pathological changes in temporal lobe epilepsy, perhaps because of the protective effects of 28K calbindin, which is enriched in this region of the hippocampus (Sloviter et al., 1991), and because of the extensive presence of neurotrophic factors in these neurons. Other protective mechanisms are undoubtedly involved. For example, the presence of glutamate dehydrogenase in astrocytes in these hippocampal regions (Aoki et al., 1987) may offer pyramidal neurons a measure of resistance to glutaminergic neurotoxicity. Knowledge of the neuroanatomical distribution of neurotrophic factors and regulatory enzyme systems is essential to our understanding of pathological processes that tend to occur (or not occur) in specific brain sites.

## C. Systems Analysis

### 1. Extrapyramidal motor system

One central neural system that has received considerable attention in the area of neurotrophic factor research is the extrapyramidal motor system. This system includes a broad spectrum of forebrain, brainstem, and spinal cord structures that participate in the temporal and spatial regulation of the three great motor systems of the body: the somatic (striated voluntary muscles), the limbic/autonomic (involuntary muscles of the gut, cardiovascular system, glands), and the endocrine (pituitary regulated) (Fallon and Loughlin, 1987). These monumental tasks are undertaken by a significant mass of the central nervous system, especially the cortex, basal ganglia, and limbic system. Therefore, we will focus on select subsystems. The subsystems, however, comprise key elements in our study of the etiology of and treatments for Parkinson's disease and Huntington's disease.

Fig. 4 outlines some fundamental connections between a few regions involved in the extrapyramidal motor system. Projections from pyramidal neurons in the deep layers of cortex reach the caudate putamen (one sector of the striatum), which then projects to the globus pallidus (one sector of the pallidum) and substantia nigra (another sector of the pallidum also containing dendrites of dopaminergic neurons). The globus pallidus projects to the entopeduncular nucleus (or internal segment of the globus pallidus). The entopeduncular nucleus and substantia nigra pars reticulata project, in turn, to thalamic nuclei, which send a return projection to cortex, completing a major cortico–subcortical loop. Descending projections from the substantia nigra pars reticulata to the deep superior colliculus and pedunculopontine nucleus ultimately lead to descending projections to motoneurons of the brainstem and spinal cord. These so-called cortico–striato–pallidal output projections are composed of numerous parallel and convergent channels that contain topographically as well as nontopographically organized systems that program motor behavior (for review, see Fallon and Loughlin, 1987). One pathway that appears to facilitate focused throughput in this system is the ascending dopaminergic projection from the substantia nigra to the caudate putamen, that is, the nigrostriatal pathway. The cardinal pathologic finding in Parkinson's disease is degeneration of this pathway. In Huntington's disease, on the other hand, the cardinal pathologic finding is the degeneration of striatal neurons that have descending projections to the globus pallidus and substantia nigra. One major thrust in neurotrophic factor research is determining therapeutic strategies to reverse these progressive pathologic changes and to develop appropriate transplantation techniques to replace the degenerated neural systems. Understanding the distribution of neurotrophic factors in these brain regions is one of the critical first steps in developing therapeutic strategies for these extrapyramidal disorders.

### 2. Striatum

Examination of the distributions of neurotrophic factors in the striatal and pallidal/nigral regions reveals some interesting patterns (Fig. 5). First, the caudate

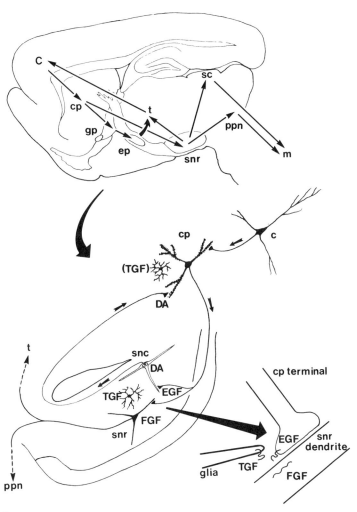

**FIGURE 4**

Some basic connections of the extrapyramidal motor systems. *Top,* Parasagittal section of the rat brain showing cortico-subcortical loops, with projections from cortex (c) to caudate putamen (cp, or dorsal striatum), then to globus pallidus (gp), substantia nigra (snr), and entopeduncular nucleus (ep). ep and snr project to thalamus (t), which returns a projection to cortex; snr also projects to the superior colliculus (sc) and pedunculopontine nucleus (ppn), which project to motoneuronal regions (m) of the brainstem and spinal cord. *Bottom,* Expanded view of some of these connections in the substantia nigra. The dopamine neurons (DA) of the substantia nigra pars compacta (snc) are shown projecting to and receiving projections from the cp. GABAergic neurons of the substantia nigra pars reticulata (snr) are depicted containing aFGF (FGF). cp terminals of striato-nigral axons contain EGF and local glia contain TGFα precursor (TGF). *Bottom right,* Magnified view of these neurotrophic factors in the snr. See text for discussion.

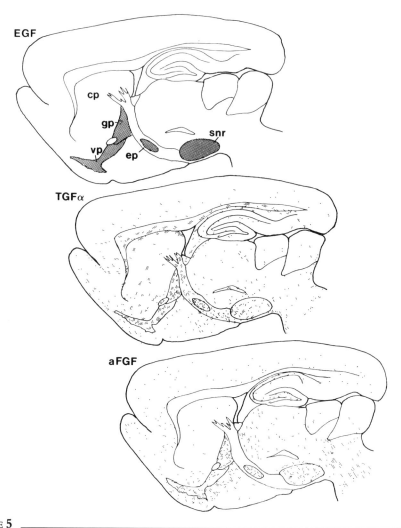

**FIGURE 5**

Parasagittal sections of brain illustrating localization of EGF in axons in pallidal brain regions (*top*), TGFα precursor in a subpopulation of astrocytes throughout the brain, but especially in pallidal regions (*middle*), and aFGF in neuronal cell bodies throughout the brain, again concentrated in pallidal regions (*bottom*).

putamen, which contains nigral dopaminergic terminals that degenerate in Parkinson's disease and striatal neurons that degenerate in Huntington's disease, does not contain an impressive array of known neurotrophic factors (Fig. 5). NGF, NGF mRNA, and NGF low molecular weight receptor mRNA are found in the striatum (Shelton and Reichardt, 1986; Mobley et al., 1989; Senut et al., 1990), which may serve to act neurotrophically for local cholinergic interneurons, but these levels are also quite low. There is a report of the presence of a striatal-derived neuronotrophic factor (Dal Toso et al., 1988) and IGF-I (Noguchi et al., 1987) in the striatum, but the combined levels of known neurotrophic factors appear to be rather low in the striatum, especially in comparison with other extrapyramidal brain areas. Further, it is not clear how noncholinergic striatal neurons, which constitute over 95% of striatal neurons, are influenced by neurotrophic factors. For example, anterograde transport to, or retrograde axonal transport of neurotrophic messengers from, other extrapyramidal structures may provide access to neurotrophic factors. In contrast, a dramatic increase of TGFα-precursor immunoreactivity in astrocytes in the transplant site is seen following fetal midbrain suspension transplants into an adult dopamine-denervated striatum (Loughlin et al., 1989). Intrastriatal infusions of TGFα, which decrease lesion-induced behavioral deficits, also increase TGFα, (S. E. Loughlin and J. H. Fallon, unpublished observations). These findings suggest that TGFα, which binds to the EGF receptor, may be an important neurotrophic factor in Parkinson's disease therapy.

A potentially important manifestation of neurotrophic factor heterogeneity in the striatum relates to the progression of neurodegeneration in Huntington's disease. This disease presents gradients of degeneration of striatal neurons, that is, there is an early loss of striatal enkephalin projections to the external globus pallidus and a later loss of striatal substance P projections to the internal globus pallidus, with relative sparing of enkephalin projections to the substantia nigra pars compacta (Reiner et al., 1988). These findings suggest differences in the susceptibility of matrix and patch neurons of the striatum, perhaps because of differences in neurotrophic factor availability as well as intrinsic lethal cellular programs.

### 3. Pallidum/Substantia Nigra

The pallidum consists of the globus pallidus, ventral pallidum, entopeduncular nucleus (internal segment of globus pallidus in primates), and substantia nigra pars reticulata. These structures are separated in the brain by incursions of the internal capsule, cerebral peduncle, and medial forebrain bundle, but they have similar cell types and some common histochemical features. One interesting feature of these pallidal structures is the unique distribution of EGF-like immunoreactivity (Figs. 5 and 6) (Fallon et al., 1984; Alheid and Heimer, 1988) that is present in striatal neuronal processes and terminals (J. H. Fallon and S. E. Loughlin, unpublished observations) in these regions (Fig. 6). Note the highly restrictive distribution of EGF and the complex geometry and physical separation of the pallidal structures in which EGF is localized. Such a distribution makes it difficult for nonneuroanatomical techniques (such as whole brain assays or incomplete dissections) to detect EGF in the brain. In addition to EGF in axons and terminals, neuronal cell bodies in these pallidal structures contain aFGF immunoreactivity (Figs. 5 and 7a) and numerous astrocytes contain TFGα-precursor immunoreactivity (Figs. 5 and 7b) (Fallon et al., 1990,1988). IGF-I has also been reported to be localized in the globus pallidus (Noguchi et al., 1987). Thus, the pallidal regions appear, at present, to be particularly enriched in a multitude of neurotrophic factors.

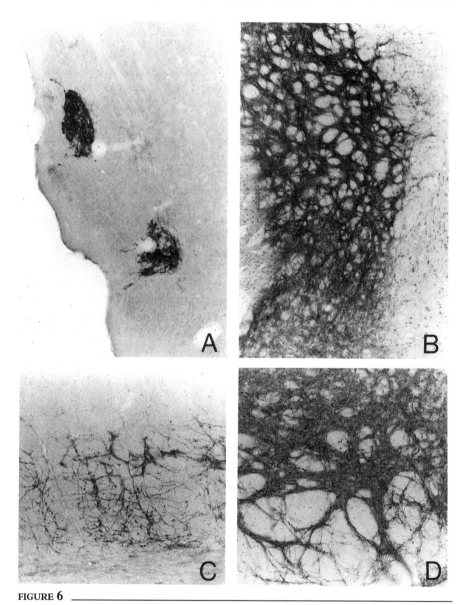

**FIGURE 6**

EGF-like immunoreactivity (DAB–immunoperoxidase technique) in pallidal regions of the brain. (A) Islands of Calleja sector of ventral pallidum. (B) Globus pallidus and downward extension into the ventral pallidum. (C) Entopeduncular nucleus. (D) Substantia nigra pars reticulata. (Photo micrographs courtesy of J. Fallon and K. Seroogy.)

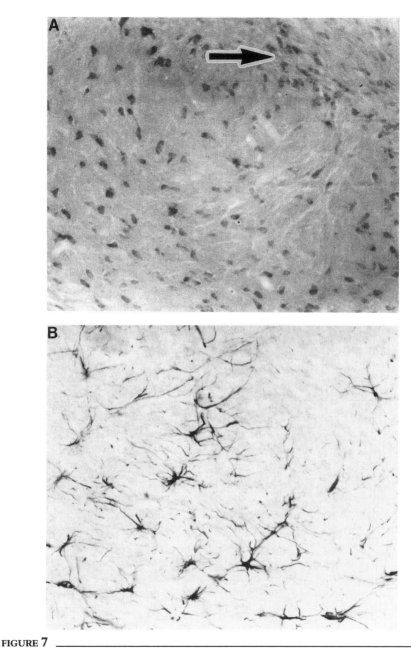

**FIGURE 7**

Neurotrophic factor immunoreactivities (DAB–immunoperoxidase technique) in the substantia nigra pars reticulata. (A) aFGF in neuronal cell bodies in pars reticulata and in a portion of the pars compacta (arrow). (B) TGFα precursor in astrocytes.

One pallidal structure of particular interest in disorders of the extrapyramidal motor systems in the substantia nigra, which is composed of the pallidal-like pars reticulata and the overlying pars compacta, which contains dopaminergic neurons projecting to the forebrain and dendrites that invade the pars reticulata (Fig. 4). What is the neurotrophic milieu of these cells? As discussed earlier, and as illustrated in Fig. 4, EGF-like immunoreactivity is present in striatal fiber and terminals known to project to dopaminergic dendrites and GABAergic neurons in the pars reticulata. Many of the large GABAergic neurons in this area contain aFGF. TGF$\alpha$-precursor is present in a subpopulation of astrocytes in the pars reticulata. In vitro studies have shown that EGF and bFGF stimulate cell proliferation and dopamine uptake in nigral cultures, but this effect depends on the presence of glia (Knusel et al., 1990). The morphological differentiation of embryonic dopaminergic neurons also depends on the presence of astrocytes (Lieth et al., 1989). The compartmentalization of neurotrophic factors in different neuronal and glial elements in the substantia nigra (Fig. 5) and the receptor binding characteristics of EGF and TGF$\alpha$ to the EGF receptor (Chapters 11 and 12), and aFGF and bFGF to FGF receptors (Chapters 9 and 10) suggest a number of plausible interactions in the substantia nigra. One working hypothesis is that a "sequential" neuronal–glial interaction (Fig. 2) mediates neurotrophic activity after damage. For example, FGFs (without a consensus secretory signal) are released from nigral neurons on injury. Increases in FGF activity have been reported following lesions in other regions of the central nervous system (Nieto-Sampedro et al., 1988b); bFGF has been shown to be retrogradely transported in the nigrostriatal pathway (Ferguson et al., 1990). In response to injury (and in aged animals), EGF-receptor immunoreactivity appears in reactive astrocytes (Nieto-Sampedro et al., 1988a). FGF receptors are present on neurons (Type 1) and glia (Type 2) (Chapter 10) and FGF binding is known to inhibit EGF binding in other tissues (Chapter 9). Thus, the neurotrophic effects of FGF may be mediated through regulation of astrocytic TGF$\alpha$ synthesis, release, and binding to local EGF (and, perhaps, independent TGF$\alpha$) receptors on glia and neurons. By inhibiting further EGF binding, FGF may regulate continued neurotrophic effects using autocrine, paracrine, and juxtacrine mechanisms.

Dopaminergic neurons also may be regulated tonically by long-distance axonal transport mechanisms. These mechanisms include anterograde and retrograde transport of neurotrophic factors in any of a number of structures interconnecting with the substantia nigra, for example, anterograde transport of EGF from the striatum. Other factors could be transported from the striatum by anterograde or retrograde mechanisms. A partially characterized (15-kDa) striatal-derived neuronotrophic factor has been reported to be neurotrophic for dopaminergic neurons (Dal Toso et al., 1988). BDNF also has neurotrophic effects on these neurons (Zigmond and Stricker, 1989; Knusel et al., 1990; Hyman et al., 1991) but its source in vivo is still in question, since BDNF mRNA is not detectable in striatum (Maisonpierre et al., 1990). Opioid peptides also are known to possess neurotrophic activity (Chapter 20) and the substantia nigra receives dense dynorphin and enkephalin inputs from the striatum (see Fallon and Ciofi, 1990, for review).

In conclusion, fiber pathways, local neurons, and glia of the substantia nigra contain a mosaic of resident neurotrophic factors. The anatomical localization of these factors provides a substrate for complex neurotrophic activity in the intact and damaged substantia nigra.

## VI. SUMMARY

The central nervous system possesses the molecular and cellular machinery to synthesize, release, and respond to a wide variety of heterogeneously distributed neurotrophic factors. The unique morphology of neural tissue provides a substrate for long-distance transport of neurotrophic factors, as well as for local interactions between neurons, glia, the vasculature, and cerebrospinal fluid. A regional analysis of the central nervous system reveals that one brain area, for example, the hippocampus, may contain members of several families of neurotrophic factors. The distribution of the neurotrophins (NGF, BDNF, NT-3), the fibroblast growth factors (aFGF, bFGF, and others), IGF-1, and TGFα reveals that both distinct and overlapping neurotrophic activity may occur for local neurons and glia, as well as for input and output systems. A systems analysis of the central nervous system reveals that the extrapyramidal motor system contains system-specific neurotrophic factors such as pallidal EGF, as well as a rich assortment of other neurotrophic factors that are present also in other systems. Different levels of the extrapyramidal motor system have been found to contain a spectrum of known neurotrophic factors. For example, the pallidum is enriched in the neurotrophic factors studied to date, whereas the striatum is relatively lacking in such factors. An analysis of neurotrophic factors in the substantia nigra reveals a differential distribution of EGF, aFGF, and TGFα in axons, cell bodies, and astrocytes. These factors may subserve a sequential type of neurotrophic activity linking neurons and glia in the intact and damaged extrapyramidal motor system.

A major function of the nervous system is to process and store information. Many of the changes associated with such activity are similar to those that occur in ontogeny. Phenomena such as long-term potentiation, learning, and adaptation are associated with measurable changes in metabolic activity, ion fluxes, receptor/transmitter/enzyme expression, synapse formation, spine growth, and glial proliferation. Thus, a continuum of processes can be mediated by many of the same messengers, including neurotransmitters and neurotrophic factors.

## ACKNOWLEDGMENTS

This work was supported by NIH grant NS 15321 to J. H. Fallon and NIH grant NS 26761 to S. E. Loughlin, as well as by a Bud Corbin Neuroscience Award and grants from the APDA SCC. We wish to thank Pippa Jones for typing the manuscript.

## REFERENCES

Alheid, G. F., and Heimer, L. (1988). New perspectives in basal forebrain organization of special relevance for neuropsychiatric disorders: The striatopallidal, amygdaloid, and corticopetal components of substantia innominata. *Neurosci.* **27**, 1–39.

Anden, N. -E., Fuxe, K., Hamberger, B., and Hokfelt, T. (1966). A quantitative study on the nigro-neostriatal dopamine neuron system in the rat. *Acta Physiol. Scand.* **67**, 306–312.

Ankelesaria, M., Teixido, J., Laiho, M., Pierce, J. H., Greenberger, J. S., and Massague, J. (1990). Cell–cell adhesion mediated by binding of membrane-anchored transforming growth factor alpha to epidermal growth factor receptors promotes cell proliferation. *Proc. Natl. Acad. Sci. U.S.A.* **87**, 3289–3293.

Aoki, C., Milner, T. A., Sheu, K-F. R., Blass, J. P., and Pickel, V. M. (1987). Regional distribution of astrocytes with intense immunoreactivity for glutamate dehydrogenase in rat brain: Implications for neuron–glia interactions in glutamate transmission. *J. Neurosci.* **7**, 2214–2231.

Banting, F. G., and Best, C. H. (1922). The internal secretion of the pancreas. *J. Lab. Clin. Med.* **7**, 251–251.

Burgess, A. W. (1989). The origins of growth factors: Tethelin lost. *Trends Biochem. Sci.*

Butcher, L. L., and Woolf, N. J. (1989). Neurotrophic agents may exacerbate the pathologic cascade of Alzheimer's disease. *Neurobiol. Aging* **10**, 557–570.

Christmas, P., Callaway, J., Fallon, J., Jones, J., and Haigler, H. T. (1991). Selective secretion of annexin 1, a protein without a signal sequence, by the human prostate gland. *J. Biol. Chem.* **266**, 2499–2507.

Clevenger, C. V., Altman, S. W., and Prystowsky, M. B. (1991). Requirement of nuclear prolactin for interleukin-2-stimulated proliferation of T lymphocytes. *Science* **253**, 77–79.

Cohen, S. (1962). Isolation of a mouse submaxillary protein accelerating incisor eruption and eyelid opening in the newborn animal. *J. Biol. Chem.* **237**, 1555–1562.

Dal Toso, R., Giorgi, O., Soranzo, C., Kirschner, G., Ferrari, G., Favaron, M., Benvegnu, D., Presti, D., Vicini, S., Toffano, G., Azzone, G. F., and Leon, A. (1988). Development and survival of neurons in dissociated fetal mesencephalic serum-free cell cultures. I. Effects of cell density and of an adult mammalian striatal-derived neuronotrophic factor (SDNF). *J. Neurosci.* **8**, 733–745.

De Felipe, J., and Jones, E. G. (1991). "Cajal's Degeneration and Regeneration of the Nervous System." Oxford University Press, New York.

Ernfors, P., Wetmore, C., Olson, L., and Persson, H. (1990). Identification of cells in rat brain and peripheral tissues expressing mRNA for members of the nerve growth factor family. *Neuron* **5**, 511–526.

Fallon, J. H., and Ciofi, P. (1990). Dynorphin-containing neurons. In "Handbook of Chemical Neuroanatomy" (A. Bjorklund, T. Hokfelt, and M. J. Kuhar, eds.), Vol. 9, pp. 1–130. Elsevier, Amsterdam.

Fallon, J. H., and Loughlin, S. E. (1982). Monoamine innervation of the forebrain: Forebrain collateralization. *Brain Res. Bull.* **9**, 295–307.

Fallon, J. H., and Loughlin, S. E., Annis, C. E. (1988). Growth factors in the CNS. *Int. Narc. Res. C.* (Abs.)

Fallon, J. H., and Loughlin, S. E. (1987). Monamine innervation of cerebral cortex and a theory of the role of monoamines in cerebral cortex and basal ganglia. In "Cerebral Cortex" (E. G. Jones and A. Peters, eds.), Vol. 8, pp. 41–127. Plenum, New York.

Fallon, J. H., Seroogy, K. B., Loughlin, S. E., Morrison, R. S., Bradshaw, R. A., Knauer, D. J., and Cunningham, D. J. (1984). Epidermal growth factor immunoreactive material in the central nervous system: Location and development. *Science* **224**, 1107–1109.

Fallon, J. H., Annis, C. M., Gentry, L. E., Twardzik, D. R., and Loughlin, S. E. (1990). Localization of cells containing transforming growth factor alpha precursor immunoreactivity in the basal ganglia of the adult rat brain. *Growth Factor* **2**, 241–250.

Fallon, J. H., DiSalvo, J., Loughlin, S. E., Gimenez-Gallego, G., Seroogy, K., Bradshaw, R. A., Morrison, R. S., Ciofi, P., and Thomas, K. A. (1991). Localization of acidic fibroblast growth factor within the mouse brain using biochemical and immunocytochemical techniques. *Growth Factors,* in press.

Ferguson, I. A., Schweitzer, J. B., and Johnson, E. M., Jr. (1990). Basic fibroblast growth factor: Receptor-mediated internalization, metabolism, and anterograde axonal transport in retinal ganglion cells. *J. Neurosci.* **10,** 2176–2189.

Gospodarowicz, D. (1974). Localization of a fibroblast factor and its effect alone and with hydrocortisone on 3T3 cell growth. *Nature (London)* **249,** 123–127.

Hamburger, V. (1934). The effects of wing bud extirpation on the development of the central nervous system in chick embryos. *J. Exp. Zool.* **68,** 449–494.

Hofer, M., Pagliushi, S. R., Hohn, A., Leibrock, J., and Barde, Y-A. (1990). Regional distribution of brain-derived neurotrophic factor mRNA in the adult mouse brain. *EMBO J.* **9,** 2459–2464.

Hoffman, R. S. (1940). The growth activating effect of extracts of adult and embryonic tissues of the rat on fibroblast colonies in culture. *Growth* **4,** 361–376.

Hyman, C., Hofer, M., Barde, Y-A., Juhasz, M., Yancopoulos, G. D., Squinto, S. P., and Lindsay, R. M. (1991). BDNF is a neurotrophic factor for dopaminergic neurons of the substantia nigra. *Nature (London)* **350,** 230–232.

Janet, T., Miehe, M., Pettman, B., Labourdette, G., and Sensenbrenner, M. (1987). Ultrastructural localization of fibroblast growth factor in neurons of rat brain. *Neurosci. Lett.* **80,** 153–157.

Knusel, B., and Hefti, F. (1988). Development of cholinergic pedunculopontine neurons in vitro: Comparison with cholinergic septal cells and response to nerve growth factor, ciliary neuronotrophic factor, and retinoic acid. *J. Neurosci. Res.* **21,** 365–375.

Knusel, B., Michel, P. P., Schwaber, J. S., and Hefti, F. (1990). Selective and nonselective stimulation of central cholinergic and dopaminergic in vitro by nerve growth factor, basic fibroblast growth factor, epidermal growth factor, insulin, and the insulin-like growth factors I and II. *J. Neurosci.* **10,** 558–570.

Lauterborn, J. C., Isackson, P. J., and Gall, C. M. (1991). Nerve growth factor mRNA-containing cells are distributed within regions of cholinergic neurons in the rat basal forebrain. *J. Comp. Neurol.* **306,** 439–446.

Levi-Montalcini, R., and Hamburger, V. (1953). A diffusable agent of mouse sarcoma, producing hyperplasia of sympathetic ganglia and hyperneurotization of viscera in the chick embryo. *J. Exp. Zool.* **123,** 233–287.

Lieth, E., Towle, A. C., and Lauder, J. M. (1989). Neuronal–glial interactions: Quantitation of astrocyte influences on development of catecholamine neurons. *Neurochem. Res.* **14,** 979–985.

Lorente de Nó, R. (1934). Studies on the structure of the cerebral cortex. II. Continuation of the study of the ammonic system. *J. Psychol. Neurol.* **46,** 113–177.

Loughlin, S. E., Annis, C. M., Twardzik, D. R., Lee, D. C., Gentry, L., and Fallon, J. H. (1989). Growth factors in opioid rich brain regions: Distribution and response to intrastriatal transplants. *Adv. Biosci.* **75,** 403–406.

Maisonpierre, P. C., Belluscio, L., Friedman, B., Alderson, R. F., Wiegand, S. J., Furth, M. E., Lindsay, R. M., and Yancopoulos, G. D. (1990). NT-3, BDNF, and NGF in the developing rat nervous system: Parallel as well as reciprocal patterns of expression. *Neuron* **5,** 501–509.

Mobley, W. C., Woo, J. E., Edwards, R. H., Riopelle, R. J., Longo, F. M., Weskamp, G., Otten, U., Valletta, J. S., and Johnston, M. V. (1989). Developmental regulation of nerve growth factor and its receptor in the rat caudate–putamen. *Neuron* **3,** 655–664.

Nieto-Sampedro, M., Gomez-Pincilla, F., Knauer, D. J., and Broderick, J. T. (1988a). Epidermal growth factor receptor immunoreactivity in rat brain astrocytes. Response to injury. *Neurosci. Lett.* **91,** 276–282.

Nieto-Sampedro, M., Lim, R., Hicklin, D. J., and Cotman, C. W. (1988b). Early release of glia maturation factor and acidic fibroblast growth factor after rat brain injury. *Neurosci. Lett.* **86,** 361–365.

Noguchi, T., Kurata, L. M., and Sugisaki, T. (1987). Presence of a somatomedin-C-immunoreactive substance in the central nervous system: Immunohistochemical mapping studies. *J. Neuroencrinol.* **46**, 277–282.

Pettmann, B., Labourdette, G., Weibel, M., and Sensenbrenner, M. (1986). The brain fibroblast growth factor (FGF) is localized in neurons. *Neurosci. Lett.* **68**, 175–180.

Phillips, H. S., Hains, J. M., Laramee, G. R., Rosenthal, A., and Winslow, J. W. (1990). Widespread expression of BDNF but not NT3 by target areas of basal forebrain cholinergic neurons. *Science* **250**, 290–294.

Ramón y Cajal, S. (1882). La retine des vertebres. *La Cellule* **9**, 121–246.

Ramón y Cajal, S. (1891). "Histologie du Systeme Nerveux de L'Homme et des Vertebres," Vol. I.

Rapraeger, A. C., Krufka, A., and Olwin, B. B. (1991). Requirement of heparan sulfate for bFGF-mediated fibroblast growth and myoblast differentiation. *Science* **252**, 1705–1708.

Reiner, A., Albin, R. L., Anderson, K. D., D'Amato, C. J., Penney, J. B., and Young, A. B. (1988). Differential loss of striatal projection neurons in Huntington's disease. *Proc. Natl. Acad. Sci. U.S.A.* **85**, 5733–5737.

Senut, M-C., Lamour, Y., Lee, J., Brachet, P., and Dicou, E. (1990). Neuronal localization of the nerve growth factor precursor-like immunoreactivity in the rat brain. *Int. J. Dev. Neurosci.* **8**, 65–80.

Shelton, D. L., and Reichardt, L. F. (1986). Studies on the expression of the beta nerve growth factor (NGF) gene in the central nervous system: Level and regional distribution of NGF mRNA suggest that NGF functions as a trophic factor for several distinct populations of neurons. *Proc. Natl. Acad. Sci. U.S.A.* **83**, 2714–2718.

Sloviter, R. S., Sollas, A. L., Barbaro, N. M., and Laxer, K. D. (1991). Calcium-binding protein (calbindin-D28k) and parvalbumin immunocytochemistry in the normal and epileptic human hippocampus. *J. Comp. Neurol.* **308**, 381–396.

Trowell, O. A., Chir, B., and Willmer, E. N. (1939). Growth of tissues *in vitro*. IV. The effects of some tissue extracts on the growth of periosteal fibroblasts. *J. Exp. Biol.* **16**, 60–70.

Wetmore, C., Ernfors, P., Persson, H., and Olson, L. (1990). Localization of brain-derived neurotrophic factor mRNA to neurons in the brain by *in situ* hybridization. *Exp. Neurol.* **109**, 141–152.

Whitehouse, P. J., Price, D. L., Clark, A. W., Coyle, J. D., and DeLong, M. R. (1981). Alzheimer's disease: Evidence for selective loss of cholinergic neurones from the nucleus basalis. *Ann. Neurol.* **10**, 122–126.

Wilcock, G. K., Esiri, M. M., Bowen, D. M., and Smith, C. (1983). The nucleus basalis in Alzheimer's disease: Cell counts and cortical biochemistry. *Neuropathol. Appl. Neurobiol.* **9**, 175–179.

Wilcox, B. J., and Unnerstall, J. R. (1991). Expression of acidic fibroblast growth factor mRNA in the developing and adult rat brain. *Neuron* **6**, 397–409.

Wong, S. T., Winchell, L. F., McCune, B. K., Earp, H. S., Teixido, J., Massague, J., Herman, B., and Lee, D. C. (1989). The TGF-alpha precursor expressed on the cell surface bids to the EGF receptor on adjacent cells, leading to signal transduction. *Cell* **56**, 495–506.

Zigmond, M. J., and Stricker, E. M. (1989). Animal models of Parkinson's using selective neurotoxins: Clinical and basic implications. *Int. Rev. Neurobiol.* **31**, 1–79.

# Neurotrophic Factors: What Are They and What Are They Doing?

Franz Hefti, Timothy L. Denton, Beat Knusel, and Paul A. Lapchak

## I. HISTORY

Two seminal discoveries underlie the conception of the field of neurotrophic factor study: developmental neuronal death and nerve growth factor (NGF). The concept that survival of neurons during development is influenced by signals provided by their target tissues was proposed and firmly established by the now classical studies of Viktor Hamburger (Hamburger, 1934). Since Viktor Hamburger was a student of Hans Speeman, in whose laboratory the induction process responsible for neural organization of amphibian embryos was discovered, the roots of the field of neurotrophic factor study can be traced to 19th century European reductionistic embryology, which created the concept of "developmental mechanics" and represents at least part of the conceptual origin of modern molecular biology. The discovery of NGF has been narrated many times (e.g., Levi-Montalcini, 1975). It was based on an incidental observation by Bruecker that mouse tumor tissue implanted in chick embryos strongly stimulated innervation of the tumor by host sensory neurons (Bruecker, 1948). An important advance was made by Rita Levi-Montalcini, who showed that a soluble factor was responsible for the stimulation of neuronal growth (Levi-Montalcini and Hamburger, 1953). The serendipitous observation, made by Seymour Cohen, that a crude nuclease preparation made from snake venom was a more concentrated source of the factor than the original tumor made possible the isolation of the factor, which was characterized as a protein and named "nerve growth factor" (Cohen et al., 1954).

The three decades that followed the discovery of NGF saw the establishment of its crucial role in the development of sensory and sympathetic neurons of the peripheral nervous system (for review, see Thoenen and Barde, 1980). In the late

1970s, circumstantial evidence was obtained for the existence of other molecules with actions on neurons comparable to those of NGF, and the terms "neurotrophic" and "neuronotrophic factors" appeared frequently in the literature. The gene encoding NGF was cloned and characterized (Scott et al., 1983). Initial findings were published that showed that NGF affects the development of central nervous system (CNS) neurons, in particular of cholinergic neurons of the basal forebrain (Schwab et al., 1979; Honegger and Lenoir, 1982; Gnahn et al., 1983). The role of NGF in the CNS then was expanded to the adult brain by the demonstration that NGF prevents degenerative changes of forebrain cholinergic cells after experimental injury; the associated speculation that NGF mechanisms play a role in neurodegenerative diseases was made (Hefti, 1983; Hefti et al., 1984). During recent years, other neurotrophic factors have been isolated and characterized, including brain-derived neurotrophic factor (BDNF; Barde et al., 1982) and ciliary neurotrophic factor (CNTF; Barbin et al., 1984). The key discovery by Barde and Thoenen that BDNF is structurally related to NGF led to the discovery of neurotrophin-3 (NT-3; Ernfors et al., 1990; Hohn et al., 1990; Jones and Reichardt, 1990; Maisonpierre et al., 1990; Rosenthal et al., 1990) and neurotrophin-4 and -5 (NT-4, NT-5; Hallbook et al., 1991; Berkemeier et al., 1991) establishing the existence of a family of NGF-related molecules now called "neurotrophins" (Hohn et al., 1990). Also during the past decade, several previously known growth factors were found to regulate neuronal survival in vitro and, in consequence, are considered neurotrophic factors. This group of molecules includes the fibroblast growth factors (FGFs; Morrison et al., 1986; Walicke et al., 1986); epidermal growth factor (EGF; Morrison et al., 1987), and the insulin-like growth factors (IGFs; Recio-Pinto et al., 1986). Given these findings, the field of neurotrophic factors has attracted many new investigators and now represents a major area of brain research. Concepts and tools are available for the exploration of the role of these factors in neuronal development, including adult function and aging, as well as their role in pathophysiology and their possible practical use in developing pharmacologic strategies to treat pathologic conditions.

## II. DEFINITION

There is no generally accepted definition for the term "neurotrophic factor." A narrow definition proposed by Barde (1989) restricts the use of the term to the historical meaning: proteins that are involved in the regulation of survival of neurons during development. In contrast, the program of the annual meeting of the Society for Neurochemistry includes a vast number of molecules under the heading "trophic agents." The restrictive definition of neurotrophic factors as developmental survival-promoting agents has been a powerful hypothesis for the guidance of research on neuronal development. Still, even NGF, the best studied neurotrophic factor, has been shown only in the peripheral nervous system to act clearly as a survival-promoting factor, and seems to be involved in nonneuronal functions as well. Therefore, the presently available body of data suggests a more inclusive definition that would include CNTF, EGF, FGF, the IGFs, and other molecules for which an action as target-derived factor has not been demonstrated conclusively and which are discussed in the chapters of this volume. For practical reasons, to avoid an all-inclusive definition, we think a definition of trophic factors should exclude membrane-bound molecules that affect adhesion and survival of neurons as well as nonprotein compounds such as thyroid hormones, steroids, and gangliosides. Therefore, we suggest the following definition. *Neurotrophic factors are endogenous soluble proteins regulating survival, growth, morphological plasticity, or synthesis of proteins for differentiated functions of neurons.* This definition combines aspects of a functional and physical nature and is compatible with the view that growth factors are multifunctional (Sporn and Roberts, 1988). Table 1 shows a list of characterized proteins that fulfill this proposed definition of neurotrophic factors. Our definition

TABLE 1

**List of characterized proteins exhibiting neurotrophic activities**

| Growth factor | References[a] |
|---|---|
| Proteins initially characterized as neurotrophic factors | |
| Nerve growth factor (NGF) | Thoenen et al., 1987 |
| | Whittemore and Seiger, 1987 |
| | Hefti et al., 1989 |
| Brain-derived neurotrophic factor (BDNF) | Barde et al., 1982 |
| | Leibrock et al., 1989 |
| Neurotrophin-3 (NT-3) | Ernfors et al., 1990 |
| | Hohn et al., 1990 |
| | Maisonpierre et al., 1990 |
| | Rosenthal et al., 1990 |
| Neurotrophin-4 (NT-4) | Hallbrook et al., 1991 |
| Neurotrophin-5 (NT-5) | Berkemeier et al., 1991 |
| Ciliary neurotrophic factor (CNTF) | Lin et al., 1989 |
| | Stöckli et al., 1989 |
| Heparin-binding neurotrophic factor (HBNF) | Kovesdi et al., 1990 |
| Growth factors with neurotrophic activity | |
| Basic fibroblast growth factor (bFGF) | Morrison et al., 1986 |
| | Walicke, 1988 |
| Acidic fibroblast growth factor (aFGF) | Walicke, 1988 |
| Insulin-like growth factors (IGFs), insulin | Aizenman et al., 1986 |
| | Baskin et al., 1987 |
| Epidermal growth factor (EGF) | Fallon et al., 1984 |
| | Morrison et al., 1987 |
| Transforming growth factor α (TGFα) | Derynck, 1988 |
| | Fallon et al., 1990 |
| Interleukin 1 | Spranger et al., 1990 |
| Interleukin 3 | Kamegai, 1990 |
| Interleukin 6 | Hama et al., 1989 |
| Protease nexin I and II | Monard, 1987 |
| | Oltersdorf et al., 1989 |
| | Whitson et al., 1989 |
| Cholinergic neuronal differentiation factor | Yamamori et al., 1989 |

[a]References given refer to recent reviews or recent key publications.

and others reflect current concepts and may help organize subfields of research. The value of such definitions, however, is limited, since the terminology will be dictated ultimately by findings produced by the reductionistic molecular analysis of physiologic phenomena.

## III. PHYSIOLOGICAL FUNCTIONS OF NEUROTROPHIC FACTORS

This volume contains detailed summaries of our knowledge of most of the presently known neurotrophic factors, and demonstrates the rapid progress made in recent years and the large number of new findings. Since the field is in a phase of rapid expansion, it seems premature to attempt to reconcile all existing data into a single coherent body of knowledge. Therefore, rather than attempt to present a global summary and concept of neurotrophic factor function, this chapter attempts to identify key issues to be resolved by future research. Unavoidably, the presentation is biased by our own scientific interests and areas of expertise. Often, the discussion is limited to NGF, given the large amount of data available for this prototypical molecule. When possible, key publications or reviews are cited in support of this overview; the reader is referred to the other chapters for more complete citations.

### A. Synthesis and Release

Control of neuronal survival and growth by neurotrophic factors requires synthesis and release by producer cells and distribution to receptors on responsive neurons. Many key questions regarding mechanisms and regulation of these processes remain unanswered. The neurotrophins (NGF, BDNF, NT-3, NT-4, NT-5) and EGF have been shown to be synthesized as larger precursor molecules with signal peptides typical of secretory proteins (Carpenter and Cohen, 1990). Precursors of NGF and EGF are processed by specific proteases that can remain associated with the proteins that exhibit neurotrophic activity and, with other proteins, form high molecular weight complexes (Jongstra-Bilen et al., 1989). Such complexes exist for NGF in the salivary glands and the glands of the reproductive system (Harper and Thoenen, 1980). The high concentrations of the 7S complex that occurs in salivary glands allowed Cohen and collaborators to purify the active βNGF molecule. For many years, the 7S complex isolated from mouse salivary glands represented the only significant source for the preparation of βNGF. At this time, complex forms of NGF have not been demonstrated in typical peripheral target tissues of NGF-responsive cells, nor in the brain, and it remains to be seen whether such complexes represent a common feature of neurotrophic factors. In contrast, other neurotrophic factors, including CNTF and IL-1, are synthesized as full length proteins without signal sequences for release, like typical cytoplasmic proteins (Le and Vilcek, 1987; Dinarello, 1988; Lin et al., 1989; Stöckli et al., 1989). bFGF has been demonstrated to be released from stimulated fibrosarcoma cells by an unconventional mechanism (Kaudel et al., 1991). Physiological release mechanisms for some other proteins with neurotrophic proteins remain to be established. It seems possible that cytoplasmic proteins with neurotrophic actions are released only after injury and cell death to become functionally important as wound healing factors, in the same manner that platelet-derived growth factor is released from platelets.

Initial findings on the regulation of neurotrophic factor synthesis and release have been published during recent years. The best studied case is the regulation of NGF synthesis in the sciatic nerve after axonal injury. Injury of this nerve is followed by infiltration of macrophages that produce IL-1. IL-1 in turn stimulates the production of NGF by Schwann cells. There is a rapid elevation of both NGF mRNA and NGF levels (Lindholm et al., 1987). In contrast to this situation in the peripheral nervous system, NGF synthesis in the septo-hippocampal system is not affected by the transection of axons of NGF-responsive cells (Korsching et al., 1986). However, IL-1, which stimulates NGF synthesis in cultured astrocytes, elevates hippocampal NGF mRNA when administered to the brain (Spranger et al., 1990). Interestingly, there is a high degree of co-localization of IL-1 and NGF mRNA in the adult brain, consistent with the possibility that IL-1 is involved in the regulation of NGF synthesis (Spranger et al., 1990). The mechanism by which IL-1 stimulates NGF synthesis in the sciatic nerve remains to be established. It has been reported that, in cultured sciatic nerve fibroblasts, the elevation of NGF mRNA induced by IL-1 is caused by both increased transcription and stabilization of this mRNA (Lindholm et al., 1988).

An understanding of the regulatory elements of the genes encoding neurotrophic factors represents an important part of the mechanistic analysis of the regulation of their synthesis. In the human, NGF mRNA is synthesized from a single-copy gene located on chromosome 1 (Francke et al., 1983). The regulatory elements for transcription of the mouse NGF gene seem to be located in a 5-kb fragment located upstream from the transcriptional start site, since fusion of this sequence with a growth hormone reporter gene and insertion into transgenic mice resulted in expression of the reporter gene in areas normally expressing NGF (Alexander et al., 1989). The detailed analysis of trophic factor genes and their regulatory elements is likely to provide key elements for the understanding of the regulation of synthesis and release of neurotrophic factors.

The recent molecular genetic analysis of trophic factors has resulted in the production of recombinant proteins in sufficient quantities for detailed analysis of their biologic and pharmacologic actions. The practical implications of the availability of such proteins cannot be overestimated, since the limited availability of purified factors, and associated problems with purity and stability, often have limited investigative progress severely. bFGF, the IGFs, and a number of the ILs have become available during the past few years. In contrast, sufficient quantities of recombinant NGF were produced only recently (Barnett et al., 1990; Knusel et al., 1990a). Also, it must be emphasized that recombinant proteins must be characterized in detail before they are used in biologic studies, since the consequences of minor changes in sequence on the biologic activities are poorly understood. For example, human recombinant NGF is approximately three times more potent that purified murine NGF on cultured rat cholinergic neurons or chick sensory neurons (Knusel et al., 1990a).

A key question to be addressed for all neurotrophic factors is: What is the anatomical level at which the synthesis and release of neurotrophic factors is regulated? At the grossest level, regulating developmental neuronal survival by limiting the amounts of a neurotrophic factor, it may be sufficient that neurotrophic factor release by target cells occurs constitutively and is uniform for all cells of a target tissue, resulting in a homogeneous distribution throughout this tissue. There is reason to believe that individual target cells could regulate their neurotrophic factor release independently. In transgenic mice expressing an insulin promotor–NGF fusion gene, NGF production from the transgene by individual pancreatic β cells and their individual sympathetic innervation was found to be heterogeneous, suggesting that the NGF released was acting in the immediate vicinity of NGF-producing cells (Edwards et al., 1989). Finally, it is conceivable that release of neurotrophic factors could be regulated at a subcellular level and could vary among different parts of a producer cell. Such local release of neurotrophic factors could be involved in synaptic rearrangements during functional plasticity and adaptation to injury. Although not based on experimental evidence, the speculation of local stimulus-regulated neurotrophic factor release seems not unreasonable, since mechanisms for the release of neurotransmitters are localized at special sites of the cell surface. The question of diffusion of neurotrophic factors is part of the question of anatomical level of regulation of synthesis and release. Rate and distance of diffusion of neurotrophic factors is not clear. Local regulation of release at the subcellular level would be made impossible by diffusion of trophic molecules beyond the immediate vicinity of the site of release. Alternatively, diffusion over several millimeters would be compatible with regulation of neurotrophic factor release at the level of a peripheral organ or an entire brain area. It seems interesting to note that similar critical questions remain unanswered for the diffusion of some neurotransmitters. It is unclear whether neurotransmitter molecules can diffuse in sufficient quantities to stimulate receptors outside the area of the synapse at which they are released.

## B. Receptors and Transducing Mechanisms

Receptors for different growth factors have been shown to contain intracellular domains with the activity or structure of protein kinases; the regulation of their activities is believed to be a primary event in the actions of the stimulated receptor. It was shown for the EGF receptor that binding of EGF to the extracellular ligand-binding domain activates a cytoplasmic domain with tyrosine kinase function that itself is phosphorylated on activation (Carpenter and Cohen, 1990). The receptors for insulin and IGF-I also have been shown to carry tyrosine-specific protein kinases in their cytoplasmic domains that share significant homology

(Ullrich et al., 1986). Further, there is evidence from immunoblot studies with antiphosphotyrosine antibodies that bFGF also stimulates tyrosine phosphorylation (Coughlin et al., 1988), two putative mouse embryonic neuronal receptors for bFGF have been found that contain a sequence for a protein tyrosine kinase (Reid et al., 1990).

The very recent time has seen rapid progress in the understanding of the transduction mechanism of NGF and the more recently discovered members of the neurotrophin family. At least two kinds of proteins are involved in the formation of functional receptors for neurotrophins. These are the low affinity NGF receptor protein (p75NGFR; Chao et al., 1986; Radeke et al., 1987) and products of *trk*-related proto-oncogenes (Hempstead et al., 1991; Kaplan et al., 1991; Squinto et al, 1990,1991; Soppet et al., 1991; Cordon-Cardo et al., 1991; Lamballe et al., 1991). The *trk* gene products, but not the low affinity NGF receptor protein, exhibit protein kinase activity (Martin-Zanca et al., 1989). Studies on various cells expressing *trk*-type transgenes suggest that individual *trk* receptors bind to and stimulate tyrosine phosphorylation of different subsets of neurotrophins. *Trk* binds to NGF but not BDNF, *trk*B binds BDNF but not NGF (Cordon-Cardo et al., 1991; Klein et al., 1991). NT-3 is capable of interacting with *trk, trk*B, and with *trk*C (Cordon-Cardo et al., 1991; Lamballe et al., 1991), and these multiple interactions may allow NT-3 to act on populations of neurons expressing different *trk* gene products. NT-5 activates *trk* and *trk*B (Berkemeier et al., 1991). Interestingly, it seems possible that the low affinity NGF receptor protein is also part of the BDNF receptor (Rodriguez-Tebar et al., 1990), which suggests that accessory proteins may mediate the specificity of the high affinity receptors for both molecules. This hypothesis is, however, not supported by the mismatch between the widespread distribution of BDNF (Phillips et al., 1990) and the very limited distribution of low affinity NGF receptor protein (Koh et al., 1989a) in the adult rat brain. These findings suggest the existence of low affinity NGF receptor homologs that await discovery.

Progress in the study of the mechanisms of action, as well as the functions, of neurotrophic factors could be accelerated by the availability of specific inhibitors. Recently, alkaloid-like compounds of microbial origin, K-252a and K-252b, have been shown to inhibit the actions of NGF selectively in several responsive cellular systems. These compounds are very potent inhibitors of different protein kinases but do not interfere with the binding of NGF to its receptor (Koizumi et al., 1988; Nakanishi et al., 1988). K-252a and K-252b reversibly and selectively inhibit the NGF-induced morphologic transformation of proliferating PC12 pheochromocytoma cells into neuron-like cells; they also inhibit the NGF-stimulated, but not the bFGF- or EGF-stimulated, phosphorylation of selected proteins (Hashimoto, 1988; Koizumi et al., 1988). K-252a interferes with NGF actions on nerve cells of the mammalian peripheral nervous system in primary cultures (Matsuda et al., 1988); NGF actions on forebrain cholinergic neurons were inhibited by K-252a and K-252b (Knusel and Hefti, 1991). K-252b exhibits very low toxicity and could become a useful tool to inhibit NGF actions in vivo. In general, specific inhibitors for the various neurotrophic proteins would allow us to identify their actions and to differentiate among the effects produced by the various factors, both during development and in the adult brain.

The synthesis and disposition of receptors represents another level of potential regulation of neurotrophic factor mechanisms. Interestingly, the expression of NGF receptor seems to be up-regulated by its own ligand. In various culture systems, and in vivo, addition of NGF to NGF-responsive cells elevates levels of NGF receptors (Higgins et al., 1989; Lindsay et al., 1990). Similarly, EGF stimulates the expression of its own receptor in an epidermoid carcinoma cell line (Clark et al., 1985). Such regulation by positive feedback is contrary to the widely accepted concept, based on findings obtained in neurotransmitter systems, that there is a negative correlation between ligand release and receptor density. If the positive correlation between neurotrophic factor release and receptor synthesis is found to be a typical feature, this positive regulation of receptors must be attributed a key functional role. This mechanism may assure the rapid fixation of newly established specific synaptic connections between afferent and target cell during development or in adult plasticity.

## C. Cellular Actions of Neurotrophic Factors

Neurotrophic factors stimulate a multitude of cellular effects at the structural and functional level that result in the promotion of survival and differentiation of responsive neurons. Such actions include stimulation of the synthesis of structural proteins, stimulation of general metabolic processes, and stimulation of mechanisms necessary for neurotransmitter synthesis and release. These multiple actions of trophic factors are probably best characterized in pheochromocytoma PC12 cells that, when exposed to NGF or bFGF, cease to proliferate and, over the course of several days, develop properties reminiscent of sympathetic neurons. They extend neurites, become electrically excitable, and can form synapses with muscle cells in culture (Greene and Tischler, 1976). Early events induced within minutes by NGF include hydrolysis of phosphoinositides, changes in phosphorylation of proteins, increase in nutrient uptake, changes in surface morphology, and induction of early genes. Late events include the induction of several neuron-specific genes, including transmitter-related enzymes and proteins involved in the formation of ion channels (see Pollock et al., 1990, for references). Despite the multitude of actions described, no consensus has emerged on how the different observations relate to each other and to the biologic effects of NGF and bFGF. Also, it is not clear whether the known elements represent a small or large part of the elements forming the entire mechanism, that ultimately remain to be discovered.

The cellular actions of trophic factors on primary nerve cells are characterized very poorly. bFGF promotes survival and neurite outgrowth of a variety of neurons in culture and stimulates transmitter-specific enzymes in brain cholinergic and dopaminergic neurons (Walicke et al., 1986; Knusel et al., 1990b). NGF supports the survival and neurite outgrowth of sympathetic neurons and induces the expression of the transmitter-specific enzymes tyrosine hydroxylase and dopamine β-hydroxylase, of Tα1 α-tubulin, and of SCG10, a membrane-bound protein abundant in developing neurons (see Thoenen and Barde, 1980, for review; Stein et al., 1988; Mathew and Miller, 1990). In cultured cholinergic neurons of the basal forebrain, NGF stimulates survival and neurite growth, induces the expression of choline acetyltransferase at the transcriptional level, and stimulates activities of acetylcholinesterase and the high-affinity choline uptake system (Hartikka and Hefti, 1988a,b; Lorenzi et al., 1992). Levels of various neuropeptides are elevated by NGF in sensory neurons (Otten and Lorez, 1983). BDNF supports the survival of subpopulations of sensory neurons and retinal ganglion cells, and stimulates the activity of transmitter-specific functions in central cholinergic and dopaminergic neurons (Johnson et al., 1986; Alderson et al., 1990; Knusel et al., 1990b). IL-3 and IL-6 have been reported to stimulate choline acetyltransferase activity in forebrain cholinergic neurons (Hama et al., 1989; Kamegai, 1990). When compared with the list of the many known actions of NGF on PC12 cells, these findings illustrate the very rudimentary nature of the knowledge of neurotrophic factor actions on brain cells at the cellular level. In the next few years, research is likely to bring a detailed description of the molecular and structural changes induced by neurotrophic factors in responsive populations of neurons. A detailed understanding of their cellular actions is necessary to develop hypotheses about the functional role of neurotrophic factors during development and in the adult brain, as is attempted in the next sections of this chapter.

## D. Are Neurotrophic Factors Identical to Target-Derived Retrograde Messengers?

In the peripheral sympathetic nervous system, NGF is synthesized in target tissues; levels of both NGF mRNA and NGF itself correlate with the density of sympathetic innervation (Heumann et al., 1984; Shelton and Reichardt, 1984). Sympathetic neurons synthesize NGF receptors and the receptor protein is transported anterogradely to the terminals (Johnson et al., 1987). As discussed earlier, the transducing mechanism mediating the actions of NGF remains unclear. However, it is clear that NGF is internalized and transported retrogradely with its receptor (Hendry et al., 1974; Johnson et al., 1987). The retrograde transport is argued to be functionally important, because interference with axonal transport by axotomy or local administration of microtubule-disrupting drugs results in cell death, and the effects of systemic administration of such drugs and of axotomy can be reversed by administration of NGF (Thoenen and Barde, 1980, for review). These findings suggest that NGF, or, alternatively, a signal other than NGF that is generated after binding of NGF to receptors on terminals, is retrogradely transported to the cell bodies where it elicits specific changes in gene expression. These findings form the basis of the general concept that neurotrophic factors produced and released by target tissue control survival and growth of innervating neurons. The concept is supported by more recent findings on the role of NGF in the development and function of the cholinergic septo-hippocampal system (see Thoenen et al., 1987;

Whittemore and Seiger, 1987; Hefti et al., 1989, for reviews). There, NGF is selectively synthesized by hippocampal target cells, and NGF receptors are expressed and anterogradely transported by cholinergic afferent neurons. As in the peripheral nervous system, after binding, NGF and NGF receptors are retrogradely transported by the cholinergic cells; these processes result in the stimulation of cellular survival and the functioning of cholinergic neurons.

The concept of target-derived retrogradely acting neurotrophic factors has been a strong guiding force in the search for novel factors during the past decade. Many investigators have attempted to purify factors from target tissue by using cultured afferent neurons as an in vitro assay system. Surprisingly, this approach does not seem to have been as successful as anticipated. Only CNTF was purified from the target tissue of parasympathetic ciliary cells (Barbin et al., 1984). BDNF was purified from pig brain based on its ability to promote survival of peripheral sensory neurons of chick (Barde et al., 1982). Many other attempts to purify target-derived growth factors from brain tissue failed or resulted in the characterization of known protein growth factors. The reason for the lack of success may be the inherent difficulty in purifying low-abundance proteins or, alternatively, the very limited number of existing factors.

Although other molecules with neurotrophic actions, including EGF, bFGF, and the IGFs, have been known for many years, retrograde transport of these proteins by neurons has not been demonstrated convincingly yet. A recent study designed to establish retrograde transport for bFGF showed that bFGF was transported anterogradely by retinal ganglion cells (Ferguson et al., 1990). The lack of retrograde transport does not preclude a role as a target-derived messenger, since a secondary signal generated by the receptor-transducing mechanism at the terminals could be transported to the cell bodies (see Chapter 3). Alternatively, bFGF and other neurotrophic molecules may represent local signals. The classical neurotrophic factor, NGF, seems to have a functional role in several nonneuronal systems, in addition to its role as target-derived signal in the nervous system. The most abundant sources of NGF are tissues not directly associated with NGF-responsive neuronal populations and include salivary glands and glands of the male reproductive system (see Harper and Thoenen, 1980, for review). In addition, NGF has been reported to act on mast cells and granulocytes, suggesting a role in immunoregulation (Aloe, 1988; Matsuda et al., 1988). Also, mRNAs encoding BDNF and, in particular, NT-3 have been detected in several peripheral tissues, suggesting nonneuronal roles of these proteins (Hohn et al., 1990; Maisonpierre et al., 1990; Rosenthal et al., 1990). These findings support the concepts that (1) growth factors are multifunctional; (2) the role as target-derived retrograde messenger is one among several functions of proteins with neurotrophic actions, and (3) this role may not be shared by all growth factors involved in development and differentiation of neurons. Characterization of the temporal and spatial aspects of synthesis and release of neurotrophic molecules in the brain and their correlation with the expression of functional receptors represents a key task for the future. Such studies will result in the identification of specific roles of the factors during development and in adulthood.

## E. Are Neurotrophic Factors Regulators of Developmental Neuronal Survival?

A role for NGF and BDNF as regulators of neuronal survival during development has been well established for responsive populations of peripheral neurons. For NGF, this role is demonstrated most dramatically by the administration of antibodies against NGF to neonatal mice or rats, resulting in an irreversible degeneration of sympathetic ganglia. Conversely, the treatment of newborn mice or rats with exogenous NGF prevents the naturally occurring cell death of sympathetic neurons in development. Survival of peripheral sensory neurons is not affected by injections of NGF or anti-NGF in neonatal rats. However, transplacental transfer of antibodies from pregnant rats immunized against NGF, or direct injection of the embryos in utero with anti-NGF, results in drastic numeric reduction of neurons in sensory ganglia (Johnson et al., 1980). Similarly, injections of BDNF into quail embryos prevents the normally occurring neuronal death in sensory ganglia (Hofer and Barde, 1988). Also, injections of BDNF into embryonic chicks, in which migrated cells from the neural crest are experimentally separated from the crest, attenuate the death of sensory neurons normally induced by the lesion (Kalcheim et al., 1987).

The situation is less clear in the brain. Despite the demonstration that NGF promotes survival of forebrain cholinergic neurons in vitro, a role as survival factor has not been demonstrated in vivo. No published reports suggest developmental death of part of the population of cholinergic neurons. Infusions of antibodies against NGF reduce the amount of choline acetyltransferase protein (Vantini et al., 1989). However, because of the intrinsic complexity of the brain, these findings do not substantiate clearly that the number of surviving cholinergic neurons is reduced. Developmental neuronal death does not seem to be a general feature in the mammalian brain and is difficult to establish because, in contrast to the peripheral nervous system, populations of neurons do not exist as isolated ganglia but typically are intermingled with other neurons (McBride et al., 1990). Further, small "shrunken" neurons surviving after periods of cell death cannot easily be distinguished from nonneuronal cells. The lack of positive demonstration of developmental neuronal death in the brain suggests the possibility that the physiologic function of neurotrophic factors during brain development may be different from their function in the development of peripheral sympathetic neurons and, further, that regulation of developmental death is one of several different functions of neurotrophic factors.

The most frequently stated hypothesis explaining the regulation of developmental neuronal death predicts that competition of individual nerve cells for limited amounts of trophic factors results in the selection of those neurons receiving the trophic factor in sufficient quantities (Barde, 1989). Accordingly, existence in limited amounts can be seen as a prerequisite for a protein to be considered a neurotrophic factor. However, tissue concentrations of any active molecule do not provide, necessarily, an accurate indication of biologically effective concentrations. As discussed earlier, there is virtually no information on diffusion rates and distances for neurotrophic proteins, parameters that are likely to affect their biologic effectiveness. For example, most measurements of tissue concentrations do not distinguish between intracellular and extracellular pools. It also must be emphasized that alternative hypotheses for the regulation of developmental neuronal death do not require necessarily that neurotrophic factors occur in limiting amounts. It has been suggested that growing neurons compete for restricted sites on targets that provide access to trophic factors and that the synthesis of trophic factor does not represent a limiting factor (Oppenheim, 1989). The situation is further complicated by recent findings of Harper and Davies (1990), whose detailed analysis of neuronal death in the mouse trigeminal ganglion suggests that regional differences in NGF synthesis in target tissues maintain differences in cell number initially present in subpopulations within these ganglia, rather than select cells from a uniform original population. The occurrence and mechanisms of neurotrophic factor-regulated cell death are likely to differ among different populations of neurons. These arguments stress the necessity of obtaining a detailed analysis of the interactions of individual neurotrophic factors with specific neuronal populations.

## F. Are Neurotrophic Factors Developmental Differentiation Factors?

Many of the actions of trophic factors on brain neurons are compatible with a role as differentiation-inducing factors, rather than as regulators of survival. NGF enhances neurite extension, and tyrosine hydroxylase expression, of developing sympathetic neurons independent of survival, since differentiation is promoted under experimental conditions in which the number of sympathetic neurons remains unchanged (Hefti et al., 1982). Similarly, NGF can stimulate the expression of choline acetyltransferase by cholinergic neurons in the absence of survival-promoting action (Hartikka and Hefti, 1988b). Effects of most of the other neurotrophic proteins listed in Table 1 are not clear because of the scarcity of data. However, our own findings suggest that bFGF and EGF also can stimulate biochemical differentiation of responsive neurons in the absence of a regulatory action on survival, since these factors seem to increase tyrosine hydroxylase expression by cultured dopaminergic neurons without affecting their survival. The following discussion distinguishes morphologic and biochemical differentiation. Although these processes may be related and regulated by the same mechanisms, it is possible that their regulation is independent and mediated by different trophic molecules.

Despite the existence of a considerable body of data obtained on morphologic differentiation in cell cultures, the exact role of neurotrophic factors during morphologic differentiation in vivo is understood very poorly. Cell culture studies typically show a general stimulation of neurite outgrowth, as shown with NGF and forebrain cholinergic neurons in cultures of dissociated cells, explant slice cultures, and cultures of reaggregating cells (Hartikka and Hefti, 1988a; Hatanaka et al., 1988; Hsiang et al., 1989; Gahwiler et al., 1990). Since, in vivo, NGF is available in the target tissue, but not in the area of the cell bodies or dendrites, the biologic meaning of the cell culture findings remains unclear. Earlier findings suggested a role of trophic factors as chemotactic molecules. NGF released by target cells was proposed to act as a chemoattractant for sympathetic and sensory axons, based on cell culture studies showing that local NGF administration can direct neurite growth to the NGF source (Gunderson and Barrett, 1979). However, no comparable findings have been published since that suggest a similar role for other neurotrophic factors. Moreover, Harper et al. (1990) showed that, in the chick embryo, developing sensory neurons lack NGF receptors at the stage when their fibers are growing toward their targets and, further, that NGF synthesis in the target field starts only at the time of arrival of these fibers. These findings seem to exclude a role for NGF of chemotactic factor in the establishment of chick sensory projections.

An alternative hypothesis for the developmental function of neurotrophic factors is derived from earlier work by Campenot, who showed that local administration of NGF to cultured sympathetic neurons selectively stimulates growth of those neurites that are exposed to NGF (Campenot, 1982a,b, 1987). These findings suggest the possibility that the local availability of NGF stimulates growth of axons near putative target cells. Such stimulation could be a prerequisite for the establishment of specific synaptic connections between trophic factor-synthesizing and -responsive cells. The temporal expression of NGF and NGF receptors in the cholinergic septo-hippocampal system can be considered support for this speculation. NGF expression in the hippocampal target tissue is maximal during the time of maximum expansion of the cholinergic input in this area (Large et al., 1986). Simultaneously, NGF receptor expression by cholinergic input neurons is maximal (Koh and Loy, 1989). It is interesting to compare this proposed mechanism of trophic factors as local regulators of axonal growth with the original trophic factor hypothesis. Although, according to the evidence we have cited, neurotrophic factors do not act as chemotactic factors, the local stimulation of axonal growth will also produce a similar effect of maximal growth at the sites of highest concentration of NGF.

Most of the neurotrophic molecules listed in Table 1 have been shown to regulate transmitter-related biochemical differentiation of responsive neurons. NGF, BDNF, bFGF, IGF-I, IL-3, and IL-6 all elevate choline acetyltransferase (ChAT) activity in forebrain cholinergic neurons (Hama et al., 1989; Kamegai et al., 1990; Knusel et al., 1990b, 1991). EGF elevates dopamine uptake by mesencephalic dopaminergic neurons (Knusel et al., 1990b). CNTF and the cholinergic differentiation factor (CNDF) induce the expression of cholinergic traits in normally adrenergic sympathetic neurons (Yamamori et al., 1989). Like the stimulation of morphologic differentiation, the mechanisms mediating the biochemical differentiation are understood very poorly. It seems that the stimulatory action of NGF on forebrain cholinergic neurons is regulated at the transcriptional level, since addition of NGF to cultured cholinergic neurons increases the abundance of a species of ChAT mRNA. The recent cloning of the ChAT gene may reveal the regulatory elements involved in the mediation of this action of NGF (Lorenzi et al., 1992). In the case of cholinergic neurons, it is anticipated that trophic factors will also induce the expression of other proteins involved in transmitter synthesis and release, including yet uncharacterized proteins that mediate the specific high affinity choline uptake that is believed to be the rate-limiting step in acetylcholine synthesis. The molecular characterization of such effects remains a major goal in the characterization of the biologic actions of all neurotrophic factors.

## G. Selectivity of Neurotrophic Factors and Temporal Windows of Responsiveness during Development

The established developmental roles for neurotrophic factors include the regulation of cell death and of specific mechanisms during morphologic growth and biochemical differentiation. Further, most of the neurotrophic factors listed in Table 1, including NGF (Cattaneo and McKay, 1990), also seem stimulate the proliferation of specific cell populations (see Roher, 1990, for review; Cattaneo and McKay, 1990). Given this multitude of different actions of neurotrophic molecules, the questions arises of how specificity of individual actions is achieved.

Is there a selective neurotrophic factor for every major population of neurons? The relatively selective effect of NGF on peripheral sympathetic and forebrain cholinergic neurons suggests the existence of other factors with a similar degree of selectivity for other populations of neurons. The discovery of BDNF, NT-3, NT-4 and NT-5, suggests the existence of yet additional members of this family and supports the view that there is a large number of neurotrophic factors that selectively affect their target populations. The discovery of several novel members of the FGF family of proteins (Hebert et al., 1990) lends additional support to this concept.

In conflict with the hypothesis that there is a large number of neurotrophic factors with a high degree of specificity, FGF and the IGFs seem to have a much lower degree of selectivity and affect a larger number of neuronal populations than does NGF. Further, single populations of neurons can respond to more than one neurotrophic factor, as illustrated by the response of forebrain cholinergic neurons to NGF, BDNF, and bFGF (Alderson et al., 1990; Knusel et al., 1991). Based on these findings, it can be speculated that the number of neurotrophic factors is relatively small and that the exact temporal and spatial regulation of neurotrophic factor release and receptor synthesis and disposition is sufficient to determine the selectivity of their actions. This speculation implies the existence of specific temporal windows of responsiveness of a given neuronal population to various trophic factors. Support for this notion is derived from findings with peripheral sensory neurons and central cholinergic neurons that respond at different times to NGF and BDNF (Alderson et al., 1990; Knusel et al., 1991). Presently available findings do not permit us to distinguish between the two possibilities of many selective neurotrophic factors or a few neurotrophic factors acting on selectively responsive neurons. In summary, the findings obtained to date suggest that particular neurotrophic factors may behave very differently and that both highly selective and nonselective neurotrophic factors exist.

## H. Function of Neurotrophic Factors in the Adult Nervous System

That many neurotrophic factors are synthesized in the adult brain suggests that these molecules are functionally important in the adult brain as well. However, the precise nature of this functional role is not clear. The fact that the cell bodies undergo degenerative changes and, at least some of them, die after transection of their axonal connection with their target areas led to the generally held belief that target-derived neurotrophic factors are necessary for survival of responsive cells. In particular, this view is supported by the finding that axonal transection of septo-hippocampal cholinergic neurons results in atrophic changes and death of cholinergic cell bodies (Hefti, 1986). However, recent findings by Sofroniew et al. (1990) do not support this view, since complete destruction of the hippocampal cell bodies without transection of the cholinergic axons failed to result in death of most of the septal cholinergic cell bodies, although many of them apparently atrophied. It cannot be ruled out that remaining glial cells in hippocampal tissue, or other cells in the vicinity of cholinergic cell bodies begin to synthesize NGF or other growth factors to compensate for the hippocampal tissue.

Alternative speculations about the function of neurotrophic factors in the adult organism can be derived from earlier studies of NGF and adult sympathetic neurons. Adult sympathetic neurons are less vulnerable to NGF deprivation than are developmental sympathetic neurons; single injections of antibodies to NGF result only in transitory degenerative changes (Goedert et al., 1978). In adult mice, injections of anti-NGF antibodies result in a transitory disappearance of sympathetic terminals in various tissues. The density of these terminals is enhanced by injections of NGF (Bjerre et al., 1975a,b). Based on these findings, it can be speculated that endogenous NGF may be involved in the regulation of synaptic rearrangements in the adult nervous system. This speculation seems particularly attractive in light of the fact that many neurotrophic factors are expressed in the adult hippocampus. This structure seems to be of key importance in memory processes (Squire and Davis, 1981), suggesting the possibility that hippocampal neurons have a high intrinsic plasticity that is regulated by neurotrophic factors. Endogenous neurotrophic factors may stimulate neurite growth in local areas with elevated synaptic activity. The report that seizures elevate hippocampal NGF and BDNF expression (Gall and Isackson, 1989; Zafra et al., 1990) and that, in cultures of sympathetic neurons, local NGF administration stimulates growth of those neurites exposed to NGF (Campenot 1982a,b, 1987) tend to support this concept. The study of the functional role of neurotrophic factors in the adult brain appears to be a particularly fruitful area of investigation that has the potential to lead to better understanding of the mechanisms involved in the functions of the brain, including memory and cognition.

## I. Neurotrophic Factors and Aging

Only very tentative evidence is available on the performance of neurotrophic factor-related mechanisms in the aged nervous system; this information is limited to NGF. No studies have been carried out so far to assess NGF-related functions in sympathetic and sensory systems during aging, nor is it known whether expression of NGF and its receptors is maintained at a constant level during the entire life-span, or whether there are age-related declines in some of these functions. In the brain, levels of both NGF and its mRNA were found to be diminished in the brain of aged rats (Lärkfors et al., 1987); there is a decrease in the intensity of NGF-receptor immunoreactivity in the basal forebrain of both aged rats and aged humans (Koh et al., 1989b; Hefti and Mash, 1989). Impaired individuals were selected, from a group of rats, based on poor performance in a swimming maze paradigm. In the forebrain of these impaired animals, the cell body size of cholinergic neurons was found to be smaller than in nonimpaired aged rat. Intraventricular infusion of NGF for 2 weeks was found to partially reverse both the morphologic atrophy of cholinergic neurons and the behavioral deficits (Fischer et al., 1989). These findings suggest the possibility that neurotrophic factor-related mechanisms are altered during the aging process and that limited availability of, or reduced sensitivity to, neurotrophic factors may be partly responsible for age-related declines in neuronal and behavioral functions. Neuronal degenerative changes are not uniform throughout the aged brain and many neuronal populations retain normal plasticity (McNeill, 1983). It can be speculated that those neurons that retain plasticity also retain their responsiveness to neurotrophic factors.

## IV. PHARMACOLOGICAL EXPLOITATION OF NEUROTROPHIC FACTORS

Currently used drugs and most pharmacologic compounds being developed to treat disorders of the nervous system influence mechanisms related to neuronal impulse flow and synaptic transmission. They do not affect the structural features of the central nervous system and there are no known clinically effective compounds that are able to promote regeneration, plasticity, and maintenance of the structural integrity of selected neuronal systems reliably. Elucidation of neurotrophic factor mechanisms may lead to a new structurally oriented neuropharmacology. NGF, because of its pronounced and rather selective trophic action on cholinergic neurons of the basal forebrain, is presently being considered as an experimental treatment in Alzheimer's disease (Hefti and Schneider, 1989) and may have other therapeutic applications (Snider and Johnson, 1989). NGF may become paradigmatic for therapeutic applications of other neurotrophic factors. Initial clinical studies with neurotrophic factors will likely be carried out by administration of the protein at the site of desired action. Applications involving the brain present the difficulty of circumventing the blood–brain barrier. Demonstration of efficacy by using infusion devices hopefully will stimulate research into alternative methods of administration, including slow-releasing intracerebral implants (Powell et al., 1990) and grafting of genetically modified cells that selectively secrete a desired neurotrophic factor (Rosenberg et al., 1988). Given the manifold systems affected by neurotrophic factors, one must direct awareness to the possibility that the pharmacologic use of growth factors and trophic agents may, in some cases, be detrimental to neuronal function, contrary to the beneficial role implied by the term "trophic." Such detrimental actions must be considered and ruled out before clinically using trophic molecules.

The search for specific agonists and antagonists of neurotrophic factors presents a major task for pharmacology. Some initial reports shed light on the possibility of producing active fragments of NGF or other neurotrophic factors. Longo et al. (1990) reported the production of peptide fragments that inhibit biologic actions of NGF. Fragments of EGF, bFGF, IGF-I, and IGF-II appear to retain activity (Komiriya et al., 1984; Ballard et al., 1987; Baird et al., 1988; Konishi et al., 1989). A synthetic peptide analog of TGFα (transforming growth factor) was shown to specifically inhibit EGF and TGFα growth stimulatory actions on a responsive cell line (Eppstein et al., 1989). Although effective concentrations of the active peptides were often much higher than those of the native proteins, these reports demonstrate the feasibility of an approach that attempts to reduce trophic factors to smaller molecules that exert the same actions on the specific receptors. Studies with peptide fragments and modified peptides will provide a basis for molecular modeling studies attempting to replace peptides with nonpeptide effectors.

The search for low molecular weight molecules that affect endogenous neurotrophic factor mechanisms represents another possible method of pharmacologically manipulating these systems. Various compounds stimulate the synthesis and secretion of NGF by cultured cells (Shinoda et al., 1990). Clovanemagnolol, a newly isolated neoligan, supports neuronal survival in culture (Fukuyama et al., 1990). Tricyano-aminopropene mimics NGF in supporting the survival of cultured sympathetic neurons, and also increases the rate of sciatic nerve regeneration (Paul et al., 1990). A recent report states that specific long-chain fatty alcohols support the survival of cortical neurons and, after systemic administration, protect septal cholinergic neurons from lesion-induced degeneration (Borg et al., 1990). NGF expression in the brain seems to be influenced by steroid hormones (Katoh-Semba et al., 1990), indicating another avenue for possible pharmacologic manipulation of central neurotrophic factor expression. Further, such compounds will become useful as tools in establishing neurotrophic factor mechanisms.

## V. CONCLUSIONS

The field of neurotrophic factor research has developed from studies of nerve growth factor and its involvement in the regulation of neuronal survival during development. The field now represents a rapidly growing area of research that is likely to provide significant insight into the mechanisms of nervous system development and function. During development, neurotrophic factors regulate morphologic and functional differentiation in addition to their role as regulators of neuronal survival. In the adult nervous system, neurotrophic factors appear to play a part in mechanisms of structural plasticity. Pharmacologic exploitation of neurotrophic factor mechanisms is likely to provide experimental tools to analyze mechanisms of development and plasticity and may yield therapeutically useful molecules.

**REFERENCES**

Alderson, R. F., Alterman, A. L., Barde, Y.-A., and Lindsay, R. M. (1990). Brain-derived neurotrophic factor increases survival and differentiates functions of rat septal cholinergic neurons in culture. *Neuron* **5,** 297–306.

Alexander, J. M., Hsu, D., Penchuk, L., and Heinrich, G. (1989). Cell-specific and developmental regulation of a nerve growth factor–human growth hormone fusion gene in transgenic mice. *Neuron* **3,** 133–139.

Aloe, L. (1988). The effect of nerve growth factor and its antibodies on mast cells in vivo. *J. Neuroimmunol.* **18,** 1–12.

Aizenman, Y., Weischsel, Jr. M. E., and De Vellis, J. (1986). Changes in insulin and transferrin requirements of pure brain neuronal cultures during embryonic development. *Proc. Natl. Acad. Sci. U.S.A.* **83,** 2263–2266.

Baird, A., Schubert, D., Ling, N., and Guillemin, R. (1988). Receptor- and heparin-binding domains of basic fibroblast growth factor. *Proc. Natl. Acad. Sci. U.S.A.* **85,** 2324–2328.

Ballard, F. J., Francis, G. L., Ross, M., Bagley, C. J., May, B., and Wallace, J. C. (1987). Natural and synthetic forms of insulin-like growth factor-1 (IGF-1) and the potent derivative, destripeptide IGF-1: Biological activities and receptor binding. *Biochem. Biophys. Res. Commun.* **149,** 398–404.

Barbin, G., Selak, I., Manthorpe, M., and Varon, S. (1984). Use of central neuronal cultures for the detection of neuronotrophic agents. *Neurosci.* **12,** 33–43.

Barde, Y. A. (1989). Trophic factors and neuronal survival. *Neuron* **2,** 1525–1534.

Barde, Y. A., Edgar, D., and Thoenen, H. (1982). Purification of a new neurotrophic factor from mammalian brain. *EMBO J.* **1,** 549–553.

Barnett, J., Baecker, P., Routledge-Ward, C., Bursztyn-Pettegrew, H., Chow, J., Nguyen, B., and Bach, C. (1990). Human β nerve growth factor obtained from a baculovirus expression system has potent in vitro and in vivo neurotrophic activity. *Exp. Neurol.* **110,** 11–24.

Baskin, D. G., Figlewicz, D. P., Woods, S. C., Porte, D., and Dorsa, D. M. (1987). Insulin in the brain. *Ann. Rev. Physiol.* **49,** 335–347.

Berkemeier, L. R., Winslow, J. W., Kaplan, D. R., Nikolics, K., Goeddel, D. V., and Rosenthal, A. (1991). Neurotrophin-5: A novel neurotrophic factor that activates trk and trkB. *Neuron,* **7,** 857–866.

Bjerre, B., Bjorklund, A., Mobley, W., and Rosengren, E. (1975a). Short- and long-term effects of nerve growth factor on the sympathetic nervous system in the adult mouse. *Brain Res.* **94,** 263–277.

Bjerre, B., Wiklund, L., and Edwards, D. C. (1975b). A study of the de- and regenerative changes in the sympathetic nervous system of the adult mouse after treatment with the antiserum to nerve growth factor. *Brain Res.* **92,** 257–278.

Borg, J., Kesslak, P. J., and Cotman, C. W. (1990). Peripheral administration of a long-chain fatty alcohol promotes septal cholinergic neurons survival after fimbria–fornix transection. *Brain Res.* **518,** 295–298.

Bruecker, E. D. (1948). Implantation of tumors in the hind limb field of the embryonic chick and the developmental response of the lumbosacral nervous system. *Anat. Rec.* **102,** 369–389.

Campenot, R. B. (1982a). Development of sympathetic neurons in compartmentalized cultures. I. Local control of neurite growth by nerve growth factor. *Dev. Biol.* **93,** 1–12.

Campenot, R. B. (1982b). Development of sympathetic neurons in compartmentalized cultures. II. Local control of neurite survival by nerve growth factor. *Dev. Biol.* **93,** 13–21.

Campenot, R. B. (1987). Local control of neurite sprouting in cultured sympathetic neurons by nerve growth factor. *Dev. Brain Res.* **37,**293–301.

Carpenter, G., and Cohen, S. (1990). Epidermal growth factor. *J. Biol. Chem.* **265,** 7709–7712.

Cattaneo, E., and McKay, R. (1990). Proliferation and differentiation of neuronal stem cells regulated by nerve growth factor. *Nature (London)* **347,** 762–765.

Chao, M. V., Bothwell, M. A., Ross, H. Koprowski, A. A., Lanahan, C. R., Buck, R. S., and Sehgal, A. (1986). Gene transfer and molecular cloning of the human NGF receptor. *Science* **232,** 518–521.

Clark, A. J. L., Ishii, S. Richert, N., Merlino, G. T., and Pastan, I. (1985). Epidermal growth factor regulates the expression of its own receptor. *Proc. Natl. Acad. Sci. U.S.A.* **82,** 8374–8378.

Cohen, S., Levi-Montalcini, R., and Hamburger, V. (1954). A nerve growth-stimulating factor isolated from snake venom. *Proc. Natl. Acad. Sci. U.S.A.* **40,** 1014–1018.

Cordon-Cardo, C., Tapley, P., Jing, S., Nandure, V., O'Rourke, E., Lamballe, F., Kovary, K., Klein, R., Jones, K. R., Reichardt, L. F., and Barbacid, M. (1991). The trk tyrosine protein kinase mediates the mitogenic properties of nerve growth factor and neurotrophin-3. *Cell* **66,** 173–183.

Coughlin, S. R., Barr, P. J., Cousens, L. S., Fretto, L. J., and Williams, L. T. (1988). Acidic and basic fibroblast growth factors stimulate tyrosine kinase activity in vivo. *J. Biol. Chem.,* **263,** 988–993.

Dinarello, S. A. (1988). Biology of interleukin 1. *FASEB J.* **2,** 108–115.

Derynck, R. (1988). Transforming growth factor alpha. *Cell* **54,** 593–595.

Edwards, R. H., Rutter, W. J., and Hanahan, D. (1989). Directed expression of NGF to pancreatic β cells in transgenic mice leads to selective hyperinnervation of the islets. *Cell* **58,** 161–170.

Eppstein, D. A., Marsh, Y. V., Schryver, B. B., and Bertics, P. J. (1989). Inhibition of epidermal growth factor/transforming growth factor-alpha-stimulated cell growth by a synthetic peptide. *J. Cell Physiol.* **141,** 420–430.

Ernfors, P., Wetmore, C., Olson, L., and Persson, H. (1990). Identification of cells in rat brain and peripheral tissues expressing mRNA for members of the nerve growth factor family. *Neuron* **5,** 511–526.

Fallon, J. H., Seroogy, K. B., Loughlin, S. E., Morrison, R. S., Bradshaw, R. A., Knauer, D. J. et al. (1984). Epidermal growth factor immunoreactive material in the central nervous system: Location and development. *Science* **224,** 1107–1109.

Fallon, J. H., Annis, C. M., Gentry, L. E., Twardzik, D. R., and Loughlin, S. E. (1990). Localization of cells containing transforming growth factor-alpha precursor immunoreactivity in the basal ganglia of the adult rat brain. *Growth Factors* **2,** 241–250.

Ferguson, I. A., Schweitzer, J. B., and Johnson, E. M., Jr. (1990). Basic fibroblast growth factor: Receptor-mediated internalization, metabolism, and anterograde axonal transport in retinal ganglion cells. *J. Neurosci.* **10,** 2176–2189.

Fischer, W., Gage, F. H., and Bjorklund, A. (1989). Degenerative changes in forebrain cholinergic nuclei correlate with cognitive impairments in aged rats. *J. Neurosci.* **1,** 33–45.

Francke, U. DeMartinville, B., Coussens, L., and Ullrich, A. (1983). The human gene for the beta-subunit of nerve growth factor is located on the proximal short arm of chromosome 1. *Science* **22,** 1248–1250.

Fukuyama, Y., Otoshi, Y., and Kodama, M. (1990). Structure of clovanemagnolol, a novel neurotrophic sesquiterpene-neolignan from *Magnolia obovata*. *Tetrahedron Lett.* **31,** 4477–1480.

Gahwiler, B. H., Rietschin, L., Knopfel, T., and Enz, A. (1990). Continuous presence of nerve growth factor is required for maintenance of cholinergic septal neurons in organotypic slice cultures. *Neuroscience* **36,** 27–31.

Gall, C. M., and Isackson, P. J. (1989). Limbic seizures increase neuronal production of messenger RNA for nerve growth factor. *Science* **245,** 758–761.

Gnahn, H., Hefti, F., Heumann, R., Schwab, M., and Thoenen, H. (1983). NGF-mediated increase of choline acetyltransferase (ChAT) in the neonatal forebrain: Evidence for a physiological role of NGF in the brain? *Dev. Brain Res.* **9,** 45–52.

Goedert, M., Otten, U., and Thoenen, H. (1978). Biochemical effects of antibodies against nerve growth factor on developing and differentiated sympathetic ganglia. *Brain Res.* **148,** 264–268.

Greene, L. A., and Tischler, A. (1976). Establishment of a noradrenergic clonal line of rat adrenal pheochromocytoma cells which respond to nerve growth factor. *Proc. Natl. Acad. Sci. U.S.A.* **73,** 2424–2428.

Gundersen, R. W., and Barrett, J. N. (1979). Neuronal chemotaxis: Chick dorsal root axons turn toward high concentrations of nerve growth factor. *Science* **206,** 1079–1080.

Hallbook, F., Ibanez, C. F., and Persson, H. (1991). Evolutionary studies of nerve growth factor family reveal a new member abundantly expressed in xenopus ovary. *Neuron* **6,** 845–858.

Hama, T., Miyamoto, M., Tsukui, H., Nishio, C., and Hatanaka, H. (1989). Interleukin-6 as a neurotrophic factor for promoting the survival of cultured basal forebrain cholinergic neurons from postnatal rats. *Neurosci. Lett.* **104,** 340–344.

Hamburger, V. (1934). The effects of wing bud extirpation on the development of the central nervous system in chick embryos. *J. Exp. Zool.* **68,** 449–494.

Harper, G. P., and Thoenen, H. (1980). Nerve growth factor: Biological significance, measurement, and distribution. *J. Neurochem.* **34,** 5–16.

Harper, S. and Davies, A. M. (1990). NGF mRNA expression in developing cutaneous epithelium related to innervation density. *Development* **110,** 515–519.

Hartikka, J., and Hefti, F. (1988a). Comparison of nerve growth factor's effects on development of septum, striatum, and nucleus basalis cholinergic neurons in vitro. *J. Neurosci. Res.* **21,** 352–364.

Hartikka, J., and Hefti, F. (1988b). Development of septal cholinergic neurons in culture: Plating density and glial cells modulate effects of NGF on survival, fiber growth, and expression of transmitter-specific enzymes. *J. Neurosci.* **8,** 2967–2985.

Hashimoto, S. (1988). K-252a, a potent protein kinase inhibitor, blocks nerve growth factor-induced neurite outgrowth and changes in the phosphorylation of proteins in PC12h cells. *J. Cell Biol.* **107,** 1531–1539.

Hashimoto, S., and Hagino, A. (1989). Blockage of nerve growth factor action in PC12h cells by staurosporine, a potent protein kinase inhibitor. *J. Neurochem.* **53,** 1675–1685.

Hatanaka, H., Tsukui, H., and Nihonmatsu, I. (1988). Developmental change in the nerve growth factor action from induction of choline acetyltransferase to promotion of cell survival in cultured basal forebrain cholinergic neurons from postnatal rats. *Dev. Brain Res.* **39,** 88–95.

Hebert, J. M., Basilico, C., Goldfarb, M., Haub, O., and Martin, G. R. (1990). Isolation of cDNAs encoding four mouse FGF family members and characterization of their expression patterns during embryogenesis. *Dev. Biol.* **138,** 454–463.

Hefti, F., Gnahn, H., Schwab, M. E., and Thoenen, H. (1982). Induction of tyrosine hydroxylase by nerve growth factor and by elevated K+ concentrations in cultures of dissociated sympathetic neurons. *J. Neurosci.* **2,** 1554–1566.

Hefti, F. (1983). Alzheimer's disease caused by a lack of nerve growth factor? *Ann. Neurol.* **13,** 109–110.

Hefti, F., David, A., and Hartikka, J. (1984). Chronic intraventricular injections of nerve growth factor elevate hippocampal choline acetyltransferase activity in adult rats with partial septo-hippocampal lesions. *Brain Res.* **293,** 311–317.

Hefti, F. (1986). Nerve growth factor (NGF) promotes survival of septal cholinergic neurons after fimbrial transections. *J. Neurosci.* **6,** 2155–2162.

Hefti, F., and Mash, D. C. (1989). Localization of nerve growth factor receptors in the normal human brain and in Alzheimer's disease. *Neurobiol. Aging* **10,** 75–87.

Hefti, F., and Schneider, L. S. (1989). Rationale for the planned clinical trials with nerve growth factor in Alzheimer's disease. *Psych. Devel.* **4,** 297–315.

Hefti, F., Hartikka, J., and Knusel, B. (1989). Function of neurotrophic factors in the adult and aging brain and their possible use in the treatment of neurodegenerative diseases. *Neurobiol. Aging* **10,** 515–533.

Hempstead, B. L., Martin-Zanca, D., Kaplan, D. R., Parada, L. F., and Chao, M. V. (1991). High-affinity

NGF binding requires coexpression of the trk proto-oncogene and the low-affinity NGF receptor. *Nature* **344**, 339–341.

Hendry, I. A., and Iversen, L. L. (1973). Reduction in the concentration of nerve growth factor in mice after sialectomy and castration. *Nature (London)* **243**, 500–504.

Hendry, I. A., Stockel, K., and Thoenen, H. (1974). The retrograde axonal transport of nerve growth factor. *Brain Res.* **68**, 103–121.

Heumann, T., Schwab, M., and Thoenen, H. (1981). A second messenger required for nerve growth factor biological activity? *Nature (London)* **292**, 838–840.

Heumann, R., Korsching, S., Scott, J., and Thoenen, H. (1984). Relationship between levels of nerve growth factor (NGF) and its messenger RNA in sympathetic ganglia and peripheral target tissues. *EMBO J.* **3**, 3183–3189.

Higgins, G. A., and Mufson, E. J. (1989). NGF receptor gene expression is decreased in the nucleus basalis in Alzheimer's disease. *Exp. Neurol.* **106**, 222–236.

Higgins, G. A., Koh, S., Chen, K. S., and Gage, F. H. (1989). Nerve growth factor (NGF) induction of NGF receptor gene expression and cholinergic neuronal hypertrophy within the basal forebrain of the adult rat. *Neuron* **3**, 247–256.

Hofer, M. M., and Barde, Y. A. (1988). Brain-derived neurotrophic factor prevents neuronal death in vivo. *Nature (London)* **331**, 261–262.

Hohn, A., Leibrock, J., Bailey, K., and Barde, Y. A. (1990). Identification and characterization of a novel member of the nerve growth factor/brain-derived neurotrophic factor family. *Nature (London)* **344**, 339–341.

Honegger, P., and Lenoir, D. (1982). Nerve growth factor (NGF) stimulation of cholinergic telencephalic neurons in aggregating cell cultures. *Dev. Brain Res.* **3**, 229–239.

Hsiang, J., Heller, A., Hoffmann, P. C., Mobley, W. C., and Wainer, B. H. (1989). The effects of nerve growth factor on the development of septal cholinergic neurons in reaggregate cell cultures. *Neurosci.* **29**, 209–223.

Johnson, E. M., Groin, P. D., Jr., Brandeis, L. D., and Pearson, J. (1980). Dorsal root ganglion neurons are destroyed by exposure in utero to maternal antibody to nerve growth factor. *Science* **210**, 916–918.

Johnson, E. M., Taniuchi, M., Clark, H. B., Springer, J. E., Koh, S., Tayrien, M., and Loy, R. (1987). Demonstration of the retrograde transport of nerve growth factor receptor in the peripheral and central nervous system. *J. Neurosci.* **7**, 923–929.

Johnson, J. E., Barde, Y. -A., Schwab, M., and Thoenen, H. (1986). Brain-derived neurotrophic factor supports the survival of cultured rat retinal ganglion cells. *J. Neurosci.* **6**, 3031–3038.

Jones, K. R., and Reichardt, L. F. (1990). Molecular cloning of a human gene that is a member of the nerve growth factor family. *Proc. Natl. Acad. Sci. U.S.A.* **87**, 8060–8064.

Jongstra-Bilen, J., Coblentz, L., and Shooter, E. M. (1989). The in vitro processing of the NGF precursors by the γ-subunit of the 7S NGF complex. *Mol. Brain Res.* **5**, 159–169.

Kalcheim, C., Barde, Y. A., Thoenen, H., and LeDouarin, N. M. (1987). In vivo effect of brain-derived neurotrophic factor on the survival of developing dorsal root ganglion cells. *EMBO J.* **6**, 2811–2813.

Kamegai, M., Niijima, K., Kunishita, T., Nishizawa, M., Ogawa, M., Araki, M., Ueki, A., Konishi, Y., and Tabira, T. (1990). Interleukin 3 as a trophic factor for central cholinergic neurons in vitro and in vivo. *Neuron* **2**, 429–436.

Kandel, J., Bossy-Wetzel, E., Radvanyi, F., Klagsburn, M., Folkman, J. and Hanahan, D. (1991). Neovascularization is associated with a switch to the export of bFGF in the multistep development of fibrosarcoma. *Cell* **66**, 1095–1104.

Kaplan, D. R., Hempstead, B. L., Martin-Zanca, D., Chao, M. V., and Parada, L. F. (1991). The trk proto-oncogene product: A signal transducing receptor for nerve growth factor. *Science* **525**, 554–558.

Kaplan, D. R., Martin-Zanca, D., and Parada, L. F. (1991). Tyrosine phosphorylation and tyrosine kinase activity of the trk proto-oncogene product induced by NGF. *Nature* **350**, 158–160.

Katoh-Semba, R., Semba, R., Kashiwamata, S., and Kato, K. (1990). Influences of neonatal and adult exposures to testosterone on the levels of the β-subunit of nerve growth factor in the neural tissues of mice. *Brain Res.* **522,** 112–117.

Klein, R., Nanduri, V., Jing, S., Lamballe, F., Tapley, P., Bryant, S., Cordon-Cardo, C., Jones, K. R., Reichardt, L. F., Barbacid, M. (1991). The trkB tyrosine protein kinase is a receptor for brain-derived neurotrophic factor and neurotrophin-3. *Cell* **66,** 395–404.

Knusel, B., and Hefti, F. (1991). K-252b is a selective and nontoxic inhibitor of nerve growth factor actions on cultured brain neurons. *J. Neurochem.* **57,** 955–962.

Knusel, B., Burton, L. E., Longo, F. M., Mobley, W. C., Koliatsos, V. E., Price, D. L., and Hefti, F. (1990a). Trophic actions of recombinant human nerve growth factor on cultured rat embryonic CNS cells. *Exp. Neurol.* **110,** 274–283.

Knusel, B., Michel, P. P., Schwaber, J. S., and Hefti, F. (1990b). Selective and nonselective stimulation of central cholingeric and dopaminergic development in vitro by nerve growth factor, basic fibroblast growth factor, epidermal growth factor, insulin, and the insulin-like growth factors I and II. *J. Neurosci.* **10,** 558–570.

Knusel, B., Winslow, J. W., Rosenthal, A., Burton, L. E., Seid, D. P., Nikolics, K., and Hefti, F. (1991). Promotion of central cholinergic and dopaminergic neuron differentiation by brain-derived neurotrophic factor but not neurotrophin-3. *Proc. Natl. Acad. Sci. U.S.A.* **88,** 961–965.

Koh, S., and Loy, R. (1989). Localization and development of nerve growth factor-sensitive rat basal forebrain neurons and their afferent projections to hippocampus and neocortex. *J. Neurosci.* **9,** 2999–3018.

Koh, S., Oyler, G. A., and Higgins, G. A. (1989a). Localization of nerve growth factor receptor messenger RNA and protein in the adult rat brain. *Exp. Neurol.* **106,** 209–221.

Koh, S., Chang, P., Collier, T. J., and Loy, R. (1989b). Loss of NGF receptor immunoreactivity in basal forebrain neurons of aged rats: Correlation with spatial memory impairment. *Brain Res.* **498,** 397–404.

Koizumi, S., Contreras, M. L., Matsuda, Y., Hama, T., Lazarovici, P., and Guroff, G. (1988). K-252a: A specific inhibitor of the action of nerve growth factor on PC12 cells. *J. Neurosci.* **8,** 715–721.

Komiriya, A., Hortsch, M., Meyers, C., Smith, M., Kanety, H., and Schlessinger, J. (1984). Biologically active synthetic fragments of epidermal growth factor: Localization of a major receptor-binding region. *Proc. Natl. Acad. Sci. U.S.A.* **81,** 1351–1355.

Konishi, Y., Kotts, C. E., Bullock, L. D., Tou, J. S., and Johnson, D. A. (1989). Fragments of bovine insulin-like growth factors I and II stimulate proliferation of rat L6 myoblast cells. *Biochem.* **28,** 8872–8877.

Korsching, S., Heumann, R., Thoenen, H., and Hefti, F. (1986). Cholinergic denervation of the rat hippocampus by fimbrial transection leads to a transient accumulation of nerve growth factor (NGF) without change in mRNA (NGF) content. *Neurosci. Lett.* **66,** 175–180.

Kovesdi, I., Fairhurst, J. L., Kretschmer, P. J., and Bohlen, P. (1990). Heparin-binding neurotrophic factor (HBNF) and MK, members of a new family of homologous, developmentally regulated proteins. *Biochem. Biophys. Res. Commun.* **172,** 850–854.

Lamballe, F., Klein, R., and Barbacid, M. (1991). TrkC, a new member of the trk family of tyrosine protein kinases, is a receptor for neurotrophin-3. *Cell* **66,** 967–979.

Lärkfors, L., Ebendal, T., Whittemore, S. R., Persson, H., Hoffer, B., and Olson, L. (1987). Decreased level of nerve growth factor (NGF) and its messenger RNA in the aged rat brain. *Mol. Brain Res.* **3,** 55–60.

Large, T. H., Bodary, S. C., Clegg, D. O., Weskamp, G., Otten, U., and Reichardt, L. F. (1986). Nerve growth factor gene expression in the developing rat brain. *Science* **234,** 352–355.

Le, T., and Vilcek, J. (1987). Tumor necrosis factor and interleukin-1: Cytokines with multiple overlapping biological activities. *Lab-Invest.* **56,** 234–248.

Leibrock, J., Lottspeich, F., Hohn, A., Hofer, M., Hengerer, B., Masiakowski, P., Thoenen, H. et al. (1989). Molecular cloning and expression of brain-derived neurotrophic factor. *Nature* **341,** 149–152.

Levi-Montalcini, R. (1975). NGF: An uncharted route. *In* "The Neurosciences: Paths of Discovery" (F.

G. Woolen, J. P. Swazey, and G. Adelman, eds.), pp. 245–265. MIT Press, Cambridge, Massachusetts.

Levi-Montalcini, R., and Hamburger, V. (1953). A diffusible agent of mouse sarcoma, producing hyperplasia of sympathetic ganglia and hyperneurotization of viscera in the chick embryo. *J. Exp. Zool.* **123,** 233–287.

Lin, L-F., H., Mismer, D., Lile, J. D., Armes, L. G., Butler, E. T., III, Vannice, J. L., and Collins, F. (1989). Purification, cloning, and expression of ciliary neurotrophic factor (CNTF). *Science* **246,** 1023–1025.

Lindholm, D., Heumann, R., Meyer, M., and Thoenen, H. (1987). Interleukin 1 regulates synthesis of nerve growth factor in nonneuronal cells of rat sciatic nerve. *Nature (London)* **230,** 658–661.

Lindholm, D., Heumann, R., Hengerer, B. and Thoenen, H. (1988). Interleukin 1 increases stability and transcription of mRNA encoding nerve growth factor in cultured rat fibroblasts. *J. Biol. Chem.* **263,** 16348–16351.

Lindsay, R. M., Shooter, E. M., Radeke, J. J., Misko, T. P., Dechant, G., Thoenen, H., and Lindholm, D. (1990). Nerve growth factor regulates expression of the nerve growth factor receptor gene in adult sensory neurons. *Eur. J. Neurosci.* **2,** 389–396.

Longo, F. M., Vu, T. K., and Mobley, W. C. (1990). The in vitro biological effect of nerve growth factor is inhibited by synthetic peptides. *Cell Reg.* **1,** 189–195.

Lorenzi, M. V. Hefti, F., Knusel, B., and Strauss, W. L. (1992). A developmental change in choline acetyltransferase gene transcription: A potential role for nerve growth factor. *Neurosci. Lett.,* in press.

McBride, R. L., Feringa, E. R., Garver, M. K., and Williams, J. K., Jr. (1990). Retrograde transport of fluoro-gold in corticospinal and rubrospinal neurons 10 and 20 weeks after T-9 spinal cord transection. *Exp. Neurol.* **108,** 83–85.

McNeill, T. H. (1983). Neural structure and aging. *Rev. Biol. Res. Aging* **1,** 163–178.

Maisonpierre, P. C., Belluscio, L., Friedman, B., Alderson, R. F., Wiegand, S. J., and Furth, R. F. (1990). NT-3, BDNF, and NGF in the developing rat nervous system: Parallel as well as reciprocal patterns of expression. *Neuron* **5,** 501–509.

Martin-Zanca, D., Oskam, R., Mitra, G., Copeland, T., and Barbacid, M. (1989). Molecular and biochemical characterization of the human trk proto-oncogene. *Mol. Cell. Biol.* **9,** 24–33.

Mathew, T. C., and Miller, F. D. (1990). Increased expression to T alpha 1 alpha-tubulin mRNA during collateral and NGF-induced sprouting of sympathetic neurons. *Dev. Biol.* **141,** 84–92.

Matsuda, H., Coughlin, M. D., Bienenstock, J., and Denburg, J. A. (1988). Nerve growth factor promotes human hemopoietic colony growth and differentiation. *Proc. Natl. Acad. Sci. U.S.A.* **85,** 6508–6512.

Monard, D. (1987). Role of protease inhibition in cellular migration and neuritic growth. *Biochem. Pharmacol.* **36,** 1389–1392.

Morrison, R. S. (1987). Fibroblast growth factors: Potential neurotrophic agents in the central nervous system. *J. Neurosci. Res.* **17,** 99–101.

Morrison, R. S., Sharma, A., DeVellis, J., and Bradshaw, R. A. (1986). Basic fibroblast growth factor supports the survival of cerebral cortical neurons in primary culture. *Proc. Natl. Acad. Sci. U.S.A.* **83,** 7537–7541.

Morrison, R. S., Kornblum, H. I., Leslie, F. M., and Bradshaw, R. A. (1987). Trophic stimulation of cultured neurons from neonatal rat brain by epidermal growth factor. *Science* **238,** 72–75.

Nakanishi, S., Yamada, K., Kase, H., Nakamura, S. and Nonomura, Y. (1988). K-252a, a novel microbial product, inhibits smooth muscle myosin light chain kinase. *J. Biol. Chem.* **263,** 6215–6219.

Oltersdorf, T. et al. (1989). The secreted form of the Alzheimer's amyloid precursor protein with the Kunitz domain is protease nexin-II. *Nature,* **341,** 144–147.

Oppenheim, R. W. (1989). The neurotrophic theory and naturally occurring mononeuron death. *Trends Neurosci.* **12,** 252–255.

Otten, U., and Lorez, H. P. (1983). Nerve growth factor increases substance P, cholecystokinin, and vasoactive intestinal polypeptide immunoactivities in primary sensory neurons of newborn rats. *Neurosci. Lett.* **34,** 153–158.

Paul, J. W., Quach, T. T., Duchemin, A.-M., Schrier, B. K., and DaVanzo, J. P. (1990). 1,1,3-Tricyano-2-amino-1-propene (Triap): A small molecule which mimics or potentiates nerve growth factor. *Dev. Brain Res.* **55,** 21–27.

Phillips, H. S., Hains, J. M., Laramee, G. R., Rosenthal, A., and Winslow, J. W. (1990). Widespread expression of BDNF but not NT3 by target areas of basal forebrain cholinergic neurons. *Science* **250,** 290–292.

Pollock, J. D., Krempin, M., and Rudy, B. (1990). Differential effects of NGF, FGF, EGF, cAMP, and dexamethasone on neurite outgrowth and sodium channel expression in PC12 cells. *J. Neurosci.* **10,** 2626–2637.

Powell, E. M., Sobarzo, M. R., and Saltzman, W. M. (1990). Controlled release of nerve growth factor from a polymeric implant. *Brain Res.* **515,** 309–311.

Radeke, M. J., Misko, T. P., Hsu, C, Herzenberg, L. A., and Shooter, E. M. (1987). Gene transfer and molecular cloning of the rat nerve growth factor receptor: A new class of receptors. *Nature (London)* **325,** 593–597.

Recio-Pinto, E., Rechler, M. M., and Ishii, D. N. (1986). Effects of insulin, insulin-like growth factor-II, and nerve growth factor on neurite formation and survival in cultured sympathetic and sensory neurons. *J. Neurosci.* **6(5),** 1211–1219.

Reid, H. H., Wilkes, A. F., and Bernard, O. (1990). Two forms of the basic fibroblast growth factor receptor-like mRNA are expressed in the developing mouse brain. *Proc. Natl. Acad. Sci. U.S.A.* **87,** 1596–1600.

Rodriguez-Tebar, A., Dechant, G., and Barde, Y. A. (1990). Binding of brain-derived neurotrophic factor to the nerve growth factor receptor. *Neuron* **4,** 487–492.

Rohrer, H. (1990). The role of growth factors in the control of neurogenesis. *Eur. J. Neurosci.* **2,** 1005–1015.

Rosenberg, M. B., Friedmann, T., Robertson, R. C., Tuszynski, M., Wolff, J. A., Breakefield, X. O., and Gage, F. H. (1988). Grafting genetically modified cells to the damaged brain: Restorative effects of NGF expression. *Science* **242,** 1575–1578.

Rosenberg, P. A. (1988). Catecholamine toxicity in cerebral cortex in dissociated cell culture. *J. Neurosci.* **8,** 2887–2894.

Rosenthal, A., Goeddel, D. V., Nguyen, T., Lewis, M., Shih, A., Laramee, G. R., Nikolics, K., and Winslow, J. W. (1990). Primary structure and biological activity of a novel human neurotrophic factor. *Neuron* **4,** 767–773.

Rydel, R. E., and Greene, L. A. (1987). Acidic and basic fibroblast growth factors promote stable neurite outgrowth and neuronal differentiation in cultures of PC12 cells. *J. Neurosci.* **7,** 3639–3653.

Schwab, M. E., Otten, U., Agid, Y., and Thoenen, H. (1979). Nerve growth factor (NGF) in the rat CNS: Absence of specific retrograde axonal transport and tyrosine hydroxylase induction in locus coeruleus and substantia nigra. *Brain Res.* **168,** 473–483.

Scott, J., Selby, M., Urdea, M., Quiroga, M., Bell, G. I., and Rutter, W. J. (1983). Isolation and nucleotide sequence of a cDNA encoding the precursor of mouse nerve growth factor. *Nature (London)* **302,** 538–540.

Seeley, P. J., Keith, C. H., Shelanski, M. L., and Greene, L. A. (1983). Pressure microinjection of nerve growth factor and anti-nerve growth factor into the nucleus and cytoplasm: Lack of effects on neurite outgrowth from pheochromocytoma cells. *J. Neurosci.* **3,** 1488–1494.

Shelton, D. L. and Reichardt, L. F. (1984). Expression of the nerve growth factor gene correlates with the density of sympathetic innervation in effector organs. *Proc. Natl. Acad. Sci. U.S.A.* **81,** 7951–7955.

Shinoda, I., Furukawa, Y., and Furukawa, S. (1990). Stimulation of nerve growth factor synthesis/secretion by propentofylline in cultured mouse astroglial cells. *Biochem. Pharmacol.* **39,** 1813–1816.

Snider, W. D., and Johnson, E. M., Jr. (1989). Neurotrophic molecules. *Ann. Neurol.* **26,** 489–506.

Sofroniew, M., Galletly, N., Isacson, O., and Svendsen, C. (1990). Survival of adult basal forebrain cholinergic neurons after loss of target neurons. *Science* **247,** 338–342.

Soppet, D., Escandon, E., Maragos, J., Kaplan, D. R., Hunter, T., Nikolics, K., and Parada, L. F. (1991). The neurotrophic factors BDNF and NT-3 are ligands for the trkB tyrosine kinase receptor. *Cell* **65,** 895–903.

Squinto, S. P., Aldrich, T. H., Lindsay, R. M., Morrissey, D. M., Panayotatos, N., Bianco, S. M., Furth, M. E., and Yancopoulos, G. D. (1990). Identification of functional receptors for ciliary neurotrophic factor on neuronal cell lines and primary neurons. *Neuron* **5**, 757–766.

Squinto, S. P., Stitt, T. N., Aldrich, T. H., Davis, S. M., Bianco, S. M., Rdziejewski, C., Glass, D. J., Masiakowski, P., Furth, M. E., Valenzuela, D. M., DiStefano, P. S., and Yancopoulos, G. D. (1991). trkB encodes a functional receptor for brain-derived neurotrophic factor and neurotrophin-3 but not NGF. *Cell* **65**, 885–893.

Sporn, M. B., and Roberts, A. B. (1988). Peptide growth factors are multifunctional. *Nature (London)* **332**, 217–219.

Spranger, M., Lindholm, D., Bandtlow, C., Heumann, R., Gnahn, H., Naher-Noe, M., and Thoenen, H. (1990). Regulation of nerve growth factor (NGF) synthesis in the rat central nervous system: Comparison between the effects of interleukin-1 and various growth factors in astrocyte cultures and in vivo. *Eur. J. Neurosci.* **2**, 69–76.

Squire, L. R., and Davis, H. P. (1981). The pharmacology of memory: A neurobiological perspective. *Ann. Rev. Pharmacol. Toxicol.* **21**, 323–356.

Stein, R., Mori, N., Matthews, K., Lo, L. -C., and Anderson, D. J. (1988). The NGF-inducible SCG10 mRNA encodes a novel membrane-bound protein present in growth cones and abundant in developing neurons. *Neuron* **1**, 463–476.

Stöckli, K. A., Lottspeich, F., Sendtner, M., Masiakowski, P., Carroll, P., Gotz, R., Lindholm, D., and Thoenen, H. (1989). Molecular cloning, expression, and regional distribution of rat ciliary neurotrophic factor. *Nature (London)* **342**, 920–923.

Thoenen, H., and Barde, Y. A. (1980). Physiology of nerve growth factor. *Physiol. Rev.* **60**, 1284–1335.

Thoenen, H. Bandtlow, C., and Heumann, R. (1987). The physiological function of nerve growth factor in the central nervous system: Comparison with the periphery. *Rev. Physiol. Biochem. Pharmacol.* **109**, 145–178.

Ullrich, A., Gray, A., Tam, A. W., Yang-Feng, T., Tsubokawa, M., Collins, C., Henzel, W., Le Bon, T., Kathuria, S., Chen, E., Jacobs, S., Francke, U., Ramachandran, J., and Fujita-Yamaguchi, Y. (1986). Insulin-like growth factor I receptor primary structure: Comparison with insulin receptor suggests structural determinant that defines functional specificity. *EMBO J.* **5**, 2503–2512.

Vantini, G., Schiavo, N., DiMartino, A., Polato, P., Triban, C., Callegaro, L., Toffano, G., and Leon, A. (1989). Evidence for a physiological role of nerve growth factor in the central nervous system of neonatal rats. *Neuron* **3**, 267–273.

Walicke, P., Varon, S., and Manthorpe, M. (1986). Purification of a human red blood cell protein supporting the survival of cultured CNS neurons, and its identification as catalase. *J. Neurosci.* **6**, 1114–1121.

Walicke, P. A. (1988). Basic and acidic fibroblast growth factors have trophic effects on neurons from multiple CNS regions. *J. Neurosci.* **8**, 2618–2627.

Whitson, J. S., Selkoe, D. J., and Cotman, C. W. (1989). Amyloid beta protein enhances the survival of hippocampal neurons in vitro. *Science* **243**, 1488–1490.

Whittemore, S. R., and Seiger, A. (1987). The expression, localization, and functional significance of beta-nerve growth factor in the central nervous system. *Brain Res. Rev.* **12**, 439–464.

Whittemore, S. R., Larkfors, L., Ebendal, T., Holets, V. R., Ericsson, A., and Persson, H. (1987). Increased beta-nerve growth factor messenger RNA and protein levels in neonatal rat hippocampus following specific cholinergic lesions. *J. Neurosci.* **7**, 244–251.

Yamamori, T., Fukada, K., Aebersold, R., Korsching, S., Fann, M. J., and Patterson, P. H. (1989). The cholinergic neuronal differentiation factor from heart cells is identical to leukemia inhibitory factor. *Science* **246**, 1412–1416.

Zafra, F., Hengerer, B., Leibrock, J., Thoenen, H., and Lindholm, D. (1990). Activity dependent regulation of BDNF and NGF mRNAs in the rat hippocampus is mediated by non-NMDA glutamate receptors. *EMBO J.* **9**, 3545–3550.

# Synergy, Retrograde Transport, and Cell Death

Ian A. Hendry and Michael F. Crouch

## I. INTRODUCTION

Many molecules are required for the developmental survival of neurons. The ones that will be of major interest in this chapter are the ones that provide the innervating neurons with information about the target region. Two possibly related forms of cell death occur during development. Normal histogenic or ontogenic cell death occurs in practically every neuronal system examined. The other form of death occurs after removal of the target tissue prior to innervation or after some other disturbance in the connection between the target tissue and its innervating neurons, for example, after axotomy during the critical period when neuron–target tissue connections are being established. It is likely that the mechanisms for both types of cell death are virtually the same; hence, they will be considered together. The developmental control of this cell death is the thrust of this chapter.

## II. CELL DEATH

### A. Target-Controlled Neuronal Survival

Talk of neurotrophic factors usually stimulates thinking about the developmental stage at which neurons innervate their target. At this stage, a large proportion, usually more than 50%, of the total neuronal population dies (Hamburger and Levi-Montalcini, 1949). In addition, if the target tissue is removed prior to innervation, nearly all the neurons that are destined to supply it will die at the time that innervation would have taken place (Hamburger, 1934). An artificial increase in the size of the target field (Hamburger, 1939) or a reduction in the number of neurons innervating a target of the same size leads to an increase in the extent of neuronal survival (Pilar et al., 1980), showing that the ratio of innervation to target size determines the number of surviving neurons, not a change in the production of a trophic factor. Thus, the target tissue provides a factor that is essential for neuronal survival.

Although the final number of neurons in any nerve center seems to depend on the size of the available target, regional variations can occur because of differential neuroblast proliferation, for example, in the chick embryo spinal cord (Oppenheim et al., 1989).

Protein and RNA synthesis are required in the embryo, for both naturally occurring and induced cell death (Oppenheim et al., 1990). In addition, the death of cultured neurons after the removal of nerve growth factor (NGF) is prevented by inhibitors of RNA and protein synthesis (Martin et al., 1988). Thus, neuronal cell death is an active process that requires biosynthetic events. It seems likely that neurotrophic factors have a dual function of stimulating genes that promote survival and differentiation, as well as suppressing genes that would kill the cell (Oppenheim et al., 1990). The finding that interferon (IFN) retards cell death in cultures of sympathetic neurons after NGF withdrawal is intriguing, and suggestive of a role for $2'$, $5'$-oligoadenosine synthetase in the process of neuronal rescue (Chang et al., 1990). Since the product of this enzyme can activate an RNAse, IFN may interrupt the "death program" by causing the degradation of mRNA critical for this program.

## B. In Vivo Survival Factors

### 1. Retrophins

The simplest mechanism by which the target tissue can control the innervating neurons is for the target cell to make a neurotrophic molecule that is available only in limited amounts. This retrogradely transported neurotrophin (retrophin) (Hendry and Hill, 1980) is taken up by the nerve terminals or growth cones, and retrogradely transported to the neuronal perikarya, where it acts to promote the survival of the neuron that has transported it (Fig. 1).

Although this mechanism is the simplest way, it is not the only way to get a message from the periphery. An in-depth analysis of other mechanisms that may lead to the chromatolytic signal has been made previously (Cragg, 1970).

### 2. Neurotrophic agents and factors that rescue neurons in vivo

For a potential survival factor identified in culture to be seen as a viable neurotrophic factor, it must be able to promote neuronal survival in vivo. Many neurotrophic factors have been shown to have effects in vivo, but the possibility always exists that the effects of an agent are pharmacologic, that is, that the agent mimics the action of the endogenous factor, often at very high concentrations, rather than being the physiologic factor itself. The physiologic factor will need to be present in limited amounts to achieve the desired control of survival.

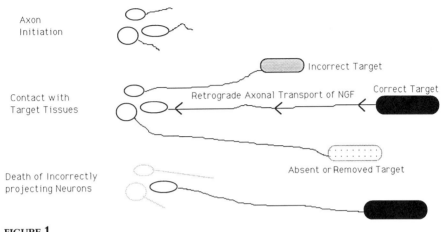

FIGURE 1

The neurotrophic theory of cell death based on the actions of NGF suggests that target tissues release the neurotrophic factor, which is taken up by receptor-mediated endocytosis and transported back to the cell body, where it has its effect on the nucleus to cause survival.

The classical studies of NGF have shown that this molecule can cause the survival of sympathetic (Levi-Montalcini, 1965) and sensory neurons (Levi-Montalcini and Angeletti, 1968). This evidence sets the model for the in vivo investigation of other putative survival factors that are discussed more fully in other chapters. NGF causes the survival of sympathetic neurons (Hendry and Campbell, 1976) and can rescue them after axotomy (Hendry, 1975) and 6-hydroxydopamine treatment (Levi-Montalcini et al., 1975).

Several members of the fibroblast growth factor (FGF) family may exhibit neurotrophic activity (Delli-Bovi et al., 1988; Basilico et al., 1989). Polyclonal antibodies to basic FGF (bFGF) have been shown to react to a higher molecular weight molecule that appears to cause the survival of ciliary neurons in culture (Grothe et al., 1990).

Basic FGF is found in the adrenal medulla chromaffin granules (Westermann et al., 1990). Gel foams soaked in bFGF can rescue adrenal preganglionic neurons after electrolytic destruction of the adrenal medulla (Blottner et al., 1989a,b).

Acidic FGF (aFGF) enhances nerve regeneration in sciatic nerve (Cordeiro et al., 1989) and retinal ganglion cell processes (Lipton et al., 1988). In addition, aFGF stimulates adrenal chromaffin cells to proliferate and extend neurites but fails to cause long-term survival (Claude et al., 1988).

Ciliary neurotrophic factor (CNTF) can rescue adrenal preganglionic neurons after electrolytic destruction of the adrenal medulla (Blottner et al., 1989a,b) and facial neurons after axotomy (Sendtner et al., 1990).

Muscle-derived cholinergic differentiating factor reverses naturally occurring cell death of motoneurons (Oppenheim et al., 1988), but not that of sensory or sympathetic preganglionic neurons in vivo (McManaman et al., 1990).

Brain-derived neurotrophic factor (BDNF) has been isolated from pig brain (Barde et al., 1987) and its sequence has been determined (Hohn et al., 1990); it is a member of the NGF protein family. BDNF binds to the low affinity NGF receptor (Rodriguez-Tébar et al., 1990). Messenger RNA encoding BDNF has been demonstrated in adult mouse brain (Hofer et al., 1990). It has a different spectrum of activity than NGF and prevents the ontogenic death of nodose neurons (Hofer and Barde, 1988).

Insulin-like growth factor-I (IGF-I) causes an enhanced regeneration of the rat sciatic nerve after a freezing lesion (Kanje et al., 1989; Sjöberg and Kanje, 1989).

### 3. Control of availability

If a neurotrophic factor is to have in vivo physiologic relevance, there must be a mechanism for controlling its availability to neurons. This control may be via competition between neurons for a limited amount of factor, an effect that will be exacerbated if the neurons remove the trophic factor released from the target to control its availability. Neurotrophic factors for sympathetic and parasympathetic neurons are elevated in the rat ventricle after chemical denervation (Kanakis et al., 1985) and growth factors for ciliary neurons are elevated in skeletal muscle after denervation (Hill and Bennett, 1983). The effects of 6-hydroxydopamine and colchicine on levels of NGF in the sympathetic ganglia and target tissue strongly suggest that the control of NGF levels in the target is through its removal by retrograde axonal transport in the innervating neurons (Korsching and Thoenen, 1985).

The availability of trophic factors may be regulated by other mechanisms such as binding to the substrate, as has been proposed for FGF (Eccleston et al., 1985; Rogelj et al., 1989). If this is the case, then the control of availability is regulated by the composition of the target tissue itself and not by the innervating neurons.

## C. Trans-Synaptic Influences

One of the requirements for neuronal survival appears to be the presence of appropriate presynaptic transmission. Naturally occurring cell death is enhanced in sympathetic and parasympathetic ganglia after blockade of ganglionic neurotransmission with pempidine (Maderdrut et al., 1988).

Conversely, neuronal survival is enhanced by blockade of postsynaptic transmission. Sympathetic preganglionic neuron cell death is reduced by treatment with NGF (Oppenheim et al., 1982a) and hemicholinium (Oppenheim et al., 1982b), perhaps because of the enlargement of the sympathetic ganglia and the supply of the preganglionic requirement for neurotrophic factors by the hypertrophied target. Blockade of activity of the target with curare leads to an increase in survival of motoneurons (Oppenheim, 1981, 1984). On the other hand, increase in activity by electrical stimulation of the limb muscles in the chick embryo increases cell death (Oppenheim and Núñez, 1982).

Many factors are involved in the regulation of cell death that may act synergistically. One attractive model for developmental control is the interaction between presynaptic and postsynaptic factors. For example, in the developing sympathetic nervous system, both presynaptic and target tissue influences control the final number of neurons (Black et al., 1972; Hendry, 1973, 1975). The number of embryonic chick spinal motoneurons that survive during the period of naturally occurring cell death is influenced by factors from both the target tissue (Hamburger, 1934, 1975; Oppenheim, 1981) and an intact descending afferent system (Okado and Oppenheim, 1984). Two distinct factors have been isolated from muscle and spinal cord that clearly promote motoneuron survival; there is synergy between these factors (Dohrmann et al., 1987).

## D. Antibodies

The only way to define a role for a neurotrophic factor positively is to block its activity in vivo and demonstrate a developmental change. The only effective blocking agents presently available are antibodies. Data has been obtained by antibody administering in vivo to developing animals.

Antibodies to NGF destroy the sympathetic nervous system in neonatal rats (Levi-Montalcini and Angeletti, 1966), but need to be present earlier to have an effect on the sensory neurons (Johnson et al., 1980). Experiments in which mothers have been made hyperimmune have established clearly the essential nature of NGF in the development of the sensory and sympathetic systems (Gorin and Johnson, 1980a). NGF is required for in vivo survival of dorsal root ganglia (DRG) in the adult rat and rabbit (Gorin and Johnson, 1980b; Johnson et al., 1982). These experiments were long term, however, compared with those in culture (Lindsay, 1988), and these hyperimmune animals may have antibodies that cross react with other members of the NGF family such as BDNF.

A monoclonal antibody to aFGF that blocks its biologic activity in vitro has been generated (Watters et al., 1989). This antibody impairs the normal development of the murine parasympathetic nervous system (Hendry et al., 1988), indicating that aFGF or an antigenically related species is important for parasympathetic neuronal development.

## III. AXOPLASMIC TRANSPORT

For the target to have a direct influence only on the neuron innervating it, there must be a specific retrograde message carried via the axon from the target to neuron nucleus. Despite the logic of the retrophin model, the retrograde axonal transport of only very few putative neurotrophic molecules has been described.

## A. Specific Transport

NGF binds to specific receptors, is internalized at nerve terminals, and transported with its receptor to the cell body in coated vesicles that eventually fuse with lysosomes, where the NGF is degraded (Schwab, 1977). It is unlikely that the internalization of NGF is the active intracellular message, since intracellular NGF itself does not have any effects (Rohrer et al., 1982) and intracellular antibodies do not affect the cellular response to NGF (Heumann et al. 1981).

The receptor for NGF is also retrogradely transported by neurons, both in the periphery and in the central nervous system, as shown by the intracellular localization of the antibody to the receptor (Johnson et al., 1987). NGF receptors are transiently expressed on motoneurons during development in newborn rats and NGF is retrogradely transported by these neurons (Yan et al., 1988). Since these receptors are not capable of mediating any traditional neurotrophic effects of survival, retrograde axonal transport of NGF in neurons is not necessarily synonymous with the ability of NGF to exert trophic activity. There is one report of NGF retrograde transport by ciliary neurons (Max et al., 1978), but subsequent experiments have shown only nonspecific transport (Hendry and Belford, 1991). NGF is transported by neurons of the chick DRG during the critical phase of their development (Brunso-Bechtold and Hamburger, 1979).

IGF-I, which is important in the regulation of peripheral nerve regeneration (Linnemann and Bock, 1986), has been reported to be retrogradely transported in the sciatic nerve (Hansson et al., 1986, 1987). However, investigators have failed to detect the high affinity uptake and retrograde axonal transport of IGFs by motoneuron processes in situ (Caroni and Grandes, 1990).

The physiologic significance of the specific retrograde axonal transport of NGF or, indeed, any other neurotrophic molecule is not resolved. The possible purposes of the transport of NGF include:

1. a means of getting NGF to the cell body, where it or a breakdown product could exert its trophic effect.
2. a means of getting the receptor to the perikaryon so it can act as the trophic message.
3. merely reflecting the presence of receptors capable of internalizing NGF, since an appropriate second messenger, which is subsequently transported to the cell body, may be generated at the nerve terminal. (Any second messenger that must reach the cell body by its own retrograde transport would need to be very stable.)
4. enabling a short-lived or unstable second messenger to reach the cell body from the periphery since the receptor–NGF complex in the transported vesicle is presumably continuously capable of generating a second messenger.
5. providing a means to regulate the levels of NGF in the target and serving as a mechanism to remove NGF from the terminal.
6. inhibiting active cell death induced by another molecule or inhibiting the retrograde axonal transport of such another molecule.

## B. Nonspecific Transport

Most of the molecules that are retrogradely transported (e.g., horseradish peroxidase, FGF, bovine serum albumin, cytochrome C) are only transported to a minor degree, presumably as a result of pinocytosis of extracellular fluid in the region of the nerve terminal (Heuser and Reese, 1973) into coated vesicles (Zacks and Saito, 1969). This uptake depends on the activity of the nerve (Edds, 1950; Soinila and Eränkö, 1983), unlike the receptor-mediated uptake of NGF, which was not seen to change after alterations in the firing pattern of sympathetic neurons (Lees et al., 1981; Stöckel et al., 1978).

Although aFGF is a good candidate for parasympathetic neurotrophic factor (Watters et al., 1989), specific retrograde axonal transport of FGF by ciliary neurons has not been demonstrated (Hendry and Belford, 1991). bFGF also was not retrogradely transported in the adult rat sciatic nerve, nor from iris to trigeminal or superior cervical ganglion (Ferguson, 1991). This suggests either that these neurons do not express significant numbers of receptors for FGF or that the injected FGF is in some way unavailable for uptake by the nerve terminals.

FGF is, however, anterogradely transported; after injection into the vitreous body of the eye, radioactivity was detected in retinal ganglion cell projections. The transport was saturable, blocked by wheat germ agglutinin and heparin, and augmented by heparinase (Ferguson, 1991).

FGF, when injected into the eye, remains bound, but when soluble heparin or heparinase is injected at the same time the FGF is lost from the eye (Ferguson, 1991). The anterograde transport of FGF seen in retinal ganglion cells is augmented by heparinase and reduced by heparin (Ferguson, 1991). Thus the bound FGF is released differently by these two procedures, being more available to receptors when released by heparinase and less available when released by heparin.

## C. Blockade of Axoplasmic Transport

Blockade of retrograde axonal transport has two results: one is an increase of growth factors in peripheral tissues and the other is cell death during the critical period of innervation. The former suggests control of neurotrophic factor concentration by the innervating neurons. Evidence conflicts, however, on whether there are direct effects on growth factor production in the target. The expression of the NGF gene correlates with the density of sympathetic innervation in effector organs (Korsching and Thoenen, 1983; Shelton and Reichardt, 1984). NGF mRNA is expressed in the developing rat brain, suggesting a functional role for NGF in the development of the central nervous system (CNS) (Large et al., 1986). Neurotrophic factors for sympathetic and parasympathetic neurons are elevated in the rat ventricle after chemical denervation (Kanakis et al., 1985), but NGF mRNA levels during development are not altered by sympathectomy (Clegg et al., 1989). Growth factors for ciliary neurons are elevated in skeletal muscle after denervation (Hill and Bennett, 1983). Changes in target levels of growth factors are reflected in a change in the innervation pattern of the target tissue. When the sympathetic system is removed, there is a compensatory increase in the extent of sensory innervation (Hill et al., 1988; Kessler et al., 1983a,b). The simplest explanation is that NGF normally is removed from the target by retrograde transport and the neurons do not themselves affect NGF mRNA production.

Axotomy-induced cell death and the death that occurs with blockade of axoplasmic transport are probably caused by the same mechanism and demonstrate that some molecule must be physically transported to exert the neurotrophic effect. Colchicine can be used to block axoplasmic transport and will mimic the effects of axotomy in ciliary neurons (Pilar and Landmesser, 1972). Vinblastine (Chen et al., 1977; Johnson, 1978) which blocks axoplasmic transport, and 6-hydroxydopamine (Levi-Montalcini et al., 1975), which interferes with axoplasmic transport by destruction of the nerve terminals, both destroy the sympathetic nervous system when administered to neonatal rats; this sympathectomy can be reversed by NGF. This demonstrates that NGF can affect cell survival by direct action on the cell bodies, not only by retrograde transport from the terminals. Circulating growth factors can be taken up by nerve terminals and transported to the cell body. For example, $I^{125}$-labeled NGF accumulates in the sympathetic ganglia after systemic injection; this accumulation can be blocked by axotomy (Schmidt et al., 1983; Angeletti et al., 1972). Thus, effects of systemic administration may still be exerted by the transported neurotrophic factors, but the effects of NGF on 6-hydroxydopamine- or vinblastine-treated neurons cannot be explained by terminal uptake.

## IV. INTERACTIONS BETWEEN NEURONALLY ACTIVE MOLECULES

### A. Synergy

Many factors in conditioned medium promote the survival of motoneurons (Calof and Reichardt, 1984; Dohrmann et al., 1986, 1987; Smith et al., 1986), neurite extension (Pannese, 1976; Henderson et al., 1981; Calof and Reichardt, 1984), and choline acetyltransferase induction (Giller et al., 1973; Giess and Weber, 1984; Smith et al., 1986; McManaman et al., 1988; Martinou et al., 1989a,b). Neurons require multiple factors for their development and survival. When these factors are tested in tissue culture, they are traditionally grown in serum-containing medium or in chemically defined medium to which the potential survival factor is added. Then the issue is whether the factors present in the serum, the defined medium, or the added factors are the survival factors. For example, the serum requirement for neurons in culture can be fulfilled by medium containing high levels of, among other things, insulin (Bottenstein et al., 1980). This result could be interpreted to mean that insulin is a survival factor for neurons, perhaps by cross-reaction with IGF-I receptors. Experiments of neuronal survival in vitro usually examine the effects of single agonists at high concentrations, but in vivo interactions between many molecules at low concentrations is probably much more relevant. Thus, in our understanding of neuronal development, we must examine interactions between molecules at low concentrations since these interactions may better define the physiologic role of the molecules.

### 1. Division

Few studies have been undertaken in the search for factors causing neuronal division. If a survival factor is acting during the time of cell division, there will be an increase in the number of dividing cells (Hendry, 1977). Thus, in many cases it is not possible to distinguish between factors causing division and those causing survival. Sympathetic cell division is regulated by insulin, bFGF, and epidermal growth factor (EGF); no other mitogen tested, nor NGF itself, had an effect (DiCicco-Bloom et al., 1990). It is interesting that, in these experiments, it was shown that sympathetic neuroblast mitosis can take place in neurons with neurites (DiCicco-Bloom et al., 1990).

### 2. Survival factors

*a. NGF*  NGF is the most extensively studied of the neuronal survival factors and has been shown to have many interactions with other molecules. The role of NGF in adult DRG survival suggests that the critical messengers for neuronal survival are not needed throughout the life of the neuron. Adult neurons respond to NGF but do not require it. In cultures of adult DRG, insulin is not required for survival but the survival is improved by polylysine and/or laminin on the substrate (Lindsay, 1988). Interferons ($\alpha$, $\beta$, $\gamma$) prevent sympathetic neuronal death after NGF deprivation for a short time, but cannot replace NGF (Chang et al., 1990). Sensory and sympathetic neurons grown in culture die when NGF is withdrawn; this death is prevented by the $Ca^{+2}$-blocking drug flunarizine at doses considerably higher than those required to block the voltage-dependent $Ca^{2+}$ channels. In addition, flunarizine also prevented the cell death that occurs in newborn rats after axotomy (Rich and Hollowell, 1990).

*b. FGF*  FGF causes parasympathetic and sensory neuronal survival (Belford et al., 1989) but its interactions have been described on many cell types. The FGF family at present includes seven related heparin-binding proteins: aFGF (Jaye et al., 1986), bFGF (Abraham et al., 1986a,b), int-2 (Dickson and Peters, 1987), hst/kFGF (Yoshida et al., 1987; Delli-Bovi et al., 1988), FGF-5 (Zhan et al., 1988), FGF-6 (Marics et al., 1989), and KGF (Finch et al., 1989). The primary sequences of aFGF and bFGF are 50% homologous (Esch et al., 1985a,b). Basic FGF is a mitogenic growth factor that is synthesized by many cell types including fibroblasts and endothelial cells (Klagsbrun and Vlodavsky, 1988) and stored in the extracellular matrix, where it is associated with heparin-containing molecules that regulate its activity and may control its release to interact with its target cells (Flaumenhaft et al., 1989; Globus et al., 1989; Presta et al., 1989). Heparin potentiates aFGF action by prolonging its biologic half-life (Damon et al., 1989). Heparin-like molecules mediate the binding of bFGF to the extracellular matrix and heparinase destroys this binding (Bashkin et al., 1989). The basement membrane of the eye has substantial binding capacity for aFGF and bFGF (Jeanny et al., 1987). Heparin-like substances account for the bulk of the binding of FGF to the substrate in cultures of bovine capillary endothelial cells; less than 10% of the $^{125}$I-labeled FGF binds directly to the tissue high affinity receptors. This heparin binding of FGF may act as a source for subsequent release to bind to the receptors, providing a mechanism to sequester FGF so it is not available to the nearby cells (Probstmeier et al., 1989).

Cultured bovine capillary endothelial cells synthesize heparin sulfate proteoglycans (HSPG) that are both secreted into the culture medium and deposited in the extracellular matrix. Both forms bind bFGF, which can be readily released by heparinase or plasmin. Plasmin releases FGF bound to the partly degraded HSPG whereas heparinase releases FGF alone (Saksela and Rifkin, 1990). FGF bound to the heparin is less sensitive to proteolytic degradation (Saksela et al., 1988). The plasminogen–plasmin system works predominantly at cell-to-cell or cell-to-substrate contact areas (Dennis et al., 1989). Close contact is required in many cases when differentiation of one cell type is mediated by another. Thus, the degradation of the HSPG core may be a mechanism for controlling the release of FGF from the extracellular matrix during target innervation. Wheat germ agglutinin blocks the binding of bFGF to its high affinity receptors (Feige and Baird, 1988). In vivo regional changes in the levels of sulfated polysaccharides also may alter the effectiveness of FGF.

*c. Retinoids* The retinoids constitute a large family of natural and synthetic compounds that possess vitamin A activity or structural homology to retinol (Goodman, 1984). Their effects on cellular proliferation, differentiation, and positional specification during development have led to intense interest in these compounds as both putative morphogens and possible therapeutic agents for neoplastic processes. Retinoic acid (RA) alone has no effect on neuronal survival but in the presence of aFGF has a pronounced dose-related effect on survival (Narayanan and Narayanan, 1982). There is also an additional increase in the activity of FGF in combination with both heparin and RA (Fig. 2).

It is possible that synergistic actions such as these with RA, glycosaminoglycans, and FGF may be responsible for the correct outgrowth, for example, of sensory neurons into the limb.

The RA receptor belongs to the superfamily of nuclear receptors that may affect gene expression (Petkovich et al., 1987). RA increases the number of NGF receptors in neuroblastoma cells and alters their morphology (Haskell et al., 1987). RA may act by increasing the numbers of FGF receptors on ciliary neurons. Modulation by RA of receptor number or function is well described (Rees et al., 1979; Jetten, 1980, 1982; Heath et al., 1981; Saito et al., 1982; Komura et al., 1986; Haskell et al., 1987; Harper, 1988; Yung et al., 1989). EGF binding capacity to fetal lung is increased 2.5-fold by RA through an increase in receptor numbers. Prostaglandin

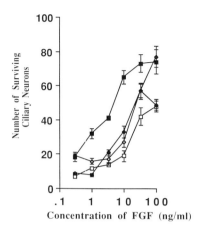

FIGURE 2

Effect of retinoic acid and heparin FGF promoted ciliary neuronal survival. Neurons were cultured as described previously (Bonyhady et al., 1980) in the presence of a serial dilution of FGF alone (□), with retinoic acid (◊), with heparin (♦), or with both (■). Values are means and SEM (vertical bars) of 4–6 experiments.

2 secretion is increased 6.4-fold in the presence of both RA and EGF (Oberg and Carpenter, 1989). RA specifically increases the number of EGF receptors in various fibroblast and epidermal cell lines (Jetten, 1980; Komura et al., 1986). However, in ME180 cells, RA decreases the number of EGF receptors when grown in 10% fetal calf serum (Zheng and Goldsmith, 1990). One neurotrophic factor—purpurin—is known to be a retinol-binding growth factor (Schubert et al., 1987), but no synergy for this system has been described.

For some years it has been known that posteriorly located chick wing bud mesenchyme (termed the zone of polarizing activity or ZPA), when grafted anteriorly on the wing bud, produces mirror-image symmetrical duplication of the host digit pattern (Saunders and Gasseling, 1968). Homoplastic and xenoplastic combinations have shown the duplicated appendages to be always of host type (wing or leg) and species, demonstrating that the grafted mesenchyme does not contribute to the supernumerary structure but induces anteroposterior polarity in the host (Balcuns Maccabe et al., 1973). Interest in RA as the natural morphogen was generated by the remarkable observation that RA applied to the anterior chick limb bud mimics the action of the ZPA by inducing additional digits in mirror-image symmetry to those of the host (Tickle et al., 1982; Summerbell, 1983). The identification of gradients of retinoids in the chick limb bud has implicated these molecules further as endogenous morphogens (Maden and Summerbell, 1986; Maden et al., 1988). Biologically active RA has been isolated in limb bud at concentrations similar to those of exogenous RA required to induce formation of extra digits (Thaller and Eichele, 1987). Significantly, RA was enriched some 2.5-fold in the posterior quarter of the limb bud. Additionally, it has been shown that release of radiolabeled RA or its analogs from implanted ion-exchange beads forms a steady-state concentration gradient across the wing bud (Eichele et al., 1985; Tickle et al., 1985).

To innervate the digits correctly, the neurons may respond to changing levels of RA in the presence of a constant level of neurotrophic factor. This may be one of the factors guiding the neurons to appropriate digits.

***d. bFGF and IGF-I*** The trophic effects of bFGF and IGF-I on fetal hypothalamic cells were additive (Torres-Aleman et al., 1990). On cells transformed with SV40, however, bFGF and IGF-I showed no additive effects, despite being added at submaximal concentrations (Torres-Aleman et al., 1990). This result shows that many of the synergistic and coordinate interactions between growth factors may reflect the constitutive level of some of the second messengers involved in the response, which would be altered under the differing culture conditions used in the investigation.

***e. TGFβ*** TGFβ (transforming growth factor β) is a potent survival factor for motoneurons in culture (Martinou et al., 1990), but only in the presence of serum and 36 m$M$ KCl (Giess and Weber, 1984) or in culture on lysed astrocytes. These experiments were carried out using relatively high concentrations of factor, which may mask more subtle synergistic effects. This effect only occurred at low cell densities. In high density cultures, TGFβ had no effect, demonstrating that its beneficial effect could be replaced by cell–cell contact. Thus, cell density per se or other factors released by dense cultures (Manthorpe et al., 1983), may affect the internal milieu of the cell and alter its survival requirements.

***f. Phorbol esters*** Phorbol esters stimulate protein kinase C (PKC) activity and have shown synergy with other factors to cause survival. IGF-I or high concentrations of insulin alone had little effect on the survival of embryonic day 8 chick ciliary neurons in culture. Similarly, the PKC activator phorbol dibutyrate (PdBu) had only a minor survival-promoting activity. In combination, however, IGF-I and PdBu were highly synergistic. In a similar fashion, FGF-induced neurotrophic activity was greatly enhanced by PdBu, and less so by IGF-I. When added alone, FGF-induced cell survival required the presence of 1% serum. However, when FGF or IGF-I was present with PdBu, the serum requirement was eliminated (Fig. 3). That is, these agonist combinations apparently could induce the total neurotrophic second messenger requirement for these cells (Crouch and Hendry, 1991). These results suggest that IGF-I may be a candidate parasympathetic neurotrophic molecule, and highlight the possibility that combinations of growth factors, rather than individual molecules, may dictate parasympathetic nervous system development in vivo.

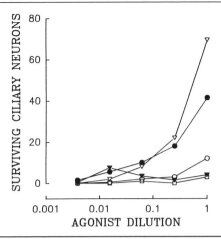

**FIGURE 3**

Effect of FGF, IGF-I, and phorbol dibutyrate (PdBu) on ciliary neuronal survival. Neuronal cultures in serum-free Dulbecco's medium were cultured in the presence of 3-fold serial dilutions from a maximal concentration of FGF of 5.5 nM (▲), of IGF-I of 68 nM (○), of PdBu of 0.75 μM (□), of IGF-I of 68 nM with PdBu constant at 10 nM 2l) and of IGF-I of 68 nM and FGF of 5.5 nM (Δ). Values are duplicate determinations of surviving neurons.

### 3. Axon elongation

Neurite outgrowth, like neuronal survival, depends on multiple factors that tend to act in concert. Two main components act on nerve fibers: substrate-bound molecules and soluble molecules. These components must interact at the level of the cytoskeleton to generate motility at the growth cone. The soluble molecules can act locally at the nerve terminal and NGF acts directly on the nerve fibers to stabilize their ramification (Campenot, 1982a,b). NGF (Campenot, 1977) and IGF-I (Caroni and Grandes, 1990) do not need to be internalized and retrogradely transported to induce local sprouting. Neurite outgrowth can take place on both substrate-bound and cell-surface molecules (Tomaselli and Reichardt, 1988; Tomaselli et al., 1988a,b); these two molecular interactions are distinct (Tomaselli et al., 1986). NGF action on neurite outgrowth in PC12 cells is modified by sulfated glycosaminoglycans (Damon et al., 1988).

The concept that neuronal survival and neurite outgrowth are different manifestations of the same anabolic mechanism (Varon and Bunge, 1978) is not likely to be true. NGF and BDNF enhance axonal regeneration but are not required for the survival of adult sensory neurons (Lindsay, 1988). In fact, adult sensory neurons grown in Bottenstein chemically defined medium (Bottenstein et al., 1979, 1980) do not even require insulin for neurite extension and do not require any survival factors for short-term survival (Lindsay, 1988). This result suggests the presence of internalized NGF or a stable second messenger required by the long transport time from the periphery. NGF has neuritogenic but not survival-promoting effects on afferent auditory neurons (Lefebvre et al., 1990). Purine analongs reversibly suppress NGF-dependent neurite outgrowth in PC12 cells and neurons, but have no effect on NGF-promoted neuronal survival of both sensory and sympathetic ganglia (Greene et al., 1990).

A local response to target-secreted growth factor may explain the intramuscular nerve branching of motoneurons seen after blocking nerve activity with bungarotoxin and curare (Dahm and Landmesser, 1988).

An HSPG that promotes neurite outgrowth has been purified by immunological techniques (Reichardt and Kelly, 1983) using a monoclonal antibody shown to block the biologic activity of the factor (Matthew and Patterson, 1983).

Interleukin-2 promotes sympathetic but not sensory neurite extension (Haugen and Letourneau, 1990). This action requires the addition of insulin, phosphocreatine, selenite, transferrin, and pyruvate.

Insulin clearly is involved in neurite outgrowth, but only in combination with other molecules. High insulin concentrations can cross-occupy IGF receptors but both insulin and IGF can enhance neurite formation and tubulin mRNA levels (Mill et al., 1985; Fernyhough and Ishii, 1987; Fernyhough et al., 1989). through occupancy of their own receptors (Recio-Pinto and Ishii, 1988). Insulin and IGF-I stabilize tubulin mRNA to increase the relative synthesis of tubulin proteins. In view of the homology between insulin and NGF (Frazier et al., 1972), the insulin effect may be due to cross-reaction on the receptors. Neither insulin nor IGF-II, however, competes for high affinity NGF receptors on SH-SY5Y cells (Recio-Pinto and Ishii, 1984). AntiNGF does not inhibit the neurite outgrowth-promoting effects of insulin and IGF-II, but anti-insulin antibodies do (Recio-Pinto and Ishii, 1984; Recio-Pinto et al., 1984). Insulin and IGF-II are required to permit binding of NGF and neurite formation by human neuroblastoma cells (Recio-Pinto et al., 1984).

Chick motoneurons will grow neurites in the presence of a laminin-coated substrate, E19 chick muscle protein extract, and serum. Under these conditions, addition of IGF-I or IGF-II in the 0.1–10 n$M$ range or insulin at ≥ 20 n$M$ greatly enhances the neurite outgrowth response (Caroni and Grandes, 1990). IGFs alone do not support neurite sprouting. Insulin, IGF-I, or IGF-II with NGF effects neurite formation and survival of cultured sympathetic and sensory neurons (Recio-Pinto et al., 1986). Neurite formation in the sympathetic system is modified by NGF, insulin, and the tumor promoter receptors (Ishii et al., 1985).

Thus, it appears that successful neurite outgrowth requires the interactions of many molecules to achieve the required selectivity of innervation.

### 4. Phenotypic determination or transmitter plasticity

Several proteins influence neurotransmitter traits in sympathetic neurons (Fukada, 1985, Kessler et al., 1986; Saadat et al., 1989). A 45-kDa glycoprotein, sympathetic cholinergic differentiating factor, and leukemia inhibitory factor are the same molecule (Yamamori et al., 1989), which increases choline acetyltransferase activity and decreases tyrosine hydroxylase (TH) activity (Fukada, 1985). A 22-kDa protein isolated from rat sciatic nerve, the ciliary neutrophic factor (Manthorpe et al., 1986; Lin et al., 1989; Stöckli et al., 1989), has a similar effect (Saadat et al., 1989). Partially purified extracts of brain also induce cholinergic traits in sympathetic neurons (Kessler et al., 1986); this effect is facilitated by plasma membranes and membrane-bound factors (Lee et al., 1990). There is no obvious pattern of transmitter induction in sympathetic neurons. Depolarization (Black et al., 1971; Black and Geen, 1973; Goodman et al., 1974; Walicke et al., 1977; Kessler et al., 1981, 1983a; Kessler and Black, 1982; Kessler, 1986), cAMP analogs (MacKay and Iversen, 1972; Goodman et al., 1974; Keen and McLean, 1974), and nonneuronal cells (Patterson and Chun, 1974; Kessler et al., 1983b; Kessler, 1984, 1986); all affect transmitter expression in different ways and result in diverse transmitter combinations, suggesting that co-localized transmitters are regulated independently. Substance P levels are increased in sympathetic ganglia by interleukin-1 and by stimulated splenocytes (Jonakait and Schotland, 1990).

Depolarization stimulates choline acetyltransferase in mouse spinal cord cultures (Ishida and Deguchi, 1983) but not in rat purified motoneurons (Martinou et al., 1989a,b), suggesting a synergistic component from nonmotoneurons.

The synthesis of complex glycosphingolipids occurs in PC12 cells with NGF or forskolin activation. The sequence of activation altered the amount of fucose incorporated into the glycolipids. This result suggests that the timing of exposure to neurotrophic factors may alter the developmental profile of sympathetic neurons (Schwarting et al., 1990).

It is clear that, depending on the interactions between factors, multiple agonists in different concentrations or acting sequentially in different orders can lead to the development of many different phenotypes.

### 5. Interactions with other cell types

Many of the factors that cause survival of neurons also have actions on nonneuronal cells. This characteristic is often reflected in the apparently inappropriate names for the factors, for example, fibroblast growth factor. It is apparent that the same molecule in several distinct systems, probably using the same receptor and the same second messenger system, has a different functional role. For example, the "pure neurotrophic factor" NGF interacts synergistically with interleukin-5 and granulocyte–macrophage colony stimulating factor (GM-CSF) to promote basophilic differentiation of Miles leukemic ANET cells (Tsuda et al., 1990). The danger is that the name itself will influence our thinking and preclude the discovery of the real functions of the molecule.

## V. MECHANISMS OF INTERACTION

Many of the experiments carried out with growth factors show the results of the activation of their second messengers; depending on the other activating factors present, the factors may result in varied effects. These effects may or may not have any physiologic significance, especially when very high doses of factors that are not normally present in the environment of the cells are used. Instead, the results allow us to determine the capabilities of the cell.

## A. Synergy between Second Messengers

The interaction between second messenger systems is extremely complex, so only the more obvious examples will be discussed here.

### 1. Tyrosine kinases and GTP binding proteins

The activity of several of the growth factors with receptors that contain tyrosine kinase activity (for example, EGF, FGF, and insulin) and of compounds that activate protein kinase C (for example, phorbol esters) are inhibited by pertussis toxin (Crouch et al., 1990). This evidence suggests that the actions of these molecules may be mediated also in part by GTP-binding proteins. The GTP-binding proteins $G_i$ and $G_o$ are both phosphorylated in phospholipid vesicles on activation of incorporated insulin receptors, suggesting that the G proteins and insulin receptor interact in the lipid environment (Krupinski et al., 1988). The inactive GDP-bound form of the G protein is phosphorylated preferentially.

Cells responding to FGF bind it to cell-surface receptors that possess intrinsic tyrosine kinase activity (Coughlin et al., 1988; Friesel et al., 1989). Two polypeptides binding to FGF have been identified by cross-linking (Friesel et al., 1986; Neufeld and Gospodarowicz, 1986; Kan et al., 1988; Walicke et al., 1989). More recently, two distinct high affinity receptors for FGF have been cloned (Sunderland, 1978). These are the *flg* and *bek* human genes; when expressed in 3T3 cells, both gene products appear to bind aFGF and bFGF with equal high affinity.

### 2. Phosphorylation

Phosphorylation is a mechanism by which several factors may converge on one transduction pathway. NGF has been shown to be involved in the regulation of the phosphorylation of many proteins, but which of these may be involved in the signaling cascade and which of these are spurious is not clear. NGF increases the activities of many protein kinases, including an endogenous 250-kDa cytoskeletal protein, a microtubule-associated protein (MAP-2) (Landreth et al., 1990), cAMP-dependent and NGF-dependent kinases that both phosphorylate MAP-2 (Sano et al., 1990), stathmin (Doye et al., 1990), and $Ca^{2+}$/phospholipid-dependent protein kinases (Cremins et al., 1986). NGF down-regulates calmodulin-dependent protein kinase III in PC12 cells via a cAMP-dependent protein kinase (Brady et al., 1990). NGF increases the activity of calpactin I and endonexin II in PC12 cells (Schlaepfer and Haigler, 1990). Although NGF and FGF regulate neurite outgrowth and gene expression via protein kinase C- and cAMP-independent mechanisms (Damon et al., 1990), NGF-directed neurite outgrowth is inhibited by sphingosine, an inhibitor of protein kinase C (Hall et al., 1988).

The relationship between the low affinity (fast) and high affinity (slow) receptors for NGF still is not resolved; it is not yet established whether the high-affinity receptor for NGF is linked to a kinase. Activation of PC12 cells with insulin and NGF produces an overlapping pattern of phosphorylated proteins (Halegoua and Patrick, 1980).

Cytoplasmic sequences of NGF receptor are required for high affinity binding (Hempstead et al., 1990); this requirement may reflect internalization (Stach and Wagner, 1982; Olender et al., 1981) or the interaction with an accessory protein required for high affinity binding (Hempstead et al., 1990). NGF is sequestered after binding to embryonic sensory neurons (Olender et al., 1981). There is some controversy about the existence of nuclear receptors for NGF. NGF receptors are found in the plasma membrane and nuclei of embryonic DRG neurons (Andres et al., 1977).

Protein I is phosphorylated at multiple sites by cAMP- and $Ca^{2+}$-dependent protein kinases with a different spectrum of phosphorylation (Huttner et al., 1981). In the superior cervical ganglion, the phosphorylation of protein I is regulated by dopamine and depolarizing agents (Nestler and Greengard, 1980).

### 3. Substrate requirement for neurotrophic action

Neurons appear to require contact with an appropriate substrate to survive. Many studies have been done on the substrate requirement for neuronal survival and neurite outgrowth. The molecules on the extracellular matrix and nonneuronal cell surfaces required for outgrowth differ (Ard et al., 1987). It seems likely that most of the actions of the substrate molecules are through specific receptors, presumably with their own second messenger systems. Parasympathetic motoneurons interact with both Schwann cell-derived neurite promoting factors and laminin during development (Tomaselli and Reichardt, 1988). Retinal ganglion cells and ciliary neurons lose their responsiveness to laminin while their axons are innervating the targets (Landmesser and Pilar, 1974a,b; Cohen et al., 1986, 1987; Tomaselli and Reichardt, 1988), suggesting that the presence of laminin receptors on both central and peripherl neurons may be regulated by target contact. Laminin has been shown to be elevated in vivo during the regeneration of the sciatic nerve (Kuecherer-Ehret et al., 1990). Chick motoneurons respond to a myotube-conditioned medium, and laminin is essential for the substrate-binding neurite-promoting activity (Calof and Reichardt, 1985). NGF and laminin can act in synergy to promote neurite initiation and extension from the trigeminal V motor nucleus of early chick embryos (Heaton, 1989), but fibronectin and muscle-derived conditioned medium cannot (Heaton et al., 1990). When fibronectin interacts with heparin, there is a structural change in its NH-terminal domain that may give rise to a matrix-driven translocation (Khan et al., 1988). L1 and N-cadherin bind Schwann cells or astrocytes and neurons to enable good neurite outgrowth to occur (Bixby et al., 1987, 1988; Neugebauer et al., 1988).

The integrins are receptors for laminin, collagen type IV, and fibronectin (Tomaselli et al., 1987, 1988a,b). The binding of these factors is regulated by local factors, most probably intracellular second messengers (Tomaselli and Reichardt, 1988). Integrin receptors interact with the cytoskeleton in a linkage that uses the cytoskeleton-associated protein talin (Horwitz et al., 1986; Burn et al., 1988). The interaction between the integrins, talin, and the matrix may be inhibited by tyrosine phosphorylation of the b1 subunit cytoplasmic domain of the fibronectin receptor (Hirst et al., 1986; Buck and Horwitz, 1987).

## B. Induction of Receptors

Receptor production also can be induced, leading to possible synergistic effects. For example, an increase in $K^+$ induces the formation of the high affinity NGF receptor on chick DRG cells (Ennulat and Stach, 1987).

## C. Multifactorial Response

The interaction between calcium influx and phosphorylation is another potential site at which many of the receptor-mediated cascades can converge. PKC can be activated fully at low $Ca^{2+}$ concentrations if specific diglycerides and phospholipids are present (Kishimoto et al., 1989). Depolarization ($Ca^{2+}$ influx) leads to phosphorylation of synapsin Ia and Ib, protein IIIa and IIIb (Wang et al., 1988), and 87- and 49-kDa proteins (Wang et al., 1988). $Ca^{2+}$ influx into nerve terminals leads to activation of $Ca^{2+}$/calmodulin kinases I and II, PKC, and some phosphatases (Wang et al., 1988). Synapsin I is a neuron-specific phosphoprotein that can interact with small synaptic vesicles and F-actin (Benfenati et al., 1989). In PC12 cells it is phosphorylated at a unique site when activated with NGF (Romano et al., 1987); the phosphorylation results in formation of F-actin bundles (Bahler and Greengard, 1987). This protein is likely to be of major importance in transmitter secretion and formation of synaptic connections. The antitrypanosomal and antifilarial drug suramin weakly inhibits $Ca^{2+}$ uptake by the endoplasmic reticulum of permeabilized cells and strongly inhibits $Ca^{2+}$ induced by $IP_3$ and GTP from nonmitochondrial stores in Swiss 3T3 fibroblasts (Ghosh et al., 1988; Seewald et al., 1989), but fails to block $Ca^{2+}$ release by arachidonic acid (Powis et al., 1990; Seewald et al., 1990), whereas heparin only blocks $Ca^{2+}$ release induced by $IP_3$ (Ghosh et al., 1988; Kobayashi et al., 1988; Seewald et al., 1989). Heparin may inhibit $IP_3$-induced $Ca^{2+}$ release by binding to a receptor for $IP_3$ on the endoplasmic reticulum $Ca^{2+}$-release channel (Worley et al., 1987; Ghosh et al., 1988). cAMP-dependent kinase but not PKC or $Ca^{2+}$/calmodulin-dependent protein kinase phosphorylation of $IP_3$ receptor decreases its release of calcium (Supattapone et al., 1988). The GTP releaseable pool of $Ca^{2+}$ appears to be different from the pool released by $IP_3$ (Dawson, 1985; Ghosh et al., 1988) and GTP may induce the translocation of $Ca^{2+}$ from an $IP_3$-insensitive to an $IP_3$-sensitive pool (Mullaney et al., 1987).

The increase in internal $Ca^{2+}$ caused by opening of calcium channels during a prolonged depolarization may activate many metabolic processes. Entry of $Ca^{2+}$ through L-type channels induces c-*fos* expression in PC12 cells (Morgan and Curran, 1986). Long-term depolarization of rat sympathetic neurons by levels of $K^+ > 20$ m$M$ depressed acetylcholine synthesis and increased catecholamine synthesis (Walicke et al., 1977; Walicke and Patterson, 1981). These effects oppose those of conditioned medium, suggesting that $Ca^{2+}$ can have opposite effects on sympathetic neurons. Further, the effects of cardiac extracts on TH induction were not caused by changes in $Ca^{2+}$ (Hill et al., 1980). These effects of calcium are mediated by $Ca^{2+}$ entry through L-channels (Vidal et al., 1989). TH expression is stimulated by $Ca^{2+}$ entry into sympathetic neurons while choline acetyltransferase induction was blocked (Vidal et al., 1989), but TH induction by elevated $K^+$ was unaffected by the removal of external $Ca^{2+}$ (Hefti et al., 1982; Kessler, 1985). In the absence of external $Ca^{2+}$, depolarization by high $K^+$ does not cause any change in internal $Ca^{2+}$ in rat sympathetic neurons (Koelle et al., 1982). The blockade of choline acetyltransferase induction was unaffected by calmidazolium, a calmodulin inhibitor (Vidal et al., 1989), but TH and c-*fos* induction by depolarization were blocked by another calmodulin inhibitor, trifluoperazine (Hefti et al., 1982; Morgan and Curran, 1986).

The beta-adrenergic receptor is only phosphorylated in vivo when it has bound the appropriate ligand (Benovic et al., 1986). The beta-adrenergic receptor kinase can translocate after activation by beta agonists and prostaglandin E1 to act on multiple adenylate cyclase-coupled receptors (Strasser et al., 1986). The beta-adrenergic receptor kinase is part of a multigene family (Benovic et al., 1989a,b) and is inhibited by several polyanions including heparin (Benovic et al., 1989a,b). The $IP_3$ receptor has been isolated (Chadwick et al., 1990) and has a size of 224 kDa. Heparin has been shown to block $Ca^{2+}$ release in smooth muscle (Kobayashi et al., 1988, 1989) and $IP_3$-dependent gating of $Ca^{2+}$-channel conductance (Ehrlich and Watras, 1988), as well as block the binding of $IP_3$ to its receptor (Chadwick et al., 1990).

## D. Gene Induction

NGF is involved in the induction of many genes. Multiple mechanisms can achieve this regulation (Cho et al., 1989). NGF alters the mRNA for the alpha subunit of $G_s$, the adenylate cyclase-activating protein (Tjaden et al., 1990); c-*fos* (Milbrandt, 1986); and a gene in PC12 cells for NGFI-B that is rapidly and transiently expressed. NGFI-B is a 61 kDa protein with structural homology to the glucocorticoid receptor gene family (Milbrandt, 1988). NGF regulates the TH gene in PC12 cells via a fat-specific element (FSE) that binds the *fos–jun* transcription factor complex (Gizang-Ginsberg and Ziff, 1990). NGF and EGF both induce rapid changes in the level of proto-oncogenes in PC12 cells (Greenberg et al., 1985).

TGF stimulates the transcription of genes encoding extracellular matrix proteins and beta-integrin (Keski-Oja et al., 1988; Rossi et al., 1988).

Increases in intraneuronal free calcium result in a rapid transient induction of the *fos* and *jun* proto-oncogenes (Morgan and Curran, 1988). $Ca^{2+}$ inducibility of human c-*fos* has been localized to a 60-bp sequence distinct from that involved in transcriptional control by growth factors and containing the TGACGTTT motif (Sheng et al., 1988; Fisch et al., 1989). A similar motif in the mouse c-*fos* promoter binds a putative transcription factor (Gilman et al., 1988). The TH promoter contains the GACGTCA motif (Harrington et al., 1987). It is thus possible that TH and c-*fos* are independently regulated by $Ca^{2+}$ through the same mechanism.

## VI. OTHER MECHANISMS TO ACHIEVE A RETROGRADE MESSAGE

If the neurotrophic molecule itself is not the second messenger, then there must be translocation of some other messenger as the signal from the terminals in the target tissue to the nucleus.

Two potential types of neurotrophic molecule have either a short-lived second messenger or a long-lived second messenger. The method to transfer the message will differ greatly for each. If the molecule required to deliver the message to the nucleus is long lived, then the transport of the second messenger is sufficient. Otherwise there is a need to transport the message-generating system. The latter appears to be the case for NGF. The two possibilities are illustrated in Fig. 4.

### A. G-Protein Translocation

One of the major arguments against aFGF and bFGF as neurotrophic factors is the lack of an appropriate secretory sequence (Esch et al., 1985a,b), making their regulated release by the target tissue unlikely. If FGF is not this target-derived neurotrophic factor, then one of the other members of the FGF gene family may be (Huebner et al., 1988; Zhan et al., 1988; Brookes et al., 1989). In this case, the protein would have to share an antigenic epitope with aFGF and may itself be retrogradely transported. If, in fact, aFGF is an essential neurotrophic factor for the ciliary neurons, then the retrograde message must be conveyed by a mechanism other than the retrograde transport of FGF itself. It is not essential, however, for the neurotrophic factor itself to be retrogradely transported, but clearly some other signal must reach the neuronal cell body. A likely alternative is the retrograde axonal transport of one of the second messengers. It has been shown that in platelets there is a translocation of G protein from the cell membrane to the cytoskeleton (Crouch et al., 1989). It has also been shown that activation of *BALB*/c 3T3 fibroblasts results in the translocation of the alpha subunit of $G_i$ from the membrane to the nuclear

---

FIGURE 4

Scheme for the two different types of retrograde neurotrophic factors. (A) Classical retrophin. The neurotrophic factor is taken up by receptor-mediated endocytosis at the nerve terminal into receptor-coated vesicles that are transported back to the cell body where the labile second messenger generated by the receptor–factor complex can reach the nucleus. (B) Nontransported retrophin. The growth factor–receptor complex at the nerve terminal initiates a cascade that results in the generation of a stable second messenger that is capable of being retrogradely transported to reach the nucleus undegraded.

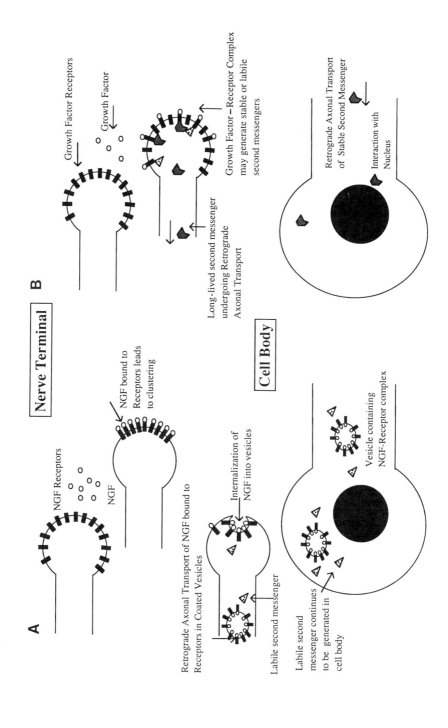

chromatin (Crouch, 1991). A similar phenomenon could result in the retrograde axonal transport of the neurotrophic message for FGF.

## B. Phosphorylated Proteins

Any of the phosphorylation reactions may generate an appropriate second messenger to transport to the cell body. Further, phosphorylation reactions can provide a mechanism for the interaction of growth factors at the terminal to generate specific alterations in potential retrograde messengers (Greengard, 1987).

## C. Myristoylation–Demyristoylation

Alterations in the hydrophobicity of molecules may come about by addition or cleavage of a lipid component, resulting in a molecule suitable for retrograde transport. The 87 kDa protein phosphorylated by PKC seems to be the same as a 68 kDa protein myristoylated in response to lipopolysaccharide in macrophages (Aderem et al., 1988). This myristoylation could allow the targeting of the protein to the membrane, where it is more readily phosphorylated by PKC; this would prime the cells for a subsequent activation. The 87 kDa protein has been localized in brain, where it is phosphorylated in response to phorbol esters and depolarization (Wang et al., 1988, 1989). This protein appears to be myristoylated in murine frontal cortex cells (Aderem et al., 1988).

GAP-43 appears in DRG cells and in the dorsal horn in the spinal cord of neonatal rats following a peripheral nerve injury (Woolf et al., 1990).

Myristoylated alanine-rich C-kinase substrate (MARCKS) (Stumpo et al., 1989) is an 87 kDa protein that is a major specific substrate for PKC in rat brain (Ouimet et al., 1990). MARCKS is translocated after phosphorylation in isolated nerve terminals (Wang et al., 1989), an event associated with its release from membranes into the cytosol (Narayanan and Narayanan, 1981).

## VII. CONCLUSIONS

The second messenger systems allow a complexity of neuronal activation that is appropriate for their development. It is unlikely, in most systems, that the simple model of the retrogradely transported neurotrophic molecule is the explanation for cell death and survival. There is a greater opportunity to benefit from the complexity of the external environment of the terminal region if interactions can occur between active molecules. The final signal conveyed to the cell body may reflect the type of second messenger used by the factor (see Fig. 4). If this messenger is ephemeral, then there is a need to take its mechanism of production to the cell body, so the transport of the factor–receptor complex is required. On the other hand, if a long-lived messenger is generated, then only it must be conveyed to the cell body. This may be the more usual method for target tissue–neuron communication.

## REFERENCES

Abraham, J. A., Mergia, A., Whang, J. L., Tumulo, A., Friedman, J. Hjerrild, K. A., Gospodarowicz, D., and Fiddes, J. C. (1986a). Nucleotide sequence of a bovine clone encoding the angiogenic protein, basic fibroblast growth factor. *Science* **233**, 545–548.

Abraham, J. A., Whang, J. L., Tumulo, A., Mergia, A., Friedman, J., Gospodarowicz, D., and Fiddes, J. C. (1986b). Human basic fibroblast growth factor: Nucleotide sequence and genomic organization. *EMBO J.* **5**, 2523–2528.

Aderem, A. A., Albert, K. A., Keum, M. M., Wang, J. K., Greengard, P., and Cohn, Z. A. (1988). Stimulus-dependent myristoylation of a major substrate for protein kinase C. *Nature (London)* **332**, 362–364.

Andres, R. Y., Jeng, I., and Bradshaw, R. A. (1977). Nerve growth factor receptors: Identification of distinct classes in plasma membranes and nuclei of embryonic dorsal root neurons. *Proc. Natl. Acad. Sci. U.S.A.* **74**, 2785–2789.

Angeletti, R. H., Angeletti, P. U., and Levi-Montalcini, R. (1972). Selective accumulation of ($^{125}$I)-labelled nerve growth factor in sympathetic ganglia. *Brain Res.* **46**, 421–425.

Ard, M. D., Bunge, R. P., and Bunge, M. B. (1987). Comparison of the Schwann cell surface and Schwann cell extracellular matrix as promotors of neurite growth. *J. Neurocytol.* **16**, 539–555.

Bahler, M., and Greengard, P. (1987). Synapsin I bundles F-actin in a phosphorylation-dependent manner. *Nature (London)* **326**, 704–707.

Balcuns Maccabe, A., Gasseling, M. T., and Saunders, J. W. (1973). Spatiotemporal distribution of mechanisms that control outgrowth and anteroposterior polarization of the limb bud in the chick embryo. *Mech. Ageing Dev.* **2**, 1–12.

Barde, Y. A., Davies, A. M., Johnson, J. E., Lindsay, R. M., and Thoenen, H. (1987). Brain derived neurotrophic factor. *Prog. Brain Res.* **71**, 185–189.

Bashkin, P., Doctrow, S., Klagsbrun, M., Svahn, C. M., Folkman, J., and Vlodavsky, I. (1989). Basic fibroblast growth factor binds to subendothelial extracellular matrix and is released by heparitinase and heparin-like molecules. *Biochemistry* **28**, 1737–1743.

Basilico, C., Newman, K. M., Curatola, A. M., Talarico, D., Mansukhani, A., Velcich, A., and Delli-Bovi, P. (1989). Expression and activation of the K-*fgf* oncogene. *Ann. N.Y. Acad. Sci.* **567**, 95–103.

Belford, D. A., Godovac-Zimmermann, J., Simpson, R. J., and Hendry, I. A. (1989). Identification of the parasympathetic neurotrophic factor. *Neurosci. Lett. (suppl.)* **34**, S58.

Benfenati, F., Valtorta, F., Bahler, M., and Greengard, P. (1989). Synapsin I, a neuron-specific phosphoprotein interacting with small synaptic vesicles and F-actin. *Cell Biol. Int. Rep.* **13**, 1007–1021.

Benovic, J. L., Strasser, R. H., Caron, M. G., and Lefkowitz, R. J. (1986). Beta-adrenergic receptor kinase: identification of a novel protein kinase that phosphorylates the agonist-occupied form of the receptor. *Proc. Natl. Acad. Sci. U.S.A.* **83**, 2797–2801.

Benovic, J. L., DeBlasi, A., Stone, W. C., Caron, M. G., and Lefkowitz, R. J. (1989a). Beta-adrenergic receptor kinase: Primary structure delineates a multigene family. *Science* **246**, 235–240.

Benovic, J. L., Stone, W. C., Caron, M. G., and Lefkowitz, R. J. (1989b). Inhibition of the beta-adrenergic receptor kinase by polyanions. *J. Biol. Chem.* **264**, 6707–6710.

Bixby, J. L., Pratt, R. S., Lilien, J., and Reichardt, L. F. (1987). Neurite outgrowth on muscle cell surfaces involves extracellular matrix receptors as well as $Ca^{2+}$-dependent and -independent cell adhesion molecules. *Proc. Natl. Acad. Sci. U.S.A.* **84**, 2555–2559.

Bixby, J. L., Lilien, J., and Reichardt, L. F. (1988). Identification of the major proteins that promote neuronal process outgrowth on Schwann cells in vitro. *J. Cell Biol.* **107,** 353–361.
Black, I. B., and Geen, S. C. (1973). Trans-synaptic regulation of adrenergic neuron development: Inhibition by ganglionic blockade. *Brain Res.* **63,** 291–302.
Black, I. B., Hendry, I. A., and Iversen, L. L. (1971). Trans-synaptic regulation of growth and development of adrenergic neurons in a mouse sympathetic ganglion. *Brain Res.* **34,** 229–240.
Black, I. B., Hendry, I. A., and Iversen, L. L. (1972). Effects of surgical decentralization and nerve growth factor on the maturation of adrenergic neurons in a mouse sympathetic ganglion. *J. Neurochem.* **19,** 1367–1377.
Blottner, D., Brüggemann, W., and Unsicker, K. (1989a). Ciliary neurotrophic factor supports target-deprived preganglionic sympathetic spinal cord neurons. *Neurosci. Lett.* **105,** 316–320.
Blottner, D., Westermann, R., Grothe, C., Bohlen, P., and Unsicker, K. (1989b). Basic fibroblast growth factor in the adrenal gland. *Eur. J. Neurosci.* **1,** 471–478.
Bonyhady, R. E., Hendry, I. A., Hill, C. E., and McLennan, I. S. (1980). Characterization of a cardiac muscle factor required for the survival of cultured parasympathetic neurons. *Neurosci. Lett.* **18,** 197–201.
Bottenstein, J. E., Hayashi, I., Hutchings, S., Masui, H., Mather, J. P., McClure, D. B., Ohasa, S., Rizzino, A., Sato, G., Serrero, G., Wolfe, R. A., and Wu, R. (1979). The growth of cells in serum-free hormone-supplemented media. *Methods Enzymol.* **58,** 94–109.
Bottenstein, J. E., Skaper, S. D., Varon, S., and Sato, G. H. (1980). Selective survival of neurons from chick embryo sensory ganglionic dissociates utilizing serum-free supplemented medium. *Exp. Cell Res.* **125,** 183–190.
Brady, M. J., Nairn, A. C., Wagner, J. A., and Palfrey, H. C. (1990). Nerve growth factor-induced down-regulation of calmodulin-dependent protein kinase III in PC12 cells involves cyclic AMP-dependent protein kinase. *J. Neurochem.* **54,** 1034–1039.
Brookes, S., Smith, R., Thurlow, J., Dickson, C., and Peters, G. (1989). The mouse homologue of hst/k-FGF: Sequence, genome organization, and location relative to int-2. *Nucleic Acids Res.* **17,** 4037–4045.
Brunso-Bechtold, J. K., and Hamburger, V. (1979). Retrograde transport of nerve growth factor in chicken embryo. *Proc. Natl. Acad. Sci. U.S.A.* **76,** 1494–1496.
Buck, C. A., and Horwitz, A. F. (1987). Integrin, a transmembrane glycoprotein complex mediating cell–substratum adhesion. *J. Cell Sci. Suppl.* **8,** 231–250.
Burn, P., Kupfer, A., and Singer, S. J. (1988). Dynamic membrane–cytoskeletal interactions: Specific association of integrin and talin arises in vivo after phorbol ester treatment of peripheral blood lymphocytes. *Proc. Natl. Acad. Sci. U.S.A.* **85,** 497–501.
Calof, A. L., and Reichardt, L. F. (1984). Motoneurons purified by cell sorting respond to two distinct activities in myotube-conditioned medium. *Dev. Biol.* **106,** 194–210.
Calof, A. L., and Reichardt, L. F. (1985). Response of purified chick motoneurons to myotube-conditioned medium: Laminin is essential for the substratum-binding, neurite outgrowth-promoting activity. *Neurosci. Lett.* **59,** 183–189.
Campenot, R. B. (1977). Local control of neurite development by nerve growth factor. *Proc. Natl. Acad. Sci. U.S.A.* **74,** 4516–4519.
Campenot, R. B. (1982a). Development of sympathetic neurons in compartmentalized cultures. II. Local control of neurite survival by nerve growth factor. *Dev. Biol.* **93,** 13–21.
Campenot, R. B. (1982b). Development of sympathetic neurons in compartmentalized cultures. I. Local control of neurite growth by nerve growth factor. *Dev. Biol.* **93,** 1–12.
Caroni, P., and Grandes, P. (1990). Nerve sprouting in innervated adult skeletal muscle induced by exposure to elevated levels of insulin-like growth factors. *J. Cell Biol.* **110,** 1307–1317.
Chadwick, C. C., Saito, A., and Fleischer, S. (1990). Isolation and characterization of the inositol trisphosphate receptor from smooth muscle. *Proc. Natl. Acad. Sci. U.S.A.* **87,** 2132–2136.
Chang, J. Y., Martin, D. P., and Johnson, E. M. (1990). Interferon suppresses sympathetic neuronal cell death caused by nerve growth factor deprivation. *J. Neurochem.* **55,** 436–445.
Chen, M. G., Chen, J. S., Calissano, P., and Levi-Montalcini, R. (1977). Nerve growth factor prevents

vinblastine destructive effects on sympathetic ganglia in newborn mice. *Proc. Natl. Acad. Sci. U.S.A.* **74**, 5559–5563.

Cho, K. O., Skarnes, W. C., Minsk, B., Palmieri, S., Jackson-Grusby, L., and Wagner, J. A. (1989). Nerve growth factor regulates gene expression by several distinct mechanisms. *Mol. Cell. Biol.* **9**, 135–143.

Claude, P., Parada, I. M., Gordon, K. A., D'Amore, P. A., and Wagner, J. A. (1988). Acidic fibroblast growth factor stimulates adrenal chromaffin cells to proliferate and to extend neurites, but is not a long-term survival factor. *Neuron* **1**, 783–790.

Clegg, D. O., Large, T. H., Bodary, S. C., and Reichardt, L. F. (1989). Regulation of nerve growth factor mRNA levels in developing rat heart ventricle is not altered by sympathectomy. *Dev. Biol.* **134**, 30–37.

Cohen, J., Burne, J. F., Winter, J., and Bartlett, P. F. (1986). Retinal ganglion cells lose response to laminin with maturation. *Nature (London)* **322**, 465–467.

Cohen, J., Burne, J. F., McKinlay, C., and Winter, J. (1987). The role of laminin and the laminin/fibronectin receptor complex in the outgrowth of retinal ganglion cell axons. *Dev. Biol.* **122**, 407–418.

Cordeiro, P. G., Seckel, B. R., Lipton, S. A., D'Amore, P. A., Wagner, J., and Madison, R. (1989). Acidic fibroblast growth factor enhances peripheral nerve regeneration in vivo. *Plast. Reconstr. Surg.* **83**, 1013–1019.

Coughlin, S. R., Barr, P. J., Cousens, L. S., Fretto, L. J., and Williams, L. T. (1988). Acidic and basic fibroblast growth factors stimulate tyrosine kinase activity in vivo. *J. Biol. Chem.* **263**, 988–993.

Cragg, B. G. (1970). What is the signal for chromatolysis? *Brain Res.* **23**, 1–21.

Cremins, J., Wagner, J. A., and Halegoua, S. (1986). Nerve growth factor action is mediated by cyclic AMP-1 and $Ca^{2+}$/phospholipid-dependent protein kinases. *J. Cell Biol.* **103**, 887–893.

Crouch, M. F. (1991). Growth factor-induced cell division is paralled by translocation of $G_i\alpha$ to the nucleus. *FASEB J.* **5**, 200–206.

Crouch, M. F., and Hendry, I. A. (1991). Co-activation of insulin-like growth factor-I receptors and protein kinase C results in parasympathetic neuronal survival. *J. Neurosci. Res.* **28**, 115–120.

Crouch, M. F., Winegar, D. A., and Lapetina, E. G. (1989). Epinephrine induces changes in the subcellular distribution of the inhibitory GTP-binding protein Gi alpha-2 and a 38-kDa phosphorylated protein in the human platelet. *Proc. Natl. Acad. Sci. U.S.A.* **86**, 1776–1780.

Crouch, M. F., Belford, D. A., Milburn, P. J., and Hendry, I. A. (1990). Pertussis toxin inhibits EGF-, phorbol ester-, and insulin-stimulated DNA synthesis in BALB/c3T3 cells: Evidence for post receptor activation of Gi alpha. *Biochem. Biophys. Res. Commun.* **167**, 1369–1376.

Dahm, L. M., and Landmesser, L. T. (1988). The regulation of intramuscular nerve branching during normal development and following activity blockade. *Dev. Biol.* **130**, 621–644.

Damon, D. H., D'Amore, P. A., and Wagner, J. A. (1988). Sulfated glycosaminoglycans modify growth factor-induced neurite outgrowth in PC12 cells. *J. Cell Physiol.* **135**, 293–300.

Damon, D. H., Lobb, R. R., D'Amore, P. A., and Wagner, J. A. (1989). Heparin potentiates the action of acidic fibroblast growth factor by prolonging its biological half-life. *J. Cell Physiol.* **138**, 221–226.

Damon, D. H., D'Amore, P. A., and Wagner, J. A. (1990). Nerve growth factor and fibroblast growth factor regulate neurite outgrowth and gene expression in PC12 cells via both protein kinase C- and cAMP-independent mechanisms. *J. Cell Biol.* **110**, 1333–1339.

Dawson, A. P. (1985). GTP enhances inositol trisphosphate-stimulated $Ca^{2+}$ release from rat liver microsomes. *FEBS Lett.* **185**, 147–150.

Delli-Bovi, P. Curatola, A. M., Newman, K. M., Sato, Y., and Moscatelli, D. (1988). Processing secretion and biological properties of a novel growth factor of the fibroblast growth factor family with oncogenic potential. *Mol. Cell. Biol.* **8**, 2933–2941.

Dennis, P. A., Saksela, O., Harpel, P., and Rifkin, D. B. (1989). Alpha 2-macroglobulin is a binding protein for basic fibroblast growth factor. *J. Biol. Chem.* **264**, 7210–7216.

DiCicco-Bloom, E., Townes-Anderson, E., and Black, I. B. (1990). Neuroblast mitosis in dissociated culture: Regulation and relationship to differentiation. *J. Cell Biol.* **110**, 2073–2086.

Dickson, C., and Peters, G. (1987). Potential oncogene product related to growth factors. *Nature (London)* **326**, 833–830.

Dohrmann, U., Edgar, D., Sendtner, M., and Thoenen, H. (1986). Muscle-derived factors that support survival and promote fiber outgrowth from embryonic chick spinal motor neurons in culture. *Dev. Biol.* **118**, 209–221.

Dohrmann, U., Edgar, D., and Thoenen, H. (1987). Distinct neurotrophic factors from skeletal muscle and the central nervous system interact synergistically to support the survival of cultured embryonic spinal motor neurons. *Dev. Biol.* **124**, 145–152.

Doye, V., Boutterin, M. C., and Sobel, A. (1990). Phosphorylation of stathmin and other proteins related to nerve growth factor-induced regulation of PC12 cells. *J. Biol. Chem.* **265**, 11650–11655.

Eccleston, P. A., Gunton, D. J., and Silberberg, D. H. (1985). Requirements for brain cell attachment, survival, and growth in serum-free medium: Effects of extracellular matrix, epidermal growth factor, and fibroblast growth factor. *Dev. Neurosci.* **7(5–6)**, 308–322.

Edds, M. V. (1950). Collateral regeneration of residual motor axons in partially denervated muscles. *J. Exp. Zool.* **113**, 517–552.

Ehrlich, B. E., and Watras, J. (1988). Inositol 1,4,5-triphosphate activates a channel from smooth muscle sarcoplasmic reticulum. *Nature (London)* **336**, 583–586.

Eichele, G., Tickle, C., and Alberts, B. M. (1985). Studies on the mechanism of retinoid-induced pattern duplications in the early chick limb bud: Temporal and spatial aspects. *J. Cell Biol.* **101**, 1913–1920.

Ennulat, D. J., and Stach, R. W. (1987). Induction of the high-affinity nerve growth factor receptor on embryonic chicken sensory nerve cells by elevated potassium. *Neurochem. Res.* **12**, 839–850.

Esch, F., Baird, A., Ling, N., Ueno, N., Hill, F., Denoroy, L., Klepper, R., Gospodarowicz, D., Böhlen, P., and Guillemin, R. (1985a). Primary structure of bovine pituitary basic fibroblast growth factor (FGF) and comparison with the amino-terminal sequence of bovine brain acidic FGF. *Proc. Natl. Acad. Sci. U.S.A.* **82**, 6507–6511.

Esch, F., Ueno, N., Baird, A., Hill, F., Denoroy, L., Ling, N., Gospodarowicz, D., and Guillemin, R. (1985b). Primary structure of bovine brain acidic fibroblast growth factor (FGF). *Biochem. Biophys. Res. Commun.* **133**, 554–562.

Feige, J. J., and Baird, A. (1988). Glycosylation of the basic fibroblast growth factor receptor. The contribution of carbohydrate to receptor function. *J. Biol. Chem.* **263**, 14023–14029.

Ferguson, I. A., Schweitzer, J. B., and Johnson, E. M. (1991). Basic fibroblast growth factor: Receptor-mediated internalization, metabolism, and anterograde axonal transport in retinal ganglion cells. *J. Neurosci.* **10**, 2176–2189.

Fernyhough, P., and Ishii, D. N. (1987). Nerve growth factor modulates tubulin transcript levels in pheochromocytoma PC12 cells. *Neurochem. Res.* **12**, 891–899.

Fernyhough, P., Mill, J. F., Roberts, J. L., and Ishii, D. N. (1989). Stabilization of tubulin mRNAs by insulin and insulin-like growth factor I during neurite formation. *Brain Res. Mol. Brain Res.* **6**, 109–120.

Finch, P. W., Rubin, J. S., Miki, T., Ron, D., and Aaronson, S. A. (1989). Human KGF is FGF-related with properties of a paracrine effector of epithelial cell growth. *Science* **245**, 752–755.

Fisch, T. M., Prywes, R., Simon, M. C., and Roeder, R. G. (1989). Multiple sequence elements in the c-*fos* promoter mediate induction by cAMP. *Genes Dev.* **3**, 198–211.

Flaumenhaft, R., Moscatelli, D., Saksela, O., and Rifkin, D. B. (1989). Role of extracellular matrix in the action of basic fibroblast growth factor: Matrix as a source of growth factor for long-term stimulation of plasminogen activator production and DNA synthesis. *J. Cell Physiol.* **140**, 75–81.

Frazier, W. A., Angeletti, R. H., and Bradshaw, R. A. (1972). Nerve growth factor and insulin. *Science* **176**, 482–488.

Friesel, R., Burgess, W. H., Mehlman, T., and Maciag, T. (1986). The characterization of the receptor for endothelial cell growth factor by covalent ligand attachment. *J. Biol. Chem.* **261**, 7581–7584.

Friesel, R., Burgess, W. H., and Maciag, T. (1989). Heparin-binding growth factor 1 stimulates tyrosine phosphorylation in NIH 3T3 cells. *Mol. Cell. Biol.* **9**, 1857–1865.

Fukada, K. (1985). Purification and partial characterization of a cholinergic neuronal differentiation factor. *Proc. Natl. Acad. Sci. U.S.A.* **82**, 8795–8799.

Ghosh, T. K., Eis, P. S., Mullaney, J. M., Ebert, C. L., and Gill, D. L. (1988). Competitive, reversible, and potent antagonism of inositol 1,4,5-trisphosphate-activated calcium release by heparin. *J. Biol. Chem.* **263**, 11075–11079.

Giess, M. C., and Weber, M. (1984). Acetylcholine metabolism in rat spinal cord cultures: Regulation by a factor involved in the determination of the neurotransmitter phenotype of sympathetic neurons. *J. Neurosci.* **4**, 1442–1452.

Giller, E. L., Schrier, B. K., Shainburg, A., Fisk, H. R., and Nelson, P. G. (1973). Choline acetyltransferase activity is increased in combined cultures of spinal cord and muscle cells from mice. *Science* **182**, 588–589.

Gilman, M. Z., Berkowitz, L. A., Feramisco, J. R., Franza, B. R., Graham, R. M., Riabowol, K. T., and Ryan, W. A. (1988). Intracellular mediators of c-*fos* induction. *Cold Spring Harbor Symp. Quant. Biol.* **53**, 761–767.

Gizang-Ginsberg, E., and Ziff, E. B. (1990). Nerve growth factor regulates tyrosine hydroxylase gene transcription through a nucleoprotein complex that contains c-*Fos*. *Genes Dev.* **4**, 477–491.

Globus, R. K., Plouet, J., and Gospodarowicz, D. (1989). Cultured bovine bone cells synthesize basic fibroblast growth factor and store it in their extracellular matrix. *Endocrinology* **124**, 1539–1547.

Goodman, D. S. (1984). Vitamin A and retinoids in health and disease. *N. Engl. J. Med.* **310**, 1023–1031.

Goodman, R., Oesch, F., and Thoenen, H. (1974). Changes in enzyme patterns produced by high potassium concentration and dibutyryl cyclic AMP in organ cultures of sympathetic ganglia. *J. Neurochem.* **23**, 369–378.

Gorin, P. D., and Johnson, E. M. (1980a). Effects of exposure to nerve growth factor antibodies on the developing nervous system of the rat: An experimental autoimmune approach. *Dev. Biol.* **80**, 313–323.

Gorin, P. D., and Johnson, E. M. (1980b). Effects of long-term nerve growth factor deprivation on the nervous system of the adult rat: An experimental autoimmune approach. *Brain Res.* **198**, 27–42.

Greenberg, M. E., Greene, L. A., and Ziff, E. B. (1985). Nerve growth factor and epidermal growth factor induce rapid transient changes in proto-oncogene transcription in PC12 cells. *J. Biol. Chem.* **260**, 14101–14110.

Greene, L. A., Volonté, C., and Chalazonitis, A. (1990). Purine analogs inhibit nerve growth factor-promoted neurite outgrowth by sympathetic and sensory neurons. *J. Neurosci.* **10**, 1479–1485.

Greengard, P. (1987). Neuronal phosphoproteins. Mediators of signal transduction. *Mol. Neurobiol.* **1**, 81–119.

Grothe, C., Zachmann, K., Unsicker, K., and Westermann, R. (1990). High molecular weight forms of basic fibroblast growth factor recognized by a new anti-bFGF antibody. *FEBS Lett.* **260**, 35–38.

Halegoua, S., and Patrick, J. (1980). Nerve growth factor mediates phosphorylation of specific proteins. *Cell* **22**, 571–581.

Hall, F. L., Fernyhough, P., Ishii, D. N., and Vulliet, P. R. (1988). Suppression of nerve growth factor-directed neurite outgrowth in PC12 cells by sphingosine, an inhibitor of protein kinase C. *J. Biol. Chem.* **263**, 4460–4466.

Hamburger, V. (1934). The effects of wing bud extirpation on the development of the central nervous system in chick embryos. *J. Exp. Zool.* **68**, 449–494.

Hamburger, V. (1939). Motor and sensory hyperplasia following limb bud transplantation in chick embryos. *Physiol. Zool.* **13**, 268–284.

Hamburger, V. (1975). Cell death in the development of the lateral motor column of the chick embryo. *J. Comp. Neurol.* **160**, 535–546.

Hamburger, V., and Levi-Montalcini, R. (1949). Proliferation, differentiation, and degeneration in the spinal ganglia of the chick embryo under normal and experimental conditions. *J. Exp. Zool.* **111**, 457–502.

Hansson, H. A., Dahlin, L. B., Danielsen, N., Fryklund, L., Nachemson, A. K., Polleryd, P., Rozell, B., Skottner, A., Stemme, S., and Lundborg, G. (1986). Evidence indicating trophic importance of IGF-I in regenerating peripheral nerves. *Acta Physiol. Scand.* **126**, 609–614.

Hansson, H. A., Rozell, B., and Skottner, A. (1987). Rapid axoplasmic transport of insulin-like growth factor I in the sciatic nerve of adult rats. *Cell Tissue Res.* **247,** 241–247.

Harper, R. A. (1988). Specificity in the synergism between retinoic acid and EGF on the growth of adult human skin fibroblasts. *Exp. Cell Res.* **178,** 254–263.

Harrington, C. A., Lewis, E. J., Krzemien, D., and Chikaraishi, D. M. (1987). Identification and cell type specificity of the tyrosine hydroxylase gene promoter. *Nucleic Acids Res.* **15,** 2363–2384.

Haskell, B. E., Stach, R. W., Werrbach-Perez, K., and Perez-Polo, J. R. (1987). Effect of retinoic acid on nerve growth factor receptors. *Cell Tissue Res.* **247,** 67–73.

Haugen, P. K., and Letourneau, P. C. (1990). Interleukin-2 enhances chick and rat sympathetic, but not sensory, neurite outgrowth. *J. Neurosci. Res.* **25,** 443–452.

Heath, J., Bell, S., and Rees, A. R. (1981). Appearance of functional insulin receptors during the differentiation of embryonal carcinoma cells. *J. Cell Biol.* **91,** 293–297.

Heaton, M. B. (1989). Influence of laminin on the responsiveness of early chick embryo neural tube neurons to nerve growth factor. *J. Neurosci. Res.* **22,** 390–396.

Heaton, M. B., Paiva, M., and Swanson, D. (1990). Comparative responsiveness of early chick neural tube neurons to muscle-conditioned medium, laminin, NGF, and fibronectin. *Brain Res. Dev. Brain Res.* **52,** 113–119.

Hefti, F., Gnahn, H., Schwab, M. E., and Thoenen, H. (1982). Induction of tyrosine hydroxylase by nerve growth factor and by elevated $K^+$ concentrations in cultures of dissociated sympathetic neurons. *J. Neurosci.* **2,** 1554–1566.

Hempstead, B. L., Patil, N., Thiel, B., and Chao, M. V. (1990). Deletion of cytoplasmic sequences of the nerve growth factor receptor leads to loss of high-affinity ligand binding. *J. Biol. Chem.* **265,** 9595–9598.

Henderson, C. E., Huchet, M., and Changeux, J. P. (1981). Neurite outgrowth from embryonic chicken spinal neurons is promoted by media conditioned by muscle cells. *Proc. Natl. Acad. Sci. U.S.A.* **78,** 2625–2629.

Hendry, I. A. (1973). Trans-synaptic regulation of tyrosine hydroxylase activity in a developing mouse sympathetic ganglion: Effects of nerve growth factor (NGF), NGF-antiserum and pempidine. *Brain Res.* **56,** 313–320.

Hendry, I. A. (1975). The response of adrenergic neurones to axotomy and nerve growth factor. *Brain Res.* **94,** 87–97.

Hendry, I. A. (1977). Cell division in the developing sympathetic nervous system. *J. Neurocytol.* **6,** 299–309.

Hendry, I. A., and Belford, D. A. (1991). Lack of retrograde axonal transport of the heparin binding growth factors by chick ciliary neurons. *Int. J. Dev. Neurosci.* **9,** 243–250.

Hendry, I. A., and Campbell, J. (1976). Morphometric analysis of rat superior cervical ganglion after axotomy and nerve growth factor treatment. *J. Neurocytol.* **5,** 351–360.

Hendry, I. A., and Hill, C. E. (1980). Retrograde axonal transport of target tissue-derived macromolecules. *Nature (London)* **287(5783),** 647–649.

Hendry, I. A., Hill, C. E., Belford, D. A., and Watters, D. J. (1988). A monoclonal antibody to a parasympathetic neurotrophic factor causes immunoparasympathectomy in mice. *Brain Res.* **475,** 160–163.

Heumann, R., Schwab, M., and Thoenen, H. (1981). A second messenger required for nerve growth factor biological activity? *Nature (London)* **292,** 838–840.

Heuser, J. E., and Reese, T. S. (1973). Evidence for recycling of synaptic vesicle membrane during transmitter release at the frog neuromuscular junctions. *J. Cell Biol.* **57,** 315–344.

Hill, C. E., Hendry, I. A., and McLennan, I. S. (1980). Development of cholinergic neurons in cultures of rat superior cervical ganglia. Role of calcium and macromolecules. *Neurosci.* **5,** 1027–1032.

Hill, C. E., Jelinek, H., Hendry, I. A., McLennan, I. S., and Rush, R. A. (1988). Destruction by anti-NGF of autonomic, sudomotor neurones and subsequent hyperinnervation of the foot pad by sensory fibres. *J. Neurosci. Res.* **19,** 474–482.

Hill, M. A., and Bennett, M. R. (1983). Cholinergic growth factor from skeletal muscle elevated following denervation. *Neurosci. Lett.* **35,** 31–35.

Hirst, R., Horwitz, A., Buck, C., and Rohrschneider, L. (1986). Phosphorylation of the fibronectin receptor complex in cells transformed by oncogenes that encode tyrosine kinases. *Proc. Natl. Acad. Sci. U.S.A.* **83**, 6470–6474.

Hofer, M., and Barde, Y. A. (1988). Brain-derived neurotrophic factor prevents neuronal death in vivo. *Nature (London)* **331**, 261–262.

Hofer, M., Pagliusi, S. R., Hohn, A., Leibrock, J., and Barde, Y. A. (1990). Regional distribution of brain-derived neurotrophic factor mRNA in the adult mouse brain. *EMBO J.* **9**, 2459–2464.

Hohn, A., Leibrock, J., Bailey, K. A., and Barde, Y. A. (1990). Identification and characterization of a novel member of the nerve growth factor/brain-derived neurotrophic factor family. *Nature (London)* **344**, 339–341.

Horwitz, A., Duggan, K., Buck, C., Beckerle, M. C., and Burridge, K. (1986). Interaction of plasma membrane fibronectin receptor with talin—a transmembrane linkage. *Nature (London)* **320**, 531–533.

Huebner, K., Ferrari, A. C., Delli Bovi, P., Croce, C. M., and Basilico, C. (1988). The FGF-related oncogene, K-FGF, maps to human chromosome region 11q13, possibly near int-2. *Oncogene Res.* **3**, 263–270.

Huttner, W. B., DeGennaro, L. J., and Greengard, P. (1981). Differential phosphorylation of multiple sites in purified protein I by cyclic AMP-dependent and calcium-dependent protein kinases. *J. Biol. Chem.* **256**, 1482–1488.

Ishida, I., and Deguchi, T. (1983). Effect of depolarizing agents on choline acetyltransferase and acetylcholinesterase activities in primary cell cultures of spinal cord. *J. Neurosci.* **3**, 1818–1823.

Ishii, D. N., Recio-Pinto, E., Spinelli, W., Mill, J. F., and Sonnenfeld, K. H. (1985). Neurite formation modulated by nerve growth factor, insulin, and tumor promoter receptors. *Int. J. Neurosci.* **26**, 109–127.

Jaye, M., Howk, R., Burgess, W., Ricca, G. A., Chiu, I. M., Ravera, M. W., O'Brien, S. J., Modi, W. S., Maciag, T., and Drohan, W. N. (1986). Human endothelial cell growth factor: Cloning, nucleotide sequence, and chromosome localization. *Science* **233**, 541–549.

Jeanny, J. C., Fayein, N., Moenner, M., Chevallier, B., Barritault, D., and Courtois, Y. (1987). Specific fixation of bovine brain and retinal acidic and basic fibroblast growth factors to mouse embryonic eye basement membranes. *Exp. Cell Res.* **171**, 63–75.

Jetten, A. M. (1980). Retinoids specifically enhance the number of epidermal growth factor receptors. *Nature (London)* **284**, 626–629.

Jetten, A. M. (1982). Effects of retinoic acid on the binding and mitogenic activity of epidermal growth factor. *J. Cell Physiol.* **110**, 235–240.

Johnson, E. M. (1978). Destruction of the sympathetic nervous system in neonatal rats and hamsters by vinblastine: Prevention by concomitant administration of nerve growth factor. *Brain Res.* **141**, 105–118.

Johnson, E. M., Gorin, P. D., Brandeis, L. D., and Pearson, J. (1980). Dorsal root ganglion neurons are destroyed by exposure in utero to maternal antibody to nerve growth factor. *Science* **210(4472)**, 916–918.

Johnson, E. M., Gorin, P. D., Osborne, P. A., Rydel, R. E., and Pearson, J. (1982). Effects of autoimmune NGF deprivation in the adult rabbit and offspring. *Brain Res.* **240**, 131–140.

Johnson, E. M., Taniuchi, M., Clark, H. B., Springer, J. E., Koh, S., Tayrien, M. W., and Loy, R. (1987). Demonstration of the retrograde transport of nerve growth factor receptor in the peripheral and central nervous system. *J. Neurosci.* **7**, 923–929.

Jonakait, G. M., and Schotland, S. (1990). Conditioned medium from activated splenocytes increases substance P in sympathetic ganglia. *J. Neurosci. Res.* **26**, 24–30.

Kan, M., DiSorbo, D., Hou, J., Hoshi, H., Mansson, P. E., and McKeehan, W. L. (1988). High and low affinity binding of heparin-binding growth factor to a 130-kDa receptor correlates with stimulation and inhibition of growth of a differentiated human hepatoma cell. *J. Biol. Chem.* **263**, 11306–11313.

Kanakis, S. J., Hill, C. E., Hendry, I. A., and Watters, D. J. (1985). Sympathetic neuronal survival factors change after denervation. *Brain Res.* **352**, 197–202.

Kanje, M., Skottner, A., Sjöberg, J., and Lundborg, G. (1989). Insulin-like growth factor I (IGF-I) stimulates regeneration of the rat sciatic nerve. *Brain Res.* **486**, 396–398.

Keen, P., and McLean, W. G. (1974). Effect of dibutyryl-cyclic AMP and dexamethasone on noradrenaline synthesis in isolated superior cervical ganglia. *J. Neurochem.* **22**, 5–10.

Keski-Oja, J., Raghow, R., Sawdey, M., Loskutoff, D. J., Postlethwaite, A. E., Kang, A. H., and Moses, H. L. (1988). Regulation of mRNAs for type-1 plasminogen activator inhibitor, fibronectin, and type I procollagen by transforming growth factor-beta. Divergent responses in lung fibroblasts and carcinoma cells. *J. Biol. Chem.* **263**, 3111–3115.

Kessler, J. A. (1984). Non-neuronal cell conditioned medium stimulates peptidergic expression in sympathetic and sensory neurons in vitro. *Dev. Biol.* **106**, 61–69.

Kessler, J. A. (1985). Differential regulation of peptide and catecholamine characters in cultured sympathetic neurons. *Neurosci.* **15**, 827–839.

Kessler, J. A. (1986). Differential regulation of cholinergic and peptidergic development in the rat striatum in culture. *Dev. Biol.* **113**, 77–89.

Kessler, J. A., and Black, I. B. (1982). Regulation of substance P in adult rat sympathetic ganglia. *Brain Res.* **234**, 182–187.

Kessler, J. A., Adler, J. E., Bohn, M. C., and Black, I. B. (1981). Substance P in principal sympathetic neurons: Regulation by impulse activity. *Science* **214**, 335–336.

Kessler, J. A., Adler, J. E., Bell, W. O., and Black, I. B. (1983a). Substance P and somatostatin metabolism in sympathetic and special sensory ganglia in vitro. *Neurosci.* **9**, 309–318.

Kessler, J. A., Adler, J. E., and Black, I. B. (1983b). Substance P and somatostatin regulate sympathetic noradrenergic function. *Science* **221(4615)**, 1059–1061.

Kessler, J. A., Bell, W. O., and Black, I. B. (1983c). Interactions between the sympathetic and sensory innervation of the iris. *J. Neurosci.* **3**, 1301–1307.

Kessler, J. A., Conn, G., and Hatcher, V. B. (1986). Isolated plasma membranes regulate neurotransmitter expression and facilitate effects of a soluble brain cholinergic factor. *Proc. Natl. Acad. Sci. U.S.A.* **83**, 3528–3532.

Khan, M. Y., Jaikaria, N. S., Frenz, D. A., Villanueva, G., and Newman, S. A. (1988). Structural changes in the $NH_2$-terminal domain of fibronectin upon interaction with heparin. Relationship to matrix-driven translocation. *J. Biol. Chem.* **263**, 11314–11318.

Kishimoto, A., Kikkawa, U., Ogita, K, Shearman, M. S., and Nishizuka, Y. (1989). The protein kinase C family in the brain: Heterogeneity and its implications. *Ann. N.Y. Acad. Sci.* **568**, 181–186.

Klagsbrun, M., and Vlodavsky, I. (1988). Biosynthesis and storage of basic fibroblast growth factor (BFGF) by endothelial cells: Implication for the mechanism of action of angiogenesis. *Prog. Clin. Biol. Res.* **266**, 55–61.

Kobayashi, S., Somlyo, A. V., and Somlyo, A. P. (1988). Heparin inhibits the inositol 1,4,5-triphosphate-dependent, but not the independent, calcium release induced by guanine nucleotide in vascular smooth muscle. *Biochem. Biophys. Res. Commun.* **153**, 625–631.

Kobayashi, S., Kitazawa, T., Somlyo, A. V., and Somlyo, A. P. (1989). Cytosolic heparin inhibits muscarinic and $\alpha$-adrenergic $Ca^{2+}$ release in smooth muscle. Physiological role of inositol 1,4,5-trisphosphate in pharmacomechanical coupling. *J. Biol. Chem.* **264**, 17997–18004.

Koelle, G. B., Ruch, G. A., Rickard, K. K., and Sanville, U. J. (1982). Regeneration of cholinesterases in the stellate and normal and denervated superior cervical ganglion of the cat following inactivation by sarin. *J. Neurochem.* **38**, 1695–1698.

Komura, H., Wakimoto, H., Chen, C. F., Terakawa, N., Aono, T., Tanizawa, O., and Matsumoto, K. (1986). Retinoic acid enhances cell responses to epidermal growth factor in mouse mammary gland in culture. *Endocrinology* **118**, 1530–1536.

Korsching, S., and Thoenen, H. (1983). Nerve growth factor in sympathetic ganglia and corresponding target organs of the rat: Correlation with density of sympathetic innervation. *Proc. Natl. Acad. Sci. U.S.A.* **80**, 3513–3516.

Korsching, S., and Thoenen, H. (1985). Treatment with 6-hydroxydopamine and colchicine decreases nerve growth factor levels in sympathetic ganglia and increases them in the corresponding target tissues. *J. Neurosci.* **5**, 1058–1061.

Krupinski, J., Rajaram, R., Lakonishok, M., Benovic, J. L., and Cerione, R. A. (1988). Insulin-dependent phosphorylation of GTP-binding proteins in phospholipid vesicles. *J. Biol. Chem.* **263,** 12333–12341.

Kuecherer-Ehret, A., Graeber, M. B., Edgar, D., Thoenen, H., and Kreutzberg, G. W. (1990). Immunoelectron microscopic localization of laminin in normal and regenerating mouse sciatic nerve. *J. Neurocytol.* **19,** 101–109.

Landmesser, L., and Pilar, G. (1974a). Synaptic transmission and cell death during normal ganglionic development. *J. Physiol. (London)* **241,** 737–749.

Landmesser, L., and Pilar, G. (1974b). Synapse formation during embryogenesis on ganglion cells lacking a periphery. *J. Physiol. (London)* **241,** 715–736.

Landreth, G. E., Smith, D. S., McCabe, C., and Gittinger, C. (1990). Characterization of a nerve growth factor-stimulated protein kinase in PC12 cells which phosphorylates microtubule-associated protein 2 and pp250. *J. Neurochem.* **55,** 514–523.

Large, T. H., Bodary, S. C., Clegg, D. O., Weskamp, G., Otten, U., and Reichardt, L. F. (1986). Nerve growth factor gene expression in the developing rat brain. *Science* **234,** 352–355.

Lee, J. M., Adler, J. E., and Black, I. B. (1990). Regulation of neurotransmitter expression by a membrane-derived factor. *Exp. Neurol.* **108,** 109–113.

Lees, G., Chubb, I., Freeman, C., Geffen, L., and Rush, R. A. (1981). Effect of nerve activity on transport of nerve growth factor and dopamine beta-hydroxylase antibodies in sympathetic neurons. *Brain Res.* **214,** 186–189.

Lefebvre, P. P., Leprince, P., Weber, T., Rigo, J. M., Delree, P., and Moonen, G. (1990). Neuronotrophic effect of developing otic vesicle on cochleo-vestibular neurons: Evidence for nerve growth factor involvement. *Brain Res.* **507,** 254–260.

Levi-Montalcini, R. (1965). Growth regulation of sympathetic nerve cells. *Arch. Ital. Biol.* **103,** 832–846.

Levi-Montalcini, R., and Angeletti, P. U. (1966). Immunosympathectomy. *Pharmacol. Rev.* **18,** 619–628.

Levi-Montalcini, R., Aloe, L., Mugnaini, E., Oesch, F., and Thoenen, H. (1975). Nerve growth factor induces volume increase and enhances tyrosine hydroxylase synthesis in chemically axotomized sympathetic ganglia of newborn rats. *Proc. Natl. Acad. Sci. U.S.A.* **72,** 595–599.

Lin, L. F., Mismer, D., Lile, J. D., Armes, L. G., Butler, E. T., Vannice, J. L., and Collins, F. (1989). Purification, cloning, and expression of ciliary neurotrophic factor (CNTF). *Science* **246,** 1023–1025.

Lindsay, R. M. (1988). Nerve growth factors (NGF, BDNF) enhance axonal regeneration but are not required for survival of adult sensory neurons. *J. Neurosci.* **8,** 2394–2405.

Linnemann, D., and Bock, E. (1986). Developmental study of detergent solubility and polypeptide composition of the neural cell adhesion molecule. *Dev. Neurosci.* **8,** 24–30.

Lipton, S. A., Wagner, J. A., Madison, R. D., and D'Amore, P. A. (1988). Acidic fibroblast growth factor enhances regeneration of processes by postnatal mammalian retinal ganglion cells in culture. *Proc. Natl. Acad. Sci. U.S.A.* **85,** 2388–2392.

MacKay, A. V., and Iversen, L. L. (1972). Increased tyrosine hydroxylase activity of sympathetic ganglia cultured in the presence of dibutyryl cyclic AMP. *Brain Res.* **48,** 424–426.

McManaman, J. L., Crawford, F. G., Stewart, S. S., and Appel, S. H. (1988). Purification of a skeletal muscle polypeptide which stimulates choline acetyltransferase activity in cultured spinal cord neurons. *J. Biol. Chem.* **263,** 5890–5897.

McManaman, J. L., Oppenheim, R. W., Prevette, D., and Marchetti, D. (1990). Rescue of motoneurons from cell death by a purified skeletal muscle polypeptide: Effects of the ChAT development factor, CDF. *Neuron* **4,** 891–898.

Maden, M., and Summerbell, D. (1986). Retinoic acid-binding protein in the chick limb bud: Identification at developmental stages and binding affinities of various retinoids. *J. Embryol. Exp. Morphol.* **97,** 239–250.

Maden, M., Ong, D. E., Summerbell, D., and Chytil, F. (1988). Spatial distribution of cellular protein binding to retinoic acid in the chick limb bud. *Nature (London)* **335,** 733–735.

Maderdrut, J. L., Oppenheim, R. W., and Prevette, D. (1988). Enhancement of naturally occurring cell death in the sympathetic and parasympathetic ganglia of the chicken embryo following blockade of ganglionic transmission. *Brain Res.* **444**, 189–194.

Manthorpe, M., Luyten, W., Longo, F. M., and Varon, S. (1983). Endogenous and exogenous factors support neuronal survival and choline acetyltransferase activity in embryonic spinal cord cultures. *Brain Res.* **267**, 57–66.

Manthorpe, M., Skaper, S. D., Williams, L. R., and Varon, S. (1986). Purification of adult rat sciatic nerve ciliary neuronotrophic factor. *Brain Res.* **367(1–2)**, 282–286.

Marics, I., Adelaide, J., Raybaud, F., Mattei, M. -G., Coulier, F., Planche, J., De Lapeyriere, O., and Birnbaum, D. (1989). Characterization of the HST-related FGF-6 gene, a new member of the fibroblast growth factor gene family. *Oncogene* **4**, 335–340.

Martin, D. P., Schmidt, R. E., DiStefano, P. S., Lowry, O. H., Carter, J. G., and Johnson, E. M. (1988). Inhibitors of protein synthesis and RNA synthesis prevent neuronal death caused by nerve growth factor deprivation. *J. Cell Biol.* **106**, 829–844.

Martinou, J. C., Bierer, F., Van Thai, A. L., and Weber, M. J. (1989a). Influence of the culture substratum on the expression of choline acetyltransferase activity in purified motoneurons from rat embryos. *Dev. Brain Res.* **47**, 251–262.

Martinou, J. C., Van Thai, A. L., Cassar, G., Roubinet, F., and Weber, M. J. (1989b). Characterization of two factors enhancing choline acetyltransferase activity in cultures of purified rat motoneurons. *J. Neurosci.* **9**, 3645–3656.

Martinou, J. C., Van Thai, A. L., Valette, A., and Weber, M. J. (1990). Transforming growth factor beta 1 is a potent survival factor for rat embryo motoneurons in culture. *Brain Res. Dev. Brain Res.* **52**, 175–181.

Matthew, W. D., and Patterson, P. H. (1983). The production of a monoclonal antibody that blocks the action of a neurite outgrowth-promoting factor. *Cold Spring Harbor Symp. Quant. Biol.* **48**, 625–631.

Max, S. R., Schwab, M., Dumas, M., and Thoenen, H. (1978). Retrograde axonal transport of nerve growth factor in the ciliary ganglion of the chick and the rat. *Brain Res.* **159**, 411–415.

Milbrandt, J. (1986). Nerve growth factor rapidly induces c-*fos* mRNA in PC12 rat pheochromocytoma cells. *Proc. Natl. Acad. Sci. U.S.A.* **83**, 4789–4793.

Milbrandt, J. (1988). Nerve growth factor induces a gene homologous to the glucocorticoid receptor gene. *Neuron* **1**, 183–188.

Mill, J. F., Chao, M. V., and Ishii, D. N. (1985). Insulin, insulin-like growth factor II, and nerve growth factor effects on tubulin mRNA levels and neurite formation. *Proc. Natl. Acad. Sci. U.S.A.* **82**, 7126–7130.

Morgan, J. I., and Curran, T. (1986). Role of ion flux in the control of c-*fos* expression. *Nature (London)* **322**, 552–555.

Morgan, J. I., and Curran, T. (1988). Calcium as a modulator of the immediate-early gene cascade in neurons. *Cell Calcium* **9**, 303–311.

Mullaney, J. M., Chueh, S. H., Ghosh, T. K., and Gill, D. L. (1987). Intracellular calcium uptake activated by GTP. Evidence for a possible guanine nucleotide-induced transmembrane conveyance of intracellular calcium. *J. Biol. Chem.* **262**, 13865–13872.

Narayanan, C. H., and Narayanan, Y. (1982). Abnormal differentiation of selected nuclear centers in the brain of a duck embryo associated with partial duplication of the primitive streak. *J. Morphol.* **172**, 287–297.

Narayanan, Y., and Narayanan, C. H. (1981). Ultrastructural and histochemical observations in the developing iris musculature in the chick. *J. Embryol. Exp. Morphol.* **62**, 117–127.

Nestler, E. J., and Greengard, P. (1980). Dopamine and depolarizing agents regulate the state of phosphorylation of protein I in the mammalian superior cervical sympathetic ganglion. *Proc. Natl. Acad. Sci. U.S.A.* **77**, 7479–7483.

Neufeld, G., and Gospodarowicz, D. (1986). Basic and acidic fibroblast growth factors interact with the same cell surface receptors. *J. Biol. Chem.* **261**, 5631–5637.

Neugebauer, K. M., Tomaselli, K. J., Lilien, J., and Reichardt, L. F. (1988). N-cadherin, NCAM, and integrins promote retinal neurite outgrowth on astrocytes in vitro. *J. Cell Biol.* **107**, 1177–1187.

Oberg, K. C., and Carpenter, G. (1989). EGF-induced PGE2 release is synergistically enhanced in retinoic acid treated fetal rat lung cells. *Biochem. Biophys. Res. Commun.* **162**, 1515–1521.

Okado, N., and Oppenheim, R. W. (1984). Cell death of motoneurons in the chick embryo spinal cord. IX. The loss of motoneurons following removal of afferent inputs. *J. Neurosci.* **4**, 1639–1652.

Olender, E. J., Wagner, B. J., and Stach, R. (1981). Sequestration of $^{125}$I-labeled beta nerve growth factor by embryonic sensory neurons. *J. Neurochem.* **37**, 436–442.

Oppenheim, R. W. (1981). Cell death of motoneurons in the chick embryo spinal cord. V. Evidence on the role of cell death and neuromuscular function in the formation of specific peripheral connections. *J. Neurosci.* **1**, 141–151.

Oppenheim, R. W. (1984). Cell death of motoneurons in the chick embryo spinal cord. VIII. Motoneurons prevented from dying in the embryo persist after hatching. *Dev. Biol.* **101**, 35–39.

Oppenheim, R. W., and Núñez, R. (1982). Electrical stimulation of hindlimb increases neuronal cell death in chick embryo. *Nature (London)* **295**, 57–59.

Oppenheim, R. W., Maderdrut, J. L., and Wells, D. J. (1982a). Cell death of motoneurons in the chick embryo spinal cord. VI. Reduction of naturally occurring cell death in the thoracolumbar column of Terni by nerve growth factor. *J. Comp. Neurol.* **210**, 174–189.

Oppenheim, R. W., Maderdrut, J. L., and Wells, D. J. (1982b). Reduction of naturally occurring cell death in the thoraco-lumbar preganglionic cell column of the chick embryo by nerve growth factor and hemicholinium-3. *Brain Res.* **255**, 134–139.

Oppenheim, R. W., Haverkamp, L. J., Prevette, D., McManaman, J. L., and Appel, S. H. (1988). Reduction of naturally occurring motoneuron death in vivo by a target-derived neurotrophic factor. *Science* **240**, 919–922.

Oppenheim, R. W., Cole, T., and Prevette, D. (1989). Early regional variations in motoneuron numbers arise by differential proliferation in the chick embryo spinal cord. *Dev. Biol.* **133**, 468–474.

Oppenheim, R. W., Prevette, D., Tytell, M., and Homma, S. (1990). Naturally occurring and induced neuronal death in the chick embryo in vivo requires protein and RNA synthesis: Evidence for the role of cell death genes. *Dev. Biol.* **138**, 104–113.

Ouimet, C. C., Wang, J. K., Walaas, S. I., Albert, K. A., and Greengard, P. (1990). Localization of the MARCKS (87 kDa) protein, a major specific substrate for protein kinase C, in rat brain. *J. Neurosci.* **10**, 1683–1698.

Pannese, E. (1976). An electron microscope study of cell degeneration in chick embryo spinal ganglia by a sialic acid-binding receptor. *Neuropathol. Appl. Neurobiol.* **2**, 247–267.

Patterson, P. H., and Chun, L. L. Y. (1974). The influence of nonneuronal cells on catecholamine and acetylcholine synthesis and accumulation in cultures of dissociated sympathetic neurons. *Proc. Natl. Acad. Sci. U.S.A.* **71**, 3607–3610.

Petkovich, M., Brand, N. J., Krust, A., and Chambon, P. (1987). A human retinoic acid receptor which belongs to the family of nuclear receptors. *Nature (London)* **330**, 444–450.

Pilar, G., and Landmesser, L. (1972). Axotomy mimicked by localized colchicine application. *Science* **177**, 1116–1118.

Pilar, G., Landmesser, L., and Burstein, L. (1980). Competition for survival among developing ciliary ganglion cells. *J. Neurophysiol.* **43**, 233–254.

Powis, G., Seewald, M. J., Sehgal, I., Iaizzo, P. A., and Olsen, R. A. (1990). Platelet-derived growth factor stimulates nonmitochondrial $Ca^{2+}$ uptake and inhibits mitogen-induced $Ca^{2+}$ signaling in Swiss 3T3 fibroblasts. *J. Biol. Chem.* **265**, 10266–10273.

Presta, M., Maier, J. A. M., Rusnati, M., and Ragnotti, G. (1989). Basic fibroblast growth factor is released from endothelial extracellular matrix in a biologically active form. *J. Cell Physiol.* **140**, 68–74.

Probstmeier, R., Kühn, K., and Schachner, M. (1989). Binding properties of the neural cell adhesion molecule to different components of the extracellular matrix. *J. Neurochem.* **53**, 1794–1801.

Recio-Pinto, E., and Ishii, D. N. (1984). Effects of insulin, insulin-like growth factor-II, and nerve growth factor on neurite outgrowth in cultured human neuroblastoma cells. *Brain Res.* **302**, 323–334.

Recio-Pinto, E., and Ishii, D. N. (1988). Insulin and insulin-like growth factor receptors regulating neurite formation in cultured human neuroblastoma cells. *J. Neurosci. Res.* **19,** 312–320.

Recio-Pinto, E., Lang, F. F., and Ishii, D. N. (1984). Insulin and insulin-like growth factor II permit nerve growth factor binding and the neurite formation response in cultured human neuroblastoma cells. *Proc. Natl. Acad. Sci. U.S.A.* **81,** 2562–2566.

Recio-Pinto, E., Rechler, M. M., and Ishii, D. N. (1986). Effect of insulin, insulin-like growth factor II, and nerve growth factor on neurite formation and survival in cultured sympathetic and sensory neurons. *J. Neurosci.* **6,** 1211–1219.

Rees, A. R., Adamson, E. D., and Graham, C. F. (1979). Epidermal growth factor receptors increase during differentiation of embryonal carcinoma cells. *Nature (London)* **281,** 309–311.

Reichardt, L. F., and Kelly, R. B. (1983). A molecular description of nerve terminal function. *Ann. Rev. Biochem.* **52,** 871–926.

Rich, K. M., and Hollowell, J. P. (1990). Flunarizine protects neurons from death after axotomy or NGF deprivation. *Science* **248,** 1419–1421.

Rodriguez-Tébar, A., Dechant, G., and Barde, Y. A. (1990). Binding of brain-derived neurotrophic factor to the nerve growth factor receptor. *Neuron* **4,** 487–492.

Rogelj, S., Klagsbrun, M., Atzmon, R., Kurokawa, M., Haimovitz, A., Fuks, Z., and Vlodavsky, I. (1989). Basic fibroblast growth factor is an extracellular matrix component required for supporting the proliferation of vascular endothelial cells and the differentiation of PC12 cells. *J. Cell Biol.* **109,** 823–831.

Rohrer, H. Schäfer, T., Korsching, S., and Thoenen, H. (1982). Internalization of nerve growth factor by pheochromocytoma PC12 cells: Absence of transfer to the nucleus. *J. Neurosci.* **2,** 687–697.

Romano, C., Nichols, R. A., and Greengard, P. (1987). Synapsin I in PC12 cells. II. Evidence for regulation by NGF of phosphorylation at a novel site. *J. Neurosci.* **7,** 1300–1306.

Rossi, P., Karsenty, G., Roberts, A. B., Roche, N. S., Sporn, M. B., and de Crombrugghe, B. (1988). A nuclear factor 1 binding site mediates the transcriptional activation of a type I collagen promoter by transforming growth factor-beta. *Cell* **52,** 405–414.

Saadat, S., Sendtner, M., and Rohrer, H. (1989). Ciliary neurotrophic factor induces cholinergic differentiation of rat sympathetic neurons in culture. *J. Cell Biol.* **108,** 1807–1816.

Saito, M., Ueno, I., and Egawa, K. (1982). Effect of retinoic acid and 12-O-tetradecanoyl phorbol-13-acetate on the binding of epidermal growth factor to its cellular receptors. *Biochim. Biophys. Acta* **717,** 301–304.

Saksela, O., and Rifkin, D. B. (1990). Release of basic fibroblast growth factor–heparan sulfate complexes from endothelial cells by plasminogen activator-mediated proteolytic activity. *J. Cell Biol.* **110,** 767–775.

Saksela, O., Moscatelli, D., Sommer, A., and Rifkin, D. B. (1988). Endothelial cell-derived heparan sulfate binds basic fibroblast growth factor and protects it from proteolytic degradation. *J. Cell Biol.* **107,** 743–751.

Sano, M., Nishiyama, K., and Kitajima, S. (1990). A nerve growth factor-dependent protein kinase that phosphorylates microtubule-associated proteins in vitro: Possible involvement of its activity in the outgrowth of neurites from PC12 cells. *J. Neurochem.* **55,** 427–435.

Saunders, J. W., and Gasseling, M. T. (1968). Ectodermal–mesenchymal interactions in the origins of limb symmetry. *In* "Epithelial–Mesenchymal Interactions" (R. Fleischmayer and R. E. Billingham, eds.), pp. 78–97. Williams and Wilkins, Baltimore.

Schlaepfer, D. D., and Haigler, H. T. (1990). Expression of annexins as a function of cellular growth state. *J. Cell Biol.* **111,** 229–238.

Schmidt, R. E., Modert, C. W., Yip, H. K., and Johnson, E. M. (1983). Retrograde axonal transport of intravenously administered $^{125}$I-nerve growth factor in rats with streptozotocin-induced diabetes. *Diabetes* **32,** 654–663.

Schubert, D., LaCorbiere, M., and Esch, F. (1987). A chick neural retinal adhesion and survival molecule is a retinol-binding protein. *J. Cell Biol.* **102,** 2295–2301.

Schwab, M. E. (1977). Ultrastructural localization of a nerve growth factor–horseradish peroxidase

(NGF-HRP) coupling product after retrograde axonal transport in adrenergic neurons. *Brain Res.* **130,** 190–196.
Schwarting, G. A., Tischler, A. S., and Donahue, S. R. (1990). Fucosylation of glycolipids in PC12 cells is dependent on the sequence of nerve growth factor treatment and adenylate cyclase activation. *Dev. Neurosci.* **12,** 159–171.
Seewald, M. J., Schlager, J. J., Olsen, R. A., Melder, D. C., and Powis, G. (1989). High molecular weight dextran sulfate inhibits intracellular $Ca^{2+}$ release and decreases growth factor-induced increases in intracellular free $Ca^{2+}$ in Swiss 3T3 fibroblasts. *Cancer Commun.* **1,** 151–156.
Seewald, M. J., Olsen, R. A., and Powis, G. (1990). Suramin blocks intracellular $Ca^{2+}$ release and growth factor-induced increases in cytoplasmic free $Ca^{2+}$ concentration. *Cancer Lett.* **49,** 107–113.
Sendtner, M., Kreutzberg, G. W., and Thoenen, H. (1990). Ciliary neurotrophic factor prevents the degeneration of motor neurons after axotomy. *Nature (London)* **345,** 440–441.
Shelton, D. L., and Reichardt, L. F. (1984). Expression of the beta-nerve growth factor gene correlates with the density of sympathetic innervation in effector organs. *Proc. Natl. Acad. Sci. U.S.A.* **81,** 7951–7955.
Sheng, M., Dougan, S. T., McFadden, G., and Greenberg, M. E. (1988). Calcium and growth factor pathways of c-*fos* transcriptional activation require distinct upstream regulatory sequences. *Mol. Cell. Biol.* **8,** 2787–2796.
Sjöberg, J., and Kanje, M. (1989). Insulin-like growth factor (IGF-I) as a stimulator of regeneration in the freeze-injured rat sciatic nerve. *Brain Res.* **485,** 102–108.
Smith, R. G., Vaca, K., McManaman, J. L., and Appel, S. H. (1986). Selective effects of skeletal muscle extract fractions on motoneuron development in vitro. *J. Neurosci.* **6,** 439–447.
Soinila, S., and Eränkö, O. (1983). Increase in the number of non-neuronal cells in superior cervical ganglia of developing rats after contralateral ganglionectomy. *Neurosci.* **9,** 911–915.
Stach, R., and Wagner, B. J. (1982). Sequestration requirements for the degradation of $^{125}$I-labeled beta nerve growth factor bound to embryonic sensory neurons. *J. Neurosci. Res.* **7,** 403–411.
Stöckel, K., Dumas, M., and Thoenen, H. (1978). Uptake and subsequent retrograde axonal transport of nerve growth factor (NGF) are not influenced by neuronal activity. *Neurosci. Lett.* **10,** 61–64.
Stöckli, K. A., Lottspeich, F., Sendtner, M., Masiakowski, P., Carroll, P., Götz, R., Lindholm, D., and Thoenen, H. (1989). Molecular cloning, expression, and regional distribution of rat ciliary neurotrophic factor. *Nature (London)* **342,** 920–923.
Strasser, R. H., Benovic, J. L., Caron, M. G., and Lefkowitz, R. J. (1986). Beta-agonist- and prostaglandin E1-induced translocation of the beta-adrenergic receptor kinase: Evidence that the kinase may act on multiple adenylate cyclase-coupled receptors. *Proc. Natl. Acad. Sci. U.S.A.* **83,** 6362–6366.
Stumpo, D. J., Graff, J. M., Albert, K. A., Greengard, P., and Blackshear, P. J. (1989). Nucleotide sequence of a cDNA for the bovine myristoylated alanine-rich C kinase substrate (MARCKS). *Nucleic Acids Res.* **17,** 3987–3988.
Summerbell, D. (1983). The effect of local application of retinoic acid to the anterior margin of the developing chick limb. *J. Embryol. Exp. Morphol.* **78,** 269–289.
Sunderland, S. (1978). "Nerves and Nerve Injuries." Churchill Livingstone, Edinburgh.
Supattapone, S., Danoff, S. K., Theibert, A., Joseph, S. K., Steiner, J., and Snyder, S. H. (1988). Cyclic AMP-depenent phosphorylation of a brain inositol trisphosphate receptor decreases its release of calcium. *Proc. Natl. Acad. Sci. U.S.A.* **85,** 8747–8750.
Thaller, C., and Eichele, G. (1987). Identification and spatial distribution of retinoids in the developing chick limb bud. *Nature (London)* **327,** 625–628.
Tickle, C., Alberts, B., Wolpert, L., and Lee, J. (1982). Local application of retinoic acid to the limb bond mimics the action of the polarizing region. *Nature (London)* **296,** 564–566.
Tickle, C., Lee, J., and Eichele, G. (1985). A quantitative analysis of the effect of all-trans-retinoic acid on the pattern of chick wing development. *Dev. Biol.* **109,** 82–95.
Tjaden, G., Aguanno, A., Kumar, R., Benincasa, D., Gubits, R. M., Yu, H., and Dolan, K. P. (1990).

Density-dependent nerve growth factor regulation of Gs-alpha RNA in pheochromocytoma 12 cells. *Mol. Cell. Biol.* **10**, 3277–3279.

Tomaselli, K. J., and Reichardt, L. F. (1988). Peripheral motoneuron interactions with laminin and Schwann cell-derived neurite-promoting molecules: Developmental regulation of laminin receptor function. *J. Neurosci. Res.* **21**, 275–285.

Tomaselli, K. J., Reichardt, L. F., and Bixby, J. L. (1986). Distinct molecular interactions mediate neuronal process outgrowth on non-neuronal cell surfaces and extracellular matrices. *J. Cell Biol.* **103**, 2659–2672.

Tomaselli, K. J., Damsky, C. H., and Reichardt, L. F. (1987). Interactions of a neuronal cell line (PC12) with laminin, collagen IV, and fibronectin: Identification of integrin-related glycoproteins involved in attachment and process outgrowth. *J. Cell Biol.* **105**, 2347–2358.

Tomaselli, K. J., Damsky, C. H., and Reichardt, L. F. (1988a). Purification and characterization of mammalian integrins expressed by a rat neuronal cell line (PC12): Evidence that they function as α/β heterodimeric receptors for laminin and type IV collagen. *J. Cell Biol.* **107**, 1241–1252.

Tomaselli, K. J., Neugebauer, K. M., Bixby, J. L., Lilien, and Reichardt, L. F. (1988b). N-cadherin and integrins: Two receptor systems that mediate neuronal process outgrowth on astrocyte surfaces. *Neuron* **1**, 33–43.

Torres-Aleman, I., Naftolin, F., and Robbins, R. J. (1990). Trophic effects of basic fibroblast growth factor on fetal rat hypothalamic cells: Interactions with insulin-like growth factor I. *Brain Res. Dev. Brain Res.* **52**, 253–257.

Tsuda, T., Switzer, J., Bienenstock, J., and Denburg, J. A. (1990). Interactions of hemopoietic cytokines on differentiation of HL-60 cells. Nerve growth factor is a basophilic lineage-specific co-factor. *Int. Arch. Allergy Appl. Immunol.* **91**, 15–21.

Varon, S., and Bunge, R. P. (1978). Trophic mechanisms in the peripheral nervous system. *Ann. Rev. Neurosci.* **1**, 327–361.

Vidal, S., Raynaud, B., and Weber, M. J. (1989). The role of $Ca^{2+}$ channels of the L-type in neurotransmitter plasticity of cultured sympathetic neurons. *Brain Res. Mol. Brain Res.* **6**, 187–196.

Walicke, P. A., and Patterson, P. H. (1981). On the role of $Ca^{2+}$ in the transmitter choice made by cultured sympathetic neurons. *J. Neurosci.* **1**, 343–350.

Walicke, P. A., Campenot, R. B., and Patterson, P. H. (1977). Determination of transmitter function by neuronal activity. *Proc. Natl. Acad. Sci. U.S.A.* **74**, 5767–5771.

Walicke, P. A., Feige, J. J., and Baird, A. (1989). Characterization of the neuronal receptor for basic fibroblast growth factor and comparison to receptors on mesenchymal cells. *J. Biol. Chem.* **264**, 4120–4126.

Wang, J. K., Walaas, S. I., and Greengard, P. (1988). Protein phosphorylation in nerve terminals: Comparison of calcium/calmodulin-dependent and calcium/diacylglycerol-dependent systems. *J. Neurosci.* **8**, 281–288.

Wang, J. K., Walaas, S. I., Sihra, T. S., Aderem, A., and Greengard, P. (1989). Phosphorylation and associated translocation of the 87-kDa protein, a major protein kinase C substrate, in isolated nerve terminals. *Proc. Natl. Acad. Sci. U.S.A.* **86**, 2253–2256.

Watters, D. J., Belford, D. A., Hill, C. E., and Hendry, I. A. (1989). Monoclonal antibody that inhibits biological activity of a mammalian ciliary neurotrophic factor. *J. Neurosci. Res.* **22**, 60–64.

Westermann, R., Johannsen, M., Unsicker, K., and Grothe, C. (1990). Basic fibroblast growth factor (bFGF) immunoreactivity is present in chromaffin granules. *J. Neurochem.* **55**, 285–292.

Woolf, C. J., Reynolds, M. L., Molander, C., O'Brien, C., Lindsay, R. M., and Benowitz, L. I. (1990). The growth-associated protein GAP-43 appears in dorsal root ganglion cells and in the dorsal horn of the rat spinal cord following peripheral nerve injury. *Neurosci.* **34**, 465–478.

Worley, P. F., Baraban, J. M., Supattapone, S., Wilson, V. W., and Snyder, S. H. (1987). Characterization of inositol triphosphate receptor binding in brain. *J. Biol. Chem.* **262**, 12132–12136.

Yamamori, T., Fukada, K., Aebersold, R., Korsching, S., Fann, M. -J., and Patterson, P. H. (1989). The cholinergic neuronal differentiation factor from heart cells is identical to leukemia inhibitory factor. *Science* **246**, 1412–1416.

Yan, Q., Snider, W. D., Pinzone, J. J., and Johnson, E. M. (1988). Retrograde transport of nerve growth factor (NGF) in motoneurons of developing rats: Assessment of potential neurotrophic effects. *Neuron* **1**, 335–343.

Yoshida, T., Miyagawa, K., Odagiri, H., Sakamoto, H., Little, P. F. R., Terada, M., and Sugimura, T. (1987). Genomic sequence of *hst*, a transforming gene encoding a protein homologous to fibroblast growth factors and the int-2-encoded protein. *Proc. Natl. Acad. Sci. U.S.A.* **84**, 7305–7309.

Yung, W. K. A., Lotan, R., Lee, P., Lotan, D., and Steck, P. A. (1989). Modulation of growth and epidermal growth factor receptor activity by retinoic acid in human glioma cells. *Cancer Res.* **49**, 1014–1019.

Zacks, S. I., and Saito, A. (1969). Uptake of exogenous horseradish peroxidase by coated vesicles in mouse neuromuscular junctions. *J. Histochem. Cytochem.* **17**, 161–170.

Zhan, X., Bates, B., Hu, X. G., and Goldfarb, M. (1988). The human FGF-5 oncogene encodes a novel protein related to fibroblast growth factors. *Mol. Cell. Biol.* **8**, 3487–3495.

Zheng, Z. -S., and Goldsmith, L. A. (1990). Modulation of epidermal growth factor receptors by retinoic acid in ME180 cells. *Cancer Res.* **50**, 1201–1205.

# 4 Primary Response Gene Expression in the Nervous System

Alaric T. Arenander and Harvey R. Herschman

## I. INTRODUCTION

The brain is the most complex of our organs, both structurally and functionally. Although changes in brain organization and function are most extensive and pronounced during early developmental periods, neurons and neural networks change throughout the entire life-span. Both neural development and adult neural plasticity are dependent on environmental signals that activate coordinated programs of differential gene expression, leading to short- and long-term changes in brain structure and function. What are the critical causal alterations in gene expression in response to extracellular signals that are responsible for these subsequent alterations in brain function?

Neurons, like all nucleated eukaryotic cells, exhibit ligand-induced transcription-dependent changes in cellular phenotype. In the brain, the complex array of changing environmental signals includes growth factors, hormones, neurotransmitters, ions, and metabolic products. Intracellular signal transduction initiated by ligand–receptor interaction leads, in many cases, to activation of various second messenger signaling elements that can subsequently alter gene expression. In the past decade, a third intracellular "messenger system" has been described: a newly discovered group of ligand-induced, rapidly transcribed genes whose protein products, in many cases, act to directly coordinate subsequent differential gene expression that results in alternative phenotypic responses. How do the products of these primary response genes participate in the intracellular processes that enable a complex and variable array of extracellular signaling agents to activate specific physiologic responses for each cell in a diverse multicellular system? This chapter discusses possible roles of the ligand-inducible primary response genes in the development and function of the brain.

In the past 10 years, a new group of genes has been discovered. These genes are characterized by two primary attributes: (1) rapid and, in many cases, transient expression following presentation of an extracellular stimulus and (2) transcriptional activation that is independent of protein synthesis (for review, see Morgan and Curran, 1989, 1991; Herschman, 1991). These genes have been referred to as "immediate early" genes (Lau and Nathans, 1985; Curran and Morgan, 1987) or "early response" genes (Sukhatme et al., 1987). We prefer the term "primary response" genes, because of precedence (Yamamoto and Alberts, 1976) and because, in many cases, their identification and cloning has been based on the characteristic that their induction can occur with no intervening protein synthesis, requiring only the modification of pre-existing transcription factors.

## II. *fos* AND *jun* GENES AS PROTOTYPICAL PRIMARY RESPONSE GENES

### A. c-*fos* and c-*jun*

The *fos* gene (c-*fos* or *fos*) is the normal cellular gene whose retroviral derivative, v-*fos*, was first identified as the oncogene associated with a radiation-induced murine osteosarcoma virus (Curran and Teich, 1982; Curran et al., 1983). In the past 8 years, considerable research has been carried out on c-*fos* in an attempt to understand its regulation and its roles in both normal and abnormal cell growth and differentiation. In most cell types examined, the basal levels of both c-*fos* message and FOS protein are low. A wide range of extracellular signals, acting on a diversity of cell types, is able to induce the rapid and transient accumulation of c-*fos* mRNA (for review, see Herschman, 1991; Morgan and Curran, 1991). Increased transcription can often be detected within 5 min (Greenberg and Ziff, 1984; Greenberg et al., 1985). Cytoplasmic levels of c-*fos* message often peak within 30 min and return to baseline values within 60 min (Müller et al., 1984).

The transient nature of c-*fos* expression appears to be the consequence of a combination of (1) destabilizing features in both the 3' untranslated and the open reading frame regions of the c-*fos* mRNA that result in rapid degradation and (2) an autorepression loop in which newly synthesized FOS protein represses transcription from its own promoter. The FOS protein is also relatively unstable; its half-life is about 2 hr (Curran et al., 1984; Müller et al., 1984). Induction of *fos* expression by extracellular ligands such as growth factors is independent of protein synthesis; ligand-induced c-*fos* transcription is not blocked by cycloheximide. Accumulation of *fos* mRNA must, therefore, be caused by activation of pre-existing transcription factors that undergo ligand-dependent post translational modification following activation of second messenger pathways, leading to transactivation of the *fos* promoter and *fos* transcription. Depending on the inducing ligand, FOS protein undergoes considerable post translational modification, usually serine and threonine phosphorylation (Curran et al., 1984; Barber and Verma, 1987). The c-*fos* gene encodes a 62-kDa nuclear protein that contains three important domains: a transcription-regulating region, a "leucine zipper" (Landschulz et al., 1988; see also Kouzarides and Ziff, 1988; Ransone et al., 1989; Schuermann et al., 1989), and a contiguous DNA-binding region (Nakabeppu and Nathans, 1989; Risse et al., 1989; Abate et al., 1990). Thus, FOS is a member of the bZIP family of proteins (basic region–leucine zipper arrangement). The leucine zipper motif is a series of leucine residues (seven for FOS) that are repeated to align the leucines on the same surface of an amphipathic α helix. Dimerization occurs between similar leucine zipper regions. As a result, FOS can form heterodimers with many other zipper-possessing proteins.

A protein known as p39 is routinely observed in immunoprecipitates of FOS isolated from cells by using anti-FOS antibodies. p39 was subsequently identified as JUN, the product of the c-*jun* gene. c-*jun*, the cellular homolog of another avian retrovirus oncogene (Rauscher et al., 1988a), is also a primary response gene whose transcription is rapidly and transiently induced by a variety of ligands (Quantin and Breathnach, 1988; Ryder and Nathans, 1988; Ryseck et al., 1988). Like FOS, the JUN protein has a leucine zipper domain (Nakabeppu et al., 1988; Cohen et al., 1989). FOS:JUN heterodimers form as a result of interaction between their leucine zipper domains. FOS and JUN are components of a transcriptional activation complex, AP-1 (Lee et al., 1987; Rauscher et al., 1988b), that influences the transcription of various secondary response target genes that possess 5′ *cis*-acting nucleotide sequences termed AP-1 sites or TREs (TPA-responsive-elements).

FOS will not form homodimers. The ability of FOS to activate target genes is dependent on its ability to form heterodimers with other leucine zipper-containing proteins such as JUN (Nakabeppu et al., 1988; Cohen et al., 1989). In contrast, JUN forms homodimers that are functional in transcriptional regulation. FOS:JUN heterodimers exhibit higher affinity for AP-1 sites than do JUN:JUN homodimers (Rauscher et al., 1988c). Alternative dimerization of leucine zipper transcription factors to form hetero- and homodimers with differing affinities and specificities for AP-1 sites and distinct transactivation properties may be one mechanism by which cells achieve altered responsiveness to different ligands or by which distinct cell types can exhibit differential responsiveness to the same ligand.

### B. *fos* and *jun* Gene Families

Antibodies originally raised against FOS are able to precipitate a number of FOS-related antigens (FRAs; Franza et al., 1987; Cohen and Curran, 1988; Cohen et al., 1989; Matsui et al., 1990; Nishina et al., 1990). FRA-1 and FRA-2 possess several regions of amino acid sequence similarity to FOS.

Like c-*fos*, they exhibit protein synthesis-independent ligand-induced transcriptional activation; they are primary response genes. *fosB* cDNA was cloned by screening with a c-*fos* probe (Zerial et al., 1989); it is also a primary response gene from this family. The time courses of induction of FRA-1, FRA-2, and *fosB* mRNA accumulation are, however, delayed significantly compared with FOS. Many of the original studies on FOS induction kinetics and distribution, both in cultured cells and in vivo, were performed with antibodies that do not distinguish among the protein products of the various members of the *fos* gene family. Most previous studies, therefore, probably were detecting not only FOS but, depending on the time of analysis following stimulation, FOS-like immunostaining (FLI) caused by FRA-1, FRA-2, and other protein products from this gene family.

Like FOS, JUN is encoded by a member of a family of genes whose products have structural similarity. These include the *junB* (Ryder et al., 1988) and *junD* (Hirai et al., 1989; Ryder et al., 1989) genes. Although the various FOS proteins cannot form heterodimers or homodimers, they can form heterodimers with the various JUN proteins (Nakabeppu et al., 1988; Cohen et al., 1989; Zerial et al., 1989). The members of the *jun* family can form homodimers as well as heterodimers with FOS and with other members of the *jun* and *fos* families. Because of the varying kinetics of expression among the members of the *fos* and *jun* families, a variety of AP-1-binding entities, differing in time after induction, thus can be formed (see Sonnenberg et al., 1989a). To date, 15 distinct AP-1-binding dimers have been demonstrated. The varying composition of the AP-1 nuclear complex following ligand stimulation is likely to alter the specificity and transregulatory activity of this DNA-binding complex for different secondary target genes in cells.

### C. c-*fos* Induction Pathways

Serum and purified growth factors, acting as mitogenic agents on 3T3 cells, were among the first ligands shown to stimulate c-*fos* induction (Greenberg and Ziff, 1984; Müller et al., 1984; Kruijer et al., 1985). Initially, it was thought that induced *fos* expression might be associated with either a specific set of extracellular signals (e.g., growth factors) or a specific type of phenotypic response (e.g., mitogenesis). However, the c-*fos* gene exhibits surprisingly little ligand, cell, or response specificity. Part of the reason for the nearly ubiquitous inducibility of *fos*, by a variety of ligands in many cell systems, lies in the structure of the *fos* promoter. A variety of second messenger cascades leads to induction of *fos* mRNA by transactivation of one or another of several regulatory acting sequences in the *fos* promoter. The primary site of transcriptional activation for a number of ligands is the 20-bp serum response element (SRE), also called the dyad symmetry element (DSE) (for a recent review, see Treisman, 1990). A diverse set of ligands that includes tetradecanoyl phorbol acetate (TPA), insulin, epidermal growth factor (EGF), and nerve growth factor (NGF) can activate separate, and in some cases additive, secondary messenger pathways that activate transcription factors that regulate c-*fos* transcription via the 14-bp inner core of the SRE. A protein termed the serum response factor, or p67$^{SRF}$ (Gilman et al., 1986; Treisman, 1986, 1987; Greenberg et al., 1987; Prywes and Roeder, 1987), has been shown to dimerize, bind to another component, p62$^{TCF}$ (Shaw et al., 1989), and recognize the c-*fos* SRE (see Schröter et al., 1990). SRE consensus sequences are present in the promoters of several other primary response genes encoding nuclear proteins (Rollins et al.,

1988; Tsai-Morris et al., 1988; Ryseck et al., 1989; Cortner and Farnham, 1990). The *fos* promoter also contains a cAMP-responsive element (CRE) (Gilman et al., 1986; Fisch et al., 1987, 1989; Sheng et al., 1988, 1990; Berkowitz et al., 1989). Ligands that activate adenylate cyclase induce an elevated level of cAMP, which then activates cyclic AMP-dependent protein kinase (PKA) (for review, see Montminy et al., 1990). The CRE-binding protein (CREB) is a substrate for PKA. PKA phosphorylation of CREB activates the transcription factor capacity of CREB and results in CRE-mediated transcription from the c-*fos* gene. Depolarizing stimuli associated with increase in intracellular $Ca^{2+}$ also induce *fos* via the CRE (see Sheng et al., 1990). Like FOS and JUN, CREB is encoded by one of a family of genes whose protein products have related structures and functions. The members of this gene family are known as activating transcription factors (ATFs) (Hai et al., 1989). Although the CREB/ATF genes are not primary response genes, their encoded proteins contain leucine zipper domains related to those of FOS and JUN. Thus, ligand-induced cAMP-mediated phosphorylation of CREB/ATF proteins in the nucleus by PKA appears to be an alternative mechanism which a subclass of extracellular signals rapidly can activate transcription of the c-*fos* gene.

### D. Inhibitory Control of *fos*

In the absence of growth factors (or serum), accumulation of *fos* mRNA is inhibited at the levels of both transcriptional initiation and elongation. Both processes are necessary for maintaining a low basal level of *fos* mRNA (Lamb et al., 1990). The FOS protein can autogenically repress transcription from the *fos* promoter, apparently by interacting with the transcription complex at the SRE (Lucibello et al., 1989; Rivera et al., 1990; Guis et al., 1991). FRA-1 can substitute for FOS in repressing c-*fos* transcription. Since FRA-1 usually is delayed in its expression following ligand induction and has a more extended half-life, it may be responsible for the lengthy refractory period of *fos* transcription observed following induction.

The secondary inhibitory element regulating c-*fos* transcription is the *fos* intragenic regulatory element (FIRE), located at the end of the first exon of the c-*fos* gene (Lamb et al., 1990). This 21-bp element binds an uncharacterized protein complex that blocks transcriptional elongation, resulting in truncated (first exon) *fos* mRNA (Bonnieu et al., 1989). Quiescent cells, either treated with cycloheximide (Fort et al., 1987) or untreated (Bonnieu et al., 1989), exhibit a significant level of this truncated *fos* transcription. Co-injection of SRE and FIRE oligonucleotides into quiescent rat fibroblasts to compete with endogenous DNA sequences for putative binding factors induced c-*fos* within 60 min. Injection of either oligonucleotide alone does not lead to *fos* induction. These data suggest that repression in unstimulated cells requires factors acting at both the SRE and the FIRE elements in the c-*fos* gene. Mutated FIRE sequences are ineffective and SRE sequences injected prior to serum stimulation block c-*fos* expression. The factors binding at the SRE and FIRE sequences appear to be distinct.

## E. Cross-Talk

Initially, the intracellular transduction pathways activated by various ligands were thought to be separate processes leading to gene expression. In many cases, signaling pathways do appear to function in reasonably separate and independent modes. However, reports of pathway interaction or "cross-talk" have added substantial complexity to the analysis of ligand induction of primary response genes.

The consensus sequence of CREs differs from that of TREs by one nucleotide (CRE:-TGAGCTCA-; TRE:-TGACTCA-) (Comb et al., 1986; Montminy et al., 1986). Although, in general, transcriptional activation by TPA or forskolin is mediated via promoter TRE and CRE sequences, respectively (Sassone-Corsi et al., 1990), recent experiments suggest that considerable cross-talk may exist in these signal transduction pathways. Gel-retardation studies with pure FOS and JUN (rather than nuclear extracts) and co-transfection experiments with combinations of reporter genes containing either CRE or TRE sequences along with *jun* and *fos* expression vectors demonstrate that, under these transfection conditions, (1) JUN can transactivate CRE sequences, (2) FOS can enhance JUN binding and activation of CRE promoter constructs, and (3) forskolin dramatically enhances the JUN/AP-1 influence on CRE sequences. How or why these two pathways converge on transcription is still not clear.

Additional examples of partial interaction or convergence can be found at nearly every level of the transduction cascade (see, for example, Yamamoto et al., 1988; Conzalez et al., 1989; Rozengurt and Sinnett-Smith, 1990; Sassone-Corsi et al., 1990; Yang-Yen et al., 1990), including elevation of second messengers ($Ca^{2+}$, cAMP, phospholipids), kinase activation (PKC, PKA, growth factor receptor kinases), kinase-phosphorylation of substrates, heterodimer combinations (JUN:FOS, JUN:CREB; vida supra, previous text), and *cis*-acting response elements (TREs, CREs). Thus, signal transduction pathways responsive to environmental stimulation display intimate coupling at several nodes linking receptor–ligand interaction to gene activation. In many cases described to date, the interaction has an additive or even synergistic effect. A notable exception is the ability of FOS:JUNB complexes to inhibit AP-1 transcriptional activation (Chiu et al., 1989; Schütte, 1989).

In summary, several themes emerge from the research carried out on the c-*fos* and c-*jun* gene families, the regulation of the transcription of these genes, and the characterization of their gene products. One central theme is that cells possess the ability to produce multiple nuclear factors whose levels of expression and/or levels of posttranslational activation can be modulated by ligand activation of various intracellular signaling pathways. The shared structural motifs for dimerization among these transcription factors adds another layer of combinatorial control of subsequent gene expression. Finally, primary response genes themselves possess a complex array of regulatory sequences in their 5' flanking regions. As a result, cooperative interactions between transcription factors may exert a hierarchical control over primary response gene transcription.

## III. OTHER PRIMARY RESPONSE GENES

The induction of primary response genes appears to be a central component of cellular adaptive responses to environmental signals. It is estimated that there are between 100 and 300 primary response genes (Almendral et al., 1988). In addition to the *fos* and *jun* gene families, a number of other primary response genes also appear to function as part of the third messenger system for ligand-induced long-term alterations in cellular phenotype. We will describe briefly some of the best characterized primary response genes whose expression has been demonstrated in the nervous system. For a comprehensive description of the full range of primary response genes, see the review by Herschman (1991).

### A. *egr1*

The murine *egr1* gene encodes a 533-amino-acid protein containing three potential zinc fingers. The *egr1* cDNA was cloned as a primary response gene from serum-treated 3T3 cells (Sukhatme et al., 1987). *egr1* is also known as *krox24* (Lemaire et al., 1988), TIS8 (Lim et al., 1987), and *zif268* (Christy et al., 1988). *egr1* protein is subject to phosphorylation and glycosylation, and binds to a specific consensus sequence found in its own promoter region as well as in those of several other primary response genes and secondary response genes (Cao et al., 1990; Lemaire et al., 1990; McMahon et al., 1990). The *egr1* protein can function as a sequence-specific *trans*-activating factor (Lemaire et al., 1990). NGFI-A encodes the 508-amino-acid rat homolog of *egr1* (Milbrandt, 1987; Changelian et al., 1989); gene 225 is the human *egr1* homolog (Wright et al., 1990).

### B. *krox20*

The *krox20* gene encodes another murine zinc-finger protein with sequence similarity to the *Drosophila* kruppel homeobox gene (Chaiver et al., 1988, 1989). The *egr2* gene is the human homolog of *krox20* (Joseph et al., 1988). *krox20/egr2* and *egr1/krox24* share sequence similarities, particularly in their zinc-finger domains, and can bind to the same DNA promoter sequence to *trans*-activate reporter gene constructs (Chavier et al., 1990). However, the two genes are expressed with distinct developmental and cell-type specific patterns (Lemaire et al., 1988).

### C. *nur77*

The *nur77/N10* primary response gene also was cloned from serum-treated 3T3 cells (Hazel et al., 1988; Ryseck et al., 1989).The cDNA for this gene was also cloned as a TPA-induced primary response gene, TIS1, in 3T3 cells (Varnum et al., 1989b). The rat homolog, cloned as an NGF-inducible gene (Milbrandt, 1988; Watson and Milbrandt, 1989), is known as NGFI-B. *nur77*/NGFI-B encodes a zinc finger-containing nuclear protein. Sequence analysis identified it as a member of the ligand-inducible steroid receptor superfamily that includes the estrogen, glucocorticoid, vitamin D, and retinoic acid receptors (Evans, 1988). In recent work, NGFI-B is observed to undergo rapid posttranslational modification with distinct subcellular distribution. A highly phosphorylated species of NGFI-B is found predominantly in the cytoplasm; the underphorphorylated species is localized in the nucleus (Fahrner et al., 1990).

### D. scip

*scip* (suppressed cAMP inducible POU) is a POU-domain gene restricted primarily to the nervous system, in particular, developing glia (Monuki et al., 1989, 1990). It appears that *scip* may play several different, but important, regulatory roles in the development of the nervous system. Both central and peripheral glia express this cAMP-inducible protein. The putative DNA-binding domain of *scip* is closely related to that of members of the POU transcription factor family. *scip* is a primary response gene, based on (1) its ligand-induced, protein synthesis-independent, rapid (< 1 hr) induction and (2) its ability to be superinduced by forskolin in the presence of cycloheximide. However, *scip* differs from most other known or putative nuclear primary response genes (e.g., c-*fos*) because its expression is not transiently induced, but remains expressed in the continued presence of ligand.

## IV. IN VITRO EXPRESSION OF PRIMARY RESPONSE GENES IN NEURAL SYSTEMS

### A. Neuronal Cells

The rate PC12 pheochromocytoma cell line is the most widely employed cell line for the study of growth factor control of neuronal growth and differentiation (Greene and Tischler, 1982). Neurobiologic studies of primary response genes began with the discovery that NGF rapidly and transiently induced c-*fos* mRNA in PC12 cells (Curran and Morgan, 1985; Greenberg et al., 1985; Kruijer et al., 1985). The NGF-mediated expression of c-*fos* suggested that primary response genes may play a key role in neuronal differentiation. However, subsequent studies showed that almost every growth factor analyzed rapidly induced c-*fos* expression in PC12 cells, regardless of the eventual phenotypic response. For example, FGF also induced both neurite outgrowth and c-*fos* expression in PC12 cells. But EGF, which stimulates PC12 proliferation, also induces c-*fos* expression. Obviously, c-*fos* cannot be a unique mediator of ligand-specific phenotypic response. Other peptides and growth factors, including platelet-derived growth factor (PDGF) (Kruijer et al., 1985) and IL-6 (Satoh et al., 1988), have been reported to induce *fos* expression in PC12 cells. Growth factor induction of *fos* in PC12 cells appears to be dependent on receptor kinase-mediated phosphorylation of pre-existing transcription factors. The *fos* promoter contains two adjacent regulatory elements that mediate NGF induction via two distinct nuclear protein complexes (Visvader et al., 1988). One of these *cis*-acting sequences is the SRE, described previously, that mediates c-*fos* induction by several growth factors in 3T3 cells. Recall that NGFI-A and NGFI-B were initially cloned as NGF-inducible primary response genes. (Milbrandt, 1987, 1988). Subsequent studies demonstrated that primary response genes other than c-*fos*, including NGFI-A/TIS8 and NGFI-B/TIS1, could also be induced in PC12 cells by EGF and FGF, as well as by NGF (Kujubu et al., 1987).

Neuronal depolarization also induces c-*fos* expression in PC12 cells. Elevated $K^+$ or rapid $Ca^{2+}$-channel activation, leading to $Ca^{2+}$ influx, induces c-*fos* expression. Treatment either with the $Ca^{2+}$ agonist Bay K-8644 or with barium can lead to rapid appearance of *fos* mRNA (Curran and Morgan, 1986; Morgan and Curran, 1986; Sheng et al., 1988). Membrane depolarization caused by neurotransmitter binding also elicits a marked rise in *fos* mRNA levels. Stimulation of acetylcholine receptors on PC12 cells by nicotine rapidly induces *fos* transcription (Greenberg et al., 1986). Growth factor-mediated c-*fos* induction is not $Ca^{2+}$ dependent. The studies using depolarization, $Ca^{2+}$ flux, and neurotransmitters suggest that $Ca^{2+}$ influx through voltage sensitive $Ca^{2+}$ channels activates a separate intracellular signaling pathway that leads to the rapid and transient induction of c-*fos*. Subsequent studies demonstrated that depolarization (Kujubu et al., 1987) and acetylcholine agonists (Altin et al., 1991) can also induce the expression of NGFI-A/TIS8 and NGFI-B/TIS1 in PC12 cells. Treatment of PC12 cells with sodium butyrate to induce a chromaffin cell-like phenotype rapidly induces *fos* mRNA, but induces *jun* mRNA only after a significant delay (Naranjo et al., 1990). The set of posttranslationally modified FOS proteins differs from the pattern seen after NGF treatment. Studies of human adrenal pheochromocytomas show constitutive high levels of c-*fos* and c-*myc* mRNA levels compared with undetectable mRNA levels in bovine adrenal medulla tissue (Goto et al., 1989).

Cultured cerebellar granule cells also have been used to study the expression of primary response genes. Szekely and co-workers (1987, 1989) have reported that NMDA-type (*N*-methyl-D-aspartate) glutamate stimulation of cultured granule cells induces c-*fos* mRNA and FOS protein accumulation.

Adrenal medullary chromaffin cells function as part of the sympatho-adrenal system and respond to systemic (e.g., angiotensin) and local (e.g., acetylcholine) signals by coordinately induced expression of specific genes (for review, see Unsicker, 1989). Cultures of tyrosine hydroxylase (TH)-positive chromaffin cells can be induced by both types of signals (nicotine and angiotensin) to express both c-*fos* mRNA and FLI (Stachowiak et al., 1990). Angiotensin-mediated induction of *fos* is protein kinase C (PKC) dependent in these cells. Thus, the PKC dependent angiotensin-mediated increase in TH, phenylethanolamine *N*-methyltransferase, and proenkephalin mRNAs may be coordinated by the expression of *fos*. Nicotine appeared to induce FRAs as well as FOS, and to increase the extent of posttranslational modification of FOS. Depolarization by veratridine and forskolin treatment also increases FLI in these cells. In summary, a variety of extracellular signals can induce c-*fos* expression in specific neuronal cell populations in culture.

## B. Glial Cells

A number of glial cell populations can be cultured and used to study the expression of primary response genes and their roles in cell growth and development. These include primary cultures of nearly homogeneous populations of astrocytes, oligodendrocytes, and progenitor cells (see Arenander and de Vellis, 1989). Primary rat neocortical astrocytes can be obtained in large numbers, passed as secondary cultures, and examined for primary response gene induction under a variety of conditions. A wide variety of agents is capable of inducing many primary response genes in these cultures (Arenander et al., 1989a,b,c; Condorelli et al., 1989). For example, the peptide mitogens EGF and FGF induce accumulation of mRNAs from the *nur77*/TIS1, TIS7/PC4, *egr1*/TIS8, TIS11, TIS21, c-*fos*, c-*jun*, and *junB* genes. TPA also acts as a mitogen for these cells and induces the same set of primary response genes. Induction of primary response genes in astrocytes, however, is not associated uniquely with any specific phenotypic response. Thus, forskolin, which increases cAMP and lead to morphologic differentiation, also induces the same set of primary response genes. Treatment of cells with combinations of ligands often results in additive levels of induction of these genes, suggesting the presence of separate independent ligand-inducible pathways that converge at the transcriptional response. For example, co-treatment of astrocytes with TPA and EGF and/or FGF gives additive levels of several primary response gene mRNAs (Arenander et al., 1989a). These changes are correlated with the additive effects on cell proliferation. Oligodendrocytes also respond to extracellular signals by expressing a number of primary response genes (A.T. Arenander, unpublished data).

Primary response gene expression also has been examined in two different clones of rat glioma cells. C6 glioma cells can express primary response genes when stimulated by a number of different agents, including insulin, insulin-like growth factor, hydrocortisone, and isoproterenol (Mocchetti et al., 1989; Arenander et al., 1991). F-98 rat glioma cells treated with sodium butyrate stop proliferating and morphologically and biochemically differentiate. Rapid and transient induction of *fos* occurs in sodium butyrate-treated F-98 cells followed by increases in *sis*, fibronectin, and collagen mRNAs (Tang et al., 1990), suggesting that *fos* may participate in sodium butyrate-induced glial differentiation. Treatment of NG108-15 neuroblastoma–glioma hybrid cells by adenosine agonists induces the expression of c-*fos*, but not of c-*jun*, mRNA (Gubits et al., 1990). The kinetics of c-*fos* induction in the neuron–glia hybrids were somewhat slower than in 3T3 cells. Specific adenosine antagonists were capable of inhibiting the increase in c-*fos* mRNA.

Schwann cell cultures have been examined for the expression of the primary response gene *scip*. *scip* is not expressed in Schwann cells cultured in the absence of neuronal cells. However, treatment of Schwann cells with forskolin, which raises intracellular cAMP by activating adenyl cyclase (Monuki et al., 1989), readily induces *scip* mRNA with rapid and prolonged kinetics. Subsequently, at about 20 hr, one sees expression of myelin differentiation markers such as MBP and PO. In normal development, expression of *scip* mRNA also precedes the onset of these myelin structural genes (MBP, PLP, and PO) in maturing oligodendrocytes and Schwann cells. Interestingly, constitutively high basal levels of *c-jun* mRNA are downregulated rapidly (about 30 min) and completely in response to cAMP activation in Schwann cells. The inverse pattern of expression of these two primary response genes in response to cAMP may be significant in the combinatorial mechanisms by which primary response genes are thought to control gene expression.

*Tst-1* is apparently the same gene as *scip* (He et al., 1989). *scip/Tst-1* is expressed in neurons of the embryonic brain and peripheral nervous system, especially in rapidly proliferating progenitor cells of the ventricular zone of the neural tube. Lemke and his co-workers (Monuki et al., 1989, 1990) suggest that *scip* may play two different roles in brain development. *scip* may serve (1) as a transiently appearing nuclear factor regulating proliferation and lineage decisions of O-2A progenitor cells in the brain and proliferation of Schwann cells in the peripheral nervous system (PNS) and (2) as a stably expressed transcription regulator in terminally differentiated cells. Thus, *scip* represents a member of a new family of primary response genes that could serve to regulate glial and neuronal growth and to determine cell fate.

## V. IN VIVO EXPRESSION OF PRIMARY RESPONSE GENES

The observation that depolarization induces *c-fos* expression in PC12 cells encouraged researchers to examine the expression of this gene in the central nervous system (CNS) in response to a variety of stimulation paradigms. In this section, we describe the current status of primary response gene induction in the CNS following perturbation, in studies ranging from pharmacologic induction of seizures to the effects of growth factors. Most in vivo studies have examined the patterns of *c-fos* mRNA and FLI. Stimulation of the brain by pharmacologic, electrical, or mechanical means induces the rapid, and in many cases transient, expression of *c-fos* mRNA in anatomic and cell-type specific patterns of neural expression (see Dragunow and Faull, 1989). In only a few instances have other primary response genes been studied.

Surprisingly little work on primary response gene expression has been carried out during periods of dramatic cellular growth and development. Most of the studies on expression of these genes have been conducted in the adult animal. Only a few studies survey the patterns of primary response gene expression in the developing brain. Ruppert and Wille (1987) showed that *c-fos* induction in dissociated cerebellum cells was age related. Caubet (1989) later showed a progressive restriction of *c-fos* mRNA in the mouse nervous system. Initially, *c-fos* was found to be widely distributed. Over time, however, c-*fos* expression became limited to specific regions. Differential patterns of expression of c-*jun*, *junB*, and *krox20* mRNA have been reported (Wilkinson et al., 1989a,b). c-*jun* is highly localized within the CNS, in particular in the forebrain ventricular zone of proliferating progenitor cells. *junB* transcripts were not reported in CNS. In contrast, a segment-specific pattern of *krox20* mRNA was observed in the embryonic hindbrain (Wilkinson et al., 1989a). With time, *krox20* expression became further restricted to hindbrain nuclei. Glia of spinal and cranial ganglia also express this zinc-finger gene.

Most studies with whole animals have identified cells expressing c-*fos* mRNA or FLI in the brain as neuronal. In light of the cell culture data demonstrating the relative ease with which primary response gene induction can be observed in cultured glial cells, the lack of responsiveness of the neuroglial elements in the intact CNS is striking. To date, only heat shock (Dragunow et al., 1989c), ischemia (Herrera and Robertson, 1989), and focal lesions impinging on white matter tracts (Dragunow and Robertson, 1988; Dragunow et al., 1990b,c) have been reported to induce FLI in glial cells. The reason for this dramatic "insensitivity" of glial cells to CNS stimulations, such as massive electrically and chemically induced seizures and lesion-induced hemispheric changes similar to spreading depression, is not known. A trivial explanation is that glial cells are capable of expressing primary response genes under these conditions, but that this response remains undetected due to lack of proper cell identification or to inadequate levels of primary response gene induction. We think it more likely that, under most conditions in the intact nervous system, extracellular signals capable of inducing primary response genes are not presented to glial cells and/or the control of primary response gene expression in these glia is tightly regulated.

Before beginning a detailed discussion of induced expression of primary response genes in the nervous system, we will discuss methodology issues associated with the in vivo perturbation studies of primary response genes. The reagents used in detecting primary response gene induction influence the interpretation of the findings. For example, many studies use a rabbit IgG antibody raised against a synthetic antigen, M-peptide, of the FOS protein (Curran et al., 1985). This small peptide is from a region of FOS that is highly conserved within the *fos* gene family. Thus, many studies use antisera that cross-react with FOS and the FRAs. These studies measure FLI rather than FOS immunoreactivity. The lack of specificity of anti-FOS antisera initially led to an oversimplified view of the complex nature of induced expression of the *fos* gene family members in response to stimulation. FOS levels usually decline within several hours; the long-lasting FLI is due to the retarded expression of one or more FRAs.

Many studies use animals under anesthesia. A problem inherent in models of epilepsy, kindling, and pain is that results obtained in anesthetized animals may be altered substantially when compared with awake preparations, complicating the analysis of mechanisms underlying each condition. Similar reservations are present in the mapping of FOS expression: anesthetics may modify neural responsiveness significantly. For example, anesthetic drugs inhibit stimulus-induced FLI in spinal cord neurons; a gradient of increased suppression is observed in cells from superficial to deeper laminae (Presley et al., 1990). Thus, evoked FLI expression in anesthetized animals may reflect an underestimation of activated neuronal subpopulations in the circuit because of differential sensitivity of cells to anesthetics. In particular, polysynaptically activated cells may be suppressed strongly by such agents.

The influence of uncontrolled experimental variables results in two confounding outcomes: higher basal levels of primary response gene expression and false positive patterns of activation. Nonspecific or stress-associated perturbations may include the effects of (1) exploratory behavior, (2) animal handling, and (3) experimental procedures. For example, several studies have reported high basal neuronal *fos* expression (Dragunow and Robertson, 1988; Insel, 1990) and FLI elevation in a variety of brain areas following treatment of animals with several stressors (Ceccatelli et al., 1989). Even gentle mechanical stimulation of the hindlimb can increase FLI in dorsal horn neurons, albeit not to the same extent or levels observed following strong noxious stimuli (Bullitt, 1989). The "stress" produced by handling animals can induce c-*fos* in neurons, particularly in cortical cells (Sharp et al., 1989b). Similarly, c-*fos* is induced in dentate granule cells, hippocampal pyramidal cells, primary olfactory cortex, and various neocortical areas following intraperitoneal injections of saline. This expression can be blocked by diazepam (Nakajima et al., 1989). Stress associated with ear clipping of "control" animals for identification purposes results in substantial FLI induction, resulting in a pattern similar to that observed following induction of seizures (Morgan et al., 1987). More recent experiments have been designed carefully to prevent such variables from influencing the pattern of *fos* expression (Chang et al., 1988; Graybiel et al., 1990; Gibbs et al., 1990; Presley et al., 1990).

## A. Seizures

Treatment with pentylenetetrazol (PTZ) provides an intense global CNS stimulation. Animals so treated exhibit seizures within minutes. c-*fos*, c-*jun*, *junB*, *egr1*, and NGFI-B mRNAs are rapidly and transiently induced in the brains of these animals after the onset of the seizure (Morgan et al., 1987; Saffen et al., 1988; Sonnenberg et al., 1989b), Sukhatme et al., 1988; Watson and Milbrandt, 1989). A distinct temporal sequence of primary response gene expression at specific anatomical sites is observed. Initially, the FLI is located in limbic structures that can exhibit high basal levels of FLI. These include the dentate granule cells, the pyriform cortex, and the amygdala. Within a period of several hours after stimulation, FLI appears prominently in the neocortex, hippocampal pyramidal cells, basal ganglia, various thalamic and hypothalamic nuclei, and the locus coeruleus. FLI in these studies has been shown to be composed of temporally staggered expression of *fos* family members—FOS appears first, followed by FRAs (FRA-35K and FRA-46K; Sonnenberg et al., 1989a,b; Morgan and Curran, 1989)—not unlike the temporal sequence observed in PC12 cells (Cohen and Curran, 1988). This shift in primary response gene protein composition must alter the composition of the AP-1 transcription complex. In fact, AP-1 DNA-binding capacity increases rapidly and is sustained for about 10 hr following the onset of PTZ seizures (Sonnenberg et al., 1989a,b), suggesting a shift from FOS to FRA dominance of the AP-1 complex. This programmed alteration in AP-1 composition may, in turn, alter the selection and extent of secondary response gene expression (vida infra).

Induction of seizures by other means also induces *fos*. Electrocortical shock (ECS), another paradigm of seizure induction, gives patterns of primary response gene expression similar to those of the PTZ model (Sonnenberg et al., 1989c). Induction of seizures by a vitamin $B_6$ antagonist also induced c-*fos* mRNA and protein expression (Mizuno et al., 1989). Mice undergoing ethanol withdrawal spontaneously exhibit seizures likened to kindling. In animals with this type of seizure, transient (24-hr) induction of c-*fos* mRNA in the hippocampus and neocortex is observed (Dave et al., 1989). Ethanol withdrawal without seizures does not lead to induction of *fos*. Finally, recurrent limbic seizures induced by lesions to the dentate gyrus hilus rapidly increase c-*fos* mRNA in the hippocampus (White and Gall, 1987). The increase in *fos* was sensitive to the anticonvulsant pentobarbital. The induction of *fos* may be important in the differential neuropeptide gene expression in the entorhinal cortex after seizures.

In an attempt to clarify the mechanisms responsible for seizure-induced FLI in the CNS, pharmacologic experiments were carried out with glutamate receptor agonists and blockers (Sonnenberg et al., 1989c). Treatment of animals with NMDA produces a rapid, marked, and transient elevation of *fos* mRNA. The NMDA antagonists MK-801 and APV greatly attenuate PTZ stimulation of *fos*. Kainate treatment gives a rapid, but sustained, increase in *fos* mRNA that is not blocked by NMDA antagonists (see also, Le Gal La Salle, 1988; Popovici et al., 1988). Immunoblot studies demonstrate that both convulsants, NMDA and kainic acid, generate the appearance of a series of antigens in brain extracts: initially FOS, followed by FRA-35K and FRA-46K. MK-801 also reduces KCl-induced hemispheric spreading depression and expression of FLI in cortical neurons (Herrera and Robertson, 1990).

Following PTZ or ECS treatment, *fos* induction in the brain remains refractory for many hours (Morgan et al., 1987; Winston et al., 1990). The mechanism for this may be the autoregulatory negative feedback control FOS exerts on its own promoter (Guis et al., 1991). The inhibition induced by an acute ECS seizure is reset within 18 hr; at this point a second seizure can evoke a complete *fos* response. In contrast, chronic daily seizure induction leads to a progressive and profound suppression of c-*fos* and c-*jun* mRNA expression and FOS expression (Winston et al., 1990). Basal FOS levels also were reduced in the chronic animals. Possible explanations for this chronic suppressed state include (1) long-term effects of FRAs on secondary response genes whose products are, in turn, capable of regulating c-*fos* expression and (2) a change in the intracellular pathways that couple seizure activity and gene expression. Whatever the mechanism, chronic seizures can induce an adaptive response that includes a long lasting desensitization of primary response gene expression and a consequent alteration in the level and molecular composition of the AP-1 complex that is likely to influence its role on secondary response gene activity.

### B. Kindling and Primary Response Gene Expression

Kindling is a widely used animal model of human epilepsy. The expression of primary response genes during the generation of the kindled state has been postulated to contribute to the molecular mechanisms that underlie the changes in neuronal structure and function that occur in these animals (Dragunow et al., 1989a). The kindled state appears after 1–2 weeks of repeated subconvulsant electrical stimulation of limbic structures. Over this period of time, the brain is irreversibly altered so low-level electrical stimuli will now precipitate a convulsive state. Once established, the kindled state will last throughout the animal's life. In the initial period of kindling, both c-*fos* mRNA (Shin et al., 1990) and FLI (Dragunow and Robertson, 1987a) are elevated in dentate granule cells and hippocampal pyramidal cells. Thus, c-*fos* expression has been proposed as a causal component of the process of kindling. However, the same pattern of c-*fos* expression is found in both naive and kindled animals (Shin et al., 1990). Moreover, there is no alteration in the basal FOS level in kindled animals. As was noted in the PTZ paradigm, a minimal level of stimulus intensity and/or duration is necessary for *fos* induction. Dragunow et al. (1989b) have extensively reviewed the literature on c-*fos* expression in kindled animals.

### C. Long-Term Potentiation

Long-term potentiation (LTP) in the rat hippocampal formation is a well-known paradigm for stimulus-evoked long-lasting changes in neuronal excitability in response to subsequent challenge (Kennedy, 1989). Unlike kindling, LTP does not involve seizure induction. The stimuli that lead to hippocampal LTP also induce FLI. However, the correlation does not hold across a range of excitory input; levels of stimulation can be sufficiently low to yield no obvious change in c-*fos*, but still produce LTP. These findings argue against a causal relationship between *fos* expression and LTP (Douglas et al., 1988; Dragunow et al., 1989a,b). Cole et al. (1989) demonstrated that NGFI-B/*zif268*/*egr1*/TIS8 is also induced by LTP. Moreover, the correlation between LTP and NGFI-B expression cannot be dissociated by dose–response studies. However, induction of LTP, but not NGFI-A expression, can be blocked by the activation of a neuronal inhibitory input (Wisden et al., 1990). This result would suggest that induction of NGFI-A may be necessary, but is not sufficient, for the establishment of LTP. Moreover, actinomycin D administration did not prevent the establishment or maintenance of the early phases of LTP over the initial 3-hr period, suggesting that the induction of primary response gene expression is not necessary for the early stages of LTP (Otani et al., 1989).

However, the expression of primary response genes such as c-*fos* and NGFI-B was not analyzed in these experiments.

### D. Mapping of Neuronal Networks

Functional mapping of neural networks has, until recently, depended on either electrophysiologic measurements or measurements of 2-deoxyglucose (2DG) uptake after stimulation. In the latter case, the relative intensity of 2DG associated with neurons is thought to represent the relative degree of neuronal activity in response to afferent input (Sokoloff, 1989). The examination of primary response gene expression patterns in response to various forms of CNS stimulation now offers an additional mode of analysis of neuronal connectivity. However, the pattern of activated neurons observed and the interpretation of the data can be greatly influenced by a variety of experimental variables, including (1) the extent and intensity of stimulation, (2) the method of detection, (3) the degree of control over nonspecific stimuli (sensory/stress effects), (4) the state of arousal of the animal (awake or anesthetized), and (5) the influence of biorhythms. Primary response gene analysis allows mapping of functionally related neural elements, activated by a specific stimulus at the single cell level, in extensive regions of tissue. However, it is possible that not all activated (excited or inhibited) cells in a network display the activation of a common primary response gene product; some neurons may be "restricted" with respect to expression of specific primary response genes.

Both 2DG and primary response gene analysis provide estimates of the degree of cell activity. However, 2DG measures overall metabolism, whereas primary response gene analysis, such as FLI, measures the extent and degree of coupling of intracellular signaling pathways to gene expression. Each method comes with advantages and disadvantages. Anatomical resolution, specificity of target detection, signal to noise level, basal compared with induced levels, and temporal resolution are all important variables. 2DG methods are inherently low resolution; one cannot determine the number or types of cells that account for the increase in metabolic rate in a positive region. Primary response gene studies have the distinct advantages of (1) cellular or subcellular resolution, (2) low basal levels of expression and, usually, all-or-none evoked nuclear expression, (3) the choice of several messages or proteins to detect, and (4) the ability to be highly specific (e.g., *fos* only) or relatively broad (e.g., exploiting the *fos* family cross-reactivity of many sera) in the choice of probes.

Electrophysiologic recording techniques provide excellent time and spatial resolution, but are restricted by the lack of population response information. Functional changes in populations of cells can be determined easily by primary response gene mapping at the mRNA or protein level. Mapping of a particular neural circuit by several primary response genes at different times may reveal a more comprehensive and dynamic picture of activated neural networks. Thus, primary response gene mapping presents a complementary source of information concerning the existence of neural circuits, their functions, and their temporal responses to stimuli.

### E. Sensory Stimulation

Physiologic sensory stimulation has been used to induce primary response genes in the CNS. Stimuli used in these studies have been both relatively gentle and painless (nonnociceptive) and painful (nociceptive). Hunt et al. (1987) reported the first example of sensory stimulation induction of FLI in dorsal horn neurons of the spinal cord.

Studies using stimulation of the motor/sensory cortex to map out polysynaptic corticocerebellar connectivity have been carried out. Cortical stimulation results in FLI appearing in selected regions of the cerebellum and pontine nuclei (Sagar et al., 1988). Stimulation produced a clear map of cerebellar somatotopic organization (Sharp et al., 1989c). Patches of cerebellar granule cells and, to a lesser extent, Purkinje cells displayed FLI. FLI thus provided cellular resolution of the entire cerebellum, in contrast to electrophysiologic and 2DG methods.

The mechanisms by which responses to sensory stimuli are incorporated into short-term (minutes to hours) and long-term memory storage (hours to days) have been examined recently at a cellular and molecular level (for review, see Kandel, 1989). One fundamental distinction between these two types of processes is that mechanisms responsible for long-term information storage in response to sensory stimulation display a critical window of sensitivity to disruption. A period of 60 to 90 min following stimulus onset has been described in most biological systems, from invertebrates to humans, as a critical interval during which the early consolidation phase of long-term storage can be disrupted. Early studies showed that a short period of cycloheximide treatment during and after learning interfered with long-term, but not short-term, memory, suggesting that protein synthesis was a necessary event for the transition.

Analysis of the consequences of sensory stimulation in the invertebrate *Aplysia* has also suggested that new protein and mRNA synthesis are required for long-term memory, but not short-term memory (Montarolo et al., 1986). A cascade of events in the consolidation process has been delineated using an in vitro model system composed of a two-neuron monosynaptic reflex circuit from this animal. A sensitizing stimulus or the application of the corresponding neurotransmitter, serotonin (5-HT), elevates cAMP. Increased cAMP levels result in the activation of protein kinase A and the phosphorylation of at least 17 specific substrate proteins defined on two-dimensional gels (Sweatt and Kandel, 1989). Although long-term memory leads to a persistent increase in the same set of phosphorylated proteins, their sustained elevation in the phosphorylated state is blocked by inhibitors of protein and mRNA synthesis, unlike the short-term increase. These data suggest that the stimulus must induce a transcription- and translation-dependent response that is required for the establishment of the long-term memory and the persistently phosphorylated forms of these proteins. In addition, a set of proteins whose rates of synthesis are increased or decreased during the initiation phase of this long-term memory have been described (Barzilai et al., 1989); expression from these genes has many of the properties of the induction of the primary response genes elicited in other systems. These "early" proteins in the 5-HT response appear as several waves of rapidly and transiently expressed proteins, whose accumulation is transcription dependent. The induction of these rapidly synthesized mRNAs appears to be mediated by the activation of the cAMP intracellular cascade, the second messenger pathway for 5-HT (Schacher et al., 1988); transient gene expression regulated by cAMP appears necessary for long-term facilitation.

Recent studies suggest that activation of protein kinase A, as a result of the 5-HT-induced elevation of cAMP, leads to the phosphorylation and activation of CREB-like transcription factors in the sensory neuron nucleus in this *Aplysia* model (Dash et al., 1990). Evidence for this conclusion comes from two types of experiments: (1) extracts of these sensory neurons contain components that bind specifically to mammalian CRE sequences and (2) microinjected CRE oligonucleotides block long-term, but not short-term, facilitation in a sequence-specific and dose-dependent manner, presumably by competing with CREs of target genes for the binding of PKA-activated CREB-like transcription factors. Thus, cellular changes underlying long-term memory, such as changes in synaptic morphology and physiology, require a set of molecular events that now includes the synthesis of new messages and proteins from genes that have the characteristics of primary response genes.

## F. Pain Systems

Various modes of noxious stimulation, pain pathway stimulation, and pharmacologic intervention have been examined to monitor the activity and functional connectivity of nociceptive neurons. Using FLI, induced FOS expression could be detected in specific spinal cord neuron populations, consistent with those identified by previous anatomic and electrophysiologic studies of nociceptive input territories and nociceptive neurons (Hunt et al., 1987; Bullitt, 1989; Menétrey et al., 1989). Basal FLI levels are very low. FLI appears within 1 hr, peaks at 2 hr, and remains elevated for 8–10 hr before returning to basal levels by 16 hr. This protracted course suggests that FRAs as well as FOS are being detected in these experiments.

A more recent study has examined the influence of the narcotic analgesic morphine on noxious stimulus-evoked FLI in spinal neurons in awake rats (Presley et al., 1990; Gogas et al., 1991). This study uses a continuous stimulation model of tonic pain elicited by subcutaneous application of formalin. Injection of systemic morphine prior to the onset of the persistent pain stimulus produced a dose-dependent naloxone-reversible suppression of FLI expression. The extent of opiate receptor-mediated suppression of FLI was lamina-dependent; the superficial lamina was most resistant to blocking and the deeper laminae were markedly inhibited. The data suggest that nociceptive information processing in the spinal cord neuronal populations is regulated differentially. Other studies, using noxious thermal stimulation, demonstrated that intravenous morphine suppressed FLI in both superficial and deep spinal neurons in a dose- and naloxone-dependent fashion. In contrast, the NMDA antagonist ketamine had no effect, suggesting that morphine is not affecting *fos* expression through NMDA receptor-mediated influx of $Ca^{2+}$ in spinal neurons (Tölle et al., 1990). Thus, NMDA-gated channels do not seem necessary for c-*fos* expression in nociceptive spinal neurons. Since ketamine at the same dose blocks behavioral response to pain, the effects of NMDA blockage may lie at supraspinal levels. The studies demonstrating that NMDA antagonists ketamine and MK-801 can suppress FLI expression in neocortical neurons following mechanical damage (Dragunow et al., 1990a; Herrera and Robertson, 1989, 1990) is consistent with this view. Recent studies show that a µ-selective opioid agonist injected intracerebroventricularly produces a lamina-selective, dose-dependent, naloxone-reversible inhibition of both formalin-induced pain behavior and *fos* expression in spinal neurons (Gogas et al., 1991). The effects of morphine are considered to be a combination of both direct (at the level of the spinal neurons) and indirect (via supraspinal centers) events.

In contrast to the work in spinal cord neurons, in which morphine injection in unstimulated rats had no effect on basal FLI, acute morphine injection elicits rapid and transient induction of c-*fos* mRNA and FLI in the rat caudate putamen (Chang et al., 1988). Induction appears to be via opiate receptor activation, since *fos* mRNA elevation was completely abolished by naloxone. The expression of *fos* may function as part of the intracellular cascade that links opiate receptors to secondary target gene expression, in particular to the suppression of the proenkephalin gene (see Chang et al., 1988). The authors suggest that FOS and FRAs, by binding to promoter regions of secondary response genes, may couple short-term changes induced by opiates with long-term changes, such as altered gene regulation, associated with opiate addiction.

Addictive drugs exert long lasting influences on brain function. Dopaminergic and other monoaminergic systems often are altered substantially. It is possible that psychomoter stimulants such as amphetamine and cocaine induce drug-specific molecular responses in neural populations by activating specific patterns of primary response genes. Rats injected with either amphetamine or cocaine exhibit rapid and widespread induction of c-*fos* in the sensorimotor and limbic striatum. The induction is sensitive to dopamine receptor blockage (Graybiel et al., 1990). However, there were substantial differences in the anatomic distribution patterns of *fos* mRNA in the striatum in response to these two agents. Amphetamine induced *fos* primarily in the striosomes, whereas cocaine induced a more homogeneous pattern of expression across both striosomes and in the striosome matrix compartment. The two drugs differed pharmacologically as well. Although D1 receptor antagonists blocked *fos* induction by either drug, reserpine depletion of monoamines resulted in suppression of c-*fos* induction by cocaine, but did not suppress amphetamine induction of c-*fos*. These results are consistent with previous work showing that D1 receptor activation could induce FLI in the striatum (Robertson et al., 1989). These experiments suggest that dopamine-regulated primary response gene expression in the striatum may play an important role in the long-term effects of addictive drugs on neural function.

## G. Lesions

Lesions to the CNS elicit a wide range of cell-type specific responses; glial proliferation and scar formation are notable examples (Lindsay, 1986). Primary response gene expression in response to lesions has been studied widely using a number of model systems in the CNS and PNS. Many of the changes in primary response gene expression evoked by lesions are found in neuroglial cells: astrocytes and oligodendrocytes in the CNS and Schwann cells in the nerve fibers of the PNS. In the CNS, although the lesions are usually limited in size, the cells throughout the entire ipsilateral cortex express *fos* (Dragunow and Robertson, 1988; Herrera and Robertson, 1989; Sharp et al., 1989a). The initial response is considered to be predominantly in neuronal cells and to be a result of a process similar to spreading depression that can sweep across the hemisphere. Similar results are observed following application of KCl; this agent induces both spreading depression and FLI expression (Herrera and Robertson, 1990). A more localized and protracted response is observed subsequently for cells identified as glial near the site of lesion (Dragunow and Robertson, 1988; Dragunow et al., 1990b).

Peripheral nerve injury sets in motion a number of short-term and long-term cellular responses, including changes in gene expression and trophic effects. Several studies have examined the expression of primary response genes following nerve lesion. Transection of the infraorbital nerve, which innervates the rat whiskers and supplies somatotropic information to thalamic neurons and cortical "barrels," evokes FLI in trigeminal sensory neurons and facial motoneurons (innervating whisker muscles) located in the brainstem (Sharp et al., 1989a). FLI expression was evident in most neurons for several days, and could be seen elevated up to 2 weeks. Most remarkable was the downregulation of *fos* expression in the contralateral neocortex barrel fields that map each of the denervated whiskers. The change in cortical neuronal FLI over several days may be a response to lack of sensory activity or to reduced levels of thalamic trophic influences.

Preganglion denervation of the cervical sympathetic trunk leads to induction of FLI within 1 hr in both satellite and Schwann cells, but not in principal neuronal cells (Roivainen and Koistinaho, 1990). Maximal staining was observed at 1 day and remained high for 8 days before returning to basal levels at 14 days. Thus, PNS nerve transection leads to long-term activation of *fos*.

*scip* mRNA expression correlates with transient Schwann cell proliferation in sciatic nerve transection experiments (Monuki et al., 1989, 1990). *scip* expression strongly correlates with the onset, duration, and extent of Schwann cell division following transection. *scip* is not expressed in either the preceding premyelinating cell state or, following transection-induced proliferation, in the quiescent, terminally differentiating, myelinating state. In contrast, NGF receptor mRNA is present in Schwann cells prior to transection but is absent when proliferation ceases, whereas MBP mRNA could be detected in the quiescent state following transection but not prior to the surgery. These data suggest that *scip* may play a role in the induction of the other genes. In co-transfection assays of cultured Schwann cells, *scip* functions as a specific transcriptional repressor of expression from MBP, PO, or NGF receptor promoter/CAT constructs. *scip* may thus exert a dual function, simultaneously signaling genomic programs for cell division and for repression of cell differentiation.

Sciatic nerve lesions cause a rapid increase in both c-*fos* and c-*jun* mRNAs (Hengerer et al., 1990). NGF mRNA transcription was also induced, with delayed kinetics relative to c-*fos* and c-*jun*. These studies also showed that FOS can act on an intronic AP-1 site of the NGF gene. Thus, *fos* is considered to play a role in NGF upregulation in Schwann cells of lesioned sciatic nerves. These data are discussed more extensively in the section on secondary response genes (see Section VIB).

## H. Neuroactive Agents

Activation of various receptor systems in vivo by specific agonists permits selective anatomic mapping of brain areas as measured by *fos* mRNA or protein. The adenosine receptor modulates cAMP levels and $Ca^{2+}$-dependent processes in the caudate putamen. Intraperitoneal injection of caffeine results in motor hyperactivity or tonic-clonic seizures, depending on the dose. Injection of caffeine also transiently induces c-*fos* mRNA. The expression is dose dependent and the pattern of activation is highly selective (Nakajima et al., 1989). c-*fos* induction in response to caffeine is found specifically in the caudate putamen and not in the regions commonly reported during other seizure paradigms. Pharmacologic modulation of the adenosine receptor reversed *fos* expression induced by caffeine. Interestingly, treatment of animals with RO-5-4864, a peripherally active benzodiazepine that is proconvulsant in rodents (Weissman et al., 1983), had no effect on saline-injected controls but dramatically potentiated *fos* induction by subconvulsive doses of caffeine. Although RO-5-4864 has been found to potentiate NGF-stimulated *fos* induction (Curran and Morgan, 1985) and the induction of other primary response genes in PC12 cells (Kujubu et al., 1987), it also potentiates TPA induction of primary response genes in cultured glia (Arenander et al., 1989c). Thus, part of the potentiation observed in vivo could be glial in origin.

Adrenergic receptors mediate changes in c-*fos* mRNA expression in the brain (Gubits et al., 1989). Intraperitoneal injection of drug vehicle leads to a transient increase in the levels of *fos* message that could be blocked partially by the β-receptor antagonist propanolol. This "stress"-induced increase was potentiated by the α2-receptor antagonist yohimbine. Adrenergic agonists are powerful inducers of c-*fos* and other primary response genes in astrocytes in culture (Arenander et al., 1989a). It will be of great interest to determine the cell types that respond to adrenergic agonists in the intact brain. Primary response gene expression in the CNS in response to adrenergic stimulation also suggests a possible mechanism for the high basal levels of FLI seen associated with stress-induced animals.

Cell survival and growth following brain injury depend on changes in gene expression. The neocortical damage-associated increase in FLI in neuronal cells may represent a mechanism which primary response gene induction regulates expression of secondary response genes necessary for proper response to injury (Sharp et al., 1989a,b). Neurotrophic factors, in particular NGF, are thought to modulate neuronal response to injury (Thoenen and Barde, 1980; Nieto-Sampedro et al., 1982). In addition, NGF can induce primary response genes in neurons (Curran and Morgan, 1985; Kujubu et al., 1987). Thus, NGF-induced primary response gene products may participate in coordinating secondary response gene expression and consequent phenotypic response to focal cortical injury. The injection of NGF into neocortical sites induces a pattern of FLI not unlike the distribution seen with cortical lesions (Sharp et al., 1989b) or spreading depression (Herrera and Robertson, 1989). Intracerebroventricular injection of NGF in mice leads to c-*fos* mRNA expression predominantly in the neocortex, in contrast to the increases in c-*fos* mRNA levels in septum and hippocampus, but not neocortex, elicited by arginine, vasopressin, or oxytocin (Giri et al., 1990).

## I. Neuroendocrine Systems

The hypothalamus contains a number of integrated neuroendocrine networks responsible for maintaining various aspects of physiologic homeostasis. Information concerning bodily processes impinges on these circuits, both in the form of neural input and as circulating agents. Many circulating factors control hypothalamic function by regulating cell-specific gene expression necessary for adaptive responses to changing neuroendocrine conditions. For example, changes in water balance are processed by the paraventricular and supraoptic nuclei. Dehydration studies demonstrate the induction of FLI in these cell populations (Sagar et al., 1988; Gibbs et al., 1990), suggesting that primary response gene induction can mark cell-specific neural activation in the hypothalamus in response to distinct environmental cues.

Reproductive control relies on rhythmic fluctuations in circulating estradiol levels. Target neurons located in the preoptic area (POA), ventral medial nuclei (VMN), and arcuate nucleus respond to changes in circulating estradiol levels with changes in specific gene expression (Mobbs and Pfaff, 1988; Pfaff and Schwartz-Giblin, 1988). Estradiol injected into ovariectomized rats increases FLI in the POA, the VMN, and the medial amygdala (Insel, 1990). Significant changes in FLI occur only after 12 hr, peaking at 24–48 hr. The anatomical pattern is specific; no changes in FLI are observed in the hippocampus or dentate or pyriform cortex. Control PTZ-induced seizures produced the opposite distribution of FLI expression.

It is unclear whether the changes in FLI elicited by estradiol in the CNS are direct or indirect. A direct effect of estradiol on c-*fos* transcription is potentially possible, since the *fos* promoter contains a putative estrogen responsive element overlapping the SRE (Greenberg et al., 1987). However, the slow time course argues for an indirect effect. In contrast, estradiol injections lead to rapid ($< 1$ hr) induction of *fos* in uterine epithelial cells (Loose-Mitchell et al., 1988; Weisz and Bresciani, 1988; Gibbs et al., 1990). Insel (1990) suggest that estradiol stimulates CNS expression and release of opiates that, in turn, induce c-*fos*. A previous report of rapid (< 30 min) induction of *fos* mRNA in the hippocampus in response to estradiol may be a result of stress-associated brain activation (Cattaneo and Maggi, 1990). Finally, another carefully designed study of sex steroids and *fos* expression in the hypothalamus reports no changes in *fos* mRNA or FLI following estradiol injection (Gibbs et al., 1990). Cell type- and steroid-specific dose-dependent increases in uterine c-*fos* mRNA were demonstrated in similarly treated animals, and the methodology was capable of detecting basal as well as PTZ- and dehydration-induced FLI in the expected neuronal cell populations.

Indirect induction of c-*fos* was also found in a recent study investigating the role of primary response genes in mediating gonadal steroid activation of luteinizing hormone-releasing hormone (LHRH) neurons and the subsequent predicted luteinizing hormone surge. Immunocytochemically identified LHRH neurons located within the POA and the anterior hypothalamus expressed FLI 6 hr after combined estrogen–progesterone treatment in immature females, but not in males (Hoffman et al., 1990). Since LHRH neurons do not possess nuclear estrogen receptors (Shivers et al., 1983), it seems likely that the effect of estrogen–progesterone injection must be indirect, presumably through local hypothalamic neural circuits. Thus, based on FLI mapping studies, it appears that estradiol elicits a rapid direct effect on uterine target tissue and a delayed indirect effect on hypothalamic target cells.

The adreno-hypothalamo-pituitary axis is a key neuroendocrine system, responsible for control of metabolic processes and stress responses. The adrenal gland secretes glucocorticoids, which function in part as negative feedback signals to the CRF/vasopressin neurons in the parvicellular paraventricular nucleus of the hypothalamus. Adrenalectomy results in a prompt increase in glucose metabolism (Kadekaro et al., 1988), electrical activity (Kasia and Yamashita, 1988), and specific gene expression (see Jingami et al., 1985; Swanson and Simmons, 1989) in these cells. Recent studies show a close correlation between induction of neuronal FLI and associated neuroendocrine activity following adrenalectomy (Jacobson et al., 1990); dramatic, long lasting increases in FLI in CRF/vasopressin neurons of the parvicellular paraventricular nucleus occur following adrenalectomy. The induction of FLI was blocked and reversed by corticosterone replacement. Extrahypothalamic sites did not show altered FLI expression after surgery. The "stressless" removal of glucocorticoids, supplied in the drinking water as replacement therapy, gave an induction of FLI similar to acute adrenalectomy. Thus, *fos* family members serve as markers of neurons known to undergo significant phenotypic changes as a result of removal of feedback signals from the adrenal gland.

## J. Biological Rhythms

Light increases c-*fos* and FLI expression in the rat suprachiasmatic nucleus (SCN) (Rea, 1989; Rusak et al., 1990). More recent studies also show that c-*fos* is induced in the SCN of the hamster following photic stimulation (Kornhauser et al., 1990). These primary response gene mapping experiments show that c-*fos* mRNA is elevated specifically in the neuronal subpopulation of the SCN that receive retinohypothalamic afferents conveying environmental light cues. Even a 5-min exposure of light was sufficient to induce both the rapid transient expression of c-*fos* message and a shift in the circadian pacemaker. These data suggest that, following a short photic stimulus, a series of early molecular events including gene expression occur in specific neuronal population in the SCN, leading to long-term changes in cellular mechanisms that are potentially responsible for circadian rhythms and entrainment of behavior (Rosbash and Hall, 1989). The link between *fos* expression and entrainment mechanisms was strengthened further by the findings that both c-*fos* induction and phase shifting of behavior were dependent on the circadian phase and that the photic thresholds for both processes were correlated (Kornhauser et al., 1990). The data suggest that *fos* may be a molecular effector in the photic entrainment pathway in mammals.

Trans-synaptic c-*fos* induction is also observed in the rat pineal gland (Carter, 1990). The pineal synthesizes and releases a number of indole compounds that mediate profound physiologic effects. The biologic activity of the pineal follows a strict circadian pattern, regulated by adrenergic control of the sympathetic nervous system and coordinated by the circadian pacemaker in the SCN. As part of the diurnal variation in pineal metabolism and the production of melatonin, the onset of darkness evokes a rapid and transient increase of c-*fos* mRNA. The induction of c-*fos* in the pineal can be mimicked by injection of isoproterenol, an adrenergic agonist, and blocked by removal of the superior cervical ganglion, which innervates the gland and is the source of photic information. Thus, *fos* may participate in transcription-dependent regulation of pineal circadian rhythms. The circadian expression of melatonin requires sympathetic innervation; trans-synaptic activation of the cAMP intracellular pathway rapidly induces expression of serotonin N-acetyltransferase, the rate-limiting enzyme in the melatonin biosynthetic pathway. Just as FOS has been suggested to couple β-adrenergic stimulation of nerve growth factor synthesis in glial cells (Mocchetti et al., 1989), FOS may also mediate the circadian pattern of pineal biosynthetic pathways.

## VI. SECONDARY RESPONSE GENES AS TARGETS OF PRIMARY RESPONSE GENES

It seems likely that those primary response genes that encode nuclear proteins may play roles in regulating the expression of secondary response genes whose products are involved in subsequent expression of altered cellular phenotypes. Direct evidence for a *trans*-acting role of FOS in transcription has been demonstrated for several secondary response genes; these include the genes encoding adipocyte P2 protein (Distel et al., 1987), collagen (Setoyama et al., 1986), transin (Kerr et al., 1988; Machida et al., 1989), collagenase (Schönthal et al., 1988), proenkephalin (Sonnenberg et al., 1989c), NGF (Hengerer et al., 1990), and tyrosine hydroxylase (Gizang-Ginsberg and Ziff, 1990). Next we briefly summarize the induction mechanisms for several secondary response genes that are dependent on growth factor or stimulus-induced primary response gene expression in cells of the nervous system.

## A. Proenkephalin

The proenkephalin (PENK) gene is expressed in select populations of CNS neurons and its expression is regulated by a variety of stimuli. Following seizure induction, PENK mRNA accumulates in dentate granule cells (White and Gall, 1987; Sonnenberg et al., 1989c), but only subsequent to the peak accumulation of mRNAs for c-*fos*, c-*jun*, and *junB*. The time courses and anatomical coincidence of c-*fos*, c-*jun*, *junB*, and PENK induction in PTZ-stimulated animals suggests that PENK may be a primary response gene whose expression is dependent on increased AP-1 DNA-binding activity. The presence in the PENK promoter of an AP-1 consensus DNA-binding site that binds FOS:JUN heterodimers is consistent with this hypothesis (Sonnenberg et al., 1989c). Finally, transient transfection experiments demonstrate that JUN and FOS synergistically transactivate the PENK promoter (Sonnenberg et al., 1989c). The data suggest that induction of PENK in the nervous system is likely to be dependent on prior expression of the primary response genes that encode the inducible AP-1 transcription factor complex.

## B. Nerve Growth Factor

Glial cells, under the appropriate conditions, synthesize and release NGF (for review, see Arenander and deVellis, 1989). Two model systems have been used to study ligand-mediated induction of NGF mRNA. Both experimental approaches suggest that FOS plays a crucial role in the expression of NGF. In C6 glioma cells, NGF is regulated by β-adrenergic stimulation. The β-adrenergic agonist isoproterenol dramatically increases cAMP concentrations (de Vellis and Brooker, 1974). Rapid and transient expression of c-*fos* mRNA and FLI follow (Arenander et al., 1989a; Mocchetti et al., 1990). β-Adrenergic stimulation of C6 cells also leads, after several hours, to increases in NGF mRNA (Dal Toso et al., 1987, 1988), NGF synthesis, and release of NGF into the culture medium (Schwartz et al., 1977). Cycloheximide inhibits the isoproterenol-mediated appearance of both FOS and NGF mRNA, suggesting that synthesis of one or more primary response gene products may play a part in NGF mRNA transcription. In this model, the intracellular cascade stimulated by isoproterenol most likely involves the phosphorylation/activation of CREB, which stimulates *fos* transcription. Elevated FOS, in turn, would lead to increased AP-1 activity and subsequent NGF transcription.

The second model system for NGF induction in glial cells involves Schwann cells of the sciatic nerve. Sciatic nerve transection results in NGF mRNA accumulation and synthesis of NGF in Schwann cells (Heumann et al., 1987a,b). c-*fos* and c-*jun* mRNA are expressed prior to the appearance of NGF mRNA (Hengerer et al., 1990). The delay in NGF induction (after 4 hr) and the ability of cycloheximide to partially suppress NGF message accumulation suggest that the NGF gene is a secondary response target gene of primary response genes (Mocchetti et al., 1989; Hengerer et al., 1990). Co-transfection experiments in murine fibroblasts, employing chimeric genes in which the FOS protein is expressed under the control of the metallothionein promoter and the NGF promoter regulates the chloramphenicol acetyltransferase reporter gene, suggest that elevated FOS expression is required for activation of the promoter of the NGF gene. The NGF gene contains nine potential AP-1 regulatory elements. Footprinting assays showed only one AP-1 site, located in the first intron of the NGF gene, binds a protein complex after FOS induction. Deletion or point mutation of this AP-1 site blocked transactivation of the NGF promoter by FOS. Thus, induction of NGF in Schwann cells following nerve transection is probably a result of rapid FOS induction by trauma-associated environmental signals. A rapid transient signaling pathway that includes primary response gene expression apparently results in a long lasting phenotypic response that includes NGF synthesis.

## C. Tyrosine Hydroxylase

Tyrosine hydroxylase is the rate-limiting enzyme in the catecholamine biosynthetic pathway. The transcription of TH is activated in PC12 cells by treatment with EGF (Lewis and Chikaraishi, 1987), cAMP (Tank et al., 1986; Lewis et al., 1987), glucocorticoids (Tank et al., 1986; Lewis et al., 1987), and NGF (Gizang-Ginsberg and Ziff, 1990). NGF-induced expression of TH is preceded by induction of FOS and JUN (Gizang-Ginsberg and Ziff, 1990). Moreover, cycloheximide inhibits TH message induction by NGF, suggesting that prior synthesis of primary response gene products is essential for subsequent TH expression. A specific region of the TH promoter, termed the TH-FSE element, appears to be responsible for transcriptional activation by NGF. The TH-FSE sequence, which contains an AP-1 site, is required for NGF induction of the TH promoter in transfection experiments. This sequence also binds both in vitro synthesized FOS:JUN complexes and nuclear proteins from NGF-treated PC12 cells that react with anti-FOS antibodies. Finally, deletion of this sequence from reporter constructs abolishes the ability of NGF to transactivate the TH promoter.

The kinetics of appearance of the AP-1 complex induced in PC12 cells by NGF initially correlate with TH transcriptional activation. However, the AP-1 complex continues to accumulate long after TH transcription has declined. Members of the *fos* family display different patterns of induction following NGF treatment. FOS peaks early, before the FRAs. Appearance of FOS, rather than the *fos*-related antigens most closely correlates with NGF-induced TH transcription. At later times following NGF treatment, FRAs predominate and may substitute for FOS in the AP-1 complex, leading to a reduced level of TH transcription. These data suggest that NGF-induced FOS and JUN proteins, and perhaps other members of the *fos* and *jun* families, play causal roles in the subsequent induction of TH expression in PC12 cells. NGF induction of TH in PC12 cells is likely to be regulated in part by at least three primary response gene products (FOS, JUN, and a FRA), participating in a nuclear AP-1 complex whose composition and function varies with time after exposure to NGF.

### D. Transin

Neuronal morphogenesis requires neurite outgrowth and cell migration. In order for cell processes such as growth cones to invade new territory, proteases are released to degrade the intercellular matrix (Matrisian and Hogan, 1990). Transin, the rat homolog of human stromelysin, degrades proteoglycans, fibronectin, laminin, and collagen (Chin et al., 1985). Transin mRNA is induced by EGF, TPA, PDGF, or IL-1 in rodent fibroblasts (Matrisian et al., 1985, 1986a,b). Both induction kinetics and dependence on protein synthesis characterize transin as a secondary response gene, suggesting that one or more primary response gene products are necessary for its expression. PDGF-mediated induction of transin is FOS dependent, whereas EGF induction of transin is FOS independent in fibroblast cell lines (Kerr et al., 1988).

NGF induces transin expression in PC12 cells (Machida et al., 1989). Transin mRNA accumulation occurs subsequent to the NGF-induced peaks in c-*fos* and c-*jun* messages. Moreover, NGF-mediated transin induction can be blocked by cycloheximide, suggesting that prior induction of primary response genes is necessary for this transcriptional response. These results suggest that NGF may induce transin transactivation by a mechanism that involves primary response gene products such as FOS. Other primary response gene products are also candidates, since c-*fos* is only one of many nuclear primary response genes induced by NGF in PC12 cells (Kujubu et al., 1987; Millbrandt, 1988, Arenander et al., 1989a,b,c; Bartel et al., 1989). Unlike the other examples of secondary response genes presented previously, there is no direct evidence yet for FOS as a mediator of NGF induction of transin in PC12 cells. However, the correlated time course of induction of *fos* and of transin, the dependence of transin on protein synthesis, and previous reports of FOS-dependent transin induction in other cell types make it likely that FOS is involved in the NGF-induced expression of this gene as well.

### VII. SPECIFICITY OF PRIMARY RESPONSE GENE EXPRESSION

Most investigators initially expected that primary response genes would function as specific determinants of ligand-mediated phenotypic change. The results with c-*fos* were, therefore, perplexing; c-*fos*, and most of the other primary response genes, could be induced by a diverse range of ligands in almost every cell type examined. These findings questioned whether primary response genes could function as determinants of distinct phenotypic responses. Accumulated evidence now suggests that, although some of the primary response genes may serve individually to activate crucial switches for phenotypic alteration, specificity is most likely to be encoded by combinatorial control mechanisms resulting from differential use (both qualitatively and quantitatively) of the primary response gene products. The protein products of a number of primary response genes, differentially interacting via a number of processes, appear to generate alternative sets of nuclear protein complexes that selectively regulate the distinctive readout of genomic programs.

## A. Quantitative Parameters

The precise encoding of environmental information may begin as a function of quantitative and qualitative expression of primary response genes. Variation in the onset, duration, cessation, and degree of message transcription, translation, and posttranslational modification of various primary response genes could yield a very complex time-dependent mixture of numerous nuclear factors interacting at secondary response gene promoters. Differential patterns of expression of many primary response gene mRNAs have been shown for both glia and neurons (Kujubu et al., 1987; Arenander et al., 1989a,b,c; Bartel et al., 1989; Altin et al., 1991). In PC12 cells, KCl induces a stronger response of c-*fos* and NGFI-B than does either NGF or EGF. However, the *egr1* and c-*jun* genes exhibit an opposite pattern of sensitivity (NGF > EGF >> KCl; Bartel et al., 1989). Experiments with rat astrocytes show that NGFI-B, TIS7, TIS11, and c-*fos*, but not *egr1*, are more strongly induced by TPA and EGF than by FGF (Arenander et al., 1989b). Astrocytes display different primary response gene induction kinetics, depending on the adrenergic receptor agonist treatment (Arenander et al., 1989c). The majority of primary response genes exhibit more rapid onset of induction in response to an α agonist than to a β agonist. A mixed receptor agonist, norepinephrine, gave an entirely different pattern of induction. These findings suggest that the coupling of extracellular signals to gene promoters is subject to a wide range of regulation. Such differential control among intracellular pathways and nuclear factors suggests that quantitative distinctions in inducibility of primary response genes in response to alternative ligands may be important in determining secondary response gene expression and subsequent cellular phenotype.

Since the extracellular environment is a complex and variable mixture of many ligands and other environmental cues, simultaneous modulation of several intracellular pathways is almost certainly more the rule than the exception. Published reports support the notion that simultaneous activation of cells by multiple ligands can quantitatively alter primary response gene expression patterns (Arenander et al., 1989a,c). Both additive and synergistic interactions are observed. For example, NGFI-B/TIS1, *egr1*/TIS8, TIS11, TIS21, and c-*fos* all demonstrate at least additive levels of message induction when astrocytes are co-treated with TPA and EGF or FGF. Within this group of primary response genes, NGFI-B/TIS1 and TIS11 also show differential responsiveness to each of the ligands alone (Arenander et al., 1989a). Potentiation of primary response gene induction is also primary response gene selective. Astrocytes treated with the benzodiazepine RO-5-4864 exhibit no induction of primary response gene mRNAs. However, this ligand dramatically superinduces NGFI-B/TIS1, TIS7, TIS11, and c-*fos*, but not *egr1*/TIS8 (Arenander et al., 1989c). Neurotransmitters also elicit gene-specific differences in the induction of primary response genes. The addition of lithium to cultures of astrocytes treated with carbachol dramatically potentiated the accumulated of mRNA for TIS11 and c-*fos*, but not NGFI-B, TIS7, and TIS21 (Arenander et al., 1989a). These data suggest that exposure to complex mixtures of ligand effectors can have profoundly different effects on the various primary response genes, with corresponding distinctly differing long-term consequences resulting from distinct subsets of induced secondary response gene expression.

## B. Cell-Type Restriction of Primary Response Gene Expression

The first example of "restriction" or "extinguishment" of primary response gene expression was observed in PC12 cells. Although the NGFI-A/TIS8, TIS7/PC4, NGFI-B/TIS1, and TIS11 genes could all be induced—albeit to varying degrees—in PC12 cells in response to growth factors, tumor promoters, and depolarization, no mRNA for the TIS10 gene could be detected in these cells, in response to any inducer. In contrast, TIS10, which encodes a protein with substantial sequence similarity to prostaglandin endoperoxide synthase (Kujubu et al., 1991), is inducible by growth factors and tumor promoters in other cell lines (Kujubu et al., 1987; Lim et al., 1987, 1989; Arenander et al., 1989c). It appears that developmentally regulated processes have "restricted" or "extinguished" the TIS10 gene so that no extracellular signal can elicit expression of this gene in PC12 cells. The zinc-finger transcription factor *egr2/krox20*, which is highly inducible in 3T3 cells (Chavier et al., 1988, 1989) and in human fibroblasts (Joseph et al., 1988), also was demonstrated subsequently to show restriction in PC12 cells; no inducer could elicit expression of this gene (Joseph et al., 1988). Cell-type restriction of primary response gene expression has been shown subsequently to be a frequent phenomenon. Thus, for example, the TIS1/*nur77*/NGFI-B gene cannot be induced in human neutrophils or in a murine myeloid cell line by either TPA or the hematopoietic growth factor GM-CSF (granulocyte macrophage colony stimulating factor), despite robust induction of other primary response genes such as TIS8/*egr1*/NGFI-A and TIS11 (Varnum et al., 1989a). By using in situ hybridization, we have observed restricted expression of many primary response genes in glial progenitor cell populations; cell-type specific restriction of primary response gene expression appears to occur in cells of the nervous system. It appears that developmentally determined processes restrict, in a cell lineage-dependent fashion, the subsets of primary response genes that can be expressed in individual cells in response to extracellular signals. This process is likely to play a pivotal role in determining which secondary response genes can be activated in response to extracellular signals, therefore limiting the phenotypic responses of distinct cell types following stimulation by a common ligand.

# REFERENCES

Abate, C., Rauscher, F. J., Gentz, R., and Curran, T. (1990). Expression and purification of the leucine zipper and the DNA-binding domains of *fos* and *jun*: Both *fos* and *jun* directly contact DNA. *Proc. Natl. Acad. Sci.U.S.A.* **87,** 1032–1036.

Almendral, J. M., Sommer, D., Macdonald-Bravo, H., Burckhardt, J., Perera, J., and Bravo, R. (1988). Complexity of the early genetic response to growth factors in mouse fibroblasts. *Mol. Cell. Biol.* **8,** 2140–2148.

Altin, J. G., Kujubu, D., Raffioni, S., Eveleth, D. D., Herschman, H. R., and Bradshaw, R. A. (1991). Differential induction of primary-response (TIS) genes in PC12 pheochromocytoma cells and the unresponsive variant PC12nnr5*. *J. Biol. Chem.* **266,** 5401–5406.

Arenander, A. T., and de Vellis, J. (1989). Development of the nervous system. In "Basic Neurochemistry: Molecular, Cellular, and Medical Aspects" (G. J. Siegel, R. W. Albers, B. W. Agranoff, and P. Molinoff, eds.), 4th Ed., pp. 479–506. Raven Press, New York.

Arenander, A. T., de Vellis, J., and Herschman, H. R. (1989a). Induction of c-*fos* and TIS genes in cultures rat astrocytes by neuro-transmitters. *J. Neurosci. Res.* **24,** 107–114.

Arenander, A. T., Lim, R. W., Varnum, B. C., Cole, R., de Vellis, J., and Herschman, H. R. (1989b). TIS gene expression in cultured rat astrocytes: Induction by mitogens and stellation agents. *J. Neurosci. Res.* **23,** 247–256.

Arenander, A. T., Lim, R. W., Varnum, B. C., Cole, R., de Vellis, J., and Herschman, H. R. (1989c). TIS gene expression in cultured rat astrocytes: Multiple pathways of induction by mitogens. *J. Neurosci. Res.* **23,** 257–265.

Arenander, A. T., Cheng, J., and de Vellis, J. (1991). Early events in the hormonal regulation of glial gene expression: Early response genes. In "Molecular Biology and Physiology of Insulin and Insulin-like Growth Factors" (M. Raizada and Derek Le Rorth, eds.), Plenum Press, New York. pp. 335–350.

Barber, J. R., and Verma, I. M. (1987). Modification of *fos* proteins: Phosphorylation of c-*fos*, but not v-*fos*, is stimulated by 12-tetradecanoyl-phorbol-13-acetate and serum. *Mol. Cell. Biol.* **7,** 2201–2211.

Bartel, D., Sheng, M., Lau, L., and Greenberg, L. (1989). Growth factors and membrane depolarization activate distinct programs of early response gene expression: Dissociation of *fos* and *jun* induction. *Genes Dev.* **3,** 304–313.

Barzilai, A., Kennedy, T. E., Sweatt, J. D., and Kandel, E. R. (1989). 5-HT modulates protein synthesis and the expression of specific proteins during long-term facilitation in *Aplysia* sensory neurons. *Neuron* **2,** 1577–1586.

Berkowitz, L. A., Riabowol, K. T., and Gilman, M. Z. (1989). Multiple sequence elements of a single functional class are required for cyclic AMP responsiveness of the mouse c-*fos* promoter. *Mol. Cell. Biol.* **9,** 4272–4281.

Bonnieu, A., Rech, J., Jeanteur, P., and Fort, P. (1989). Requirements for c-*fos* mRNA down regulation in growth stimulated murine cells. *Oncogene* **4,** 881–888.

Bullitt, E. (1989). Induction of c-*fos*-like protein within the lumbar spinal cord and thalamus of the rat following peripheral stimulation. *Brain Res.* **493,** 391–397.

Cao, X., Koski, R. A., Gashler, A., McKiernan, M., Morris, C. F., Gaffney, R., Hay, R. V., and Sukhatme, V. P. (1990). Identification and characterization of the *egr1* gene product, a DNA-binding zinc finger protein induced by differentiation and growth signals. *Mol. Cell. Biol.* **10,** 1931–1939.

Carter, D. A. (1990). Temporally defined induction of c-*fos* in the rat pineal. *Biochem. Biophys. Res. Commun.* **166,** 589–594.

Cattaneo, E., and Maggi, A. (1990). c-*fos* induction by estrogen in specific rat brain areas. *Eur. J. Pharmacol. Mol. Pharmacol.* **188,** 153–159.

Caubet, J. F. (1989). c-*fos* proto-oncogene expression in the nervous system during mouse development. *J. Cell Biol.* **9,** 2269–2272.

Ceccatelli, S., Villar, M. J., Goldstein, M., and Hökfelt, T. (1989). Expression of c-*fos* immunoreactivity

in transmitter-characterized neurons after stress. *Proc. Natl. Acad. Sci. U.S.A.* **86,** 9569–9573.
Chang, S. L., Squinto, S. P., and Harlan, R. E. (1988). Morphine activation of c-*fos* expression in rat brain. *Biochem. Biophys. Res. Commun.* **157,** 698–704.
Changelian, P. S., Feng, P., King, T. C., and Milbrandt, J. (1989). Structure of the NGFI-A gene and detection of upstream sequences responsible for its transcriptional induction by nerve growth factor. *Proc. Natl. Acad. Sci. U.S.A.* **86,** 377–381.
Chavier, P., Zerial, M., Lemaire, P., Almendral, J., Bravo, R., and Charnay, P. (1988). A gene encoding a protein with zinc fingers is activated during $G_0/G_1$ transition in cultured cells. *EMBO J.* **7,** 29–35.
Chavier, P., Janssen-Timmen, U., Mattei, M.-G., Zerial, M., Bravo, R., and Charnay, P. (1989). Structure, chromosome location, and expression of the mouse zinc finger gene *krox20*: Multiple gene products and coregulation with the photo-oncogene c-*fos*. *Mol. Cell. Biol.* **9,** 787–797.
Chavier, P., Vesque, C., Galliot, B., Vigneron, M., Dollé, P., Duboule, D., and Charnay, P. (1990). The segment-specific gene Krox-20 encodes a transcription factor with binding sites in the promoter region of the Hox-1.4 gene. *EMBO J.* **9,** 1209–1218.
Chin, J. R., Murphy, G., and Werb, Z. (1985). Stromelysin, a connective tissue-degrading metalloendopeptidase secreted by stimulated rabbit synovial fibroblasts in parallel with collagenase. Biosynthesis, isolation, characterization, and substrates. *J. Biol. Chem.* **260,** 12367–12376.
Chiu, R., Angel, P., and Karin, M. (1989). *junB* differs in its biological properties from, and is a negative regulator of c-*jun*. *Cell* **59,** 979–986.
Christy, B. A., Lau, L. F., and Nathans, D. (1988). A gene activated in mouse 3T3 cells by serum growth factors encodes a protein with "zinc finger" sequences. *Proc. Natl. Acad. Sci. U.S.A.* **85,** 7857–7861.
Cohen, D. R., and Curran, T. (1988). *fra1*: A serum-inducible cellular immediate-early gene that encodes a *fos*-related antigen. *Mol. Cell. Biol.* **8,** 2063–2069.
Cohen, D. R., Ferreira, P. C. P., Gentz, R., Franza, B. R., Jr., and Curran, T. (1989). The product of a *fos*-related gene, *fra1*, binds cooperatively to the AP-1 site with *jun*: Transcription factor AP-1 is comprised of multiple protein complexes. *Genes Dev.* **3,** 173–184.
Cole, A. J., Saffen, D. W., Baraban, J. M., and Worley, P. F. (1989). Rapid increase of an immediate-early gene messenger RNA in hippocampal neurons by synaptic NMDA receptor activation. *Nature (London)* **340,** 474–476.
Comb, M. C., Birnberg, N. C., Seasholtz, A., Herbert, E., and Goodman, H. M. (1986). A cyclic AMP- and phorbol ester-inducible DNA element. *Nature (London)* **323,** 353–356.
Condorelli, D., Kaczmarek, L., Nicoletti, F., Arcidiacono, P., Dell'Albani, P., Ingrao, F., Magri, G., Malaguarneara, L., Avola, R., Messina, A., and Giuffrida-Stella, A. M. (1989). Induction of proto-oncogene *fos* by extracellular signals in primary glial cell cultures. *J. Neurosci. Res.* **23,** 234–239.
Cortner, J., and Farnham, P. J. (1990). Identification of the serum-responsive transcription initiation site of the zinc finger gene *krox20*. *Mol. Cell. Biol.* **10,** 3799–3791.
Curran, T., and Morgan, J. I. (1985). Superinduction of the *fos* gene by nerve growth factor in the presence of peripherally active benzodiazepines. *Science* **229,** 1265–1268.
Curran, T., and Morgan, J. I. (1986). Barium modulates c-*fos* expression and post-translational modification. *Proc. Natl. Acad. Sci. U.S.A.* **83,** 8521–8524.
Curran, T., and Morgan, J. I. (1987). Memories of *fos*. *BioEssays* **7,** 255–258.
Curran, T., and Teich, N. M. (1982). Candidate product of the FBJ murine osteosarcoma virus oncogene: Characterization of a 55,000 dalton phosphoprotein. *J. Virol.* **42,** 114–122.
Curran, T., MacConnell, W. P., van Straaten, F., and Verma, I. M. (1983). Structure of the FMJ murine osteosarcoma virus genome: Molecular cloning of its associated helper virus and the cellular homolog of the v-*fos* gene from mouse and human cells. *Mol. Cell. Biol.* **3,** 914–921.
Curran, T., Miller, A. D., Zokas, L., and Verma, I. M. (1984). Viral and cellular *fos* proteins: A comparative analysis. *Cell* **36,** 259–268.

Curran, T., Van Beveren, C., Ling, N., and Verma, I. M. (1985). Viral and cellular *fos* proteins are complexed with a 39,000 dalton cellular protein. *Mol. Cell. Biol.* **5**, 167–172.
Dal Toso, R., De Bernardi, M. A., Costa, E., and Mocchetti, I. (1987). Beta-adrenergic receptor of NGF-mRNA content in rat C6-2B glioma cells. *Neuropharmacol.* **26**, 1783–1786.
Dal Toso, R., DeBernardi, M. A., Brooker, G., Costa, E., and Mocchetti, I. (1988). Beta adrenergic and prostaglandin receptor activation increases nerve growth factor mRNA content in C6-2B rat astrocytoma cells. *J. Pharmacol. Exp. Therap.* **246**, 1190–1193.
Dash, P. K., Hochner, B., and Kandel, E. R. (1990). Injection of the cAMP-responsive element into the nucleus of *Aplysia* sensory neurons blocks long-term facilitation. *Nature (London)* **345**, 718–721.
Dave, R. J., Tabakoff, B., and Hoffman, P. L. (1989). Ethanol withdrawal seizures produce increased *c-fos* mRNA in mouse brain. *Mol. Pharmacol.* **37**, 367–371.
de Vellis, J., and Brooker, G. (1974). Reversal of catecholamine refractoriness by inhibitors of RNA and protein synthesis. *Science* **186**, 1221–1223.
Distel, R. J., Ro, H.-S., Rosen, B. S., Groves, D. L., and Spiegelman, B. M. (1987). Nucleoprotein complexes that regulate gene expression in adipocyte differentiation: Direct participation of *c-fos*. *Cell* **49**, 835–844.
Douglas, R. M., Dragunow, M., and Robertson, H. A. (1988). High-frequency discharge of dentate granule cells, but not long-term potentiation, induces *c-fos* protein. *Mol. Brain Res.* **4**, 259–262.
Dragunow, M., and Faull, R. (1989). The use of *c-fos* as a metabolic marker in neuronal pathway tracing. *J. Neurosci. Meth.* **29**, 261–265.
Dragunow, M., and Robertson, H. A. (1987a). Kindling stimulation induces *c-fos* protein(s) in granule cells of the rat dentate gyrus. *Nature (London)* **329**, 441–442.
Dragunow, M., and Robertson, H. A. (1987b). Generalized seizures induce *c-fos* protein(s) in mammalian neurons. *Neurosci. Lett.* **82**, 157–161.
Dragunow, M., and Robertson, H. A. (1988). Brain injury induces *c-fos* protein(s) in nerve and glial-like cells in adult mammalian brain. *Brain Res.* **455**, 295–299.
Dragunow, M., Abraham, W. C., Goulding, M., Mason, S. E., Robertson, H. A., and Faull, R. L. M. (1989a). Long-term potentiation and the induction of *c-fos* mRNA proteins in the dentate gyrus of unanaesthetized rats. *Neurosci. Lett.* **101**, 274–280.
Dragunow, M., Currie, R. W., Faull, R. L., Robertson, H. A., and Jansen, K. (1989b). Immediate-early genes, kindling and long-term potentiation. *Neurosci. Biobehav. Rev.* **13**, 301–313.
Dragunow, M., Currie, R. W., Robertson, H. A., and Faull, R. (1989c). Heat shock induces *c-fos* protein-like immunoreactivity in glial cells in adult rat brain. *Exp. Neurol.* **106**, 105–109.
Dragunow, M., Faull, R. L. M., and Jansen, K. L. R. (1990a). MK-801, an antagonist of NMDA receptors, inhibits injury-induced *c-fos* protein accumulation in rat brain. *Neurosci. Lett.* **109**, 128–133.
Dragunow, M., Goulding, M., Faull, R. L., Ralph, R., Mee, E., and Frith, R. (1990b). Induction of *c-fos* mRNA and protein in neurons and glia after traumatic brain injury: Pharmacological characterization. *Exp. Neurol.* **107**, 236–248.
Dragunow, M., Robertson, G. S., Faull, R. L. M., Robertson, H. A., and Jansen, K. (1990c). D2 dopamine receptor antagonists induce *fos* and related proteins in rat striatal neurons. *Neurosci.* **37**, 287–294.
Evans, R. M. (1988). The steroid and thyroid hormone receptor superfamily. *Science* **240**, 889–895.
Fahrner, T. J., Carroll, S. L., and Milbrandt, J. (1990). The NGFI-B protein, an inducible member of the thyroid/steroid receptor family, is rapidly modified posttranslationally. *Mol. Cell. Biol.* **10**,6454–6549.
Fisch, T. M., Prywes, R., and Roeder, R. G. (1987). *c-fos* sequences necessary for basal expression and induction by epidermal growth factor, 12-O-tetradecanoyl phorbol-13-acetate, and the calcium ionophore. *Mol. Cell. Biol.* **7**, 3490–3502.
Fisch, T. M., Prywes, R., Simon, M. C., and Roeder, R. G. (1989). Multiple sequence elements in the *c-fos* promoter mediate induction by cAMP. *Genes Dev.* **3**, 198–211.
Fort, P., Rech, J., Piechaczyk, M., Bonnieu, A., Jeanteur, P., and Blanchard, J. M. (1987). Regulation

of c-fos gene expression in hamster fibroblasts: Initiation and elongation of transcription and mRNA degradation. *Nucleic Acid Res.* **15**, 5657–5667.

Franza, B. R., Sambucetti, L. C., Cohen, D. R., and Curran, T. (1987). Analysis of fos protein complexes and fos-related antigens by high-resolution two-dimensional gel electrophoresis. *Oncogene* **1**, 213–221.

Frisch, S. M., and Ruley, H. E. (1987). Transcription from the stromelysin promoter is induced by interleukin-1 and repressed by dexamethasone. *J. Biol. Chem.* **262**, 16300–16304.

Gibbs, R. B., Mobbs, C. V., and Pfaff, D. W. (1990). Sex steroids and fos expression in rat brain and uterus. *Mol. Cell. Neurosci.* **1**, 29–40.

Gilman, M. Z., Wilson, R. N., and Weinberg, R. A. (1986). Multiple protein binding sites in the 5'-flanking region regulate c-fos express ion. *Mol. Cell. Biol.* **6**, 4305–4315.

Giri, P. R., Dave, J. R., Tabakoff, B., and Hoffman, P. L. (1990). Arginine vasopressin induces the expression of c-fos in the mouse septum and hippocampus. *Mol. Brain Res.* **7**, 131–137.

Gizang-Ginsberg, E., and Ziff, E. B. (1990). Nerve growth factor regulates tyrosine hydroxylase gene transcription through a nucleoprotein complex that contain c-fos. *Genes Dev.* **4**, 477–491.

Gogas, K. R., Presley, R. W., Levine, J. D., and Basbaum, A. I. (1991). The anti-nociceptive action of supraspinal opioids results from an increase in descending inhibitory control: Correlation of nociceptive behavior and c-fos expression. *Neurosci.*

Gonzalez, G. A., Yamamoto, K. K., Fischer, W. H., Darr, D., Menzel, P., Biggs, W., III, Vale, W. W., and Montminy, M. R. (1989). A cluster of phosphorylation sites on the cyclic AMP-regulated nuclear factor CREB predicted by TIS sequence. *Nature (London)* **337**, 749–752.

Goto, K., Ogo, A., Yanase, T., Haji, M., Ohashi, M., and Nawata, H. (1989). Expression of c-fos and c-myc proto-oncogenes in human adrenal pheochromocytomas. *J. Clin. Endocr. Metab.* **70**, 353–357.

Graybiel, A. M., Moratalla, R., and Robertson, H. A. (1990). Amphetamine and cocaine induce drug-specific activation of the c-fos gene in striosome-matrix compartments and limbic subdivisions of the striatum. *Proc. Natl. Acad. Sci. U.S.A.* **87**, 6912–6916.

Greenberg, M., and Ziff, E. (1984). Stimulation of 3T3 cells induces transcription of the c-fos proto-oncogene. *Nature (London)* **331**, 433–438.

Greenberg, M., Green, L., and Ziff, E. (1985). Nerve growth factor and epidermal growth factor induce rapid transient changes in proto-oncogene transcription in PC-12 cells. *J. Biol. Chem.* **260**, 14101–14110.

Greenberg, M. E., Hermanowski, A. L., and Ziff, E. B. (1986). Effect of protein synthesis inhibition on growth factor activation of c-fos, c-myc, and actin gene transcription. *Mol. Cell. Biol.* **6**, 1050–1057.

Greenberg, M. E., Siegfried, Z., and Ziff, E. B. (1987). Mutation of the c-fos dyad symmetry element inhibits serum inducibility *in vivo* and nuclear regulatory factor binding *in vitro*. *Mol. Cell Biol.* **7**, 1217–1225.

Greene, L., and Tischler, A. (1982). PC-12 pheochromocytoma cultures in neurobiological research. *Adv. Cell. Neurobiol.* **3**, 373–414.

Gubits, R. M., and Fairhurst, J. L. (1988). c-fos mRNA levels are increased by the cellular stressors, heat shock, and sodium arsenite. *Oncogene* **3**, 163–168.

Gubits, R., Smith, T., Fairhurst, J., and Yu, H. (1989). Adrenergic receptors mediate c-fos mRNA levels in brain. *Mol. Brain Res.* **6**, 29–45.

Gubits, R. M., Wollack, J. B., Yu, H., and Liu, W.-K. (1990). Activation of adenosine receptors induces c-fos, but not c-jun, expression in neuron–glia hybrids and fibroblasts. *Mol. Brain. Res.* **8**, 275–281.

Guis, D., Cao, X., Rauscher, F. J., III, Cohen, D. R., Curran, T., and Sukhatme, V. P. (1991). Transcriptional activation and repression by fos are independent functions: The C terminus represses immediate-early gene expression via CArG elements. *Mol. Cell. Biol.* **10**, 4243–4255.

Hai, T. W., Liu, F., Coudos, W. J., and Green, M. R. (1989). Transcription factor ATF cDNA clones: An extensive family of leucine zipper proteins able to selective form DNA-binding heterodimers. *Genes Dev.* **3**, 2083–2090.

Hazel, T. G., Nathans, D., and Lau, L. F. (1988). A gene inducible by serum growth factors encodes a member of the steroid and thyroid hormone receptor superfamily. *Proc. Natl. Acad. Sci. U.S.A.* **85,** 8444–8448.

He, X., Treacy, M. N., Simmons, D. M., Ingraham, H. A., Swanson, L. W., and Rosenfeld, M. G. (1989). Expression of a large family of POU-domain regulatory genes in mammalian brain development. *Nature (London)* **340,** 35–42.

Hengerer, B., Lindholm, D., Heumann, R., Ruther, U., Wagner, E. F., and Thoenen, H. (1990). Lesion-induced increase in nerve growth factor mRNA is mediated by c-*fos*. *Proc. Natl. Acad. Sci. U.S.A.* **87,** 3899–3903.

Herrera, D. G., and Robertson, H. A. (1989). Unilateral induction of c-*fos* protein in cortex following cortical devascularization. *Brain Res.* **503,** 205–213.

Herrera, D. G., and Robertson, H. A. (1990). Application of potassium chloride to the brain surface induces the c-*fos* proto-oncogene: Reversal by MK-801. *Brain Res.* **510,** 166–170.

Herschman, H. R. (1991). Primary response genes induced by growth factors and tumor promoters. *In* "Annual Reviews of Biochemistry" (C. Richardson, ed.), Volume 60, pp. 281–319. Annual Reviews, Palo Alto, California.

Heumann, R., Korsching, S., Bandtlow, C., and Thoenen, H. (1987a). Changes of nerve growth factor synthesis in nonneuronal cells in response to sciatic nerve transection. *J. Cell Biol.* **104,** 1623–1631.

Heumann, R., Lindholm, D., Bandtlow, C., Meyer, M., Radeke, M. J., Misko, T. P., Shooter, E., and Thoenen, H. (1987b). Differential regulation of mRNA encoding nerve growth factor and its receptor in rat sciatic nerve during development, degeneration, and regeneration: Role of macrophages. *Proc. Natl. Acad. Sci. U.S.A.* **84,** 8735–8739.

Hirai, S.-I., Ryseck, R.-P., Mechta, F., Bravo, R., and Yaniv, M. (1989). Characterization of *junD*: A new member of the *jun* proto-oncogene family. *EMBO J.* **8,** 1433–1439.

Hoffman, G. E., Lee, W. S., Attardi, B., Yann, V., and Fitzsimmons, M. D. (1990). Leutinizing hormone-releasing hormone neurons express c-*fos* antigen after steroid activation. *Endocrinology* **126,** 1736–1741.

Hunt, S. P., Pini, A., and Evan, G. (1987). Induction of c-*fos*-like protein in spinal cord neurons following sensory stimulation. *Nature (London)* **328,** 632–634.

Insel, T. R. (1990). Regional induction of c-*fos*-like protein in rat brain after estradiol administration. *Endocrinology* **126,** 1264–1849.

Jacobson, L., Sharp, F. R., and Dallman, M. F. (1990). Induction of *fos*-like immunoreactivity in hypothalamic CRF neurons after adrenalectomy in the rat. *Endocrinology* **126,** 1709–1719.

Jingami, H., Matsukura, S., Numa, S., and Imura, H. (1985). Effects of adrenalectomy and dexamethasone administration on the level of preprocortisotropin-releasing factor mRNA in the hypothalamus and adrenocorticotropin/beta-lipotropin precursor mRNA in the pituitary in rats. *Endocrinology* **117,** 1314–1331.

Joseph, L. J., LeBeau, M. M., Jamieson, G. A., Acharya, S., Shows, T. B., Rowley, J. D., and Sukhatme, V. P. (1988). Molecular cloning, sequencing, and mapping of *egr2*, a human early growth response gene encoding a protein with "zinc-binding finger" structure. *Proc. Natl. Acad. Sci. U.S.A.* **85,** 7164–7168.

Kadekaro, M., Ito, M., and Gross, P. M. (1988). Local cerebral glucose utilization is increased in acutely adrenalectomized rats. *Neuroendocrinol.* **47,** 329–338.

Kandel, E. R. (1989). Genes, nerve cells, and the remembrance of things past. *J. Neuropsychiatry* **1,** 103–125.

Kasai, M., and Yamashita, H. (1988). Inhibition by cortisol of neurons in the paraventricular nucleus of the hypothalamus in adrenalectomized rats, an *in vitro* study. *Neurosci. Letts.* **91,** 59–67.

Kennedy, M. B. (1989). Regulation of synaptic transmission in the central nervous system: Long-term potentiation. *Cell* **59,** 777–787.

Kerr, L., Holt, J., and Matrisian, L. (1988). Growth factors regulate transin gene expression by c-*fos* dependent and c-*fos* independent pathways. *Science* **242,** 1424–1427.

Kornhauser, J. M., Nelson, D. E., Mayo, K. E., and Takahashi, J. S. (1990). Photic and circadian regulation of c-*fos* gene expression in the hamster suprachiasmatic nucleus. *Neuron* **5,** 127–134.

Kouzarides, T., and Ziff, E. (1988). The role of the leucine zipper in the *fos–jun* interaction. *Nature (London)* **336**, 646–651.
Kruijer, W., Schubert, D., and Verma, I. M. (1985). Induction of the proto-oncogene *fos* by nerve growth factor. *Proc. Natl. Acad. Sci. U.S.A.* **82**, 7330–7334.
Kujubu, D. A., Lim, R. W., Varnum, B. C., and Herschman, H. R. (1987). Induction of transiently expressed genes in PC-12 pheochromocytoma cells. *Oncogene* **1**, 257–262.
Kujubu, D. A., Fletcher, B. S., Yarnum, B. C., Lim, R. W., and Herschman, H. R. (1991) TIS10, a phorbol ester tumor promoter-inducible mRNA from Swiss 3T3 cells, encodes a novel prostaglandin synthase/cycloxygenase homologue. *J. Biol. Chem.* **266**, 12866–12872.
Lamb, N. J. C., Fernandez, A., Tourkine, N., Jeanteur, P., and Blanchard, J.-M. (1990). Demonstration in living cells of an intragenic negative regulatory element within the rodent c-*fos* gene. *Cell* **61**, 485–496.
Landschulz, W. H., Johnson, P. F., and McKnight, S. L. (1988). The leucine zipper: A hypothetical structure common to a new class of DNA binding proteins. *Science* **240**, 1759–1764.
Lau, L. F., and Nathans, D. (1985). Identification of a set of genes expressed during the $G_0/G_1$ transition of cultured mouse cells. *EMBO J.* **4**, 3145–3151.
Le Gal La Salle, G. (1988). Long-lasting and sequential increase of c-*fos* oncoprotein expression in kainic acid-induced status epilepticus. *Neurosci. Lett.* **88**, 127–130.
Lee, W. P., Mitchell, P., and Tjian, R. (1987). Purified transcription factor AP-1 interacts with TPA-inducible enhancer elements. *Cell* **49**, 741–752.
Lemaire, P., Revelant, O., Bravo, R., and Charnay, P. (1988). Two mouse genes encoding potential transcription factors with identical DNA-binding domains are activated by growth factors in cultured cells. *Proc. Natl. Acad. Sci. U.S.A.* **85**, 4691–4695.
Lemaire, P., Vesque, C., Schmitt, J., Stunnenberg, H., Frank, R., and Charnay, P. (1990). The serum-inducible mouse gene *krox24* encodes a sequence-specific transcriptional activator. *Mol. Cell. Biol.* **10**, 3456–3467.
Lewis, E. J., and Chikaraishi, D. M. (1987). Regulated expression of the tyrosine hydroxylase gene by epidermal growth factor. *Mol. Cell. Biol.* **7**, 3332–3336.
Lewis, E. J., Harrington, C. A., and Chikaraishi, D. M. (1987). Transcriptional regulation of the tyrosine hydroxylase gene by glucocorticoid and cyclic AMP. *Proc. Natl. Acad. Sci. U.S.A.* **84**, 3550–3554.
Lim, R., Varnum, B. C., and Herschman, H. R. (1987). Cloning of sequences induced as a primary response following mitogen treatment of density arrested Swiss 3T3 cells. *Oncogene* **1**, 263–270.
Lim, R. W., Varnum, B. C., O'Brien, T. G., and Herschman, H. R. (1989). Induction of tumor promoter inducible genes in murine 3T3 cell lines and TPA non-proliferative 3T3 variants can occur through both protein kinase C-dependent and independent pathways. *Mol. Cell. Biol.* **9**, 1790–1793.
Lindsay, R. M. (1986). Reactive gliosis. *In* "Astrocytes" (S. Federoff and A. Vernadakis, eds.), Vol. 3, pp. 231–262. Academic Press, New York.
Loose-Mitchell, D. S., Chiapetta, C., and Stancel, G. M. (1988). Estrogen regulation of c-*fos* messenger ribonucleic acid. *Mol. Endocrinol.* **2**, 946–956.
Lucibello, F. C., Lowag, C., Neuberg, M., and Müller, R. (1989). Trans-repression of the mouse c-*fos* promoter: A novel mechanism for *fos*-mediated trans-regulation. *Cell* **59**, 999–1007.
Machida, C. M., Rodland, K., Matrisian, L., Magun, B. E., and Ciment, G. (1989). NGF induction of the gene encoding the protease transin accompanies neuronal differentiation in PC12 cells. *Neuron* **2**, 1587–1596.
McMahon, A. P., Champion, J. E., McMahon, J. A., and Sukhatme, V. P. (1990). Developmental expression of the putative transcription factor *Egr-1* suggests that *Egr-1* and c-*fos* are co-regulated in some tissues. *Development* **108**, 281–287.
Matrisian, L. M., and Hogan, B. L. M. (1990). Growth factor-regulated proteases and extracellular matrix remodeling during mammalian development. *Curr. Top. Dev. Biol.* **24**, 219–259.
Matrisian, L. M., Glaischenhaus, N., Gesnel, M. C., and Breathnach, R. (1985). Epidermal growth factor and oncogenes induce transcription of the same cellular mRNA in rat fibroblasts. *EMBO J.* **4**, 1435–1440.

Matrisian, L. M., Bowden, G. T., Krieg, P., Furstenberger, G., Briand, J. P., LeRoy, P., and Breathnach, R. (1986a). The mRNA coding for the secreted protease transin is expressed more abundantly in malignant than in benign tumors. *Proc. Natl. Acad. Sci. U.S.A.* **83**, 9413–9417.

Matrisian, L. M., LeRoy, P., Ruhlmann, C., Gesnel, M. C., and Breathnach, R. (1986b). Isolation of the oncogene and epidermal growth factor-induced transin gene: Complex control in rat fibroblasts. *Mol. Cell. Biol.* **6**, 1679–1686.

Matsui, M., Nomura, N., and Ishizaki, R. (1990). Isolation of human *fos*-related genes and their expression during monocyte-macrophage differentiation. *Oncogene* **5**, 249–255.

Ménétrey, D., Gannon, A., Levine, J. D., and Basbaum, A. I. (1989). Expression of c-*fos* protein in interneurons and projection neurons of the rat spinal cord in response to noxious somatic, articular, and visceral stimulation. *J. Comp. Neurol.* **285**, 177–195.

Milbrandt, J. (1987). A nerve growth factor-induced gene encodes a possible transcriptional regulatory factor. *Science* **238**, 797–799.

Milbrandt, J. (1988). Nerve growth factor induces a gene homologous to the glucocorticoid receptor gene. *Neuron* **1**, 183–188.

Mizuno, A., Mizobuchi, Y., Ishibashi, Y., and Matsuda, M. (1989). c-*fos* mRNA induction under vitamin B6 agonist-induced seizure. *Neurosci. Lett.* **98**, 272–275.

Mobbs, C. V., and Pfaff, D. W. (1988). Estradiol-regulated neuronal plasticity. *Curr. Top. Membrane Transport* **31**, 191–215.

Mocchetti, I., De Bernardi, M. A., Szekely, A. M., Alho, H., Brooker, G., and Costa, E. (1989). Regulation of nerve growth factor biosynthesis by beta-adrenergic receptor activation in astrocytoma cells: A potential role of c-*fos* protein. *Proc. Natl. Acad. Sci. U.S.A.* **86**, 3891–3895.

Montarolo, P. G., Goelet, P., Castellucci, V. F., Morgan, J., Kandel, E. R., and Schacher, S. (1986). A critical period for macromolecular synthesis in long-term heterosynaptic facilitation in *Aplysia*. *Science* **234**, 1249–1254.

Montminy, M. R., Sevarino, K. A., Wagner, J. A., Mandel, G., and Goodman, R. H. (1986). Identification of a cyclic-AMP-responsive element within the rat somatostatin gene. *Proc. Natl. Acad. Sci. U.S.A.* **83**, 6682–6686.

Montminy, M. R., Gonzaliz, G. A., and Yamamoto, K. K. (1990). Regulation of cAMP-inducible genes by CREB. *Trends Neurosci.* **13**, 184–188.

Monuki, E. S., Weinmaster, G., Kuhn, R., and Lemke, G. (1989). *scip*: A glial POU domain gene regulated by cyclic AMP. *Neuron* **3**, 783–793.

Monuki, E. S., Kuhn, R., Weinmaster, G., Trapp, B. D., and Lemke, G. (1990). Expression and activity of the POU transcription factor *scip*. *Science* **249**, 1300–1303.

Morgan, J., and Curran, T. (1986). Role of ion flux in the control of c-*fos* expression. *Nature (London)* **322**, 552–555.

Morgan, J. I., and Curran, T. (1989). Stimulus-transcription coupling in neurons: Role of cellular immediate-early genes. *Trends Neurosci.* **12**, 459–462.

Morgan, J. I., and Curran, T. (1991). Stimulus-transcription coupling in the nervous system: Involvement of the inducible proto-oncogenes *fos* and *jun*. *Ann. Rev. Neurosci.* **14**, 421–451.

Morgan, J. I., Cohen, D. R., Hempstead, J. L., and Curran, T. (1987). Mapping patterns of c-*fos* expression in the central nervous system after seizure. *Science* **237**, 192–197.

Müller, R., Bravo, R., and Burckhardt, J. (1984). Induction of c-*fos* gene and protein by growth factors precedes activation of c-*myc*. *Nature (London)* **312**, 716–720.

Nakabeppu, Y., and Nathans, D. (1989). The basic region of *fos* mediates specific DNA binding. *EMBO J.* **8**, 3833–3841.

Nakabeppu, Y., Ryder, K., and Nathans, D. (1988). DNA binding activities of three murine *jun* proteins: Stimulation by *fos*. *Cell* **55**, 907–915.

Nakajima, T., Daval, J.-L., Morgan, P. F., Post, R. M., and Marangos, P. J. (1989). Adenosinergic modulation of caffeine-induced c-*fos* mRNA expression in mouse brain. *Brain Res.* **501**, 307–314.

Naranjo, J. R., Mellström, B., Auwerx, J., Mollinedo, F., and Sassone-Corsi, P. (1990). Unusual c-*fos* induction upon chromaffin PC12 differentiation by sodium butyrate: Loss of *fos* autoregulatory function. *Nucleic Acids Res.* **18**, 3605–3610.

Nieto-Sampedro, M., Lewis, E. R., Cotman, C. W., Manthorpe, M., Skaper, S. D., Barbin, G., Longo, F. M., and Varon, S. (1982). Brain injury causes a time-dependent increase in neuronotrophic activity at the lesion site. *Science* **217**, 860–861.

Nishina, H., Sato, H., Suzuki, T., Sato, N., and Iba, H. (1990). Isolation and characterization of *fra-2*: A new member of the *fos* gene family. *Proc. Natl. Acad. Sci. U.S.A.* **87**, 3619–3623.

Otani, S., Marshall, C. J., Tate, W. P., Goddard, G. V., and Abraham, W. C. (1989). Maintenance of long-term potentiation in rat dentate gyrus requires protein synthesis but not messenger RNA synthesis immediately post-tetanization. *Neurosci.* **28**, 519–526.

Pfaff, D. W., and Schwartz-Giblin, S. (1988). Cellular mechanisms of female reproductive behaviors. In "The Physiology of Reproduction" (E. Knobil, J. Neill, et al., eds.) pp. 1487–1501. Raven Press, New York.

Popovici, T., Barbin, G., and Ben Ari, Y. (1988). Kainic acid-induced seizures increase c-*fos*-like protein in the hippocampus. *Eur. J. Pharmacol.* **150**, 405–406.

Presley, R. W., Ménétrey, D., Levine, J. D., and Basbaum, A. I. (1990). Systemic morphine suppresses noxious stimulus-evoked *fos* protein like immunoreactivity in the rat spinal cord. *J. Neurosci.* **10**, 323–335.

Prywes, R., and Roeder, R. G. (1987). Purification of the c-*fos* enhancer binding protein. *Mol. Cell. Biol.* **7**, 3482–3489.

Quantin, B., and Breathnach, R. (1988). Epidermal growth factor stimulates transcription of the c-*jun* proto-oncogene in rat fibroblasts. *Nature (London)* **334**, 538–539.

Ransone, L. J., Visvader, J., Sassone-Corsi, P., and Verma, I. M. (1989). *fos–jun* interaction: Mutational analysis of the leucine zipper domain of both proteins. *Genes Dev.* **3**, 770–781.

Rauscher, F. J., III, Cohen, D. R., Curran, T., Bos, T. J., Vogt, D., Bohmann, R., Tjian, R., and Franza, B. R., Jr. (1988a). *fos*-associated protein p39 is the product of the *jun* proto-oncogene. *Science* **240**, 1010–1016.

Rauscher, F. J., III, Sambucetti, L. C., Curran, T., Distel, R. J., and Spiegelman, B. M. (1988b). Common DNA binding site for *fos* protein complexes and transcription factor AP-1. *Cell* **52**, 471–480.

Rauscher, F. J., III, Voulalas, P. J., Franza, B. R., Jr., and Curran, T. (1988c). *fos* and *jun* bind cooperatively to the AP-1 site: Reconstitution *in vitro*. *Genes Dev.* **2**, 1687–1699.

Rea, M. A. (1989). Light increases *fos*-related protein immunoreactivity in the rate suprachiasmatic nuclei. *Brain Res. Bull.* **23**, 577–581.

Risse, G., Jooss, K., Neuberg, M., Bruller, H.-J., and Müller, R. (1989). Asymmetrical recognition of the palindromic AP-1 binding site (TRE) by *fos* protein complexes. *EMBO J.* **8**, 3825–3832.

Rivera, V. M., Sheng, M., and Greenberg, M. E. (1990). The inner core of the serum response element mediates both the rapid induction and subsequent repression of c-*fos* transcription following serum stimulation. *Genes Dev.* **4**, 255–268.

Robertson, G., Herrera, D., Dragunow, M., and Robertson, H. (1989). L-DOPA activates c-*fos* in the striatum ipsilateral to a 6-hydroxydopamine lesion of the substantia nigra. *Eur. J. Pharmacol.* **159**, 99–100.

Roivainen, R., and Koistinaho, J. (1990). Decentralization induces long-term c-*fos* protein-like immunoreactivity in non-neuronal cells in the rat superior cervical ganglion. *Neurosci. Lett.* **119**, 105–108.

Rollins, B. J., Morrison, E. D., and Stiles, C. D. (1988). Cloning and expression of JE, a gene inducible by platelet-derived growth factor and whose product has cytokine-like properties. *Proc. Natl. Acad. Sci. U.S.A.* **85**, 3738–3742.

Rosbash, M., and Hall, J. C. (1989). The molecular biology of circadian rhythms. *Neuron* **3**, 387–398.

Rozengurt, E., and Sinnett-Smith, J. (1990). Bombesin stimulation of fibroblast mitogenesis: Specific receptors, signal transduction and early events. *Phil. Trans. R. Soc. Lond. (Biol.)* **327**, 209–221.

Ruppert, C., and Wille, W. (1987). Proto-oncogene c-*fos* is highly induced by disruption of neonatal but not of mature brain tissue. *Mol. Brain. Res.* **2**, 51–56.

Rusak, B., Robertson, H. A., Wisden, W., and Hunt, S. P. (1990). Light pulses that shift rhythms induce gene expression in the suprachiasmatic nucleus. *Science* **248**, 1237–1240.

Ryder, K., and Nathans, D. (1988). Induction of proto-oncogene c-*jun* by serum growth factors. *Proc. Natl. Acad. Sci. U.S.A.* **85,** 8464–8467.

Ryder, K., Lau, L. F., and Nathans, D. (1988). A gene activated by growth factors is related to the oncogene v-*jun*. *Proc. Natl. Acad. Sci. U.S.A.* **85,** 1487–1491.

Ryder, K., Lanahan, A., Perez-Albuerne, E., and Nathans, D. (1989). *jun*-D: A third member of the *jun* gene family. *Proc. Natl. Acad. Sci. U.S.A.* **86,** 1500–1503.

Ryseck, R. P., Hirai, S. I., Yaniv, M., and Bravo, R. (1988). Transcriptional activation of c-*jun* during the $G_0/G_1$ transition in mouse fibroblasts. *Nature (London)* **334,** 535–537.

Ryseck, R. P., MacDonald-Bravo, H., Mattei, M.-G., Ruppert, S., and Bravo, R. (1989). Structure, mapping, and expression of a growth factor inducible gene encoding a putative hormonal binding receptor. *EMBO J.* **8,** 3327–3335.

Saffen, D. W., Cole, A. J., Worley, P. F., Christy, B. A., Ryder, K., and Baraban, J. M. (1988). Convulsant-induced increase in transcription factor messenger RNAs in rat brain. *Proc. Natl. Acad. Sci. U.S.A.* **85,** 7795–7799.

Sagar, S. M., Sharp, F. R., and Curran, T. (1988). Expression of c-*fos* protein in brain: Metabolic mapping at the cellular level. *Science* **240,** 1328–1331.

Sassone-Corsi, P., Ransone, L. J., and Verma, I. M. (1990). Cross-talk in signal transduction: TPA-inducible factor *jun*/AP-1 activates cAMP-responsive enhancer elements. *Oncogene* **5,** 427–431.

Satoh, T., Nakamura, S., Taga, T., Matsuda, T., Hirano, T., Kishimoto, T., and Kaziro, Y. (1988). Induction of neuronal differentiation in PC12 cells by B-cell stimulatory factor/interleukin-6. *Mol. Cell. Biol.* **8,** 3546–3549.

Schacher, S., Castellucci, V. F., and Kandel, E. R. (1988). cAMP evokes long-term facilitation in *Aplysia* sensory neurons that requires new protein synthesis. *Science* **240,** 1667–1669.

Schönthal, A., Herrlich, P., Rahmsdorf, H. J., and Ponta, H. (1988). Requirement for *fos* gene expression in the transcriptional activation of collagenase by other oncogenes and phorbol esters. *Cell* **54,** 325–334.

Schröter, H., Mueller, C. G. F., Meese, K., and Nordheim, A. (1990). Synergism in ternary complex formation between the dimeric glycoprotein $p67^{srf}$, polypeptide $p62^{tcf}$ and the c-*fos* serum response element. *EMBO J.* **4,** 1123–1130.

Schuermann, M., Neuberg, M., Hunter, J. B., Jenuwein, T., Ryseck, R. P., Bravo, R., and Müller, R. (1989). The leucine repeat motif in *fos* protein mediates complex formation with *jun*/AP-1 and is required for transformation. *Cell* **56,** 507–516.

Schütte, J., Viallet, J., Nau, M., Segal, S., Fedorko, J., and Minna, J. (1989). *jun*-B inhibits and c-*fos* stimulates the transforming and trans-activating activities of c-*jun*. *Cell* **59,** 987–997.

Schwartz, J. P., Chuang, D.-M., and Costa, E. (1977). Increase in nerve growth factor content of C6 glioma cells by the activation of a β-adrenergic receptor. *Brain Res.* **137,** 369–375.

Setoyama, C., Frunzio, R., Liau, G., Mudryj, M., and de Crombrugghe, B. (1986). Transcription activation encoded by the v-*fos* gene. *Proc. Natl. Acad. Sci. U.S.A.* **83,** 3213–3217.

Sharp, F. R., Gonzalez, M. F., Hisanaga, K., Mobley, W. C., and Sagar, S. M. (1989a). Induction of the c-*fos* gene product in rat forebrain following cortical lesions and NGF injections. *Neurosci. Lett.* **100,** 117–122.

Sharp, F. R., Gonzalez, M. F., Sharp, J. W., and Sagar, S. M. (1989b). c-*fos* expression and ($^{14}$C)-2-deoxyglucose uptake in the caudal cerebellum of the rat during motor/sensory cortex stimulation. *J. Comp. Neurol.* **284,** 621–636.

Sharp, F. R., Griffith, J., Gonzalez, M. F., and Sagar, S. M. (1989c). Trigeminal nerve section induces *fos*-like immunoreactivity (FLI) in brainstem and decreases FLI in sensory cortex. *Mol. Brain Res.* **6,** 217–220.

Shaw, P. E., Schroter, H., and Nordheim, A. (1989). The ability of a ternary complex to form over the serum response element correlates with serum inducibility of the c-*fos* promoter. *Cell* **56,**563–572.

Sheng, M., Dougan, S. T., McFadden, G., and Greenberg, M. E. (1988). Calcium and growth factor pathways of c-*fos* transcriptional activation require distinct upstream regulatory sequences. *Mol. Cell Biol.* **8,** 2787–2796.

Sheng, M., McFadden, G., and Greenberg, M. E. (1990). Membrane depolarization and calcium induce c-fos transcription via phosphorylation of transcription factor CREB. *Neuron* **4,** 571–582.

Shin, C., McNamara, J. O., Morgan, J. I., Curran, T., and Cohen, D. R. (1990). Induction of c-fos mRNA expression by after discharge in the hippocampus of naive and kindled rats. *J. Neurochem.* **55,** 1050–1055.

Shivers, B. D., Harlan, R. E., Morrell, J. I., and Pfaff, D. W. (1983). Absence of oestradiol concentration in cell nuclei of LHRH-immunoreactive neurones. *Nature (London)* **304,** 345–347.

Sokoloff, L. (1989). Circular and energy metabolism of the brain. *In* "Basic Neurochemistry: Molecular, Cellular, and Medical Aspects" (G. J. Siegel, R. W. Albers, F. Agranof, and P. Molinoff, eds.), 4th ed., pp. 565–590. New York, Raven Press.

Sonnenberg, J. L., Macgregor-Leon, P. F., Curran, T., and Morgan, J. I. (1989a). Dynamic alterations occur in the levels and composition of transcription factor AP-1 complexes after seizure. *Neuron* **3,** 359–365.

Sonnenberg, J. L., Mitchelmore, C., Macgregory-Leon, P. F., Hempstead, J., Morgan, J. I., and Curran, T. (1989b). Glutamate receptor agonists increase the expression of *fos*, Fra, and AP-1 DNA binding activity in the mammalian brain. *J. Neurosci. Res.* **24,** 72–80.

Sonnenberg, J. L., Rauscher, F. J., Morgan, J. I., and Curran, T. (1989c). Regulation of proenkephalin by proto-oncogene *fos* and *jun*. *Science* **246,** 1622–1625.

Stachowiak, M. K., Sar, M., Tuominen, R. K., Jiang, H.-K., Iadarola, M. J., Poisner, A. M., and Hong, J. S. (1990). Stimulation of adrenal medullary cells *in vivo* and *in vitro* induces expression of c-fos proto-oncogene. *Oncogene Res.* **5,** 69–73.

Sukhatme, V. P., Kartha, S., Toback, F. G., Taub, R., Hoover, R. G., and Tsai-Morris, C. (1987). A novel early growth response gene rapidly induced by fibroblast, epithelial cell and lymphocyte mitogens. *Oncogene Res.* **1,** 343–355.

Sukhatme, V. P., Cao, X., Chang, L. C., Tsai-Morris, C., Stamenkovich, D., Ferre, P. C., Cohen, D. R., Edwards, S. A., Shows, T. B., Curran, T., LeBeau, M. M., and Adamson, E. D. (1988). A zinc finger-encoding gene co-regulated with c-fos during growth and differentiation, and after cellular depolarization. *Cell* **53,** 37–43.

Swanson, L. W., and Simmons, D. M. (1989). Differential steroid hormone and neural influences on peptide mRNA in CRH cells of the paraventricular nucleus: A hybridization histochemical study in the rat. *J. Comp. Neurol.* **285,** 413–425.

Sweatt, J. D., and Kandel, E. R. (1989. Persistent and transcriptionally dependent increase in protein phosphorylation in long-term facilitation of *Aplysia*. *Nature (London)* **339,** 51–54.

Szekely, A., Barbaccia, M., and Costa, E. (1987). Activation of specific glutamate receptor subtypes increases c-fos proto-oncogene expression in primary cultures of neonatal rat cerebellar granule cells. *Neuropharmacol.* **26,** 1779–1782.

Szekely, A. M., Barbaccia, M. L., Alho, H., and Costa, E. (1989). In primary cultures of cerebellar granule cells the activation of N-methyl-D-aspartate-sensitive glutamate receptors induces c-fos mRNA expression. *Mol. Pharmacol.* **35,** 401–408.

Tang, S.-J., Ko, L.-W., Lee, Y.-H. W., and Wang, F.-F. (1990). Induction of *fos* and *sis* proto-oncogene and genes of the extracellular matrix proteins during butyrate induced glioma differentiation. *Biochem. Biophys. Acta* **1048,** 59–65.

Tank, A. W., Currella, P., and Ham, L. (1986). Induction of mRNA for tyrosine hydroxylase by cyclic AMP and glucocorticoids in a rat pheochromocytoma cell line: Evidence for the regulation of tyrosine hydroxylase synthesis by multiple mechanisms in cells exposed to elevated levels of both inducing agents. *Mol. Pharmacol.* **30,** 497–503.

Thoenen, H., and Barde, Y.-A. (1980). Physiology of nerve growth factor. *Physiol. Rev.* **60,** 1284–1335.

Tölle, T. R., Castro-Lopes, J. M., Coimbra, A., and Zieglgansberger, W. (1990). Opiates modify induction of c-fos proto-oncogene in the spinal cord of the rat following noxious stimulation. *Neurosci. Lett.* **111,** 46–51.

Treisman, R. (1986). Identification of a protein-binding site that mediates transcriptional response to the c-fos gene to serum factors. *Cell* **46,** 567–574.

Treisman, R. (1987). Identification and purification of a polypeptide that binds to the c-*fos* serum response element. *EMBO J.* **6,** 2711–2717.
Treisman, R. (1990). The SRE: A growth factor responsive transcriptional regulator. *In* "Seminars in Cancer Biology, Transcription Factors, Differentiation and Cancer" (N. C. Jones, ed.), pp. 47–58, Saunders, London.
Tsai-Morris, C.-H., Cao, X., and Sukhatme, V. P. (1988). 5' flanking sequence and genomic structure of *Egr-1*, a murine mitogen-inducible zinc finger-encoding gene. *Nucleic Acids Res.* **16,** 8835–8846.
Unsicker, K., Seidl, K., and Hofmann, H. D. (1989). The neuro-endocrine ambiguity of sympathoadrenal cells. *Int. J. Dev. Neurosci.* **7,** 413–417.
Varnum, B. C., Lim, R. W., Kaufman, S. E., Gasson, J. C., Greenberger, J. S., and Herschman, H. R. (1989a). Granulocyte-macrophage colony-stimulating factor induces a unique pattern of primary response TIS genes in both proliferating and terminally differentiated myeloid cells. *Mol. Cell. Biol.* **9,** 3580–3583.
Varnum, B. C., Lim, R. W., Sukhatme, V. P., and Herschman, H. R. (1989b). Nucleotide sequence of a cDNA encoding TIS11, a message induced in Swiss 3T3 cells by the tumor promoter tetradecanoyl phorbol acetate. *Oncogene* **4,** 119–120.
Visvader, J., Sassone-Corsi, P., and Verma, I. (1988). Two adjacent promoter elements mediate nerve growth factor activation of the c-*fos* gene and bind distinct nuclear complexes. *Proc. Natl. Acad. Sci. U.S.A.* **85,** 9474–9478.
Watson, M. A., and Milbrandt, J. (1989). The NGFI-B gene, a transcriptionally inducible member of the steroid receptor gene superfamily: Genomic structure and expression in rat brain after seizure. *Mol. Cell. Biol.* **9,** 4213–4219.
Weissman, B. A., Cott, J., Paul, S. M., and Skolnick, P. (1983). Ro5-4864: A potent benzodiazepine convulsant. *Eur. J. Pharmacol.* **90,** 149–150.
Weisz, A., and Bresciani, F. (1988). Estrogen induces expression of c-*fos* and c-*myc* proto-oncogenes in rat uterus. *Mol. Endocrinol.* **2,** 816–824.
White, J. D., and Gall, C. M. (1987). Differential regulation of neuropeptide and proto-oncogene mRNA content in the hippocampus following recurrent seizures. *Mol. Brain Res.* **3,** 21–29.
Wilkinson, D., Bhatt, S., Chavrier, P., Bravo, R., and Charnay, P. (1989a). Segment-specific expression of a zinc-finger gene in the developing nervous system of the mouse. *Nature (London)* **337,** 461–465.
Wilkinson, D., Bhatt, S., Ryseck, R.-P., and Bravo, R. (1989b). Tissue-specific expression of c-*jun* and *junB* during organogenesis in the mouse. *Development* **106,** 465–471.
Winston, S. M., Hayward, M. D., Nestler, E. J., and Duman, R. S. (1990). Chronic electroconvulsive seizures down-regulate expression of the immediate-early genes c-*fos* and c-*jun* in rat cerebral cortex. *J. Neurochem.* **54,** 1920–1925.
Wisden, W., Errington, M. L., Williams, S., Dunnett, S. B., and Waters, C. (1990). Differential expression of immediate early genes in the hippocampus and spinal cord. *Neuron* **4,** 603–614.
Wright, J. J., Gunter, K. C., Mitsuya, H., Irving, S. G., Kelly, K., and Siebenlist, U. (1990). Expression of a zinc finger gene in HTLV-I- and HTLV-II-transformed cells. *Science* **248,** 588–591.
Yamamoto, K., and Alberts, B. (1976). Steroid receptors: Elements for modulation of eukaryotic transcription. *Ann. Rev. Biochem.* **45,** 721–746.
Yamamoto, K. K., Gonzalez, G. A., Biggs, W. H., III, and Montminy, M. R. (1988). Phosphorylation-induced binding and transcriptional efficacy of nuclear factor CREB. *Nature (London)* **334,** 494–498.
Yang-Yen, J.-F., Chambard, J.-C., Sun, Y.-L., Smeal, T., Schmidt, T. J., Drouin, J., and Karin, M. (1990). Transcriptional interference between c-*jun* and the glucocorticoid receptor: Mutual inhibition of DNA binding due to direct protein-protein interaction. *Cell* **62,** 1205–1215.
Zerial, M., Toschi, L., Ryseck, R. P., Schuermann, M., Müller, R., and Bravo, R. (1989). The product of a novel growth factor activated gene, *fosB*, interacts with *jun* proteins enhancing their DNA binding activity. *EMBO J.* **8,** 805–813.

# 5  Nerve Growth Factor and Related Substances: Structure and Mechanism of Action

Joseph G. Altin and Ralph A. Bradshaw

## I. INTRODUCTION

Nerve growth factor (NGF) was the first neurotrophic factor to be identified (Levi-Montalcini and Hamburger, 1951). It was also the first polypeptide growth factor to be identified as such and occupies, therefore, a unique place in both neural and cell biology. It has come to be recognized as a prototype for at least one class of neurotrophic substances, that is, those that support neuronal maintenance and survival through retrograde transport mechanisms (retrophins) (Hendry et al., 1974). Its interaction with and importance for the development of sympathetic neurons was recognized early through the use of antibodies (Levi-Montalcini and Booker, 1960) and its relationship to other endocrine-like substances was suggested through sequential comparisons to insulin and insulin-related substances (Frazier et al., 1972). As subsequent research has shown, NGF is not limited to peripheral nervous targets (Longo and Mobley, 1991); its structural relationship, and therefore its evolutionary relationship, to insulin, as revealed by recent three-dimensional studies (McDonald et al., 1991), is probably questionable. Nonetheless, both of these observations are of major importance in the history of neurobiology and the study of polypeptide growth factors.

In this chapter, the structural properties of NGF, and of the recently identified sequentially similar neurotrophins, are described. In addition, mechanistic properties of the hormone, primarily as they have been derived from studies with the neuroparadigm, the PC12 cell, as well as properties and characteristics of NGF receptors are outlined. A more detailed account of aspects of biosynthesis of the hormone and its distribution and function in the central nervous system are found in the ensuing chapters. Other reviews on NGF, which provide further detail in certain areas, include Server and Shooter (1977), Bradshaw (1978), Greene and Shooter (1980), Thoenen and Barde (1980), and Halegoua et al. (1991).

## II. NGF STRUCTURE

### A. Sequences of NGF and Associated Subunits

Although NGF activity was first identified in two sarcoma tissues (Levi-Montalcini and Hamburger, 1951) and in certain snake venoms (Cohen 1959), the vast majority of the chemical and physical properties of NGF have been deduced from the NGF obtained from adult male mouse submandibular glands (Cohen, 1960). The active principle of this preparation is quite similar to the NGF that occurs in a variety of other tissues; however, other aspects are apparently unique. This is exemplified by the fact that mouse NGF can be isolated in two distinct forms. The first, referred to as 7S NGF (designating its sedimentation coefficient), is a high molecular weight complex containing two copies each of three types of polypeptide chains (Varon et al., 1967). Designated by the Greek letters $\alpha$, $\beta$, and $\gamma$, these subunits readily can be dissociated and reassembled, under appropriate conditions. The presence of 1–2 gram atoms of zinc is important both for stability and reassociation (Pattison and Dunn, 1975). The second preparation of NGF, commonly referred to, particularly in the early literature, as 2.5S NGF (Bocchini and Angeletti, 1969), shows a greater degree of proteolytic modification at both termini but is biologically and immunologically indistinguishable from the $\beta$ subunit (either as it occurs in the 7S complex or in disassociated form). The 7S complex appears to be unique in the mouse submandibular gland, presumably because the $\alpha$ and $\gamma$ subunits are not synthesized elsewhere. They actually represent members of an interesting glandular kallikrein family that is extensively expressed in this tissue (Richards et al., 1989). The properties and characteristics of the $\beta$ subunit and the two glandular kallikreins, the $\alpha$ and $\gamma$ subunits, will be described in the ensuing two sections.

#### 1. βNGF

The mouse $\beta$NGF subunit, either as derived from the 7S complex or isolated directly, occurs as a tightly associated noncovalent dimer (Angeletti et al., 1971). The longest mature form of the sequence isolated contains 118 amino acids (Angeletti and Bradshaw, 1971). The protein commences with amino terminal serine and terminates with carboxy terminal arginine. In preparations of 2.5S NGF, the initial 8 residues are deleted from approximately half of the polypeptide chains (Angeletti et al., 1973). Similarly, the carboxy terminal arginine residue also is removed, although generally on a smaller percent of subunits. The sequence of the molecule, as determined from protein chemical methods, is shown in Fig. 1 (Angeletti and Bradshaw, 1971). The molecule contains three interchain disulfide bonds and is quite basic in character, with an observed pI of 9.3.

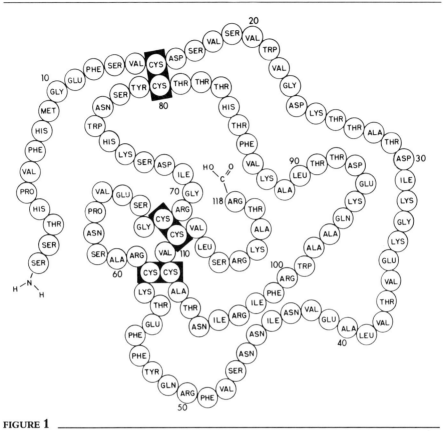

**FIGURE 1**

Amino acid sequence of the primary subunit of 2.5S NGF from mouse submaxillary gland. Reproduced from Angeletti and Bradshaw (1971).

Not unexpectedly, the predicted protein sequence from the corresponding cDNA of mouse NGF (Scott et al., 1983) revealed that the mature form of NGF was excised from a precursor structure containing residues. Most of these are removed from the amino terminus; only two residues are cut from the carboxy terminus. The amino terminal serine of the mature protein is immediately preceded by two basic residues, suggesting a typical dibasic processing site. The carboxy terminal dipeptide results from the cleavage of an Arg–Arg sequence (the dipeptide removed is Arg–Gly). These features are illustrated in Fig. 2. A putative role for the γ subunit in the removal of the Arg–Gly dipeptide is described in the ensuing section.

The sequences of the mature form of NGF from human, guinea pig, rat, bovine, chicken, and cobra have been determined from corresponding cDNA or genomic clones (Ullrich et al., 1983; Ebendal et al., 1986; Meier et al., 1986; Wion et al., 1986; Selby et al., 1987; Whittemore et al., 1988; Schwarz et al., 1989). In

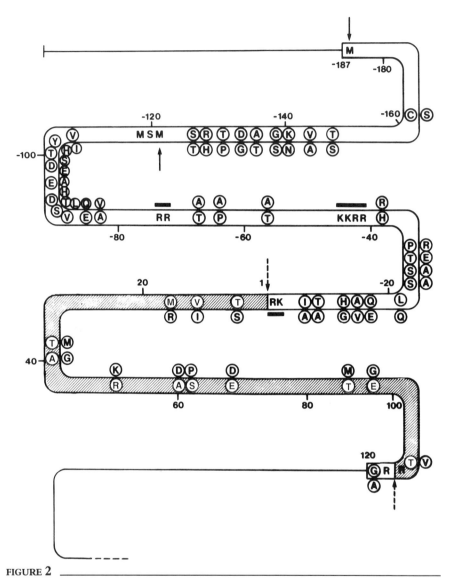

FIGURE 2

Precursor sequences of mouse and human βNGF. Residues indicated in circles mark the differences (inside the bar, mouse; outside the bar, human) between the two molecules. The shaded area corresponds to the mature sequence. Solid arrows indicate alternate interior methionine sites; broken arrows mark the minimum precursor processing sites to generate the native molecule. Solid horizontal bars indicate dibasic sequences (that are real or potential processing sites). Data from Ullrich et al. (1983) and Scott et al. (1983).

addition, partial sequences of viper, *Xenopus,* and salmon NGF also have been reported (Hallbook et al., 1991) (see Fig. 3). As is usually the case, there is a substantial amount of residue conservation between species, as well as a number of variable regions. The conserved residues are likely to contribute directly to receptor interactions or to maintaining essential three-dimensional structures (see subsequent text).

Some additional insight into active site structures comes from various chemical modification experiments. In one study (Frazier et al., 1973), it was shown that $Trp_{21}$ is accessible to solvent and is not required for biologic activity whereas $Trp_{99}$ and $Trp_{76}$ are buried partially and completely, respectively, and their modification leads to a loss of biologic function. All three residues are conserved in all species to date. Similar studies support the importance of the histidines as positions 77 and 86 (Dunbar et al., 1984). However, only the former is found in all species, suggesting that the latter residue does not participate directly in receptor interactions. Modification of βNGF with 1-guanyl-3,5-dimethylpyrazole and cyclohexanedione indicates that lysine and arginine residues do not directly participate in receptor interactions; complete modification of the latter residues is necessary before biologic activity is lost, whereas alteration of the lysine residues is without effect (Silverman, 1978). These observations suggest, as has been observed in other systems, that receptor interactions are contributed probably primarily by van der Waals contacts and generally do not involved residues that are readily modified post translationally. The advent of the three-dimensional structure (see subsequent text) should be a major advance in defining the receptor interaction sites of NGF.

### 2. α and γNGF

In the mouse 7S complex, the βNGF dimer is associated with two molecules of α subunit and two molecules of γ subunit. The latter interaction depends on the presence of the carboxy terminal arginine of the β subunit; thus, γ subunits are presumably not associated with complexes in which the carboxy terminal arginine is missing. This is consistent with observations that some high molecular weight complexes show lower sedimentation coefficients (Server and Shooter, 1977). The interaction of α and γ subunits with the β subunits is noncovalent and is affected by both pH and concentration. Both α–β and β–γ complexes can be formed and detected by ultracentrifugation (Silverman and Bradshaw, 1982). These show markedly different pH stability profiles, indicating rather different sites of interaction with the β subunit. There is no evidence to suggest that the α and γ subunits form significant contracts. The formation of covalent bonds between the β subunit polypeptides and the α and γ subunits (thus stabilizing the interactions) results in the inactivation of the β subunit, presumably because it prevents receptor interactions by steric hindrance. Covalent cross-linking of the β dimers does not have any effect on biologic activity (Stach and Shooter, 1974; Pulliam et al., 1975). Neither α nor γ subunits show any biologic activity with neuronal cells.

In addition to their specific interactions with the β subunit, the γ subunits also display catalytic activity toward peptide substrates formed by arginine residues

```
            -10                  -1   1                10
hNGF     P F N R T H R S K R   S S S H P I F H R G E F S V C D S
gpNGF    S V N R T H R S K R   S S T H P V F H M G E F S V C D S
rNGF     S F N R T H R S K R   S S T H P V F H M G E F S V C D S
mNGF     P F N R T H R S K R   S S T H P V F H M G E F S V C D S
bNGF             H R S K R     S S S H P V F H R G E F S V C D S
chNGF    S L N R T A R T K R   - T A H P V L H R G E F S V C D S
sNGF     S L N R N I R A K R   - E D H P V H N L G E H S V C D S
mNT3         N R T S P R R K R - Y A E H K S H R G E Y S V C D S
mBDNF    A A N M S M R V R R   - - H S D P A R R G E L S V C D S
chBDNF                   R     - - H S D P A R R G E L S V C D S
xBDNF                    R     - - H S D P A R R G E L S V C D S
xNT4     P A N K T S R L K R (ASGSD) S V S L S R R G E L S V C D S
                                ● ●

              20                30                40
hNGF     V S V W V - - G D K T T A T D I K G K E V M V L G E V N
gpNGF    V S V W V - - A D K T T A T D I K G K E V T V L A E V N
rNGF     V S V W V - - G D K T T A T D I K G K E V T V L G E V N
mNGF     V S V W V - - G D K T T A T D I K G K E V T V L A E V N
bNGF     I S V W V - - G D K T T A T D I K G K E V M V L G E V N
chNGF    V S M W V - - G D K T T A T D I K G K E V T V L G E V N
sNGF     V S A W V - - T - K T T A T D I K G N T V T V M E N V N
mNT3     E S L W V - - T D K S S A I D I R G H Q V T V L G E I K
mBDNF    I S E W V T A A D K K T A V D M S G G T V T V L E K V P
chBDNF   T S E W V T A A E K K T A V D M S G A T V T V L E K V P
xBDNF    I S E W V T A A N K K T A V D M S G G T V T V L E K V P
xNT4     V N V W V - - T D K R T A V D D R G K I V T V M S E I Q
            ●   ● ●       ● ●   ● ●                         ● ●

              50                60                70
hNGF     I N N S V F K Q Y F F E T K C R D P N P V D S G C R G
gpNGF    V N N N V F K Q Y F F E T K C R D P S P V E S G C R G
rNGF     I N N S V F K Q Y F F E T K C R A P N P V E S G C R G
mNGF     I N N S V F R Q Y F F E T K C R A S N P V E S G C R G
bNGF     I N N S V F K Q Y F F E T K C R D P N P V D S G C R G
chNGF    I N N V F K Q Y F F E T K C R D P R P V S S G C R G
sNGF     L D N K V Y K Q Y F F E T K C K N P N P E P S G C R G
vNGF                             K C K N P S P V S G G C R G
xNGF                             K C R D P K P V S S G C R G
fNGF                             T C R G A R A G S S G C L G
mNT3     T G N S P V K Q Y F Y E T R C K E A R P V K N G C R G
chNT3                            R C K E A K P V K N G C R G
fNT3                             K C R T A K P F K S G C R G
yNT3                             R C K E S K P G K N G C R G
mBDNF    V S K G Q L K Q Y F Y E T K C N P M G Y T K E G C R G
chBDNF   V P K G Q L K Q Y F Y E T K C N P K G Y T K E G C R G
xBDNF    V S K G Q L K Q Y F Y E T K C N P M G Y M K E G C R G
vBDNF                            K C S T K G Y A K E G C R G
fBDNF                            K C N P M G Y T K E G C R G
yBDNF                            K C N P K G F T N E G C R G
xNT4     T L T G P L K Q Y F F E T K C N P S G S T T R G C R G
vNT4                             K C N P A G G T V G G C R G
           ●   ●         ●
```

**FIGURE 3**

Comparison of amino acid sequences of the mature regions of NGF, NT3, BDNE, and NT4 from different species. Amino acid sequences are shown for human NGF (hNGF) (Ullrich et al., 1983), guinea pig NGF (gpNGF) (Schwarz et al., 1989), rat NGF (rNGF) (Whittemore et al., 1988), mouse NGF (mNGF) (Scott et al., 1983; Ullrich et al., 1983), bovine NGF (bNGF) (Meier et al., 1986), chicken NGF (chNGF) (Ebendal et al., 1986; Meier et al., 1986; Wion et al., 1986), cobra NGF (sNGF)

```
                      80                   90                  100
hNGF    I D S K H W N S Y C T T T H T F V K A L T M D G - K Q A
gpNGF   I D S K H W N S Y C T T T H T F V K A L T T D N - K Q A
rNGF    I D S K H W N S Y C T T T H T F V K A L T T D D - K Q A
mNGF    I D S K H W N S Y C T T T H T F V K A L T T D E - K Q A
bNGF    I D A K H W N S Y C T T T H T F V K A L T M D G - K Q A
chNGF   I D A K H W N S Y C T T T H T F V K A L T M E G - K Q A
sNGF    I D S S H W N S Y C T E T D T F I K A L T M E G - N Q A
vNGF    I D A K H W N S Y C T T T D T F V R A L T M E G - N Q A
xNGF    I D A K H W N S Y C T T T H T F V R A L T M E G - K Q A
fNGF    I D G R H W N S Y C T N S H T F V R A L T S F K - D L V
mNT3    I D D K H W N S Q C K T S Q T Y V R A L T S E N N K L V
chNT3   I D D K H W N S Q C K T S Q T Y V R A L T S E N N K L V
fNT3    I D D K H W N S Q C K T S Q T Y V R A L T Q D R T - S V
yNT3    I D D K H W N S Q C K T S Q T Y V R A L S K E N N K Y V
mBDNF   I D K R H W N S Q C R T T Q S Y V R A L T M D S K K R I
chBDNF  I D K R H W N S Q C R T T Q S Y V R A L T M D N K K R V
xBDNF   I E K R Y W N S Q C R T T Q S Y V R A F T M D S K K K V
vBDNF   I D K R Y W N S Q C R T T Q S Y V R A L T M D N K K R I
fBDNF   I D K R H Y N S Q C R T T Q S Y V R A L T M D N K K K I
yBDNF   I D K K H W N S Q C R T S Q S Y V R A L T M D S R K K I
xNT4    V D K K Q W I S E C K A K Q S Y V R A L T I D A N K L V
vNT4    V D R R H W I S E C K A K Q S Y V R A L T M D S D K I V
          •           • • •           • •
```

```
                         110              120
hNGF    A W R F I R I D T A C V C V L S R K A V R R A
gpNGF   A W R F I R I D T A C V C V L N R K A A R R G
rNGF    A W R F I R I D T A C V C V L S R K A A R R G
mNGF    A W R F I R I D T A C V C V L S R K A T R R G
bNGF    A W R F I R I D T A C V C V L S R K T G Q R A
chNGF   A W R F I R I D T A C V C V L S R K S G R P
sNGF    S W R F I R I E T A C V C V I T K K G N
mNT3    G W R W I R I D T S C V C A L S R K I G R T
mBDNF   G W R F I R I D T S C V C T L T I K R G R
chBDNF  G W R F I R I D T S C V C T L - - - - - -
xBDNF   G W R F I R I D T S C V C T L - - - - - -
xNT4    G W R W I R I D T A C V C T L L S R T G R T
                        •       • •
```

(Selby et al., 1987), viper NGF (vNGF) (Hallbook et al., 1991), *Xenopus* NGF (xNGF) (Hallbook et al., 1991), salmon NGF (fNGF) (Hallbook et al., 1991), mammalian NT3 (mNT3) (Ernfors et al., 1990; Hohn et al., 1990; Kaisho et al., 1990; Maisonpierre et al., 1990; Rosenthal et al., 1990), chicken NT3 (chNT3) (Hallbook et al., 1991), salmon NT3 (fNT3) (Hallbook et al., 1991), ray NT3 (yNT3) (Hallbook et al., 1991), mammalian BDNF (mBDNF) (Leibrock et al., 1989; Hohn et al., 1990; Maisonpierre et al., 1990; Rosenthal et al., 1990), chicken BDNF (chBDNF) (Isackson et al., 1991), *Xenopus* BDNF (xBDNF) (Isackson et al., 1991), viper BDNF (vBDNF) (Hallbook et al., 1991), salmon BDNF (fBDNF) (Hallbook et al., 1991), ray BDNF (yBDNF) (Hallbook et al., 1991), *Xenopus* NTA (xNT4) (Hallbook et al., 1991), and viper NT4 (vNT4) (Hallbook et al., 1991). Absolutely conserved amino acid residues are boxed. Residues differing among NGF, BDNF, NT3, and NT4, but conserved in species comparisons of each factor are indicated by filled circles.

(Greene et al., 1969). Chemical modification and other studies suggested that the enzyme was a member of the serine protease family. This was subsequently confirmed by determination of the complete amino acid sequence (Thomas et al., 1981b), which revealed that the protein contained all the features associated with enzymes of this family and, in addition, was sequentially most closely related to the kallikrein family. The protein as isolated contains internal cleavage sites (introduced by limited proteolysis posttranslationally) and one site of N-linked glycosylation (Thomas et al., 1981a). The origin of the proteolytic "nicks" may be in part autolytic but the expression of recombinant subunits suggests that this is only partially the case (Blaber et al., 1990). Sequence analysis of a full-length cDNA clone (Ullrich et al., 1984) also revealed that the γ subunit, as well as other members of this family, is produced in a pre-pro-form that is presumably activated in the usual fashion. In addition, sequence analysis established that tetrapeptide is removed quantitatively from the mature protein at one of the internal sites of processing.

The identification of the α subunit as another member of the glandular kallikrein class was not apparent until a partial amino acid sequence and the corresponding cDNA clone were available (Isackson and Bradshaw, 1984; Isackson et al., 1984). Although the molecule has greater than 80% identity to the γ subunits, it possesses no catalytic activity with any peptide or ester substrate examined. The reasons for this are apparent on inspection of the amino acid sequence. First, the normal activation site, characterized by an Arg–Ile/Val bond, has been altered to produce an Arg–Gln bond. The deletion of the adjacent four amino acids undoubtedly further impairs any normal activation. Thus the α subunit exits as a quasi-zymogen. In addition, several other mutations, including the substitution of histidine for glycine at a position adjacent to the active-site serine, a residue unaltered in all serine proteases studied to date, would probably not be well tolerated catalytically. Thus, the α subunit has undergone modifications, possibly to allow more specific interactions with the β subunit, at the expense of its catalytic "heritage." There are no other known biologic functions of the α subunit. As does the γ subunit, it contains one glycosylation site and undergoes at least one internal site of proteolytic modification in the mature form. In this case, the modification clearly is not introduced autolytically.

In addition to the α and γ subunits, the mouse submandibular gland produces a considerable number of other glandular kallikreins, all homologous to the α and γ subunits (Richards et al., 1989). Only one of these, renal kallikrein, appears to be expressed in any tissues other than the submandibular gland. The reasons for this are obscure. At least one of these kallikreins apparently is responsible for the removal of the octapeptide from the amino terminus of the β subunit (see previous text) (Fahnestock et al., 1991) and a fourth member of the family is specifically associated with epidermal growth factor (EGF), which also occurs in this tissue (Blaber et al., 1987). Designated epidermal growth factor binding protein (EGF-BP), it shares many similar structural and functional properties with γNGF, including association with the mature EGF protein through a carboxy terminal arginine residue. The two proteins differ only in 35 positions (of 237), yet are absolutely specific for their associations with the precursors and mature forms of their respective growth factors. The preparation of chimeric proteins in which the amino terminal, central, and carboxy terminal segments of these two kallikreins (γNGF and EGF-BP) have been interchanged has revealed additional sites of interaction between the kallikreins and the hormones that occur outside the active site (M. Blaber, P. J. Isackson, and R. A. Bradshaw, unpublished observations).

## B. Sequences of Other Neurotrophins

Although NGF was tentatively identified as a distantly related member of the insulin family (Frazier et al., 1972) only recently have much more closely related homologs been identified. The first of these, brain derived neurotrophic factor (BDNF), was isolated from pig brain and shown to act on sensory neurons unresponsive to NGF (Barde et al., 1982). Its similar sequence and, therefore, evolutionary relationship were not apparent until the structure was determined (Leibrock et al., 1989; Isackson et al., 1991). A comparison of the mature sequences of BDNF molecules from a number of species is shown in Fig. 3. As in the NGF molecules from different species, there are both conserved and variable regions. These similarities and differences take on greater significance with the realization that the various NGF-related substances have some common receptor interactions as well as some distinct binding sites. Through the use of PCR (polymerase chain reaction) and other sophisticated cloning techniques, additional NGF-related substances have been identified (Hohn et al., 1990; Maisonpierre et al., 1990; Rosenthal et al., 1990; Hallbook et al., 1991). These have been designated neurotrophin 3 (NT-3) and neurotrophin 4 (NT-4); the entire family is now appropriately referred to by the more general name of neurotrophin. (Accordingly, NGF would be equivalent to neurotrophin 1 and BDNF to neurotrophin 2, although neither original name is likely to disappear from the literature in the near future.) Like BDNF, NT-3 shows a distinct pattern of target cell responsiveness with some overlap with both NGF and BDNF. For all the neurotrophins, the similarities in sequence, including apparent conservation of the disulfide bonds, suggest that the molecules will have similar three-dimensional structures. As for NGF, a more detailed analysis of the conserved structures of the neurotrophins will be possible with the completion of the three-dimensional analysis of NGF.

## C. Three-Dimensional Properties

The first crystallization of βNGF was reported by Wlodawer et al. (1975). However, a high resolution structure for the molecule has only been obtained recently (McDonald et al., 1991). Raman spectroscopy previously had indicated a high degree of beta structure, an indication that NGF and the insulin-related substances were unlikely to share large portions of common three-dimensional structure (Williams et al., 1982). The model that has been obtained from X-ray data confirms this prediction, indicating the presence of seven β strands in three antiparallel pairs.

At either end of the elongated molecule are found various loop structures. In some cases, they show close conservation among species in one neurotrophin type with significant variations between families. In other cases, there is strong conservation in two or more neurotrophin types. These similarities and differences presumably account for the receptor-binding properties observed. The structure is clearly quite unique and does not correspond to any other growth factor or hormone previously studied.

## III. RECEPTORS

### A. Properties

As do other polypeptide hormones and growth factors, NGF exerts its responses through initial complexing with specific cell-surface receptors. Although assuming a variety of different forms and structures, molecules with this function invariably are transmembrane proteins that are organized into extracellular, transmembrane, and intracellular domains. The extracellular portion, which contains the ligand binding site, is usually heavily glycosylated and is often rich in disulfide bonds, whereas the intracellular domain often contains a defined signaling unit, most commonly a protein tyrosine kinase (Yarden and Ullrich, 1989). Occasionally the receptor/signaling unit is composed of more than one polypeptide, like, for example, hormones using adenyl cyclase effector systems. The transmembrane segment is usually an α helix. Receptor proteins comprise one to many such segments; the receptors for polypeptide growth factors tend to contain a single pass through the membrane.

The initial characterization of the NGF receptor involved the determination of equilibrium and kinetic binding constants using radiolabeled NGF tracers (Banerjee et al., 1973; Herrup and Shooter, 1973; Frazier et al., 1974). These studies defined both high and low affinity binding constants on a variety of responsive cells. In a definitive study, Sutter et al. (1979) identified a small number of high affinity receptors with association constants of $10^{-11}M$ and a much larger number of sites with binding constants approximately 2 orders of magnitude smaller. The differences were related entirely to the dissociation rate and gave rise to a nomenclature of "slow" and "fast" receptor types that corresponded to the high and low affinity forms, respectively. Importantly, it was shown that the high affinity form of the receptor solely is responsible for biologic activity. Similar studies with rat PC12 cells, described in detail in the next section, also showed the same two receptor types, albeit the difference between the high and low affinity receptors is somewhat smaller.

The first molecular characterization of the NGF receptor arose from hydrodynamic studies by Costrini et al. (1979). In this study, solubilized receptors from sympathetic neurons gave a measured molecular weight of 135,000. Covalent cross-linking of similar preparations showed equal amounts of labeled receptors with molecular masses of 143 and 112 kDa (Massague et al., 1981). Proteolytic mapping suggested that the lower molecular weight form was derived from the higher molecular weight form. Applications of the same technology to PC12 cells also indicated the presence of two receptor forms with molecular masses of 158 and 110 kDa (Massague et al., 1982). These molecular sizes were assumed to include one β polypeptide, suggesting that the masses of the high molecular weight receptors in sympathetic neurons and PC12 cells were 130 and 145 kDa, respectively.

Similar experiments using different cross-linking reagents identified a different class of NGF receptors on human melanoma cells (Grob et al., 1983). These studies demonstrated an 80-kDa species with no corresponding high molecular weight form. (A 200-kDa form was observed in some experiments but was attributed to dimerization of the 80-kDa species.) Hosang and Shooter (1985,1987), using the protocol of Massague et al. (1981), subsequently showed that the lower molecular weight species in PC12 cells apparently corresponded to the low affinity receptor (which was deemed to be analogous to the 80-kDa species of the human melanoma cells) and the high molecular weight form was judged to correspond to the high affinity form.

Molecular cloning of the low affinity receptor from PC12 cells and from human melanoma cells (Johnson et al., 1986; Radeke et al., 1987) demonstrated that these receptors corresponded to proteins of <50 kDa that showed a level of sequence similarity anticipated for rat and human species differences. Both molecules were deemed to be glycosylated with both $N$- and $O$-linked carbohydrate, accounting, at least in part, for the greater observed mass on SDS polyacrylamide gel electrophoresis. These molecules contain clearly recognizable extracellular, transmembrane, and intracellular domains. However, the last does not show any recognizable signaling entity, that is, it is devoid of a tyrosine kinase-like sequence. The absence of such a moiety, and the identification of the high molecular weight form as the high affinity receptor, led to the speculation that the NGF receptor formed a complex with an "accessory protein" that would account for the higher molecular weight (Chao, 1990). An alternative model suggested the existence of an entirely separate protein that corresponded to the high affinity form of the receptor, a view supported by the results of Kouchalakos and Bradshaw (1986) arising from direct receptor isolation experiments using sympathetic neurons.

The absence of a tyrosine kinase-like domain in the cloned receptors was further confounded by the observations that NGF, in PC12 cells, induced ligand-dependent protein tyrosine phosphorylations (Maher, 1988; Miyasaka et al., 1991) (see subsequent text). This evidence suggested that the "accessory protein" might be a tyrosine kinase, perhaps of a *src*-like nature, a view supported by the putative molecular mass of the protein, calculated from the difference between the high and low molecular size forms (approximately 60 kDa). However, these observations also were consistent with the existence of an independent entity that contained a tyrosine kinase. To some degree, the identification of the proto-oncogene *trk* as a ligand-binding entity in PC12 cells and in some responsive NGF neurons has resolved this issue. However, ambiguities, as described here, remain.

## B. *trk* as an NGF Receptor

The *trk* oncogene originally was identified from colon cancer cells as a protein containing a tropomyosin domain linked to a receptor-like tyrosine kinase (Martin-Zanca et al., 1986). The identification of the corresponding proto-oncogene (Martin-Zanca et al., 1989) revealed a structure consistent with a polypeptide growth factor receptor. Interestingly, its distribution was found to be limited to selective neurons of the peripheral nervous system (Martin-Zanca et al., 1990). The observation that *trk* autophosphorylation could be stimulated by NGF provided the first direct evidence that the ligand for *trk* was indeed NGF (Kaplan et al., 1991a). Subsequently, a number of groups provided evidence that (1) *trk* could bind NGF directly; (2) PC12 cells contained *trk*, and introduction of the *trk* proto-oncogene into PC12nnr5 cells, an NGF-unresponsive mutant of the PC12 cell line which fails to express *trk*, rescues their responsiveness to NGF (Loeb et al., 1991) and (3) *trk* transfected into oocytes could be stimulated by NGF (to cause germinal vessel breakdown), providing evidence that *trk* could function directly as a receptor for NGF (Kaplan et al., 1991b; Klein et al., 1991; Nebreda et al., 1991). As a result, two new models for NGF receptor structure have arisen (that really are modifications of the old models). In one case, *trk* acts alone to signal NGF responses, in the other, the low molecular size receptor (now abbreviated LNGFR) and *trk* form a complex (Bothwell, 1991). At present, these two models have not been reconciled. It is still controversial whether *trk* by itself binds NGF with high or low affinity. Klein et al. (1991) suggest that *trk* alone is the high affinity form of the NGF receptor in, for example, PC12 cells, a perspective that is consistent with binding studies and covalent cross-linking experiments (Hosang and Shooter, 1985,1987). In support of the view that *trk* acts alone, NGF is a potent mitogen in fibroblasts that express the *trk* proto-oncogene product and are devoid of LNGFR (Cordon-Carlo et al., 1991). Additionally, Weskamp and Reichardt (1991) have reported immunologic data that suggests the existence of a class of receptors that do not have LNGFR epitopes.

However, substantial evidence also supports the opposite view, that is, that LNGFR is involved in NGF signaling and that it and *trk* form a complex with NGF bound between them (Bothwell, 1991). First, direct measurements suggest that high affinity receptors can only be formed in the presence of both molecules (Hempstead et al., 1991). Second, NGF–EGF chimeric receptors suggest that PC12 cells do not respond unless the transmembrane and cytoplasmic domains of the LNGFR are present in these cells (Yan et al., 1991). Third, Berg et al. (1991) have shown that mutants of LNGFR in the cytoplasmic domain can prevent NGF-induced tyrosine phosphorylations. Independently, Eveleth and Bradshaw (unpublished observations) have raised antipeptide antibodies to LNGFR and have shown that these reagents block NGF responses in PC12 cells, in apparent contradiction to the findings of Weskamp and Reichardt (1991). Interestingly, if this model proves to be correct, it actually could be viewed as a variation on the "accessory protein" model, even though the molecular weight of *trk* is considerably different than that anticipated from the cross-linking experiments.

In related studies, it has been shown that *trk* B, a proto-oncogene closely related to *trk* that is also found in various nervous system cells, binds specifically to BDNF (Klein et al., 1991b; Soppet et al., 1991; Squinto et al., 1991). Previous binding experiments (Rodriguez-Tebar et al., 1990) have shown that BDNF and NGF (and now NT-3) are capable of binding the low molecular weight NGF receptor with essentially equal affinities. Thus, BDNF may also act solely through *trk* B or as a complex with LNGFR, analogous to the models for NGF.

Regardless of the resolution of the receptor model for NGF (and the other neurotrophins), it is clear that these entities can produce a tyrosine phosphorylation signal as their initial response to the binding of ligand. It is, at present, unclear whether this is the sole initial signal generated at the plasma membrane. It has been suggested that LNGFR contains a consensus G protein binding sequence at its carboxy terminus (in the cytoplasmic domain) (E. Shooter, personal communication) that could act either in concert with or independently of the tyrosine phosphorylation signals generated by *trk* (or related protein tyrosine kinase receptors). These and the various secondary responses of NGF, primarily in PC12 cells, that lead to the biologic response to this hormone, are described in the ensuing section.

## IV. MECHANISM OF ACTION OF NGF

### A. General Responses

#### 1. Use of PC12 cells as a model system

Because of the complexity of cell types that exists in the mammalian nervous system, studies on the mechanism of action of NGF hitherto have relied heavily on cell culture models, in particular the rat pheochromocytoma cell line PC12, which has been an invaluable tool for studying neuronal development. Much of the biochemical work on the mechanism of action of NGF (and more recently FGF) has used these cells, which are a tumor cell line derived from the adrenal medulla of the rat. The cells grow in culture as round chromaffin-like cells, with a cell-cycle time of ~48 hr; the addition of NGF or FGF induces the differentiation of these cells into ones that morphologically and biochemically resemble sympathetic neurons (Greene and Tischler, 1976,1982; Togari et al., 1985; Rydel and Greene, 1987; Schubert et al., 1987). Although the molecular mechanisms underlying these ligand-induced changes are not yet fully understood, the morphologic response, namely neurite outgrowth, forms the basis for a bioassay for differentiation, and has been a convenient marker for dissecting the mechanism of intracellular signaling by NGF in many recent studies.

Clearly, some differences between differentiated PC12 cells and central and peripheral neurons exist. PC12 cells are transformed cells that, no doubt, respond differently than normal neurons in many respects. They generally are grown in an environment that is quite distinct from that which exists for neurons in the nervous system, a milieu that normally would include the presence of other specific growth factors, as well as specific interactions with other cells. However, despite these differences, PC12 cells clearly have proven to be a useful in vitro system for dissecting the mechanistic pathways of the neurotrophic factors. A detailed description of these activations in PC12 cells, particularly as they relate to NGF, follows.

### 2. Biochemical changes induced by NGF

*a. Overview*  The induction of differentiation of PC12 cells by NGF is accompanied by a number of biochemical responses similar to those induced by mitogens in other systems. For example, the addition of NGF to PC12 cells has been reported to induce rapid alterations in phospholipid metabolism (Lakshmanan, 1978; Traynor et al., 1982; Contreras and Guroff, 1987; Chan et al., 1989; Altin and Bradshaw, 1990), changes in ion fluxes across the plasma membrane (Boonstra et al., 1983; Morgan and Curran, 1986; Pandiella-Alonso et al., 1986), the phosphorylation of specific proteins (Blenis and Erikson, 1986; Cremins et al., 1986; Landreth and Williams, 1987; Rowland et al., 1987; Mutoh et al., 1988; Vulliet et al., 1989), and alteration in the activity of numerous enzymes including ornithine decarboxylase (Greene and McGuire, 1978) and cyclic AMP- and $Ca^{2+}$/phospholipid-dependent protein kinases (Blenis and Erickson, 1986; Cremins et al., 1986; Hama et al., 1986,1987; Rowland et al., 1987). In addition, specific primary response genes such as c-*fos* (Curran and Morgan, 1985; Greenberg et al., 1985; Kruijer et al., 1985; Milbrandt, 1986) and c-*jun* (Wu et al., 1989), as well as a variety of other genes (Greene and Rein, 1977; Greene and Tischler, 1982; Kujubu et al., 1987; Leonard et al., 1987), also are induced, with variable timetables. Although many of these events may be the direct result of the action of NGF in these cells, not all may be necessary for committing the cells to undergo the morphologic transformation that subsequently occurs. The ensuing sections describe those intracellular signals generated during the early stages of the action of NGF that may be important in mediating the induction of neurite outgrowth.

*b. Protein synthesis*  The action of NGF is accompanied by the synthesis of many cytoplasmic and nuclear proteins including ornithine decarboxylase (Greene and McGuire, 1978; Fujii et al., 1982; Tiercy and Shooter, 1986), tubulin and neurofilament proteins that may be involved in neurite extension (Brugg and Matus, 1988; Drubin et al., 1988; Lindenbaum et al., 1988; Ikenaka et al., 1990), and proteins for binding $Ca^{2+}$ (Masiakowski and Shooter, 1988), as well as the synthesis and release of various neurotransmitters (see e.g., Greene and Tischler, 1982). The induction of neurite outgrowth in PC12 cells that have not been exposed to NGF previously occurs after a lag period of 12–24 hr, and is blocked by inhibitors of transcription. In contrast, the induction of neurite outgrowth following exposure to NGF is more rapid, and is unaffected by inhibitors of transcription (Burstein and Greene, 1978). This evidence has led to suggestions that ongoing protein translation is required for the neurite outgrowth response (Greene and Shooter, 1980), and is consistent with the observation that there are transcription-dependent increases in the rates of synthesis of several cytoplasmic and nuclear proteins within the first few hours of NGF treatment (Tiercy and Shooter, 1986). In enucleated PC12 cells, NGF can still initiate neurite outgrowth during the first 24 hr, but this induction is less efficient, and thereafter the neurites themselves are less stable than those formed by intact PC12 cells (Nichols et al., 1989). This evidence suggests that transcription is essential for continued neurite growth and neurite maintenance, but may be required only partly in the neurite initiation or induction process.

*c. Internalization of NGF* It has been established that the binding of NGF to neurons is followed by internalization of the NGF–receptor complex by receptor-mediated endocytosis. In responsive neurons, this is followed by retrograde transport of the complex to the cell body (Hendry et al., 1974; Stoeckel et al., 1974; Brunso-Bechtold and Hamburger, 1979; Taniuchi and Johnson, 1985; Johnson and Taniuchi, 1987). Since neurons can interact with NGF at the presynaptic membrane, this process may play an important role in intracellular signaling. Internalization of NGF occurs exclusively through the high affinity receptors (Bernd and Greene, 1984; Green et al., 1986). Moreover, internalization of the NGF–

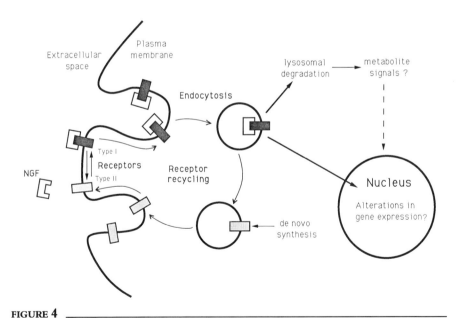

**FIGURE 4**

Internalization of the NGF–receptor complex in NGF-responsive cells. The binding of NGF to the high-affinity NGF receptor is followed by an internalization of the NGF–receptor complex. The internalized NGF is degraded in the lysosomes, a process that can generate potential intracellular signals. In addition, a significant proportion of the internalized NGF escapes degradation and either becomes associated with the nucleus (or other organelle) or recycles with the internalized receptors to re-emerge with the newly synthesized pool of receptors.

receptor complex appear to be associated with an interconversion of the low and the high affinity forms of the receptor (Eveleth and Bradshaw, 1988; Buxser et al., 1990). On internalization, NGF is degraded rapidly in the lysosomes, but a significant proportion either finds its way to other compartments or re-emerges at the cell surface (see Fig. 4; Eveleth and Bradshaw, 1988). That internalization may constitute an important part of the pathway for intracellular signaling by NGF is suggested by the observation that a mutant cell line PC12nnr5 binds NGF normally, but the cells neither internalize nor respond to NGF (Green et al., 1986; D. D. Eveleth and R. A. Bradshaw, unpublished observations). These cells lack the oncogene *trk,* which also binds to NGF (see previous text) (L. Greene, unpublished observations), indicating that internalization depends on the presence of this protein in PC12 cells. Whether or not a cell can bind and internalize NGF in the absence of the low molecular weight receptor is unknown (see previous text). Also consistent with this role for *trk* are the observations that introduction of the human LNGFR cDNA into a medulloblastoma cell line, which does not lead to the internalization of NGF, results in the expression of both high and low affinity NGF binding and induction of early transcriptional effects like *c-fos,* but not induction of differentiation (Pleasure et al., 1990).

The role of receptor-mediated internalization in the action of NGF is unclear. One possibility is that metabolites of NGF, produced by lysosomal degradation, may serve as intracellular signals for the action of NGF. Interestingly, it is reported that high affinity NGF binding sites do exist in the nucleus (Andres et al., 1977; Yankner and Shooter, 1979; Rakowicz-Szulczynska et al., 1988). One study reports that NGF binds to chromatin proteins in association with the DNA, and inhibits ribosomal RNA synthesis and cell proliferation in certain tumor cells (Rakowicz-Szulczynska and Koprowski, 1989). The significance of these findings is presently unknown, but they do indicate a potential contribution to the mechanism of intracellular signaling by NGF that escapes degradation and recycling. However, since the direct introduction of NGF into the free cytoplasm of PC12 cells by infusion techniques does not induce differentiation of the cells (Heumann et al., 1981), any intracellular role for NGF also apparently requires receptor binding and internalization.

## B. Initial Signal Transduction

### 1. Tyrosine phosphorylation

The finding that the *trk* protein, which contains a tyrosine kinase, is an NGF receptor suggests that an early event in the signaling cascade activated by NGF may involve protein tyrosine phosphorylation. Although at least one early study in PC12 cells suggested that NGF does not cause tyrosine phosphorylation of proteins (Boonstra et al., 1985), subsequent evidence indicated that the binding of NGF to PC12 cells was indeed associated with such events (Maher, 1988,1989; Schanen-King et al., 1991). It was found that protein tyrosine phosphorylation by NGF is induced rapidly, reaching a maximal level within 2–5 min of NGF addition, and then declines. Protein tyrosine phosphorylation by NGF is inhibited or blocked by MTA (5'-methylthioadenosine) and by the protein kinase inhibitor K-252a, two agents that inhibit morphologic differentiation of PC12 cells by NGF (Seeley et al., 1984; Hashimoto, 1988; Koizumi et al., 1988). However, MTA did not affect protein tyrosine phosphorylation by EGF (a known inducer of protein tyrosine phosphorylation), suggesting that the EGF- and NGF-stimulated protein tyrosine kinases have distinct properties (Maher, 1988). Inhibition of NGF-induced tyrosine phosphorylation of the phosphoproteins pp40, pp42, and MAP kinase by K-252a and by another related protein kinase inhibitor, staurosporine, also have been reported (Miyasaka et al., 1991). However, vanadate, an inhibitor of protein tyrosine phosphatase that also leads to increases in tyrosine phosphorylation of proteins, does not induce neurite outgrowth in PC12 cells, suggesting that protein tyrosine phosphorylation is not sufficient to induce morphologic differentiation (Maher, 1989). Interestingly, recent studies have shown that, in PC12 cells, NGF stimulates the phosphorylation of phospholipase C-γ on tyrosine and serine (Kim et al., 1991; Vetter et al., 1991), probably involving an association of the *trk* NGF receptor with phospholipase C-γ1 (Ohmichi et al., 1991). This provides at least one mechanism for stimulating phospholipid hydrolysis and activating other signaling pathways involving such entities as $Ca^{2+}$ flux and protein kinase C (see subsequent text).

Also consistent with a role for protein tyrosine phosphorylation in the action of NGF are the observations that transfection of the *src* oncogene (which encodes the cellular protein tyrosine kinase $pp60^{v\text{-}src}$) into PC12 cells results in morphologic differentiation in the absence of NGF (Alema et al., 1985). Differentiation of PC12 cells by *src* is not inhibited in K-252a, suggesting that if $pp60^{v\text{-}src}$ is involved in the action of NGF, the inhibitor must be acting at a site closer to the NGF receptor in the signaling pathway (Rausch et al., 1989). This notion is supported by the finding (Eveleth et al., 1989) that introduction of the *src* gene also can lead to the morphologic differentiation of PC12nnr5 cells, a variant of the PC12 cell line that binds NGF but lacks the *trk* receptor. That *src,* or a *src*-related tyrosine kinase, is involved in the action of NGF is further suggested by the finding that microinjection of a monoclonal antibody to $pp60^{v\text{-}src}$ into fused PC12 cells inhibits neurite induction by NGF (Kremer et al., 1989). The relationship of the *src*-like tyrosine kinase involved to the tyrosine kinase of *trk,* which is apparently associated with the NGF receptor, is not yet known. It is noteworthy, however, that $pp60^{v\text{-}src}$ has been shown to be localized in the growth cones of PC12 cells (Sobue and Kanda, 1988) and that protein tyrosine kinases have been detected in growth cone glycoproteins from rat brain (Cheng and Sahyoun, 1990). Moreover, $pp60^{c\text{-}src}$ is regulated by tyrosine kinase phosphorylation (Okada and Nakagawa, 1989), providing a mechanism for possible activation by *trk* (Hempstead et al., 1991; Kaplan et al., 1991a,b).

### 2. Cyclic nucleotides

Evidence of the effect of NGF on the levels of cyclic AMP, a potential second messenger in the action of NGF, has been contradictory. NGF-induced increases in the intracellular levels of cyclic AMP in PC12 cells and responsive neurons have been reported in some studies (Schubert et al., 1978) but not in others (Frazier et al., 1973; Hatanaka et al., 1978; Laasberg et al., 1988). It has also been reported that NGF causes a rapid activation of adenylate cyclase and a simultaneous decrease in the activity of the high affinity GTPase, with no significant change in the level of cyclic AMP (Golubeva et al., 1989). Interestingly, stimulation of PC12 cells and cultured sympathetic and sensory neurons with membrane-permeable analogs of cyclic AMP, or with agents known to increase intracellular levels of cyclic AMP, can mimic at least partially the effect of NGF (Frazier et al., 1973) and can act synergistically with NGF in inducing neurite outgrowth (Richter-Landsberg and Jastorff, 1986; Rydel and Greene, 1988a,b). Also, in PC12 cells, both cyclic AMP and NGF modulate the synthesis of a similar set of proteins (Garrels and Schubert, 1979; Mann et al., 1989) and elicit a similar pattern of protein phosphorylation and expression of primary response genes, including *c-fos* (Halegoua and Patrick, 1980; Wong et al., 1980; Curran and Morgan, 1985; Milbrandt, 1986).

Despite these similarities, significant differences between the effects of NGF and cyclic AMP in PC12 cells exist. These include an inability of cyclic AMP to induce "priming" (Gunning et al., 1981a,b; Greene et al., 1982; Heumann et al., 1983), or to stimulate the induction of a high molecular weight microtubule-associated protein (Greene et al., 1983). More importantly, the induction of neurite outgrowth by NGF is not inhibited by the presence of cyclic AMP antagonists (Braumann et al., 1986; Richter-Landsberg and Jastorff, 1986). Also, a PC12 cell-derived mutant deficient in cyclic AMP-dependent protein kinase A forms neurites in response to NGF but fails to respond to dibutyryl cyclic AMP (Glowacka and Wagner, 1990). These findings suggest that the induction of morphologic differentiation of PC12 cells by NGF does not require cyclic AMP; hence, the induction of neurite outgrowth by cyclic AMP occurs by a different mechanism than that of NGF. Nonetheless, increases in cyclic AMP may participate in the induction of certain responses by NGF. In PC12 cells, the stimulation of tyrosine hydroxylase (Cremins et al., 1986), GAP-43 expression (Costello et al., 1990), and down-regulation of $Ca^{2+}$/calmodulin-dependent kinase III (Brady et al., 1990) have been found to occur through a cyclic AMP-dependent mechanism (but see Doherty et al., 1987). It is reported that NGF increases cyclic GMP levels in PC12 cells (Laasberg et al., 1988); however, the significance of this effect is presently unknown.

In summary, NGF may act to stimulate cAMP production transiently but probably acts indirectly. cAMP clearly is not required to manifest the full NGF response, although it can act independently to mimic some of the actions of NGF.

## C. Signal Transmission

### 1. Overview

Although all the immediate responses that arise as a consequence of NGF binding to its functional receptor may not be known, a variety of responses clearly arise secondarily. These include, but are not necessarily limited to, phospholipid production and hydrolysis, activation of protein kinase C, and stimulation of *ras* and *src*. In concert or independently, these (and possibly other) entities transmit the signal transduced at the plasma membrane to the nuclear compartment, ultimately altering gene expression and the phenotypic profile (and response) of the cell. The effects of NGF on gene expression will be described in Section IVD. The following sections describe the signal transmission response to NGF.

### 2. Involvement of p21*ras*

The p21*ras* family of proteins plays a key role in signal transduction in many cells (reviewed by Barbacid, 1987; Nishimura and Sekiya, 1987; Santos and Nebreda, 1989). *ras* is a 21-kDa membrane-bound GTP-binding protein with the ability to hydrolyze GTP; it is active when bound to GTP and inactive when bound to GDP. That *ras* may play a role in the action of NGF was suggested by the finding that neuronal differentiation of PC12 cells, similar to that induced by NGF, can be induced by the introduction of *ras* transforming proteins into the cells (Bar-Sagi and Feramisco, 1985; Guerrero et al., 1986). Also the levels of Kirsten c-*ras* and Harvey c-*ras* mRNA and the activity of p21*ras* are increased following treatment of PC12 cells with NGF (Curran and Morgan, 1985; Qiu and Green, 1991). Other studies suggested that the levels of expression of a v-Ha-*ras* gene transfected into PC12 cells correlate well with the expression of neuron-associated properties (Sugimoto et al., 1988), although cells expressing the proto-oncogenic form of *ras* differentiate only if guanosine-5'-O-3-thiotriphosphate is co-introduced into these cells (Satoh et al., 1987). This evidence is consistent with the observation that introduction of *ras* proteins into cultured embryonic neurons mimics the biologic activity of neurotrophic factors in vitro (Borasio et al., 1989). Further support for the involvement of *ras* in the action of NGF comes from the observation that microinjection into fused PC12 cells of a monoclonal antibody Y13-259, which inhibits the function of all rat p21*ras* species by inhibiting GDP/GTP exchange and by blocking the interaction of *ras* with the GTPase-activating protein (GAP) (Furth et al., 1982; Hattori et al., 1987; Srivastava et al., 1989), inhibits the induction of neurite outgrowth by NGF and bFGF (Hagag et al., 1986; Altin et al., 1991a; Kremer et al., 1991).

Unlike the introduction of oncogenic *ras* proteins into PC12 cells (which induces differentiation), the microinjection of the protein into NIH 3T3 cells leads to their transformation (Stacey and Kung, 1984). On the other hand, microinjection of the Y13-259 antibody can induce a reversal of the transformed phenotype (Kung et al., 1986) and can inhibit the initiation of DNA synthesis by serum as well as the proliferation of certain types of tumor cells (Mulcahy et al., 1985; Stacey et al., 1987). These findings support a role for *ras* in the control of cellular proliferation. These two apparently contrasting effects of *ras* (inducing proliferation in some cell types and differentiation in others) will not be resolved until there is a more precise understanding of the role of the protein in signal transduction.

Despite its probable role in the action of NGF, it is not yet clear how *ras* is activated by NGF in cells. The GTPase activity of the protein is increased some

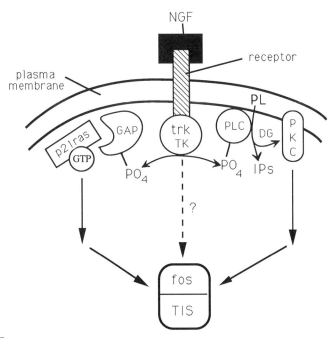

**FIGURE 5**

Signal transduction by the *trk* NGF receptor (or receptor complex). Signaling can occur through tyrosine phosphorylation and activation of phospholipase C (PLC), which can stimulate hydrolysis of phospholipids (PL) that generate inositol phosphates (IPs) and 1,2-diacylglycerol (DG), leading to the activation of protein kinase C (PKC). As well, signaling probably occurs through the tyrosine phosphorylation of the GTPase-activating protein GAP which regulates the GTPase activity of p21*ras*. The direct tyrosine phosphorylation of other signaling proteins by *trk* is also possible. These pathways are likely to play an important role in the induction of primary response genes like *fos* and other TPA-inducible sequences (TIS) by NGF.

50-fold in the presence of GAP, and there is evidence that GAP can be phosphorylated by tyrosine kinases and by protein kinase C (Molloy et al., 1989; Downward et al., 1990), suggesting a mechanism by which receptors can regulate *ras* function (see Fig. 5). Thus, activation of the *trk* tyrosine kinase following the binding of NGF in PC12 cells could lead to an activation of the GTPase activity of *ras*, presumably through a mechanism involving GAP, as has been proposed for other systems (Vogel et al., 1988; Molloy et al., 1989; Santos and Nebreda, 1989). Evidence supporting this view has been presented recently (Qiu and Green, 1991). It is also unclear whether *ras*, or perhaps even GAP, can transduce signals directly to the nucleus, or whether other effectors are involved (Cales et al., 1988; Garrett et al., 1989; McCormick 1989). In some systems it is known that the tyrosine phosphorylation of *ras* GAP is accompanied by an association of the complex with two unidentified molecules of 62 and 190 kDa that are themselves phosphorylated on tyrosine (reviewed by Cantley et al., 1991). In PC12 cells, *ras*-induced differentiation occurs through a mechanism that involves the proto-oncogenes *fos* and *jun* (Sassone-Corsi et al., 1989; Wu et al., 1989). Interestingly, one report suggests that in T-lymphocytes *ras* can be activated by protein kinase C (Downward et al., 1990). Other studies suggest that protein kinase C may be involved in the action of *ras* but in some cases, including NIH 3T3 cells (Maly et al., 1989) and Swiss 3T3 cells (Lloyd et al., 1989), p21*ras* may act through a protein kinase C-independent mechanism. Nonetheless, it seems well established that *ras*-transformed cells have elevated levels of 1,2-diacylglycerol (Lacal et al., 1987; Huang et al., 1988; Kato et al., 1988; Lacal, 1990); and there is evidence that *ras* may be involved in the regulation of phospholipase activity (Bar-Sagi and Feramisco, 1986; Bar-Sagi et al., 1988; Montgomery et al., 1988; but see Yu et al., 1988). *ras*-mediated induction of DNA synthesis in NIH 3T3 cells requires the activity of phospholipase C (Smith et al., 1990). Although a clear picture of *ras* involvement in the mechanism of NGF (or any other factor) has not emerged from these observations, it seems that *ras* plays a direct and probably indispensable role in the differentiation process (but see Qiu et al., 1991).

### 3. Stimulation of phospholipid metabolism by NGF

*a. Phosphoinositide hydrolysis* The action of many hormones that interact with receptors on the plasma membrane is associated with the hydrolysis of phosphatidylinositol 4,5-bisphosphate ($PIP_2$), which produces two second messengers, inositol 1,4,5-triphosphate ($IP_3$) and 1,2-diacylglyerol (DG) (see, e.g., Berridge, 1984,1987; Catt et al., 1991). $PIP_2$ is synthesized from phosphatidylinositol by the action of kinases (Carpenter and Cantley, 1990), and receptor-mediated hydrolysis is catalyzed by a phospholipase (phospholipase C) thought to be coupled to the receptor through a GTP-binding protein (G protein) (see Fig. 6) (Chambard et al., 1987; Boyer et al., 1989; Fain, 1990; Smrcka et al., 1991). Although the primary substrate for this mechanism generally is thought to be $PIP_2$, in some systems phosphatidylinositol 4-phosphate (PIP), phosphatidylinositol (PI), and phosphatidylinositol 3,4,5-triphosphate, as well as other phospholipids such as phosphatidylcholine, can be hydrolyzed by different enzymes (Whitman et al., 1988; Pelech and Vance, 1989; Billah and Anthes, 1990; Exton, 1990).

The 1,4,5 isomer of $IP_3$ is a messenger for the mobilization of $Ca^{2+}$ from intracellular stores (probably the endoplasmic reticulum), and can undergo phosphorylation or dephosphorylation to produce a host of other metabolites or potential second messengers (for reviews, see Berridge and Irvine, 1989; Bansal and Majerus, 1990; Downes and Macphee, 1990; Catt et al., 1991). There is evidence that $IP_3$ (and possibly $IP_4$) may be involved in stimulating $Ca^{2+}$ influx across the plasma membrane (for reviews, see Altin and Bygrave, 1988a; Boynton et al., 1990; Putney, 1990). The question of how the stimulation of $Ca^{2+}$ influx into the cell is regulated is still subject to speculation, although there is evidence that $Ca^{2+}$ enters the cell through a pathway that connects the extracellular space to the endoplasmic reticulum and that $Ca^{2+}$ then is released from the endoplasmic reticulum in a "pulsatile" or an oscillatory fashion (Berridge and Galione, 1988; Berridge, 1990; Peterson and Wakui, 1990). The other immediate product of phospholipid hydrolysis is DG, which activates a $Ca^{2+}$/phospholipid-dependent protein kinase, protein kinase C (Nishizuka, 1986,1988). $Ca^{2+}$ and protein kinase C are known to alter the activity of numerous enzymes and, in many instances, these changes can account for the induction of the physiologic response to the ligand (see Fig. 6).

The phospholipase C that hydrolyzes $PIP_2$ to $IP_3$ and DG can be any member of what seems to be a large family of enzymes; at least five immunologically distinct forms of the enzyme have been isolated from a variety of mammalian tissues (Rhee et al., 1989; Meldrum et al., 1991). These forms differ in their molecular mass, isoelectric point, pH optima, and calcium dependency, indicating the existence of multiple isozymes. Three phospholipase C isozymes of 150, 145, and 85 kDa recently purified from bovine brain were found to be specific for phosphatidylinositol (PI) and the polyphosphoinositides, but did not hydrolyze other phospholipids such as phosphatidylcholine or phosphatidylethanolamine. Hydrolysis of both PI and $PIP_2$ by the three enzymes is $Ca^{2+}$-dependent; however, at low $Ca^{2+}$ concentr-

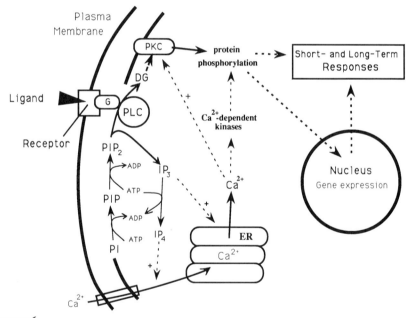

FIGURE 6

Receptor-mediated hydrolysis of phosphoinositides. Receptors linked to the hydrolysis of phosphatidylinositol-4,5-bisphosphate ($PIP_2$) are often coupled to a phospholipase C (PLC) through a GTP-binding protein (or G-protein). The hydrolysis generates two second messengers, inositol 1,4,5-trisphosphate ($IP_3$) and 1,2-diacylglycerol (DG). $IP_3$ is the message involved in the mobilization of $Ca^{2+}$ from the endoplasmic reticulum (ER), whereas DG and $Ca^{2+}$ activate protein kinase C (PKC). $Ca^{2+}$ (through the activation of $Ca^{2+}$-sensitive kinases) and PKC can lead to phosphorylation of and alteration in the activity of intracellular enzymes; in many instances this can account for the induction of the physiological response by the ligand.

ation, $PIP_2$ is the preferred substrate for all three enzymes (Rhee et al., 1989). It is unclear why there are so many phospholipase C isozymes; the stimulation of a given receptor may result in the activation of a specific phospholipase C isozyme, perhaps with a specificity for a particular pool of phospholipid (Martin, 1991). Interestingly, phospholipase C-γ is a substrate for the platelet-derived growth factor (PDGF) and EGF receptors (Margolis et al., 1989; Meisenhelder et al., 1989; Wahl et al., 1989; Anderson et al., 1990). It is reported that in PC12 cells, NGF stimulates the phosphorylation of phospholipase C-γ (Miyasaka et al., 1991; Ohmichi et al., 1991; Vetter et al., 1991), leading to its activation (Nishibe et al., 1990).

Stimulation of inositol phospholipid metabolism is apparently an early event in the signal transduction cascade activated by NGF, as suggested by the observations that NGF stimulates the incorporation of $^{32}PO_4$ into phosphatidylinositol in rat superior cervical ganglia (Lakshmanan, 1978) and into phosphatidic acid, and, to a lesser extent, into other phospholipids such as phosphatidylcholine in PC12 cells (Traynor et al., 1982). Subsequently, it was reported that NGF stimulates a rapid increase in inositol phospholipid turnover (Traynor, 1984; Contreras and Guroff, 1987). In contrast to these studies, Pandiella-Alonso et al. (1986) could not detect any change in the level of $IP_3$.

Altin and Bradshaw (1990) also found that NGF (and bFGF) induced significant production of inositol phosphates in PC12 cells; however, this increase was found to be much smaller than that induced by the muscarinic agonist carbachol, which is reported to stimulate phosphoinositide hydrolysis and the mobilization of intracellular $Ca^{2+}$ (Pozzan et al., 1986; Vicentini et al., 1986). Furthermore, the major phospholipid product of NGF induction is not $IP_3$ but inositol phosphate (which is ineffective in mobilizing intracellular $Ca^{2+}$), an observation that is consistent with other reports that $Ca^{2+}$ flux changes induced by NGF in PC12 cells are small (Pandiella-Alonso et al., 1986; Lazarovici et al., 1989) and often escape detection (Landreth et al., 1980; van Calker et al., 1989).

It is unclear whether the relatively small stimulation of phosphoinositide hydrolysis and mobilization of cellular $Ca^{2+}$ induced by NGF in PC12 cells is necessary for mediating neurite outgrowth. The phosphatidylinositol cycle is purported to be activated in spreading BHK cells (Breuer and Wagener, 1989) and has been implicated in the assembly of actin filaments and their anchoring to cellular membranes in other systems (Burn, 1988; Lassing and Lindberg, 1988; Goldschmidt-Clermont et al., 1990). On the other hand, $Ca^{2+}$ appears to be required for the attachment of PC12 cells to substrate (Schubert and Whitlock, 1977, Schubert et al., 1978), an event that is necessary for the induction of neurite regeneration by NGF (Greene and Tischler, 1982). It may be relevant that sustained increases in $Ca^{2+}$ are reported to be involved in the cholinergic differentiation of PC12 cells by glioma-conditioned medium and retinoic acid (Mizuno et al., 1989), that $Ca^{2+}$ is thought to control neurite outgrowth in cultured chick dorsal root ganglia by regulating the stability of actin filaments (Lankford and Letourneau, 1989), and that $Ca^{2+}$ has been implicated in the regulation of growth cone behavior (Kater and Mills, 1991). In addition, NGF is reported to stimulate $Ca^{2+}$ uptake by PC12 cells (Nikodijevic and Guroff, 1991) and to induce the expression of both $Na^+$ and $Ca^{2+}$ channels during their differentiation (Garber et al., 1989; Pollock et al., 1990; Rausch et al., 1990). Interestingly, the stimulation of $Ca^{2+}$ influx by the $Ca^{2+}$-channel agonist Bay K-8644 potentiates neurite outgrowth in response to NGF in superior cervical ganglia (Rogers and Hendry, 1990). It is possible, therefore, that alterations in phosphoinositide hydrolysis, including changes in $Ca^{2+}$ flux and the consequent $Ca^{2+}$-dependent phosphorylation of proteins, may participate in the action of NGF in the elaboration of neurites. However, because carbachol, an agent that induces a considerable increase in inositol phosphates and intracellular $Ca^{2+}$ (Pozzan et al., 1986; Vicentini et al., 1986; Altin and Bradshaw, 1990), does not itself induce neurite outgrowth in PC12 cells (Schubert et al., 1978), it is unlikely that an increase in intracellular $Ca^{2+}$ alone is sufficient to elicit the morphologic response to NGF. It seems that some other signal(s), generated from the binding of NGF to its receptor, also must be required for this action.

As part of the phosphoinositide hydrolysis cycle, NGF (and bFGF) also induces the production of DG, resulting in the activation of protein kinase C. Studies by Heasley and Johnson (1989) showed that NGF rapidly stimulates significant production of DG in PC12 cells. Similarly, Altin and Bradshaw (1990) found that the addition of NGF to PC12 cells increases the mass level of DG 2-fold within the first 1–2 min, rising to 2- to 3-fold by 15 min, and then decreasing slightly by 30 min. This increase was dependent on the presence of extracellular $Ca^{2+}$ and was inhibited by both phenylarsine oxide (an inhibitor of receptor-mediated endocytosis) and the methyltransferase inhibitor 5′-deoxy-5′-methylthioadenosine. However, as described next, DG also arises from the cleavage of other phospholipids. The effect of DG on protein kinase C is also described.

***b. Phosphatidylcholine*** Using $^{32}P$ NMR, Miccheli et al. (1989) reported that the action of NGF in PC12 cells is accompanied by rapid changes in the levels of phosphatidylcholine and phosphatidylethanolamine, suggesting that NGF also stimulates the metabolism of these phospholipids. The breakdown of phosphatidylcholine (the major phospholipid of eukaryotic cell plasma membrane, constituting ~20% of total lipids) has been reported to occur following receptor-mediated stimulation of a number of different cell types, including hepatocytes and T-lymphocytes (Bocckino et al., 1985; Saltiel et al., 1987; Polverino and Barritt, 1988; Rosoff et al., 1988). This breakdown can occur through the activation of phospholipase C, phospholipase D, or phospholipase $A_2$ (Bocckino et al., 1987; Balsinde et al., 1988; Shukla and Halenda, 1991). The mechanism by which these enzymes are activated is not entirely clear. In some cells, receptor activation of these enzymes may occur through a G protein, analogous to the mechanism for the hydrolysis of inositol phospholipids (Gilman, 1987; Casey and Gilman, 1988; Fain et al., 1988). In others, activation may occur secondary to phosphoinositide hydrolysis, resulting from the increase in intracellular $Ca^{2+}$ or the activation of protein kinase C.

Since each class of phospholipid has a characteristic fatty acid profile, it is possible to determine the sources of the DG that is produced by the action of NGF in PC12 cells by determining the molecular species profile of the fatty acids present in the phospholipids of these cells and comparing these with the DG species profile produced by stimulation with the growth factor (see e.g., Pessin and Raben, 1989). Such an analysis in PC12 cells indicated that total cellular phosphatidylinositol is composed primarily (70%) of sn-1-stearoyl-2-arachidonoyl glycerol (SAG), whereas total cellular phosphatidylcholine is composed of a number of different species, of which SAG represents only 5%. Further, a comparison of the nature of the fatty acid moieties on the DGs produced by the action of NGF and carbachol in PC12 cells showed that the DGs generated by 1 min after NGF (or bFGF) stimulation appear to arise primarily from hydrolysis of phosphoinositides, whereas after 1-min stimulation by carbachol there is an essentially equal contribution of the DG from the hydrolysis of phosphoinositides and phosphatidylcholine. In contrast, after 15 min of stimulation, the DGs in all cases are generated from a combination of phosphatidylinositol and phosphatidylcholine hydrolysis, with a predominance of the latter (Pessin et al., 1991). These findings suggest that NGF only transiently stimulates $PIP_2$ hydrolysis and has a much greater effect on phosphatidylcholine cleavage. The significance of this finding is not yet known.

*c. Glycosyl-phosphatidylinositol* Another mechanism by which DG can be produced in certain cells is by cleavage of glycosyl-phosphatidylinositol (GPI) by action of a GPI-specific phospholipase C or D, perhaps through an interaction involving a specific G protein as described for the action of insulin in cultured BC3H1 myocytes (Saltiel et al., 1987). The hydrolysis of free GPI on the cytoplasmic side of the plasma membrane would result in the production of DG and inositol phosphate glycan (another potential second messenger). A similar mechanism has been proposed for the release of GPI-anchored proteins such as heparin sulfate proteoglycan, alkaline phosphatase, or lipoprotein lipase (Saltiel and Cuatrecasas, 1988). Chan et al. (1989) have suggested that NGF also stimulates the hydrolysis of GPI in PC12 cells. NGF stimulates the production of labeled DG in cells treated with [$^3$H]myristate, but not in cells treated with [$^3$H]arachidonate, suggesting that DG is rich in myristic acid (Chan et al., 1989). It is noteworthy that the action of interleukin-1 in T-lymphocytes (Rosoff et al., 1988) and of insulin in BC3H1 cells (Saltiel et al., 1987) is similar to that of NGF in PC12 cells in that the effects are not accompanied by large increases in inositol 1,4,5-triphosphate or by changes in $Ca^{2+}$ flux. A recent study suggests that an NGF-induced hydrolysis of GPI in cochleovestibular ganglia is mediated by the low affinity NGF receptor and exerts a proliferative effect in these cells (Represa et al., 1991).

*d. Production of arachidonic acid and eicosanoids* The production of DG is often accompanied by the generation of arachidonic acid through the action of a DG-sensitive lipase (reviewed in Berridge, 1984). Arachidonic acid also can be generated directly through the action of phospholipase $A_2$ on membrane phospholipids such as phosphatidylinositol and phosphatidylcholine (Lapetina, 1982). 'ncreased levels of arachidonic acid in cells can be associated with the activation of the arachidonate cascade, which leads to the production of eicosanoids—both prostaglandins and leukotrines (Needleman et al., 1986). These metabolites of arachidonic acid can serve both as intracellular messengers (Bevan and Wood, 1987) and as intercellular hormone-like substances (see, e.g., Altin and Bygrave, 1988b).

Radioimmunoassay of PC12 cell-conditioned medium from PC12 cell lysates indicates that these cells produce and release increased amounts of prostaglandin (PGE) in response to NGF (DeGeorge et al., 1988). This eicosanoid response occurs within minutes of exposure to NGF, and is probably associated with the production of arachidonic acid through a mechanism independent of phosphoinositide turnover (Fink and Guroff, 1990). The possibility that eicosanoids play a role in the action of NGF is suggested by the observation that induction of neurite outgrowth in PC12 cells by NGF is inhibited by agents such as mepacrine and 4-bromophenylbromide (inhibitors of phospholipase $A_2$) and by some inhibitors of lipoxygenase metabolism (DeGeorge et al., 1988). It remains to be shown, however, that the observed effects of these inhibitors on NGF-induced neurite outgrowth is not simply due to the toxicity of the agents or some nonspecific effect. Interestingly, like DG, arachidonic acid is an activator of protein kinase C (Sekiguichi et al., 1988; Shearman et al., 1989; Das, 1991) and may play a role in this action. In addition, lipoxygenase products recently have been shown to inhibit the activity of *ras* GAP, thereby activating p21*ras* or switching it to the "on" state (Han et al., 1991). Thus, in addition to the possible tyrosine phosphorylation of GAP described earlier, this provides a mechanism for possible regulation of p21*ras* activity by NGF. However, since the addition of arachidonic acid directly to PC12 cells does not stimulate neurite outgrowth (J. G. Altin and R. A. Bradshaw, unpublished observations), a role for arachidonic acid metabolism in the induction of this response remains to be demonstrated.

### 4. Role of protein kinase C

Numerous reports indicate that the action of NGF in PC12 cells is associated with the activation of protein kinase C, a result consistent with the effects of NGF on phospholipid metabolism and the production of DG described earlier. Protein kinase C activity has been reported to increase 3- to 4-fold within the first hour of exposing PC12 cells to NGF (Cremins et al., 1986; Hama et al., 1986,1987). Despite the fact that protein kinase C has been shown to phosphorylate many proteins in PC12 cells, its role in the induction of neurite outgrowth by NGF has been the subject of much controversy. This is probably not only because of differences in the techniques used in different laboratories (including, perhaps, the particular source or phenotypic variability of the PC12 cells), but also perhaps because of a lack of specificity of some of the inhibitors such as sphingosine, K-252a, and staurosporine (see, e.g., Ruegg and Burgess et al., 1989; Lowe et al., 1990; Sohal and Cornell, 1990), and possibly because of the technique for downregulation of the enzyme (Cooper et al., 1989) that is currently in use.

Several studies support a role for protein kinase C in the action of NGF. Tetradecanoyl phorbol acetate (TPA), an activator of protein kinase C, potentiates neurite induction of NGF in PC12 cells (Burstein et al., 1982) and induces neurite outgrowth in central nervous system neurons (Moskal and Morrison, 1987). In agreement with these observations, sphingosine (a reversible inhibitor of protein kinase C) blocks NGF-induced neurite outgrowth in PC12 cells (Hall et al., 1988). Also, the methyltransferase inhibitor MTA, which has been reported to inhibit a number of responses induced by NGF including the induction of neurite outgrowth (Seeley et al., 1984; Acheson and Thoenen, 1987) and the rapid redistribution of F-actin (Paves et al., 1988,1990), also inhibits the production of DG by this growth factor (Altin and Bradshaw, 1990). However, neither TPA alone nor stimulation of the cells with carbachol (which stimulates the production of DG) can induce neurite outgrowth in PC12 cells. Further, the elimination of protein kinase C in PC12 cells by prolonged treatment with phorbol esters (Matthies et al., 1987) has been reported not to inhibit neurite induction by NGF (Reinhold and Neet, 1989). These latter findings suggest that the activation of protein kinase C is neither necessary nor sufficient for induction of morphologic differentiation of the cells by NGF.

Apart from the associated technical difficulties, another possible explanation for the seemingly conflicting results is the existence of multiple forms of protein kinase C. Protein kinase C originally was discovered as the receptor for phorbol esters (a class of tumor promoters). It is a serine/threonine protein kinase, and only a few of its substrates have been identified (Nishizuka, 1986; Huang, 1990). At least seven subspecies of protein kinase C have been identified in different tissues (Nishizuka, 1988; Parker et al., 1989). Four subspecies ($\alpha$, $\beta1$, $\beta2$, and $\gamma$) emerged from the initial screening of a variety of cDNA libraries. Protein kinase C isozymes I, II, and III belong to a family of closely related enzymes whose activities are dependent on $Ca^{2+}$, phospholipid and DG; they have an overall sequence homology greater than 70% and are encoded by separate genes (Nishizuka, 1988). These isoforms account for the majority of the $Ca^{2+}$/DG/phospholipid-dependent histone kinase activity in the extracts of mammalian brain. A novel protein kinase C activity, PKC$\epsilon$, has been purified from brain (Konno et al., 1989; Akita et al., 1990; Hagiwara et al., 1990). This enzyme can be activated in the absence of $Ca^{2+}$ and has been shown to have different substrate specificity than the other three types. Recent evidence suggests that the protein kinase C isozymes also may be regulated differently by different divalent cations such as $Ca^{2+}$, different phospholipids, different fatty acids such as arachidonic acid, by TPA, and by DGs of different fatty acid composition (Sekiguchi et al., 1988; Huang et al., 1989; Huang and Huang, 1990; Pelosin et al., 1990). These observations, albeit in different cell types and in vitro, suggest that specific protein kinase C isozymes may be activated by agonists under conditions in which a particular species of divalent cation such as $Ca^{2+}$ or a particular species of phospholipid, fatty acid, or DG may exist.

In an attempt to clarify more specifically the role of protein kinase C in neurite outgrowth, Altin et al. (1992) studied the effects of NGF on PC12 cells microinjected with a monoclonal antibody to protein kinase C (monoclonal antibody PKC-1.9). This antibody was raised against highly purified rat brain protein kinase

C, and is purported to inhibit strongly the activity of all protein kinase C isozymes in vitro and in vivo without affecting either the cAMP-dependent protein kinase or the $Ca^{2+}$/calmodulin-dependent kinase (Mochly-Rosen and Koshland, 1987,1988). Microinjection of the PKC-1.9 antibody significantly inhibited neurite outgrowth by NGF when scored 24 hr after microinjection of the antibody and exposure of the cells to the growth factor (Altin et al., 1992). In addition, introduction of the antibody was associated with a significant increase in cell loss, suggesting that protein kinase C may be important in mediating cell attachment or the survival-promoting effects of the growth factor, supporting the observation that TPA promotes the survival of chick sensory and sympathetic neurons in culture (Bhave et al., 1990). These findings suggest that protein kinase C is involved at least partly in mediating both neurite outgrowth and the health-promoting effects of NGF.

### 5. Protein phosphorylation

Protein phosphorylation is clearly a recurring theme in growth factor mechanisms. NGF induces the phosphorylation of many proteins in PC12 cells, including cytosolic and nuclear proteins, synapsin I (Romano et al., 1987), various neurofilament proteins (Sihag et al., 1988; Aletta et al., 1989), cytoskeletal proteins (Landreth and Williams, 1987), and ribosomal protein S6 (Hashimoto and Hagino, 1990; Mutoh et al., 1992). Mechanisms for the phosphorylation of such proteins by NGF could involve phosphorylation by the NGF receptor tyrosine kinase *trk*, by protein kinase C, and by the $Ca^{2+}$/calmodulin-dependent kinases and cyclic AMP-dependent protein kinases. In addition, it has been established that NGF leads to the activation of numerous other protein kinases including an NGF-sensitive kinase (Togari and Guroff, 1985), a serine protein kinase activity (referred to as S6 kinase) (Blenis and Erickson, 1986; Mutoh et al., 1988), N-kinase (Rowland et al., 1987; Volante et al., 1989; Rowland-Gagne and Greene, 1990; Volante and Greene, 1990a), a proline-directed serine/threonine protein kinase (Vulliet et al., 1989), and microtubule-associated protein kinases (Gotoh et al., 1990; Miyasaka et al., 1990; Sano et al., 1990) and the serine/threonine protein kinase B-raf (Oshima et al., 1991). The mechanisms by which these kinases are activated are unclear and often seem to involve more than one signal pathway.

Evidence suggesting that protein kinases may be involved in the action of NGF also follows from studies using various inhibitors. For instance, K-252a, a protein kinase inhibitor that inhibits neurite outgrowth and other responses in PC12 cells, also inhibits the NGF-induced phosphorylation of a number of proteins including tyrosine kinases, NSP100, MAP2/pp250 kinase, and kemptide kinase (Hashimoto, 1988; Koizumi et al., 1988; Lazarovici et al., 1989; Smith et al., 1989). A recent study suggests that K-252a inhibits the NGF-induced tyrosine phosphorylation and activation of the *trk* NGF receptor (Berg et al., 1992). A related inhibitor, staurosporine, inhibits neurite induction under some conditions, but is also capable of mimicking some actions of NGF in PC12 cells (Hashimoto and Hagino, 1989a,b; Tischler et al., 1990,1991). Interestingly, the induction of neurite outgrowth in PC12 cells and in the sympathetic and sensory neurons is inhibited by purine analogs such as

6-thioguanine; this inhibition correlates with an ability of the agents to inhibit protein kinase N (Volante et al., 1989; Greene et al., 1990). Purine analogs also inhibit the induction of ornithine decarboxylase under some conditions, but apparently have no effect on NGF-promoted neuronal survival or the induction of c-*fos* mRNA expression, suggesting that the agents may be specific only for some actions of NGF (Volante et al., 1989; Greene et al., 1990; Volante and Greene, 1990b).

The observation that most of the intracellular protein kinases reported to be activated by NGF are activated also by EGF, an agent that does not promote neurite outgrowth, casts some uncertainty on the role of these kinases in NGF action. They may, however, be necessary but not sufficient in these processes. Recently, there has been considerable interest in protein kinases that are specifically activated by NGF, since these kinases may play a significant role in the neurite induction process. A recent study has shown that a high molecular weight microtubule-associated protein kinase is phosphorylated and activated specifically by agents such as NGF and bFGF that induce neurite outgrowth, but not by EGF (Tsao et al., 1990). It remains to be shown whether a direct activation of this or similar kinase is sufficient to elicit the rapid morphologic changes, such as membrane ruffling (Bar-Sagi and Feramisco, 1985), the rapid reorganization of F-actin and microfilaments (Paves et al., 1988,1990), and the regulation of growth cone shape and morphology (Connolly et al., 1987), that are seen seconds after the addition of NGF to cells. Clearly, NGF can act on multiple signaling pathways that have both divergent and convergent elements. An analysis of the particular role of each of these elements, in both the short- and long-term actions of NGF, no doubt will be required to give a better understanding of the mechanism of action of the growth factor.

## D. Induction of Primary Response Genes

The various chemical changes that are transduced by NGF and lead to characteristic changes ultimately manifest themselves in the nucleus as alterations in gene expression. The first such responses are by genes that are affected by pre-existing transcription regulators and, therefore, can be detected in the absence of protein synthesis. These are identified as primary response or immediate early genes and often encode proteins that are themselves transcription factors (Herschman, 1989; see also Chapter 4).

One such set of genes was identified as TPA-inducible sequences (TIS) that were induced by mitogens in 3T3 fibroblasts in the presence of cycloheximide (Lim et al., 1987). Since TPA activates protein kinase C, an enzyme that can be activated physiologically by $Ca^{2+}$ and DG (Nishizuka, 1986), the TIS genes represent a set of genes that responds to growth factors that can potentially stimulate the hydrolysis of membrane phospholipids leading to the production of DG. That NGF, as well as TPA and elevated $K^+$ (which appear to potentiate the action of NGF in inducing neurite outgrowth), also induces TIS genes (with the exception of TIS10) in PC12 cells (Kujubu et al., 1987) is thus consistent with this hypothesis. These genes may, therefore, also be involved in the early signaling mechanism(s) of neurotrophic factors. However, EGF, which is mitogenic for PC12 cells and does not induce neurite outgrowth (Schubert et al., 1978), also induces TIS genes (Kujubu et al., 1987). Nonetheless, many of these genes have been identified independently as ones induced by NGF; they may be necessary but not sufficient for its response.

In a study on the induction of TIS genes in PC12 cells by agents that stimulate, or fail to stimulate, neurite outgrowth, Altin et al. (1991b) compared the expression of three of these genes—TIS1 (also designated NGFI-A), TIS8 (also designated NGFI-B), and TIS21—in PC12 cells following simulation by NGF, bFGF, carbachol, and TPA. They showed that, although the level of expression of TIS8 and TIS21 is similar for all three ligands (as well as for TPA), the induction of TIS1 is slight after stimulation by NGF and TPA, is only just detectable after stimulation by bFGF, but is strong after stimulation by carbachol (Altin et al., 1991b), suggesting that at least this gene is induced in a ligand-specific manner, a conclusion supported by similar observations for other ligands in other systems (Arenander et al., 1989; Herschman, 1989; Lim et al., 1989). Similar experiments with PC12nnr5 cells suggested that bFGF, but not NGF, can induce TIS gene expression in these cells, albeit the cells do not respond to this ligand either. As well as supporting previous findings that induction of the TIS genes is not sufficient to permit neurite outgrowth in PC12 cells (Kujubu et al., 1987), these observations provide evidence for differences in the signaling mechanisms of NGF and bFGF. In addition, the observed differences in the induction of the genes by NGF, bFGF, carbachol, and TPA may provide a mechanism of eliciting the ligand-specific differences in the response that is ultimately manifested in these cells. Since the promoter for the c-*fos* gene (and possibly those for other TIS genes) contains elements responsive to $Ca^{2+}$ (Sheng and Greenberg, 1990; Sheng et al., 1990), the observed differences in gene expression may be explained in terms of the differing abilities of the ligands NGF and carbachol to induce increases in intracellular $Ca^{2+}$ (see, e.g., Trejo and Brown, 1991).

Altin et al. (1991b) also showed that both NGF and bFGF still can induce neurite outgrowth and elicit a significant TIS gene response in PC12 cells in which protein kinase C had been down-regulated by prolonged treatment with TPA. Although the TIS gene response to NGF is inhibited about 59% by the down-regulation, the TIS gene responses to carbachol and TPA are effectively blocked under these conditions. This suggests that, unlike the induction of the genes by carbachol (and TPA), which occurs apparently through the activation of protein kinase C alone, induction of the genes by NGF (and probably bFGF) can occur both by a protein kinase C-dependent and by a protein kinase C-independent mechanism. This conclusion is consistent with other studies on NGF-dependent induction of genes in PC12 cells. For example, the early response genes d2 (TIS8) and d5 (Cho et al., 1989), the NGF-induced stimulation of the ornithine decarboxylase gene (Damon et al., 1990; Trotta et al., 1990), and the induction of c-*fos* (Sigmund et al., 1990) also can be induced in PC12 cells in a protein kinase C-independent fashion. In two of these studies, it was shown that morphologic differentiation of the cells still could occur in protein kinase C down-regulation

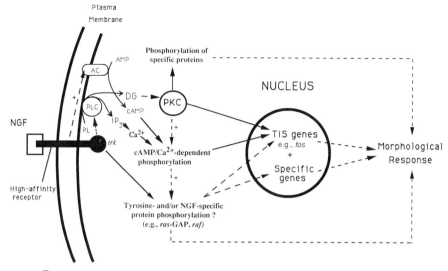

**FIGURE 7**

Pathways of immediate early gene induction by NGF in PC12 cells. NGF can elicit induction of immediate early genes through activation of both protein kinase C-dependent and protein kinase C-independent mechanisms. The tyrosine kinase associated with the high-affinity NGF receptor can lead to the phosphorylation and the activation of phospholipase C-γ (PLC) which can promote phosphoinositide hydrolysis (PL), the generation of 1,2-diacylglycerol (DG), and the activation of protein kinase C (PKC). As well, it is possible that the *trk* tyrosine kinase associated with the high-affinity NGF receptor can, by interacting with other proteins, such as *ras*-GAP and pp74$^{c\text{-}raf}$, transduce an intracellular signal by a mechanism independent of protein kinase C.

cells (Damon et al., 1990; Sigmund et al., 1990). Sigmund et al. (1990) also showed that c-*fos* responses to NGF, but not to bFGF, were reduced in protein kinase C down-regulated cells, suggesting that, as for the induction of TIS genes, a protein kinase C-dependent pathway for NGF may operate parallel to a protein kinase C-independent one.

The observation that NGF (and bFGF), unlike carbachol, can elicit the expression of TIS and other early response genes in PC12 cells in which protein kinase C is down-regulated by prolonged treatment with TPA suggests that both growth factors have other pathways for nuclear signaling available, perhaps involving the phosphorylation and activation of the serine/tyrosine kinase B-raf (Oshima et al., 1991) (see Fig. 7). An alternative explanation is that the growth factors can activate a protein kinase C isozyme that is not sensitive to down-regulation by TPA. However, differences in the pathways of nuclear signaling are likely to exist as a consequence of the basic differences in the structures of the receptors for NGF, bFGF, and carbachol. The receptor for carbachol does not have any obvious intrinsic tyrosine kinase activity, but it has been shown that the receptor for bFGF (Lee et al., 1989) and, more recently, the receptor for NGF (see previous text) do have a tyrosine kinase in their intracellular domain. These structures clearly activate other pathways and at least some of these can result in nuclear signaling as shown by the TIS gene responses, and by the finding that the *src* proteins can activate promoters including c-*fos* in other cells (Fujii et al., 1989). Such pathways are separately unavailable to carbachol, hence, the inhibition of carbachol-induced TIS gene responses in protein kinase C down-regulated cells.

One TIS gene of particular note is c-*fos*, which encodes a highly phosphorylated protein of 380 amino acids that localizes in the nucleus where it interacts with members of the *jun* family of proteins. The FOS and JUN proteins make up the activator protein-1 (AP-1) transcriptional complex which regulates the transcription of genes possessing AP-1 promoters. Thus, such genes are important in regulating the transcription of other genes important in the control of cell proliferation and possibly of cell differentiation (Chiu et al., 1988; Sheng and Greenberg, 1990).

Induction of c-*fos* may constitute a potential signal for differentiation of PC12 cells by NGF because its transcription is rapidly and transiently increased (50- to 100-fold) within minutes of adding NGF (or bFGF) to PC12 cells (Curran and Morgan, 1985; Milbrandt, 1986; Sheng et al., 1988). Moreover, increased AP-1 DNA-binding activity has been found in nuclear extracts of *ras*-infected PC12 cells, suggesting a possible involvement of FOS and JUN in the action of RAS (Sassone-Corsi et al., 1989; Wu et al., 1989). The possibility that FOS is an effector downstream of RAS is also supported by studies in other cell types, for example, NIH 3T3 and rat 208F fibroblast cells, in which the microinjection of the transforming *ras* protein causes a rapid induction of c-*fos* (Stacey et al., 1987; Riabowol et al., 1988a) and the expression of antisense c-*fos* mRNA reverses the transforming effects of the *ras* oncogene (Ledwith et al., 1990). Thus, the induction of c-*fos* may constitute an essential part of the signaling mechanism for the induction of neurite outgrowth by NGF although increased c-*fos* expression is not observed when neurite outgrowth is induced by transfection of a mouse N-*ras* gene into a subline of PC12 (Guerrero et al., 1986,1988), and overexpression of the c-*fos* and *fra-1* genes in PC12 cells apparently inhibits neurite induction by NGF (Ito et al., 1989,1990).

In recent studies, to test the involvement of c-*fos* in neurite induction by NGF, affinity-purified antibodies that immunoprecipitate FOS and other FOS-related and associated antigens including FRA1 and JUN were microinjected into PC12 cells. Surprisingly, these antibodies resulted in a significant potentiation of neurite induction during the first 24 hr of exposure to either NGF or bFGF (Altin et al., 1991c), suggesting that, in PC12 cells, induction of the *fos* protein is not necessary and, in fact, delays or interferes with neurite induction by NGF. In contrast to the effect of the anti-FOS antibodies on neurite induction by NGF, the microinjection of the antibodies into PC12 cells inhibited the stimulation of DNA synthesis by serum, suggesting that, in PC12 cells, FOS is involved in a pathway, the major effect of which culminates in cellular proliferation. These later findings are consistent with the observation that the antibodies also inhibit the stimulation of DNA synthesis by serum in rat fibroblasts (Riabowol et al., 1988b; Vosatka et al., 1989).

It is reported that an initial effect of NGF on PC12 cells is to stimulate a cycle of cell division (Greene and Tischler, 1982). It is possible, therefore, that the induction of c-*fos* by NGF constitutes a necessary step in that process. Recent work suggests that expression of the N-*ras* gene in a subline of PC12 cells inhibits the induction of c-*fos* by NGF and bFGF (Thomson et al., 1990). Since the pathway of neurite induction by NGF is thought to involve *ras* (Hagag et al., 1986; Altin et al., 1991a), an action of NGF may be to block or suppress c-*fos* expression some time after its initial induction. This action, as well as the elimination of any constitutively expressed c-*fos* by autoregulatory mechanisms, may be necessary for "priming" in these cells, which occurs in the early phase of the NGF and bFGF response before the stimulation of any neurite outgrowth (Greene and Tischler, 1982; Rydel and Greene, 1987). Evidence that induction of c-*fos* by NGF may play a role in other aspects of the NGF response in PC12 cells is suggested by the finding that induction of c-*fos* is required for expression of the tyrosine hydroxylase gene (Gizang-Ginsberg and Ziff, 1990), as well as by the observation that microinjection of anti-FOS antibodies inhibits cell attachment and cell survival (Altin et al., 1991c).

## V. SUMMARY AND CONCLUSIONS

It is clear that NGF, and probably all growth factors, can activate a cascade of events known to be involved in intracellular signaling and presumably involved in some aspect of the action of the growth factor on cells. However, there is still considerable uncertainty about which signaling pathways are important in the induction of morphologic differentiation induced by NGF. Some studies suggest that there may be considerable redundancy in the signaling mechanisms that are activated by this growth factor. Such redundancies, if they exist, may contribute to complicating the task of identifying the important signals. A number of areas require further investigation before a clear picture of the mechanism by which NGF elicits its biologic effects is available.

The nature of the NGF receptor involved in high affinity NGF binding, an event that is necessary for neurite induction, is still unresolved. The *trk* protein appears to be important in high affinity NGF binding but may require the presence of the LNGFR. Further, many of the important substrates of the *trk* tyrosine kinase remain to be identified. Certain growth factors, acting through tyrosine kinase receptors, can induce an association of molecules such as *ras* GAP, phosphatidylinositol 3-kinase, $pp60^{c-src}$, phospholipase C-$\gamma$, and $pp74^{c-raf}$ (Morrison et al., 1988; Whitman et al., 1988; Cantley et al., 1991; Koch et al., 1991). Such events appear to involve a recruitment of molecules from the cytoplasm to the plasma membrane, and may be important in many aspects of intracellular signaling including the regulation of ion transport and the generation of important second messengers, such as inositol phosphate glycan (Saltiel and Cuatrecasas, 1988) or particular isomers of inositol phosphate (Whitman et al., 1988; Bansal and Majerus, 1990). In addition, the recruitment of molecules from the cytoplasm may play a role in

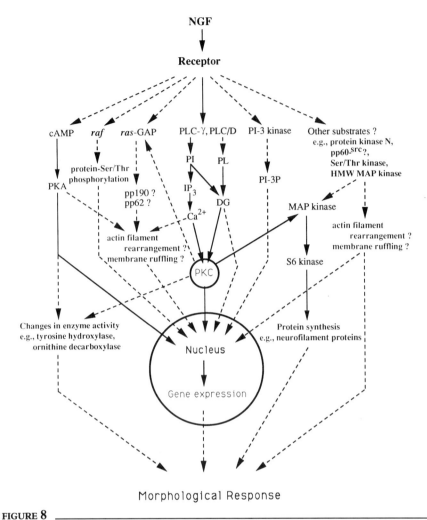

**FIGURE 8**
Summary of the signaling pathways activated by NGF and some of the responses that are induced.

regulating the clustering and internalization of receptors, ligands, and associated molecules, as well as the anchoring of certain cytoskeletal elements to the cell membranes (Cantley et al., 1991). The extent to which these events are involved in the action of NGF is still uncertain.

Clearly, there seems to be abundant potential for NGF to activate intracellular signaling mechanisms (Fig. 8). This is illustrated also by the seemingly different ways in which the growth factor can elicit the induction of primary response genes. An area of considerable importance is that of deciphering those signals that are crucial for induction of the ultimate neurotrophic response to the growth factor. It is possible that signals that do not seem to be important in the short term (<1 hr; e.g., phosphinositide hydrolysis and mobilization of cellular $Ca^{2+}$) may, in fact, be important in the long term (>6 hr), as suggested by the fact that in PC12 cells, NGF must be present for a minimum period of time (12–24 hr) before any subsequent morphologic differentiation can occur. Another area for future study is to establish which, if any, of these signals, including perhaps the induction of some of the TIS genes, are redundant. The possibility of redundancy is suggested, for example, by the finding that induction of c-*fos* is not required for the induction of neurite outgrowth by NGF. The real possibility exists, however, that such signals still may play an unidentified role in the response of the individual neurons that constitute the mammalian nervous system to the growth factor. This possibility will, no doubt, be the focus of much attention in the future.

The elucidation of the mechanism of NGF in PC12 cells, although certain to be informative, still will leave the all-important issues of NGF action in peripheral and central neurons unexplained. The spectrum of these responses is described in part in the two ensuing chapters, in which our understanding of the synthesis and expression of NGF and the other neurotrophins is discussed. Here, too, many questions remain unanswered.

## ACKNOWLEDGMENTS

The portions of this work emanating from R. A. Bradshaw's laboratory were supported in part by U.S. Public Health Service research grant NS19964 and American Cancer Society grant BE44. We are particularly grateful to Judy Brown for her assistance in preparing the manuscript. We also wish to thank Paul Isackson and Tom Blundell for sharing data with us prior to publication.

## REFERENCES

Acheson, A., and Thoenen, H. (1987). Both short- and long-term effects of nerve growth factor on tyrosine hydroxylase in calf adrenal chromaffin cells are blocked by S-adenosylhomocysteine hydrolase inhibitors. *J. Neurochem.* **48,** 1416–1424.

Akita, Y., Ohno, S., Konno, Y., Yano, A., and Suzuki, K. (1990). Expression and properties of two distinct classes of the phorbol ester receptor family, four conventional protein kinase C types, and a novel protein kinase C. *J. Biol. Chem.* **265,** 354–362.

Alema, S., Casalbore, P., Agostini, E., and Tato, F. (1985). Differentiation of PC12 phaeochromocytoma cells induced by v-*src* oncogene. *Nature (London)* **316,** 557–559.

Aletta, J. M., Shelanski, M. L., and Greene, L. A. (1989). Phosphorylation of the peripherin 58-kDa neuronal intermediate filament protein. Regulation by nerve growth factor and other agents. *J. Biol. Chem.* **264,** 4619–4627.

Altin, J. G., and Bradshaw, R. A. (1990). Production of 1,2-diacylglycerol in PC12 cells by nerve growth factor and basic fibroblast growth factor. *J. Neurochem.* **54,** 1666–1676.

Altin, J. G., and Bygrave, F. L. (1988a). Second messengers and the regulation of $Ca^{2+}$ fluxes by $Ca^{2+}$-mobilizing agonists in rat liver. *Biol. Rev.* **63,** 551–611.

Altin, J. G., and Bygrave, F. L. (1988b). Non-parenchymal cells as mediators of physiological responses in liver. *Mol. Cell. Biochem.* **83**, 3–14.

Altin, J. G., Wetts, R., and Bradshaw, R. A. (1991a). Microinjection of a p21ras antibody into PC12 cells inhibits neurite outgrowth induced by nerve growth factor and basic fibroblast growth factor. *Growth Factors* **4**, 145–155.

Altin, J. G., Kujubu, D. A., Raffioni, S., Eveleth, D. D., Herschman, H. R., and Bradshaw, R. A. (1991b). Differential induction of primary-response (TIS) genes in PC12 pheochromocytoma cells and the unresponsive variants PC12nnr5. *J. Biol. Chem.* **266**, 5401–5406.

Altin, J. G., Wetts, R., Riabowol, K. T., and Bradshaw, R. A. (1992). Testing the in vivo role of c-*fos* and protein kinase C on the induction of neurite outgrowth in PC12 cells. *Mol. Biol. Cell* **3**, 323–333.

Anderson, D., Koch, C. A., Grey, L., Ellis, C., Moran, M. F., and Pawson, T. (1990). Binding of SH2 domains of phospholipase $C_\gamma 1$, GAP, and Src to activated growth factor receptors. *Science* **250**, 979–981.

Andres, R. Y., Jeng, I., and Bradshaw, R. A. (1977). Nerve growth factor receptors: Identification of distinct classes in plasma membranes and nuclei of embryonic dorsal root neurons. *Proc. Natl. Acad. Sci. U.S.A.* **74**, 2785–2789.

Angeletti, R. H., and Bradshaw, R. A. (1971). The amino acid sequence of 2.5S mouse submaxillary gland nerve growth factor. *Proc. Natl. Acad. Sci. U.S.A.* **68**, 2417–2420.

Angeletti, R. H., Bradshaw, R. A., and Wade, R. D. (1971). Subunit structure and amino acid composition of mouse submaxillary gland nerve growth factor. *Biochemistry* **10**, 463–469.

Angeletti, R. H., Hermodsen, M. A., and Bradshaw, R. A. (1973). The amino acid sequence of mouse 2.5S nerve growth factor. II. Isolation and characterization of the thermolytic and peptic peptides and the complete covalent structure. *Biochemistry* **12**, 90–100.

Arenander, A. T., Lim, R. W., Varnum, B. C., Cole, R., de Vellis, J., and Herschman, H. R. (1989). TIS gene expression in cultured rat astrocytes: Multiple pathways of induction by mitogens. *J. Neurosci. Res.* **23**, 257–265.

Balsinde, J., Diez, E., and Mollinedo, F. (1988). Phosphatidylinositol-specific phospholipase D. A pathway for generation of a second messenger. *Biochem. Biophys. Res. Commun.* **154**, 502–508.

Banerjee, S. P., Snyder, S. H., Cuatrecasas, P., and Greene, L. A. (1973). Binding of nerve growth factor receptor in sympathetic ganglia. *Proc. Natl. Acad. Sci. U.S.A.* **70**, 2519–2523.

Bansal, V. S., and Majerus, P. W. (1990). Phosphatidylinositol-derived precursors and signals. *Ann. Rev. Cell Biol.* **6**, 41–67.

Barbacid, M. (1987). *ras* genes. *Ann. Rev. Biochem.* **56**, 779–827.

Barde, Y.-A., Edgar, D., and Thoenen, H. (1982). Purification of a new neurotrophic factor from mammalian brain. *EMBO J.* **1**, 549–553.

Bar-Sagi, D., and Feramisco, J. R. (1985). Microinjection of the *ras* oncogene protein into PC12 cells induces morphological differentiation. *Cell* **42**, 841–848.

Bar-Sagi, D., and Feramisco, J. R. (1986). Induction of membrane ruffling and fluid-phase pinocytosis in quiescent fibroblasts by *ras* proteins. *Science* **233**, 1061–1068.

Bar-Sagi, D., Suhan, J. P., McCormick, F., and Feramisco, J. R. (1988). Localization of phospholipase $A_2$ in normal and *ras*-transformed cells. *J. Cell Biol.* **106**, 1649–1658.

Berg, M. M., Steinberg, D. W., Hempstead, B. L., and Chao, M. V. (1991). The low-affinity p75 nerve growth factor (NGF) receptor mediates NGF-induced tyrosine phosphorylation. *Proc. Natl. Acad. Sci. U.S.A.* **88**, 7106–7110.

Berg, M. M., Sternberg, D. W., Parada, L. F., and Chao, M. V. (1992). K-252a inhibits nerve growth factor-induced *trk* proto-oncogene tyrosine phosphorylation and kinase activity. *J. Biol. Chem.* **267**, 13–16.

Bernd, P., and Greene, L. A. (1984). Association of [125]I-nerve growth factor with PC12 pheochromocytoma cells: Evidence for internalization via high-affinity receptors only and for long term regulation by nerve growth factor of both high and low-affinity receptors. *J. Biol. Chem.* **259**, 15509–15514.

Berridge, M. J. (1984). Inositol trisphosphate and diacylglycerol as second messengers. *Biochem. J.* **220**, 345–360.

Berridge, M. J. (1987). Inositol trisphosphate and diacylglycerol: Two interacting second messengers. *Ann. Rev. Biochem.* **56,** 159–193.

Berridge, M. J. (1990). Calcium oscillations. *J. Biol. Chem.* **265,** 9583–9586.

Berridge, M. J., and Galione, A. (1988). Cytosolic calcium oscillators. *FASEB J.* **2,** 3074–3082.

Berridge, M. J., and Irvine, R. F. (1989). Inositol phosphates and cell signalling. *Nature (London)* **341,** 197–205.

Bevan, S., and Wood, J. N. (1987). Arachidonic-acid metabolites as second messengers. *Nature (London)* **328,** 20.

Bhave, S. V., Malhotra, R. K., Wakade, T. D., and Wakade, A. R. (1990). Survival of chick embryonic sensory neurons in culture is supported by phorbol esters. *J. Neurochem.* **54,** 627–632.

Billah, M. M., and Anthes, J. C. (1990). The regulation and cellular functions of phosphatidylcholine hydrolysis. *Biochem. J.* **269,** 281–291.

Blaber, M., Isackson, P. J., and Bradshaw, R. A. (1987). A complete amino acid sequence for the major epidermal growth factor binding protein in the male mouse submandibular gland. *Biochemistry* **26,** 6742–6749.

Blaber, M., Isackson, P. J., and Bradshaw, R. A. (1990). The characterization of recombinant mouse glandular kallikreins from *E. coli*. *Proteins: Structure, Function Genetics* **7,** 280–290.

Blenis, J., and Erikson, R. L. (1986). Regulation of protein kinase activities in PC12 pheochromocytoma cells. *EMBO J.* **5,** 3441–3447.

Bocchini, V., and Angeletti, P. U. (1969). The nerve growth factor: Purification as a 30,000-molecular weight protein. *Proc. Natl. Acad. Sci. U.S.A.* **64,** 787–794.

Bocckino, S. B., Blackmore, P. F., and Exton, J. H. (1985). Stimulation of 1,2-diacylglycerol accumulation in hepatocytes by vasopressin, epinephrine, and angiotensin II. *J. Biol. Chem.* **260,** 14201–14207.

Bocckino, S. B., Blackmore, P. F., Wilson, P. B., and Exton, J. H. (1987). Phosphatidate accumulation in hormone-treated hepatocytes via a phospholipase D mechanism. *J. Biol. Chem.* **262,** 15309–15315.

Boonstra, J., Moolenaar, W. H., Harrison, P. H., Moed, P., van der Saag, P. T., and de Laat, S. W. (1983). Ionic responses and growth stimulation induced by nerve growth factor and epidermal growth factor in rat pheochromocytoma (PC12) cells. *J. Cell Biol.* **97,** 92–98.

Boonstra, J., van der Saag, P. T., Feijen, A., Bisschop, A., and de Laat, S. (1985). Epidermal growth factor, but not nerve growth factor, stimulates tyrosine-specific protein-kinase activity in pheochromocytoma (PC12) plasma membranes. *Biochimie* **67,** 1177–1183.

Borasio, G. D., John, J., Wittinghofer, A., Barde, Y.-A., Sendtner, M., and Heumann, R. (1989). *ras*p21 protein promotes survival and fiber outgrowth of cultured embryonic neurons. *Neuron* **2,** 1087–1096.

Bothwell, M. (1991). Keeping track of neurotrophin receptors. *Cell* **65,** 915–918.

Boyer, J. L., Hepler, J. R., and Harden, T. K. (1989). Hormone and growth factor receptor-mediated regulation of phospholipase C activity. *Trends Pharmacol. Sci.* **10,** 360–364.

Boynton, A. L., Dean, N. M., and Hill, T. D. (1990). Inositol 1,3,4,5-tetrakisphosphate and regulation of intracellular calcium. *Biochem. Pharmac.* **40,** 1933–1939.

Bradshaw, R. A. (1978). Nerve growth factor. *Ann. Rev. Biochem.* **47,** 191–216.

Brady, M. J., Nairn, A. C., Wagner, J. A., and Palfrey, H. C. (1990). Nerve growth factor-induced down-regulation of calmodulin-dependent protein kinase III in PC12 cells involves cyclic AMP-dependent protein kinase. *J. Neurochem.* **54,** 1034–1039.

Braumann, T., Jastorff, B., and Richter-Landsberg, C. (1986). Fate of cyclic nucleotides in PC12 cell cultures: Uptake, metabolism, and effects of metabolites on nerve growth factor-induced neurite outgrowth. *J. Neurochem.* **47,** 912–919.

Breuer, D., and Wagener, C. (1989). Activation of the phosphatidylinositol cycle in spreading cells. *Exp. Cell Res.* **182,** 659–663.

Brugg, B., and Matus, A. (1988). PC12 cells express juvenile microtubule-associated proteins during nerve growth factor-induced neurite outgrowth. *J. Cell Biol.* **107,** 643–650.

Brunso-Bechtold, J. K., and Hamburger, V. (1979). Retrograde transport of nerve growth factor in chicken embryo. *Proc. Natl. Acad. Sci. U.S.A.* **76,** 1494–1496.

Burn, P. (1988). Phosphatidylinositol cycle and its possible involvement in the regulation of cytoskeleton—membrane interactions. *J. Cell. Biochem.* **36**, 15–24.

Burstein, D. E., and Greene, L. A. (1978). Evidence for RNA synthesis-dependent and -independent pathways for stimulation of neurite outgrowth by nerve growth factor. *Proc. Natl. Acad. Sci. U.S.A.* **75**, 6059–6063.

Burstein, D. E., Blumberg, P. M., and Greene, L. A. (1982). Nerve growth factor-induced neuronal differentiation of PC12 pheochromocytoma cells: Lack of inhibition by a tumor promoter. *Brain Res.* **247**, 115–119.

Buxser, S., Decker, D., and Ruppel, P. (1990). Relationship among types of nerve growth factor receptors on PC12 cells. *J. Biol. Chem.* **265**, 12701–12710.

Cales, C., Hancock, J. F., Marshall, C. J., and Hall, A. (1988). The cytoplasmic protein GAP is implicated as the target for regulation by the *ras* gene product. *Nature (London)* **332**, 548–550.

Cantley, L. C., Auger, K. R., Carpenter, C., Duckworth, B., Graziani, A., Kapeller, R., and Soltoff, S. (1991). Oncogenes and signal transduction. *Cell* **64**, 281–302.

Carpenter, C. L., and Cantley, L. C. (1990). Phosphoinositide kinases. *Biochem.* **29**, 11147–11156.

Casey, P. J., and Gilman, A. G. (1988). G protein involvement in receptor-effector coupling. *J. Biol. Chem.* **263**, 2577–2580.

Catt, K. J., Hunyady, L., and Balla, T. (1991). Second messengers derived from inositol lipids. *J. Bioenergetics Biomembranes* **23**, 7–27.

Chambard, J. C., Paris, S., L'Allemain, G., and Pouyssegur, J. (1987). Two growth factor signalling pathways in fibroblasts distinguished by pertussis toxin. *Nature (London)* **326**, 800–803.

Chan, B. L., Chao, M. V., and Saltiel, A. R. (1989). Nerve growth factor stimulates the hydrolysis of glycosyl-phosphatidylinositol in PC-12 cells: A mechanism of protein kinase C regulation. *Proc. Natl. Acad. Sci. U.S.A.* **86**, 1756–1760.

Chao, M. V. (1990). Nerve growth factor. *In* "Peptide Growth Factors and Their Receptors" (M. B. Sporn and A. B. Roberts, eds.), Vol. II, pp. 135–165. Springer-Verlag, Berlin.

Cheng, N., and Sahyoun, N. (1990). Neuronal tyrosine phosphorylation in growth cone glycoproteins. *J. Biol. Chem.* **265**, 2417–2420.

Chiu, R., Boyle, W. J., Meek, J., Smeal, T., Hunter, T., and Karin, M. (1988). The c-*fos* protein interacts with c-*jun*/AP-1 to stimulate transcription of AP-1 responsive genes. *Cell* **54**, 541–552.

Cho, K.-O., Skarnes, W. C., Minsk, B., Palmieri, S., Jackson-Grusby, L., and Wagner, J. A. (1989). Nerve growth factor regulates gene expression by several distinct mechanisms. *Mol. Cell. Biol.* **9**, 135–143.

Cohen, S. (1959). Purification and metabolic effects of a nerve growth-promoting protein from snake venom. *J. Biol. Chem.* **234**, 1129–1137.

Cohen, S. (1960). Purification of a nerve-growth promoting protein from the mouse salivary gland and its neurocytotoxic antiserum. *Proc. Natl. Acad. Sci. U.S.A.* **46**, 302–311.

Connolly, J. L., Seeley, P. J., and Greene, L. A. (1987). Rapid regulation of neuronal growth cone shape and surface morphology by nerve growth factor. *Neurochem. Res.* **12**, 861–868.

Contreras, M. L., and Guroff, G. (1987). Calcium-dependent nerve growth factor-stimulated hydrolysis of phosphoinositides in PC12 cells. *J. Neurochem.* **48**, 1466–1472.

Cooper, D. R., Watson, J. E., Acevedo-Duncan, M., Pollet, R. J., Standaert, M. L., and Farese, R. V. (1989). Retention of specific protein kinase C isozymes following chronic phorbol ester treatment in BC3H-1 myocytes. *Biochem. Biophys. Res. Commun.* **161**, 327–334.

Cordon-Carlo, C., Tapley, P., Jing, S., Nanduri, V., O'Rourke, E., Lamballe, F., Kovary, K., Klein, R., Jones, K. R., Reichardt, L. F., and Barbacid, M. (1991). The *trk* tyrosine protein kinase mediates the mitogenic properties of nerve growth factor and neurotrophin 3. *Cell* **66**, 173–183.

Costello, B., Meymandi, A., and Freeman, J. A. (1990). Factors influencing GAP-43 gene expression in PC12 pheochromocytoma cells. *J. Neurosci.* **10**, 1398–1406.

Costrini, N. V., Kogan, M., Kukreja, K., and Bradshaw, R. A. (1979). Physical properties of the detergent-extracted nerve growth factor receptor of sympathetic ganglia. *J. Biol. Chem.* **254**, 11242–11246.

Cremins, J., Wagner, J. A., and Halegoua, S. (1986). Nerve growth factor action is mediated by cyclic AMP- and $Ca^{2+}$/phospholipid-dependent protein kinases. *J. Cell. Biol.* **103**, 887–893.

Curran, T., and Morgan, J. I. (1985). Superinduction of c-*fos* by nerve growth factor in the presence of peripherally active benzodiazepines. *Science* **229**, 1265–1268.

Damon, D. H., D'Amore, P. A., and Wagner, J. A. (1990). Nerve growth factor and fibroblast growth factor regulate neurite outgrowth and gene expression in PC12 cells via both protein kinase C- and cAMP-independent mechanisms. *J. Cell Biol.* **110**, 1333–1339.

Das, U. N. (1991). Arachidonic acid as a mediator of some of the actions of phorbolmyristate acetate, a tumor promotor and inducer of differentiation. *Prostaglandins Leukotrienes Essential Fatty Acids* **42**, 241–244.

DeGeorge, J. J., Walenga, R., and Carbonetto, S. (1988). Nerve growth factor rapidly stimulates arachidonate metabolism in PC12 cells: Potential involvement in nerve fiber growth. *J. Neurosci. Res.* **21**, 323–332.

Doherty, P., Mann, D. A., and Walsh, F. S. (1987). Cholera toxin and dibutyryl cyclic AMP inhibit the expression of neurofilament protein induced by nerve growth factor in cultures of naive and primed PC12 cells. *J. Neurochem.* **49**, 1676–1687.

Downes, C. P., and Macphee, C. H. (1990). *myo*-Inositol metabolites as cellular signals. *Eur. J. Biochem.* **193**, 1–18.

Downward, J., Graves, J. D., Warne, P. H., Rayter, S., and Cantrell, D. A. (1990). Stimulation of p21*ras* upon T-cell activation. *Nature (London)* **346**, 719–723.

Drubin, D., Kobayashi, S., Kellogg, D., and Kirschner, M. (1988). Regulation of microtubule protein levels during cellular morphogenesis in nerve growth factor-treated PC12 cells. *J. Cell Biol.* **106**, 1583–1591.

Dunbar, J. C., Tregear, G. W., and Bradshaw, R. A. (1984). Histidine residue modification inhibits binding of murine β nerve growth factor to its receptor. *J. Prot. Chem.* **3**, 349–356.

Ebendal, T., Larhammar, D., and Persson, H. (1986). Structure and expression of the chicken β-nerve growth factor. *EMBO J.* **5**, 1483–1487.

Ernfors, P., Ibanez, C. F., Ebendal, T., Olson, L., and Persson, H. (1990). Molecular cloning and neurotrophic activities of a protein with structural similarities to β-nerve growth factor: Developmental and topographical expression in the brain. *Proc. Natl. Acad. Sci. U.S.A.* **87**, 5454–5458.

Eveleth, D. D., and Bradshaw, R. A. (1988). Internalization and cycling of nerve growth factor in PC12 cells: Interconversion of type II (fast) and type I (slow) nerve growth factor receptors. *Neuron* **1**, 929–936.

Eveleth, D. D., Hanecak, R., Fox, G. M., Fan, H., and Bradshaw, R. A. (1989). v-*src* genes stimulate neurite outgrowth in pheochromocytoma (PC12) variants unresponsive to neurotrophic factors. *J. Neurosci. Res.* **24**, 67–71.

Exton, J. H. (1990). Signaling through phosphatidylcholine breakdown. *J. Biol. Chem.* **265**, 1–4.

Fahnestock, M., Woo, J. E., Lopez, G. A., Snow, J., Walz, D. A., Arici, M. J., and Mobley, W. C. (1991). βNGF endopeptidase: Structure and activity of a kallikrein encoded by the gene *m*GK-22. *Biochemistry* **30**, 3443, 3450.

Fain, J. N. (1990). Regulation of phosphoinositide-specific phospholipase C. *Biochim. Biophys. Acta* **1053**, 81–88.

Fain, J. N., Wallace, M. A., and Wojcikiewicz, R. J. H. (1988). Evidence for involvement of guanine nucleotide-binding regulatory proteins in the activation of phospholipases by hormones. *FASEB J.* **2**, 2569–2574.

Fink, D. W., Jr., and Guroff, G. (1990). Nerve growth factor stimulation of arachidonic acid release from PC12 cells: Independence from phosphoinositide turnover. *J. Neurochem.* **55**, 1716–1726.

Frazier, W. A., Angeletti, R. H., and Bradshaw, R. A. (1972). Nerve growth factor and insulin. *Science* **176**, 482–488.

Frazier, W. A., Angeletti, R. H., Sherman, R., and Bradshaw, R. A. (1973). The topography of mouse 2.5S nerve growth factor: The reactivity of tyrosine and trytophan. *Biochemistry* **12**, 3281–3293.

Frazier, W. A., Boyd, L. F., Pulliam, M. W., Szutowiez, A., and Bradshaw, R. A., (1974). Properties and

specificity of binding sites for $^{125}$I-nerve growth factor in embryonic heart and brain. *J. Biol. Chem.* **249,** 5918–5923.
Fujii, D. K., Massoglia, S. L., Savion, N. and Gospodarowicz, D. (1982). Neurite outgrowth and protein synthesis by PC12 cells as a function of substratum and nerve growth factor. *J. Neuroscience* **2,** 1157–1175.
Fujii, M., Shalloway, D., and Verma, I. M. (1989). Gene regulation by tyrosine kinases: src protein activates various promoters, including c-*fos. Mol. Cell. Biol.* **9,** 2493–2499.
Furth, M. E., Davis, L. J., Fleurdelys, B., and Scolnick, E. M. (1982). Monoclonal antibodies to the p21 products of the transforming gene of Harvey murine sarcoma virus and of the cellular *ras* gene family. *J. Virol.* **43,** 294–304.
Garber, S. S., Hoshi, T., and Aldrich, R. W. (1989). Regulation of ionic currents in pheochromocytoma cells by nerve growth factor and dexamethasone. *J. Neurosci.* **9,** 3976–3987.
Garrels, J. I., and Schubert, D. (1979). Modulation of protein synthesis by nerve growth factor. *J. Biol. Chem.* **254,** 7978–7985.
Garrett, M. D., Self, A. J., van Oers, C., and Hall, A. (1989). Identification of distinct cytoplasmic targets for *ras*/R-*ras* and *rho* regulatory proteins. *J. Biol. Chem.* **264,** 10–13.
Gilman, A. G. (1987). G Proteins: Transducers of receptor-generated signals. *Ann. Rev. Biochem.* **56,** 615–649.
Gizang-Ginsberg, E., and Ziff, E. B. (1990). Nerve growth factor regulates tyrosine hydroxylase gene transcription through a nucleoprotein complex that contains c-*fos. Genes Dev.* **4,** 477–491.
Glowacka, D., and Wagner, J. A. (1990). Role of the cAMP-dependent protein kinase and protein kinase C in regulating the morphological differentiation of PC12 cells. *J. Neurosci. Res.* **25,** 453–462.
Goldschmidt-Clermont, P. J., Machesky, L. M., Baldassare, J. J., and Pollard, T. D. (1990). The actin-binding protein profilin binds to $PIP_2$ and inhibits its hydrolysis by phospholipase C. *Science* **247,** 1575–1578.
Golubeva, E. E., Posypanova, G. A., Kondratyev, A. D., Melnik, E. I., and Severin, E. S. (1989). The influence of nerve growth factor on the activities of adenylate cyclase and high-affinity GTPase in pheochromocytoma PC12 cells. *FEBS Lett.* **247,** 232–234.
Gotoh, Y., Nishida, E., Yamashita, T., Hoshi, M., Kawakami, M., and Sakai, H. (1990). Microtubule-associated-protein (MAP) kinase activated by nerve growth factor and epidermal growth factor in PC12 cells. Identity with the mitogen-activated MAP kinase of fibroblastic cells. *Eur. J. Biochem.* **193,** 661–669.
Green, S. H., Rydel, R. E., Connolly, J. L., and Greene, L. A. (1986). PC12 cell mutants that possess low- but not high-affinity nerve growth factor receptors neither respond to nor internalize nerve growth factor. *J. Cell. Biol.* **102,** 830–843.
Greenberg, M. E., Greene, L. A., and Ziff, E. B. (1985). Nerve growth factor and epidermal growth factor induce rapid transient changes in proto-oncogene transcription in PC12 cells. *J. Biol. Chem.* **260,** 14101–14110.
Greene, L. A., and McGuire, J. C. (1978). Induction of ornithine decarboxylase by nerve growth factor dissociated from effects on survival and neurite outgrowth. *Nature (London)* **276,** 191–194.
Greene, L. A., and Rein, G. (1977). Release, storage, and uptake of catecholamines by a clonal cell line of nerve growth factor (NGF) responsive pheochromocytoma cells. *Brain Res.* **129,** 247–263.
Greene, L. A., and Shooter, E. M. (1980). The nerve growth factor: Biochemistry, synthesis, and mechanism of action. *Ann. Rev. Neurosci.* **3,** 353–402.
Greene, L. A., and Tischler, A. S. (1976). Establishment of a noradrenergic clonal line of rat adrenal pheochromocytoma cells which respond to nerve growth factor. *Proc. Natl. Acad. Sci. U.S.A.* **73,** 2424–2428.
Greene, L. A., and Tischler, A. S. (1982). PC12 pheochromocytoma cultures in neurobiological research. *Adv. Cell. Neurobiol.* **3,** 373–414.
Greene, L. A., Shooter, E. M., and Varon, S. (1969). Subunit interaction and enzymatic activity of mouse nerve growth factor. *Biochemistry* **8,** 3735–3741.
Grob, P. M., Berlot, C. H., and Bothwell, M. (1983). Affinity labeling and partial purification of nerve

growth factor receptors from rat pehochromocytoma and human melanoma cells. *Proc. Natl. Acad. Sci. U.S.A.* **80,** 6819–6823.

Guerrero, I., Wong, H., Pellicer, A., and Burstein, D. E. (1986). Activated N-ras gene induces neuronal differentiation of PC12 rat pheochromocytoma cells. *J. Cell. Physiol.* **129,** 71–76.

Guerrero, I., Pellicer, A., and Burstein, D. E. (1988). Dissociation of c-*fos* from ODC expression and neuronal differentiation in a PC12 subline stably transfected with an inducible N-ras oncogene. *Biochem. Biophys. Res. Commun.* **150,** 1185–1192.

Gunning, P. W., Landreth, G. W., Bothwell, M. A., and Shooter, E. M. (1981a). Differential and synergistic actions of nerve growth factor and cyclic AMP in PC12 cells. *J. Cell Biol.* **89,** 240–245.

Gunning, P. W., Letourneau, P. C., Landreth, G. E., and Shooter, E. M. (1981b). The action of nerve growth factor and dibutyryl adenosine cyclic 3'–5'-monophosphate on rat pheochromocytoma reveals distinct stages in the mechanisms underlying neurite outgrowth. *J. Neurosci.* **1,** 1085–1095.

Hagag, N., Halegoua, S., and Viola, M. (1986). Inhibition of growth factor-induced differentiation of PC12 cells by microinjection of antibody to *ras* p21. *Nature (London)* **319,** 680–682.

Hagiwara, M., Uchida, C., Usuda, N., Nagata, T., and Hidaka, H. (1990). ε-Related protein kinase C in nuclei of nerve cells. *Biochem. Biophys. Res. Commun.* **168,** 161–168.

Halegoua, S., and Patrick, J. (1980). Nerve growth factor mediates phosphorylation of specific proteins. *Cell* **22,** 571–581.

Halegoua, S., Armstrong, R. C., and Kreme, N. E. (1991). Dissecting the mode of action of a neuronal growth factor. *Curr. Topics Microbiol. Immunol.* **165,** 119–170.

Hall, F. L., Fernyhough, P., Ishii, D. N., and Vulliet, P. R. (1988). Suppression of nerve growth factor-directed neurite outgrowth in PC12 cells by sphingosine, an inhibitor of protein kinase C. *J. Biol. Chem.* **263,** 4460–4466.

Hallbook, F., Ibanez, C. F., and Persson, H. (1991). Evolutionary studies of the nerve growth factor family reveal a novel member abundantly expressed in *Xenopus* ovary. *Neuron* **6,** 845–858.

Hama, T., Huang, K. P., and Guroff, G. (1986). Protein kinase C as a component of a nerve growth factor-sensitive phosphorylation system in PC12 cells. *Proc. Natl. Acad. Sci. U.S.A.* **83,** 2353–2357.

Hama, T., Huang, F. L., Yoshida, Y., Huang, K-P., and Guroff, G. (1987). Nerve growth factor-induced changes in protein kinase C levels and activity in PC12 cells. *Soc. Neurosci. Abstr.* **13(1),** 552.

Han, J-W., McCormick, F., and Macara, I. G. (19910. Regulation of Ras-GAP and the neurofibromatosis-1 gene product by eicosanoids. *Science* **252,** 576–579.

Hashimoto, S. (1988). K-252a, a potent protein kinase inhibitor, blocks nerve growth factor-induced neurite outgrowth and changes in the phosphorylation of proteins in PC12h cells. *J. Cell Biol.* **107,** 1531–1539.

Hashimoto, S., and Hagino, A. (1989a). Blockage of nerve growth factor action in PC12h cells by staurosporine, a potent protein kinase inhibitor. *J. Neurochem.* **53,** 1675–1685.

Hashimoto, S., and Hagino, A. (1989b). Staurosporine-induced neurite outgrowth in PC12h cells. *Exp. Cell Res.* **184,** 351–359.

Hashimoto, S., and Hagino, A. (1990). Nerve growth factor-induced transient increase in the phosphorylation of ribosomal protein S6 mediated through a mechanism independent of cyclic AMP-dependent protein kinase and protein kinase C. *J. Neurochem.* **55,** 970–980.

Hatanaka, H., Otten, U., and Thoenen, H. (1978). Nerve growth factor-mediated selective induction of ornithine decarboxylase in rat pheochromocytoma: A cyclic AMP-independent process. *FEBS Lett.* **92,** 313–316.

Hattori, S., Clanton, D. J., Satoh, T., Nakamura, S., Kaziro, Y., Kawakita, M., and Shih, T. Y. (1987). Neutralizing monoclonal antibody against *ras* oncogene product p21 which impairs guanine nucleotide exchange. *Mol. Cell. Biol.* **7,** 1999–2002.

Heasley, L. E., and Johnson, G. L. (1989). Regulation of protein kinase C by nerve growth factor, epidermal growth factor, and phorbol esters in PC12 pheochromocytoma cells. *J. Biol. Chem.* **264,** 8646–8652.

Hempstead, B. L., Martin-Zanca, D., Kaplan, D. R., Parada, L. F., and Chao, M. V. (1991). High-affinity NGF binding requires coexpression of the *trk* proto-oncogene and the low-affinity NGF receptor. *Nature (London)* **350,** 678–683.

Hendry, I. A., Stockel, K., Thoenen, H., and Iversen, L. L. (1974). The retrograde axonal transport of nerve growth factor. *Brain Res.* **68,** 103–121.

Herrup, K., and Shooter, E. M. (1973). Properties of the β nerve growth factor receptor of avian dorsal root ganglia. *Proc. Natl. Acad. Sci. U.S.A.* **70,** 3884–3888.

Herschman, H. R. (1989). Extracellular signals, transcriptional responses and cellular specificity. *Trends Biochem. Sci.* **14,** 455–458.

Heumann, R., Schwab, M., and Thoenen, H. (1981). A second messenger required for nerve growth factor biological activity? *Nature (London)* **282,** 838–840.

Heumann, R., Kachel, V., and Thoenen, H. (1983). Relationship between NGF-mediated volume increase and "priming effect" in fast and slow reacting clones of PC12 pheochromocytoma cells. *Exp. Cell Res.* **145,** 179–190.

Hohn, A., Leibrock, J., Bailey, K., and Barde, Y.-A. (1990). Identification and characterization of a novel member of the nerve growth factor/brain-derived neurotrophic factor family. *Nature (London)* **344,** 339–341.

Hosang, M., and Shooter, E. M. (1985). Molecular characteristics of nerve growth factor receptors on PC12 cells. *J. Biol. Chem.* **260,** 655–662.

Hosang, M., and Shooter, E. M. (1987). The internalization of nerve growth factor by high affinity receptors on pheochromocytoma PC12 cells. *EMBO J.* **6,** 1197–1201.

Huang, F. L., Yoshida, Y., Cunha-Melo, J. R., Beaven, M. A., and Huang, K.-P. (1989). Differential down-regulation of protein kinase C isozymes. *J. Biol. Chem.* **264,** 4238–4243.

Huang, K.-P. (1990). Role of protein kinase C in cellular regulation. *Biofactors* **2,** 171–178.

Huang, K.-P., and Huang, F. L. (1990). Differential sensitivity of protein kinase C isozymes to phospholipid-induced inactivation. *J. Biol. Chem.* **265,** 738–744.

Huang, M., Chida, K., Kamata, N., Nose, K., Kato, M., Homma, Y., Takenawa, T., and Kuroki, T. (1988). Enhancement of inositol phospholipid metabolism and activation of protein kinase C in *ras*-transformed rat fibroblasts. *J. Biol. Chem.* **263,** 17975–17980.

Ikenaka, K., Nakahira, K., Takayama, C., Wada, K., Hatanaka, H., and Mikoshiba, K. (1990). Nerve growth factor rapidly induces expression of the 68-kDa neurofilament gene by posttranscriptional modification in PC12h-R cells. *J. Biol. Chem.* **265,** 19782–19785.

Isackson, P. J., and Bradshaw, R. A. (1984). The α-subunit of mouse 7S nerve growth factor is an inactive serine protease. *J. Biol. Chem.* **259,** 5380–5383.

Isackson, P. J., Ullrich, A., and Bradshaw, R. A. (1984). Mouse 7S nerve growth factor: Complete sequence of a cDNA coding for the α-subunit precursor and its relationship to serine proteases. *Biochemistry* **23,** 5997–6002.

Isackson, P. J., Towner, M. D., and Huntsman, M. M. (1991). Comparison of mammalian, chicken and Xenopus brain derived neurotrophic factor coding sequences. *FEBS Lett.* **285,** 260–264.

Ito, E., Sonnenberg, J. L., and Naraynan, R. (1989). Nerve growth factor-induced differentiation in PC12 cells is blocked by *fos* oncogene. *Oncogene* **4,** 1193–1199.

Ito, E., Sweterlitsch, L. A., Tran, P. B-V., Rauscher, F. J., III, and Narayanan, R. (1990). Inhibition of PC-12 cell differentiation by the immediate early gene *fra*-1. *Oncogene* **5,** 1755–1760.

Johnson, D., Lanahan, A., Buck, C. R., Sehgal, A., Morgan, C., Mercer, E., Bothwell, M., and Chao, M. (1986). Expression and structure of the human NGF receptor. *Cell* **47,** 545–554.

Johnson, E. M., and Taniuchi, M. (1987). Nerve growth factor (NGF) receptors in the central nervous system. *Biochem. Pharmac.* **36,** 4189–4195.

Kaisho, Y., Yoshimura, K., and Nakahama, K. (1990). Cloning and expression of a cDNA encoding a novel human neurotrophic factor. *FEBS Lett.* **266,** 187–191.

Kaplan, D. R., Martin-Zanca, D., and Parada, L. F. (1991a). Tyrosine phosphorylation and tyrosine kinase activity of the *trk* proto-oncogene product induced by NGF. *Nature (London)* **350,** 158–160.

Kaplan, D. R., Hempstead, B. L., Martin-Zanca, D., Chao, M. V., and Parada, L. F. (1991b). The *trk*

proto-oncogene product: A signal transducing receptor for nerve growth factor. *Science* **252**, 554–558.
Kater, S. B., and Mills, L. R. (1991). Regulation of growth cone behaviour by calcium. *J. Neurosci.* **11**, 891–899.
Kato, H., Kawai, S., and Takenawa, T. (1988). Disappearance of diacylglycerol kinase translocation in ras-transformed cells. *Biochem. Biophys. Res. Commun.* **154**, 959–966.
Kim, U-H., Fink, D., Jr., Kim, H. S., Park, D. J., Contreras, M. L., Guroff, G. and Rhee, S. G. (1991). Nerve growth factor stimulates phosphorylation of phospholipase C-γ in PC12 cells. *J. Biol. Chem.* **266**, 1359–1362.
Klein, R., Jing, S., Nanduri, V., O'Rourke, E., and Barbacid, M. (1991a). The *trk* proto-oncogene encodes a receptor for nerve growth factor. *Cell* **65**, 189–197.
Klein, R., Nanduri, V., Jing, S., Lamballe, F., Tapley, P., Bryant, S., Cordon-Carlo, C., Jones, K. R., Reichardt, L. F., and Barbacid, M. (1991b). The *trk* B tyrosine kinase is a receptor for brain-derived neurotrophic factor and neurotrophin 3. *Cell* **66**, 395–403.
Koch, C. A., Anderson, D., Moran, M. F., Ellis, C., and Pawson, T. (1991). SH2 and SH3 domains: Elements that control interactions of cytoplasmic signaling proteins. *Science* **252**, 668–674.
Koizumi, S., Contreras, M. L., Matsuda, Y., Hama, T., Lazarovici, P., and Guroff, G. (1988). K-252a: A specific inhibitor of the action of nerve growth factor on PC12 cells. *J. Neurosci.* **8**, 715–721.
Konno, Y., Ohno, S., Akita, Y., Kawasaki, H., and Suzuki, K. (1989). Enzymatic properties of a novel phorbol ester receptor/protein kinase, nPKC. *J. Biochem.* **106**, 673–678.
Kouchalakos, R. N., and Bradshaw, R. A. (1986). Nerve growth factor receptor from rabbit sympathetic ganglia membranes: Relationship between subforms. *J. Biol. Chem.* **261**, 16054–16059.
Kremer, N. E., Brugge, J. S., and Halegoua, S. (1989). Src and ras related antigens are required for neuronal differentiation of PC12 cells by NGF and bFGF. *Soc. Neurosci. Abstr.* **15(1)**, 866.
Kremer, N. E., D'Arcangelo, G., Thomas, S. M., DeMarco, M., Brugge, J. S., and Halegouna, S. (1991). Signal transduction by nerve growth factor and fibroblast growth factor in PC12 cells requires a sequence of Src and Ras actions. *J. Cell Biol.* **115**, 809–819.
Kruijer, W., Schubert, D. and Verma, I. M. (1985). Induction of the proto-oncogene *fos* by nerve growth factor. *Proc. Natl. Acad. Sci. U.S.A.* **82**, 7330–7334.
Kujubu, D. A., Lim, R. W., Varnum, B. C., and Herschman, H. R. (1987). Induction of transiently expressed genes in PC12 pheochromocytoma cells. *Oncogene* **1**, 257–262.
Kung, H-F., Smith, M. R., Bekesi, E., Manne, V., and Stacey, D. W. (1986). Reversal of transformed phenotype by monoclonal antibodies against Ha-*ras* p21 proteins. *Exp. Cell Res.* **162**, 363–371.
Laasberg, T., Pihlak, A., Neuman, T., Paves, H., and Saarma, M. (1988). Nerve growth factor increases the cyclic GMP level and activates the cyclic GMP phosphodiesterase in PC12 cells. *FEBS Lett.* **239**, 367–370.
Lacal, J. C. (1990). Diacylglycerol production in *Xenopus laevis* oocytes after microinjection of p21$^{ras}$ proteins is a consequence of activation of phosphatidylcholine metabolism. *Mol. Cell. Biol.* **10**, 333–340.
Lacal, J. C., Moscat, J., and Aaronson, S. A. (1987). Novel source of 1,2-diacylglycerol elevated in cells transformed by Ha-*ras* oncogene. *Nature (London)* **330**, 269–272.
Lakshmanan, J. (1978). Nerve growth factor induced turnover of phosphatidylinositol in rat superior cervical ganglia. *Biochem. Biophys. Res. Commun.* **82**, 767–775.
Landreth, G. E., and Williams, L. K. (1987). Nerve growth factor stimulates the phosphorylation of a 250-kDa cytoskeletal protein in cell-free extracts of PC12 cells. *Neurochem. Res.* **12**, 943–950.
Landreth, G., Cohen, P., and Shooter, E. M. (1980). $Ca^{2+}$ transmembrane fluxes and nerve growth factor action on a clonal cell line of rat pheochromocytoma. *Nature (London)* **283**, 202–204.
Lankford, K. L., and Letourneau, P. C. (1989). Evidence that calcium may control neurite outgrowth by regulating the stability of actin filaments. *J. Cell Biol.* **109**, 1229–1243.
Lapetina, E. G. (1982). Regulation of arachidonic acid production: Role of phospholipases C and $A_2$. *Trends Pharmaceut. Sci.* **7**, 115–118.
Lassing, I., and Lindberg, U. (1988). Specificity of the interaction between phosphatidylinositol 4,5-bisphosphate and the profilin: Actin complex. *J. Cell. Biochem.* **37**, 255–267.

Lazarovici, P., Levi, B-Z., Lelkes, P. I., Koizumi, S., Fujita, K., Matsuda, Y., Ozato, K., and Guroff, G. (1989). K-252a inhibits the increase in c-*fos* transcription and the increase in intracellular calcium produced by nerve growth factor in PC12 cells. *J. Neurosci. Res.* **23**, 1–8.

Ledwith, B. J., Manam, S., Kraynak, A. R., Nichols, W. W., and Bradley, M. O. (1990). Antisense-*fos* RNA causes partial reversion of the transformed phenotypes induced by the c-Ha-*ras* oncogene. *Mol. Cell. Biol.* **10**, 1545–1555.

Lee, P. L., Johnson, D. E., Cousens, L. S., Fried, V. A., and Williams, L. T. (1989). Purification and complementary DNA cloning of a receptor for basic fibroblast growth factor. *Science* **245**, 57–60.

Leibrock, J., Lottspeich, F., Hohn, A., Hifer, M., Hengerer, B., Masiakowski, P., Thoenen, H., and Barde, Y.-A. (1989). Molecular cloning and expression of brain-derived neurotrophic factor. *Nature (London)* **341**, 149–152.

Leonard, D. G. B., Ziff, E. B., and Greene, L. A. (1987). Identification and characterization of mRNAs regulated by nerve growth factor in PC12 cells. *Mol. Cell. Biol.* **7**, 3156–3167.

Levi-Montalcini, R., and Booker, B. (1960). Destruction of the sympathetic ganglia in mammals by an anti-serum to a nerve growth protein. *Proc. Natl. Acad. Sci. U.S.A.* **46**, 384–391.

Levi-Montalcini, R., and Hamburger, V. (1951). Selective growth stimulating effects of mouse sarcoma on the sensory and sympathetic nervous system of the chick embryo. *J. Exp. Zool.* **116**, 321–363.

Lim, R. W., Varnum, B. C., and Herschman, H. R. (1987). Cloning of tetradecanoyl phorbol ester-induced "primary response" sequences and their expression in density-arrested Swiss 3T3 cells and a TPA non-proliferative variant. *Oncogene* **1**, 263–270.

Lim, R. W., Varnum, B. C., O'Brien, T. G., and Herschman, H. R. (1989). Induction of tumor promotor-inducible genes in murine 3T3 cell lines and tetradecanoyl phorbol acetate non-proliferative 3T3 variants can occur through protein kinase C-dependent and -independent pathways. *Mol. Cell. Biol.* **9**, 1790–1793.

Lindenbaum, M. H., Carbonetto, S., Grosveld, F., Flavell, D., and Mushynski, W. E. (1988). Transcriptional and post-transcriptional effects of nerve growth factor on expression of the three neurofilament subunits in PC-12 cells. *J. Biol. Chem.* **263**, 5662–5667.

Lloyd, A. C., Paterson, H. F., Morris, J. D. H., Hall, A., and Marshall, C. J. (1989). p21H-*ras*-induced morphological transformation and increases in c-*myc* expression are independent of functional protein kinase C. *EMBO J.* **8**, 1099–1104.

Loeb, D. M., Maragos, J., Martin-Zanca, D., Chao, M. V., Parada, L. F., and Greene, L. A. (1991). The *trk* proto-oncogene rescues NGF responsiveness in mutant NGF-responsive PC12 cell lines. *Cell* **66**, 961–966.

Longo, F. M., and Mobley, W. C. (1991). Nerve growth factor: Studies addressing its expression and actions in the central nervous system. In "Growth Factors and Alzheimer's Disease" (E. Hefti, P. Brachet, B. Will, and Y. Christen, eds.), pp. 39–60. Springer-Verlag, Berlin.

Lowe, J. H. N., Huang, C.-L., and Ives, H. E. (1990). Sphingosine differentially inhibits activation of the $Na^+/H^+$ exchanger by phorbol esters and growth factors. *J. Biol. Chem.* **265**, 7188–7194.

McCormick, F. (1989). ras GTPase activating protein: Signal transmitter and signal terminator. *Cell* **56**, 5–8.

McDonald, N. Q., Lapatto, R., Murray-Rust, J., Gunning, J., Wlodawer, A., and Blundell, T. L. (1991). A new protein fold revealed by a 2.3 Å resolution crystal structure of nerve growth factor. *Nature (London)* in press.

Maher, P. A. (1988). Nerve growth factor induces protein-tyrosine phosphorylation. *Proc. Natl. Acad. Sci. U.S.A.* **85**, 6788–6791.

Maher, P. A. (1989). Role of protein tyrosine phosphorylation in the NGF response. *J. Neurosci. Res.* **24**, 29–37.

Maisonpierre, P. C., Bellluscio, L., Squinto, S., Ip, N. Y., Furth, M. E., Lindsay, R. M., and Yancopoulos, G. D. (1990). Neurotrophin-3: A neurotrophic factor related to NGF and BDNF. *Science* **247**, 1446–1451.

Maly, K., Uberall, F., Loferer, H., Doppler, W., Oberhuber, H., Groner, B., and Grunicke, H. H. (1989). Ha-*ras* activates the $Na^+/H^+$ antiporter by a protein kinase C-independent mechanism. *J. Biol. Chem.* **264**, 11839–11842.

Mann, D. A., Doherty, P., and Walsh, F. S. (1989). Increased intracellular cyclic AMP differentially modulates nerve growth factor induction of three neuronal recognition molecules involved in neurite outgrowth. *J. Neurochem.* **53,** 1581–1588.

Margolis, B., Rhee, S. G., Felder, S., Mervic, M., Lyall, R., Levitzki, A., Ullrich, A., Zilberstein, A., and Schlessinger, J. (1989). EGF induces tyrosine phosphorylation of phospholipase C-II: A potential mechanism for EGF receptor signaling. *Cell* **57,** 1101–1107.

Martin, T. F. J. (1991). Receptor regulation of phosphoinositidase C. *Pharmac. Ther.* **49,** 329–345.

Martin-Zanca, D., Hughes, S. H., and Barbacid, M. (1986). A human oncogene formed by the fusion of truncated tropomyosin and protein tyrosine kinase sequences. *Nature (London)* **319,** 743–748.

Martin-Zanca, D., Oskam, R., Mitra, G., Copeland, T., and Barbacid, M. (1989). Molecular and biochemical characterization of the human *trk* proto-oncogene. *Mol. Cell. Biol.* **9,** 24–33.

Martin-Zanca, D., Barbacid, M., and Parada, L. F. (1990). Expression of the *trk* proto-oncogene is restricted to the sensory cranial and spinal ganglia of neural crest origin in mouse development. *Genes Dev.* **4,** 683–694.

Masiakowski, P., and Shooter, E. M. (1988). Nerve growth factor induces the genes for two proteins related to a family of calcium-binding proteins in PC12 cells. *Proc. Natl. Acad. Sci. U.S.A.* **85,** 1277–1281.

Massague, J., Guillette, B. J., Czech, M. P., Morgan, C. J., and Bradshaw, R. A. (1981). Identification of a nerve growth factor receptor protein in sympathetic ganglia membranes by affinity-labeling. *J. Biol. Chem.* **256,** 9419–9424.

Massague, J., Buxser, S., Johnson, G. L., and Czech, M. (1982). Affinity labeling of a nerve growth factor receptor component on rat pheochromocytoma (PC12) cells. *Biochim. Biophys. Acta* **693,** 205–212.

Matthies, H. J. G., Palfrey, H. C., Hirning, L. D., and Miller, R. J. (1987). Down regulation of protein kinase C in neuronal cells: Effects on neurotransmitter release. *J. Neurosci.* **7,** 1198–1206.

Meier, R., Becker-Andre, M., Gotz, R., Heumann, R., Shaw, A., and Thoenen, H. (1986). Molecular cloning of bovine and chick nerve growth factor (NGF): Delineation of conserved and unconserved domains and their relationship to the biological activity and antigeneity of NGF. *EMBO J.* **5,** 1489–1493.

Meisenhelder, J., Such, P-G., Rhee, S. G., and Hunter, T. (1989). Phospholipase C-γ is a substrate for the PDGF and EGF receptor protein-tyrosine kinases in vivo and in vitro. *Cell* **57,** 1109–1122.

Meldrum, E., Parker, P. J., and Carozzi, A. (1991). The PtdIns-PLC superfamily and signal transduction. *Biochim. Biophys. Acta* **1092,** 49–71.

Miccheli, A., Salvatore, A. M., Delfini, M., Conti, F., and Calissano, P. (1989). A $^{31}$P NMR study of the NGF action on PC12 cell phospholipid metabolism. *Neurosci. Res. Commun.* **4,** 33–39.

Milbrandt, J. (1986). Nerve growth factor rapidly induces c-*fos* mRNA in PC12 rat pheochromocytoma cells. *Proc. Natl. Acad. Sci. U.S.A.* **83,** 4789–4793.

Miyasaka, T., Chao, M. V., Sherline, P., and Saltiel, A. R. (1990). Nerve growth factor stimulates a protein kinase in PC-12 cells that phosphorylates microtubule-associated protein-2. *J. Biol. Chem.* **265,** 4730–4735.

Miyasaka, T., Sternberg, D. W., Miyasaka, J., Sherline, P., and Saltiel, A. R. (1991). Nerve growth factor stimulates protein tyrosine phosphorylation in PC-12 pheochromocytoma cells. *Proc. Natl. Acad. Sci. U.S.A.* **88,** 2653–2657.

Mizuno, N., Matsuoka, I., and Kurihara, K. (1989). Possible involvements of intracellular $Ca^{2+}$ and $Ca^{2+}$-dependent protein phosphorylation in cholinergic differentiation of clonal rat pheochromocytoma cells (PC12) induced by glioma-conditioned medium and retinoic acid. *Dev. Brain Res.* **50,** 1–10.

Mochly-Rosen, D., and Koshland, D. E., Jr. (1987). Domain structure and phosphorylation of protein kinase C. *J. Biol. Chem.* **262,** 2291–2297.

Mochly-Rosen, D., and Koshland, D. E., Jr. (1988). A general procedure for screening inhibitory antibodies: Application for identifying anti-protein kinase C antibodies. *Anal. Biochem.* **170,** 31–37.

Molloy, C. J., Bottaro, D. P., Fleming, T. P., Marshall, M. S., Gibbs, J. B., and Aaronson, S. A. (1989). PDGF induction of tyrosine phosphorylation of GTPase activating protein. *Nature (London)* **342**, 711–714.

Montgomery, G. W. G., Jagger, B. A., and Bailey, P. D. (1988). Normal cellular Ha-ras p21 protein causes local disruption of bilayer phospholipid. *Biochemistry* **27**, 4391–4395.

Morgan, J. I., and Curran, T. (1986). Role of ion flux in the control of c-*fos* expression. *Nature (London)* **322**, 552–555.

Morrison, D. K., Kaplan, D. R., Rapp, U., and Roberts, T. M. (1988). Signal transduction from membrane to cytoplasm: Growth factors and membrane-bound oncogene products increase Raf-1 phosphorylation and associated protein kinase activity. *Proc. Natl. Acad. Sci. U.S.A.* **85**, 8855–8859.

Moskal, J. R., and Morrison, R. S. (1987). Signal transduction mechanisms that promote neurite outgrowth in cultures of CNS neurones. *Soc. Neurosci. Abstr.* **13(1)**, 1605.

Mulcahy, L. S., Smith, M. R., and Stacey, D. W. (1985). Requirement for *ras* proto-oncogene function during serum-stimulated growth of NIH 3T3 cells. *Nature (London)* **313**, 241–243.

Mutoh, T., Rudkin, B. B., Koizumi, S., and Guroff, G. (1988). Nerve growth factor, a differentiating agent, and epidermal growth factor, a mitogen, increase the activities of different S6 kinases in PC12 cells. *J. Biol. Chem.* **263**, 15853–15856.

Mutoh, T., Rudkin, B. B., and Guroff, G. (1992). Differential responses of the phosphorylation of ribosomal protein S6 to nerve growth factor and epidermal growth factor in PC12 cells. *J. Neurochem.* **58**, 58, 175–185.

Nebreda, A. R., Martin-Zanca, D., Kaplan, D. R., Parada, L. F., and Santos, E. (1991). Induction by NGF of meiotic maturation of *Xenopus* oocytes expressing the *trk* proto-oncogene product. *Science* **252**, 558–561.

Needleman, P., Turk, J., Jakschik, B. A., Morrison, A. R., and Lefkowitz, J. B. (1986). Arachidonic acid metabolism. *Ann. Rev. Biochem.* **55**, 69–102.

Nichols, R. A., Chandler, C. E., and Shooter, E. M. (1989). Enucleation of the rat pheochromocytoma clonal cell line, PC12: Effect on neurite outgrowth. *J. Cell. Physiol.* **141**, 301–309.

Nikodijevic, B., and Guroff, G. (1991). Nerve growth factor-induced increase in calcium uptake by PC12 cells. *J. Neurosci. Res.* **28**, 192–199.

Nishibe, S., Wahl, M. I., Hernandez-Sotomayor, S. M. T., Tonks, N. K., Rhee, S. G., and Carpenter, G. (1990). Increase of the catalytic activity of phospholipase C-γ1 by tyrosine phosphorylation. *Science* **250**, 1253–1255.

Nishimura, S., and Sekiya, T. (1987). Human cancer and cellular oncogenes. *Biochem. J.* **243**, 313–327.

Nishizuka, Y. (1986). Studies and perspectives of protein kinase C. *Science* **233**, 305–312.

Nishizuka, Y. (1988). The molecular heterogeneity of protein kinase C and its implications for cellular regulation. *Nature (London)* **334**, 661–665.

Ohmichi, M., Decker, S. J., Pang, L., and Saltiel, A. R. (1991). Nerve growth factor binds to the 140 kd *trk* proto-oncogene product and stimulates its association with the *src* homology domain of phospholipase Cγ1. *Biochem. Biophys. Res. Comm.* **179**, 217–223.

Okada, M., and Nakagawa, H. (1989). A protein tyrosine kinase involved in regulation of pp60$^{c-src}$ function. *J. Biol. Chem.* **264**, 20886–20893.

Oshima, M., Sitanandam, G., Rapp, U. R., and Guroff, G. (1991). The phosphorylation and activation of B-raf in PC12 cells stimulated by nerve growth factor. *J. Biol. Chem.* **266**, 23753–23760.

Pandiella-Alonso, A., Malgaroli, A., Vicentini, L. M., and Meldolesi, J. (1986). Early rise of cytosolic $Ca^{2+}$ induced by NGF in PC12 and chromaffin cells. *FEBS Lett.* **208**, 48–51.

Parker, P. J., Kour, G., Marais, R. M., Mitchell, F., Pears, C., Schapp, D., Stabel, S., and Webster, C. (1989). Protein kinase C—A family affair. *Mol. Cell. Endocrin.* **65**, 1–11.

Pattison, S. E., and Dunn, M. F. (1975). On the relationship of zinc ion to the structure and function of the 7S NGF protein. *Biochemistry* **14**, 2733–2739.

Paves, H., Neuman, T., Metsis, M., and Saarma, M. (1988). Nerve growth factor induces rapid redistribution of F-actin in PC12 cells. *FEBS Lett.* **235**, 141–143.

Paves, H., Neuman, T., Metsis, M., and Saarma, M. (1990). Nerve growth factor-induced rapid

reorganization of microfilaments in PC12 cells: Possible roles of different second messenger systems. *Exp. Cell Res.* **186,** 218–226.
Pelech, S. L., and Vance, D. E. (1989). Signal transduction via phosphatidylcholine cycles. *Trends Biochem. Sci.* **14,** 28–30.
Pelosin, J-M., Keramidas, M., Souvignet, C., and Chambaz, E. M. (1990). Differential inhibition of protein kinase C subtypes. *Biochem. Biophys. Res. Commun.* **169,** 1040–1048.
Pessin, M. S., and Raben, D. M. (1989). Molecular species analysis of 1,2-diglycerides stimulated by α-thrombin in cultured fibroblasts. *J. Biol. Chem.* **264,** 8729–8738.
Pessin, M. S., Altin, J. G., Jarpe, M., Tansley, F., Bradshaw, R. A., and Raben, D. M. (1991). Carbachol stimulates a different phospholipid metabolism than nerve growth factor and basic fibroblast growth factor in PC12 cells. *Cell Regulation* **2,** 383–390.
Petersen, O. H., and Wakui, M. (1990). Oscillating intracellular $Ca^{2+}$ signals evokes by activation of receptors linked to inositol lipid hydrolysis: Mechanism of generation. *J. Membrane Biol.* **118,** 93–105.
Pleasure, S. J., Reddy, U. R., Venkatakrishnan, G., Roy, A. K., Chen, J., Ross, A. H., Trojanowski, J. Q., Pleasure, D. E., and Lee, V. M-Y. (1990). Introduction of nerve growth factor (NGF) receptors into a medulloblastoma cell line results in expression of high- and low-affinity NGF receptors but not NGF-mediated differentiation. *Proc. Natl. Acad. Sci. U.S.A.* **87,** 8496–8500.
Pollock, J. D., Krempin, M., and Rudy, B. (1990). Differential effects of NGF, FGF, EGF, cAMP, and dexamethasone on neurite outgrowth and sodium channel expression in PC12 cells. *J. Neurosci.* **10,** 2626–2637.
Polverino, A. J., and Barritt, G. J. (1988). On the source of the vasopressin-induced increases in diacylglycerol in hepatocytes. *Biochim. Biophys. Acta* **970,** 75–82.
Pozzan, T., Di Virgilio, F., Vicentini, L. M., and Meldolesi, J. (1986). Activation of muscarinic receptors in PC12 cells. Stimulation of $Ca^{2+}$ influx and redistribution. *Biochem. J.* **234,** 547–553.
Pulliam, M. W., Boyd, L. F., Baglan, N. C., and Bradshaw, R. A. (1975). Specific binding of covalently cross-linked mouse nerve growth factor to responsive peripheral neurons. *Biochem. Biophys. Res. Comm.* **67,** 1281–1289.
Putney, J. W., Jr. (1990). Capacitative calcium entry revisited. *Cell Calcium* **11,** 611–624.
Qiu, M-S., and Green, S. H. (1991). NGF and EGF rapidly activate p21$^{ras}$ in PC12 cells by distinct, convergent pathways involving tyrosine phosphorylation. *Neuron* **7,** 937–946.
Qiu, M-S., Pitts, A. F., Winters, T. R., and Green, S. H. (1991). *ras* Isoprenylation is require for *ras*-induced but not for NGF-induced neuronal differentiation of PC12 cells. *J. Cell. Biol.* **115,** 795–808.
Radeke, M. J., Misko, T. P., Hsu, C., Herzenberg, L. A., and Shooter, E. M. (1987). Gene transfer and molecular cloning of the rat nerve growth factor receptor. *Nature (London)* **325,** 593–597.
Rakowicz-Szulczynska, E. M., and Koprowski, H. (1989). Antagonistic effect of PDGF and NGF on transcription of ribosomal DNA and tumor cell proliferation. *Biochem. Biophys. Res. Commun.* **163,** 649–656.
Rakowicz-Szulczynska, E. M., Herlyn, M., and Koprowski, H. (1988). Nerve growth factor receptors in chromatin of melanoma cells, proliferating melanocytes, and colorectal carcinoma cells *in vitro*. *Cancer Res.* **48,** 7200–7206.
Rausch, D. M., Dickens, G., Doll, S., Fujita, K., Koizumi, S., Rudkin, B. B., Tocco, M., Eiden, L. E., and Guroff, G. (1989). Differentiation of PC12 cells with v-src: Comparison with nerve growth factor. *J. Neurosci. Res.* **24,** 49–58.
Rausch, D. M., Lewis, D. L., Barker, J. L., and Eiden, L. E. (1990). Functional expression of dihydropyridine-insensitive calcium channels during PC12 cell differentiation by nerve growth factor (NGF), oncogenic ras, or src tyrosine kinase. *Cell. Mol. Neurobiol.* **10,** 237–255.
Reinhold, D. S., and Neet, K. E. (1989). The lack of a role for protein kinase C in neurite extension and in the induction of ornithine decarboxylase by nerve growth factor in PC12 cells. *J. Biol. Chem.* **264,** 3538–3544.
Represa, J., Avila, M. A., Miner, C., Giraldez, F., Romero, G., Clemente, R., Mato, J. M., and

Varela-Nieto, I. (1991). Glycosyl-phosphatidylinositol/inositol phosphoglycan: A signaling system for the low-affinity nerve growth factor receptor. *Proc. Natl. Acad. Sci. USA* **88**, 8016–8019.

Rhee, S. G., Such, P.-G., Ryu, S-H., and Lee, S. Y. (1989). Studies of inositol phospholipid-specific phospholipid C. *Science* **244**, 546–550.

Riabowol, K. T., Fink, J. S., Gilman, M. Z., Walsh, D. A., Goodman, R. H., and Feramisco, J. R. (1988a). The catalytic subunit of cAMP-dependent protein kinase induces expression of genes containing cAMP-responsive enhancer elements. *Nature (London)* **336**, 83–86.

Riabowol, K. T., Vosatka, R. J., Ziff, E. B., Lamb, N. J., and Feramisco, J. R. (1988b). Microinjection of *fos*-specific antibodies blocks DNA synthesis in fibroblast cells. *Mol. Cell. Biol.* **8**, 1670–1676.

Richards, R. I., Coghlan, J. P., Digby, M., Drinkwater, C. C., Lloyd, C., Lyons, I., and Zhang, X.-Y. (1989). Molecular biology of the glandular kallikrein genes of mouse and man. In "The Kallikrein–Kinin System in Health and Disease" (H. Fritz, I. Schmidt, and G. Dietze, eds.), pp. 215–225. Limbach-Verlag, Braunschweig.

Richter-Landsberg, C., and Jastorff, B. (1986). The role of cAMP in nerve growth factor-promoted neurite outgrowth in PC12 cells. *J. Cell Biol.* **102**, 821–829.

Rodriguez-Tebar, A., Dechant, G., and Barde, Y.-A. (1990). Binding of brain-derived neurotrophic factor to the nerve growth factor receptor. *Neuron* **4**, 487–492.

Rogers, M., and Hendry, I. (1990). Involvement of dihydropyridine-sensitive calcium channels in nerve growth factor-dependent neurite outgrowth by sympathetic neurons. *J. Neurosci.* **26**, 447–454.

Romano, C., Nichols, R. A., and Greengard, P. (1987). Synapsin 1 in PC12 cells. 11. Evidence for regulation by NGF of phosphorylation at a novel site. *J. Neurosci.* **7**, 1300–1306.

Rosenthal, A., Goeddel, D. V., Nguyen, T., Lewis, M., Shih, A., Laramee, G. R., Nikolics, K., and Winslow, J. W., (1990). Primary structure and biological activity of a novel human neurotrophic factor. *Neuron* **4**, 767–773.

Rosoff, P. M., Savage, N., and Dinarello, C. A. (1988). Interleukin-1 stimulates diacylglycerol production in T lymphocytes by a novel mechanism. *Cell* **54**, 73–81.

Rowland, E. A., Muller, T. H., Goldstein, M., and Greene, L. A. (1987). Cell-free detection and characterization of a novel nerve growth factor-activated protein kinase in PC12 cells. *J. Biol. Chem.* **262**, 7504–7513.

Rowland-Gagne, E., and Greene, L. A. (1990). Multiple pathways of N-kinase activation in PC12 cells. *J. Neurochem.* **54**, 424–433.

Ruegg, U. T., and Burgess, G. M. (1989). Staurosporine, K252 and UCN-01: Potent but nonspecific inhibitors of protein kinases. *Trends Pharmacol. Sci.* **10**, 218–220.

Rydel, R. E., and Greene, L. A. (1987). Acidic and basic fibroblast growth factors promote stable neurite outgrowth and neuronal differentiation in cultures of PC12 cells. *J. Neurosci.* **7**, 3639–3653.

Rydel, R. E., and Greene, L. A. (1988a). 8-Substituted cAMP analogs can replace the NGF requirement of cultured rat sympathetic and sensory neurons: Evidence for parallel neurotrophic pathways. In "Neuronal Plasticity and Trophic Factors" (G. Biggio, P. F. Spano, G. Toffano, S. H. Appel, and G. L. Gessa, eds.), pp. 73–85. Liviana Press, Padova.

Rydel, R. E., and Greene, L. A. (1988b). cAMP analogs promote survival and neurite outgrowth in cultures of rat sympathetic and sensory neurons independently of nerve growth factor. *Proc. Natl. Acad. Sci. U.S.A.* **85**, 1257–1261.

Saltiel, A. R., and Cuatrecasas, P. (1988). In search of a second messenger for insulin. *Am. J. Physiol.* **255**, C1–C11.

Saltiel, A. R., Sherline, P., and Fox, J. A. (1987). Insulin-stimulated diacylglycerol production results from the hydrolysis of a novel phosphatidylinositol glycan. *J. Biol. Chem.* **262**, 1116–1121.

Sano, M., Nishiyama, K., and Kitajima, S. (1990). A nerve growth factor-dependent protein kinase that phosphorylates microtubule-associated proteins in vitro: Possible involvement of its activity in the outgrowth of neurites from PC12 cells. *J. Neurochem.* **55**, 427–435.

Santos, E., and Nebreda, A. R. (1989). Structural and functional properties of *ras* proteins. *FASEB J.* **3**, 2151–2163.

Sassone-Corsi, P., Der, C. J., and Verma, I. M. (1989). *ras*-Induced neuronal differentiation of PC12 cells: Possible involvement of *fos* and *jun*. *Mol. Cell. Biol.* **9**, 3174–3183.

Satoh, T., Nakamura, S., and Kaziro, Y. (1987). Induction of neurite formation in PC12 cells by microinjection of proto-oncogenic Ha-*ras* protein preincubated with guanosine-5'-O-(3-thiotriphosphate). *Mol. Cell. Biol.* **7,** 4553–4556.

Schanen-King, C., Nel, A., Williams, L. K., and Landreth, G. (1991). Nerve growth factor stimulates the tyrosine phosphorylation of MAP2 kinase in PC12 cells. *Neuron* **6,** 915–922.

Schubert, D., and Whitlock, C. (1977). Alteration of cellular adhesion by nerve growth factor. *Proc. Natl. Acad. Sci. U.S.A.* **74,** 4055–4058.

Schubert, D., LaCorbiere, M., Whitlock, C., and Stallcup, W. (1978). Alterations in the surface properties of cells responsive to nerve growth factor. *Nature (London)* **273,** 718–723.

Schubert, D., Ling, N., and Baird, A. (1987). Multiple influences of a heparin-binding growth factor on neuronal development. *J. Cell Biol.* **104,** 635–643.

Schwarz, M. A., Fisher, D., Bradshaw, R. A., and Isackson, P. J. (1989). Isolation and sequence of a cDNA clone of β-nerve growth factor from the guinea pig prostate gland. *J. Neurochem.* **52,** 1203–1209.

Scott, J., Selby, M., Urdea, M., Quiroga, M., Bell, G. I., and Rutter, W. J. (1983). Isolation and nucleotide sequence of a cDNA encoding the precursor of mouse nerve growth factor. *Nature (London)* **302,** 538–590.

Seeley, P. J., Rukenstein, A., Connolly, J. L., and Greene, L. A. (1984). Differential inhibition of nerve growth factor and epidermal growth factor effects on the PC12 pheochromocytoma line. *J. Cell Biol.* **98,** 417–426.

Sekiguchi, K., Tsukuda, M., Ase, K., Kikkawa, U., and Nishizuka, Y. (1988). Mode of activation and kinetic properties of three distinct forms of protein kinase C from rat brain. *J. Biochem.* **103,** 759–765.

Selby, M. J., Edwards, R. H., and Rutter, W. J. (1987). Cobra nerve growth factor: Structure and evolutionary comparison. *J. Neurosci. Res.* **18,** 293–298.

Server, A. C., and Shooter, E. M. (1977). Nerve growth factor. *Adv. Prot. Chem.* **31,** 339–370.

Shearman, M. S., Naor, Z., Sekiguchi, K., Kishimoto, A., and Nishizuka, Y. (1989). Selective activation of the γ-subspecies of protein kinase C from bovine cerebellum by arachidonic acid and its lipoxygenase metabolites. *FEBS Lett.* **243,** 177–182.

Sheng, M., and Greenberg, M. E. (1990). The regulation and function of c-*fos* and other immediate early genes in the nervous system. *Neuron* **4,** 477–485.

Sheng, M., Dougan, S. T., McFadden, G., and Greenberg, M. E. (1988). Calcium and growth factor pathways of c-*fos* transcriptional activation require distinct upstream regulatory sequences. *Mol. Cell. Biol.* **8,** 2787–2796.

Sheng, M., McFadden, G., and Greenberg, M. E. (1990). Membrane depolarization and calcium induce c-*fos* transcription via phosphorylation of transcription factor CREB. *Neuron* **4,** 571–582.

Shukla, S. D., and Halenda, S. P. (1991). Phospholipase D in cell signalling and its relationship to phospholipase C. *Life Sciences* **48,** 851–866.

Sigmund, O., Naor, Z., Anderson, D. J., and Stein, R. (1990). Effect of nerve growth factor and fibroblast growth factor on SCG10 and c-*fos* expression and neurite outgrowth in protein kinase C-depleted PC12 cells. *J. Biol. Chem.* **265,** 2257–2261.

Sihag, R. K., Jeng, A. Y., and Nixon, R. A. (1988). Phosphorylation of neurofilament proteins by protein kinase C. *FEBS Lett.* **233,** 181–185.

Silverman, R. E. (1978). Interactions within the mouse nerve growth factor complex. Ph.D. Thesis. Washington University, St. Louis, Missouri.

Silverman, R. E., and Bradshaw, R. A. (1982). Nerve growth factor: Subunit interactions in the mouse submaxillary gland 7S complex. *J. Neurosci. Res.* **8,** 127–136.

Smith, D. S., King, C. S., Pearson, E., Gittinger, C. K., and Landreth, G. E. (1989). Selective inhibition of nerve growth factor-stimulated protein kinases by K-252a and 5'-S-methyladenosine in PC12 cells. *J. Neurochem.* **53,** 800–806.

Smith, M. R., Liu, Y-L., Kim, H., Rhee, S. G., and Kung, H-F. (1990). Inhibition of serum- and ras-stimulated DNA synthesis by antibodies to phospholipase C. *Science* **247,** 1074–1077.

Smrcka, A. V., Helper, J. R., Brown, K. O., and Sternweis, P. C. (1991). Regulation of pholphosphoinositide-specific phospholipase C activity by purified $G_q$. *Science* **251**, 804–807.
Sobue, K., and Kanda, K. (1988). Localization of pp60$^{c\text{-}src}$ in growth cone of PC12 cell. *Biochem. Biophys. Res. Commun.* **157**, 1383–1389.
Sohal, P. S., and Cornell, R. B. (1990). Sphingosine inhibits the activity of rat liver CTP: Phosphocholine cytidyltransferase. *J. Biol. Chem.* **265**, 11746–11750.
Soppet, D., Escandon, E., Maragos, J., Middlemas, D. S., Reid, S. W., Blair, J., Burton, L. E., Stauton, B. R., Kaplan, D. R., Hunter, T., Nikolies, K., and Parada, L. F. (1991). The neurotrophic factors brain derived neurotrophic factor and neurotrophin-3 are ligands for the *trk* B tyrosine kinase receptor. *Cell* **65**, 895–903.
Squinto, S., Stitt, T. N., Aldrich, T. H., Davis, S., Bianco, S. M., Radziejewski, C., Glass, D. J., Masiakowski, P., Furth, M. E., Valenzuela, D. M., DiStefano, P. S., and Yancopoulos, G. D. (1991). *trk* B encodes a functional receptor for brain-derived neurotrophic factor and neurotrophin-3 but not nerve growth factor. *Cell* **65**, 885–893.
Srivastava, S. K., Di Donato, A., and Lacal, J. C. (1989). H-*ras* mutants lacking the epitope for the neutralizing monoclonal antibody Y13-259 show decreased biological activity and are deficient in GTPase-activating protein interaction. *Mol. Cell. Biol.* **9**, 1779–1783.
Stacey, D. W., and Kung, H-F. (1984). Transformation of NIH 3T3 cells by microinjection of Ha-*ras* p21 protein. *Nature (London)* **310**, 508–511.
Stacey, D. W., DeGudicibus, S. R., and Smith, M. R. (1987). Cellular *ras* activity and tumor cell proliferation. *Exp. Cell Res.* **171**, 232–242.
Stach, R. E., and Shooter, E. M. (1974). The biological activity of cross-linked β nerve growth factor protein. *J. Biol. Chem.* **249**, 6668–6674.
Stockel, J., Paravicini, U., and Thoenen, H. (1974). Specificity of the retrograde axonal transport of nerve growth factor. *Brain Res.* **76**, 413–421.
Sugimoto, Y., Noda, M., Kitayama, H., and Ikawa, Y. (1988). Possible involvement of two signaling pathways in induction of neuron-associated properties by v-Ha-*ras* gene in PC12 cells. *J. Biol. Chem.* **263**, 12102–12108.
Sutter, A., Riopelle, R. J., Harris-Warrick, R. M., and Shooter, E. M. (1979). NGF receptor characterization of two distinct classes of binding sites on chick embryo sensory ganglia cells. *J. Biol. Chem.* **254**, 5972–5982.
Taniuchi, M., and Johnson, E. M., Jr. (1985). Characterization of the binding properties and retrograde axonal transport of a monoclonal antibody directed against the rat nerve growth factor receptor. *J. Cell Biol.* **101**, 1100–1106.
Thoenen, H., and Barde, Y.-A. (1980). Physiology of nerve growth factor. *Physiol. Rev.* **60**, 1284–1335.
Thomas, K. A., Silverman, R.E., Jeng, Z., Baglan, N. C., and Bradshaw, R. A. (1981a). Electrophoretic heterogeneity and polypeptide chain structure of the γ-subunit of mouse submaxillary gland 7S nerve growth factor. *J. Biol. Chem.* **256**, 9147–9155.
Thomas, K. A., Baglan, N. C., and Bradshaw, R. A. (1981b). The amino acid sequence of the γ-subunit of mouse submaxillary gland 7S nerve growth factor. *J. Biol. Chem.* **256**, 9156–9166.
Thomson, T. M., Green, S. H., Trotta, R. J., Burstein, D. E., and Pellicer, A. (1990). Oncogene N-*ras* mediates selective inhibition of c-*fos* induction by nerve growth factor and basic fibroblast growth factor in a PC12 cell line. *Mol. Cell. Biol.* **10**, 1556–1563.
Tiercy, J-M., and Shooter, E. M. (1986). Early changes in the synthesis of nuclear and cytoplasmic proteins are induced by nerve growth factor in differentiating rat PC12 cells. *J. Cell Biol.* **103**, 2367–2378.
Tischler, A. S., Ruzicka, L. A., and Perlman, R. L. (1990). Mimicry and inhibition of nerve growth factor effects: Interactions of staurosporine, forskolin, and K252a in PC12 cells and normal rat chromaffin cells in vitro. *J. Neurochem.* **55**, 1159–1165.
Tischler, A. S., Ruzicka, L. A., and Dobner, P. R. (1991). A protein kinase inhibitor, staurosporine, mimics nerve growth factor induction of neurotensin/neuromedin N gene expression. *J. Biol. Chem.* **266**, 1141–1146.

Togari, A., and Guroff, G. (1985). Partial purification and characterization of a nerve growth factor-sensitive kinase and its substrate from PC12 cells. *J. Biol. Chem.* **260**, 3804–3811.

Togari, A., Dickens, G., Kuzuya, H., and Guroff, G. (1985). The effect of fibroblast growth factor on PC12 cells. *J. Neurosci.* **5**, 307–316.

Traynor, A. E. (1984). The relationship between neurite extension and phospholipid metabolism in PC12 cells. *Dev. Brain Res.* **14**, 205–210.

Traynor, A. E., Schubert, D., and Allen, W. R. (1982). Alterations of lipid metabolism in response to nerve growth factor. *J. Neurochem.* **39**, 1677–1683.

Trejo, J., and Brown, J. H. (1991). c-*fos* and c-*jun* are induced by muscarinic receptor activation of protein kinase C but are differentially regulated by intracellular calcium. *J. Biol. Chem.* **266**, 7876–7882.

Trotta, R. J., Thomson, T. M., Lacal, J.-C., Pellicer, A., and Burstein, D. E. (1990). Potentiation of oncogenic N-ras-induced neurite outgrowth and ornithine decarboxylase activity by phorbol dibutyrate and protein kinase inhibitor H-8. *J. Cell. Physiol.* **143**, 68–78.

Tsao, H., Aletta, J. M., and Greene, L. A. (1990). Nerve growth factor and fibroblast growth factor selectively activate a protein kinase that phosphorylates high molecular weight microtubule-associated proteins. Detection, partial purification, and characterization in PC12 cells. *J. Biol. Chem.* **265**, 15471–15480.

Ullrich, A., Gray, A., Berman, C., and Dull, T. J. (1983). Human beta nerve growth factor sequence is highly homologous to that of mouse. *Nature (London)* **303**, 821–825.

Ullrich, A., Gray, A., Wood, W. I., Hayflick, J., and Seeburg, P. H. (1984). Isolation of a cDNA clone coding for the γ-subunit of mouse nerve growth factor using a high-stringency selection procedure. *DNA* **3**, 387–391.

van Calker, D., Takahata, K., and Heumann, R. (1989). Nerve growth factor potentiates the hormone-stimulated intracellular accumulation of inositol phosphates and $Ca^{2+}$ in rat PC12 pheochromocytoma cells: Comparison with the effect of epidermal growth factor. *J. Neurochem.* **52**, 38–45.

Varon, S., Nomura, J., and Shooter, E. M. (1967). The isolation of the mouse nerve growth factor protein in a high molecular weight form. *Biochemistry* **6**, 2202–2209.

Vetter, M. L., Martin-Zanca, D., Parada, L. F., Bishop, J. M., and Kaplan, D. R. (1991). Nerve growth factor rapidly stimulates tyrosine phosphorylation of phospholipase C-γ1 by a kinase activity associated with the product of the *trk* proto-oncogene. *Proc. Natl. Acad. Sci. U.S.A.* **88**, 5650–5654.

Vicentini, L. M., Ambrosini, A., Di Virgilio, F., Meldolesi, J., and Pozzan, T. (1986). Activation of muscarinic receptors in PC12 cells. Correlation between cytosolic $Ca^{2+}$ rise and phosphoinositide hydrolysis. *Biochem. J.* **234**, 555–562.

Vogel, U. S., Dixon, R. A. F., Schaber, M. D., Diehl, R. E., Marshall, M. S., Scolnick, E. M., Sigal, I. S., and Gibbs, J. B. (1988). Cloning of bovine GAP and its interaction with oncogenic *ras* p21. *Nature (London)* **335**, 90–93.

Volante, C., and Greene, L. A. (1990a). Nerve growth factor (NGF) responses by non-neuronal cells: Detection by assay of a novel NGF-activated protein kinase. *Growth Factors* **2**, 321–331.

Volante, C., and Greene, L. A. (1990b). Induction of ornithine decarboxylase by nerve growth factor in PC12 cells: Dissection by purine analogues. *J. Biol. Chem.* **265**, 11050–11055.

Volante, C., Rukenstein, A., Loeb, D. M., and Greene, L. A. (1989). Differential inhibition of nerve growth factor responses by purine analogues: Correlation with inhibition of a nerve growth factor-activated protein kinase. *J. Cell Biol.* **109**, 2395–2403.

Vosatka, R. J., Hermanowski-Vosatka, A., Metz, R., and Ziff, E. B. (1989). Dynamic interactions of c-fos protein in serum-stimulated 3T3 cells. *J. Cell. Physiol.* **138**, 493–502.

Vulliet, P. R., Hall, F. L., Mitchell, J. P., and Hardie, D. G. (1989). Identification of a novel proline-directed serine/threonine protein kinase in rat pheochromocytoma. *J. Biol. Chem.* **264**, 16292–16298.

Wahl, M. I., Olashaw, N. E., Nishibe, S., Rhee, S. G., Pledger, W. J., and Carpenter, G. (1989). Platelet-derived growth factor induces rapid and sustained tyrosine phosphorylation of phospholipase C-γ in quiescent BALB/c 3T3 cells. *Mol. Cell. Biol.* **9**, 2934–2943.

Weskamp, G. and Reichardt, L. F. (1991). Evidence the biological activity of NGF is mediated through a novel subclass of high affinity receptors. *Neuron* **6,** 649–663.

Whitman, M., Downes, C. P., Keeler, M., Keller, T., and Cantley, L. (1988). Type I phosphatidylinositol kinase makes a novel inositol phospholipid, phosphatidylinositol-3-phosphate. *Nature (London)* **332,** 644–646.

Whittemore, S. R., Friedman, P. L., Larhammar, D., Persson, H. J., Gonzalez-Carvajal, M., and Holets, V. R. (1988). Rat beta-nerve growth factor sequence and site of synthesis in the adult hippocampus. *J. Neurosci. Res.* **20,** 403–410.

Williams, R., Gaber, B., and Gunning, J. (1982). Raman spectroscopic determination of the secondary structure of crystalline nerve growth factor. *J. Biol. Chem.* **257,** 13321–13323.

Wion, D., Perret, C., Frechin, N., Keller, A., Behar, G., Brachet, P., and Auffray, C. (1986). Molecular cloning of the avian β-nerve growth factor gene: Transcription in brain. *FEBS lett.* **203,** 82–86.

Wlodawer, A., Hodgson, K. D., and Shooter, E. M. (1975). Crystallization of nerve growth factor from mouse submaxillary glands. *Proc. Natl. Acad. Sci. U.S.A.* **72,** 777–779.

Wong, Y. M., Tolson, N. W, and Guroff, G. (1980). Increased phosphorylation of specific nuclear proteins in superior cervical ganglia and PC12 cells in response to nerve growth factor. *J. Biol. Chem.* **255,** 10481–10492.

Wu, B-Y., Fodor, E. J. B., Edwards, R. H., and Rutter, W. J. (1989). Nerve growth factor induces the proto-oncogene c-*jun* in PC12 cells. *J. Biol. Chem.* **264,** 9000–9003.

Yarden, Y., and Ullrich, A. (1989). Molecular analysis of signal transduction by growth factors. *Perspectives Biochem.* 132–138.

Yan, H., Schlessinger, J., and Chao, M. V. (1991). Chimeric NGF–EGF receptors define domains responsible for neuronal differentiation. *Science* **252,** 561–563.

Yankner, B. A., and Shooter, E. M. (1979). Nerve growth factor in the nucleus: Interaction with receptors on the nuclear membrane. *Proc. Natl. Acad. Sci. U.S.A.* **76,** 1269–1273.

Yu, C-L., Tsai, M-H., and Stacey, D. W. (1988). Cellular *ras* activity and phospholipid metabolism. *Cell* **52,** 63–71.

# 6 Regulation of Nerve Growth Factor Expression

Robert H. Edwards

Nerve growth factor (NGF), like other growth factors, acts as a signal between the cell that produces it and the cell that responds. The nature of this signal depends critically on the amount, location, and timing of NGF expression. Based largely on these features of its expression, the NGF signal appears to fall into at least three recognizable patterns. First, cells in the nervous system that receive input from NGF-responsive neurons produce NGF, presumably to control their innervation. Second, neural supporting cells synthesize NGF after nerve injury, possibly to assist regeneration. Third, NGF and its receptor are expressed in a number of settings that suggest a role in such diverse biologic processes as early neural development and spermatogenesis.

Two complementary approaches have illuminated the role of NGF in the nervous system: observation of the response to NGF and analysis of its expression. The administration of NGF has dramatic effects on neuronal properties both in vivo and in vitro, including cell survival, process outgrowth, and neurotransmitter expression. These observations have suggested that endogenous NGF acts similarly during development, and have tended to emphasize the role of NGF as a maintenance factor. The analysis of endogenous NGF expression has supported some of the predicted functions and refined or eliminated others. However, limitations in the ability both to study endogenous NGF expression and to manipulate its expression in relevant target cells have greatly restricted our understanding of its exact role in vivo.

NGF appears to act as a target-derived factor that regulates the structure of specific neural connections. The low levels of endogenous NGF expression in both the central and peripheral nervous systems preclude direct study of its regulation and release, forcing a reliance on mRNA determination during development and correlations with the density of innervation to derive hypotheses about its function.

In addition, previous studies have relied necessarily on systemic administration to show the effects of NGF deprivation or excess and thus have neglected the local effects that are more relevant to the function of NGF in the target. Since NGF-responsive neurons in the peripheral nervous system do not form precise connections with their targets, these experimental limitations have not been considered critical for the neurotrophic hypothesis. However, in the central nervous system, local release of NGF by target cells that are themselves neurons, as well as its effect on connections that are classical synapses, clearly requires an explanation that is more than a simple analogy of peripheral systems. Thus, understanding the true scope of NGF action in the nervous system depends on further characterization of its regulation and on the ability to manipulate NGF levels in relevant cells in vivo.

The postulated role of NGF in other settings lacks the support of observed responses to a change in the level of NGF and derives almost entirely from circumstantial evidence of regulated expression. Injury of peripheral nerve activates the expression of NGF and its receptor by supporting cells, suggesting a role for NGF in regeneration. However, to what extent NGF participates in the re-establishment of connections remains unclear. The expression of NGF receptor in early neural development also suggests a role in such events as differentiation and axon outgrowth, but increasing or reducing NGF expression at these times to determine the true nature of the signal has not been possible. Similarly, the expression of NGF and NGF receptor in different cells of the testis, as well as in a host of cells during nonneural development, suggests a variety of functions that remain difficult to confirm.

This review addresses the regulation of NGF expression in situations in which its function is well understood as well as in settings in which its role is characterized poorly. In many cases, the pattern of regulation provides major clues to function; understanding NGF regulation in more detail will yield important insights. In all cases, the ability to manipulate NGF expression in relevant cells eventually will lead to a definite assignment of specific functions and, in some cases, may be necessary simply to reveal the extent of NGF action.

## I. EXPRESSION IN THE TARGET

### A. Biologic Regulation

#### 1. Tissue distribution

Target tissues innervated by NGF-responsive neurons express extremely low levels of NGF. Initially, investigators used a two-site radioimmunoassay to quantitate NGF in extracts from various peripheral targets (Korsching and Thoenen, 1983). Although the presence of NGF in a tissue did not necessarily indicate synthesis at that site and NGF that was secreted by a target was retrogradely transported by responsive neurons, the level of NGF in vivo generally correlated with the level determined from explanted tissue that was cultured in vitro. Availability of the NGF cDNA permitted the first unambiguous determination of the sites of synthesis in vivo (Scott et al., 1983; Ullrich et al., 1983). Different target tissues contain different amounts of NGF mRNA, but even the most highly expressing tissues contain no more than one transcript per 10–100 cells, on average (Shelton and Reichardt, 1984).

*a. Peripheral nervous system: sympathetic targets* In the peripheral nervous system, the level of expression generally parallels the density of sympathetic innervation as measured by tyrosine hydroxylase activity (Korsching and Thoenen, 1983; Shelton and Reichardt, 1984). These data have been interpreted as supporting the neurotrophic hypothesis that target tissues produce factors such as NGF to support the survival (and differentiation) of specific neurons that supply them. Presumably, NGF production serves to regulate the appropriate extent of innervation for the various targets; the heart is highly regulated by sympathetic input and expresses relatively high levels of NGF mRNA, but the liver requires little neural regulation and expresses virtually no NGF. However, resolution of the cellular sites of synthesis in sympathetic targets rarely has been possible. Analysis of NGF mRNA by in situ hybridization shows a signal in the dilator muscle of the iris, but this signal is very weak and the heavily innervated iris expresses more NGF than do most peripheral targets (Bandtlow et al., 1987). In addition, $^{35}$S-labeled probes often do not permit localization at the level of individual cells in tissue sections.

Several exceptions to the general correlation between NGF production and innervation include the mouse submaxillary gland, the snake venom gland, the guinea pig prostate, and the placenta. The mouse submaxillary gland expresses particularly high levels of NGF, around 0.1% of total mRNA in the male and 0.01% in the female (Ishii and Shooter, 1975). However, the mouse gland receives no excess innervation relative to other peripheral targets (Korsching et al., 1985). This result can probably be explained by the polarity of the glandular epithelium, whereby NGF is secreted from the apical surface into the gland lumen in a highly regulated manner (Wallace and Partlow, 1976). Thus, responsive neurons with their terminals under the basement membrane would not have access to the high levels of NGF; measurement of retrogradely transported NGF in the mouse superior cervical ganglion shows no dramatic elevations (Korsching et al., 1985). The polarity of secretion into the gland clearly sequesters NGF from the neural supply, and has significant implications for the secretion of NGF from similarly polarized central targets that are themselves neurons. Since the high levels of NGF in the mouse submaxillary gland have no clear effect on local NGF-responsive neurons, and since the homologous glands in other mammals do not show such elevated NGF expression, regulation in this tissue will not be discussed further in the context of a target-derived neurotrophic factor. The snake venom gland and guinea pig prostate probably represent other anomalous sources of large amounts of NGF that may be related to the specific function of their secretions (Hogue et al., 1976; Harper et al., 1979). However, placental tissue from several species expresses amounts of NGF comparable to those in heavily innervated sympathetic targets

(Edwards et al., 1986). A biologic role for placental NGF has not been determined, but NGF may support early neural development as a systemic hormone rather than in a local paracrine manner like that of target-derived NGF. NGF expression in the testis and in early neural development also appears to represent exceptions to the general correlation between NGF level and innervation, but the levels are not increased dramatically relative to those seen in targets of innervation (Ayer-Le-Lievre et al., 1988b).

*b. Peripheral nervous system: sensory targets*   In the peripheral nervous system, NGF supports the survival of sensory as well as sympathetic neurons (Gorin and Johnson, 1979). Sensory neurons are considerably less abundant than sympathetic neurons and the absence of a single reliable marker for them has not allowed a simple correlation between NGF expression and sensory innervation in most target tissues. Sensory neurons also derive trophic support from their central processes, which may further complicate the correlation with NGF expression in peripheral targets. However, the rodent whisker pad receives extensive sensory innervation from trigeminal neurons, and expresses very high levels of NGF mRNA (Davies et al., 1987). In situ hybridization shows NGF transcripts in the whisker pad epithelia, and considerably fewer in surrounding mesenchyme (see Section I,B,1,e). Further, the ophthalmic, maxillary, and mandibular branches of the trigeminal nerve that supplies the sensory innervation of the head show levels of NGF mRNA that correlate with the extent of sensory supply (highest for the maxillary, moderate for mandibular, lowest for ophthalmic) (Harper and Davies, 1990). Thus, the correlation between NGF expression and extent of innervation also appears to hold for peripheral sensory processes, at least in one region. Surprisingly, the three divisions of the trigeminal nerve appear to undergo roughly the same extent of developmental cell death, despite the different concentrations of NGF in the three targets and the differences in final density of innervation (Harper and Davies, 1990). Several possibilities may explain the failure to show a difference in survival in the three populations of NGF-dependent neurons. First, NGF simply may maintain a difference in cell number that existed before contact with the target. It is hard to conceive of how NGF, autonomously regulated in the peripheral target, could be adjusted so precisely to preserve exactly the same fraction of pre-existing neurons. This demonstrates a redundancy of controls in a system that probably does not require them. Second, physiologic levels of target-derived NGF simply may exceed a threshold for cell survival, and subsequent differences in innervation density are created by terminal branching. This explanation seems unlikely in light of the observed dramatic effects of NGF on cell survival, but these observations simply may reflect the use of massive amounts of exogenous NGF. Third, NGF may be involved in sensory process outgrowth, although higher levels of NGF are not achieved until the fibers contact their target. Alternatively, a co-regulated factor well may be responsible for differential outgrowth. These fundamental questions about the biology of NGF will be addressed only by manipulating its expression in target cells in vivo.

*c. Central nervous system targets* The analogy of peripheral NGF-responsive neuronal populations has helped interpret the distribution of NGF transcripts in the central nervous system. Cholinergic neurons located in the basal forebrain and corpus striatum express NGF receptors and respond to NGF administered in vivo by increasing the level of choline acetyltransferase (Gnahn et al., 1983; Mobley et al., 1985). These systems project, respectively, to the hippocampus (and cortex) and within the basal ganglia. These central targets express high levels of NGF mRNA (Shelton and Reichardt, 1986a; Korsching and Thoenen, 1988). The cerebellum, generally considered unresponsive to NGF, contains very little NGF mRNA. However, NGF transcripts appear in many brain regions in which NGF-responsive neurons have not been identified, suggesting the existence of several NGF-responsive populations, perhaps not identified previously because of restricted access to exogenous NGF or failure to assay the appropriate response. In situ hybridization clearly has revealed NGF transcripts in hippocampal pyramidal cells, small neurons in layer III of cortex, and perhaps in dentate gyrus granule cells, virtually all the targets of cholinergic input from the basal forebrain (Rennert and Heinrich, 1986; Ayer-LeLievre et al., 1988a; Spillantini et al., 1989).

## 2. Developmental expression

Study of NGF expression during development has helped delineate its role as a neurotrophic factor. In most peripheral targets, NGF mRNA rises from barely detectable levels at birth to its adult levels by 2–3 weeks after birth, then stabilizes or declines slightly (Davies et al., 1987; Selby et al., 1987a). The level of NGF protein in the target rises over the same period, then declines, presumably as it is removed by retrograde transport. Central cholinergic targets show a similar increase in NGF transcripts after birth, but in hippocampus and neocortex, NGF transcripts decline 3-fold from peak expression at 3 weeks after birth to adulthood (Large et al., 1986). In corpus striatum, NGF mRNA rises after birth, then declines to low levels (Mobley et al., 1989). At the cell bodies of NGF-responsive cholinergic neurons in the basal forebrain, NGF mRNA remains at a very low level, whereas the level of NGF protein increases 3- to 4-fold after birth, again presumably reflecting retrograde transport from the target. To discern the neurotrophic role of NGF among the large number of central pathways, some investigators have calculated the ratio of NGF protein to mRNA (Large et al., 1986). In cholinergic targets such as the neocortex and hippocampus, this ratio is low, reflecting the removal of NGF by retrograde transport. In other targets such as olfactory bulb and cerebellum, the ratio is 10- to 20-fold higher, possibly indicating the presence of NGF-responsive populations that, like the basal forebrain, have obtained the factor from a remote target. In neocortex and hippocampus, the 3-fold decline in NGF mRNA during later postnatal development, with no reduction in NGF protein, further suggests alterations in the pathway of NGF biosynthesis or in the mode of clearance (Large et al., 1986).

Detailed analysis of the time course of NGF mRNA appearance in the rodent whisker pad demonstrates very few NGF transcripts until embryonic day 12, by which time sensory fibers already have reached the target (Davies et al., 1987). Thus, target-derived NGF does not appear to play a per wet weight of tissue declines after embryonic day 13 or 14. In this predominantly sensory target, NGF expression occurs much earlier than in most peripheral targets, suggesting that one way to alter the balance of sensory and sympathetic inputs is through a change in the timing of NGF expression.

Sensory neurons depend on NGF for survival before birth, whereas sympathetic neurons have their critical period of dependence after birth. The relative restriction of NGF expression to one but not the other of these intervals thus could be expected to favor the development of one type of neuronal input. Similarly, NGF mRNA reaches a peak in the skin and cornea of the developing chick between embryonic days 8 and 12, then declines, and gradually increases after hatching in sympathetic targets (Ebendal and Persson, 1988). However, it has not been possible previously to test this hypothesis by changing the timing of NGF expression in vivo.

### 3. Regulation by neural input

To investigate the relationship between innervation and NGF expression, the iris as a prototypic sympathetic target has been studied extensively (Ebendal et al., 1980). Following explantation of the iris, the level of NGF secretion was found to increase dramatically. This result suggested a down-regulation of NGF expression by NGF-responsive innervation in vivo, from which the tissue is released in vitro. However, denervation in vivo, either nonselective (both sensory and sympathetic) or selective (sympathetic fibers only), produced only a slight increase in NGF, suggesting instead a role for trauma in the induction of NGF expression after explantation (Korsching and Thoenen, 1985) (see Section II). Since failure to remove NGF by retrograde transport may have accounted for some of the increased expression, analysis of NGF mRNA was performed; NGF mRNA was found to be elevated 10- to 20-fold in vitro (Shelton and Reichardt, 1986b). However, denervation in vivo produced no increase in NGF mRNA, indicating that the protein had to accumulate through some other mechanism, probably by failure of retrograde transport.

The neural regulation of NGF expression in the target has been further examined in vivo. Sympathectomy at birth does not alter the normal developmental expression of NGF mRNA in the heart (Clegg et al., 1989). Explantation of the rodent whisker pad also does not alter its pattern of NGF expression (Rohrer et al., 1988). Thus, the level of peripheral NGF transcripts appears to be determined by the specific target tissue, independent of the neural input, presumably to regulate the extent of that input. In the central nervous system, sectioning the cholinergic pathway to the hippocampus does not substantially influence the expression of NGF mRNA (Gasseretal, 1986; Whittemore et al., 1987). The loss of cholinergic input in Alzheimer's disease also does not appear to be accompanied by a change in cortical NGF transcripts (Goedert et al., 1986). However, this pathway represents only one of several inputs to hippocampal pyramidal neurons. In newborn mice, removal of the whisker pad reduces sensory input and results in a dramatic change in cortical cytoarchitecture, but no change in the NGF mRNA content of specifically dissected somatosensory cortex (Selby et al., 1987a).

On the other hand, recent studies suggest that neural activity may regulate NGF expression in central targets. Generalized seizures dramatically induce NGF mRNA in hippocampus and cortex (Gall and Isackson, 1989). An approximately 25-fold induction of NGF transcripts occurs in dentate gyrus granule cells within 2–3 hr following a seizure. The level of NGF mRNA in dentate gyrus granule cells then declines, but remains 2- to 12-fold elevated for more than 24 hr with continued seizure activity. NGF transcripts begin to accumulate to high levels in entorhinal cortex, piriform cortex, the amygdala, and layers II, III, and VI of neocortex 17–24 hr after the seizure (Gall and Isackson, 1989). This second wave of NGF induction presumably reflects spread of the seizure from its original site, followed by activation of early response genes and then NGF in these secondary locations. These findings have several interesting implications for both the pathogenesis of seizures and the normal regulation of NGF expression. Seizures result in the formation of somatic spines and the sprouting of mossy fiber collaterals, possibly in response to the loss of mossy cells and other hippocampal interneurons (Sutula et al., 1988). These structural changes may underlie the propensity of epileptiform cortex to generate continued seizures. In addition, these results suggest that vigorous subseizure stimulation also may activate NGF expression.

Recent observations support the regulation of NGF expression by physiologic activity. The hypothalamus of mice that fight contains dramatically increased NGF mRNA relative to controls (Spillantini et al., 1989). In addition, stimulation of hippocampal neurons in vitro with KCl depolarization elevates the mRNA for NGF and brain-derived neurotrophic factor (BDNF) more than 10-fold (Zafra et al., 1990). A variety of neurotransmitters failed to produce such an increase, whereas carbachol had a mild effect and excitatory amino acids exhibited the full effect, apparently acting through non-NMDA receptors. Although these cultures contained many cells other than neurons, it is likely that the only cells producing NGF were neurons. Further, administration of the various stimulants in vivo produced similar changes, that clearly could be separated from the effects of seizure activity (Zafra et al., 1990). These studies of NGF regulation in neurons raise several important questions. First, several agents including catecholamines also influence NGF mRNA in fibroblasts, but, as mentioned earlier, peripheral denervation in situ does not alter the level of transcripts. To demonstrate that neural activity regulates NGF mRNA in central targets and thus represents a different phenomenon, it will be important to demonstrate that the loss of excitatory input to the hippocampus in situ changes the levels of transcripts (of both NGF and BDNF), and that these changes occur in neurons and not in glia, as might occur after injury (see subsequent text). Conversely, it would be important to know that infusions of catecholamine or other neurotransmitters did not elevate NGF mRNA in various peripheral targets. However, the observations in aggressive mice strongly suggest that physiologic activity can regulate NGF transcripts. In hippocampal pyramidal neurons, the most dramatic effects on NGF mRNA occur after stimulation with excitatory amino acids, probably because these ultimately depolarize the cells to a greater extent than do other transmitters. This result suggests that the principal NGF-responsive cholinergic input to pyramidal cells does not itself provide the strongest stimulus for NGF induction. In this case, the up-regulation of NGF expression by activity might not play a strong role in the presumed stabilization of more active connections. On the other hand, endogenous cholinergic input may show a relatively stronger influence on NGF expression in situ by virtue of its location on proximal dendrites close to the axon hillock. Alternatively, NGF regulation may not act to stabilize specific inputs that are more active than other homologous inputs, but to increase cholinergic input to cells exposed to higher levels of heterologous excitatory input. Finally, it will be critical to determine the dynamic influence of altered NGF expression by target neurons on the structure of classic synapses and the physiologic relevance for their function. The control of NGF (or BDNF or NT-3) mRNA levels may well prove to be a widespread phenomenon, but it will be essential to determine its actual contribution to the structure of circuits and their plasticity.

## B. Mechanisms of Regulation

### 1. Transcription unit

*a. Alternative RNA splicing*  To determine the molecular basis for the regulated expression of NGF, the organization of the NGF transcription unit has been studied. Several NGF transcripts derive from a single gene in mouse and in humans (Edwards et al., 1986). The NGF cDNA originally isolated on the basis of oligonucleotide probes derived from amino acid sequence hybridizes to an mRNA transcript of 1.3 kb in all tissues (as well as to a second, uniformly less abundant transcript of 1.6 kb seen only in rat). The sequence predicts an NGF precursor protein of approximately 34 kDa, with the mature hormone (13 kDa) at the C terminus (Scott et al., 1983). Series of two and four basic amino acids flank the N terminus of mature NGF and also occur upstream in the prohormone. Two arginines also occur at the C terminus, followed only by a glycine residue. A series of 16 noncharged amino acids occurs approximately 70 amino acids from the N terminus of the predicted protein, and presumably acts as the signal peptide to mediate secretion. On account of the atypical internal location of this signal sequence, as well as reports of several distinct forms of the NGF complex in the mouse salivary gland (Sabouri and Young, 1986), alternative forms of NGF mRNA were sought. RNA analysis demonstrated a second, slightly shorter NGF transcript that is less abundant than the original in the mouse submaxillary gland but that predominates in more usual target tissues (Edwards et al., 1986). Only the placenta (in mouse and human) resembles the mouse submaxillary gland by containing more of the long than the short NGF transcript. cDNA cloning revealed a sequence identical to the original clone with the sole loss of 120 nucleotides from a region near the 5' end. This sequence corresponds to the second exon of the original longer transcript and contains the first AUG considered to initiate translation. Thus, translation of the shorter transcript would start at the next in-frame AUG, which is situated just upstream from the putative signal peptide and thus places the signal peptide in a more typical N-terminal location. Additional transcripts detected by cDNA cloning showed initiation of transcription in an intron in one case (or artifactual cloning of a truncated cDNA corresponding to an incompletely processes RNA intermediate) and transcription initiation upstream of the usual site in another, with a substitution of the original N-terminal residues by others of approximately the same length (Selby et al., 1987a). Both transcripts occur at very low levels in the mouse submaxillary gland and more usual target tissues.

*b. Genomic organization and chromosomal location* The mouse gene spans more than 40 kb of genomic DNA and contains four main exons. The two major transcripts initiate at the same site and thus contain the same first 33-bp exon. Restriction digests of genomic DNA indicate that this first exon is located more than 32 kb from the second; the exact distance is not determined because the various isolated phage clones do not overlap. The longer transcript contains the second 127-bp exon, whereas the shorter transcript lacks these sequences. Both contain a 124-bp third exon that is separated from the second and the fourth by approximately 4 and 6 kb, respectively, in the mouse gene. The fourth exon, shared by all transcripts, encodes the second in-frame initiation codon of the longer transcript and the first of the shorter, as well as the majority of the NGF precursor and mature NGF. Clones for the 3' end of the human NGF gene reveal a very similar structure (Ullrich et al., 1983); the gene is located on the short arm of chromosome 1 (Francke et al., 1983).

*c. Putative regulatory sequences* The sequence surrounding the 5' end of the mouse NGF gene contains several motifs characteristic of *cis*-acting regulatory regions (Selby et al., 1987a). TTAAA, located at −24 to −28 relative to the transcription start site, presumably acts like the more typical TATAA sequence to direct the initiation of transcription. AP-1-like sites occur further upstream on both strands of the 5' flanking region, but their functional significance remains unclear. The upstream region also has a high content of G and C residues, frequently arranged as (G)GGAGGG, which may represent atypical Sp1 sites (usually GGGCGG) (Selby et al., 1987a; Zheng and Heinrich, 1988). These sequences are frequently found as elements that confer the constitutive low level expression of housekeeping genes, an idea that is consistent with the expression of NGF in many innervated target tissues, although the extremely low levels and subtle variations in control clearly implicate additional regulatory mechanisms.

*d. Mechanism of transcriptional activation* The transcription factor AP-1 appears to mediate the induction of NGF mRNA by a variety of stimuli. Although sympathetic innervation does not regulate the steady-state levels of NGF mRNA in vivo, catecholamines do increase the level of NGF transcripts in L–M mouse fibroblasts in vitro (Furukawa et al., 1986). Similarly, beta-adrenergic agents stimulate the expression of NGF in C6 rat astrocytoma cells (Schwartz, 1988). This effect is preceded by an increase in intracellular cAMP, c-*fos* mRNA, and FOS protein (Mocchetti et al., 1989). Further, cycloheximide blocks both the accumulation of FOS protein and the increase in NGF mRNA. This suggests that a complex involving c-*fos* activates NGF expression after exposure to catecholamines. In L929 cells, serum and phorbol esters induce NGF transcripts; this induction can be suppressed with the protein kinase C inhibitor K-252a or cycloheximide, indicating a requirement for protein synthesis (Ebendal et al., 1980; Whittemore and Seiger, 1987; Houlgatte et al., 1988). Similarly, seizures induce c-*fos* and AP-1 expression in a variety of central neuronal populations (Morgan et al., 1987). Indeed, this result has prompted investigators to determine whether seizures subsequently induce NGF expression, and, as noted earlier, they do. Last, the induction of NGF expression in Schwann cells and perineural fibroblasts after peripheral nerve injury appears to depend on binding of the transcription factor AP-1 to a site in the first intron (Hengerer et al., 1990) (see Section II,B,1). Deletion of this AP-1 site reduced both the induction and the basal level of expression conferred by 5'sequences of the NGF gene. The induction of NGF mRNA in hippocampal neurons either in vitro or in vivo following stimulation with carbachol or kainate presumably also involves activation of transcription through occupancy of this same AP-1 site. Thus, a set of characterized early response genes appears to regulate the phasic and at least part of the tonic control of NGF expression.

Hormones can regulate the expression of NGF in the mouse submaxillary gland and in cultured cell lines, but the relevance to its expression as a neurotrophic factor in physiologic targets remains unclear. Testosterone is responsible for the 10-fold excess of male mouse submaxillary gland NGF relative to the female (Ishii and Shooter, 1975). However, in mouse L929 cells, which normally express substantial amounts of NGF, cortisone, aldosterone, and testosterone all reduce the level of NGF transcripts (Siminoski et al., 1987; Wion et al., 1987). In vivo, peripheral and central target tissues show no apparent sexual dimorphism in the expression of NGF mRNA. Other hormones, such as thyroxine or retinoic acid, may produce the observed developmental changes in NGF expression, but this relationship remains to be established (Wien et al., 1990). Further, it is important to note that myriad cultured cell lines product NGF, whereas their counterparts in vivo may not. Indeed, it is difficult to find immortalized cell lines that do not express NGF, this result in particular has complicated the analysis of NGF receptor function.

To determine whether the region flanking the 5' end of the NGF gene contains all the sequences responsible for appropriate regulation in different tissues, a construct with these sequences fused to the cDNA for growth hormone as reporter was introduced into transgenic mice and the pattern of growth hormone expression compared with that of the endogenous NGF gene (Alexander et al., 1989). The transgene conferred appropriate overexpression in the submaxillary gland, in the typical sexually dimorphic manner. Similarly, the heart, with one of the highest levels of endogenous NGF transcripts, expressed more of the transgene that did other peripheral targets. Although the level of growth hormone expression in other targets was appropriately low and thus barely detectable, these results suggest that sequences within approximately 5 kb of the genomic DNA flanking the 5' end of the mouse NGF gene contain all the elements required for appropriate tissue-specific expression. The ability to dissect *cis*-acting elements in the control of NGF gene expression requires easy access to primary cultures or physiologically appropriate immortalized cell lines. A keratinocyte cell line expresses NGF in vitro and corresponds to a physiologic target of both sympathetic and sensory innervation in vivo (Tron et al., 1990). However, as noted earlier, the relevance of many immortalized lines appears questionable, since primary cultures of central neurons and glia express NGF at levels that do not correspond to the levels seen in vivo; for example, striatal neurons secrete more NGF than hippocampal cells do (Gonzalez et al., 1990).

### 2. Biosynthesis and secretion

*a. Organization of the precursor*  The low levels of NGF produced by usual target tissues have caused a reliance on sources of anomalous expression to study NGF biosynthesis and secretion. In the mouse submaxillary gland, NGF is secreted into the saliva (Wallace and Partlow, 1976). The beta subunit of this 7S complex contains the neurotrophic activity, whereas the alpha and gamma subunits belong to the kallikrein family of serine proteases found in the salivary gland (Varon et al., 1968). The gamma subunit has protease activity, whereas the alpha subunit shows strong structural and amino acid similarity but has no proteolytic activity, apparently because of an amino acid substitution in the active site (Isackson and Bradshaw, 1984; Isackson et al., 1984). Because these proteins bind to NGF, they have alternatively been proposed to process NGF from a precursor and remain bound to their substrate or to protect NGF from proteolytic degradation after release into such protease-rich environments as saliva or other glandular secretions, such as those from the guinea pig prostate or snake venom gland. Using metabolically labeled salivary gland slices, a higher molecular size protein of 22 kDa was detected by immunoprecipitation; cleavage to mature NGF was demonstrated by both the gamma subunit and trypsin (Berger and Shooter, 1977).

The sequence of the NGF cDNA revealed that mature NGF occurred at the C terminus of a larger precursor and predicted the sites of proteolytic cleavage (see Section I,B,1,a). The availability of the cDNA also permitted expression of the precursor in vitro to determine the role of putative processing enzymes. Cell-free translation of synthetic RNA derived from transcription of an SP6 vector containing each of the two major NGF cDNAs yielded proteins of the expected sizes, approximately 34 and 27 (Edwards et al., 1988b). However, antibodies to NGF did not recognize these problems. Similarly, immunocytochemistry of the mouse submaxillary gland revealed NGF only near the apical surface of the epithelial layer, and not within the Golgi apparatus, further suggesting that the precursor located in those earlier compartments of the secretory pathway might adopt a different conformation and so fail to be recognized by antibodies raised against the mature protein (Schwab et al., 1988). To assess the potential for cleavage of the precursor by different putative processing enzymes, the precursor translated in vitro was incubated with the alpha and gamma subunits of the mouse 7S complex as well as with trypsin. All of the enzymes cleaved the precursor, but also resulted in loss of the mature NGF moiety (Edwards et al., 1988b). The production of more stable peptides from the N terminus of the precursor raises the possibility that such molecules play distinct biologic roles, but subsequent characterization of NGF cDNAs from other species has shown relatively poor conservation of the sequences between di- and tetrabasic residues in the N terminus of the precursor. Production of an antibody capable of recognizing these peptides will be necessary to assess more definitely their possible biologic functions. Several antibodies have been raised against synthetic peptides derived from the non-NGF domain of the precursor (Ebendal et al., 1989). These immunostain the appropriate cells of the mouse salivary gland, but precisely which NGF-associated peptides they recognize remains unclear. The NGF precursors produced in vitro also possibly did not fold properly, thus explaining the loss of antigenicity and the failure to yield mature NGF after processing. To pursue this possibility, the cell-free translation mixture was supplemented with dog pancreas microsomes that allow co-translational insertion of nascent secretory proteins into the lumen of a secretory compartment. The precursors synthesized under these conditions, although translocated across the endoplasmic reticulum membrane, also failed to be recognized by NGF antibodies and failed to be processed in the expected way (Edwards et al., 1988b).

***b. Heterologous expression***   To determine the pathway of NGF synthesis and secretion in cells, the cDNA has been overexpressed using a variety of systems. COS cells, which contain the SV40 T antigen and permit transient expression from a plasmid containing the SV40 origin of replication, secrete small but detectable amounts of biologically active NGF after transfection with a vector containing the NGF cDNA (Hallbook et al., 1988). This system has been used in conjunction with site-directed mutagenesis to assess the importance of specific residues in receptor binding and biologic activity (Ibanez et al., 1990). Retroviral vectors similarly confer low levels of expression on a variety of cell types in vitro, and have been used further in a model of gene therapy to supply NGF to responsive central cholinergic neurons that have been separated in vivo from their targets and, hence, their normal source of NGF (Rosenberg et al., 1988). Vaccinia virus confers transient expression on a wide variety of cultured mammalian cells, including primary lines and cells that do not permit viral replication (Edwards et al., 1988a).

A recombinant vaccinia virus expression vector containing the NGF cDNA has been used to explore the pathways of NGF synthesis and secretion in a variety of cells, including some cells that are analogous to those that express NGF in vivo (Edwards et al., 1988a). The level of expression is moderate, and permits analysis of the intracellular events in NGF synthesis, but does not produce enough protein for structural studies or therapeutic intervention. Cells infected with recombinant viruses encoding either the long or the short NGF precursor secrete biologically active NGF. Further, immunoprecipitation of metabolically labeled cells shortly after infection shows an intracellular protein of 34 kDa as well as small amounts of intracellular NGF (13 kDa) and larger amounts of secreted NGF. Pulse–chase experiments demonstrate a precursor–product relationship between the larger protein and NGF, and that the cleavage occurs inside the cell. Moreover, all the cell lines tested, including secondary skin fibroblasts, supported reasonable levels of expression and cleaved the precursor to NGF. These included the BSC40 cells used to propagate vaccinia as well as L929 cells, hamster insulinoma (HIT) cells, L8 muscle cells, Madin–Darby canine kidney cells, mouse anterior pituitary (AtT20) cells, primary smooth muscle cells, and atrial and ventricular cardiac cells. Thus, although the parathyroid gland contains predominantly a 30-kDa protein reactive with NGF antisera (Dicou et al., 1986), processing of the NGF precursor otherwise does not appear to be cell specific. This result is not surprising in light of the production of NGF by a variety of peripheral and central targets and the general parallel between NGF mRNA levels and the extent of innervation, suggesting little posttranscriptional control of NGF expression. Defective processing of the NGF precursor also does not appear to account for the clinical syndrome of familial dysautonomia, in which a variety of autonomic neurons degenerate and abnormalities of NGF in serum and fibroblasts have been detected, but in which the NGF gene is not linked to the disorder (Breakefield et al., 1984; Edwards and Rutter, 1988).

*c. Activation by proteolytic processing: importance of the precursor* The ability of NGF antisera to recognize the precursor made in vivo but not in vitro suggests that the protein made in vitro may not fold properly, even in the presence of dog pancreas microsomes. This phenomenon has been observed with other secretory proteins, and appears to be a result of incorrect redox potential in the cell-free translation mixtures (Scheele and Jacoby, 1982). Since the disturbance in folding could account for improper processing by various proteases, precursor rescued from cells infected by the recombinant vaccinia virus also was incubated with the gamma subunit of the 7S complex and with trypsin. Using this cell-expressed precursor as substrate, several enzymes released mature NGF (Edwards et al., 1988b). Trypsin performed the reaction more efficiently than did the gamma subunit, probably because the gamma subunit remained bound to the product and therefore acted in a stoichiometric manner rather than as an enzyme. Moreover, the conversion from precursor to NGF was accompanied by a dramatic increase in biologic activity (Edwards et al., 1988b). Thus, the precursor probably serves a very important function in the folding of NGF, enabling appropriate cleavage by a range of serine proteases to activate biologic function.

Unlike the products obtained from cell-free translation, recombinant vaccinia viruses encoding the two major NGF transcripts both produce a precursor of the same size. The longer transcript does appear to give rise to small amounts of some slightly larger proteins, but N-terminal sequencing of the metabolically labeled immunoprecipitated precursors obtained after infection with each virus indicates cleavage in the usual location, downstream of the common signal peptide (Edwards et al., 1988b). Translation of the N-terminal 70-amino acid extension encoded by the longer transcript may be followed by cleavage after translocation, or cellular translation simply may initiate in both cases at the common AUG located in sequences of the fourth and largest exon. Further, pulse–chase studies show equal rates of precursor cleavage and NGF secretion from both precursors. Preliminary data from infection of Schwann cells with the two vaccinia viruses shows the only major discrepancies between the two NGF precursors (see Section II,B,3). Thus, the physiologic role for the two major transcripts remains unclear, although it is conserved from mouse to human (unlike the 1.6-kb NGF transcript that appears only in rat). In addition, the precursor for brain-derived neurotrophic factor may have two potential initiation codons near the N terminus of the precursor in a position similar to that in NGF, supporting a specific physiologic role for the two forms (Hohn et al., 1990).

*d. Regulated secretory pathway* Although the recombinant vaccinia virus increases the amount of NGF and so permits its detection in a variety of cell types analogous to those that produce NGF in vivo in the peripheral nervous system, a crucial question about the regulation of NGF secretion concerns the mode of delivery by central targets that are themselves neurons, and hence polarized, and have a regulated secretory pathway. In the mouse submaxillary gland, NGF release is highly polarized into the glandular lumen, as noted earlier. Further, catecholamines dramatically induce the release of NGF into the gland, so collected mouse saliva constitutes a simple means to purify NGF (Wallace and Partlow, 1976). Infection of AtT20 cells with the recombinant viruses also demonstrates that NGF sorts into the regulated pathway of secretion (Edwards et al., 1988a). In addition, a rat insulin promoter-NGF (RIP-NGF) construct confers NGF expression on beta cells in the pancreatic islets, where immunostaining shows a distinctly granular appearance in the NGF-positive cells (Edwards et al., 1989). Thus, in addition to the recently observed regulation of NGF mRNA, activity might regulate NGF release and differences among multiple neuronal inputs may serve as a basis for competition. However, the low levels of NGF expressed by central targets currently preclude direct identification of the site of NGF release from neurons or the mode of its regulation.

If NGF sorts to the standard regulated pathway in neurons, it would enter synaptic vesicles destined for the axon terminal. However, NGF-responsive cholinergic neurons do not form a synapse at the axon terminal of NGF-producing hippocampal pyramidal or dentate gyrus granule neurons. Rather, the cholinergic neurons terminate at the proximal dendrites and cell body, suggesting that these constitute the physiologically relevant sites of NGF release. NGF delivery to responsive neurons requires either only partial sorting to synaptic vesicles, with sufficient release at nonterminal locations to support neuron survival and differentiation, or a specific pathway for targeting postsynaptic molecules to the proximal dendrites and soma. The similar size and basic pI of molecules such as NGF, BDNF, and neurotrophin-3 (as well as the basic FGFs) constitute physical characteristics that may help direct such target-derived neurotrophic factors to their appropriate cellular locations (Leibrock et al., 1989; Hohn et al., 1990). Whereas the low levels of NGF expression preclude detection of such a polarized secretory pathway in neurons, it has been possible to study NGF secretion from other polarized epithelial cell types. Madin–Darby canine kidney cells grow as a monolayer on filters and, when infected with recombinant vaccinia viruses encoding the NGF precursors, secrete most of the expressed NGF basolaterally and only small amounts apically (D. M. Currie and R. H. Edwards, unpublished data). Rat ileum supports the transepithelial transport of NGF from the gut lumen to blood, corresponding to an apical to basolateral direction (Siminuski et al., 1986). Although the data suggest that such a specific basolateral or somatodendritic secretory pathway may mediate the appropriate delivery of neurotrophic factors by central targets, this conclusion awaits demonstration, probably by increasing the levels of expression and directly visualizing the targets.

*e. Fate after secretion* Several factors may alter the fate of NGF after secretion and before uptake by a responsive neuron. As noted earlier, NGF remains bound to two other proteins in mouse saliva, the alpha and gamma subunits of the 7S complex, which may participate in processing and also protect NGF against further proteolytic degradation in the gland lumen. Distinct proteins may bind to NGF released at other sites and serve a similar function, but the low levels of NGF expression preclude their identification (Pantazis, 1983). The basic pI of NGF may also direct its binding to negatively charged proteoglycans in the extracellular matrix. Indeed, overexpression of NGF in the pancreatic islets of RIP-NGF transgenic mice leads to dramatic sympathetic hyperinnervation, but the ingrowing processes follow the basement membrane that surrounds the capillaries rather than terminate directly on the beta cells that are secreting NGF, suggesting either that the fibers cannot grow and terminate on beta cells or that released NGF binds to extracellular matrix and therefore is concentrated there rather than at the cellular site of release (Edwards et al., 1989). In addition, the NGF receptor may itself serve to modify local concentrations of NGF. The high affinity form of the receptor releases NGF very slowly and mediates the first step in retrograde transport. Thus, the receptor effectively can sequester local NGF, reduce the concentrations in surrounding areas, and consequently reduce access by other cells. This sequestration mechanism may provide the basis for distinguishing cells that survive the normal developmental program of cell death from those that do not. Heterologous neuronal populations also may compete for the limited amounts of NGF produced at the target. Chemical sympathectomy with 6-hydroxydopamine increases substance P in the iris; the increase in sensory neurotransmitter is mediated by NGF, indicating competition between different neuronal populations for a single trophic factor made in the common target (Kessler, 1985). The low affinity form of the NGF receptor, found in nonneural tissues as well as in certain neuronal populations, also may serve to modify local concentrations of NGF. In addition to its postulated role in guiding axonal regeneration (see subsequent text), NGF receptor has been identified in at least one sensory target region, the rodent whisker pad. In this tissue, receptor transcripts are expressed in mesenchyme and NGF transcripts are expressed in epithelial cells (Wyatt et al., 1990). Release of NGF from the epithelium and its subsequent binding to nonfunctional low affinity receptors in mesenchyme may account for the observed termination of many sensory fibers in mesenchyme rather than epithelium. However, the biologic significance of NGF receptors in target regions remains unclear because of the inability to manipulate the expression of these molecules in vivo.

### C. Biologic Significance of Regulated NGF Expression

The observation of regulated NGF expression in targets suggests that this regulation plays an important role in development of the nervous system. The variation in expression by different targets and its correlation with innervation density despite extraordinarily low levels in virtually all tissues indicate a highly evolved regulatory mechanism. Presumably, the density of sympathetic innervation in peripheral targets determines their function, but this remains to be demonstrated in most cases (see Section IV). The early increase in expression by sensory targets and later increase by sympathetic targets also would seem to be involved in the formation of appropriate neural inputs. In addition, the recent evidence supporting regulation of NGF expression by neural activity raises the possibility that NGF mediates the plasticity involved in a wide range of phenomena, from memory to epileptogenesis. However, the data to support the various postulated roles, particularly in the central nervous system, are entirely lacking. Alterations in a large number of molecules during development and following stimulation make it difficult to assign a specific role in the observed phenomena to NGF. Thus, it will be essential to determine whether perturbations in NGF alone suffice to produce structural and functional changes in the relevant neural circuits (see Section IV).

## II. EXPRESSION IN NEURAL SUPPORTING CELLS AFTER INJURY

### A. Observations in Peripheral Nerve

One of the earliest apparently physiologic sites of NGF synthesis to be identified was in nerve fibers of the explanted iris (Rush, 1984). This finding originally seemed controversial in light of the presumed role of NGF as a *target-derived* neurotrophic factor. However, it has become apparent that injury to peripheral nerve induces the mRNAs for NGF and NGF receptor in the nerve; this induction occurs as part of a process distinct from expression in the target.

Both sciatic nerve crush and transection increase the level of NGF transcripts just proximal and at all sites distal to the lesion, although the largest increase occurs just distal to the site of injury (Heumann et al., 1987a). The induction is biphasic: the first peak of expression is within 24 hr of injury and the second is at 3 days. The patterns of expression following crush and transection then diverge; the level of NGF mRNA falls over weeks after a crush but remains elevated after transection, suggesting that the axonal regrowth that occurs after crush but not after transection is responsible for down-regulating NGF expression.

Although the phenomenon of NGF induction by injury has been studied most extensively in the peripheral nervous system, it also may operate to some extent in the brain. Astrocytes, particularly oligodendrocytes, in culture appear to be capable of NGF expression as well (Houlgatte et al., 1989; Gonzalez et al., 1990).

### 1. Role of macrophages and cytokines

The initial wave of NGF synthesis can be reproduced in vitro by simply explanting peripheral nerve segments. However, the second induction does not occur in vitro unless macrophages are added to the culture; the factor they secrete that appears to mediate the elevation of NGF mRNA is interleukin-1 (Lindholm et al., 1987). In situ hybridization of dissociated peripheral nerve shows that the cells responsible for the induction of NGF transcripts include both Schwann cells and perineural fibroblasts (Heumann et al., 1987b). Although only small caliber sympathetic and sensory neurons appear to respond to NGF, the induction seems to occur in all supporting cells, including those that surround larger fibers from cells unresponsive to NGF (such as motoneurons) as well as those that surround smaller responsive fibers.

## B. Mechanisms of NGF Induction

### 1. Initial phase: role of early response genes

The initial activation of NGF transcripts by injury follows the induction of c-*fos* and c-*jun*. The mRNA for these proto-oncogenes appears within 1 hr, peaks at 2–3 hr, and declines over 12–14 hr (Hengerer et al., 1990). Similar to the events in hippocampal neurons after generalized seizures (see previous text), NGF follows this immediate early response, appearing at 2–4 hr and remaining elevated to 12–24 hr. The sequence of events has been reconstructed in a heterologous system in vitro. Overexpression of c-*fos* in fibroblasts under the control of the inducible metallothionein promoter is shortly followed by the induction of NGF mRNA (Hengerer et al., 1990). This induction requires the presence of an AP-1 site in the first intron of the gene. Footprinting demonstrates occupancy of this site after activation, while other possible AP-1 sites in the region flanking the 5' end of the gene show no change in bound protein (Hengerer et al., 1990).

### 2. Induction by interleukin-1: role of increased transcript stability

The mechanism by which interleukin-1 (IL-1) activates the second wave of NGF expression after nerve injury also has been investigated. IL-1 increases the stability of NGF mRNA and, to a lesser extent, the rate of transcription of the NGF gene (Lindholm et al., 1988). In the presence of actinomycin D to block RNA synthesis, IL-1 still increases the level of NGF mRNA, indicating a posttranscriptional effect on NGF transcripts, presumably at the level of stability. Cycloheximide induces NGF transcripts as well, suggesting that NGF belongs to the class of response genes, with c-*fos* and c-*jun*, that are superinduced by inhibiting protein synthesis. The 3' untranslated region of the cDNA contains a long AT-rich sequence that may be specific for molecules that mediate the inflammatory response, for example, tumor necrosis factor. The AT-rich sequence is responsible for rapid transcript degradation (Heumann, 1987), but whether this sequence is responsible for the rapid turnover of NGF mRNA has not been tested. Nuclear run-on experiments also show that IL-1 slightly increases the level of NGF transcription, although the baseline level is so low that the precise magnitude of this change is uncertain.

### 3. Translational control and fate after release

Among the various cultured cells in which NGF has been overexpressed by recombinant technology methods, glia show a unique pattern of NGF expression (D. M. Currie and R. H. Edwards, unpublished observations). After infection with the recombinant vaccinia viruses and metabolic labeling for 4 hr, immunoprecipitation demonstrates large amounts of NGF and precursor in the cell extract compared with other cells, in which most of the precursor has been processed proteolytically and secreted by this time. Pulse–chase experiments support a slower rate of secretion. Further, infection with virus encoding the longer precursor yielded 5- to 10-fold more precursor and mature NGF than the shorter precursor, despite equal amounts of mRNA. Culture conditions that favor the differentiation of Schwann cells showed larger discrepancies between the results with long and short transcript than conditions that reduce differentiation (e.g., cAMP or cholera toxin). Pulse–chase experiments indicate that these differences in protein level occur as soon as the precursors are made, indicating control at the level of translation rather than of differential stability. Infection of primary astrocyte cultures shows a similar but less striking difference between expression of the two precursors. These results suggest a novel pattern of NGF regulation in glia, particularly in Schwann cells, that may contribute to the increased NGF expression seen after injury. To assess the significance of this result further, it will be important to determine the proportion of major NGF transcripts in injured nerve. Certain populations of central astrocytes also demonstrate a discrepancy between NGF mRNA and protein when placed into culture. Whereas NGF transcripts appear at high levels, no protein is made until conditions are manipulated in a specific manner (Gonzalez et al., 1990). The slower rate of NGF release from Schwann cells may derive from the coexpression of NGF receptors (Taniuchi et al., 1986). In addition, released NGF may bind to extracellular matrix secreted by Schwann cells and be presented in this manner.

## C. Biologic Significance

Despite extensive investigation into the mechanisms underlying the observed induction of NGF and NGF receptor expression, its biologic significance remains unclear. In contrast to NGF, the receptor appears only in Schwann cells of the distal stump and remains until axonal contact is re-established. A first and critical question concerns the importance of Schwann cells for regeneration. Freeze-killing the distal stump and injecting mitomycin C to inhibit Schwann cell proliferation appear to inhibit regeneration dramatically (Hall, 1986). NGF may bind to NGF receptors at the cell surface or, possibly, during biosynthesis (Taniuchi et al., 1986). Because this receptor is low affinity, not involved in internalization of the ligand, and presumably nonfunctional, NGF will remain at the Schwann cell surface until regenerating neurons with high affinity receptors arrive (DiStefano and Johnson, 1988). Thus, locally increased concentrations of NGF will act as a temporary receding target to guide regenerating axons. Since all Schwann cells in the lesioned nerve express NGF and NGF receptors, NGF will be presented both to NGF-responsive small sensory and sympathetic fibers and to larger unresponsive fibers. Alternatively, coexpression of NGF and its receptor may conceal the presence of functional high affinity binding sites, with NGF involved in an autocrine mechanism to stimulate supporting cells. Addition of NGF to Schwann cells does increase the level of the L1 cell adhesion molecule somewhat, whereas NGF antibody reduces the amount of L1 (Seilheimer and Schachner, 1987). However, several NGF antisense oligonucleotides have failed to alter Schwann cell growth and morphology in vitro (J. M. Simmons and R. H. Edwards, unpublished observations). In addition, Schwann cells show no evidence for increased receptor turnover by pulse–chase analysis of metabolically labeled cells (J. M. Simmons and R. H. Edwards, unpublished observations), although high affinity receptors, even when clearly present, rarely represent more than 10% of the total cell NGF receptors and, thus, might escape detection.

Another way to assess the physiologic importance of NGF expression in peripheral nerve is to examine the pattern of expression during development. Although the level of NGF (and NGF receptor) mRNA per wet weight of tissue declines in the weeks following birth (Heumann et al., 1987a), this simply reflects growth of the animal, since the level does not change during this time with respect to total RNA (D. M. Currie and R. H. Edwards, unpublished observations). Thus, the regulation of NGF (and receptor) expression differs dramatically after injury and during development, and its function can be expected to differ correspondingly.

## III. EXPRESSION IN EARLY DEVELOPMENT AND NONNEURAL SETTINGS

Many brain regions express the NGF receptor early in development, suggesting a role for NGF that is distinct from its subsequent role as a target-derived neurotrophic factor (Emfors et al., 1988; Schattzmann et al., 1988; Heuer et al., 1990; Maisonpierre et al., 1990). NGF mRNA does appear between embryonic day 11 and 12, but at extremely low levels (Maisonpierre et al., 1990). Since BDNF can bind with low affinity to the NGF receptor, it is also possible that other factors in the NGF family use the same receptor to provide an entirely different set of signals at an earlier embryonic stage (Rodriguez-Tebar et al., 1990). In particular, neurotrophin-3 (NT-3) transiently rises to high levels early in brain development and may serve as a differentiation or outgrowth promoting factor in vivo (Maisonpierre et al., 1990). BDNF appears at roughly the same time, but remains at very low levels until after birth. Thus, although it is possible that trace amounts of NGF act in a previously uncharacterized manner or that NGF at these stages acts as a placentally derived systemic hormone, other molecules may prove to be the active ligands in vivo. Further definition of the pattern of growth factor expression and its correlation with developmental events will refine our assignment of function to specific factors. However, since many of the factors perform the same functions (survival, process outgrowth) in vitro, it will be particularly important to alter the levels of the various endogenous factors in vivo and observe the consequent altered patterns of differentiation and connectivity. Expression of the receptor by nonneural cells also suggests a function for members of the NGF family outside of the nervous system.

Germ cells in the adult testis express NGF identified both by in situ hybridization and by immunocytochemistry (Ayer-LeLievre et al., 1988b). Supporting Sertoli cells express the NGF receptor; this expression is modulated significantly by hormonal influences (Persson et al., 1990). Hypophysectomy, Leydig cell destruction, and blockade of the androgen receptor all up-regulate NGF mRNA; this effect is suppressed with chorionic gonadotropin and testosterone. Thus, NGF appears to mediate an interaction between germ and Sertoli cells, but the nature of the signal remains uncertain.

## IV. BIOLOGIC SIGNIFICANCE OF NGF REGULATION: STRATEGIES FOR MANIPULATION IN VIVO

To understand how the regulation of NGF expression contributes to the structure of neural circuits, one must be able to manipulate NGF expression in vivo in appropriate target cells. In particular, it is important to know the biologic significance of extremely low levels of NGF mRNA; to understand how relative differences in expression by different targets affect the distribution of neural supply; at what stages NGF acts (during process outgrowth, target contact before the critical period, in survival, or after, to control terminal branching); how upregulation by activity alters neural circuits; how NGF is released from neurons; and how different levels affect the formation of true synapses. Systemic or intraventricular NGF administration and the grafting of transfected fibroblasts have been used to elicit responses at relatively late stages of neural development (generally after birth) and, although effective in changing some behavioral parameters, do not mimic the normal development of functional connections under the control of a target-derived trophic factor (Rosenberg et al., 1988).

Using transgenic mice, it has become possible recently to alter the level of specific mRNA transcripts in particular cell populations (Hanahan, 1985). Accordingly, the known tissue specificity of insulin gene regulatory sequences has been used to increase the level of NGF mRNA expressed in one peripheral target, the islet cells of the pancreas (Edwards et al., 1989). The observed ability of many cell types to process and activate NGF from its precursor indicated that islet cells might also secrete functional NGF. The elevated NGF transcripts in beta cells give rise to amounts of NGF that can be immunostained directly in this transgenic target. Further, the overexpression leads to a drastic hyperinnervation of the islets by sympathetic fibers, with no apparent increase or decrease in the innervation of surrounding exocrine tissue. This suggests that target-derived NGF affects only the fibers that project to that target, and that relative differences in expression do not participate in a redistribution of sympathetic fibers among multiple targets. The islets of transgenic animals demonstrate a 100- to 200-fold increase in sympathetic innervation that cannot be attributed solely to altered cell survival, but certainly involves extensive terminal sprouting. Thus, the counterpart of the in vitro sprouting response to NGF probably does occur in vivo after nerves already have penetrated their target. Although the increase in NGF expression begins at least as early as embryonic day 18 in one transgenic line, sensory nerves traditionally considered NGF-responsive appear relatively unaffected, as are parasympathetic fibers. This relative unresponsiveness implicates additional factors (e.g., BDNF, competition with heterologous populations such as sympathetic neurons) in the response of these cells. In addition, the altered neural input to the islets appears to disturb the regulation of insulin secretion under stress, suggesting one physiologic reason that target-derived NGF should remain at low levels (R. H. Edwards and G. M. Grodsky, unpublished observations). A similar transgenic approach could be used to alter the level of NGF in central targets, to address the issues of NGF secretion from a neuron, the potential role of up-regulation by neural activity, and the functional consequences for central cholinergic pathways. If transgene expression begins early enough, it also will illuminate the role of the various neural growth factors (NGF, BDNF, and NT-3) in developmental events such as precursor proliferation, differentiation, and process outgrowth. The role of NGF in regeneration and nonneural development also can be addressed with this approach.

A complementary method to alter endogenous growth factor expression involves the insertional inactivation of both NGF alleles by homologous recombination in embryonic stem cells (Thomas and Capecchi, 1990). This method has very general application and does not depend on the availability of reliable tissue-specific regulatory sequences. Drastic effects early in development could be difficult to characterize and would prevent identification of subsequent roles, but precisely these less accessible phenomena currently pose major questions. A combination of the different transgenic approaches will characterize further the rapidly expanding biologic role that the analysis of expression has suggested for the family of neural growth factors.

## REFERENCES

Alexander, J. M., Hsu, D., Penchuk, L., and Heinrich, G. (1989). Cell-specific and developmental regulation of a nerve growth factor and human growth hormone fusion gene in transgenic mice. *Neuron* **3**, 133–139.

Ayer-LeLievre, C., Olson, L., Ebendal, T., Seiger, A., and Perssou, H. (1988a). Expression of the β-nerve growth factor gene in hippocampal neurons. *Science* **24**, 1339–1341.

Ayer-LeLievre, C., Olson, L., Ebendal, T., Hallbrook, F., and Persson, H. (1988b). Nerve growth factor mRNA and protein in the testis and epididymis of mouse and rat. *Proc. Natl. Acad. Sci. U.S.A.* **85**, 2628–2632.

Bandtlow, C. E., Heumann, R., Schwab, M. E., and Thoenen, H. (1987). Cellular localization of nerve growth factor synthesis by in situ hybridization. *EMBO J.* **6**, 891–899.

Berger, E. A., and Shooter, E. M. (1977). Evidence for pro-beta-nerve growth factor, a biosynthetic precursor to beta-nerve growth factor. *Proc. Natl. Acad. Sci. U.S.A.* **74**, 3647–3651.

Breakefield, X. O., Orloff, G., Castiglione, C., Coussens, L., Axelrod, F. B., and Ullrich, A. (1984). Structural gene for β-nerve growth factor not defective in familial dysautonomia. *Proc. Natl. Acad. Sci. U.S.A.* **81**, 4213–4216.

Clegg, D. O., Large, T. H., Bodary, S. C., and Reichardt, L. F. (1989). Regulation of nerve growth factor mRNA levels in developing rat heart ventricle is not altered by sympathectomy. *Dev. Biol.* **134**, 30–37.

Davies, A. M., Bandtlow, C., Heumann, R., Korsching, S., Rohrer, H., and Thoenen, H. (1987). Timing and site of nerve growth factor synthesis in developing skin in relation to innervation and expression of the receptor. *Nature (London)* **326**, 353–358.

Dicou, E., Lee, J., and Brachet, P. (1986). Synthesis of nerve growth factor mRNA and precursor protein in the thyroid and parathyroid glands of the rat. *Proc. Natl. Acad. Sci. U.S.A.* **83**, 7084–7088.

DiStefano, P. S., and Johnson, E. M. (1988). Nerve growth factor receptors on cultured rat Schwann cells. *J. Neurosci.* **8**, 231–241.

D'Mello, S. R., and Heinrich, G. (1990). Induction of nerve growth factor gene expression by 12-O-tetradecanoyl phorbol 13 acetate. *J. Neurochem.* **55**, 718–721.

Ebendal, T., and Persson, H. (1988). Detection of nerve growth factor mRNA in the developing chicken embryo. *Development* **102**, 101–106.

Ebendal, T., Olson, L., Seiger, A., and Hedlund, K.-O. (1980). Nerve growth factors in the rat iris. *Nature (London)* **286**, 25–28.

Ebendal, T., Persson, H., Larhammar, D., Lundstromer, K., and Olson, L. (1989). Characterization of antibodies to synthetic nerve growth factor (NGF) and proNGF peptides. *J. Neurosci. Res.* **22**, 223–240.

Edwards, R. H., and Rutter, W. J. (1988). Use of vaccinia virus vectors to study protein processing in human disease. *J. Clin. Invest.* **82**, 44–47.

Edwards, R. H., Selby, M. J., and Rutter, W. J. (1986). Differential RNA splicing of nerve growth factor transcripts. *Nature (London)* **319**, 784–787.

Edwards. R. H., Selby, M. J., Mobley, W. C., Weinrich, S. L., Hruby, D. E., and Rutter, W. J. (1988a). Processing and secretion of nerve growth factor: Expression in mammalian cells with a vaccinia virus vector. *Mol. Cell. Biol.* **8**, 2456–2464.

Edwards, R. H., Selby, M. J., Garcia, P. D., and Rutter, W. J. (1988b). Processing of the native nerve growth factor precursor to form biologically active NGF. *J. Biol. Chem.* **263**, 6810–6815.

Edwards, R. H., Rutter, W. J., and Hanahan, D. (1989). Directed expression of NGF to pancreatic β cells in transgenic mice leads to selective hyperinnervation of the islets. *Cell* **58,** 161–170.

Ernfors, P., Hallbook, F., Ebendal, T., Shooter, E. M., Radeke, M. J., Misko, T. P., and Persson, H. (1988). Developmental and regional expression of beta-nerve growth factor receptor mRNA in the chick and rat. *Neuron* **1,** 983–996.

Francke, U., Martinville, B., Coussens, L., and Ullrich, A. (1983). The human gene for the beta subunit of nerve growth factor is located on the proximal short arm of chromosome 1. *Science* **222,**1248–1251.

Furukawa, Y., Furukawa, S., Satoyoshi, E., and Hayashi, K. (1986). Catecholamines induce an increase in nerve growth factor content in the medium of mouse L-M cells. *J. Bio. Chem.* **261,** 6039–6047.

Gall, C. M., and Isackson, P. J. (1989). Limbic seizures increase neuronal production of messenger RNA for nerve growth factor. *Science* **245,** 758–761.

Gasser, U. E., Weskamp, G., Otten, U., and David, A. R. (1986). Time course of the elevation of nerve growth factor (NGF) content in the hippocampus and septum following lesions of the septohippocampal pathway in rats. *Brain Res.* **376,** 351–356.

Gnahn, H., Hefti, F., Heumann, R., Schwab, M. E., and Thoenen, H. (1983). NGF-mediated increase of choline acetyltransferase (ChAT) in the neonatal rat forebrain: Evidence for a physiological role of NGF in the brain? *Dev. Brain Res.* **9,** 45–52.

Goedert, M., Fine, A., Hunt, S. P., and Ullrich, A. (1986). Nerve growth factor mRNA in peripheral and central rat tissues and in the human central nervous system: Lesion effects in the rat brain and levels in Alzheimer's disease. *Mol. Brain Res.* **1,** 85–92.

Gonzalez, D., Dees, W. L., Hiney, J. K., Ojeda, S. R., and Saneto, R. P. (1990). Expression of beta-nerve growth factor in cultured cells derived from the hypothalamus and cerebral cortex. *Brain Res.* **511,** 249–258.

Gorin, P. D., and Johnson, E. M. (1979). Experimental auto-immune model of nerve growth factor deprivation: Effects on developing peripheral sympathetic neurons. *Proc. Natl. Acad. Sci. U.S.A.* **76,** 5382–5386.

Hall, S. M. (1986). The effect of inhibiting Schwann cell mitosis on the re-innervation of cellular autografts in the peripheral nervous system of the mouse. *Neuropath. Appl. Neurobiol.* **12,** 401–414.

Hallbook, F., Ebendal, T., and Persson, H. (1988). Production and characterization of biologically active recombinant beta nerve growth factor. *Mol. Cell. Biol.* **8,** 452–456.

Hallbook, F., Ayer-LeLievre, C., Ebendal, T., and Persson, H. (1990). Expression of nerve growth factor receptor mRNA during early development of the chick embryo: Emphasis on cranial ganglia. *Development* **108,** 693–704.

Hanahan, D. (1985). Heritable formation of pancreatic beta-cell tumors in transgenic mice expressing recombinant insulin/simian virus 40 oncogenes. *Nature (London)* **315,** 115–122.

Harper, G. P., Barde, Y. A., Burnstock, G., Larstairs, J. R., Dennison, M. E., Suda, K., and Vernon, C. A. (1979). Guinea pig prostate is a rich source of nerve growth factor. *Nature (London)* **279,** 160–162.

Harper, S., and Davies, A. M. (1990). NGF mRNA expression in developing cutaneous epithelium related to innervation density. *Development* **110,** 515–519.

Hengerer, B., Lindholm, D., Heumann, R., Ruther, U., Wagner, E. F., and Thoenen, H. (1990). Lesion-induced increase in nerve growth factor mRNA is mediated by c-*fos Proc. Natl. Acad. Sci. U.S.A.* **87,** 3899–3903.

Heuer, J. G., von Bartheld, C. S., Kinoshita, Y., Evers, P. C., and Bothwell, M. (1990). Alternating phases of FGF receptor and NGF receptor expression in the developing chicken nervous system. *Neuron* **5,** 283–296.

Heumann, R. (1987). Regulation of the synthesis of nerve growth factor. *J. Exp. Biol.* **132,** 133–150.

Heumann, R., Lindholm, D., Bandtlow, C., Meyer, M., Radeke, M. J., Misko, T. P., Shooter, E., and Thoenen, H. (1987a). Differential regulation of mRNA encoding nerve growth factor and its receptor in rat sciatic nerve during development, degeneration, and regeneration: Role of macrophages. *Proc. Natl. Acad. Sci. U.S.A.* **84,** 8735–8739.

Heumann, R., Korsching, S., Bandtlow, C., and Thoenen, H. (1987b). Changes of nerve growth factor synthesis in nonneuronal cells in response to sciatic nerve transection. *J. Cell Biol.* **104,** 1623–1631.

Hogue, Angeletti, R. A., Frazier, W. A., Jacobs, J. W., Niall, H. D., and Bradshaw, R. A. (1976). Purification, characterization and partial amino acid sequence of nerve growth factor from cobra venom. *Biochemistry* **15,** 26–34.

Hohn, A., Leibrock, J., Bailey, K., and Barde, Y.-A. (1990). Identification and characterization of a novel member of the nerve growth factor/brain derived neurotrophic factor family. *Nature (London)* **344,** 339–341.

Houlgatte, R., Wion, D., and Brachet, P. (1988). Serum contains a macromolecular effector promoting the synthesis of nerve growth factor (NGF) in L cells. *Biochem. Biophys. Res. Comm.* **150,** 723–730.

Houlgatte, R., Mallat, M., Brachet, P., and Prochiantz, A. (1989). Secretion of nerve growth factor in cultures of glial cells and neurons derived from different regions of the mouse brain. *J. Neurosci. Res.* **24,** 143–152.

Ibanez, C. F., Hallbook, F., Ebendal, T., and Persson, H. (1990). Structure–function studies of nerve growth factor: Functional importance of highly conserved amino acid residues. *EMBO J.* **9,** 1477–1483.

Isackson, P. J., and Bradshaw, R. A. (1984). The α-subunit of mouse 7S nerve growth factor is an inactive serine protease. *J. Biol. Chem.* **259,** 5380–5383.

Isackson, P. J., Ullrich, A., and Bradshaw, R. A. (1984). Mouse 7S nerve growth factor: Complete sequence of a cDNA coding for an α-subunit precursor and its relationship to serine proteases. *Biochemistry* **23,** 5997–6002.

Ishii, D. N., and Shooter, E. M. (1975). Regulation of nerve growth factor synthesis in mouse submaxillary glands by testosterone. *J. Neurochem.* **25,** 843–851.

Kessler, J. A. (1985). Parasympathetic, sympathetic and sensory interactions in the iris: Nerve growth factor regulates cholinergic ciliary ganglion innervation in vivo. *J. Neurosci.* **5,** 2719–2725.

Koh, S., and Loy, R. (1989). Localization and development of nerve growth factor-sensitive rat basal forebrain neurons and their afferent projections to hippocampus and neocortex. *J. Neurosci.* **9,** 2999–3018.

Korsching, S., and Thoenen, H. (1983). Nerve growth factor in sympathetic ganglia and corresponding target organs of the rat: Correlation with density of sympathetic innervation. *Proc. Natl. Acad. Sci. U.S.A.* **80,** 3513–3516.

Korsching, S., and Thoenen, H. (1985). Treatment with 6-hydroxydopamine and colchicine decreases nerve growth factor levels in sympathetic ganglia and increases them in the corresponding target tissues. *J. Neurosci.* **5,** 1058–1061.

Korsching, S., and Thoenen, H. (1988). Developmental changes of nerve growth factor levels in sympathetic ganglia and their target organs. *Dev. Biol.* **126,** 40–46.

Korsching, S., Auberger, G., Heumann, R., Scott, J., and Thoenen, H. (1985). Levels of nerve growth factor and its mRNA in the central nervous system of the rat correlate with cholinergic innervation. *EMBO J.* **4,** 1389–1393.

Large, T. H., Bodary, S. C., Clegg, D. O., Weskamp, G., Otten, U., and Reichardt, L. F. (1986). Nerve growth factor gene expression in the developing rat brain. *Science* **234,** 352–355.

Leibrock, J., Lottspeich, F., Hohn, A., Hofer, M., Hengerer, B., Masiakowski, P., Thoenen, H., and Barde, Y. A. (1989). Molecular cloning and expression of brain-derived neurotrophic factor. *Nature (London)* **341,** 149–152.

Lindholm, D., Heumann, R., Meyer, M., and Thoenen, H. (1987). Interleukin-1 regulates synthesis of nerve growth factor in non-neuronal cells of rat sciatic nerve. *Nature (London)* **330,** 658–659.

Lindholm, D., Heumann, R., Hengerer, B., and Thoenen, H. (1988). Interleukin 1 increases stability and transcription of mRNA encoding nerve growth factor in cultured rat fibroblasts. *J. Biol. Chem.* **263,** 16348–16351.

Maisonpierre, P. C., Belluscio, L., Friedman, B., Alderson, R. F., Wiegand, S. J., Furth, M. E., Lindsay,

R. M., and Yancopoulos, G. D. (1990). NT-3, BDNF and NGF in the developing rat nervous system: Parallel as well as reciprocal patterns of expression. *Neuron* **5,** 501–509.

Martinez, H. J., Dreyfus, C. F., Miller, Jonakait, G., and Black, I. B. (1985). Nerve growth factor promotes cholinergic development in brain striatal cultures. *Proc. Natl. Acad. Sci. U.S.A.* **82,** 7777–7781.

Mobley, W. C., Rutkowski, J. L., Tennekoon, G. I., Buchanan, K., and Johnston, M. V. (1985). Choline acetyltransferase activity in striatum of neonatal rats increased by nerve growth factor. *Science* **229,** 284–286.

Mobley, W. C., Woo, J. E., Edwards, R. H., Riopelle, R. J., Lougo, F. M., Weskamp, G., Otten, U., Valletta, J. S., and Johnston, M. V. (1989). Developmental regulation of nerve growth factor and its receptor in the rat caudate-putamen. *Neuron* **3,** 655–664.

Mocchetti, I., De Bernard, M. A., Szekely, A. M., Alho, H., Brooker, E., and Costa, E. (1989). Regulation of nerve growth factor biosynthesis by β-adrenergic receptor activation in astrocytoma cells: A potential role of c-Fos protein. *Proc. Natl. Acad. Sci. U.S.A.* **86,** 3891–3895.

Morgan, J. I., Cohen, D. R., Hempstead, J. L., and Curran, T. (1987). Mapping patterns of *c-fos* expression in the central nervous system after seizure. *Science* **237,** 192–197.

Pantazis, N. J. (1983). Nerve growth factor synthesized by mouse fibroblast cells in culture: Absence of alpha and gamma subunits. *Biochemistry* **22,** 4264–4271.

Persson, H., Ayer-Le Lievre, C., Soder, O., Villar, M. J., Metsis, M., Olson, L., Ritzey, M., and Hokjeit, T. (1990). Expression of beta-nerve growth factor receptor mRNA in Sertoli cells downregulated by testosterone. *Science* **247,** 704–707.

Rennert, P. D., and Heinrich, G. (1986). Nerve growth factor mRNA in brain: Localization by in situ hybridization. *Biochem. Biophys. Res. Commun.* **138,** 813–818.

Rodriguez-Tebar, A., Dechant, G., and Barde, Y.-A. (1990). Binding of brain-derived neurotrophic factor to the nerve growth factor receptor. *Neuron* **4,** 487–492.

Rohrer, H., Heumann, R., and Thoenen, H. (1988). The synthesis of nerve growth factor (NGF) in developing skin is independent of innervation. *Dev. Biol.* **128,** 240–244.

Rosenberg, M. B., Friedmann, T., Robertson, R. C., Toszyuski, M., Wolff, J. A., Breakefield, X. D., and Gage, F. H. (1988). Grafting genetically modified cells to the damaged brain: Restorative effects of NGF expression. *Science* **242,** 1575–1578.

Rush, R. A. (1984). Immunohistochemical localization of endogenous nerve growth factor. *Nature (London)* **312,** 364–367.

Saboori, A. M., and Young, M. (1986). Nerve growth factor: Biosynthetic products of the mouse salivary glands. Characterization of stable high molecular weight and 32,000 dalton nerve growth factor. *Biochemistry* **25,** 5565–5571.

Schatteman, G. C., Gibbs, L., Lanahan, A. A., Claude, P., and Bothwell, M. (1988). Expression of NGF receptor in the developing and adult primate central nervous system. *J. Neurosci.* **8,** 860–873.

Scheele, G., and Jacoby, R. (1982). Conformational changes associated with proteolytic processing of presecretory proteins allow glutathione-catalyzed formation of native disulfide bonds. *J. Biol. Chem.* **257,** 12277–12282.

Schwab, M. E., Stockel, R., and Thoenen, H. (1988). Immunocytochemical localization of nerve growth factor in the submandibular gland of adult mice by light and electron microscopy. *Cell Tissue Res.* **169,** 289–299.

Schwartz, J. P. (1988). Stimulation of nerve growth factor mRNA content in C6 glioma cells by a β-adrenergic receptor and by cyclic AMP. *Glia* **1,** 282–285.

Scott, J., Selby, M., Urdea, M., Quiroga, M., Bell, T. I., Rutter, W. J. (1983). Isolation and nucleotide sequence of a cDNA encoding the precursor of mouse nerve growth factor. *Nature (London)* **302,** 538–540.

Seilheimer, B., and Schachner, M. (1987). Regulation of neural cell adhesion molecule expression on cultured mouse Schwann cells by nerve growth factor. *EMBO J.* **6,** 1611–1616.

Selby, M. J., Edwards, R., Sharp, F., and Rutter, W. J. (1987a). Mouse nerve growth factor gene: Structure and expression. *Mol. Cell. Biol.* **7,** 3057–3064.

Selby, M. J., Edwards, R. H., and Rutter, W. J. (1987b). Cobra nerve growth factor: Structure and evolutionary comparison. *J. Neurosci. Res.* **18,** 293–298.

Shelton, D. L., and Reichardt, L. F. (1984). Expression of the beta-nerve growth factor gene correlates with the density of sympathetic innervation in effector organs. *Proc. Natl. Acad. Sci. U.S.A.* **81,** 7951–7955.

Shelton, D. L., and Reichardt, L. F. (1986a). Studies on the expression of the β nerve growth factor (NGF) gene in the central nervous system: Level and regional distribution of NGF and mRNA suggest that NGF functions as a trophic factor for several distinct populations of neurons. *Proc. Natl. Acad. Sci. U.S.A.* **83,** 2714–2718.

Shelton, D. L., and Reichardt, L. F. (1986b). Studies on the regulation of beta-nerve growth factor gene expression in the rat iris: The level of mRNA-encoding nerve growth factor is increased in irises placed in explant cultures in vitro, but not in irises deprived of sensory and sympathetic innervation in vivo. *J. Cell. Biol.* **102,** 1940–1948.

Silverman, R. E., and Bradshaw, R. A. (1982). Nerve growth factor: Subunit interactions in the mouse submaxillary gland 7S complex. *J. Neurosci. Res.* **8,** 127–136.

Siminoski, K., Gonnella, P., Bernanke, J., Owen, L., Neutra, M., and Murphy, R. A. (1986). Uptake and transepithelial transport of nerve growth factor in suckling rat ileum. *J. Cell. Biol.* **103,** 1979–1990.

Siminoski, K., Murphy, R. A., Rennert, P., and Heinrich, G. (1987). Cortisone, testosterone, and aldosterone reduce levels of nerve growth factor messenger ribonucleic acid in I-929 fibroblasts. *Endocrinology* **121,** 1432–1437.

Spillantini, M. G., Aloe, L., Alleva, E., deSimone, R., Goedert, M., and Levi-Moutalcini, R. (1989). Nerve growth factor mRNA and protein increase in hypothalamus in a mouse model of aggression. *Proc. Natl. Acad. Sci. U.S.A.* **86,** 8555–8559.

Sutula, T., He, X. X., Cavazos, J., and Scott, G. (1988). Synaptic reorganization in hippocampus induced by abnormal functional activity. *Science* **239,** 1147–1150.

Taniuchi, M., Clark, H. B., and Johnson, E. M. (1986). Induction of nerve growth factor receptor in Schwann cells after axotomy. *Proc. Natl. Acad. Sci. U.S.A.* **83,** 4094–4098.

Thomas, K. R., and Capecchi, M. R. (1990). Targeted disruption of the murine int-1 proto-oncogene resulting in severe abnormalities in midbrain and cerebellar development. *Nature (London)* **346,** 847–850.

Tron, V. A., Coughlin, M. D., Jang, D. E., Stanisz, J., and Sauder, D. N. (1990). Expression and modulation of nerve growth factor in murine keratinocytes (PAM 212). *J. Clin. Invest.* **85,** 1085–1089.

Ulrich, A., Gray, A., Berman, C., and Dull, T. J. (1983). Human beta-nerve growth factor gene sequence highly homologous to that of mouse. *Nature (London)* **303,** 821–825.

Ullrich, A., Gray, A., Wood, W. I., Haytlick, J., and Seeburg, P. H. (1984). Isolation of a cDNA clone coding for the α-subunit of mouse nerve growth factor using a high-stringency selection procedure. *DNA* **3,** 387–392.

Varon, S., Nomura, J., and Shooter, E. M. (1968). Reversible dissociation of the mouse nerve growth factor protein into different subunits. *Biochemistry* **7,** 1296–1303.

Wallace, L. J., and Partlow, L. M. (1976). Alpha-adrenergic regulation of secretion of mouse saliva rich in nerve growth factor. *Proc. Natl. Acad. Sci. U.S.A.* **73,** 4210–4214.

Whittemore, S. R., and Seiger, A. (1987). The expression, localization and functional significance of beta-nerve growth factor in the central nervous system. *Brain Res. Rev.* **12,** 439–464.

Whittemore, S. R., Ebendal, T., Lärkfors, L., Olson, L., Sieger, A., Stromberg, I., and Persson, H. (1986). Developmental and regional expression of β nerve growth factor messenger RNA and protein in rat central nervous system. *Proc. Natl. Acad. Sci. U.S.A.* **83,** 817–821.

Whittemore, S. R., Lärkfors, L., Ebendal, T., Holets, V. R., Ericsson, A., and Persson, H. (1987). Increased β-nerve growth factor messenger RNA and protein levels of neonatal rat hippocampus following specific cholinergic lesions. *J. Neurosci.* **7,** 244–251.

Wion, D., Houlgatte, R., and Brachet, P. (1986). Dexamethasone rapidly reduces the expression of the beta-NGF gene in mouse L929 cells. *Exp. Cell Res.* **162,** 562–565.

Wion, D., Houlgatte, R., Barbot, N., Barrand, P., Dicou, E., and Brachet, P. (1987). Retinoic acid increases the expression of NGF gene in mouse L cells. *Biochem. Biophys. Res. Commun.* **149,** 510–514.

Wion, D., MacGrogan, D., Houlgatte, R., and Brachet, P. (1990). Phorbol 12-myristate 13-acetate (PMA) increases the expression of the nerve growth factor gene in mouse L929 fibroblasts. *FEBS Lett.* **262,** 42–44.

Wyatt, S., Shooter, E. M., and Davies, A. M. (1990). Expression of the NGF receptor gene in sensory neurons and their cutaneous targets prior to and during innervation. *Neuron* **4,** 421–427.

Zafra, F., Hengerer, B., Leibrock, J., Thoenen, H., and Lindholm, D. (1990). Activity dependent regulation of BDNF and NGF mRNAs in the rat hippocampus is mediated by non-NMDA glutamate receptors. *EMBO J.* **9,** 3545–3550.

Zheng, M., and Heinrich, G. (1988). Structural and functional analysis of the promoter region of the nerve growth factor gene. *Mol. Brain Res.* **3,** 133–140.

# Nerve Growth Factor: Actions in the Peripheral and Central Nervous Systems

Frank M. Longo, David M. Holtzman, Mark L. Grimes, and William C. Mobley

## I. INTRODUCTION

Nerve growth factor (NGF) has an important role in the developing and mature nervous system. Herein, we review the neurotrophic actions of NGF in the peripheral and central nervous systems. NGF acts by binding to specific cell surface receptors; therefore, definition of its receptor is fundamental to understanding its biologic function. A recent discovery that considerably enhances our view of NGF actions is that p140$^{trk}$, a receptor tyrosine kinase, mediates NGF signal transduction. Other recent studies on the distribution of NGF and its receptors help elucidate the role of NGF in preventing death of developing neurons and the actions of NGF on the differentiation, maintenance, and regeneration of neurons.

The discovery that NGF is but one member of the neurotrophin gene family generates the need for an important cautionary statement. Structural relatedness of NGF to other neurotrophins creates the opportunity for cross-hybridization of nucleotide probes and for cross-reactivity of antibodies. Studies of NGF gene expression using nonspecific probes could report data influenced by the presence of other neurotrophin mRNAs. Similarly, it is possible that data from studies using NGF antibodies could reflect binding to other neurotrophins as well as to NGF. Nevertheless, several observations encourage the view that current data for the distribution and actions of NGF are largely correct. In recent studies it has been possible to validate earlier reports of the distribution of NGF mRNA in the peripheral nervous system (PNS) and the central nervous system (CNS). Also, use of monoclonal antibodies to NGF has made estimations of NGF levels less prone to error. Finally, an effect of NGF deprivation on neuronal survival has been documented in a number of studies using many different NGF antisera. Future studies using probes highly specific for NGF may identify a need to reexamine certain aspects of NGF distribution and action. Nevertheless, we believe that current data convincingly document an important role for NGF actions in the PNS and CNS.

## II. NGF ACTIONS IN THE PERIPHERAL NERVOUS SYSTEM

### A. Discovery of NGF

NGF was discovered in the course of experiments exploring the nature of the interaction between developing neural centers (i.e., ganglia or neuronal nuclei) and their innervation targets. By the late 1940s, it was known that modifying the size of the target had a marked effect on the size of the corresponding neural center. Thus, removing a limb resulted in hypoplasia of the developing neural center, whereas transplanting an extra limb produced hyperplasia (Hamburger and Levi-Montalcini, 1949). Rita Levi-Montalcini, working with Victor Hamburger, discovered that degeneration of cells was a normal event in neuronal development and that the hypoplasia of neural centers following limb bud extirpation was associated with increased degeneration and atrophy of cells in the centers that normally would supply the limb. Importantly, degeneration occurred during the same time period as it would have during normal development; in normal development, degeneration was correlated in time with the growth of neurites into the target (Hamburger and Levi-Montalcini. 1949). These studies identified the target as key to the development of neurons and suggested that the target determined the size of its neural center by regulating cell death among neurons. One possible means for explaining the influence of the target was a "metabolic exchange between the neurite and the substrate in which it grows" (Hamburger and Levi-Montalcini, 1949).

An experimental system in which an essential element in the proposed "metabolic exchange" eventually would be identified was described by Bueker in 1948. He reported the results of a histologic study in which a fragment of a mouse sarcoma was implanted into the body wall of 3-day chick embryos (Bueker, 1948). Like those of Levi-Montalcini and Hamburger, his studies were aimed at addressing the influence of the target on neural centers; he used the sarcoma to provide a relatively homogeneous target. Examination of embryos fixed 3–5 days after implantation demonstrated that sensory nerve fibers from adjacent dorsal root ganglia (DRG) entered the grafts. DRGs that innervated the grafts were enlarged by 20 to 40%. In contrast there was no apparent growth stimulation of motor neurons. Intrinsic properties of the sarcoma were thought to be responsible. When Levi-Montalcini and Hamburger repeated these experiments, they confirmed the findings of Bucker and also observed that sympathetic ganglia were enlarged and had sent their fibers into sarcoma tissue (Levi-Montalcini and Hamburger, 1951). Importantly, in these and subsequent studies, the authors noted substantial deviations from the normal pattern of embryonic development in the extent of ganglion hyperplasia, the involvement of ganglia distant from the site of the tumor, and the bizarre or excessive innervation of various viscera (Levi-Montalcini and Hamburger, 1951; Levi-Montalcini, 1987). These findings led to the hypothesis that the neoplastic cells released a soluble diffusible agent that promoted differentiation and growth of sympathetic and sensory neurons. This hypothesis was supported in experiments in which the mouse sarcoma elicited the same effects when transplanted onto the chorioallantoic membrane of chick embryos, a location that allows diffusion via the circulation but prevents direct contact between the implant and developing tissue (Levi-Montalcini and Hamburger, 1953). The factor was called the nerve growth factor. Subsequent studies showed that, when embryological day (ED) 8 sensory or sympathetic ganglia were explanted in vitro adjacent to mouse sarcoma tissue, a halo of outgrowing nerve fibers developed (Levi-Montalcini et al., 1954). The use of the fiber outgrowth response of ganglion explants as an in vitro bioassay was instrumental in the eventual purification of the NGF from two rich sources: snake venom (Cohen and Levi-Montalcini, 1956) and the male mouse submandibular gland (Cohen, 1960).

A great deal has been learned about the actions of NGF in the past 40 years. Interestingly, the most prominent features of its effects already were apparent in these early studies. Prevention of cell death among specific neuronal populations, enhanced differentiation and hypertrophy of responsive cells, and tropic effects on neurites were noted to be manifestations of NGF action.

## B. NGF and NGF Receptors in the Peripheral Nervous System

### 1. Distribution of NGF in the peripheral nervous system

Northern and dot-blot analyses of RNA have demonstrated that NGF is synthesized by the peripheral targets of sensory and sympathetic neurons (Heumann et al., 1984a; Shelton and Reichardt, 1984). The presence of NGF mRNA has recently been demonstrated using a probe specific for NGF (Maisonpierre et al., 1990). NGF mRNA was detected in fibroblasts, epithelial cells, and smooth muscle cells by in situ hybridization histochemistry. NGF mRNA also was found in supporting cells in developing nerve (Heumann et al., 1984a, 1987a,b; Shelton and Reichardt, 1984; Bandtlow et al., 1987; see also Chapter 6). At the protein level, NGF or NGF-like molecules have been studied with immunohistochemical techniques, two-site enzyme-linked immunoassays, or bioassays in which NGF-like activity was blocked by NGF antibodies. NGF-like immunoreactivity has been found in the targets of developing sensory and sympathetic neurons and in the processes and perikarya of these cells (Korsching and Thoenen, 1983; Heumann et al., 1984a; Davies et al., 1987; Finn et al., 1987). Interestingly, NGF also was found in fibers of motor nerves and in skeletal muscle of ED 15–16 mouse embryos (Finn et al., 1987).

Comparison of NGF mRNA and NGF-like protein levels in peripheral targets demonstrated that both are correlated with the degree of sympathetic innervation. The correspondence of mRNA and protein levels suggests that, in peripheral targets, NGF levels are determined predominantly by the level of NGF mRNA (Heumann et al., 1984a). High levels of NGF in sympathetic ganglia in the absence of detectable mRNA were consistent with retrograde transport of NGF from target tissue rather than with local synthesis (Heumann et al., 1984a). In support of this idea, NGF immunoreactivity was decreased in sympathetic ganglia following treatment with 6-hydroxydopamine and colchicine (Korsching and Thoenen, 1985). Studies in the mouse trigeminal sensory system indicated that NGF gene expression in the target of these neurons also is correlated with density of innervation (Davies et al., 1987).

A great deal of interest has focused on the regulation of NGF gene expression in target tissues. As indicated in Section II,C, it appears that innervating peripheral neurons exert little if any effect on gene expression in development. In support of this statement are studies of NGF expression in targets deprived of neural input. Examined in heart and skin, NGF mRNA levels were not influenced when their innervation was prevented (Rohrer et al., 1988; Clegg et al., 1989). The view that the target has a primary role in regulating its innervation gains support from these observations. The mechanisms for NGF regulation and secretion in targets, and whether or not these are different in mature subjects, are yet to be defined (see also Chapter 6).

Studies of NGF mRNA and NGF protein localization are consistent with the view that NGF is synthesized in targets of neural crest-derived sensory and sympathetic neurons and that NGF is transported retrogradely from the target to the cell bodies of these neurons. These observations provide strong support for the view that NGF is a target-derived neurotrophic factor.

At times during development and after peripheral nerve injury (see subsequent text), the target source of NGF may be cells in the developing or transected nerve (Heumann et al., 1987a,b; Johnson et al., 1988). Quantitative RNA blot hybridization studies of rat peripheral nerve detected approximately 50 fg/mg wet weight NGF mRNA, which decreased 10-fold by the third postnatal week and remained at low levels in the adult (Heumann et al., 1987a,b). Similarly, *in situ* hybridization failed to detect NGF mRNA in adult nerve but showed uniform labeling in newborn nerve (Bandtlow et al., 1987). NGF mRNA also was detected in sciatic nerve-derived Schwann cells and fibroblasts in culture. The persistence of relatively high levels of NGF mRNA in newborn nerve, a developmental stage much later than the arrival of neurites in their targets, is unexplained. NGF synthesized in nerve may supplement the target-derived source transiently or, possibly, may play some other role in the developing nerve.

### 2. NGF receptors in the peripheral nervous system

*a. Defining the NGF receptor*   NGF mediates its neurotrophic effect by binding to specific cell-surface receptors (see Chapter 5). The molecular nature of these receptors and their cellular localization have been the focus of many recent studies. NGF binds to its receptors at low affinity sites ($K_d$, $\sim 10^{-9} M$) and at high affinity sites ($K_d$, ~ $10^{-11}$ $M$; Sutter et al., 1979; Meakin and Shooter, 1991a; Weskamp and Reichardt, 1991). Dose–response curves and other biochemical data suggest that the high affinity site mediates the activity of NGF (Sutter et al., 1979; Bernd and Greene, 1984; Meakin and Shooter, 1991a). High affinity NGF binding has been demonstrated in vitro for sympathetic neurons, DRG sensory neurons, and pheochromocytoma (PC12) cells. When radiolabeled NGF is cross-linked to its receptor on PC12 cells, two receptor complexes of different size are evident; one is approximately 158 kDa and one is 100 kDa (Hosang and Shooter, 1985; Meakin and Shooter, 1991a). If the mass of an NGF dimer is subtracted from these values, the receptor proteins present in these complexes would be approximately 75 kDa and 140 kDa respectively. These molecules are referred to as p75 and p140. The gene for p75 has been cloned from human (Johnson et al., 1986), rat (Radeke et al., 1987), and chick (Ebendal et al., 1986; Escandon and Chao, 1989; Large et al., 1989; Heuer et al., 1990). The p75 gene encodes a polypeptide of approximately 400 amino acids containing one transmembrane domain, an extracellular domain featuring four cysteine repeats, and a relatively short cytoplasmic domain of approximately 150 residues. The extracellular domain binds NGF (Sehgal et al., 1989; Welcher et al., 1991). Within the cytoplasmic region, there are no readily recognizable sequence motifs that offer clues to the signal transducing mechanism, apart from one near the C terminus that could serve to interact with a G protein(s) (Myers et al., 1991).

Studies of p75 have helped clarify its role. Expression of p75 cDNA in fibroblast cell lines resulted in low affinity but not high-affinity NGF binding (Johnson et al., 1986; Radeke et al., 1987). When p75 cDNA was expressed in a mutant PC12 cell line that lacks p75 and does not bind NGF, the cells displayed both high and low affinity binding (Hempstead et al., 1989). The $K_d$ of these receptors was similar to that of those on normal PC12 cells, but the number of receptors was only about one-third of normal. The p75 transfected cells responded to NGF by induction of c-*fos* (Hempstead et al., 1989). These results suggest that p75 is required for both low and high affinity binding of NGF and that PC12 cells normally express both the p75 gene and a second gene whose product is required for high affinity binding and signal transduction. A high degree of sequence conservation points to the transmembrane and C-terminal cytoplasmic portions of p75 as critical to mediating signal transduction (Large et al., 1989). These domains likely facilitate interaction with another signal-transducing molecule. Interestingly, deletions in the cytoplasmic region of p75 prevented high affinity NGF binding (Hempstead et al., 1990) and NGF-mediated tyrosine phosphorylation of cellular proteins (Berg et al., 1991). Normally, epidermal growth factor (EGF) does not evoke neurite outgrowth from PC12 cells. However, when chimeric NGF–EGF receptors, consisting of the extracellular domain of the EGF receptor fused to the transmembrane and cytoplasmic domains of p75, were expressed in PC12 cells, EGF did induce neurite outgrowth. Chimeric receptors in which the EGF receptor transmembrane domain replaced the p75 transmembrane domain did not mediate the NGF effect (Yan et al., 1991). These observations suggest that the transmembrane and cytoplasmic domains of p75 do influence NGF binding and signal transduction.

The identity of the p140 gene product has been pursued vigorously for several years. The relevance of p140 is apparent from its specific binding to NGF under conditions in which high affinity binding is emphasized; this has suggested a role for p140 in mediating NGF actions (Hosang and Shooter, 1985). Experiments have shown that NGF induces tyrosine phosphorylation of several proteins (Maher, 1988). It was found that NGF induces rapid tyrosine phosphorylation of p140 (Kaplan et al., 1991a; Meakin and Shooter, 1991b). A substantial amount of evidence suggests that p140 is itself a tyrosine kinase and that it is encoded by *trk*, a proto-oncogene (Bothwell, 1991; Kaplan et al., 1991a; Klein et al., 1991; Meakin and Shooter, 1991a,b; Radeke and Feinstein, 1991). The discovery that p140 is the product of the *trk* proto-oncogene has enhanced our view of NGF signal transduction greatly. The *trk* gene family is a group of 140-kDa receptor tyrosine kinases; identified members are *trk*, *trkB*, and *trkC*. In situ and Northern blot hybridization studies have detected *trk* gene expression in trigeminal sensory ganglia, DRGs, and paravertebral sympathetic ganglia, all of which contain NGF-responsive neurons of neural crest origin (Martin-Zanca et al., 1990; L. Parada, personal communication). It has been demonstrated that by binding to p140$^{trk}$, NGF rapidly and specifically stimulates the tyrosine kinase activity of p140$^{trk}$ and tyrosine phosphorylation on p140$^{trk}$ (Kaplan et al., 1991a,b; Klein et al., 1991; Meakin and Shooter, 1991a,b; Nebreda et al., 1991). NGF activation of p140$^{trk}$ is critical for signal transduction. When PC12 mutant cells (PC12nnr), which express p75 but do not respond to NGF, were transfected with p140$^{trk}$, they responded morphologically to NGF at concentrations similar to those used for wild type PC12 cells (Loeb et al., 1991). In p140$^{trk}$-expressing NIH 3T3 cell, NGF acted as a mitogen and induced transient expression of c-FOS proteins (Cordon-Cardo et al., 1991). Addition of NGF to oocytes expressing p140$^{trk}$ stimulated tyrosine phosphorylation of p140$^{trk}$ and resulted in meiotic maturation.

It is now possible to attempt to define the high affinity NGF receptor and to investigate its role in signal transduction. Binding and cross-linking studies demonstrated that NGF bound with low affinity to membrane fractions prepared from COS cells and to NIH 3T3 cells expressing p140$^{trk}$ (Hempstead et al., 1991; Kaplan et al., 1991b). On the other hand, NGF bound with high affinity to the membranes of COS cells that express both p140$^{trk}$ and p75 (Hempstead et al., 1991). This has suggested that both p75 and p140 are required for high affinity binding and signal transduction. Other experiments raise the possibility that high affinity binding can occur in the absence of p75. Studies with p140$^{trk}$-expressing NIH 3T3 cells, which do not express p75, showed that the majority of NGF binding sites was of low affinity; however, several intermediate to high affinity binding sites ($K_d$, 1.0–1.2 × $10^{-10} M$) also were detected (Klein et al., 1991). Interestingly, a response to NGF requiring nanomolar concentrations was found in p140$^{trk}$-expressing oocytes; this result raises the possibility that p140$^{trk}$ may mediate NGF actions even when NGF binds with low affinity (Nebreda et al., 1991). Several models can be proposed for the high affinity signal-transducing NGF receptor. Figure 1 depicts several NGF–NGF receptor complexes. The actual composition of the high affinity signal-transducing receptor is uncertain. An important goal is to define the stoichiometry of the high affinity complex. At present, no data prove or disprove the existence of the p140$^{trk}$-containing trimolecular complexes depicted in Fig. 1D and E. Also needed is a detailed kinetic characterization of NGF binding, so the relative contributions of p140$^{trk}$ and p75 can be explored. Finally, it will be important to explore further the functional consequences of the activation of p140$^{trk}$ in the presence and the absence of p75.

***b. Distribution of NGF receptors in the peripheral nervous system*** The cloning of the p75 gene and the resulting availability of p75 probes, as well as specific antibodies,

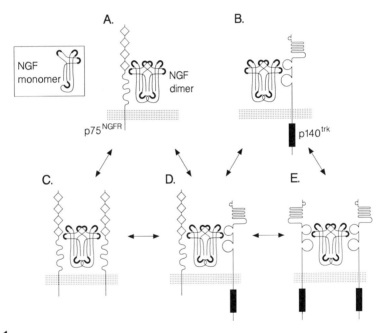

**FIGURE 1**

Interaction of NGF with its receptor. Several NGF–NGF receptor complexes are depicted. They are discussed in Sections II,B and D. Possible interactions of NGF with p75 and p140 receptors are illustrated. High affinity binding probably involves the formation of trimolecular complexes containing p140$^{trk}$, as shown in D and E. It is possible that other higher order complexes form and that they have a role in signal transduction. The bold regions in NGF indicate three β-hairpin loops (upper arms) and a reverse turn (lower arm). NGF crystallographic data (McDonald et al., 1991), inhibition of NGF activity by NGF synthetic peptides (Longo et al., 1990), and studies of NGF chimeric proteins (Ibañez et al., 1991) suggest that NGF may interact with its receptors through such surface loop regions. [Figures are adapted from the following sources: NGF, McDonald et al. (1991); p75, Mallett et al. (1990); p140, Schneider and Schweiger (1991).]

have led to several studies exploring the developmental timing and distribution of p75 gene expression. These studies have been remarkable. The distribution of p75 mRNA and protein is extremely widespread in developing animals and occurs at very early developmental stages (Yan and Johnson, 1988; Heuer et al., 1990). p75 mRNA is found in premigratory neural crest cells of the chick embryo and continues to be found in developing sensory and sympathetic ganglia, increasing in amount as the cells undergo differentiation (Heuer et al., 1990). The distribution of p75 protein in the rat is similar (Yan and Johnson, 1988). Interestingly, p75 expression is found in developing neurons derived from the neural crest and from the placode; placode-derived neurons are not known to be NGF responsive. Indeed, developing neurons in all cranial ganglia contain p75 mRNA and protein. p75 mRNA and protein also are found in developing Schwann cells (Yan and Johnson, 1988; Heuer et al., 1990). Of note, a large number of nonneural tissues was also found to express p75. p75 mRNA and protein were found in mesenchymal structures, including somites, muscle, and the tissue around hair follicles (Yan and Johnson, 1988; Heuer et al., 1990). Developing thymus also expressed p75. Expression of p75 in mesenchymal tissues was transient, decreasing in later developmental stages. Developmental down-regulation of p75 also was found in spinal cord, bursa of Fabricius, thymus (Ernfors et al., 1988), and sciatic nerve, where p75 mRNA levels were 120-fold higher at birth than in adult nerve (Heumann et al., 1987b).

Relatively few studies have detailed the distribution of high affinity NGF binding in the PNS. Nevertheless, NGF high affinity binding clearly is much more restricted than is p75 expression. Binding of $^{125}$I-labeled NGF to frozen chick sections from ED 3 to post-hatching day 3 demonstrated NGF binding with an apparent intermediate to high affinity ($K_d$, $10^{-10}M$), beginning on ED 4 in the neural crest-derived cranial and spinal sensory ganglia and paravertebral sympathetic ganglia (Raivich et al., 1985). The adrenal anlage also was labeled (Raivich et al., 1985). Transient binding also was detected in the lateral motor column and in muscle during motoneuron synapse formation, a point to be discussed in subsequent sections (Raivich et al., 1985). High affinity $^{125}$I-labeled NGF binding has been demonstrated in adult rat DRG neurons (Richardson et al., 1986), particularly those with immunoreactivity for substance P and calcitonin gene-related peptide and in which there was a high level of expression of GAP43 (Verge et al., 1989, 1990). NGF also bound with high affinity to sympathetic neurons (Richardson et al., 1986).

Studies on the identity, actions, and distribution of NGF receptor molecules emphasize the importance of p140$^{trk}$ for NGF high affinity binding and actions. As yet, a clear role for NGF has been defined only in neurons that express p140$^{trk}$. Further careful studies of p140$^{trk}$ expression in the PNS may elucidate other NGF-responsive populations. One interesting population is found in the cochleovestibular ganglion. These neurons bind and respond to NGF (Lefebvre et al., 1990; Represa et al., 1991). Whether or not they express p140$^{trk}$ is unknown. Finally, although p75 appears to contribute to signal transduction in p140$^{trk}$-expressing neurons, it is likely that p75 has functions distinct from those associated with NGF binding. Mediation of signal transduction from other neurotrophins is probable (Rodriguez-Tebar et al., 1990, 1991). Other biologic functions also may emerge for p75. Interesting recent studies in the developing kidney indicate that p75 may have a role in the morphogenesis of this organ (Sariola et al., 1991).

## C. Regulation of Developmental Cell Death by NGF

The nervous system develops as the result of progressive and regressive events. The former include neuronal proliferation, migration, and differentiation (Cowan et al., 1984). NGF actions on the differentiation of neurons will be discussed in the next section. Discovery that $p140^{trk}$, a proto-oncogene, is part of the NGF receptor and that NGF-mediated activation of $p140^{trk}$ induces mitosis in some cells (Cordon-Cardo et al., 1991) raises the possibility that NGF may act to enhance mitosis of neuronal precursors. Using a culture system for sympathetic neurons from ED 15.5 superior cervical ganglia, it was discovered that NGF had no effect on cell division (DiCicco-Bloom and Black, 1988). Examining long-term cultures of quail neural crest cells, Sieber-Blum (1991) could identify cells of the sensory neuron lineage and of the sympathoadrenal lineage. NGF treatment was found to have no significant effect on the total number of cells or on the percentage of cells that was sensory or sympathoadrenal in type (Sieber-Blum, 1991). These data argue against an effect of NGF on mitosis of sensory and sympathetic neuronal precursors, a result that conforms well to observations in vivo (Hendry, 1977). However, NGF does appear to enhance mitosis of certain neural crest-derived cells. Lillien and Claude (1985) demonstrated a mitogenic effect of NGF on adrenal chromaffin cells obtained from postnatal rats. Stemple et al. (1988) presented similar data. Interestingly, chromaffin cells differentiated into neuron-like cells on long-term exposure to NGF (Lillien and Claude, 1985). NGF also was shown to enhance mitosis among certain variants of PC12 cells, an adrenal medulla-derived cell line (Burstein and Greene, 1982). An effect of NGF on the proliferation of CNS neuroepithelial stem cells also has been reported (Cattaneo and McKay, 1990). The significance of these data for the normal development of the nervous system is uncertain. Nevertheless, the data raise the possibility that NGF may act to enhance the proliferation of certain cells, and that distinctly different responses to NGF may be recorded for different neuronal populations. Therefore it will be important to identify potential NGF-responsive embryonic populations on the basis of the presence of NGF receptors and to determine whether or not NGF influences their proliferation. Such studies may expand our view of NGF actions in the developing nervous system.

Neuron cell death is a regressive phenomenon that is widespread in the developing nervous system (Cowan et al., 1984). It serves to provide for the survival of only those neurons with functionally appropriate synaptic contacts (Cunningham, 1982; Cowen et al., 1984). Observations that bear on the nature of this process are (1) in many instances the death of neurons coincides with target innervation, (2) stereotyped morphologic patterns characterize dying neurons, and (3) neuronal degeneration appears to be an active process requiring synthesis of RNA and protein (for reviews see Martin and Johnson, 1991; Server and Mobley, 1991). Although the molecular basis of cell death in neurons remains to be elucidated, studies of NGF actions in the developing nervous system give some clues to the nature of this process and suggest experiments to elucidate further the means by which targets may act to control neuronal viability.

NGF acts as a survival factor for developing sympathetic and neural crest-derived sensory neurons. Data from studies in vivo and in vitro support this assertion. Use of NGF antibodies in vivo has provided perhaps the most compelling case. Levi-Montalcini and Booker (1960) showed that injecting NGF antiserum into postnatal mice and rats resulted in nearly complete destruction of para- and prevertebral sympathetic ganglia. Degeneration of sympathetic neurons in newborn mice treated daily with NGF antiserum was apparent under the light microscope within 2–3 days of the first injection. By the ninth postnatal day, the superior cervical ganglion was reduced in volume markedly and only a few small neurons remained. Cell counts in an animal injected for 20 days showed that only 1% of neurons was present, and that the remaining neurons were considerably smaller than normal (Levi-Montalcini and Booker, 1960). The ultrastructural manifestations of antiserum treatment were apparent within 2–12 hr and included changes in the cytoplasm and the nucleus. Nuclear changes were most marked and included disorganization of the nucleolus as well as dense clumping of chromatin (Sabatini et al., 1965; Levi-Montalcini et al., 1969; Schucker, 1972). Biochemical changes produced by antiserum injection included decreases in the ganglion in RNA and protein synthesis; in the activity of the neurotransmitter synthetic enzymes tyrosine hydroxylase, dopamine-$\beta$-hydroxylase, and DOPA decarboxylase; in the metabolism of glucose; and in impulse transmission (Goedert et al., 1978; Mobley et al., 1977). As would be expected, the target tissues of sympathetic neurons were devoid of fibers (Hamberger et al., 1965), and there was marked depletion of norepinephrine content and uptake (Mobley et al., 1977).

Degeneration of sensory neurons followed NGF antibody treatment if animals were exposed in utero, either through placental transfer of antibody from NGF-immunized mothers or by direct injection. In rats and guinea pigs exposed to NGF antibodies via placental transfer, cell loss in the dorsal root ganglion was approximately 80% (Johnson et al., 1980). Following a single in utero injection at ED 16.5 in the rat, there was a 90% decrease in the number of unmyelinated dorsal root fibers and a decrease in myelinated fibers due to fewer small diameter fibers. Accompanying biochemical changes included a large decrease in the content of substance-P- and somatostatin-like immunoreactivities in the dorsal root ganglia, the skin, and the dorsal horn of the spinal cord. A decrease in substance-P- and somatostatin-like immunoreactivity also was demonstrated for the trigeminal ganglion (Goedert et al., 1984). The NGF antiserum effect on developing sensory and sympathetic neurons appears to be caused by antibody-mediated sequestration of

NGF; thus, NGF is prevented from binding and activating its receptor (Goedert et al., 1980).

The effects of NGF administration on neuronal cell death exactly counterpose those for NGF antibodies. Hamburger and colleagues showed in the chick embryo that repeated NGF injections eliminated or markedly reduced the naturally occurring death of developing DRG sensory neurons (Hamburger at al., 1981). For sympathetic neurons of the rat superior cervical ganglion, NGF injection eliminated the normal 30% loss of neurons that occurs during the early postnatal period (Hendry and Campbell, 1976). These data argue strongly for a role for NGF in preventing the death of DRG sensory and sympathetic neurons. The targets of these neurons normally may regulate their viability by regulating the availability of NGF. To substantiate this view and to better understand how NGF modulates cell survival, it is important to place NGF actions in the context of neuronal development. Thus, it is important to know (1) when these neurons bear NGF receptors and demonstrate a response to NGF, (2) when and where the neurons first encounter NGF, (3) when developmental neuronal death occurs, (4) what the morphologic and biochemical characteristics of dying neurons are, (5) whether these characteristics are produced by NGF deprivation, and (6) whether or not changes in tissue and ganglion NGF levels are correlated with the occurrence of developmental neuronal death.

In the rat, sympathetic precursors are found in the cervical region as a column of cells along the dorsal aorta on ED 12. By ED 14, these cells have accumulated to form a primitive superior cervical ganglion (Rubin, 1985). Mitotic activity among these cells increases their number, neurons are born during the entire late embryonic period, but neurogenic mitoses are essentially complete at birth (Hendry, 1977; Rubin, 1985). Axonal outgrowth is underway in some neurons by ED12; by ED15, many axon profiles are found in the internal carotid nerve. The first sympathetic axons to enter the lacrimal gland and the eye do so on ED 15 (Rubin, 1985). The process of axonal outgrowth continues over subsequent days and one can predict that many axons will have reached their target at ED 18–19. Degeneration of superior cervical ganglion neurons is apparent by ED 19 or 20 (Wright et al., 1983). The period of neuronal cell death thus overlaps that for mitosis among neuronal precursors. Neuron number is constant between ED 19 and postnatal day (PD) 3, at which time there is a marked decrease that is correlated with a marked increase in the number of degenerating cells. Development of the superior cervical ganglion in the mouse is similar, except the schedule of events appears to be advanced by about 1 day (Fernholm, 1971; Coughlin et al., 1977; E. M. DiCicco-Bloom, personal communication).

Studies of NGF gene expression, NGF receptor gene expression, and neuronal responsiveness to NGF can be placed in the context of these developmental events. As discussed earlier, both p75 and p140$^{trk}$ are found on sympathetic neurons. p75 is present in rat sympathetic ganglia from the time of their formation (Yan and Johnson, 1988). p140$^{trk}$ mRNA has been demonstrated in the mouse by ED 17 (L.

Parada, personal communication). In fact, p140$^{trk}$ is likely to be present even earlier. Superior cervical ganglion neurons demonstrate a response to NGF at or about the time of ganglion formation. For example, embryonic mice injected with NGF at ED 12 showed a 70% increase in tyrosine hydroxylase (TH) activity at ED 14 (Kessler and Black, 1980). Interestingly, superior cervical ganglion explants or dissociates at ED 14 demonstrated NGF responsiveness for neurite growth, cell size, and induction of TH activity but showed relatively little or no dependence on NGF for survival. (Coughlin et al., 1977; Coughlin and Collins, 1985). An effect of NGF on survival could be demonstrated increasingly in sympathetic neurons from animals ED 16 and older (Coughlin and Collins, 1985). Detailed studies of NGF expression have been carried out in the mouse. NGF protein was detected at very low levels (~0.1 ng/g wet weight) at ED 13 in the submandibular gland. There was a rapid increase to ~2 ng/g over the next 4 days. NGF then decreased to about 1 ng/gm at birth and stayed at that level through PD 12. NGF levels in the superior cervical ganglion approximately mirrored those in the submandibular gland. NGF barely was detectable at ED 13 (~3 ng/g protein), increased 8-fold by ED 15, an additional several-fold by ED 17 (~100 ng/g protein), and remained at this level through the first postnatal week. The temporal correspondence of NGF levels in the submandibular gland and superior cervical ganglion is likely to reflect retrograde transport from the gland to the ganglion (Korsching and Thoenen, 1988).

Although the data are by no means complete, the following conclusions are suggested by studies of sympathetic neurons and their targets.

1. Functional NGF receptors are likely to be present on sympathetic neurons at or before the time of target contact.
2. NGF synthesis in target tissues commenced about the time that sympathetic axons first reach them.
3. Dependence on NGF for survival lags behind certain other measures of NGF responsiveness and emerges in the several days that follow initial target contact, so most neurons are dependent for survival by PD 0.
4. Onset of dependency on NGF for survival as determined by studies in vitro precedes by 2 or 3 days the period for cell death in the superior cervical ganglion. About 50% of mouse neurons depend on NGF for survival at ED 16 (Coughlin and Collins, 1985). First evidence for cell death is at ED 19 in the rat (Wright et al., 1983).

It is quite possible, then, that neonatal sympathetic neurons are predisposed to respond to perturbations in NGF supply in the target and to die if NGF levels are insufficient. The devastating effects demonstrated for NGF antiserum treatments in the neonate support the view that marked NGF dependency exists at this time. Interestingly, antiserum injections have irreversible effects only when administered before or during the period in which cell death is prominent (Otten, 1984). The timing of developmental events is thus consistent with a role for target-derived NGF in regulating survival of sympathetic neurons.

Detailed studies of NGF actions have been carried out also by Davies and colleagues in the mouse trigeminal ganglion and in its cutaneous target field, the maxillary process (Davies et al., 1987). The trigeminal ganglion can be identified at ED 9. Sensory axons first leave the ganglion at ED 9.5 and grow toward their targets. The number of fibers leaving the ganglion increases through ED 13. The epithelium of the maxillary process is contacted by the first sensory axons at ED 11. The last axons arrive just after ED 15. There is an approximate one-to-one correspondence between the number of peripherally projecting fibers leaving the ganglion and the number of neurons in the ganglion throughout development. Sensory neuron number peaks at ED 13 and, by birth (ED 19), cell death results in the loss of over 50% of neurons. Little if any further loss of neurons occurs between birth and PD 4 (Davies and Lumsden, 1984; Davies et al., 1987).

p75 and p140$^{trk}$ both are expressed in the trigeminal ganglion. p75 mRNA, which is found principally in trigeminal ganglion neurons, is detected at ED 9.5, but at very low levels. The level increased slightly by ED 10, about 3-fold between ED 10 and 11, and another 7-fold between ED 11 and ED 15, the peak of expression. When calculated on a per neuron basis, a large developmental increase in p75 mRNA occurs after ED 12; the level plateaus after its peak on ED 15 (Wyatt et al., 1990). Thus, p75 mRNA is found in neurons prior to target contact and increases markedly in the days following contact. In other studies, it was shown that p140$^{trk}$ mRNA is present in the mouse embryo at Ed 9.5; p140$^{trk}$ mRNA clearly was localized to trigeminal ganglion cells at ED 13.5 and later (Martin-Zanca et al., 1990). Davies et al. (1987) showed that a portion of neurons in ED 10 ganglia demonstrated NGF binding after 24 hr in vitro (Davies et al., 1987); Davies has shown that p140$^{trk}$ is localized to trigeminal neurons at ED 9.5 (A. Davies, personal communication). Like that of p75, p140$^{trk}$ mRNA is increased following target contact (A. Davies, personal communication). Onset of expression of p75 and p140$^{trk}$ in trigeminal neurons is correlated with NGF responsiveness. A neurite outgrowth response to NGF in vitro was first evident in ED 10 ganglia and increased to a maximal level at ED 13. NGF responsiveness was maintained into the early postnatal period. Interestingly, NGF-independent neurite growth was seen at ED 9 and 10, was diminished by ED 12, and absent thereafter (Davies and Lumsden, 1984). This finding suggests an increasing dependence on NGF for survival following target contact. In addition, it is possible that, as for sympathetic neurons, the time of onset during development for NGF responsiveness may differ from that for dependence on NGF for survival.

NGF gene expression has been examined in the maxillary process (Davies et al., 1987). NGF mRNA was found in both epithelium and mesenchyme; due to the size of these structures, NGF mRNA was more concentrated in the epithelium. NGF mRNA first was detected at ED 10.5, just 12 hr prior to first detection of the protein. Thus, NGF first was present at the time that sensory axons entered the target. NGF mRNA levels increased rapidly over the next several days, as did protein levels. Increasing NGF levels also were detected in the trigeminal ganglion, due to retrograde transport (Davies et al., 1987). NGF levels in the maxillary process fell dramatically between ED 13 and ED 15. Given the persistence of peak levels of NGF mRNA in the tissue, this decrease is likely to have been caused, at least in part, by retrograde transport of NGF. Indeed, NGF levels in the trigeminal ganglion increased during this period (Davies et al., 1987).

A review of studies examining the trigeminal ganglion and its target field allows the following conclusions.

1. Functional NGF receptors are probably present on these neurons prior to target contact.
2. NGF synthesis in target tissues commences at or near the time of target innervation.
3. In vitro, NGF-independent neurite growth is succeeded by NGF-dependent growth.
4. NGF-independent neurite growth corresponds to the period of establishing target innervation.
5. Decrease in NGF-independent growth in vitro precedes the cell death period in vivo.
6. Administration of NGF antibodies during the cell death period leads to irreversible biochemical changes (Goedert et al., 1984).

As was the case for sympathetic neurons, the timing of developmental events is consistent with a role for endogenous NGF in regulating neuronal survival.

The possibility that diffusible target-derived molecules serve as chemoattractants for developing axons is an interesting one (Tessier-Lavigne and Placzek, 1991). NGF has been shown to be neurotropic in vivo (Levi-Montalcini and Hamburger, 1951; Menesini-Chen et al., 1978) and is known to serve as a chemoattractant for regenerating axons in vitro (LeTourneau, 1978; Gunderson and Barrett, 1979, 1980). The means by which NGF orients axonal growth is unknown; its actions are accomplished over a concentration range that makes it difficult to envision how a single class of receptors (i.e., high or low affinity) could be responsible. Relative to NGF actions on developing sensory and sympathetic neurons, it is unlikely that NGF produced in targets influences the direction of axonal growth over long distances, because NGF synthesis in the target commences at about the time that axons first enter. On the other hand, local gradients of NGF within the target could direct local axonal growth and collateral sprouting (Tessier-Lavigne and Placzek, 1991).

The biochemical events induced by developmental cell death are now the subject of intense interest. Johnson and colleagues have demonstrated that the death of newborn sympathetic neurons deprived of NGF *in vitro* is an active process requiring RNA and protein synthesis (Martin et al., 1988). Studies characterizing trigeminal cell death in vitro also have been reported. Scott and Davies (1990) found that NGF deprivation resulted in the death of these neurons, an event that was prevented by inhibition of RNA and protein synthesis. These findings parallel those obtained in other developing neural systems in vivo and in vitro (Server and Mobley, 1991). These results have prompted the view that developmental neuronal cell death is an active gene-directed process. Considering the evidence presented thus far, NGF is likely to play an important role in controlling expression of this program.

Studies of NGF and cell death are at an exciting juncture. It is now possible to implicate NGF directly in naturally occurring cell death and to ask by what means NGF regulates this process. It is critical that the molecular elements of the postulated cell death program be identified and that the means for inducing this program be defined. Several candidate genes have been suggested but, as yet, none have been demonstrated to be neuron death genes (Martin and Johnson, 1991; Server and Mobley, 1991). Evidence of a role for calcium in mediating the death process has been obtained (Koike et al., 1989); data are available that document the importance of posttranscriptional events (Edwards et al., 1991). Knowledge of NGF actions and the timing of expression of NGF and its receptor in vivo should speed definition of the molecular basis of neuron death. Current data suggest that, for most neurons, target contact precedes cell death. It is uncertain whether target contact has a role in inducing the cell death program. A long-held view is that developing neurons compete for NGF in their target (Hamburger and Oppenheim, 1982). Successful competitors survive; unsuccessful ones die. The molecular events that create this competition are still undefined.

Studies in the mouse trigeminal ganglion have made note of the decrease in NGF levels in the target at the time of cell death (Davies et al., 1987). However, at this same time, NGF mRNA levels in the maxillary process are not decreasing, and the amount of NGF in the ganglion actually is increasing. When the loss of a substantial number of neurons is considered, the increase in NGF in the ganglion is even more remarkable (Davies et al., 1987). This indicates that neurons as a whole are not failing to bind or retrogradely transport NGF. Rather, certain neurons apparently are failing to bind NGF in a tissue that remains a rich NGF source. In the superior cervical ganglion also, NGF levels are stable or increasing at the time of cell death (Korsching and Thoenen, 1988). Thus, it would appear that a simple decrease in NGF levels fails to explain degeneration of either sensory or sympathetic neurons. Instead, some neurons may be gaining a greater share of NGF in the target for their use. What is the mechanism that creates the advantage enjoyed by neurons that compete successfully for NGF? One suggestion, based on studies in the trigeminal ganglion, is that NGF-mediated induction of p75, and perhaps of p140, serves to bind and sequester NGF on neurites in contact with NGF. The richer the NGF source, the greater the extent of induction. Conceivably, then, the NGF receptor may be used to enhance competition among neurons for NGF in the target. Consistent with this view are data that show that cell death is greater among superior cervical ganglion neurons with later birthdates, suggesting that an advantage is conferred on cells whose axons enter a target relatively early (Hendry, 1977).

## D. NGF and Differentiation of Peripheral Nervous System Neurons

Young neurons respond to NGF secreted from their target organ by maturing into large well-differentiated neurons. The number of ribosomes increases and the endoplasmic reticulum enlarged, signifying increased translation. The Golgi apparatus increases in prominence to accommodate routing of new secretory proteins (see Mobley et al., 1977, for review). The new proteins synthesized during maturation include neuropeptides, enzymes required for neurotransmitter biosynthesis, proteins for the plasma membrane and secretory granules, and cytoskeletal building blocks including actin, intermediate filaments, and microtubules. These global effects on differentiation are caused by activation of protein kinases, production of second messengers, and induction of gene expression. There is a brisk and transient induction of several genes, referred to as "immediate early" or primary response genes. The products of these genes activate transcription of other classes of important genes. Among the latter are many of the so-called late genes whose products contribute to neuronal form and function.

The mechanism by which NGF promotes the remarkable transformation of the immature neuron into a mature one is under intense investigation (Halegoua et al., 1991; Levi and Alema, 1991; see also Chapter 5). Since biochemistry is difficult to perform on neurons of dorsal root ganglia and sympathetic ganglia, a model system is needed. The rat adrenal medulla-derived PC12 cell line responds to treatment with NGF by ceasing mitosis and undergoing morphologic differentiation (Greene and Tischler, 1976). The cells send out neurites with growth cones and acquire the ability to generate sodium-based action potentials and to secrete catecholamines and acetylcholine (Greene and Tischler, 1976; Halegoua et al., 1991). Although under the conditions used for most experiments the survival of PC12 cells does not depend on NGF, PC12 cells are employed widely as a model system to study the short- and long-term effects of NGF on differentiation.

NGF responses can be categorized by their kinetics. The first immediate responses occur within a few minutes of NGF binding to its receptors. These include posttranslational modifications of proteins, activation of second messengers and transcriptional activation of immediate early genes. Immediate responses have been studied extensively in PC12 cells. With the discovery that $p140^{trk}$ is a receptor for NGF (Kaplan et al., 1991a,b; Klein et al., 1991; Vetter et al., 1991), experimental dissection of the events that occur at the plasma membrane is now within reach. The first signaling event in response to NGF appears to be activation of $p140^{trk}$. Ligand-stimulated autophosphorylation or cross-phosphorylation on tyrosine is associated with enhancement of the tyrosine kinase activity of $p140^{trk}$ (Meakin and Shooter, 1991a,b). Analogous to the activation of other receptor tyrosine kinases, ligand-mediated activation of $p140^{trk}$ probably creates binding sites for recruitment of a number of specific substrates from the cytosol (Cantley et al., 1991). The activated $p140^{trk}$ substrates would then act in turn to generate intracellular second messengers and transduce signals to the cell interior (Cantley et al., 1991).

Phosphorylation of phospholipase C-γ (PLC) appears to be involved in NGF action (Kim et al., 1991; Vetter et al., 1991). This enzyme activates hydrolysis of phosphoinositol bisphosphate, producing diacylglycerol and inositol triphosphate. The latter mobilizes intracellular calcium stores which, with diacylglycerol, activate protein kinase C (PKC) (Heasley and Johnson, 1989a,b; van Calker et al., 1989;

Altin and Bradshaw, 1990). PKC activation may have an important role in mediating or augmenting NGF actions. A microtubule-associated protein (MAP) kinase induced by NGF was dependent partially on PKC (Gotoh et al., 1990). A kinase that apparently is activated specifically by the NGF-induced kinase cascade (Rowland et al., 1987), and uniquely inhibited by 6-thioguanine (Volonte et al., 1989), has been designated N-kinase. For some cells, PKC has been implicated in the activation of this kinase, which is probably the same as S6 kinase (Rowland and Greene, 1990; Volonte and Greene, 1990a,b); Cantley et al., 1991). However, PKC activation was not required for an increase in ornithine decarboxylase (ODC) activity and sodium channel induction, transcription of early response genes, or neurite outgrowth (Cho et al., 1989; Damon et al., 1990; Kalman et al., 1990; Pollock et al., 1990). Thus, NGF appears to act in both PKC-dependent and -independent pathways (Clark and Lee, 1991).

The situation is similar for cAMP-dependent protein kinase A (PKA). Agents that mimic cyclic AMP (cAMP) or cause its levels to rise are capable of causing growth arrest and inducing neurite outgrowth in PC12 cells (Boonstra et al., 1987; Chijiwa et al., 1990; Pollock et al., 1990). NGF rapidly activates adenylate cyclase in PC12 cells, and a high affinity GTPase is inhibited concomitantly, possibly implicating a G protein in this affect (Golubeva et al., 1989). cAMP levels rise transiently after NGF treatment (Schubert, 1978). NGF induction of GAP43 mRNA and down-regulation of calmodulin-dependent protein kinase III activity apparently are mediated by PKA (Nairn et al., 1987; Brady et al., 1990; Costello et al., 1990). NGF increases sodium channel number in PC12 cells, in part by activating PKA but not PKC (Kalman et al., 1990). However, a selective inhibitor of PKA did not block neurite outgrowth induced by NGF (Chijiwa et al., 1990); PKA-deficient cells are capable of responding to NGF by protein phosphorylation, transcriptional induction of the immediate early gene *egr1*, ODC activity induction, and neurite outgrowth (Ginty et al., 1991). This indicates that cAMP and PKA may not be required for many aspects of NGF action, yet they may mediate some responses to NGF.

Other kinases that are distinct from PKA and PKC also are involved in NGF action. A high molecular weight MAP kinase is activated by NGF (Tsao et al., 1990) independent of PKA or PKC (Tsao and Greene, 1991). NGF-mediated activation of extracellular signal-regulated kinases (ERKs) also has been demonstrated (Boulton et al., 1991). Taylor and Landreth (1991) have reported NGF-mediated activation of a kinase that phosphorylates the C terminus of c-FOS. Induction of the protease transin was blocked by a kinase inhibitor that has no affect on PKA or PKC (Machida et al., 1991). It is not known whether these kinases are downstream in the signaling pathway, but induction of transin appears to be augmented by protein kinases A and C (Machida et al., 1991).

In addition to cAMP elevation and phosphoinositol bisphosphate hydrolysis, other second messenger systems appear to play a role in the NGF signaling cascade. NGF rapidly increases cyclic GMP (cGMP) levels and enhances cGMP phosphodiesterase activity in PC12 cells (Laasberg et al., 1988). Interest in a role for p21*ras* in NGF signal transduction was fostered when it was shown that introduction of activated *ras* caused morphologic differentiation of PC12 cells and neurotrophin-responsive embryonic neurons (Bar-Sagi and Feramisco, 1985; Guerrero et al., 1986; Smith et al., 1986; Borasio et al., 1989). Injection of antibody raised against p21*ras* inhibited NGF-induced neurite outgrowth (Altin et al., 1991). However, blockage of p21*ras* isoprenylation and membrane association did not mitigate NGF responses (Qiu et al., 1991). Possibly through an effect on *ras*-GAP (Cantley et al., 1991), NGF stimulates phospholipase $A_2$, generating arachidonic acid derivatives (DeGeorge et al., 1988; Ellis et al., 1990; Fink and Guroff, 1990). Activation of phospholipase $A_2$ may be important for neurite outgrowth (DeGeorge et al., 1988). Other lipid hydrolases are activated by NGF also, for example, the glycosyl-phosphatidylinositol/inositol phosphoglycan signaling (Chan et al., 1989). Yet another early event in signal transduction may be activation of pp60$^{c-src}$, a membrane-associated nonreceptor protein tyrosine kinase (Alema et al., 1985). Introduction of v-*src* into PC12 cells induced neurite outgrowth and NGF-inducible large external (NILE) glycoprotein (Rausch et al., 1989). Importantly, however, v-*src* did not mimic NGF actions completely (Rausch et al., 1989).

One challenge in understanding NGF signal transduction is to distinguish the unique subset of signaling events that is necessary and sufficient for the NGF response from events that mediate responses to other growth factors that do not cause neuronal differentiation of PC12 cells. The initial signal transduction events in response to NGF are likely to overlap with events that occur in response to other polypeptide growth factors such as fibroblast growth factor (FGF), EGF, insulin, and platelet-derived growth factor (PDGF) (James and Bradshaw, 1984; Cantley et al., 1991). For example, PKC is activated by a number of polypeptide growth factors. PKC phosphorylates S6 kinase, which phosphorylates ribosomal subunit S6 and activates translation of resident mRNAs (Cantley et al., 1991). EGF activates PKC, PKA, and a third 100-kDa kinase in common with NGF (Heasley and Johnson, 1989b). EGF also induces ODC and GAP43 in PC12 cells (Costello et al., 1990; Volonte and Greene, 1990a). Several immediate early genes are induced by FGF and EGF, as well as NGF (Cho et al., 1989). However, EGF does not induce neurite outgrowth in PC12 cells; in fact, EGF has no effect on microfilament organization, whereas NGF induces the redistribution of F-actin in PC12 cells (Paves et al., 1988). Phosphorylation of MAPs may affect the polymerization of microtubules, which is one step in neurite formation. EGF, FGF, and NGF can cause phosphorylation of MAPs (Rydel and Greene, 1987; Tsao et al., 1990; Miyasaka et al., 1991). However, MAP kinase activation by EGF was PKC independent, whereas that by NGF was partly PKC dependent and was induced with different kinetics (Gotoh et al., 1990). FGF can cause neurite outgrowth and an increase in sodium channel density in PC12 cells (Rydel and Greene, 1987; Pollock et al., 1990). FGF- and NGF-induced phosphorylation of TH increases its activity (Zigmond et al., 1989). Interestingly, FGF-induced phosphorylation occurs on only one of the two TH sites induced by NGF (Halegoua et al., 1991). PKC activation is responsible for phosphorylation of TH at the site induced by FGF; PKA is required for phosphorylation of the other site (Cremins et al., 1986). This evidence suggests that FGF may not activate PKA in the same manner as NGF does.

In summary, signal transduction is complex and appears to involve changes in phospholipid metabolism, nucleotide second messengers, and a cascade of tyrosine, serine, and threonine phosphorylations. Many of these signal transduction events also occur in response to other growth factors. The precise combination of events that is necessary and sufficient for initiating the NGF response has yet to be elucidated. The pharmacologic agent K-252a, which inhibits NGF action (Hashimoto, 1988; Hoizumi et al., 1988; Maher, 1988), may help define NGF-specific events (Rausch et al., 1989; Smith et al., 1989).

Transcriptional responses to the NGF signal transduction cascade occur within minutes of NGF binding. Activation of primary response genes links signal transduction with long-term changes. These genes, for example, c-*fos* and c-*jun* (perhaps as the AP-1 complex) (Kruijer et al., 1985; Curran and Franza, 1988; Wu et al., 1989), NGF I-A (*egr1*) and NGF I-B (Milbrandt, 1987, 1988) are thought to function as transcriptional activators of other genes. The expression of some of these genes appears to be short lived.

The later genes, whose expression is induced over a period of hours or days, include biosynthetic and structural proteins involved in achieving and stabilizing neuronal differentiation. Some proteins, in addition to being phosphorylated immediately, also exhibit increased expression levels at later times. These include the catecholamine biosynthetic enzyme TH, NILE glycoprotein, synapsin I, periferin, and microtubule-associated proteins MAP 1.2 and MAP 2 (Romano et al., 1987; Aletta et al., 1989; Halegoua et al., 1991). Increased gene expression is observed also for the cytoskeletal proteins actin, tubulin, MAP 3, MAP 5, and tau; intermediate filament subunits; and the plasma membrane protein thy-1 (Halegoua et al., 1991). Proteins characteristic of differentiated neurons—including neuron-specific enolase, neuropeptide Y, substance P, the enkephalin receptor, voltage-dependent type II sodium channel, and calcium channels—are induced over hours or days by NGF treatment (Lindsay and Harmar, 1989; Halegoua et al., 1991; Levi and Alema, 1991). p75 also is induced by NGF (Lindsay et al., 1990; Miller et al., 1991). The result of these changes in gene expression is morphologic and functional differentiation.

A potentially important difference between NGF actions on cells in tissue culture and on neurons in vivo is that initial fiber outgrowth of sensory and sympathetic neurons in vivo is independent of NGF (Davies et al., 1987; Korsching and Thoenen, 1988). Further, the target organ provides a point source of NGF distant from the cell body (Heumann et al., 1984a; Korsching and Thoenen, 1988). The NGF signal is communicated to the developing neuron via the axon (Korsching and Thoenen, 1983; Davies et al., 1987). The *in vitro* experiments of Campenot (1977) demonstrated that NGF applied to a chamber containing neurites is sufficient to support survival of neurons whose cell bodies were in an adjacent chamber that did not receive NGF. These data indicate that activation of NGF receptors in the neurite is very important for the NGF response and suggest that retrograde axonal transport of NGF may play a critical role in signal transduction. The retrograde transport of NGF depends on interaction of NGF with receptors at the nerve terminal (Dumas et al., 1979). The speed with which the signal returns from the growth cone is that of retrograde transport: 2–3 mm/hr (Johnson et al., 1978; Schnapp and Reese, 1989). NGF transported from the target to the cell body remains intact in the perikaryon for several hours, but there is no evidence that NGF or NGF fragments injected into the cell body can transduce the signal (Heumann et al., 1984b). It has been proposed (Misko et al., 1987; Halegoua et al., 1991) that retrograde transport of NGF occurs in membrane-bound vesicles that are moved by dynein (Schnapp and Reese, 1989) along microtubules. The vesicles could be endosomes or vesicles derived from early endosomes present in growth cones.

p75 is transported retrogradely with NGF, suggesting that it may have a role in signal transduction (Taniuchi and Johnson, 1985; Johnson et al., 1987). Of particular interest is the possibility that p140$^{trk}$ is transported retrogradely. Although this is likely, it has not yet been demonstrated. One can easily envision NGF bound to actively signaling p140$^{trk}$ receptor as it is carried in vesicles to the cell body. However reasonable this hypothesis is, it has not been proven. An alternative is that the cascade of NGF receptor-mediated events at the neurite tip modifies returning proteins or transport vesicles so they carry the signal to the cell body. Thus, an important open question is the molecular nature of the returning signal.

Internalization of NGF and its receptors has been studied in tissue culture cells. After 1 hr of NGF treatment of PC12 cells at 37°C, approximately 75% of total bound NGF was internalized (Bernd and Greene, 1984). At this time, there was no degradation of the internalized NGF (Buxser et al., 1990). Internalization apparently was mediated only via the high affinity receptor (Bernd and Greene, 1984). After internalization, some $^{125}$I-labeled NGF was associated with a Triton X-100 insoluble fraction that was shown to be the cytoskeleton (Schecter and Bothwell, 1981; Vale et al., 1985). These results may be interpreted in the context of similar data from studies of other tyrosine kinase receptors. The EGF receptor (Wiegant et al., 1986; van Bergen en Henegouwen et al., 1989) and enzymes associated with the phosphoinositide pathway, including phosphoinositide kinase, diacylglycerol kinase, and phospholipase C, become associated with the actin filament system in response to EGF (Payrastre et al., 1991). The role of this cytoskeletal sequestration of receptors and signal transducing enzymes is not known. One possibility is that so-called "signal particles" are localized to a particular domain at the cell surface to enhance actin polymerization (Lassing and Lindberg, 1985; Payrastre et al., 1991).

Figure 1 illustrates models of NGF high affinity binding and signal transduction. In one model (Fig. 1D), the high affinity NGF binding site is proposed to be composed of an NGF dimer bound to a p75-p140$^{trk}$ heterodimer (Hempstead et al., 1991). If this heterodimer complex is responsible for propagating the NGF signal to the cell body, we might expect it to be internalized in response to NGF. As mentioned earlier, retrograde transport of p75 has been demonstrated (Taniuchi and Johnson, 1985; Johnson et al., 1987). However, this finding may be explained by constitutive endocytosis and retrograde transport of p75. In an attempt to distinguish ligand-induced from constitutive internalization, Hosang and Shooter (1987) demonstrated that the cross-linked 158-kDa species, which is known to be NGF covalently bound to p140$^{trk}$ (Kaplan et al., 1991b), is internalized in PC12 cells in response to NGF, while there was no detectable ligand-induced internalization of p75. There are three interpretations of these data.

1. The NGF-p75-p140$^{trk}$ complex (Fig. 1D) may be shortlived and exist only on the plasma membrane. This interpretation would suggest that only the NGF-p140$^{trk}$ complex (Fig. 1B,E), or such complexes clustered with other p140$^{trk}$ molecules, is internalized.
2. The heterodimer (Fig. 1D) may be internalized but rapidly dissociate in endosomes. p75 then would recycle rapidly to the cell surface, as does the transferrin receptor (Klausner et al., 1983). If this model is correct, the interesting possibility arises that the NGF-p75-p140$^{trk}$ complex (Fig. 1D) may produce a different signal at the plasma membrane than NGF bound to p140$^{trk}$ following internalization (Fig. 1B,E).
3. The third possibility is that the relative abundance of p75, about 10- to 20-fold in excess of p140$^{trk}$ (Sutter et al., 1979; Vale and Shooter, 1985), makes it difficult to detect surface down-regulation of p75.

Occupancy-induced down-regulation of cell-surface receptors attenuates signal transduction in many other cell types. For example, occupied kinase-active EGF receptors were internalized 10-fold faster than empty or kinase-defective receptors (Felder et al., 1990; Honegger et al., 1990; Wiley et al., 1991). For EGF and PDGFβ receptors, tyrosine kinase activity promotes, but is not a prerequisite for, ligand-induced internalization of the ligand–receptor complex (Felder et al., 1990; Honegger et al., 1990; Sorkin et al., 1991; Wiley et al., 1991). Multiple phosphorylation sites appear to be required for EGF receptor kinase activation and for the ligand-induced internalization of the EGF–receptor complex (Helin and Beguinot, 1991; Sorkin et al., 1991).

In the cytoplasmic tail of membrane proteins, the consensus internalization signal for efficient endocytosis consists of a tight turn conformation around an aromatic residue (Collawn et al., 1990). This signal often takes the form of NPXY in Type I membrane proteins. Interestingly, this sequence is present in p140$^{trk}$ and is conserved in p140$^{trkB}$ and p140$^{trkC}$ (Lamballe et al., 1991). It is reasonable to suggest that this sequence may be important for internalization of p140$^{trk}$ receptors. Whether such internalization results in attenuation of NGF signal transduction, provides a way to return the signal-transducing receptor–NGF complex to the cell body by retrograde transport in endosomes or endosome-derived vesicles, or both, remains to be elucidated.

## E. NGF in the Adult Normal and Adult Lesioned Peripheral Nervous System

### 1. Normal Adult

In general, the survival-promoting and differentiating properties of NGF on neurons in vivo are less evident beyond the developmental period. Nevertheless, several lines of evidence suggest that NGF responsiveness is present in the adult and that NGF may have an important role in maintenance of mature neurons. High affinity NGF receptors are present on the cell bodies and fibers of adult sensory neurons (Richardson and Riopelle, 1984; Richardson et al., 1986) and retrograde transport of NGF can be demonstrated in these cells (Stockel et al., 1975; Richardson and Riopelle, 1984). Treatment of intact animals with NGF induces increased levels of substance P (Goedert et al., 1981). NGF appears not to be required for the survival of mature DRG neurons. Adult sensory neurons in vitro survived in defined serum-free medium in the absence of NGF (Lindsay, 1988; Yasuda et al., 1990). However, these cells responded to NGF by increased cell body size and increased neurite arborization (Yasuda et al., 1990). Antibody-mediated deprivation of NGF in adult rodents resulted in no detectable death of DRG neurons (Schwartz et al., 1982; Rich et al., 1984), but there was neuronal atrophy (Rich et al., 1984), reduction of axonal caliber (Gold et al., 1991), and decreased levels of substance P (Schwartz et al., 1982). Finally, axotomy, which interrupts the target-derived supply of retrogradely transported NGF, induces morphologic, biochemical, and electrophysiologic alterations. Some of these changes could be attenuated by administration of NGF to the proximal stump of transected nerve (Fitzgerald et al., 1985; Otto et al., 1987; Rich et al., 1987; Gold et al., 1991).

NGF also acts on mature sympathetic neurons, increasing their size and enhancing their axonal and dendritic arborization. NGF also increases the development of endoplasmic reticulum and the number of neurofilaments in these neurons (Angeletti et al., 1971; Bjerre et al., 1975; Ruit et al., 1990). Interestingly, there is evidence that remodeling of axonal arbors may be achieved by changing NGF levels in the target of the cell (Saffran and Crutcher, 1990). As evidence of a role for endogenous NGF, NGF antiserum administration caused a decrease in the size of sympathetic neurons and, following chronic antiserum treatment, there was a loss of 40% of these cells (Ruit et al., 1990).

### 2. Lesioned Adult

NGF also may play a role in peripheral nerve injury repair. Following sciatic nerve crush or transection, NGF mRNA levels in the proximal stump and distal nerve segment were increased (Heumann et al., 1987a,b). The level of expression first peaked within 24 hr following injury, again at 3 days, and then gradually decreased over several weeks. The fall in NGF mRNA levels corresponded to the period of reinnervation. In the absence of reinnervation (i.e., nerve transection instead of crush injury), NGF mRNA levels remained elevated (Heumann et al., 1987a,b). NGF gene expression in the injured nerve may be due, in part, to macrophage-derived interleukin-1 stimulation of NGF synthesis by nonneuronal cells (Heumann et al., 1987a,b; Lindholm et al., 1987). Further studies have shown that sciatic nerve transection causes an induction of c-*fos* and c-*jun* mRNA that precedes the increased in NGF mRNA (Hengerer et al., 1990). In vitro experiments with fibroblasts of transgenic mice containing an exogenous c-*fos* gene under the control of a metallothionein promoter showed a causal relationship between c-*fos* induction and activation of the NGF promoter. DNase I footprint studies and deletion experiments suggested that the activity of the NGF promoter was regulated by an AP-1 site in the first intron of the NGF gene (Hengerer et al., 1990). Increased NGF mRNA in the injured nerve led to increased NGF protein. Immunoassays have shown that the amount of NGF protein in the proximal stump of the transected nerve rapidly decreases and then recovers to 40–80% of control levels over the next several days (Heumann et al., 1987a). That the NGF produced in injured nerve was biologically active is indicated by in vitro studies in which nerve fragments, incubated in culture media, released factors that supported the growth of neurites from DRG and sympathetic ganglia. This activity was blocked partially by NGF antisera (Richardson and Ebendal, 1982; Windebank and Poduslo, 1986).

Increased NGF gene expression and accumulation of NGF in nerve after nerve injury suggests that local NGF synthesis may support neuronal regeneration. However, several observations conflict with this view. In a study using adult guinea pigs that were chronically deprived of NGF by prior immunization with NGF, the degree of regeneration following nerve crush was the same in sensory neurons in control and NGF-deprived animals (Rich et al., 1984). In another study, deprivation of endogenous NGF reduced postlesion collateral sprouting but had no effect on regeneration (Diamond et al., 1987). Provision of NGF to the tip of regenerating nerve for 4 weeks was shown to increase the number of myelinated axons in the regenerating distal segment, but there was no change in the total number of axons (Rich et al., 1989). By 10 weeks, extensive regeneration had occurred and no difference was observed between NGF- and control-treated subjects in the extent of this process (Hollowell et al., 1990). Evidence demonstrating an impressive effect of NGF on the number and size of axotomized neurons appears discordant with data for the extent of outgrowth of regenerating neurites. Thus, the role that is played by NGF synthesized at the tip of regenerating sensory and sympathetic neurons is uncertain. Other potential neurotrophic factors have been detected during peripheral nerve regeneration (Longo et al., 1983, 1984).

A novel suggestion for the role of NGF synthesis after nerve transection comes from in situ binding studies. Sciatic nerve transection was demonstrated to cause axonal NGF receptors to disappear and return only after completion of successful regeneration (Raivich and Kreutzberg, 1987; Raivich et al., 1991). The loss of NGF receptors is associated with a decrease in retrogradely transported NGF (Raivich et al., 1991). In contrast to the decrease in axonal NGF receptors, axotomy caused an increase in the number of NGF receptors on Schwann cells (Raivich et al., 1991). Since NGF can enhance the expression of L1, a Schwann cell surface adhesion molecule expressed after axotomy and during regeneration (Seilheimer and Schachner, 1987), it has been suggested that the primary site of NGF action during peripheral nerve regeneration is on Schwann cells, where it acts to promote remyelination of regenerated neurites (Raivich et al., 1991).

### F. Summary of NGF Actions in the Peripheral Nervous System

NGF produced in the targets of sensory and sympathetic neurons acts through specific receptors to enhance the survival, differentiation, and maintenance of these cells. Discovery that $p140^{trk}$ mediates NGF actions may facilitate the discovery of novel NGF-responsive populations. Defining the cellular substrates for $p140^{trk}$ will enhance our view of the molecular events by which NGF exerts its actions.

## III. NGF ROLE IN THE CENTRAL NERVOUS SYSTEM

In addition to playing an important role in the survival, maintenance, and differentiation of neurons in the PNS, NGF has been shown to be present in the CNS and to influence specific neuronal populations. Criteria for establishing that NGF is a neurotrophic factor in the CNS can be suggested by analysis of its actions in the PNS. These criteria would include the following.

1. Endogenous NGF is available to responsive neurons.
2. Specific NGF receptors are present to mediate NGF actions.
3. NGF shows a definite effect on the growth or differentiation of putative responsive neurons.

These points will be addressed and the actions of NGF on forebrain cholinergic neurons will be discussed. NGF effects in the lesioned CNS as well as potential NGF involvement in neurodegenerative disease also will be reviewed.

### A. Distribution of NGF and the NGF Receptor in the Central Nervous System

#### 1. NGF in the central nervous system

Several studies have described the distribution and developmental regulation of NGF and NGF mRNA in the brain and spinal cord of the rat (Korsching et al., 1985; Large et al., 1986; Shelton and Reichardt, 1986; Whittemore et al., 1986; Auberger et al., 1987; see also Chapter 6). Studies using probes specific for NGF have demonstrated NGF mRNA in the newborn and adult rat brain, with highest levels in the hippocampus, neocortex, and olfactory bulb (Maisonpierre et al., 1990). The localization of NGF gene expression in these regions has been examined by in situ hybridization histochemistry. NGF mRNA was found in hippocampal pyramidal neurons, granule and hilar cells of the dentate gyrus, neocortical neurons, and periglomerular cells of the olfactory bulb (Rennert and Heinrich, 1986; Ayer-

LeLievre et al., 1988; Bandtlow et al., 1990; Ernfors et al., 1990; Gall et al., 1991). In adult rat brain, immunoassay showed that the highest levels of NGF are found in hippocampus (1.3 ng/gm), cortex (0.75 ng/gm), and olfactory bulb (0.88 ng/gm) (Large et al., 1986). Each of these regions serves as a target of basal forebrain cholinergic neurons. In the hippocampus and neocortex, NGF mRNA and protein levels appear to be regulated coordinately during development. Both are present at low levels after birth and increase thereafter, peaking just prior to adulthood (Large et al., 1986). When injected into the adult rat hippocampus, NGF is transported retrogradely to the innervating cholinergic neurons residing in the medial septum and diagonal band. Further, the ratio of NGF protein to NGF mRNA in the hippocampus and neocortex is relatively low in comparison with that in basal forebrain (Large et al., 1986). These findings are consistent with the view that NGF production in hippocampus and cortex serves to supply NGF to basal forebrain cholinergic neurons. NGF and its mRNA also have been localized to several other CNS regions, including caudate putamen, cerebellum, hypothalamus, and spinal cord (Large et al., 1986; Whittemore et al., 1986; Mobley et al., 1989; Maisonpierre et al., 1990). Developmental regulation in these regions is distinct from that in hippocampus and neocortex (Large et al., 1986; Whittemore et al., 1986; Mobley et al., 1989; Maisonpierre et al., 1990), consistent with the view that NGF acts on other CNS neurons. Regulation of NGF gene expression in caudate putamen is discussed in a subsequent section. In the spinal cord, NGF mRNA levels are higher at birth than in adulthood (Maisonpierre et al., 1990).

Localization of NGF gene expression principally to neurons suggests that neuronal activity and neuron-to-neuron interactions may regulate NGF synthesis, release, and actions. Recent evidence to support this view is that limbic seizures produced either by electrolytic lesion of the hilus of the dentate gyrus or by intraventricular injection of kainic acid markedly increased NGF mRNA in neurons in the hippocampus and neocortex (Gall and Isackson, 1989; Gall et al., 1991). Data demonstrating that activation of non-NMDA glutamate receptors in vivo increases NGF mRNA in the rat hippocampus and cortex (Zafra et al., 1990) suggest that glutamate may have an important role in regulating NGF gene expression under physiologic conditions.

There is great interest in understanding regulation of NGF gene expression in the CNS (see Chapter 6). Neurons that project to sites of NGF synthesis may have a role; regulation of NGF expression in CNS targets by neuronal as well as nonneuronal cells is still being explored. In addition to a role for glutamate, roles for other neurotransmitters, neuropeptides, and IL-1 may emerge from this work, considering the regional differences in NGF levels and its developmental regulation, it is quite possible that the nature and timing of the mechanisms controlling the level of NGF and its release from neurons also will vary regionally.

## 2. NGF receptors in the central nervous system

As mentioned earlier, the neurotrophic activities of NGF depend on its binding to high affinity sites ($K_d$, $10^{-11}M$) on the neuronal membrane. Binding studies with iodinated murine and recombinant human NGF have revealed high affinity binding sites in a number of locations. Highest concentrations were found in forebrain cholinergic nuclei (basal forebrain and caudate putamen) and the interpenducular nucleus in the adult rodent CNS (Richardson et al., 1986; Altar et al., 1991). High affinity binding present in hippocampus, amygdala, and cortex was probably due to the presence of receptors on neurons projecting from basal forebrain (Altar et al., 1991).

Studies in vitro suggest strongly that the molecules p75 and p140$^{trk}$ are components of the high affinity NGF receptor. p140$^{trk}$ is the only member of the *trk* gene family to bind NGF (Bothwell, 1991; Klein et al., 1991; Squinto et al., 1991). Moreover, NGF binds to p140$^{trk}$ more avidly than other neurotrophins (Cordon-Cardo et al., 1991). On the other hand, p75 binds other neurotrophins; the affinity of binding NGF is similar to that for brain derived neutrophic factor (BDNF), NT-3, and probably other members of this family (Rodriguez-Tebar et al., 1990, 1991; Bothwell, 1991; Squinto et al., 1991). Predictions based on these findings are that p75 may be distributed more widely than p140$^{trk}$ and that specific responses to NGF will be found only in those neurons that express p140$^{trk}$. It is of interest, then, to know where p140$^{trk}$ and p75 molecules are found in the CNS and to determine to what extent their localization is correlated with high affinity binding and NGF responsiveness.

A number of studies have examined the expression of p75 mRNA and protein in the CNS (Buck et al., 1987; Springer et al., 1987; Yan and Johnson, 1988). In the adult rat brain, p75 mRNA and protein are present in cholinergic magnocellular neurons of the basal forebrain (Springer et al., 1987; Kiss et al., 1988; Koh and Loy, 1989; Koh et al., 1989). p75 gene expression is present in the caudate putamen in developing subjects (Gage et al., 1989; Mobley et al., 1989; Koh and Higgins, 1991). Studies on the developmental regulation of p75 in the basal forebrain and caudate putamen will be discussed further in Section III,B. A number of other neuronal populations contain p75 mRNA and protein. These include neurons in the periglomerula region of the olfactory bulb, ventral premamillary nucleus, mesencephalic trigeminal nucleus, prepositus hypoglossal nucleus, raphe nucleus, ventral cochlear nucleus, nucleus ambiguus, and cerebellar Purkinje cells (Koh et al., 1989). p75 immunoreactivity is present also in the median eminence and arcuate nucleus of the hypothalamus (Yan and Johnson, 1989). During development in the rat, p75 mRNA is very widespread in its distribution. In addition to localization over many of the same neuronal populations shown in the adult, p75 immunoreactivity, as detected using the monoclonal antibody 192-IgG, was present in spinal and brainstem motoneurons, the ganglion cell layer of the retina, the lateral geniculate nucleus, the superior olive, the external granular cell layer and deep nuclei of the cerebellum, and deep layers of the cerebral cortex (Yan and Johnson, 1988). Other structures also were immunostained, including cells in the subventricular/subependymal layer that may serve as neuronal or glial precursors (Yan and Johnson, 1988).

The pattern of *trk* expression in the CNS is being defined currently. Studies from Luis Parada and from our laboratory have shown that p140$^{trk}$ mRNA is expressed in the adult rat basal forebrain in a pattern similar or identical to that for p75 and choline acetyltransferase (ChAT) (D. Holtzman, Y. Li, J. Springer, and W. Mobley, unpublished observations). p140$^{trk}$ mRNA also is found in magnocellular neurons in the adult striatum. The distribution and size of these cells suggest that they, too, are cholinergic neurons, but this hypothesis has yet to be demonstrated conclusively. If true, p140$^{trk}$ mRNA is present in adult striatal neurons at a time when little or no p75 mRNA or protein is detected (Gage et al., 1989; D. Holtzman, Y. Li, and W. Mobley, unpublished observations). Since these cells have high affinity NGF receptors (Richardson et al., 1986) and show a robust response to NGF administered intracerebroventricularly, p140$^{trk}$ probably plays a very significant role in NGF binding and signal transduction in these neurons.

## B. NGF as a Trophic Factor for Cholinergic Neurons of the Basal Forebrain and Caudate Putamen

NGF appears to act as a trophic factor for basal forebrain and striatal (caudate putamen) cholinergic neurons. In mammals, basal forebrain cholinergic neurons appear to play a role in learning and memory (Coyle et al., 1983; Richardson and DeLong, 1988). These neurons are located in a band of continuous subcortical nuclei (medial septal, diagonal band, and nucleus basalis), and their axons supply the major cholinergic input to the hippocampus and neocortex (Mesulam et al., 1983). Cholinergic neurons produce ChAT, an enzyme that is localized to these cells selectively and catalyzes the synthesis of acetylcholine (MacIntosh, 1981). As indicated, NGF is produced in the target of basal forebrain cholinergic neurons (hippocampus and neocortex). There is a close temporal correspondence during development between NGF levels in hippocampus and neocortex and increases in basal forebrain ChAT activity (Large et al., 1986; Whittemore et al., 1986; Auberger et al., 1987). Basal forebrain cholinergic neurons bind NGF with high affinity and express p75 and p140$^{trk}$ (see previous section). The level of p75 mRNA and protein closely parallels that of ChAT, suggesting a role for NGF in coordinating their expression (J. Springer, D. Holtzman, Y. Li, and W. Mobley, unpublished observations). The intraventricular administration of NGF has been shown to effect a prominent and selective increase in ChAT activity in both the basal forebrain and its cholinergic projection areas in postnatal and adult rats (Gnahn et al., 1983; Mobley et al., 1985, 1986; Johnston et al., 1987). p75 and ChAT mRNA, as well as cholinergic neuronal size, are increased in the adult rat basal forebrain after NGF administration (Cavicchioli et al., 1989; Higgins and Mufson, 1989). Expression of other neuronal genes also is induced by NGF in these cells. The genes encoding the amyloid precursor protein and the prion protein both are increased in the developing hamster basal forebrain in vivo by the intraventricular administration of NGF (Mobley et al., 1988). NGF actions on basal forebrain cholinergic neurons have also been examined extensively in vitro (Hartikka and Hefti, 1988, 1990). More compelling evidence that endogenous NGF regulates cholinergic differentiation comes from studies in which the intraventricular administration of NGF antibodies modestly reduced ChAT activity in adult basal forebrain and its projections (Vantini et al., 1989). These data suggest that sequestration of endogenous NGF, or a cross-reacting neurotrophin, inhibited neurochemical differentiation and strengthen the argument that NGF is a trophic factor for these cells in vivo. To document conclusively that NGF is responsible for these effects, it will be necessary to block its actions using specific antibodies or antagonists. Expression in hippocampus of both BDNF and NT-3 (Maisonpierre et al., 1990; Phillips et al., 1990) developmental regulation of BDNF mRNA in hippocampus in a pattern similar to NGF (Maisonpierre et al., 1990), and evidence that BDNF acts on basal forebrain cholinergic neurons (Alderson et al., 1990) make further studies documenting the significance of NGF actions on these cells quite important. It also will be important to know whether endogenous NGF acts in normal development to enhance the survival of these neurons.

Cholinergic neurons of the caudate putamen differ from those in the basal forebrain because they are interneurons; their axons ramify entirely within the caudate putamen. These neurons appear to play a role in the control of movement (Lalley et al., 1970; Schwarz et al., 1986). Although the time course of expression during development differs from that in other brain regions, NGF mRNA and protein are present in the rat caudate putamen. Highest NGF levels are found between PD 5 and PD 20 (Mobley et al., 1989). Interestingly, these levels are only about 10% of those found in the hippocampus (Shelton and Reichardt, 1986; Weskamp and Otten, 1987). Cholinergic differentiation in the caudate putamen, as marked by induction of ChAT activity, is delayed compared with the basal forebrain, yet it parallels changes in the specific activity of high affinity NGF binding closely (Mobley et al., 1989). As in the developing basal forebrain, p75 mRNA and protein are expressed in the developing caudate putamen (Mobley et al., 1989; Koh and Higgins, 1991). However, expression of this gene is decreased in the adult; protein and mRNA levels are at or below the limit of detection by immunocytochemistry and in situ hybridization (Gage et al., 1989; Mobley et al., 1989). Because high affinity NGF binding is greatest in adults (Mobley et al., 1989), decreased p75 gene expression is quite remarkable. Interestingly, p140$^{trk}$ mRNA has been found in both postnatal and adult caudate putamen (W. C. Mobley, Y. Li, and D. M. Holtzman, unpublished observations). Data showing that these cells respond to NGF infusion suggest that NGF receptors are competent to transduce a signal. Similar to effects seen in basal forebrain, intraventricular injections of NGF in the rat have been shown to increase ChAT activity during development and in the adult (Mobley et al., 1985; Hagg et al., 1989). In addition, intrastriatally injected NGF has been shown to increase p75 mRNA, ChAT immunoreactivity, and cholinergic neuronal size in the adult rat caudate putamen (Gage et al., 1989). These observations, as a whole, suggest a role for endogenous NGF in the development as well as in the maintenance of caudate putamen cholinergic neurons and point to an important role for p140$^{trk}$ in mediating NGF signal transduction.

In summary, several lines of evidence indicate that NGF is an endogenous trophic factor for both basal forebrain and striatal cholinergic neurons. NGF is present in the target of these cells, high affinity NGF receptors are present on these cells, and expression of cholinergic markers parallels changes in NGF and NGF high affinity binding. Additionally, exogenous NGF has been shown to exert specific effects on neurochemical differentiation and on gene expression.

## C. Other Populations in the Central Nervous System

As was true in the PNS, the CNS contains many neurons that express the gene for p75. Data that clearly implicate NGF as a neurotrophic factor are available for only a few of these populations. The criteria suggested earlier for determining whether or not NGF acts as a trophic factor can and should be explored for other CNS neurons. It is likely that high affinity NGF binding and expression of $p140^{trk}$ will be correlated with NGF responsiveness. Once this has been demonstrated, more detailed studies of NGF expression can be carried out to rule on a physiologic role for NGF. For some populations, expression of p75 will not be correlated with NGF high affinity binding or with an NGF response. For these cells, p75 may serve to bind and respond to one or more of the other neurotrophins (Rodriguez-Tebar et al., 1990,1991). Alternatively, p75 on these cells may play a role distinct from signal transduction.

Spinal cord motoneurons are particularly interesting with respect to their trophic dependency. Data suggest that NGF may act on these cells.

1. High affinity NGF binding sites were present transiently in the lateral motor column and in muscle in early chick development (Raivich et al., 1985).
2. High and low affinity binding sites were present in embryonic chick spinal cord and were found to be enriched in gradient fractions containing motoneurons at ED 8 (Marchetti and McManaman, 1990).
3. There was a decline of both receptor types during chick development (Marchetti and McManaman, 1990).
4. p75 was expressed in embryonic chick motoneurons (Ernfors et al., 1988) and in rat motoneurons during embryonic and newborn periods. p75 was decreased markedly or absent by PD 10 in the rat (Yan and Johnson, 1988; Ernfors et al., 1989).
5. Radiolabeled NGF was transported retrogradely to developing motoneurons of the chick (Wayne and Heaton, 1990b) and the rat (Yan et al., 1988) but not in the adult rat (Stockel et al., 1975).
6. p75 expression was induced in mature motoneurons by sciatic nerve crush (Ernfors et al., 1989).

These findings suggest a role for NGF in early motoneuron development, but this role must be substantiated. NGF was found to increase process outgrowth from embryonic chick ventral spinal cord neurons in culture but had no effect on survival (Longo et al., 1982; Wayne and Heaton, 1990a). Arakawa et al. (1990) also failed to see an effect of NGF on survival of developing motoneurons. Yan et al. (1988) detected specific retrograde transport of NGF in the newborn rat, but no NGF effect was demonstrated on motoneuron morphology or survival following crush injury, and NGF failed to induce ChAT activity. Further studies examining p140$^{trk}$ expression and NGF responsiveness in these cells may prove valuable.

NGF may serve as a neurotrophic factor for several additional populations of CNS neurons. One of these comprises Purkinje cells in the developing cerebellum. High affinity NGF binding has been demonstrated to occur on the cell bodies and dendrites of Purkinje cells in the postnatal rat. NGF binding was not found on adult Purkinje cells (Cohen-Cory et al., 1989). Cultured embryonic Purkinje cells also bound radiolabeled NGF specifically. Survival, as well as various morphologic parameters, was enhanced in these cells when NGF treatment was combined with high potassium, aspartate, and glutamate treatment (Cohen-Cory et al., 1991). Purkinje cells are known to express p75 in developing subjects and in the adult (Yan and Johnson, 1988, 1989; Koh et al., 1989; Wanaka and Johnson, 1990). Developmental regulation of NGF binding on Purkinje cells and their response to NGF *in vitro* suggest that NGF may act as a neurotrophic factor during their differentiation. It will now be important to determine whether or not p140$^{trk}$ is expressed in these cells and to define their source of NGF. It is possible that NGF will also be shown to act on hypothalamic neurons (Yan and Johnson, 1989; Ojeda et al., 1991). An interesting finding was a marked increase in NGF mRNA in hypothalamic neurons in the preoptic and ventrolateral nucleus following intermale aggressive behavior induced by social isolation (Spillantini et al., 1989).

Studies correlating the distribution of p140$^{trk}$ with high affinity NGF binding can be used to pursue evidence of NGF responsiveness in novel populations. This approach, coupled with studies of NGF expression, may extend the list of CNS neurons for which NGF acts as a neurotrophic factor.

### D. Role of NGF in the Lesioned Central Nervous System

NGF is the prototypic neurotrophic factor. Studies of NGF actions have led to a model of neurotrophic interactions. A central feature of this model in developing systems is that elaboration of trophic factors in limiting quantities by the target allows for survival of some neurons but not others. This may occur through a competition that favors afferently projecting neurons with axonal processes bearing an appropriate level of neurotrophic factor receptors present in territories enriched in the factor. Whether or not NGF-responsive CNS neurons depend for their survival on NGF is uncertain, as yet. Neurons that respond to neurotrophic factors during development and survive may continue to depend on target-derived trophic support for their maintenance after the developmental period. Although NGF is likely to be important for the maintenance of mature neurons, the degree to which these neurons depend on target-derived trophic support and the nature of this dependence is uncertain.

The septohippocampal cholinergic system has been a particularly useful model for studying NGF in the lesioned adult CNS. Neurons from the medial septum and vertical limb of the diagonal band project to the hippocampal formation primarily via the fimbria–fornix. About 50% of these projecting neurons are cholinergic (Amaral and Kurtz, 1985). After lesion of the fimbria–fornix, a severe and relatively rapid degeneration of approximately 50% of axotomized cholinergic neurons ensues in the medial septum and vertical limb of the diagonal band (Gage et al., 1986; Hefti, 1986). In addition to losing ChAT immunostaining, these cells atrophy and eventually die (Tuszynski et al., 1990; Fischer and Bjorklund, 1991). Neuronal cell death in this model appears to be due, at least in part, to interrupted transport of NGF from the hippocampus. Importantly, intraventricularly administered NGF has been shown to prevent cholinergic neuronal shrinkage and cell death, both in fimbria–fornix-lesioned rodents and in nonhuman primates (Hefti, 1986; Williams et al., 1986; Kromer, 1987; Hagg et al., 1988; Koliatsos et al., 1990; Tuszynski et al., 1990, 1991). It is interesting to note that, following distal axotomy of hippocampal cholinergic afferents, there is neuronal atrophy without a decrease in cell number (Sofroniew et al., 1990). This may result from the ability of actively growing brain glia in the damaged CNS to up-regulate NGF synthesis markedly in certain brain regions (Lu et al., 1991). Thus, although glial cells may produce little or no NGF in the intact adult brain, glial elaboration of NGF may assume important functions in the damaged CNS. Alternatively, a loss of NGF support in the mature brain may lead to atrophied dysfunctional basal forebrain cholinergic neurons without cell death. Preliminary evidence favors the latter possibility, since NGF mRNA is decreased markedly in the hippocampal remnant following excitotoxic ablation of the hippocampal neurons (Sofroniew et al., 1991).

### E. Role of NGF in Central Nervous System Diseases

Results of experiments in which axotomized cholinergic neurons were rescued by exogenous NGF have suggested that specific neurotrophic factors may be able to reverse degeneration of injured neurons and possibly even that of those whose degeneration is due to disease or is age-related. Interestingly, in the human neurodegenerative disorders, Alzheimer's disease (AD), Huntington's disease, and certain cases of Parkinson's disease, cholinergic neurons of the basal forebrain, striatum, or both atrophy and degenerate (Oyanagi et al., 1989; for review, see Holtzman and Mobley, 1991). In AD, the shrinkage and loss of these neurons is likely to result in decreased ChAT activity and to contribute to dementia (Price, 1986). Although NGF appears to be a normal trophic factor for these cells both during development and in the mature CNS, there is currently no evidence to support a primary deficiency of NGF mRNA or protein as a cause for the dysfunction of these neurons (Goedert et al., 1986; Mobley, 1989). There is evidence for decreased p75 mRNA in the AD basal forebrain (Higgins and Mufson, 1989). There is no evidence, however, that this finding is a primary abnormality, since it occurs at a time when there is shrinkage and neuronal cell loss of p75-immunoreactive neurons (Kordower et al., 1988). p75 immunoreactivity normally is not observed on neuronal cell bodies in the postnatal human cerebral cortex. Interestingly, p75-immunoreactive cortical neurons are found in AD and in extreme advanced aging (Mufson and Kordower, 1992). Although it is unclear whether these cells are neurotrophin responsive and express a member of the *trk* gene family, these findings suggest that cortical neurons in AD exhibit plasticity in their expression of p75. Recent data concerning the neurotrophin BDNF have shown that its mRNA is decreased in the dentate gyrus in AD (Phillips et al., 1991). Again, whether this abnormality is primary or secondary and whether it contributes to cholinergic abnormalities remains to be clarified.

Although no primary abnormality of NGF or its receptor has been demonstrated in the Alzheimer brain, it is possible that NGF administration would prevent loss of basal forebrain cholinergic neurons and forestall or mitigate dementia. For example, it is conceivable that metabolic demands imposed by the disease process have increased the need for neurotrophic support without a corresponding increase in NGF production in hippocampus or cortex. Alternatively, the disease process may inhibit the activation of one or more molecular pathways important for survival and function that normally are activated by NGF. Admitting the speculative nature of these comments, it is very clear that NGF has dramatic effects on these cells in normal and lesioned subjects. Animal data that suggest that the administration of NGF to humans with AD may be beneficial are summarized here.

1. Exogenous NGF enhances the neurochemical differentiation, including ChAT activity, of basal forebrain cholinergic neurons in developing and mature animals (Gnahn et al., 1983; Mobley et al., 1986; Johnston et al., 1987).
2. Transection of the cholinergic projection to the hippocampus (fimbria–fornix) leads to the degeneration of basal forebrain neurons. This degeneration can be prevented or reversed by the intraventricular infusion of NGF in rodents and nonhuman primates (Hefti, 1986; Williams et al., 1986; Kromer et al., 1987; Hagg et al., 1988; Koliatsos et al., 1990, 1991; Tuszynski et al., 1990).
3. Intraventricular infusion of NGF in aged rats is associated with improved spatial learning and a reversal of cholinergic neuronal atrophy (Fischer et al., 1987). In addition, in a recently developed mouse model of spontaneous cholinergic neurodegeneration (Holtzman et al., 1992), intraventricular NGF can reverse neuronal atrophy (D. Holtzman, Y. Li, K. Chen, F. Gage, and W. Mobley, unpublished observations).

Among the factors that will determine whether the exogenous administration of NGF will be of benefit in AD are (1) whether or not a significant proportion of the intellectual decline in AD results from cholinergic dysfunction and (2) whether disrupting the distribution or amount of a normal target-derived agent also may disrupt normal function. Currently, there is no answer to either of these questions. A number of noncholinergic neuronal populations degenerate in the Alzheimer brain, including cells in the amygdala, hippocampus, and neocortex (Holtzman and Mobley, 1991). The degree to which dementia is caused by involvement of other neuronal populations is unclear. If their loss is critical, it is possible that even marked enhancement of cholinergic function by NGF will not produce improvement in cognition. NGF enhances sprouting from cholinergic neurons in vitro and as well as in the lesioned basal forebrain in vivo (Gage et al., 1988; Hartikka and Hefti, 1988; Tuszynski et al., 1991). It is therefore possible that NGF treatment may affect the distribution of cholinergic neurites and their pattern of innervation with undesired functional consequences. Further, noting the need at present to deliver NGF by intraventricular cannula, NGF supplied in this manner enhances growth of sympathetic perivascular fibers in rats (Saffran and Crutcher, 1990). Whether this change would have a deleterious effect is unknown. Finally, if NGF reaches the systemic circulation, it may have unwanted effects on NGF receptor-bearing cells in the immune system or the peripheral nervous system (Levi-Montalcini, 1987). This brief discussion makes clear that NGF treatment of Alzheimer patients has potential benefits as well as disadvantages. Concerns for efficacy and safety argue strongly for a clearer understanding of the pathogenesis of AD, the role of cholinergic neurons in disease symptomatology, and the actions of NGF on diseased basal forebrain cholinergic neurons. Improved methods for delivering NGF, inducing NGF gene expression, or mimicking its actions also will be important. For each of these goals, it will be essential to pursue studies in animal models of cholinergic neurodegeneration.

## REFERENCES

Alderson, R. F., Alterman, A. L., Barde, Y.-A., and Lindsay, R. M. (1990). Brain-derived neurotrophic factor increases survival and differentiated functions of rat septal cholinergic neurons in culture. *Neuron* **5**, 297–306.

Alema, S., Casalbore, P., Agostini, E., and Tato, F. (1985). Differentiation of PC12 phaeochromocytoma cells induced by v-src oncogene. *Nature (London)* **316**, 557–559.

Aletta, J. M., Shelanski, M. L., and Greene, L. A. (1989). Phosphorylation of the peripherin 58-kDa neuronal intermediate filament protein. Regulation by nerve growth factor and other agents. *J. Biol. Chem.* **264**, 4619–4627.

Altar, A. C., Burton, L. E., Bennett, G. L., and Dugich-Djordjevic, M. (1991). Recombinant human nerve growth factor is biologically active and labels novel high-affinity binding sites in rat brain. *Proc. Natl. Acad. Sci. U.S.A.* **88**, 281–285.

Altin, J. G., and Bradshaw, R. A. (1990). Production of 1,2-diacylglycerol in PC12 cells by nerve growth factor and basic fibroblast growth factor. *J. Neurochem.* **54**, 1666–1676.

Altin, J. G., Wetts, R., and Bradshaw, R. A. (1991). Microinjection of a p21ras antibody into PC12 cells inhibits neurite outgrowth induced by nerve growth factor and basic fibroblast growth factor. *Growth Factors* **4**, 145–155.

Amaral, D. G., and Kurtz, J. (1985). An anlysis of the origins of the cholinergic and noncholinergic septal projections to the hippocampal formation of the rat. *J. Comp. Neurol.* **240**, 37–59.

Angeletti, P. U., Levi-Montalcini, R., and Caramia, F. (1971). Ultrastructural changes in sympathetic neurons of newborn and adult mice treated with nerve growth factor. *J. Ultrastruct. Res.* **36**, 24–36.

Arakawa, Y., Sendtner, M., and Thoenen, H. (1990). Survival effect of ciliary neurotrophic factor

(CNTF) on chick embryonic motoneurons in culture: Comparison with other neurotrophic factors and cyctokines. *J. Neurosci.* **10,** 3507–3515.
Auberger, G., Heumann, R., Hellweg, R., Korsching, S., and Thoenen, H. I. (1987). Developmental changes of nerve growth factor and its mRNA in the rat hippocampus: Comparison with choline acetyltransferase. *Dev. Biol.* **120,** 322–328.
Ayer-LeLievre, C., Olson, L., Ebendal, T., Seiger, A., and Persson, H. (1988). Expression of the b-nerve growth factor gene in hippocampal neurons. *Science* **240,** 1339–1341.
Bandtlow, C., Heumann, R., Schwab, M. E., and Thoenen, H. (1987). Cellular localization of nerve growth factor synthesis by *in situ* hybridization. *EMBO J.* **6,** 891–899.
Bandtlow, C., Meyer, M., Lindholm, D., Spranger, M., Heumann, R., and Thoenen, H. (1990). Regional and cellular codistribution of interleukin 1b and nerve growth factor mRNA in the adult rat brain: Possible relationship to the regulation of nerve growth factor synthesis. *J. Cell Biol.* **111,** 1701–1711.
Bar-Sagi, D., and Feramisco, J. R. (1985). Microinjection of the *ras* oncogene protein into PC12 cells induces morphological differentiation. *Cell* **42,** 841–848.
Berg, M. M., Sternberg, D. W., Hempstead, B. L., and Chao, M. V. (1991). The low-affinity p75 nerve growth factor (NGF) receptor mediates NGF-induced tyrosine phosphorylation. *Proc. Natl. Acad. Sci. U.S.A.* **88,** 7106–7110.
Bernd, P., and Greene, L. A. (1984). Association of $^{125}$I-nerve growth factor with PC12 pheochromocytoma cells: Evidence for internalization via high-affinity receptors only and for long-term regulation by nerve growth factor of both high- and low-affinity receptors. *J. Biol. Chem.* **259,** 15509–15514.
Bjerre, B., Bjorklund, A., Mobley, W., and Rosengren, E. (1975). Short- and long-term effects of nerve growth factor on the sympathetic nervous system in the adult mouse. *Brain Res.* **94,** 263–277.
Boonstra, J., Mummery, C. L., Feyen, A., de Hoog, W., van der Saag, P., and de Laat, S. (1987). Epidermal growth factor receptor expression during morphological differentiation of pheochromocytoma cells, induced by nerve growth factor or dibutyryl cyclic AMP. *J. Cell Physiol.* **131,** 409–417.
Borasio, G. D., John, J., Wittinghofer, A., Barde, Y. A., Sendtner, M., and Heumann, R. (1989). ras p21 protein promotes survival and fiber outgrowth of cultured embryonic neurons. *Neuron* **2,** 1087–1096.
Bothwell, M. (1991). Keeping track of neurotrophin receptors. *Cell* **65,** 915–918.
Boulton, T. G., Nye, S. H., Robbins, D. J., Ip, N. Y., Radziejewska, E., Morgenbesser, S. D., DePinho, R. A., Panayotatos, N., Cobb, M. H., and Yancopoulos, G. D. (1991). ERKs: A family of protein-serine/threonine kinases that are activated and tyrosine phosphorylated in response to insulin and NGF. *Cell* **65,** 663–675.
Brady, M. J., Nairn, A. C., Wagner, J. A., and Palfrey, H. C. (1990). Nerve growth factor-induced down-regulation of calmodulin-dependent protein kinase III in PC12 cells involves cyclic AMP-dependent protein kinase. *J. Neurochem.* **54,** 1034–1039.
Buck, C. R., Martinez, H. J., Black, I. B., and Chao, M. V. (1987). Developmentally regulated expression of the nerve growth factor receptor gene in the periphery and brain. *Proc. Natl. Acad. Sci. U.S.A.* **84,** 3060–3063.
Bueker, E. D. (1948). Implantation of tumors in the hind limb field of the embryonic chick and the developmental response of the lumbosacral nervous system. *Anat. Rec.* **102,** 369–390.
Burstein, D. E., and Greene, L. A. (1982). Nerve growth factor has both mitogenic and anti-mitogenic activity. *Dev. Biol.* **94,** 477–482.
Buxser, S., Decker, D., and Ruppel, P. (1990). Relationship among types of nerve growth factor receptors on PC12 cells. *J. Biol. Chem.* **265,** 12701–12710.
Campenot, R. B. (1977). Local control of neurite development by nerve growth factor. *Proc. Natl. Acad. Sci. U.S.A.* **74,** 4516–4519.
Cantley, L. C., Auger, K. R., Carpenter, C., Duckworth, B., Graziani, A., Kapellar, R., and Soltoff, S. (1991). Oncogenes and signal transduction. *Cell* **64,** 281–302.

Cattaneo, E., and McKay, R. (1990). Proliferation and differentiation of neuronal stem cells regulated by nerve growth factor. *Nature (London)* **347,** 762–765.

Cavichiolli, L., Flanigan, T. P., Vantini, G., Fusco, M., Polato, P., Toffano, G., Walsh, F. S., and Leon, A. (1989). NGF amplifies expression of NGF receptor messenger RNA in forebrain cholinergic neurons of rats. *Eur. J. Neurosci.* **1,** 258–262.

Chan, B. L., Chao, M. V., and Saltiel, A. R. (1989). Nerve growth factor stimulates the hydrolysis of glycosylphosphatidylinositol in PC-12 cells: A mechanism of protein kinase C regulation. *Proc. Natl. Acad. Sci. U.S.A.* **86,** 1756–1760.

Chijiwa, T., Mishima, A., Hagiwara, M., Sano, M., Hayashi, K., Inoue, T., Naito, K., Toshioka, T., and Hidaka, H. (1990). Inhibition of forskolin-induced neurite outgrowth and protein phosphorylation by a newly synthesized selective inhibitor of cyclic AMP-dependent protein kinase, $N$-[2-($p$-bromocinnamylamino)ethyl]-5-isoquinolinesulfonamide (H-89), of PC12D pheochromocytoma cells. *J. Biol. Chem.* **265,** 5267–5272.

Cho, K. O., Skarnes, W. C., Minsk, B., Palmieri, S., Jackson, G. L., and Wagner, J. A. (1989). Nerve growth factor regulates gene expression by several distinct mechanisms. *Mol. Cell. Biol.* **9,** 135–143.

Clark, E. A., and Lee, V. M. (1991). The differential role of protein kinase C isozymes in the rapid induction of neurofilament phosphorylation by nerve growth factor and phorbol esters in PC12 cells. *J. Neurochem.* **57,** 802–810.

Clegg, D. O., Large, T. H., Bodary, S. C., and Reichardt, L. F. (1989). Regulation of nerve growth factor mRNA levels in developing rat heart ventricle is not altered by sympathectomy. *Dev. Biol.* **134,** 30–37.

Cohen, S. (1960). Purification of a nerve-growth promoting protein from the mouse salivary gland and its neuro-cytotoxic antiserum. *Proc. Natl. Acad. Sci. U.S.A.* **46,** 302–311.

Cohen, S., and Levi-Montalcini, R. (1956). A nerve growth-stimulating factor isolated from snake venom. *Proc. Natl. Acad. Sci. U.S.A.* **42,** 571–574.

Cohen-Cory, S., Dreyfus, C. F., and Black, I. B. (1989). Expression of high- and low-affinity nerve growth factor receptors by purkinje cells in the developing rat cerebellum. *Exp. Neurol.* **105,** 104–109.

Cohen-Cory, S., Dreyfus, C. F., and Black, I. B. (1991). NGF and excitatory neurotransmitters regulate survival and morphogenesis of cultured cerebellar purkinje cells. *J. Neurosci.* **11,** 462–471.

Collawn, J. F., Stangel, M., Kuhn, L. A., Esekogwu, V., Jing, S. Q., Trowbridge, I. S., and Tainer, J. A. (1990). Transferrin receptor internalization sequence YXRF implicates a tight turn as the structural recognition motif for endocytosis. *Cell* **63,** 1061–1072.

Cordon-Cardo, C., Tapley, P., Jing, S., Nanduri, V., O'Rourke, E., Lamballe, F., Kovary, K., Klein, R., Jones, K. R., Reichardt, L. F., and Barbacid, M. (1991). The *trk* tyrosine protein kinase mediates the mitogenic properties of nerve growth factor and neurotrophin-3. *Cell* **66,** 173–183.

Costello, B., Meymandi, A., and Freeman, J. A. (1990). Factors influencing GAP-43 gene expression in PC12 pheochromocytoma cells. *J. Neurosci.* **10,** 1398–1406.

Coughlin, M. D., and Collins, M. B. (1985). Nerve growth factor-independent development of embryonic mouse sympathetic neurons in dissociated cell culture. *Dev. Biol.* **110,** 392–401.

Coughlin, M. D., Boyer, D. M., and Black, I. B. (1977). Embryologic development of a mouse sympathetic ganglion *in vivo* and *in vitro*. *Proc. Natl. Acad. Sci. U.S.A.* **75,** 3438–3442.

Cowan, W. M., Fawcett, J. W., O'Leary, D. D. M., and Stanfield, B. B. (1984). Regressive events in neurogenesis. *Science* **225,** 1258–1265.

Coyle, J. T., Price, D. L., and DeLong, M. R. (1983). Alzheimer's disease: A disorder of cortical cholinergic innervation. *Science* **219,** 1184–1190.

Cremins, J., Wagner, J. A., and Halegoua, S. (1986). Nerve growth factor action is mediated by cyclic AMP- and $Ca^{+2}$/phospholipid-dependent protein kinases. *J. Cell Biol.* **103,** 887–893.

Cunningham, T. J. (1982). Naturally occurring neuron death and its regulation by developing neural pathways. *Int. Rev. Cytol.* **74,** 163–185.

Curran, T., and Franza, B. J. (1988). Fos and Jun: The AP-1 connection. *Cell* **55,** 395–397.

Damon, D. H., D'Amore, P. A., and Wagner, J. A. (1990). Nerve growth factor and fibroblast growth factor regulate neurite outgrowth and gene expression in PC12 cells via both protein kinase C- and cAMP-independent mechanisms. *J. Cell Biol.* **110**, 1333–1339.

Davies, A., and Lumsden, A. (1984). Relation of target encounter and neuronal death to nerve growth factor responsiveness in the developing mouse trigeminal ganglion. *J. Comp. Neurol.* **223**, 124–137.

Davies, A. M., Bandtlow, C., Heumann, R., Korsching, S., Rohrer, H., and Thoenen, H. (1987). Timing and site of nerve growth factor synthesis in developing skin in relation to innervation and expression of the receptor. *Nature (London)* **326**, 353–358.

DeGeorge, J. J., Walenga, R., and Carbonetto, S. (1988). Nerve growth factor rapidly stimulates arachidonate metabolism in PC12 cells: Potential involvement in nerve fiber growth. *J. Neurosci. Res.* **21**, 323–332.

Diamond, J., Coughlin, M., MacIntyre, L., Holmes, M., and Visheau, B. (1987). Evidence that endogenous b nerve growth factor is responsible for the collateral sprouting, but not the regeneration, of nociceptive axons in adult rats. *Proc. Natl. Acad. Sci. U.S.A.* **84**, 6596–6600.

DiCicco-Bloom, E., and Black, I. B. (1988). Insulin growth factors regulate the mitotic cycle in cultured rat sympathetic neuroblasts. *Proc. Natl. Acad. Sci. U.S.A.* **85**, 4066–4070.

Dumas, M., Schwab, M. E., and Thoenen, H. (1979). Retrograde axonal transport of specific macromolecules as a tool for characterizing nerve terminal membranes. *J. Neurobiol.* **10**, 179–197.

Ebendal, T., Larhammar, D., and Persson, H. (1986). Structure and expression of the chicken β-nerve growth factor. *EMBO J.* **5**, 1483–1487.

Edwards, S. N., Buckmaster, A. E., and Tolkovsky, A. M. (1991). The death programme in cultured sympathetic neurones can be suppressed at the posttranslational level by nerve growth factor, cyclic AMP, and depolarization. *J. Neurochem.* **57**, 2140–2143.

Ellis, C., Moran, M., McCormick, F., and Pawson, T. (1990). Phosphorylation of GAP and GAP-associated proteins by transforming and mitogenic tyrosine kinases. *Nature (London)* **343**, 377–381.

Ernfors, P., Hallbook, F., Ebendal, T., Shooter, E. M., Radeke, M. J., Misko, T. P., and Persson, H. (1988). Developmental and regional expression of β-nerve growth factor receptor mRNA in the chick and rat. *Neuron* **1**, 983–996.

Ernfors, P., Henschen, A., Olson, L., and Persson, H. (1989). Expression of nerve growth factor receptor mRNA is developmentally regulated and increased after axotomy in rat spinal cord motoneurons. *Neuron* **2**, 1605–1613.

Ernfors, P., Ibanez, C. F., Ebendal, T., Olson, L., and Persson, H. (1990). Molecular cloning and neurotrophic activities of a protein with structural similarities to β-nerve growth factor: Developmental and topographical expression in the brain. *Proc. Natl. Acad. Sci. U.S.A.* **87**, 5454–5458.

Escandon, E., and Chao, M. V. (1989). Developmental expression of the chicken nerve growth factor receptor gene during brain morphogenesis. *Dev. Brain Res.* **47**, 187–196.

Felder, S., Miller, K., Moehren, G., Ullrich, A., Schlessinger, J., and Hopkins, C. R. (1990). Kinase activity controls the sorting of the epidermal growth factor receptor within the multivesicular body. *Cell* **61**, 623–634.

Fernholm, M. (1971). On the development of the sympathetic chain and the adrenal medulla in the mouse. *Z. Anat. Entwickl.-Gesch.* **133**, 305–317.

Fink, D. J., and Guroff, G. (1990). Nerve growth factor stimulation of arachidonic acid release from PC12 cells: Independence from phosphoinositide turnover. *J. Neurochem.* **55**, 1716–1726.

Finn, P. J., Ferguson, I. A., Wilson, P. A., Vahaviolos, J., and Rush, R. A. (1987). Immunohistochemical evidence for the distribution of nerve growth factor in the embryonic mouse. *J. Neurocytol.* **16**, 639–647.

Fischer, W., and Bjorklund, A. (1991). Loss of AChE- and NGFr-labeling precedes neuronal death of axotomized septal-diagonal band neurons: Reversal by intraventricular NGF infusion. *Exp. Neurol.* **113**, 93–108.

Fischer, W., Wictorin, K., Bjorklund, A., Williams, L. R., Varon, S., and Gage, F. H. (1987). Amelioration of cholinergic neuron atrophy and spatial memory impairment in aged rats by nerve growth factor. *Nature (London)* **329**, 65–68.

Fitzgerald, M., Wall, P. D., Goedert, M., and Emson, P. C. (1985). Nerve growth factor counteracts the neurophysiological and neurochemical effects of chronic sciatic nerve section. *Brain Res.* **332**, 131–141.

Gage, F. H., Wictorin, K., Fischer, W., Williams, L. R., Varon, S., and Bjorklund, A. (1986). Life and death of cholinergic neurons in the septal and diagonal band following complete fimbria–fornix transection. *Neuroscience* **19**, 241–255.

Gage, F. H., Armstrong, D. M., Williams, L. R., and Varon, S. (1988). Morphologic response of axotomized septal neurons to nerve growth factor. *J. Comp. Neurol.* **269**, 147–155.

Gage, F. H., Batchelor, P., Chen, K. S., Chin, D., Higgins, G. A., Koh, S., Deputy, S., Rosenberg, M. B., Fischer, W., and Bjorklund, A. (1989). NGF receptor reexpression and NGF-mediated cholinergic neuronal hypertrophy in the damaged adult neostriatum. *Neuron* **2**, 1177–1184.

Gall, C. M., and Isackson, P. J. (1989). Limbic seizures increase neuronal production of messenger RNA for nerve growth factor. *Science* **245**, 758–761.

Gall, G., Murray, K., and Isackson, G. (1991). Kainic acid-induced seizures stimulate increased expression of nerve growth factor mRNA in rat hippocampus. *Mol. Brain Res.* **9**, 113–123.

Ginty, D. D., Glowacka, D., DeFranco, C., and Wagner, J. A. (1991). Nerve growth factor-induced neuronal differentiation after dominant represssion of both type I and type II cAMP-dependent protein kinase activities. *J. Biol. Chem.* **266**, 15325–15333.

Gnahn, H., Hefti, F., Heumann, R., Schwab, M. E., and Thoenen, H. (1983). NGF-mediated increase of choline acetyltransferase (ChAT) in the neonatal rat forebrain: Evidence for a physiological role of NGF in the brain? *Dev. Brain Res.* **9**, 45–52.

Goedert, M., Otten, U., and Thoenen, H. (1978). Biochemical effects of antibodies against nerve growth factor on developing and differentiated sympathetic ganglia. *Brain Res.* **148**, 264–268.

Goedert, M., Otten, U., Schafer, T. H., Schwab, M., and Thoenen, H. (1980). Immunosympathectomy: Lack of evidence for a complement-mediated cytotoxic mechanism. *Brain Res.* **201**, 399–409.

Goedert, M., Stoeckel, K., and Otten, U. (1981). Biological importance of the retrograde axonal transport of nerve growth factor in sensory neurons. *Proc. Natl. Acad. Sci. U.S.A.* **78**, 5895–5898.

Goedert, M., Otten, U., Hunt, S. P., Bond, A., Chapman, D., and Schlumpf, M. (1984). Biochemical and anatomical effects of antibodies against nerve growth factor on developing rat sensory ganglia. *Proc. Natl. Acad. Sci. U.S.A.* **81**, 1580–1584.

Goedert, M., Fine, A., Hunt, S. P., and Ullrich, A. (1986). Nerve growth factor mRNA in peripheral and central rat tissues and the human central nervous system: Lesion effects in the rat brain and levels in Alzheimer's disease. *Mol. Brain Res.* **1**, 85–92.

Gold, B. G., Mobley, W. C., and Matheson, S. F. (1991). Regulation of axonal caliber, neurofilament content, and nuclear localization in mature sensory neurons by nerve growth factor. *J. Neurosci.* **114**, 943–955.

Golubeva, E. E., Posypanova, G. A., Kondratyev, A. D., Melnik, E. I., and Severin, E. S. (1989). The influence of nerve growth factor on the activities of adenylate cyclase and high-affinity GTPase in pheochromocytoma PC12 cell. *FEBS Lett.* **247**, 232–234.

Gotoh, Y., Nishida, E., Yamashita, T., Hoshi, M., Kawakami, M., and Sakai, H. (1990). Microtubule-associated-protein (MAP) kinase activated by nerve growth factor and epidermal growth factor in PC12 cells. Identify with the mitogen-activated MAP kinase of fibroblastic cells. *Eur. J. Biochem.* **193**, 661–669.

Greene, L. A., and Tischler, A. S. (1976). Establishment of a noradrenergic clonal line of rat adrenal pheochromocytoma cells which respond to nerve growth factor. *Proc. Natl. Acad. Sci. U.S.A.* **73**, 2424–2428.

Guerrero, I., Wong, H., Pellicer, A., and Burstein, D. E. (1986). Activated N-*ras* gene induces neuronal differentiation of PC12 rat pheochromocytoma cells. *J. Cell Physiol.* **129**, 71–76.

Gundersen, R. W., and Barrett, J. N. (1979). Neuronal chemotaxis: Chick dorsal-root axons turn toward high concentrations of nerve growth factor. *Science* **206**, 1079–1080.

Gundersen, R. W., and Barrett, J. N. (1980). Characterization of the turning response of dorsal root neurites toward nerve growth factor. *J. Cell Biol.* **87**, 546–554.

Hagg, T., Manthorpe, M., Vahlsing, H. L., and Varon, S. (1988). Delayed treatment with nerve growth factor reverses the apparent loss of cholinergic neurons after acute brain damage. *Exp. Neurol.* **101**, 303–312.

Hagg, T., Hagg, F., Vahlsing, H. L., Manthorpe, M., and Varon, S. (1989). Nerve growth factor effects on cholinergic neurons of neostriatum and nucleus accumbens in the adult rat. *Neuroscience* **30**, 95–103.

Halegoua, S., Armstrong, R. C., and Kremer, N. E. (1991). Dissecting the mode of action of a neuronal growth factor. *Curr. Top. Microbiol. Immunol.* **165**, 119–170.

Hamberger, B., Levi-Montalcini, R., Norberg, K.-A., and Sjoquist, F. (1965). Monoamines in immunosympathectomized rats. *Int. J. Neuropharmacol.* **4**, 91–95.

Hamburger, V., and Levi-Montalcini, R. (1949). Proliferation, differentiation and degeneration in the spinal ganglia of the chick embryo under normal and experimental conditions. *J. Exp. Zool.* **111**, 457–501.

Hamburger, V., and Oppenheim, R. W. (1982). Naturally occurring neuronal death in vertebrates. *Neurosci. Comment.* **1**, 39–55.

Hamburger, V., Brunso-Bechtold, J. K., and Yip, J. W. (1981). Neuronal death in the spinal ganglia of the chick embryo and its reduction by nerve growth factor. *J. Neurosci.* **1**, 60–71.

Hartikka, J., and Hefti, F. (1988). Comparison of nerve growth factor's effects on development of septum, striatum, and nucleus basalis cholinergic neurons *in vitro*. *J. Neurosci. Res.* **21**, 352–364.

Hartikka, J., and Hefti, F. (1990). Comparison of NGF's effects on development of septum, striatum and nucleus basalis cholinergic neurons *in vitro*. *J. Neurosci. Res.* **8**, 2967–2985.

Hashimoto, S. (1988). K-252a, a potent protein kinase inhibitor, blocks nerve growth factor-induced neurite outgrowth and changes in the phosphorylation of proteins in PC12h cells. *J. Cell Biol.* **107**, 1531–1539.

Heasley, L. E., and Johnson, G. L. (1989a). Detection of nerve growth factor and epidermal growth factor-regulated protein kinases in PC12 cells with synthetic peptide substrates. *Mol. Pharmacol.* **35**, 331–338.

Heasley, L. E., and Johnson, G. L. (1989b). Regulation of protein kinase C by nerve growth factor, epidermal growth factor, and phorbol esters in PC12 pheochromocytoma cells. *J. Biol. Chem.* **264**, 8646–8652.

Hefti, F. (1986). Nerve growth factor (NGF) promotes survival of septal cholinergic neurons after fimbrial transection. *J. Neurosci.* **6**, 2155–2162.

Hefti, F., and Weiner, W. J. (1986). Nerve growth factor and Alzheimer's disease. *Ann. Neurol.* **20**, 275–281.

Helin, K., and Beguinot, L. (1991). Internalization and down-regulation of the human epidermal growth factor receptor are regulated by the carboxyl-terminal tyrosines. *J. Biol. Chem.* **266**, 8363–8368.

Hempstead, B. L., Schleifer, L. S., and Chao, M. V. (1989). Expression of functional nerve growth factor receptors after gene transfer. *Science* **243**, 373–375.

Hempstead, B. L., Patil, N., Thiel, B., and Chao, M. V. (1990). Deletion of cytoplasmic sequences of the nerve growth factor receptor leads to loss of high-affinity ligand binding. *J. Biol. Chem.* **265**, 9595–9598.

Hempstead, B. L., Martin-Zanca, D., Kaplan, D. R., Parada, L. F., and Chao, M. V. (1991). High-affinity NGF binding requires coexpression of the *trk* proto-oncogene and the low-affinity NGF receptor. *Nature (London)* **350**, 678–683.

Hendry, I. A. (1977). Cell division in the developing sympathetic nervous system. *J. Neurocytol.* **6**, 299–309.

Hendry, I. A., and Campbell, J. (1976). Morphometric analysis of rat superior cervical ganglion after axotomy and nerve growth factor treatment. *J. Neurocytol.* **5**, 351–360.

Hengerer, B., Lindholm, D., Heumann, R., Ruther, U., Wagner, E. F., and Thoenen, H. (1990). Lesion-induced increase in nerve growth factor mRNA is mediated by c-*fos*. *Proc. Natl. Acad. Sci. U.S.A.* **87**, 3899–3903.

Heuer, J. G., Fatemie-Nainie, S., Wheeler, E. F., and Bothwell, M. (1990). Structure and developmental expression of the chicken NGF receptor. *Dev. Biol.* **137**, 287–304.

Heumann, R., Korsching, S., Scott, J., and Thoenen, H. (1984a). Relationship between levels of nerve

growth factor (NGF) and its messenger RNA in sympathetic ganglia and peripheral target tissues. *EMBO J.* **3**, 3183–3189.
Heumann, R., Schwab, M., Merkl, R., and Thoenen, H. (1984b). Nerve growth factor-mediated induction of choline acetyltransferase in PC12 cells: Evaluation of the site of action of nerve growth factor and the involvement of lysosomal degradation products of nerve growth factor. *J. Neurosci.* **4**, 3039–3050.
Heumann, R., Korsching, Bandtlow, C., and Thoenen, H. (1987a). Changes of nerve growth factor synthesis in nonneuronal cells in response to sciatic nerve transection. *J. Cell Biol.* **104**, 1623–1631.
Heumann, R., Lindholm, D., Bandtlow, C., Meyer, M., Radeke, M. J., Misko, T. P., Shooter, E., and Thoenen, H. (1987b). Differential regulation of mRNA encoding nerve growth factor and its receptor in rat sciatic nerve during development, degeneration and regeneration: Role of macrophages. *Proc. Natl. Acad. Sci. U.S.A.* **84**, 8735–8739.
Higgins, G. A., and Mufson, E. (1989). NGF receptor gene expression is decreased in the nucleus basalis in Alzheimer's disease. *Exp. Neurol.* **106**, 222–236.
Hollowell, J. P., Villadiego, A., and Rich, K. M. (1990). Sciatic nerve regeneration across gaps within silicone chambers: Long-term effects of NGF and consideration of axonal branching. *Exp. Neurol.* **110**, 45–51.
Holtzman, D. M., and Mobley, W. C. (1991). Molecular studies in Alzheimer's disease. *Trends Biochem. Sci.* **16**, 140–144.
Holtzman, D. M., Li, Y., DeArmond, S. J., McKinley, M. P., Gage, F. H., Epstein, C. J., and Mobley, W. C. (1992). A mouse model of neurodegeneration: Atrophy of basal forebrain neurons in Ts 16 transplants. *Proc. Natl. Acad. Sci. U.S.A.* **89**, 1383–1387.
Honegger, A. M., Schmidt, A., Ullrich, A., and Schlessinger, J. (1990). Separate endocytic pathways of kinase-defective and -active EGF receptor mutants expressed in same cells. *J. Cell Biol.* **110**, 1541–1548.
Hosang, M., and Shooter, E. M. (1985). Molecular characteristics of nerve growth factor receptors on PC12 cells. *J. Biol. Chem.* **260**, 655–662.
Hosang, M., and Shooter, E. M. (1987). The internalization of nerve growth factor receptor by high affinity receptors on pheochromocytoma PC12 cells. *EMBO J.* **6**, 1197–1202.
Ibáñez, C. F., Ebendal, T., and Persson, H. (1991). Chimeric molecules with multiple neurotrophic activities reveal structural elements determining the specificities of NGF and BDNF. *EMBO J.* **10**, 2105–2110.
James, R., and Bradshaw, R. A. (1984). Polypeptide growth factors. *Ann. Rev. Biochem.* **53**, 259–292.
Johnson, D., Lanahan, A., Buck, C. R., Sehgal, A., Morgan, C., Mercer, E., Bothwell, M., and Chao, M. (1986). Expression and structure of the human NGF receptor. *Cell* **47**, 545–554.
Johnson, E. M., Jr., Andres, R. Y., and Bradshaw, R. A. (1978). Characterization of the retrograde transport of nerve growth factor (NGF) using high specific activity [$^{124}$I] NGF. *Brain Res.* **150**, 319–331.
Johnson, E. M., Jr., Gorin, P. D., Brandeis, L. D., and Pearson, J. (1980). Dorsal root ganglion neurons are destroyed by exposure in utero to maternal antibody to nerve growth factor. *Science* **210**, 916–918.
Johnson, E. M., Jr., Taniuchi, M., Clark, H. B., Springer, J. E., Koh, S., Tayrien, M. W., and Loy, R. (1987). Demonstration of the retrograde transport of nerve growth factor receptor in the peripheral and central nervous system. *J. Neurosci.* **7**, 923–929.
Johnson, E. M., Taniuchi, M., and DiStefano, P. S. (1988). Expression and possible function of nerve growth factor receptors on Schwann cells. *Trends Neurosci.* **11**, 299–304.
Johnston, M. V., Rutkowski, J. L., Wainer, B. H., Long, J. B., and Mobley, W. C. (1987). NGF effects on developing forebrain cholinergic neurons are regionally specific. *Neurochem. Res.* **12**, 985–994.
Kalman, D., Wong, B., Horvai, A. E., Cline, M. J., and O'Lague, P. H. (1990). Nerve growth factor acts through cAMP-dependent protein kinase to increase the number of sodium channels in PC12 cells. *Neuron* **4**, 355–366.
Kaplan, D. R., Martin-Zanca, D., and Parada, L. F. (1991a). Tyrosine phosphorylation and tyrosine

kinase activity of the *trk* proto-oncogene product induced by NGF. *Nature (London)* **350**, 158–160.

Kaplan, D. R., Hempstead, B. L., Martin-Zanca, D., Chao, M. V., and Parada, L. F. (1991b). The *trk* proto-oncogene product: A signal transducing receptor for nerve growth factor. *Science* **252**, 554–558.

Kessler, J. A., and Black, I. B. (1980). The effects of nerve growth factor (NGF) and antiserum to NGF on the development of embryonic sympathetic neurons *in vivo*. *Brain Res.* **189**, 157–168.

Kim. U.-H., Fink, D., Jr., Kim, H. S., Park, D. J., Contreras, M. L., Guroff, G., and Rhee, S. G. (1991). Nerve growth factor stimulates phosphorylation of phospholipase C-gamma in PC12 cells. *J. Biol. Chem.* **266**, 1359–1362.

Kiss, J., McGovern, J., and Patel, A. J. (1988). Immunohistochemical localization of cells containing nerve growth factor receptors in the different regions of the adult rat forebrain. *Neuroscience* **27**, 731–748.

Klausner, R. D., Ashwell, G., van Renswoude, J., Harford, J. B., and Bridges, K. R. (1983). Binding of apotransferrin to K562 cells: Explanation of the transferrin cycle. *Proc. Natl. Acad. Sci. U.S.A.* **80**, 2263–2266.

Klein, R., Jing, S., Nanduri, V., O'Rourke, E., and Barbacid, M. (1991). The *trk* proto-oncogene encodes a receptor for nerve growth factor. *Cell* **65**, 189–197.

Koh, S., and Higgins, G. A. (1991). Differential regulation of the low-affinity nerve growth factor receptor during postnatal development of the rat brain. *J. Comp. Neurol.* **313**, 494–508.

Koh, S., and Loy, R. (1989). Localization and development of nerve growth factor-sensitive rat basal forebrain neurons and their afferent projections to hippocampus and neocortex. *J. Neurosci.* **9**, 2999–3018.

Koh, S., Oyler, G. A., and Higgins, G. A. (1989). Localization of nerve growth factor receptor messenger RNA and protein in the adult rat brain. *Exp. Neurol.* **106**, 209–221.

Koike, T., Martin, D. P., and Johnson, E. M., Jr. (1989). Role of $Ca^{2+}$ channels in the ability of membrane depolarization to prevent neuronal death induced by trophic factor deprivation: Evidence that levels of internal $Ca^{2+}$ determine nerve growth factor dependence of sympathetic ganglion cells. *Proc. Natl. Acad. Sci. U.S.A.* **86**, 6421–6425.

Koizumi, S., Contreras, M. L., Matsuda, Y., Hama, T., Lazarovici, P., and Guroff, G. (1988). K-252a: A specific inhibitor of the action of nerve growth factor on PC 12 cells. *J. Neurosci.* **8**, 715–721.

Koliatsos, V. E., Nauta, H. J. W., Clatterbuck, R. E., Holtzman, D. M., Mobley, W. C., and Price, D. L. (1990). Mouse nerve growth factor prevents degeneration of axotomized basal forebrain cholinergic neurons in monkey. *J. Neurosci.* **10**, 3801–3813.

Kordower, J. H., Gash, D. M., Bothwell, M., Hersh, L., and Mufson, E. (1988). Nerve growth factor receptor and choline acetyltransferase remain colocalized in the nucleus basalis (Ch4) of Alzheimer's patients. *Neurobiol. Aging* **10**, 287–294.

Korsching, S., and Thoenen, H. (1983). Quantitative demonstration of the retrograde axonal transport of endogenous nerve growth factor. *Neurosci. Lett.* **39**, 1–4.

Korsching, S., and Thoenen, H. (1985). Treatment with 6-hydroxydopamine and colchicine decreases nerve growth factor levels in sympathetic ganglia and increases them in the corresponding target tissues. *J. Neurosci.* **5**, 1058–1061.

Korsching, S., and Thoenen, H. (1988). Developmental changes of nerve growth factor levels in sympathetic ganglia and their target organs. *Dev. Biol.* **126**, 40–46.

Korsching, S., Auburgen, G., Heumann, R., Scott, J., and Thoenen, H. (1985). Levels of nerve growth factor and its mRNA in a the central nervous system of the rat correlate with cholinergic innervation. *EMBO J.* **4**, 1389–1393.

Kromer, L. F. (1987). Nerve growth factor treatment after brain injury prevents neuronal death. *Science* **235**, 214–216.

Kruijer, W., Schubert, D., and Verma, I. M. (1985). Induction of the proto-oncogene *fos* by nerve growth factor. *Proc. Natl. Acad. Sci. U.S.A.* **82**, 7330–7334.

Laasberg, T., Pihlak, A., Neuman, T., Paves, H., and Saarma, M. (1988). Nerve growth factor increases

the cyclic GMP level and activates the cyclic GMP phosphodiesterase in PC12 cells. *FEBS Lett.* **239**, 367–370.
Lalley, P. M., Rossi, G. V., and Baker, W. W. (1970). Analysis of local cholinergic tremor mechanisms following selective neurochemical lesions. *Exp. Neurol.* **27**, 258–275.
Lamballe, F., Klein, R., and Barbacid, M. (1991). trkC, a new member of the *trk* family of tyrosine protein kinases, is a receptor for neurotrophin-3. *Cell* **66**, 967–979.
Large, T. H., Bodary, S. C., Clegg, D. O., Weskamp, G., Otten, U., and Reichardt, L. (1986). Nerve growth factor gene expression in the developing rat brain. *Science* **234**, 352–355.
Large, T. H., Weskamp, G., Helder, J. C., Radeke, M. J., Misko, T. P., Shooter, E. M., and Reichardt, L. F. (1989). Structure and developmental expression of the nerve growth factor receptor in the chicken central nervous system. *Neuron* **2**, 1123–1134.
Lassing, I., and Lindberg, U. (1985). Specific interaction between phosphatidylinositol 4,5-bisphosphate and profilactin. *Nature (London)* **314**, 472–474.
Lefebvre, P. P., Leprince, P., Weber, T., Rigo, J.-M., Delree, P., and Moonen, G. (1990). Neuronotrophic effect of developing otic vesicle on cochleo-vestibular neurons: Evidence for nerve growth factor involvement. *Brain Res.* **507**, 254–260.
LeTourneau, P. C. (1978). Chemotactic response of nerve fiber elongation to nerve growth factor. *Dev. Biol.* **66**, 183–196.
Levi, A., and Alema, S. (1991). The mechanism of action of nerve growth factor. *Ann. Rev. Pharmacol. Toxicol.* **31**, 205–228.
Levi-Montalcini, R. (1987). The nerve growth factor 35 years later. *Science* **237**, 1154–1162.
Levi-Montalcini, R., and Booker, B. (1960). Destruction of the sympathetic ganglia in mammals by an antiserum to a nerve-growth protein. *Proc. Natl. Acad. Sci. U.S.A.* **46**, 384–391.
Levi-Montalcini, R., and Hamburger, V. (1951). Selective growth stimulating effects of mouse sarcoma on the sensory and sympathetic nervous system of the chick embryo. *J. Exp. Zool.* **116**, 321–362.
Levi-Montalcini, R., and Hamburger, V. (1953). A diffusible agent of mouse sarcoma, producing hyperplasia of sympathetic ganglia and hyperneurotization of viscera in the chick embryo. *J. Exp. Zool.* **123**, 233–288.
Levi-Montalcini, R., Meyer, H., and Hamburger, V. (1954). *In vitro* experiments on the effects of mouse sarcomas 180 and 37 on the spinal and sympathetic ganglia of the chick embryo. *Cancer Res.* **14**, 49–57.
Levi-Montalcini, R., Caramia, F., and Angeletti, P. U. (1969). Alterations in the fine structure of nucleoli in sympathetic neurons following NGF-antiserum treatment. *Brain Res.* **12**, 54–73.
Lillien, L. E., and Claude, P. (1985). Nerve growth factor is a mitogen for cultured chromaffin cells. *Nature (London)* **317**, 632–634.
Lindholm, D., Heumann, R., Meyer, M., and Thoenen, H. (1987). Interleukin-1 regulates synthesis of nerve growth factor in non-neuronal cells of rat sciatic nerve. *Nature (London)* **330**, 658–659.
Lindsay, R. M. (1988). Nerve growth factors (NGF, BDNF) enhance axonal regeneration by are not required for survival of adult sensory neurons. *J. Neurosci.* **7**, 2394–2405.
Lindsay, R. M., and Harmar, A. J. (1989). Nerve growth factor regulates expression of neuropeptide genes in adult sensory neurons. *Nature (London)* **337**, 362–364.
Lindsay, R. M., Shooter, E. M., Radeke, M. J., Misko, T. P., Dechant, G., Thoenen, H., and Lindholm, D. (1990). Nerve growth factor regulates expression of the nerve growth factor receptor gene in adult sensory neurons. *Eur. J. Neurosci.* **2**, 389–396.
Loeb, D. M., Maragos, J., Martin-Zanca, D., Chao, M. V., Parada, L. F., and Greene, L. A. (1991). The *trk* proto-oncogene rescues NGF in mutant NGF-nonresponsive PC12 cell lines. *Cell* **66**, 961–966.
Longo, F. M., Manthorpe, M., and Varon, S. (19820. Spinal cord neuronotrophic factors (SCNTFs). I. Bioassay of schwannoma and other conditioned media. *Dev. Brain Res.* **3**, 277–294.
Longo, F. M., Manthorpe, M., Skaper, S. D., Lundborg, G., and Varon, S. (1983). Neuronotrophic activities accumulate *in vivo* within silicone nerve regeneration chambers. *Brain Res.* **262**, 109–116.
Longo, F. M., Hayman, E. G., Davis, G. E., Ruoslahti, E., Engvall, E., Manthorpe, M., and Varon, S.

(1984). Neurite-promoting factors and extracellular matrix components accumulating *in vivo* within nerve regeneration chambers. *Brain Res.* **309,** 105–117.

Longo, F. M., Vu, K., and Mobley, W. C. (1990). The *in vitro* biological effect of nerve growth factor is inhibited by synthetic peptides. *Cell Reg.* **1,** 189–195.

Lu, B., Yokoyama, M., Dreyfus, C. F., and Black, I. B. (1991). NGF gene expression in actively growing brain glia. *J. Neurosci.* **11,** 318–326.

McDonald, N. Q., Lapatto, R., Murray-Rust, J., Gunning, J., Wiodawer, A., and Blundell, T. L. (1991). New protein fold revealed by a 2.3-Å resolution crystal structure of nerve growth factor. *Nature (London)* **354,** 411–414.

Machida, C. M., Scott, J. D., and Ciment, G. (1991). NGF-induction of the metalloproteinase-transin/stromelysin in PC12 cells: Involvement of multiple protein kinases. *J. Cell Biol.* **114,** 1037–1048.

MacIntosh, F. C. (1981). Acetylcholine. In "Basic Neurochemistry," 3d Ed. (G. J. Siege, R. W. Albers, B. W. Agranoff, and R. Katzman, eds.), pp. 183–204. Little Brown, Massachusetts.

Maher, P. A. (1988). Nerve growth factor induces protein-tyrosine phosphorylation. *Proc. Natl. Acad. Sci. U.S.A.* **85,** 6788–6791.

Maisonpierre, P. C., Belluscio, L., Friedman, B., Alderson, R. F., Wiegand, S. J., Furth, M. E., Lindsay, R. M., and Yancopoulos, G. D. (1990). NT-3, BDNF, and NGF in the developing rat nervous system: Parallel as well as reciprocal patterns of expression. *Neuron* **5,** 501–509.

Mallett, S., Fossum, S., and Barclay, A. N. (1990). Characterization of the MRC OX40 antigen of activated CD4 positive T lymphocytes—a molecule related to nerve growth factor receptor. *EMBO J.* **9,** 1063–1068.

Marchetti, D., and McManaman, J. L. (1990). Characterization of nerve growth factor binding to embryonic rat spinal cord neurons. *J. Neurosci. Res.* **27,** 211–218.

Martin, D. P., and Johnson, E. M., Jr. (1991). Programmed cell death in the peripheral nervous system. In "Apoptosis: The Molecular Basis of Cell Death," (L. D. Tomei and F. Cope, eds.), pp. 247–261. Cold Spring Harbor Laboratory Press, Cold Spring Harbor, New York.

Martin, D. P., Schmidt, R. E., Di Stefano, P. S., Lowry, O. H., Carter, J. G., and Johnson, E. M., Jr. (1988). Inhibitors of protein synthesis and RNA synthesis prevent neuronal death caused by nerve growth factor deprivation. *J. Cell Biol.* **106,** 829–844.

Martin-Zanca, D., Barbacid, M., and Parada, L. F. (1990). Expression of the *trk* proto-oncogene is restricted to the sensory cranial and spinal ganglia of neural crest origin in mouse development. *Genes Dev.* **4,** 683–694.

Meakin, S. O., and Shooter, E. M. (1991a). Molecular investigations on the high-affinity nerve growth factor receptor. *Neuron* **6,** 153–163.

Meakin, S. O., and Shooter, E. M. (1991b). Tyrosine kinase activity coupled to the high-affinity nerve growth factor-receptor complex. *Proc. Natl. Acad. Sci. U.S.A.* **88,** 5862–5866.

Menesini-Chen, M. G., Chen, J. S., and Levi-Montalcini, R. (1978). Sympathetic nerve fibers ingrowth in the central nervous system of neonatal rodents upon intracerebral NGF injections. *Arch. Ital. Biol.* **116,** 53–84.

Mesulam, M. M., Mufson, E. J., Wainer, B. H., and Levey, A. I. (1983). Central cholinergic pathways in the rat: An overview based on an alternative nomenclature (Ch1-Ch6). *Neuroscience* **10,** 1185–1201.

Milbrandt, J. (1987). A nerve growth factor-induced gene encodes a possible transcriptional regulatory factor. *Science* **238,** 797–799.

Milbrandt, J. (1988). Nerve growth factor induces a gene homologous to the glucocorticoid receptor gene. *Neuron* **1,** 183–188.

Miller, F., Mathew, T. C., and Toma, J. G. (1991). Regulation of nerve growth factor receptor gene expression by nerve growth factor in the developing peripheral nervous system. *J. Cell Biol.* **112,** 303–312.

Misko, T. P., Radeke, M. J., and Shooter, E. M. (1987). Nerve growth factor in neuronal development and maintenance. *J. Exp. Biol.* **132,** 177–190.

Miyasaka, T., Sternberg, D. W., Miyasaka, J., Sherline, P., and Saltiel, A. R. (1991). Nerve growth factor

stimulates protein tyrosine phosphorylation in PC-12 pheochromocytoma cells. *Proc. Natl. Acad. Sci. U.S.A.* **88**, 2653–2657.
Mobley, W. C. (1989). Nerve growth factor in Alzheimer's disease: To treat or not to treat? *Neurobiol. Aging* **10**, 578–580.
Mobley, W. C., Server, A. C., Ishii, D. N., Riopelle, R. J., and Shooter, E. M. (1977). Nerve growth factor—medical progress. *N. Engl. J. Med.* **297**, 1096–1104; 1149–1158, 1211–1218.
Mobley, W. C., Rutkowski, J. L., Tennekoon, S. J., Buchanan, K., and Johnston, M. V. (1985). Choline acetyltransferase in striatum of neonatal rats increased by nerve growth factor. *Science* **229**, 284–287.
Mobley, W. C., Rutkowski, J. L., Tennekoon, G. I., Gemski, J., Buchanan, K., and Johnston, M. V. (1986). Nerve growth factor increases choline acetyltransferase activity in developing basal forebrain neurons. *Mol. Brain Res.* **1**, 53–62.
Mobley, W. C., Neve, R. L., Prusiner, S. B., McKinley, M. P. (1988). Nerve growth factor increases mRNA levels for the prion protein and the β-amyloid protein precursor in developing hamster brain. *Proc. Natl. Acad. Sci. U.S.A.* **85**, 9811–9815.
Mobley, W. C., Woo, J. E., Edwards, R. H., Riopelle, R. J., Longo, F. M., Weskamp, G., Otten, U., Valletta, J. S., and Johnston, M. V. (1989). Developmental regulation of nerve growth factor and its receptor in the rat caudate-putamen. *Neuron* **3**, 655–664.
Mufson, E. J., and Kordower, J. H. (1992). Cortical neurons express nerve growth factor receptors in advanced age and Alzheimer's disease. *Proc. Natl. Acad. Sci. U.S.A.* **89**, 569–573.
Myers, S., Dostaler, S., Weaver, D., Richardson, P., Riopelle, R., and Dow, K. (1991). A functional domain of the low affinity NGF receptor (LNGFR). *Neurosci. Abstr.* **17**, 1498.
Nairn, A. C., Nichols, R. A., Brady, M. J., and Palfrey, H. C. (1987). Nerve growth factor treatment on cAMP elevation reduces $Ca^{2+}$/calmodulin-dependent protein kinase III activity in PC12 cells. *J. Biol. Chem.* **262**, 14265–14272.
Nebreda, A. R., Martin-Zanca, D., Kaplian, D. R., Parada, L. F., and Santos, E. (1991). Induction by NGF of meiotic maturation of *Xenopus* oocytes expressing the *trk* proto-oncogene product. *Science* **252**, 558–561.
Ojeda, S. R., Hill, D. F., and Katz, K. H. (1991). The genes encoding nerve growth factor and its receptor are expressed in the developing female rat hypothalamus. *Mol. Brain Res.* **9**, 47–55.
Otten, U. (1984). Nerve growth factor and the peptidergic sensory neurons. *Trends Pharmacol. Sci.* **5**, 307–310.
Otto, D., Unsicker, K., and Grothe, C. (1987). Pharmacological effects of nerve growth factor and fibroblast growth factor applied to the transectioned sciatic nerve on neuron death in adult rat dorsal root ganglia. *Neurosci. Lett.* **83**, 156–160.
Oyanagi, K., Takahashi, H., Wakabayashi, K., and Ikuta, F. (1989). Correlative decrease of large neurons in the neostriatum and basal nucleus of Meynert in Alzheimer's disease. *Brain Res.* **504**, 354–357.
Paves, H., Neuman, T., Metsis, M., and Saarma, M. (1988). Nerve growth factor induces rapid redistribution of F-actin in PC12 cells. *FEBS Lett.* **235**, 141–143.
Payrastre, B., van Bergen en Henegouwen, P., Breton, M., den Hartigh, J., Plantavid, M., Verkleij, A. J., and Boonstra, J. (1991). Phosphoinositide kinase, diacylglycerol kinase, and phospholipase C activities associated to the cytoskeleton: Effect of epidermal growth factor. *J. Cell Biol.* **115**, 121–128.
Phillips, H. S., Hains, J. M., Laramee, G. R., Rosenthal, A., and Winslow, J. W. (1990). Widespread expression of BDNF but not NT3 by target areas of basal forebrain cholinergic neurons. *Science* **250**, 290–294.
Phillips, H. S., Hains, J. M., Armanini, M., Laramee, G. R., Johnson, S. A., and Winslow, J. W. (1991). BDNF mRNA is decreased in hippocampus of individuals with Alzheimer's disease. *Soc. Neurosci. Abstr.* **17**, 909.
Pollock, J. D., Krempin, M., and Rudy, B. (1990). Differential effects of NGF, FGF, EGF, cAMP, and dexamethasone on neurite outgrowth and sodium channel expression in PC12 cells. *J. Neurosci.* **10**, 2626–2637.

Price, D. L. (1986). New perspectives on Alzheimer's disease. *Ann. Rev. Neurosci.* **9**, 489–512.
Qiu, M. S., Pitts, A. F., Winters, T. R., and Green, S. H. (1991). ras isoprenylation is required for ras-induced but not for NGF-induced neuronal differentiation of PC12 cells. *J. Cell Biol.* **115**, 795–808.
Radeke, M. J., and Feinstein, S. C. (1991). Analytical purification of the slow, high-affinity NGF receptor: Identification of a novel 135 kd polypeptide. *Neuron* **7**, 141–150.
Radeke, M. J., Misko, T. P., Hsu, C., Herzenberg, L. A., and Shooter, E. M. (1987). Gene transfer and molecular cloning of the rat nerve growth factor receptor. *Nature (London)* **325**, 593–597.
Raivich, G., and Kreutzberg, G. W. (1987). Expression of growth factor receptors in injured nervous tissue. I. Axotomy leads to a shift in the cellular distribution of specific β-nerve growth factor binding in the injured and regenerating PNS. *J. Neurocytol.* **17**, 689–700.
Raivich, G., Zimmermann, A., and Sutter, A. (1985). The spatial and temporal pattern of bNGF receptor expression in the developing chick embryo. *EMBO J.* **4**, 637–644.
Raivich, G., Hellweg, R., and Kreutzberg, G. W. (1991). NGF receptor-mediated reduction in axonal NGF uptake and retrograde transport following sciatic nerve injury and during regeneration. *Neuron* **7**, 151–164.
Rausch, D. M., Dickens, G., Doll, S., Fujita, K., Koizumi, S., Rudkin, B. B., Tocco, M., Eiden, L. E., and Guroff, G. (1989). Differentiation of PC12 cells with v-src: comparison with nerve growth factor. *J. Neurosci. Res.* **24**, 49–58.
Rennert, P. D., and Heinrich, G. (1986). Nerve growth factor mRNA in brain: Localization by *in situ* hybridization. *Biochem. Biophys. Res. Commun.* **138**, 813–818.
Represa, J., Van De Water, T. R., and Bernd, P. (1991). Temporal pattern of nerve growth factor receptor expression in developing cochlear and vestibular ganglia in quail and mouse. *Anat. Embryol.* **184**, 421–432.
Rich, K. M., Yip, H. K., Osborne, A., Schmidt, R. E., and Johnson, E. M. (1984). Role of nerve growth factor in the adult dorsal root ganglia neuron and its response to injury. *J. Comp. Neurobiol.* **230**, 110–118.
Rich, K. M., Luszczynski, J. R., Osborne, P. A., and Johnson, E. M. (1987). Nerve growth factor protects adult sensory neurons from cell death and atrophy caused by nerve injury. *J. Neurocytol.* **16**, 261–268.
Rich, K. M., Alexander, T. D., Pryor, J. C., and Hollowell, J. P. (1989). Nerve growth factor enhances regeneration through silicone chambers. *Exp. Neurol.* **105**, 162–170.
Richardson, P. M., and Ebendal, T. (1982). Nerve growth activities in rat peripheral nerve. *Brain Res.* **246**, 57–64.
Richardson, P. M., Verge Issa, V. M. K., and Riopelle, R. J. (1986). Distribution of neuronal receptors for nerve growth factor in the rat. *J. Neurosci.* **6**, 2312–2321.
Richardson, R. T., and DeLong, M. R. (1988). A reappraisal of the functions of the nucleus basalis of Meynert. *Trends Neurosci.* **11**, 264–267.
Richardson, R. T., and Riopelle, R. J. (1984). Uptake of nerve growth factor along peripheral and spinal axons of primary sensory neurons. *J. Neurosci.* **4**, 1683–1689.
Rodriguez-Tebar, A., Dechant, G., and Barde, Y.-A. (1990). Binding of brain-derived neurotrophic factor to the nerve growth factor receptor. *Neuron* **4**, 487–492.
Rodriguez-Tebar, A., Dechant, G., and Barde, Y. A. (1991). Neurotrophins—Structural relatedness and receptor interactions. *Phil. Trans. R. Soc. London Ser. B (Biol.)* **331**, 255–258.
Rohrer, H., Heumann, R., and Thoenen, H. (1988). The synthesis of nerve growth factor (NGF) in developing skin is independent of innervation. *Dev. Biol.* **128**, 240–244.
Romano, C., Nichols, R. A., and Greengard, P. (1987). Synapsin I in PC12 cells. II. Evidence for regulation by NGF of phosphorylation at a novel site. *J. Neurosci.* **7**, 1300–1306.
Rowland, E. A., Muller, T. H., Goldstein, M., and Greene, L. A. (1987). Cell-free detection and characterization of a novel nerve growth factor-activated protein kinase in PC12 cells. *J. Biol. Chem.* **262**, 7504–7513.
Rowland, G. E., and Greene, L. A. (1990). Multiple pathways of N-kinase activation in PC12 cells. *J. Neurochem.* **54**, 423–433.

Rubin, E. (1985). Development of the rat superior cervical ganglion: Ganglion cell maturation. *J. Neurosci.* **5,** 673–684.

Ruit, K. G., Osborne, P. A., Schmidt, R. E., Johnson, E. M., and Snider, W. D. (1990). Nerve growth factor regulates sympathetic ganglion cell morphology and survival in the adult mouse. *J. Neurosci.* **10,** 2412–2419.

Rydel, R. E., and Greene, L. A. (1987). Acidic and basic fibroblast growth factors promote stable neurite outgrowth and neuronal differentiation in cultures of PC12 cells. *J. Neurosci.* **7,** 3639–3653.

Sabatini, M. T., Pellegrino de Iraldi, A., and De Roberts, E. (1965). Early effects of antiserum against the nerve growth factor on fine structure of sympathetic neurons. *Exp. Neurol.* **12,** 370–383.

Saffran, B. N., and Crutcher, K. A. (1990). NGF-induced remodeling of mature uninjured axon collaterals. *Brain Res.* **525,** 11–20.

Sariola, H., Saarma, M., Sainio, K., Arumae, U., Palgi, J., Vaahtokare, A., Thesleff, I., and Karavanov, A. (1991). Dependence of kidney morphogenesis on the expression of Nerve growth factor receptor. *Science* **254,** 571–573.

Schecter, A. L., and Bothwell, M. A. (1981). Nerve growth factor receptors on PC12 cells: Evidence for two receptor classes with differing cytoskeletal association. *Cell* **24,** 867–874.

Schnapp, B. J., and Reese, T. S. (1989). Dynein is the motor for retrograde axonal transport of organelles. *Proc. Natl. Acad. Sci. U.S.A.* **86,** 1548–1552.

Schneider, R., and Schweiger, M. (1991). A novel modular mosaic of cell adhesion motifs in the extracellular domains of the neurogenic *trk* and *trkB* tyrosine kinase receptors. *Oncogene* **6,**1807–1811.

Schubert, D. (1978). NGF-induced alterations in protein secretion and substrate-attached material of a clonal nerve cell line. *Brain Res.* **155,** 196–200.

Schucker, F. (1972). Effects of NGF-antiserum in sympathetic neurons during early postnatal development. *Exp. Neurol.* **36,** 59–78.

Schwartz, J. P., Pearson, J., and Johnson, E. M. (1982). Effects of exposure to anti-NGF on sensory neurons of adult rats and guinea pigs. *Brain Res.* **244,** 378–381.

Schwarz, M., Ikonomidou, C., Klockgether, T., Turski, L., Ellenbroek, B., and Sontag, K.-H. (1986). The role of striatal cholinergic mechanisms for the development of limb rigidity: an electromyographic study in rats. *Brain Res.* **373,** 365–372.

Scott, S. A., and Davies, A. M. (1990). Inhibition of protein synthesis prevents cell death in sensory and parasympathetic neurons deprived of neurotrophic factor *in vitro*. *J. Neurobiol.* **21,** 630–638.

Sehgal, A., Bothwell, M., and Chao, M. (1989). Gene transfer of truncated NGF receptor clones leads to cell surface expression in mouse fibroblasts. *Nucleic Acids Res.* **17,** 5623–5632.

Seilheimer, B., and Schachner, M. (1987). Regulation of neural cell adhesion molecule expression on cultured mouse Schwann cells by nerve growth factor. *EMBO J.* **6,** 1611–1616.

Server, A. C., and Mobley, W. C. (1991). Neuronal cell death and the role of apoptosis. In "Apoptosis: The Molecular Basis of Cell Death," (L. D. Tomei and F. O. Cope, eds.), pp. 263–278. Cold Spring Harbor Laboratory Press, Cold Spring Harbor, New York.

Shelton, D. L., and Reichardt, L. F. (1984). Expression of the β-nerve growth factor gene correltes with the density of sympathetic innervation in effector organs. *Proc. Natl. Acad. Sci. U.S.A.* **81,** 7951–7955.

Shelton, D. L., and Reichardt, L. F. (1986). Studies on the expression of b NGF gene in the central nervous system: Level and regional distribution of NGF mRNA suggest that NGF functions as a trophic factor for several neuronal populations. *Proc. Natl. Acad. Sci. U.S.A.* **83,** 2714–2718.

Sieber-Blum, M. (1991). Role of the neurotrophic factors BDNF and NGF in the commitment of pluripotent neural crest cells. *Neuron* **6,** 949–955.

Smith, D. S., King, C. S., Pearson, E., Gittinger, C. K., and Landreth, G. E. (1989). Selective inhibition of nerve growth factor-stimulated protein kinases by K-252a and 5′-S-methyladenosine in PC12 cells. *J. Neurochem.* **53,** 800–806.

Smith, M. R., DeGudicibus, S. J., and Stacey, D. W. (1986). Requirement for *c-ras* proteins during viral oncogene transformation. *Nature (London)* **320,** 540–543.

Sofroniew, M. V., Galletly, N. P., Isacson, O., and Svendsen, C. N. (1990). Survival of adult basal forebrain cholinergic neurons after loss of target neurons. *Science* **247**, 338–342.

Sofroniew, M. V., Cooper, J. D., Svendsen, C. N., Crossman, P., Ip, N. Y., Lindsay, R. M., Zafra, F., Castren, E., Thoenen, H., and Lindholm, D. (1991). Long term survival of septal cholinergic neurons after lesions that deplete the hippocampus of cells producing NGF or BDNF mRNA. *Soc. Neurosci. Abstr.* **17**, 221.

Sorkin, A., Waters, C., Overholser, K. A., and Carpenter, G. (1991). Multiple autophosphorylation site mutations of the epidermal growth factor receptor. Analysis of kinase activity and endocytosis. *J. Biol. Chem.* **266**, 8355–8362.

Spillantini, M. G., Aloe, L., Alleva, E., De Simone, R., Goedert, M., and Levi-Montalcini, R. (1989). Nerve growth factor mRNA and protein increase in hypothalamus in a mouse model of aggression. *Proc. Natl. Acad. Sci. U.S.A.* **86**, 8555–8559.

Springer, J. E., Koh, S., Tayrien, M. W., and Loy, R. (1987). Basal forebrain magnocellular neurons stain for nerve growth factor receptor: Correlation with cholinergic cell bodies and effects of axotomy. *J. Neurosci. Res.* **17**, 111–118.

Squinto, S. P., Stitt, T. N., Aldrich, T. H., Davis, S., Bianco, S. M., Radziewski, C., Glass, D. J., Masiakowski, P., Furth, M. E., Valenzuela, D. M., DiStefano, P. S., and Yancopoulos, G. D. (1991). *trkB* encodes a functional receptor for brain-derived neurotrophic factor and neurotrophin-3 but not nerve growth factor. *Cell* **65**, 885–893.

Stemple, D. L., Mahanthappa, N. K., and Anderson, D. J. (1988). Basic FGF induces neuronal differentiation, cell division and NGF dependence in chromaffin cells: A sequence of events in sympathetic development. *Neuron* **1**, 517–525.

Stockel, K., Schwab, M., and Thoenen, H. (1975). Specificity of retrograde transport of nerve growth factor (NGF) in sensory neurons: A biochemical and morphological study. *Brain Res.* **89**, 1–14.

Sutter, A., Riopelle, R. J., Harris, W. R. M., and Shooter, E. M. (1979). Nerve growth factor receptors. Characterization of two distinct classes of binding sites on chick embryo sensory ganglia cells. *J. Biol. Chem.* **254**, 5972–5982.

Taniuchi, M., and Johnson, E. M. J. (1985). Characterization of the binding properties and retrograde axonal transport of a monoclonal antibody directed against the rat nerve growth factor receptor. *J. Cell Biol.* **101**, 1100–1106.

Taylor, L. K., and Landreth, G. E. (1991). NGF activates a protein kinase which phosphorylates the proto-oncogene c-*fos*. *Soc. Neurosci. Abstr.* **17**, 1117.

Tessier-Lavigne, M., and Placzek, M. (1991). Target attraction: Are developing axons guided by chemotropism? *Trends Neurosci.* **14**, 303–310.

Tsao, H. S., and Greene, L. A. (1991). The roles of macromolecular synthesis and phosphorylation in the regulation of a protein kinase activity transiently stimulated by nerve growth factor. *J. Biol. Chem.* **266**, 12981–12988.

Tsao, H., Aletta, J. M., and Greene, L. A. (1990). Nerve growth factor and fibroblast growth factor selectively activate a protein kinase that phosphorylates high molecular weight microtubule-associated proteins. Detection, partial purification, and characterization in PC12 cells. *J. Biol. Chem.* **265**, 15471–15480.

Tuszynski, M. H., U, H. S, Amaral, D. G., and Gage, F. (1990). Nerve growth factor infusion in the primate brain reduces lesion-induced cholinergic neuronal degeneration. *J. Neurosci.* **10**, 3604–3614.

Tuszynski, M. H., Sang, H., Yoshida, K., and Gage, F. H. (1991). Recombinant human nerve growth factor infusions prevent cholinergic neuronal degeneration in the adult primate brain. *Ann. Neurol.* **30**, 625–636.

Vale, R. D., and Shooter, E. M. (1985). Assaying binding of nerve growth factor to cell surface receptors. *Methods Enzymol.* **109**, 21–39.

Vale, R. D., Ignatius, M. J., and Shooter, E. M. (1985). Association of nerve growth factor receptors with the triton X-100 cytoskeleton of PC12 cells. *J. Neurosci.* **5**, 2762–2770.

van Bergen en Henegouwen, P. M. P., Defize, L. H. K., de Kroon, J., van Damme, H., Verkleij, A. J., and Boonstra, J. (1989). Ligand induced association of epidermal growth factor receptor to the cytoskeleton of A431 cells. *J. Cell Biochem.* **39,** 455–465.

van Calker, D., Takahata, K., and Heumann, R. (1989). Nerve growth factor potentiates the hormone-stimulated intracellular accumulation of inositol phosphates and $Ca^{2+}$ in rat PC12 pheochromocytoma cells: Comparison with the effect of epidermal growth factor. *J. Neurochem.* **52,** 38–45.

Vantini, G., Schiavo, N., Di Martino, A., Polato, P., Triban, C., Callegaro, L., Toffano, G., and Leon, A. (1989). Evidence for a physiological role of nerve growth factor in the central nervous system of neonatal rats. *Neuron* **3,** 267–273.

Verge, V. M. K., Richardson, P. M., Benoit, R., and Riopelle, R. J. (1989). Histochemical characterization of sensory neurons with high-affinity receptors for nerve growth factor. *J. Neurocytol.* **18,** 583–591.

Verge, V. M. K., Tetzlaff, W., Richardson, P. M., and Bisby, M. A. (1990). Correlation between GAP-43 and nerve growth factor receptors in rat sensory neurons. *J. Neurosci.* **10,** 926–934.

Vetter, M. L., Martin-Zanca, D., Parada, L. F., Bishop, J. M., and Kaplan, D. R. (1991). Nerve growth factor rapidly stimulates tyrosine phosphorylation of phospholipase C-γ 1 by a kinase activity associated with the product of the *trk* protooncogene. *Proc. Natl. Acad. Sci. U.S.A.* **88,** 5650–5654.

Volonte, C., and Greene, L. A. (1990a). Induction of ornithine decarboxylase by nerve growth factor in PC12 cells: Dissection by purine analogues. *J. Biol. Chem.* **265,** 11050–11055.

Volonte, C., and Greene, L. A. (1990b). Nerve growth factor (NGF) responses by non-neuronal cells: Detection by assay of a novel NGF-activated protein kinase. *Growth Factors* **2,** 321–331.

Volonte, C., Rukenstein, A., Loeb, D. M., and Greene, L. A. (1989). Differential inhibition of nerve growth factor responses by purine analogues: Correlation with inhibition of a nerve growth factor-activated protein kinase. *J. Cell Biol.* **109,** 2395–2403.

Wanaka, A., and Johnson, E. M. (1990). Developmental study of nerve growth factor receptor mRNA expression in the postnatal rat cerebellum. *Dev. Brain Res.* **55,** 288–292.

Wayne, D. B., and Heaton, M. B. (1990a). The response of cultured trigeminal and spinal cord motoneurons to nerve growth factor. *Dev. Biol.* **138,** 473–483.

Wayne, D. B., and Heaton, M. B. (1990b). The ontogeny of specific retrograde transport of nerve growth factor by motoneurons of the brainstem and spinal cord. *Dev. Biol.* **138,** 484–498.

Welcher, A. A., Bitler, C. M., Radeke, M. J., and Shooter, E. M. (1991). Nerve growth factor binding domain of the nerve growth factor receptor. *Proc. Natl. Acad. Sci. U.S.A.* **8,** 159–163.

Weskamp, G., and Otten, U. (1987). An enzyme-linked immunoassay for nerve growth factor (NGF): A tool for studying regulatory mechanisms involved in NGF production in brain and peripheral tissues. *J. Neurochem.* **48,** 1779–1786.

Weskamp, G., and Reichardt, L. F. (1991). Evidence that biological activity of NGF is mediated through a novel subclass of high-affinity receptors. *Neuron* **6,** 649–663.

Whittemore, S. R., Ebendal, T., Larkfors, L., Olson, L., Seiger, A., Stromberg, I., and Persson, H. (1986). Developmental and regional expression of b-nerve growth factor messenger RNA and protein in the rat central nervous system. *Proc. Natl. Acad. Sci. U.S.A.* **83,** 817–821.

Wiegant, F. A., Blok, F. J., Defize, L. H., Linnemans, W. A., Verkley, A. J., and Boonstra, J. (1986). Epidermal growth factor receptors associated to cytoskeletal elements of epidermoid carcinoma (A431) cells. *J. Cell Biol.* **103,** 87–94.

Wiley, H. S., Herbst, J. J., Walsh, B. J., Lauffenburger, D. A., Rosenfeld, M. G., and Gill, G. N. (1991). The role of tyrosine kinase activity in endocytosis, compartmentation, and down-regulation of the epidermal growth factor receptor. *J. Biol. Chem.* **266,** 11083–11094.

Williams, L. R., Varon, S., Peterson, G. M., Wictorin, K., Fischer, W., Bjorklund, A., and Gage, F. H. (1986). Continuous infusion of nerve growth factor prevents basal forebrain neuronal death after fimbria–fornix transection. *Proc. Natl. Acad. Sci. U.S.A.* **83,** 9231–9235.

Windebank, A. J., and Poduslo, J. F. (1986). Neuronal growth factors produced by adult peripheral nerve after injury. *Brain Res.* **385,** 197–200.

Wright, L. L., Cunningham, T. J., and Smolen, A. J. (1983). Developmental neuron death in the rat superior cervical sympathetic ganglion: Cell counts and ultrastructure. *J. Neurocytol.* **12,** 727–738.

Wu, B. Y., Fodor, E. J., Edwards, R. H., and Rutter, W. J. (1989). Nerve growth factor induces the proto-oncogene c-*jun* in PC12 cells. *J. Biol. Chem.* **264**, 9000–9003.

Wyatt, S., Shooter, E. M., and Davies, A. M. (1990). Expression of the NGF receptor gene in sensory neurons and their cutaneous targets prior to and during innervation. *Neuron* **2**, 421–427.

Yan, H., Schlessinger, J., and Chao, M. V. (1991). Chimeric NGF–EGF receptors define domains responsible for neuronal differentiation. *Science* **252**, 561–563.

Yan, Q., and Johnson, E. M. (1988). An immunohistochemical study of nerve growth factor receptor in developing rats. *J. Neurosci.* **8**, 3481–3498.

Yan, Q., and Johnson, E. M. (1989). Immunohistochemical localization and biochemical characterization of nerve growth factor receptor in adult rat brain. *J. Comp. Neurol.* **290**, 595–598.

Yan, Q., Snider, W. D., Pinzone, J. J., and Johnson, E. M., Jr. (1988). Retrograde transport of nerve growth factor (NGF) in motoneurons of developing rats: Assessment of potential neurotrophic effects. *Neuron* **1**, 335–343.

Yasuda, T., Sobue, G., Takayuki, I., Mitsuma, T., and Takahashi, A. (1990). Nerve growth factor enhances neurite aborization of adult sensory neurons: A study in single-cell culture. *Brain Res.* **524**, 54–63.

Zafra, F., Hengerer, B., Leibrock, J., Thoenen, H., and Lindholm, D. (1990). Activity dependent regulation of BDNF and NGF mRNAs in the rat hippocampus is mediated by non-NMDA glutamate receptors. *EMBO J.* **9**, 3545–3550.

Zigmond, R. E., Schwarzschild, M. A., and Rittenhouse, A. R. (1989). Acute regulation of tyrosine hydroxylase by nerve activity and by neurotransmitters via phosphorylation. *Ann. Rev. Neurosci.* **12**, 415–461.

# Brain-Derived Neurotrophic Factor: An NGF-Related Neurotrophin

Ronald M. Lindsay

This chapter is, perhaps, one of the last ones in which almost all the published articles containing any substantial mention of brain-derived neurotrophic factor (BDNF) may be comfortably reviewed. In terms of its natural history, BDNF may be described as a molecule that has had a long gestation period, a demanding labor, but almost no childhood, before being thrust into prominence. Although several investigators have participated, the course of BDNF discovery has been charted largely by Yves-Alain Barde, currently at the Max Planck Institute for Psychiatry in Munich, with the unflinching support of Hans Thoenen. BDNF was described first in 1982, as the culmination of studies at the Biozentrum of the University of Basel, Switzerland, that started from the discovery of a distinct but NGF-like activity in the growth medium of cultured glioma cells. Purification of BDNF from pig brain, a painstaking procedure by any standard, led to the discovery of a molecule that had neuronal specificities that were both overlapping and distinct from those of nerve growth factor (NGF). Partial protein sequence information led, in 1988 and 1989, to the molecular cloning of BDNF and to the realization that BDNF and NGF were members of a gene family. Molecular cloning of BDNF has led not only to increased availability of this rare protein, by recombinant techniques, but also to the rapid discovery of neurotrophin-3 (NT-3), a third member of what has now been termed the neurotrophin family of NGF-related trophic factors. Rapidly emerging studies on the biology of BDNF suggest multiple populations of responsive CNS neurons, including important neuronal targets of neurodegenerative diseases, for example, septal cholinergic neurons, nigral dopaminergic neurons, and motor neurons. The therapeutic potential of BDNF, NGF, and NT-3 is an area of growing interest.

## I. INTRODUCTION

Although there is, as yet, little understanding at the molecular level of the mechanisms involved in the early stages of neuronal specification—creation of diverse neuronal types, extensive neuroblast proliferation, neuroblast migration, and the formation of discrete ganglia or nuclei—it is now clear that sculpting of later events in ontogeny of the vertebrate nervous system is dependent on interaction of neurons with epigenetic cues in the environment. In contrast to a precise production of the exact number of neurons required to establish the mature nervous system, it appears that vertebrate development involves an initial overproduction of neurons. Neurons in excess of those required to form either the peripheral (PNS) or central (CNS) nervous system subsequently are eliminated or pruned during phases of naturally occurring cell death (for reviews, see Hamburger and Oppenheim, 1982; Oppenheim, 1991). This phenomenon has been described for the entire vertebrate nervous system but has been studied to greatest advantage in the developing avian peripheral nervous system. Initial clues about the types of mechanisms that might regulate neuronal selection arose from limb ablation and supernumerary limb transplantation studies in the chick embryo. These studies, originating largely from the pioneering work of Victor Hamburger (Hamburger and Levi-Montalcini, 1949; Hamburger, 1958; Hollyday and Hamburger, 1976; see also, Landmesser and Pilar, 1978) and his students, have demonstrated that removal of peripheral tissue such as a limb bud, the target of developing peripheral neurons, leads to a reduction in the number of sympathetic, sensory, and motor neurons in ganglia or spinal cord segments at the level of the amputation. Conversely, transplantation of supernumerary limb buds has been shown to support survival of abnormally high numbers of neurons in adjacent motor neuron pools. These observations gave substance to the neurotrophic hypothesis and, eventually, to the study and discovery of target-derived neurotrophic factors. For a comprehensive review of the neurotrophic hypothesis and supporting studies, the reader is referred to the excellent monograph by Purves (1988) and major reviews by Levi-Montalcini and Angeletti (1966,1968), Thoenen and Barde (1980), Barde (1988,1989), and Thoenen (1991).

## II. NGF: PROTOTYPICAL NEUROTROPHIC FACTOR

The notion that peripheral tissues might influence the specificity and density of their own innervation led, both directly and indirectly, to the discovery of NGF, as reviewed by Levi-Montalcini and Angelleti (1968), Thoenen and Barde (1980), and Snider and Johnson (1989). In brief, our understanding of the neuronal biology of NGF indicates that this protein is an essential survival factor during the development of most, if not all, sympathetic neurons, a subpopulation of neural crest-derived sensory neurons (probably, for the most part, small-diameter, unmyelinated, cutaneous, nociceptive sensory neurons), and possibly cholinergic neurons of the basal forebrain (Levi-Montalcini and Angeletti, 1968; Thoenen and Barde, 1980; Johnson et al., 1986; Whittemore and Seiger, 1987; Hartikka and Hefti, 1988; Lindsay, 1988b; Snider and Johnson, 1989). In addition to being involved in the selection of neuronal populations during development (Hamburger et al., 1981), NGF is also clearly important in maintenance of the normal phenotype of mature sensory, sympathetic, and cholinergic neurons (Hefti, 1986; Lindsay, 1988b,1992). Sequestration of NGF by antibodies or disconnection of neurons from their targets leads to loss of phenotypic markers, for example, depletion of tyrosine hydroxylase (TH) activity in sympathetic neurons, loss of substance P in sensory neurons, and loss of choline acetyltransferase (ChAT) activity in septal cholinergic neurons (Levi-Montalcini and Angeletti, 1966; Jessel et al., 1979; Gorin and Johnson, 1979,1980; Johnson et al., 1980; Aloe et al., 1981; Goedert et al., 1981; Schwartz et al., 1982; Tessler et al., 1985; Hefti, 1986). Infusion of NGF has been shown to rescue or restore partially the appropriate phenotype (Fitzgerald et al., 1985; Rich et al., 1987; Williams et al., 1986; Hefti et al., 1989; Koliatsos et al., 1990; Tuszynski et al., 1990). Finally, NGF has been implicated, at least in the peripheral nervous system, as an important element in regeneration. After a crush or cut injury to the sciatic nerve, levels of NGF are up-regulated in the distal nerve stump, primarily in Schwann cells (Heumann et al., 1987a,b; Lindholm et al., 1987). Elevated levels in the stump are thought to compensate in part for loss of target-derived NGF, perhaps just to coax fibers to regenerate to their targets. After regeneration and remyelination, NGF levels return to a low level in the nerve trunk.

Our wealth of knowledge of the biology of NGF stems from two simple but important facts that allowed the isolation and characterization of this protein at an early stage. First, Levi-Montalcini and colleagues established a highly sensitive, reliable, and semiquantitative bioassay for neurotrophic factors when they discovered that NGF elicited massive neurite outgrowth from cultured explants of chick embryo paravertebral chain sympathetic ganglia or spinal sensory ganglia (dorsal root ganglia, DRG). This observation in pictorial form subsequently has adorned the covers of many textbooks, monographs, and reviews. Second, by sheer serendipity, an abundant source of NGF was discovered at an early stage, first in the venom glands of certain snakes and subsequently in the salivary glands of adult male mice. (Other relatively abundant sources of NGF have been found since in such unlikely places as the guinea pig prostate and bovine seminal plasma.) Although no obvious function has been ascribed to the high levels of NGF protein in the mouse salivary gland, this abundant source was sufficient to allow the full biochemical characterization of NGF by isolation, purification, protein chemistry, and protein sequencing, long before the advent of current techniques in gene cloning and recombinant expression of proteins.

Equally important to the biochemical characterization of NGF is the detailed characterization of those neuronal types of the peripheral and central nervous system that apparently depend on NGF for survival during development that was made possible by an abundant source of mouse NGF. Using both in vitro and in vivo approaches, it was established readily that NGF was a highly specific neurotrophic factor. Many types of neurons were found to be unresponsive to the survival-promoting effect of NGF in vitro and were refractory to NGF deprivation in vivo (reviewed by Levi-Montalcini and Angeletti, 1966,1968; Thoenen and Barde, 1980; Johnson et al., 1986; Lindsay, 1988a,b,1992). Parasympathetic neurons of the ciliary ganglion, motor neurons of the ventral spinal cord, neural placode-derived sensory neurons, and subpopulations of neural crest-derived sensory neurons emerged as neuronal populations that were unresponsive to NGF.

## III. SEARCH FOR OTHER NEUROTROPHIC FACTORS

The isolation and characterization of NGF and the production of NGF antibodies provided tools that rapidly produced firm molecular evidence in support of the target-derived neurotrophic factor hypothesis. The specificity of NGF, especially the observations that cultured motor neurons and parasympathetic neurons of the chick ciliary ganglion were refractory to NGF, initiated a painstaking search for neurotrophic molecules analogous to NGF but with distinct neuronal specificities.

## IV. DISCOVERY OF NEUROTROPHIC "ACTIVITIES" DISTINCT FROM NGF

In the early 1970s, Monard and colleagues described an activity present in the conditioned medium of a glioma cell line that induced morphologic differentiation of neuroblastoma cells (Monard et al., 1973). In the course of purification, it was demonstrated that this "glial factor" was both biologically and biochemically distinct from NGF (Monard et al., 1975). However, assay of various fractions of partially purified glial factor on explants of chick embryo DRG revealed an additional neurite-promoting activity that could not be blocked by sheep antibodies to mouse NGF. The activity derived from glioma cells elicited neurite outgrowth from DRG explants that was both qualitatively and quantitatively distinct from that elicited by NGF—less robust and predominantly of thick branched fibers (Y.-A. Barde and R.M. Lindsay, unpublished observations). To determine whether this novel activity acted directly to promote survival and neurite outgrowth of DRG neurons, it was established that glioma-cell conditioned medium supported DRG neurons in dissociated neuron-enriched cultures (Barde et al., 1978).

Wishing to draw further distinctions between NGF and this novel activity, it was established (Barde et al., 1980) that glioma-cell conditioned medium activity was more effective than NGF in promoting the survival of cultured DRG neurons from older (> ED 10) chick embryos when the neurons were cultured on a polyornithine substrate. This result was interpreted as indicating that the neurotrophic requirements of sensory neurons might change during development, first being dependent on NGF (derived perhaps from a peripheral target) and then on a factor that might be derived from glial cells in their central target, that is, the spinal cord.

However, it has been shown subsequently, both in explant cultures and in neuron-enriched cultures grown on a laminin-coated substrate, that DRG neurons of the chick respond to both NGF and BDNF as early as ED 6 in development (Lindsay et al., 1985a,b).

In first attempts to show that this novel activity found in glioma cells might be of physiologic relevance (i.e., not just a property of glioma tumor cells), it was reported that both chick and rat brain extracts and primary cultures of adult rat brain astrocytes contained an activity similar to that found in the conditioned medium of C6 glioma cells (Lindsay, 1979; Lindsay and Tarbit, 1979; Barde et al., 1980; Lindsay et al., 1982). In addition to glial cells and brain extracts, homogenates of post-hatched chick spinal cord and various embryonic chick peripheral tissues including heart, liver, and kidney were found to contain activity that promoted survival of chick DRG neurons, even in the presence of NGF-neutralizing antiserum (Lindsay and Tarbit, 1979; Lindsay and Peters, 1984a,b). Although these various activities were all distinct from NGF by immunological criteria it has never been established whether they represented a widely distributed single entity (i.e., the same molecule) or were indicative of the existence of multiple neurotrophic factors. We are now aware that subpopulations of DRG neurons are responsive to a number of non-NGF neurotrophic factors, including BDNF, NT-3, ciliary neurotrophic factor (CNTF), and possibly some members of the fibroblast growth factor family (FGFs). Many tissue extracts and cell-line conditioned media have been found to contain two or more of these factors.

## V. PURIFICATION OF BDNF FROM PORCINE BRAIN

The identification of a novel neurotrophic activity in rat brain extracts prompted the first steps of a process that finally led to the purification of BDNF from pig brain (for details, see Barde et al., 1982a,b,1983,1985a,b,1987a,b; Thoenen et al., 1982,1984a,b; Edgar and Barde, 1983; Barde and Thoenen, 1984,1985). Using chick DRG neurons to monitor purification, Barde et al. (1982a) isolated a small basic protein of 12.3 kDa that was later called brain-derived neurotrophic factor. BDNF was found to have a pI of ~10. Because it required several million-fold enrichment to achieve homogeneity, BDNF is undoubtedly a protein of extremely low abundance, a feature that is entirely consistent with the expected properties of a target-derived neurotrophic molecule. The initial purification procedure produced 1 µg BDNF from 1.5 kg pig brain. Even with no amino acid composition data or primary sequence data, the physicochemical and biologic properties of BDNF described a protein with properties that suggested it might be related to an NGF monomer, which later proved to be the case (Leibrock et al., 1989). However, many unrelated proteins are both basic and in the 12–16 kDa range, including histone proteins and fibroblast and other types of growth factors. The initial purification of BDNF called for a two-dimensional gel step and final recovery from an SDS gel (Barde et al., 1982), an effective procedure that yielded many hundred-fold purification, a single protein band by silver staining, but poor recovery of active material. Later modifications led to a higher yield of active protein (6 μg BDNF/kg brain; Lindsay et al., 1985a). Changes in some of the column chromatographic steps and the use of a C8 reverse-phase column eliminated the cumbersome problem of SDS gel electrophoresis as the final purification step, producing a yield of 1 μg BDNF/kg brain (Hofer and Barde, 1988).

BDNF purified from pig brain has been shown in DRG survival bioassays to be at least as active as NGF; saturating BDNF activity is found in the low picomolar range ($\sim$15 p$M$ or 200 pg/ml) and half maximal activity is seen at 80 pg/ml (Hofer and Barde, 1988).

## VI. NEURONAL SPECIFICITY OF BDNF

### A. Peripheral Nervous System Neurons: Distinct and Overlapping Specificity of BDNF and NGF

#### 1. Developing neurons

The purification of BDNF, albeit in small quantities, allowed further comparisons to be made between this novel factor and NGF (see reviews by Barde et al., 1983,1987a,b; Barde, 1989). In the first clear demonstration of a neuronal specificity distinct from that of NGF, BDNF was found to support survival and to elicit neurite outgrowth from neurons of the chick embryo nodose ganglion (NG), the distal sensory ganglion of the vagus or tenth cranial nerve (Lindsay et al., 1985a,b). Previous studies, both in vitro and in vivo, had established that neurons of the neural placode-derived NG were refractory to NGF (Levi-Montalcini, 1962; Lindsay, 1979; Lindsay et al., 1982; Pearson et al., 1983; Lindsay and Rohrer, 1985). In addition, it was found that, in contrast to NGF, BDNF did not support the survival of cultured sympathetic neurons. Like NGF, BDNF was found not to support survival or neurite outgrowth of chick ciliary ganglion neurons (Lindsay et al., 1985a,b), which are parasympathetic neurons that are shown to be responsive to CNTF, a neurotrophic factor identified initially in chick eye, later purified from sciatic nerve, and subsequently cloned (Lin et al., 1989; Stöckli et al., 1989). [For reviews of CNTF, see Manthorpe et al. (1989), Thoenen (1991), and Chapter 15.]

Further studies have established the broader premise that sensory neurons that respond to NGF in vitro are predominantly, if not exclusively, of neural crest origin (DRG, dorsomedial trigeminal, and jugular ganglia), whereas neural placode-derived sensory ganglia (neurons of the ventrolateral trigeminal, geniculate, vestibuloacoustic, petrosal, and nodose ganglia) are refractory to NGF (Lindsay, 1979; Davies and Lumsden, 1983; Davies and Lindsay, 1984,1985; Lindsay and Rohrer, 1985; Lindsay et al., 1985a,b; Davies et al., 1986a). In contrast to differential responsiveness to NGF, it appears that all sensory ganglia, regardless of neural crest or a neural placode origin contain BDNF-responsive neurons (Lindsay et al., 1985a,b; Davies et al., 1986a). Interestingly, the effects of NGF and BDNF in promoting the survival of ED 6–ED 12 chick DRG neurons are additive, suggesting that they may act on different, possibly functionally distinct, subpopulations of sensory neurons (Lindsay et al., 1985a,b). Further, cranial sensory neurons display a highly variable response to BDNF (Davies et al., 1986a), again suggesting a functional subpopulation of sensory neurons whose representation varies in different cranial ganglia. Another view, however, is that some sensory neurons require trophic support from both their peripheral and their central target; thus, there may be a requirement for two distinct neurotrophic factors to sustain the survival of certain sensory neurons in culture. In this context, it has been shown that sensory neurons of the trigeminal mesencephalic nucleus, proprioceptive sensory neurons that innervate the muscles of mastication, are responsive to both BDNF and a factor present in skeletal muscle (Davies et al., 1986b). At subsaturating levels, the effects of BDNF and skeletal muscle extract were found to be additive in supporting survival of these neurons. At saturating levels, however, either factor alone was equally effective, but there was no additivity. This observation has prompted the notion that the levels of each neurotrophic factor in either the peripheral or the central target of sensory neurons may be inadequate (i.e., subsaturating) for specifying and sustaining the survival of sensory neurons. Therefore, perhaps the combined effect of neurotrophic support from both peripheral and central target fields is required for the specification, selection, and maintenance of sensory neurons. Although suggestive evidence for this hypothesis has been reported (Kalcheim et al., 1987; see subsequent text), definitive evidence from an in vivo study is still lacking.

In a broad survey of the response of explants of both neural crest- and neural placode-derived chick embryo sensory ganglia to BDNF, Davies et al. (1986a) found that BDNF had little or no effect on fiber outgrowth until ED 4. Neurons of the NG, DRG, and ventrolateral trigeminal ganglion were the earliest to show a response to BDNF. However, potential effects of BDNF at earlier stages of development have been indicated from a study of the effects of BDNF on cultures of quail embryo neural crest cells. Using substance P and the HNK-1 epitope as markers, Kalcheim and Gendreau (1988), have reported that BDNF treatment produced a 1.5- to 6-fold increase in the number of HNK-1 immunopositive neuronal cells developing from cultures of neural crest cells derived from ED 3 quail embryos. Neural crest cells from different somitic levels (somites 15–26) were cultured with adjacent somites in the presence of BDNF. The greatest effect of BDNF (6-fold increase in HNK-1 positive cells) was seen with the most anterior crest cells, cells that were already postmigratory and in the process of ganglion formation. As might be expected, cultures derived from more posterior neural crest showed less of a response to BDNF, indicative of their earlier developmental stage. When comparing studies in chick and quail, it should be noted that quail development in ovo is more rapid than that of chick, and hatching occurs at 16 rather than at 21 days.

At early developmental stages, the observed effects of neurotrophic factors on cultured DRG are complicated by the fact that not only are anterior–posterior differences in the state of maturation of neurons within each ganglion possible, but there is also a spectrum of neuroblasts at different stages in the cell cycle within a single ganglion at ED 5–ED 7, as well as postmitotic neurons with birthdates that span 2–3 days. In attempts to determine the effects of NGF or BDNF on early neuronal precursor cells, as opposed to the more commonly studied effects of NGF or BDNF in sustaining survival and promoting regeneration of postmitotic neurons, Ernsberger and Rohrer (1988) isolated neuronal precursors from ED 6 chick DRG by eliminating neurons and Schwann cell precursors by immunocytolysis. DRG neuronal precursor survival initially was found to be independent of any neurotrophic factor, but survival beyond 9 hr was enhanced by either NGF or BDNF. At saturating concentrations there was no additive effect of BDNF and NGF, indicating a complete overlap in NGF and BDNF responsiveness. Interestingly, although the same proportion of cells survived with either BDNF or NGF for the first 3 days in vitro, survival in the presence of BDNF declined in the next 3 days, whereas survival of the same number of neurons continued in the presence of NGF. These results suggest that precursor cells that are initially responsive to two distinct neurotrophic factors may retain dependence on one factor while becoming refractory to the other. This notion is supported by data that showed that fewer DRG neurons are supported by BDNF at ED 9 or ED 12 than at ED 6 (Lindsay et al., 1985a,b), and that a small percentage of NG neurons are responsive to NGF at early postmitotic stages (ED 5) but not at later developmental stages (> ED 7; Lindsay and Rohrer, 1985; Lindsay et al., 1985a,b).

### 2. Adult neurons

The limited availability of BDNF has restricted its use greatly; to date, no studies have been conducted in vivo that address the role of BDNF in the maintenance of survival or phenotype of mature neurons. However, it has been shown that highly enriched cultures of adult rat DRG neurons, many of which are absolutely dependent on either NGF or BDNF during their development, can be maintained in culture in the absence of any exogenous neurotrophic support. Neither NGF or BDNF was found to enhance survival of adult rat DRG neurons, even when the neurons were grown as single cells in microwells (Lindsay, 1988b). However, both factors enhanced neurite outgrowth (Lindsay, 1988b), and NGF later was found to be necessary for expression of normal neuropeptide levels [substance P and calcitonin gene-related peptide (CGRP); Lindsay and Harmar, 1989; Lindsay et al., 1989], to induce up-regulation of the low affinity NGF receptor (Lindsay et al., 1990), and to render rat DRG neurons responsive to the excitotoxin capsaicin (Winter et al., 1988). Thus it would appear that adult sensory neurons lose their absolute dependence on target-derived neurotrophic factors for survival, but remain responsive to retrograde signals from their targets. These trophic signals still seem to be necessary to sustain normal function and phenotypic expression.

It is interesting to speculate what molecular events change during development that leads to the loss of a survival requirement for BDNF or NGF. Unpublished results (J. Garrett, C. Evison, and R.M. Lindsay) indicate that rat sensory neurons slowly lose NGF dependence over a 2-week period after birth. It has been shown by in situ hybridization that BDNF mRNA is present in some neurons of the adult rat DRG (Ernfors et al., 1990a,b). Since there are no neuronal projections and few if any synaptic contacts within the DRG, there is no rationale for BDNF acting in this locale as a target-derived factor for projecting neurons. Therefore, during maturation, DRG neurons may be induced to express BDNF which, by an autocrine, paracrine or even intracrine mechanism, is sufficient to sustain their survival in the mature state. Such a mechanism might be a general phenomenon of neuronal maturation. Whereas target dependence is an attractive hypothesis for nervous system sculpting during development, it might be a risky feature of the mature nervous system, especially in the CNS. For example, if absolute target dependence is retained in the adult, loss of one neuronal population in the mature nervous system as a consequence of injury, disease, or aging could be expected to have a transneuronal knockout effect, leading to widespread loss of other neuronal systems. The pathology of Alzheimer's disease might support this suggestion, whereas the specific loss of cells in Parkinson's disease might not. Arguably, once the appropriate number of neurons has been specified carefully during development, it would seem appropriate that mechanisms be implemented that guarantee neuronal survival in a manner that removes most of the risk associated with absolute dependence on their targets. Perhaps in the mature nervous system, local actions of neurotrophic factors (autocrine, paracrine, intracrine, or others) are sufficient to insure neuronal survival, but are inadequate to initiate full phenotypic expression, synaptic plasticity, or a regenerative response after injury. Thus the target still may play an important role in regulating these latter functions. Experiments with antisense probes may prove useful in exploring this notion in vitro and in vivo.

## B. CNS Neurons: Distinct and Overlapping Specificity of BDNF and NGF

### 1. Rat retinal ganglion cells

Based on initial studies that showed that rat brain extracts stimulated neurite outgrowth from fetal rat retinal explants in culture (Turner et al., 1983), Johnson et al. (1986) provided the first evidence that BDNF had neurotrophic actions on specific neurons of the CNS as well as of the PNS. Using retrograde cell-labeling techniques and antibodies to the cell-surface glycoprotein thy-1 as markers of the retinal ganglion cell subpopulation, it was established that BDNF greatly enhanced (> 6-fold) the survival of retinal ganglion cells in dissociated cultures of ED 17 rat retinal cells. Similar effects were found in cultures enriched 80-fold for retinal ganglion cells. This study elegantly established that the observed effects of BDNF were specific, not mediated by enhanced proliferation of ganglion cells or their precursors, and not confused by a BDNF-induced up-regulation of the thy-1 marker. NGF was found to have no effect on the survival of perinatal retinal ganglion cells (Johnson et al., 1986). In contrast, a recent report has indicated that intraocular injections of NGF can rescue retinal ganglion cells for at least 6 weeks after intracranial section of the optic nerve (Carmignoto et al., 1989). In addition to survival-promoting effects of BDNF on developing rat retinal ganglion cells, it has been shown that BDNF enhances regeneration of retinal ganglion axons in explants of adult rat retina (Thanos et al., 1989). A small effect of BDNF (25%) on the survival of adult retinal ganglion cells also was noted in this study.

### 2. Chick retinal ganglion cells

Effects of BDNF on the survival of retinal ganglion cells also has been established in cultures of embryonic chick retina. By studying the effects of BDNF on retinal cultures established from chicks of different embryonic ages, an interesting age dependency of retinal ganglion cells on BDNF was established (Rodriguez-Tébar et al., 1989). Prior to ED 6, chick retinal ganglion cells survived in culture equally well in the presence or absence of BDNF. From ED6 to ED11, the survival of retinal ganglion cells became increasingly dependent on BDNF, so by ED 11 BDNF enhanced the number of surviving thy-1-immunopositive retinal ganglion cells 3-fold. In terms of the development of retinal ganglion cells in vivo, this study indicates that retinal ganglion cells show greatest dependence on BDNF at the time when there is a peak in the number of retinal ganglion cell axons in the optic nerve. This time is immediately before the period of massive ganglion cell elimination that continues for several days after ED 11. Thus, in favor of the target-derived neurotrophic hypothesis, the maximum dependence of retinal ganglion cells on BDNF as a survival factor would appear to coincide with the period of greatest competition among these cells for a limited target.

### 3. Septal cholinergic neurons

It is now well established that NGF promotes survival and phenotypic expression of cholinergic neurons in cultures of embryonic rat basal forebrain (Hefti et al., 1985; Hartikka and Hefti, 1988; Knüsel et al., 1990). Further, intraventricular infusion of NGF has been shown both in rodents and in primates to prevent or reduce the loss of cholinergic phenotype that follows axotomy of these neurons after fimbria–fornix transection (Hefti, 1986; Williams et al., 1986; Koliatsos et al., 1990; Tuszynski et al., 1990). It has been established that BDNF has a very similar effect to NGF on cultures of ED 17 rat embryo septal cholinergic neurons (Alderson et al., 1990). At appropriate cell densities, BDNF was found to produce a 2- to 3-fold increase in the survival of cholinergic neurons as determined by histochemical staining for acetylcholinesterase (AChE) or immunocytochemical staining for ChAT. Effects of BDNF on the number of AChE- and ChAT-positive cells also were reflected in 2- to 3-fold increased levels of ChAT and AChE enzymatic activity. BDNF also produced a 3-fold increase in the number of neurons that stained positively with antibodies to the low affinity NGF receptor. Although the effects of BDNF were found to be broadly similar to those of NGF at the age studied, suggesting overlapping actions on the same neuronal population, substantial differences were found in the dose–response curves to each factor.

Interestingly, a synergistic effect of BDNF and NGF was found in terms of induction of maximal ChAT levels. Saturating levels of BDNF and NGF separately produced a maximal stimulation of 2.6- and 2.8-fold, respectively, whereas the combination of the two factors produced a more than additive effect of a 6.7-fold increase in ChAT activity. In support of this in vitro study, it has been shown subsequently by Northern blot analyses and by in situ hybridization studies that BDNF mRNA, like NGF mRNA, is abundant in the adult rat hippocampal formation (Leibrock et al., 1989; Hofer et al., 1990; Maisonpierre et al., 1990b; Phillips et al., 1990). Broadly similar effects of BDNF on basal forebrain cholinergic neurons were found later in cultures derived from slightly younger (ED 15–ED 16) rat embryos (Knüsel et al., 1991).

### 4. Dopaminergic neurons of the ventral mesencephalon

Although several reports have suggested that basic fibroblast growth factor (bFGF) and partially purified bovine brain extracts (including a striatal-derived factor; Dal Toso et al., 1988) promote the survival and differentiation of nigral dopaminergic neurons in cultures of embryonic rat ventral mesencephalon, until recently there was no evidence of a purified neurotrophic factor that acted directly on nigral dopamine neurons to enhance their survival. Now it has been established that BDNF enhances survival of dopaminergic neurons in cultures derived from the ventral mesencephalon of ED 14 rat embryos (Hyman et al., 1991). By culturing cells from this early age in an appropriate serum-free chemically defined medium, it has been possible to show that BDNF promotes a 5-fold enhancement in the number of neurons immunopositive for TH, a marker of the dopaminergic cells of the substantia nigra, after 8 days in culture. These cultures were essentially free of glial and other nonneuronal cells, indicative of a direct effect of BDNF on dopaminergic neurons. In addition to promoting the survival of dopaminergic neurons, BDNF treatment was found to protect these cells greatly from the neurotoxicity of MPTP 1-methyl-4-phenyl-1,2,3,6 tetrahydropyridine. This agent has been shown, both in humans and in nonhuman primates, to acutely induce parkinsonian symptoms (for review of MPTP, see Zigmond and Stricker, 1989). Preliminary results suggest that the protective effect of BDNF against MPTP is at the level of reducing cellular oxidative stress. This effect appears to be achieved by BDNF-induced up-regulation of enzymes involved in dissipation of free radicals and superoxide ions (Spina et al., 1992a,b).

BDNF also has been shown to produce a 2-fold increase in dopamine uptake in cultures of embryonic rat ventral mesencephalon (Knüsel et al., 1991).

## VII. BDNF PREVENTS NEURONAL DEATH IN VIVO

Perhaps the most compelling evidence in support of a physiologic role of NGF in particular, and of the target-derived neurotrophic hypothesis in general, are studies that have shown that (1) administration of exogenous NGF prevents naturally occurring developmental cell death in vivo and (2) sequestration of endogenous target-derived NGF (either by administration of heterlogous neutralizing antibodies or by exposure to maternal NGF antibodies in utero) increases neuronal cell death in vivo during development. Two studies now have shown that BDNF fulfills the criteria of a molecule involved in the regulation of naturally occurring neuronal cell death.

Implantation of a silastic membrane between the neural tube and adjacent somites results in the separation of early migrated neural crest cells of the DRG anlage from the neural tube. This barrier between the neural tube and the developing DRG results in rapid death of the DRG anlage (Kalcheim and Le Douarin, 1986), suggesting that normal development of DRG neurons is dependent on a trophic signal from the neural tube, that is, the target of the centrally projecting axons of these neurons. In support of this suggestion, it was found that implantation of silastic membranes pretreated with extracts of 3–4-day-old neural tube could rescue DRG neurons (Kalcheim and Le Douarin, 1986). Similarly, in this paradigm, pretreating silastic membranes with BDNF was found to rescue many DRG neurons (Kalcheim et al., 1987). Interestingly, the rescuing effect of BDNF required the presence of laminin. Collagen-coated BDNF-treated silastic membranes were without effect, whereas laminin-coated BDNF-treated (but not NGF-treated) membranes rescued DRG neurons in 80% of embryos examined. This requirement of laminin to mediate the effect of BDNF on early sensory neurons in ovo was in good agreement with the earlier in vitro study that showed that DRG neurons were equally sensitive to BDNF from ED 6 to ED 11 when cultured on laminin (Lindsay et al., 1985a,b) rather than polyornithine (Barde et al., 1980,1982a,b). However, it remains to be shown whether interaction of BDNF with laminin or other proteoglycans is physiologically relevant or simply a requirement of the studies carried out so far.

Whereas the study just described provided the first evidence for a possible role of BDNF in vivo, the surgical intervention required to carry out such a study raises its own problems when interpreting the results. Thus, the report by Hofer and Barde (1988) that BDNF injections into the chorioallantoic membrane of developing quail embryos prevented naturally occurring loss of both DRG neurons and NG neurons (a loss of 38% and 30%, respectively, between ED 5 and ED 12) was the first definitive evidence that BDNF fulfills one criterion for a neurotrophic factor in vivo. Thus BDNF was the first defined molecule since NGF to meet this criterion. Demonstration that neutralizing antibodies to BDNF increased neuronal death during development by sequestration of target-derived BDNF would be an alternative method of establishing physiologic relevance. This demonstration has not been possible because of limited amounts of BDNF against which to prepare antibodies, and because of the apparently poor immunogenicity of BDNF (Y.-A. Barde, unpublished observations). Poor immunogenicity may be due to the fact that mature BDNF protein is extraordinarily well conserved across species: mature BDNF protein is 100% conserved at the amino acid level in all mammals examined to date (Leibrock et al., 1989; Jones and Reichardt, 1990; Maisonpierre et al., 1990a,b, 1991; Rosenthal et al., 1990).

## VIII. MOLECULAR CLONING OF PORCINE, MOUSE, RAT, AND HUMAN BDNF

The low abundance of BDNF and the consequent problems in purifying even a few hundred micrograms of this protein made the task of obtaining any structural information extraordinarily difficult. Although recombinant DNA technology provides a remarkable way to produce rare proteins, allowing their full biochemical and biologic characterization, recombinant production can be considered only after molecular cloning of the gene of interest. Until relatively recently, the only straightforward approaches to the cloning of most genes, especially those expressed at very low levels, required either the production of a good antibody or the determination of partial primary structure. BDNF proved to be a poor immunogen, and sequencing studies were greatly hampered at early stages by a blocked N terminus. However, cleavage of purified BDNF with V8 protease, trypsin, or cyanogen bromide (under reducing conditions to cleave disulfide bridges) yielded several peptides that were purified by HPLC and found to be amenable to sequence determination by gas-phase microsequencing (Leibrock et al., 1989). Several overlapping peptides provided a suitable stretch of partial sequence to design two sets of oligonucleotides that formed the basis of a polymerase chain reaction (PCR) strategy using a pig genomic template and then a cDNA template derived from pig brain superior colliculus. Isolation and sequencing of overlapping cDNA clones finally showed that BDNF is encoded by an intronless open reading frame that yields a precursor protein of 252 amino acid residues (Leibrock et al., 1989). Mature BDNF, a protein of 120 amino acids, constitutes essentially the C-terminal half of its precursor, the N terminus of BDNF commencing after a typical canonical basic-X-basic-basic residue (Arg–Val–Arg–Arg) proteolytic cleavage site. The predicted size of mature BDNF (13 kDa) and its calculated pI of 9.99 were found to agree well with the determinations made originally with purified porcine brain BDNF. Assay of conditioned medium from transiently transfected COS cells confirmed that recombinant BDNF had biologic activity consistent with that of purified BDNF (Leibrock et al., 1989).

The most striking feature of the primary structure of BDNF is the similarity of BDNF to NGF. When compared with the sequence of NGF from six species, ranging from snake to human, mature BDNF was found to have slightly less than 50% amino acid sequence identity, including absolute conservation of six cysteine residues (Leibrock et al., 1989) that have been shown to form three disulfide bridges in NGF. Quite remarkably, the primary structure of mature BDNF has been found to be identical in all mammals from which this gene has been cloned—mouse, rat, pig, and human (Leibrock et al., 1989; Hofer et al., 1990; Maisonpierre et al.,

1990a,b,1991) [For detailed comparative analysis of the sequence of BDNF and its precursor, and comparison of its gene structure with those of NGF and NT-3, the reader is referred to original articles (Leibrock et al., 1989; Hohn et al., 1990; Maisonpierre et al., 1990a,b,1991; Rosenthal et al., 1990).]

## IX. CONSEQUENCES OF THE MOLECULAR CLONING OF BDNF

Immediate benefits of the molecular cloning of BDNF were (1) the availability of cDNA and RNA probes for quantitative Northern analysis of BDNF mRNA in tissue extracts (Leibrock et al., 1989; Maisonpierre et al., 1990a,b,1991), (2) use of similar probes to localize sites of BDNF synthesis in tissue sections (Ernfors et al., 1990a,b; Hofer et al., 1990; Phillips et al., 1990; Wetmore et al., 1990), and (3) the construction of suitable expression vectors to begin the process of producing recombinant BDNF. All these benefits have been exploited rapidly, leading to a rapid appearance of new information on the biology of BDNF.

### A. Distribution of BDNF mRNA in Brain Regions and Peripheral Tissues

#### 1. Northern analysis

BDNF mRNA is most abundant in the adult brain (Leibrock et al., 1989; Maisonpierre et al., 1990b) and, although it was first undetectable in peripheral tissues, significant levels of BDNF transcripts have been found in both rat and human heart and lung. (Maisonpierre et al., 1990a,b,1991). However, in contrast to NGF and NT-3 transcripts, which are abundant in peripheral tissues, BDNF is predominantly found in the CNS. BDNF mRNA is first detected in the rat embryo between ED 11 and ED 12, although levels of BDNF are generally low during development. The relative amounts of BDNF, NT-3, and NGF in adult rat brain were found to be quite similar, although levels of each neurotrophin in discrete brain regions vary enormously (Maisonpierre et al., 1990b). In both rat and mouse brain, highest levels of BDNF transcripts were found in hippocampus, cortex, and cerebellum. Interestingly, all three neurotrophins show their highest level of adult expression in the hippocampus (Hofer et al., 1990; Maisonpierre et al., 1990b).

The finding of high levels of BDNF mRNA in the hippocampus is consistent with the results of Alderson et al. (1990), which showed that BDNF has effects similar to those of NGF in promoting survival and phenotypic differentiation in septal cholinergic neurons. In the mouse brain (Hofer et al., 1990), it was estimated that levels of BDNF mRNA were ~ 50-fold higher than levels of NGF mRNA. It is thus likely that hippocampal BDNF has effects on multiple neurons, rather than solely cholinergic neurons projecting from the basal forebrain. Likely candidates may be intrinsic neurons, neurons projecting from entorhinal and other cortical areas, and cross-projecting hippocampal fibers. The broader implications of high levels of all three neurotrophins in the hippocampus, a structure of central importance in cognitive behavior, such as learning and memory, are that the neurotrophins may play a local role in such phenomena as long-term potentiation and synaptic remodeling, in addition to their traditional role as target-derived neurotrophic factors. In this context, it has been shown, both in cultures of hippocampal cells and in vivo, that neural activity regulates levels of both BDNF and NGF mRNA (Zafra et al., 1990). Exposure of hippocampal neurons to kainic acid induced a massive up-regulation of BDNF within 3 hr; pharmacologic analysis showed that this effect was mediated through non-NMDA glutamate receptors.

BDNF has been shown to enhance the survival of dopaminergic neurons of the developing substantia nigra and to protect these neurons from the neurotoxicity of the dopaminergic neurotoxins 6-hydroxydopamine and MPTP (Hyman et al., 1991). Consistent with this effect, BDNF mRNA has been found in the striatum of the adult mouse brain (Hofer et al., 1990), although levels in adult rat brain were barely detectable. Preliminary studies on the pattern of distribution of radiolabeled BDNF injected at different sites in the adult rat brain or into the sciatic nerve indicate specific retrograde transport of BDNF in at least all CNS and PNS neurons that have been identified to be responsive to BDNF in vitro (DiStefano et al. 1992).

That levels of BDNF transcripts are much higher in adult brain than during development (Maisonpierre et al., 1990b) may imply that BDNF is as important, if not more important, as a maintenance factor for mature neurons, as it is as a target-derived neurotrophic factor for developing neurons.

### 2. Localization of BDNF mRNA in the CNS and PNS by in situ hybridization

A number of studies have described the localization of BDNF mRNA in tissue sections of adult mouse, rat, pig, and human brain (Ernfors et al., 1990a,b; Hofer et al., 1990; Phillips et al., 1990; Wetmore et al., 1990). As predicted from Northern analysis, the highest density of cells expressing BDNF appears to be in the hippocampal formation. In the rat brain, BDNF localized intensely to neuronal cells throughout major targets of the basal forebrain cholinergic system—the hippocampus, amygdala, and neocortex (Phillips et al., 1990). In the mouse and rat hippocampal formation, BDNF localized most intensely to hilar neurons in the dentate region, pyramidal neurons in layers CA2–CA4, and granule cells of the dentate gyrus (Ernfors et al., 1990a,b; Hofer et al., 1990). CA1 pyramidal cells were labeled more weakly. Although NGF expression also is localized to cells of the dentate gyrus and neurons of the pyramidal layers, the pattern of NGF expression is sufficiently different from that of BDNF to indicate that each neurotrophic factor may be expressed in different subpopulations of hippocampal neurons (Ernfors et al., 1990a,b; Hofer et al., 1990; Phillips et al., 1990; Wetmore et al., 1990). In contrast to that of BDNF and NGF, expression of NT-3 in the hippocampus was found to be restricted to pyramidal cells of the relatively small CA2 field, the most medial regions of CA1, and dentate granule cells. Within the latter, NT-3 localized to more cells than did BDNF. Overall, the patterns of NGF, BDNF, and NT-3 mRNA localization in the adult rodent hippocampus indicate that these growth factors are synthesized by neurons. There has been no indication of expression of these rare mRNA species in nonneuronal cells. This result contrasts with the clear demonstration that adult and embryonic glial cells (and glioma cells) synthesize and secrete active NGF and other neurotrophic factors (probably NT-3) in vitro (Lindsay, 1979). It is clear that each member of the neurotrophin family has a different pattern of expression in the hippocampus (and the brain as a whole), although the present level of resolution does not rule out substantial overlap, with some neurons perhaps expressing more that one neurotrophin. At this time, the functional data with which to interpret these different patterns of expression are insufficient, but it seems likely that BDNF may have as important a role as NGF in cholinergic innervation. Perhaps there is some topographic segregation of cholinergic neurons in terms of BDNF- and NGF-expressing cells. Moreover, the broader pattern of BDNF expression in the hippocampus suggests effects on noncholinergic pathways as well.

Outside the hippocampus, BDNF mRNA was localized to scattered pyramidal cells throughout the cerebral cortex and to a large number of cells in the temporal, peripheral, and cingulate cortex (Ernfors et al., 1990a,b; Hofer et al., 1990). Strong hybridization was seen in many olfactory projection regions (Phillips et al., 1990) but not in the olfactory bulb itself, which labeled only weakly (Hofer et al., 1990). In the mouse, high levels of BDNF were found also in the pontine nuclei and preoptic area; lower levels of BDNF mRNA were noted in the hypothalamus, mammillary nuclei, the inferior and superior colliculus, and cerebellar granule cells (Hofer et al., 1990).

In the peripheral nervous system, BDNF mRNA has been localized to certain neurons of the adult rat DRG. A possible autocrine, paracrine, or intracrine function of BDNF in these neurons was speculated on earlier. In addition, it is also possible that neurons use neurotrophins in hitherto undescribed ways. Perhaps, by orthograde transport, neurons use BDNF (or NT-3 or NGF) to signal information to their targets or ensheathing cells. It is well known that an axonal signal is required for Schwann cell myelination and that innervating neurons send signals to stimulate the differentiation of peripheral tissues, especially muscle.

BDNF mRNA has been identified in human platelets (Yamamoto and Gurney, 1990). The functional significance of this result remains to be established.

## X. BINDING OF BDNF TO SPECIFIC RECEPTORS

The first study of the binding characteristics of BDNF to responsive cells demonstrated that chick embryo DRG neurons possessed two different classes of BDNF receptors—high and low affinity receptors exhibiting dissociation constants of $\sim 10^{-11}$ and $10^{-9}$ $M$, respectively (Rodriguez-Tébar and Barde, 1988). Those values are very similar to the original dissociation constants reported for NGF binding to sensory neurons (Sutter et al., 1979). However, the association and dissociation rates of BDNF were found to be much slower than NGF. On chick sensory neurons, the numbers of high affinity sites for NGF and BDNF appear to be similar. However, unlike NGF, no high affinity sites for BDNF were found on sympathetic neurons, although low affinity BDNF binding to these neurons was observed.

Studies by Rodriguez-Tébar et al. (1990) suggest that BDNF and NGF both bind to the low affinity NGF receptor (LNGFR). Until recently, the relationship between the LNGFR and functional NGF binding via the high affinity site has been a puzzle, despite the cloning of a 75-kDa glycosylated LNGFR from both rat and human (Jonnson et al., 1986; Radeke et al., 1987). Cross-linking studies with radiolabeled NGF have shown that high affinity binding of NGF results in a protein complex of higher molecular mass (~150 kDa), which compares to a complex of NGF–LNGFR that migrates at 90–100 kDa (Hosang and Shooter, 1985,1987). It was thought for some time that this higher molecular mass might be the result of an accessory protein involved in the NGF–high affinity receptor complex. However, Meakin and Shooter (1991) showed that three antibodies to different epitopes of the LNGFR effectively immunoprecipitate low but not high molecular weight NGF–receptor complexes. This results suggested that the high affinity NGF receptor might be a completely different protein than LNGFR. This suggestion was supported by the finding of phosphotyrosine residues exclusively on the high affinity NGF receptor. The detection of phosphotyrosine residues indicated that tyrosine phosphorylation might be involved early in the NGF signal transduction pathway.

From a completely different perspective, a flurry of very recent papers has established that a family of hitherto "orphan" transmembrane glycoprotein tyrosine kinase receptors, including the *trk* and *trkB* protooncogene products, may themselves or in combination with LNGFR constitute high affinity receptors for NGF, BDNF, and NT-3 (Klein et al., 1990,1991; Hempstead et al., 1991; Kaplan et al., 1991; Squinto et al., 1991). The *trk* oncogene is a human transforming gene originally isolated from a colon carcinoma biopsy (Martin-Zanca et al., 1989). Northern analysis and in situ hybridization localization of the proto-oncogene *trk* and a related tyrosine kinase receptor *trkB* (Klein et al., 1989) localized expression of these genes to neural structures and suggested a possible neuronal function for the *trk* family (Klein et al., 1989; Martin-Zanca et al., 1990). Particularly striking was the localization of *trk* proto-oncogene expression to ganglia of neural crest-derived sensory neurons (Martin-Zanca et al., 1990) and a wider distribution of *trkB* throughout the CNS and PNS (Klein et al., 1989). Given the known effects of NGF on neural crest-derived sensory neurons, the knowledge that high affinity binding of NGF appears to be associated with an NGF–receptor complex of 158 kDa that contains phosphotyrosine residues, and the fact that the *trk* proto-oncogene product is a 140-kDa transmembrane glycoprotein, several groups of investigators made the connection that the *trk* proto-oncogene family of tyrosine kinase receptors may be receptors for the neurotrophins. Kaplan et al. (1991) first demonstrated that NGF treatment of PC12 cells induced tyrosine phosphorylation and activation of *trk*. It also was established that mouse fibroblasts transfected with *trk* bound NGF with high affinity, whereas parental cells or cells transfected with *trkB* exhibited little or no binding of NGF (Klein et al., 1991). This study suggested that the product of the *trk* proto-oncogene was by itself sufficient to confer high affinity binding in the absence of the 75-kDa LNGFR. Further, cross-linking studies established that NGF was indeed binding to the *trk* protein (Klein et al., 1991). Squinto et al. (1991) have established that the *trkB* proto-oncogene encodes a functional receptor for BDNF and NT-3, but not NGF. In this study, BDNF and NT-3 but not NGF bound to both full-length and truncated forms of *trkB*. When transfected with *trkB*, PC12 cells produced neurite outgrowth on exposure to either BDNF or NT-3. Parental PC12 cells elicit neurite outgrowth in response to NGF but not to BDNF or NT-3. As for high affinity binding of NGF to *trk* (Klein et al., 1991), functional responses of *trkB* to BDNF and NT-3 were found in cells that did not express LNGFR (Squinto et al., 1991). Perhaps in contradiction to these findings, it also has been suggested from other cross-linking and equilibrium binding studies, using approaches of heterologous membrane fusion and co-transfection of LNGFR and *trk* into COS cells, that *trk* and LNGFR both constitute low affinity NGF binding sites and that both molecules are required to produce a high affinity complex (Hempstead et al., 1991). This situation would be analogous to the situation with the interleukin-2 receptor (see Waldmann, 1991).

Although the explanation of what constitutes the physiologically relevant high affinity BDNF receptor is far from complete, the tools are now available to explore whether there are multiple receptors for BDNF and what part, if any, the LNGFR plays in high affinity binding of the neurotrophin family as a whole (see Ragsdale and Woodgett, 1991). Immunocytochemical studies have shown consistently that LNGFR is found in many more cells than the few neuronal populations known to be responsive to NGF (Yan and Johnson, 1988). If, indeed, LNGFR is a common low affinity receptor for all neurotrophins, we already may have some important clues about what other neuronal types might be responsive to BDNF, NT-3, or as-yet-undiscovered members of the neurotrophin family.

## XI. NEUROTROPHIN FAMILY OF NGF-RELATED NEUROTROPHIC FACTORS—NGF, BDNF, NT-3, AND OTHERS

As mentioned earlier, the molecular cloning of BDNF has had an enormous impact on our understanding of this molecule. Northern analyses of BDNF mRNA content and in situ hybridization studies rapidly have produced a lot of new information. Recombinant BDNF now is being used in a plethora of in vitro and in vivo studies that undoubtedly will produce an avalanche of new information on the biologic activity and physiologic relevance of this neurotrophic factor.

The finding that BDNF and NGF were clearly members of a gene family suggested that other members of this family might exist. Based on a variety of PCR strategies, several groups identified a third member of the NGF family that is called neurotrophin-3 (Hohn et al., 1990; Maisonpierre et al., 1990; Rosenthal et al., 1990) and an identical protein that was termed HDNF—hippocampal-derived neurotrophic factor (Ernfors et al., 1990a,b)—and NGF-2 (Kaisho et al., 1990). Rat NT-3 shares 55% identity with both NGF and BDNF and, like them, mature NT-3 is cleaved from a much larger precursor. NT-3 has actions on neurons of the PNS and CNS and specificities that both overlap and are distinct from those of either BDNF or NGF (Hohn et al., 1990; Maisonpierre et al., 1990a).

Although elucidation of the biology of NT-3 is at an early stage, two features suggest that NT-3 functions may encompass both the classical notions of a target-derived neurotrophic factor and other roles in development and maintenance of both neurons and certain nonneural cells. First, expression of NT-3 in some regions of the nervous system (e.g., spinal cord, cerebellum, and hippocampus) is much higher during development than in the adult (Maisonpierre et al., 1990b). This result contrasts with BDNF, which is generally low in developing CNS and only reaches high levels in the adult brain. These reciprocal patterns may suggest a more major role for NT-3 as a target-derived factor during development, whereas BDNF may be more important in maintenance and regulation of mature neurons. Second, levels of NT-3 in some developing brain regions reach extraordinarily high levels compared with either NGF or BDNF. In particular, NT-3 levels are very high in the cerebellum and hippocampus during peak proliferation of the granule cells that populate these structures. NT-3 may be playing a traditional neurotrophic role, but it could be involved in regulating proliferation of granule cell precursors, either as a mitotic factor or as a factor that induces terminal differentiation. Finally, in peripheral tissues NT-3 is more abundant than either BDNF or NGF. Based on all current information, in the PNS (1) the major role of NGF may be in the development and maintenance of sympathetic neurons and nociceptive and other small unmyelinated neural crest-derived sensory neurons, (2) the major role of BDNF and NT-3 is to sculpt medium and large myelinated sensory neurons with proprioceptive and mechanoceptive modalities, and (3) BDNF and NT-3, but not NGF, are also candidates for target-derived neurotrophic factors for all the placode-derived neurons of the specialized cranial sensory ganglia. On morphologic, functional, and neurochemical criteria, there are six or more subtypes of neurons within the DRG (Rambourg et al., 1983; Lawson and Harper, 1985). Retrograde transport studies using radiolabeled neurotrophins will help define which subtypes are responsive to BDNF, NT-3, or NGF (DiStefano et al., 1992).

## XII. CLINICAL PERSPECTIVE

One of several major pathologic consequences of Alzheimer's disease is the loss of basal forebrain cholinergic neurons. The finding that NGF promotes survival of cultured basal forebrain neurons and can rescue the phenotype of septal cholinergic neurons after axotomy in the adult has raised the question of whether NGF in particular, and neurotrophic factors in general, might be used therapeutically in Alzheimer's disease (Hefti and Weiner, 1986). Although there is no evidence of a neurotrophic factor deficit in any neurodegenerative disorder, there is evidence that intraventricular infusion of NGF can restore spatial memory impairment in aged rats. Similarly, numerous experiments show that NGF can protect sensory neurons from a variety of toxic insults (for review, see Snider and Johnson, 1989), including agents used therapeutically in cancer chemotherapy (Apfel et al., 1991). Thus, pharmacologic doses of neurotrophic factors may in principle be able to protect neurons from unknown insults, be they environmental toxins, pathogens, metabolic stress, or the side effects of otherwise useful drugs.

At present no animal efficacy data are available to support the potential use of BDNF as a therapeutic agent. Such studies are in progress. However, based on the recently discovered effects of BDNF on developing basal forebrain cholinergic neurons (Alderson et al., 1990; Knüsel et al., 1991), dopaminergic neurons of the substantia nigra (Hyman et al., 1991), and cultures of ventral spinal cord cells (Wong et al. submitted unpublished observations), it is possible that BDNF may find therapeutic utility in Alzheimer's disease, Parkinson's disease, motor neuron disorders such as amyotrophic lateral sclerosis, and a variety of peripheral neuropathies. It is a long leap from the culture dish to clinical use, but the ability to produce recombinant human BDNF greatly facilitates all the intermediate steps. Even if BDNF proves effective in animal models of, for example, Parkinson's disease (Zigmond and Stricker, 1989), there will be important issues, such as intracerebral drug delivery, that will need to be overcome before clinical studies can begin. However, the diseases that may be ameliorated by BDNF or other neurotrophic factors are so debilitating that novel strategies undoubtedly will be found to speed up development and clinical application.

## XIII. CONCLUSIONS

In a recent review, Thoenen (1991) rings the death knell of the 30-year-old NGF monopoly. The remarkable NGF story has provided a molecular basis for the neurotrophic hypothesis that sprang from experimental neuroembryology. Without the fortuitous discovery of a large source of NGF in the mouse salivary gland, it is unlikely that our knowledge of neurotrophic factors would be as far advanced as it is. The isolation of the second (BDNF) and third (CNTF) neurotrophic factors has come slowly. The rapid cloning of NT-3, following the molecular cloning of BDNF, suggests that the pace is changing. Advances in molecular biology and protein chemistry will not only make it easier to speed up the purification of interesting activities, but also will provide new methods (expression cloning) to identify novel factors. Although these advances may take some of the fun out of phenomenology, they will provide easier access to novel proteins so that their biology can be elucidated in vitro and in vivo.

## ACKNOWLEDGMENTS

Much of the content of this chapter covers a review of the work of Yves Barde and a small group of excellent students and colleagues. I hope that I have done some justice to their efforts. It has been a pleasure to collaborate and share ideas with this group over the last decade or more.

For more recent interactions, I thank all my colleagues at Regeneron for their enthusiasm and their sometimes irreverent approach to neurobiology. I am grateful to Peter DiStefano for reading the manuscript and making some very useful suggestions. Eric Shooter and Len Schleifer are responsible for my being at Regeneron.

## REFERENCES

Alderson, R. F., Alterman, A. L., Barde, Y.-A., and Lindsay, R. M. (1990). Brain-derived neurotrophic factor increases survival and differentiated functions of rat septal cholinergic neurons in culture. *Neuron* **5**, 297–306.

Aloe, L., Cozzari, C., Calissano, P., and Levi-Montalcini, R. (1981). Somatic and behavioral postnatal effects of fetal injections of nerve growth factor antibodies in the rat. *Nature (London)* **291**, 413–415.

Apfel, S. C., Lipton, R. B., Arezzo, J. C., and Kessler, J. A. (1991). Nerve growth factor prevents toxic neuropathy in mice. *Ann. Neurol.* **29**, 87–90.

Arakawa, Y., Sendtner, M., and Thoenen, H. (1990). Survival effect of ciliary neurotrophic factor on chick embryonic motorneurons in culture. Comparison with other neurotrophic factors and cytokines. *J. Neurosci.* **10**, 3507–3515.

Barde, Y.-A. (1988). What if anything is a neurotrophic factor? *Trends Neurosci.* **11**, 343–346.

Barde, Y.-A. (1989). Trophic factors and neuronal survival. *Neuron* **2**, 1525–1534.

Barde, Y.-A., Lindsay, R. M., Monard, D., and Thoenen, H. (1978). New factor released by cultured glioma cells supporting survival and growth of sensory neurones. *Nature (London)* **274**, 818.

Barde, Y.-A., and Thoenen, H. (1984). Purification of a neurotrophic protein from mammalian brain. In "The Role of Cell Interactions in Early Neurogenesis" (A.-M. Duprat, A. C. Kato, and M. Weber, eds.) pp. 263–269. Plenum Press, New York.

Barde, Y.-A., and Thoenen, H. (1985). Neurotrophic factors. In "Hormones and Cell Regulation" (J. E. Dumont, B. Hamprecht, and J. Nunez, eds.) Vol. 9, 385–390. Elsevier, Amsterdam.

Barde, Y.-A., Edgar, D., and Thoenen, H. (1980). Sensory neurons in culture: Changing requirements for survival factors during embryonic development. *Proc. Natl. Acad. Sci. U.S.A.* **77**, 1199–1203.

Barde, Y.-A., Edgar, D., and Thoenen, H. (1982a). Purification of a new neurotrophic factor from mammalian brain. *EMBO J.* **1**, 549–553.

Barde, Y.-A., Edgar, D., and Thoenen, H. (1982b). Molecules involved in the regulation of neuron survival during development. In "Neuroscience Approached through Cell Culture" (S. Pfeiffer, ed.), Vol. 1, pp. 69–86. CRC Press, Boca Raton, Florida.

Barde, Y.-A., Edgar, D., and Thoenen, H. (1983). New neurotrophic factors. *Ann. Rev. Physiol.* **45**, 601–612.

Barde, Y.-A., Lindsay, R. M., and Thoenen, H. (1985a). In "Neurobiochemistry" (B. Hamprecht and V. Neuhoff, eds.), pp. 18–21. Springer Verlag, Berlin.

Barde, Y.-A., Lindsay, R. M., and Thoenen, H. (1985b). Biological characterization of a brain-derived neurotrophic factor. *Biol. Chem. Hoppe-Seyler* **366**, 324.

Barde, Y.-A., Edgar, D., and Thoenen, H. (1987a). Neurotrophic factors in the central nervous system. In "Brain Peptides" (J. Martin, M. Brownstein, and D. Krieger, eds.), pp. 240–249.

Barde, Y.-A., Davies, A. M., Johnson, J. E., Lindsay, R. M., and Thoenen, H. (1987b). Brain derived neurotrophic factor. *Prog. Brain Res.* **71**, 185–189.

Carmignoto, G., Maffei, L., Candeo, P., Cannella, R., and Comelli, C. (1989). Effect of NGF on the survival of retinal ganglion cells following optic nerve section. *J. Neurosci.* **9**, 1263–1272.

Dal Toso, R., Giorgi, O., Soranzo, C., Kirschner, G., Ferrari, G., Favaron, M., Benvegnu, D., Presti, D., Vicini, S., Toffano, G., Azzone, G. F., and Leon, A. (1988). Development and survival of neurons in dissociated fetal mesencephalic serum-free cell cultures. 1. Effects of cell density and of an adult mammalian striatal-derived neuronotrophic factor (SDNF). *J. Neurosci.* **8**, 733–745.

Davies, A. M. (1988). Role of neurotrophic factors in development. *Trends Genetics* **4**, 139–143.

Davies, A. M., and Lindsay, R. M. (1984). Neural crest-derived spinal and cranial sensory neurones are equally sensitive to NGF but differ in their response to tissue extracts. *Dev. Brain Res.* **14**, 121–127.

Davies, A. M., and Lindsay, R. M. (1985). The cranial sensory ganglia in culture: Difference in the response of placode-derived and neural crest-derived neurons to nerve growth factor. *Dev. Biol.* **111**, 62–72.

Davies, A. M., and Lumsden, A. G. S. (1983). Influence of nerve growth factor on developing dorsomedial and ventro-lateral neurons of chick and mouse trigeminal ganglia. *Int. J. Dev. Neurosci.* **1**, 171–177.

Davies, A. M., Thoenen, H., and Barde, Y.-A. (1986a). The response of chick sensory neurons to brain-derived neurotrophic factor. *J. Neurosci* **6**, 1897–1904.

Davies, A. M., Thoenen, H., and Barde, Y.-A. (1986b). Different factors from the central nervous system and the periphery regulate the survival of neurones. *Nature (London)* **319**, 497–499.

DiStefano, P. S., Friedman, B., Radziejewski, C., Alexander, C., Boland, P., Schick, C. M. Lindsay, R. M. and Weigand, S. J. (1992). The neurotrophins BDNF, NT-3 and NGF display distinct patterns of axonal transport. Neuron **8**, 983–993.

Edgar, D., and Barde, Y.-A. (1983). Neuronal growth factors. *Trends Neurosci.* **6**, 260–262.

Ernfors, P., Ibanez, C. F., Ebendal, T., Olson, L., and Persson, H. (1990a). Molecular cloning and neurotrophic activities of a protein with structural similarities to nerve growth factor: Developmental and topographical expression in the brain. *Proc. Natl. Acad. Sci. U.S.A.* **87**, 5454–5458.

Ernfors, P., Wetmore, C., Olson, L., and Persson, H. (1990b). Identification of cells in rat brain expressing mRNA for members of the nerve growth factor family. *Neuron* **5**, 511–526.

Ernsberger, U., and Rohrer, H. (1988). Neuronal precursor cells in chick dorsal rot ganglia: Differentiation and survival in vitro. *Dev. Biol.* **126**, 420–432.

Fitzgerald, M., Wall, P. D., Goedert, M., and Emson, P. C. (1985). Nerve growth factor counteracts the neurophysiological and neurochemical effects of chronic sciatic nerve section. *Brain Res.* **332**, 131–141.

Goedert, M., Stoeckel, K., and Otten, U. (1981). Biological importance of the retrograde transport of nerve growth factor in sensory neurons. *Proc. Natl. Acad. Sci. U.S.A.* **78**, 5895–5898.

Gorin, P. D., and Johnson, E. M. (1979). Experimental autoimmune model of nerve growth factor deprivation; Effect on developing peripheral sympathetic and sensory neurons. *Proc. Natl. Acad. Sci. U.S.A.* **76**, 5382–5386.

Gorin, P. D., and Johnson, E. M. (1980). Effects of long term nerve growth factor deprivation on the nervous system of the adult rat: An experimental autoimmune approach. *Brain Res.* **198**, 27–42.

Hamburger, V. (1958). Regression versus peripheral control of differentiation in motor hypoplasia. *Amer. J. Anat.* **102**, 365–410.

Hamburger, V., and Levi-Montalcini, R. (1949). Proliferation, differentiation and degeneration in the spinal ganglia of the chick embryo under normal and experimental conditions. *J. Exp. Zool.* **111**, 457–502.

Hamburger, V., and Oppenheim, R. W. (1982). Naturally-occurring neuronal death in vertebrates. *Neurosci. Comments* **1**, 38–55.

Hamburger, V., Brunso-Bechtold, J. K., and Yip, J. (1981). Neuronal death in the spinal ganglia of the chick embryo and its reduction by nerve growth factor. *J. Neurosci.* **1**, 60–71.

Hartikka, J., and Hefti, F. (1988). Development of septal cholinergic neurons in culture: Plating density and glial cells modulate effects of NGF on survival and expression of transmitter-specific enzymes. *J. Neurosci.* **8**, 2967–2985.

Hatanaka, H., Tsukui, H., and Nihonmatsu, I. (1988). Developmental change in the nerve growth factor action from induction of choline acetyltransferase to promotion of cell survival in cultured basal forebrain neurons from postnatal rat. *Dev. Brain Res.* **39**, 88–95.

Hefti, F. (1986). Nerve growth factor promotes survival of septal cholinergic neurons after fimbrial transections. *J. Neurosci.* **6**, 2155–2161.

Hefti, F., and Weiner, W. J. (1986). Nerve growth factor and Alzheimer's disease. *Annals Neurol.* **20**, 275–281.

Hefti, F., Hartikka, J., Eckenstein, F., Gnahn, H., Heumann, R., and Schwab, M. (1985). Nerve growth factor (NGF) increases choline acetyltransferase but not survival or fiber outgrowth of cultured fetal septal cholinergic neurons. *Neuroscience* **14**, 55–68.

Hefti, F., Hartikka, J., and Knüsel, B. (1989). Function of neurotrophic factors in the adult and aging brain and their possible use in the treatment of neurodegenerative diseases. *Neurobiol. Aging* **10**, 515–533.

Hempstead, B. L., Martin-Zanca, D., Kaplan, D., Parada, L., and Chao, M. V. (1991). High-affinity NGF binding requires coexpression of the *trk* proto-oncogene and the low-affinity NGF receptor. *Nature (London)* **350**, 678–683.

Heumann, R., Korsching, S., Bandtlow, C., and Thoenen, H. (1987a). Changes of nerve growth factor synthesis in nonneuronal cells in response to sciatic nerve transection. *J. Cell Biol.* **104**, 1623–1631.

Heumann, R., Lindholm, D., Bandtlow, C., Meyer, M., Radeke, M. J., Misko, T. P., Shooter, E. M., and Thoenen, H. (1987b). Differential regulation of mRNA encoding nerve growth factor and its receptor in rat sciatic nerve during development, degeneration and regeneration: Role of macrophages. *Proc. Natl. Acad. Sci. U.S.A.* **84**, 8735–8739.

Hofer, M. M., and Barde, Y.-A. (1988). Brain-derived neurotrophic factor prevents neuronal death. *Nature (London)* **331**, 261–262.

Hofer, M., Pagliusi, S. R., Hohn, A., Leibrock, J., and Barde, Y.-A. (1990). Regional distribution of brain-derived neurotrophic factor mRNA in the adult mouse brain. *EMBO J.* **9**, 2459–2464.

Hohn, A., Leibrock, J., Bailey, K., and Barde, Y.-A. (1990). Identification and characterization of a novel member of the nerve growth factor/brain-derived neurotrophic factor family. *Nature (London)* **344**, 339–341.

Hollyday, M., and Hamburger, V. (1976). Reduction of the naturally occurring motor neurons loss by enlargement of the periphery. *J. Comp. Neurol.* **170**, 311–320.

Hosang, M., and Shooter, E. M. (1985). Molecular characteristics of nerve growth factor receptors on PC12 cells. *J. Biol. Chem.* **260**, 655–662.

Hosang, M., and Shooter, E. M. (1987). The internalization of nerve growth factor by high-affinity receptors on pheochromocytoma cells. *EMBO J.* **6**, 1197–1202.

Hyman, C., Hofer, M., Barde, Y.-A., Juhasz, M., Yancopoulos, G. D., Squinto, S. P., and Lindsay, R. M. (1991). BDNF is a neurotrophic factor for dopaminergic neurons of the substantia nigra. *Nature (London)* **350**, 230–232.

Jessel, T. M., Tsunoo, A., Kanazawa, I., and Otsuka, M. (1979). Substance P depletion in the dorsal horn of rat spinal cord after section of the peripheral processes of primary sensory neurons. *Brain Res.* **168**, 247–259.

Johnson, E. M., Gorin, P. D., Brandeis, L. D., and Pearson, J. (1980). Dorsal root ganglion neurons are destroyed by exposure in utero to maternal antibody to nerve growth factor. *Science* **210**, 916–918.

Johnson, D., Lanahan, A., Buck, C. R., Sehgal, A., Morgan, C., Mercer, E., Bothwell, M., and Chao, M. V. (1986). Expression and structure of the human NGF receptor. *Cell* **47**, 545–554.

Johnson, E. M., Jr., and Yip, H. K. (1985). Central nervous system and peripheral nerve growth factor provide trophic support critical to mature sensory neuronal survival. *Nature (London)* **314**, 751–752.

Johnson, E. M., Rich, K. M., and Yip, H. K. (1986). The role of NGF in sensory neurons in vivo. *Trends Neurosci.* **9**, 33–37.

Johnson, J. E., Barde, Y.-A., Schwab, M., and Thoenen, H. (1986). Brain-derived neurotrophic factor (BDNF) supports the survival of cultured rat retinal ganglion cells. *J. Neurosci.* **6**, 3031–3038.

Jones, K. R., and Reichardt, L. F. (1990). Molecular cloning of a human gene that is member of the nerve growth factor family. *Proc. Natl. Acad. Sci. U.S.A.* **87**, 8060–8064.

Kaisho, Y., Yoshimura, K., and Nakahama, K. (1990). Cloning and expression of a cDNA encoding a novel neurotrophic factor. *FEBS Lett.* **266**, 187–191.

Kalcheim, C., and Gendreau, M. (1988). Brain-derived neurotrophic factor stimulates survival and neuronal differentiation in cultured avian neural crest. *Dev. Brain Res.* **41**, 79–86.

Kalcheim, C., Barde, Y.-A., Thoenen, H., and Le Douarin, N. M. (1987). In vivo effect of brain-derived neurotrophic factor on the survival of developing dorsal root ganglion cells. *EMBO J.* **6**, 2871–2873.

Kalcheim, C., and Le Douarin, N. M. (1986). Requirement of a neural tube signal for the differentation of neural crest cells into dorsal root ganglia. *Dev. Biol.* **116**, 451–466.

Kaplan, D. R. Martin-Zanca, D., and Parada, L. (1991). Tyrosine phosphorylation and tyrosine kinase activity of the *trk* proto-oncogene product induced by NGF. *Nature (London)* **350**, 158–160.

Klein, R., Parada, L. F., Coulier, F., and Barbacid, M. (1989). *trkB*, a novel tyrosine kinase receptor expressed during mouse neural development. *EMBO J.* **8**, 3701–3709.

Klein, R., Conway, D., Parada, L. F., and Barbacid, M. (1990). The *trkB* tyrosine kinase gene codes for a second neurogenic receptor that lacks the catalytic kinase domain. *Cell* **61**, 647–656.

Klein, R., Jing, S., Nanduri, V., O'Rourke, E., and Barbacid, M. (1991). The *trk* photo-oncogene encodes a receptor for nerve growth factor. *Cell* **65**, 189–197.

Knüsel, B., Michel, P. P., Schwaber, J. S., and Hefti, F. (1990). Selective and nonselective stimulation of central cholinergic and dopaminergic development in vitro by nerve growth factor, basic fibroblast growth factor, epidermal growth factor, insulin and the insulin-like growth factors I and II. *J. Neurosci.* **10**, 558–570.

Knüsel, B., Winslow, J. W., Rosenthal, A., Burton, L. E., Seid, D. P., Nikolics, K., and Hefti, F. (1991). Promotion of central cholinergic and dopaminergic neuron differentiation by brain-derived neurotrophic factor but not neurotrophin 3. *Proc. Natl. Acad. Sci. U.S.A.* **88**, 961–965.

Koliatsos, V. E., Nauta, H. J., Clatterbuck, R. E., Holtzman, D. M., Mobley, W. C., and Price, D. L. (1990). Mouse nerve growth factor prevents degeneration of axotomized basal forebrain cholinergic neurons in the monkey. *J. Neurosci.* **10**, 3801–3813.

Landmesser, L., and Pilar, G. (1978). Interaction between neurons and their targets during in vivo synaptogenesis. *Fed. Proc.* **37**, 2016–2022.

Lawson, S. N., and Harper, A. A. (1985). Cell types in rat dorsal ganglia: Morphological, immunocytochemical and electrophysiological analyses. *In* "Somatosensory Pathways" (W. D. Willis and L. Rowe eds.), pp. 97–103, Liss, New York.

Leibrock, J., Lottspeich, F., Hohn, A., Hengerer, B., Masiakowski, P., Thoenen, H., and Barde, Y.-A. (1989). Molecular cloning and expression of brain-derived neurotrophic factor. *Nature (London)* **341**, 149–152.

Levi-Montalcini, R. (1962). Analysis of a specific nerve growth factor and its antiserum. *Sci. Rep. 1st Super. Sanita* **2**, 345–368.

Levi-Montalcini, R., and Angeletti, P. U. (1966). Immunosympathectomy. *Pharmacol. Rev.* **18**, 619–628.

Levi-Montalcini, R., and Angeletti, P. U. (1968). Nerve growth factor. *Physiol. Rev.* **48**, 534–569.

Lin, L-F. H., Mismer, D., Lile, J. D., Armes, L. G., Butler, E. T., III, and Collins, F. (1989). Purification, cloning and expression of ciliary neurotrophic factor (CNTF). *Science* **246**, 1023–1025.

Lindholm, D., Heumann, R., and Thoenen, H. (1987). Interleukin-1 regulates synthesis of nerve growth factor in non-neuronal cells of rat sciatic nerve. *Nature (London)* **330**, 658–659.

Lindsay, R. M. (1979). Adult rat brain astrocytes support survival of both NGF-dependent and NGF-insensitive neurones. *Nature (London)* **282**, 80–82.

Lindsay, R. M. (1988a). Nerve growth factors (NGF,BDNF) enhance axonal regeneration but are not required for survival of adult sensory neurons. *J. Neurosci.* **8**, 2394–2405.

Lindsay, R. M. (1988b). The role of neurotrophic growth factors in development, maintenance and regeneration of sensory neurons. *In* "The Making of the Nervous System" (J. Parnavelas, C. D. Stern, and R. V. Stirling, eds.), pp. 148–165. Oxford University Press.

Lindsay, R. M. (1992). The role of neurotrophic factors in functional maintenance of mature sensory neurons. In "Sensory Neuron Diversity, Development, and Plasticity" (S. Scott, ed.). Oxford University Press.

Lindsay, R. M., and Harmar, A. J. (1989). Nerve growth factor regulates expression of neuropeptide genes in adult sensory neurones. *Nature (London)* **337**, 362–364.

Lindsay, R. M., and Peters, C. (1984a). Spinal cord contains neurotrophic activity for spinal sensory neurons. Late developmental appearance of a survival factor distinct from nerve growth factor. *Neuroscience* **12**, 45–51.

Lindsay, R. M., and Peters, C. (1984b). A spinal cord derived neurotrophic factor for spinal nerve sensory neurons. In "The Role of Cell Interactions in Early Neurogenesis" (A.-M. Duprat, A. C. Kato, and M. Weber, eds.). Vol. 77, pp. 299–306. Plenum Press, New York.

Lindsay, R. M., and Rohrer, H. (1985). Placodal sensory neurons in culture: Nodose ganglion neurons are unresponsive to NGF, lack NGF receptors but are supported by a liver-derived neurotrophic factor. *Dev. Biol.* **112**, 30–48.

Lindsay, R. M., and Tarbit, J. (1979). Developmentally regulated induction of neurite outgrowth from immature chick sensory neurons (DRG) by homogenates of avian or mammalian heart, liver or brain. *Neurosci. Lett.* **12**, 195–200.

Lindsay, R. M., Barber, P. C., Sherwood, M. R. C., Zimmer, J., and Raisman, G. (1982). Astrocyte cultures from adult rat brain. Derivation, characterization and neurotrophic properties of pure astroglial cells from corpus callosum. *Brain Res.* **243**, 329–343.

Lindsay, R. M., Barde, Y.-A., Davies, A. M., and Rohrer, H. (1985a). Differences and similarities in the neurotrophic requirements of sensory neurons derived from neural crest and neural placode. *J. Cell Sci. Suppl.* **3**, 115–129.

Lindsay, R. M., Thoenen, H., and Barde, Y.-A. (1985b). Placode and neural crest-derived sensory neurons are responsive at early developmental stages to brain-derived neurotrophic factor (BDNF). *Dev. Biol.* **112**, 319–328.

Lindsay, R. M., Lockett, C., Sternberg, J., and Winter, J. (1989). Neuropeptide expression in cultures of adult sensory neurons: Modulation of substance P and CGRP levels by nerve growth factor. *Neuroscience* **33**, 53–65.

Lindsay, R. M., Shooter, E. M., Radeke, M. J., Misko, T. P., Dechant, G., Thoenen, H., and Lindholm, D. (1990). Nerve growth factor regulates expression of the nerve growth factor receptor gene in adult sensory neurons. *Eur. J. Neurosci.* **2**, 389–396.

Maisonpierre, P. C., Belluscio, L., Squinto, S., Ip, N. Y., Furth, M. E., Lindsay, R. M., and Yancopoulos, G. D. (1990a). Neurotrophin-3: A neurotrophic factor related to NGF and BDNF. *Science* **247**, 1446–1451.

Maisonpierre, P. C., Belluscio, L., Friedman, B., Alderson, R., Wiegand, S. J., Furth, M. E., Lindsay, R. M., and Yancopoulos, G. (1990b). NT-3, BDNF and NGF in the developing rat nervous system: Parallel as well as reciprocal patterns of expression. *Neuron* **5**, 501–509.

Maisonpierre, P. C., Le Beau, M. M., Espinosa, R., Ip, N. Y., Belluscio, L., De La Monte, S. M., Squinto, S., Furth, M. E., and Yancopoulos, G. D. (1991). Human and rat brain-derived neurotrophic factor and neurotrophin-3: Gene structures, distributions and chromosomal localizations *Genomics* **10**, 558–568.

Manthorpe, M., Skaper, S. D., Williams, L. R., and Varon, S. (1986). Purification of adult rat sciatic nerve ciliary neurotrophic factor. *Brain Res.* **367**, 282–286.

Manthorpe, M., Ray, J., Pettman, B., and Varon, S. (1989). Ciliary neurotrophic factors. In "Nerve Growth Factors" (R. A. Rush, ed.). John Wiley, New York.

Martin-Zanca, D., Barbacid, M. and Parada, L. F. (1990). Expression of the *trk* proto-oncogene is restricted to the sensory cranial and spinal ganglia of neural crest origin in mouse development. *Genes Dev.* **4**, 683–694.

Martin-Zanca, D., Oskam, R., Mitra, G., Copeland, T., and Barbacid, M. (1989). Molecular and biochemical characteristics of the human *trk* proto-oncogene. *Mol. Cell. Biol.* **9**, 24–33.

Meakin, S. O., and Shooter, E. M. (1991). Molecular investigations on the high affinity nerve growth factor receptor. *Neuron* **6**, 153–163.

Monard, D., Solomon, F., Rentsch, M., and Gysin, R. (1973). Glia-induced morphological differentiation in neuroblastoma cells. *Proc. Natl. Acad. Sci. U.S.A.* **70**, 1894–1897.

Monard, D., Stockel, K., Goodman, R., and Thoenen, H. (1975). Distinction between nerve growth factor and glial factor. *Nature (London)* **258**, 444–445.

Oppenheim, R. W. (1991). Cell death during development of the nervous system. *Ann. Rev. Neurosci.* **14**, 453–501.

Pearson, J., Johnson, E. M., and Brandeis, L. (1983). Effects of antibodies to nerve growth factor in intrauterine development of derivatives of cranial neural crest and placode in the guinea pig. *Dev. Biol.* **96**, 32–36.

Phillips, H. S., Hains, J. M., Laramee, G. R., Rosenthal, A., and Winslow, J. W. (1990). Widespread expression of BDNF but not NT-3 by target areas of basal forebrain cholinergic neurons. *Science* **250**, 290–294.

Purves, D. (1988). A molecular basis for trophic interactions in vertebrate. *In* "Body and Brain; A Trophic Theory of Neural Connections," pp. 123–141 Harvard University Press, Cambridge, Massachusetts.

Radeke, M. J. Misko, T. P., Hsu, C., Herzenberg, L. A., and Shooter, E. M. (1987). Gene transfer and cloning of the rat nerve growth factor receptor. *Nature (London)* **325**, 593–597.

Ragsdale, C., and Woodgett, J. (1991). trking neurotrophic receptors. *Nature (London)* **350**, 660–661.

Rambourg, A., Clermont, Y., and Beaudet, A. (1983). Ultrastructural features of six types of neurons in rat dorsal root ganglia. *J. Neurocytol.* **12**, 47–66.

Rich, K. M., Yip, H. K., Osborne, P. A., Schmidt, R. E., and Johnson, E. M., Jr. (1984). Role of nerve growth factor in the adult dorsal root ganglion and its response to injury. *J. Comp. Neurol.* **230**, 110–118.

Rich, K. M., Lusyzczynski, J. R., Osborne, P. A., and Johnson, E. M. (1987). Nerve growth factor protects adult sensory neurons from death and atrophy caused by nerve injury. *J. Neurocytol.* **16**, 261–268.

Rodriguez-Tébar, A., and Barde, Y.-A. (1988). Binding characteristics of brain-derived neurotrophic factor to its receptors on neurons from the chick embryo. *J. Neurosci.* **9**, 3337–3342.

Rodriguez-Tébar, A., Jeffrey, P. L., Thoenen, H., and Barde, Y.-A. (1989). The survival of chick retinal ganglion cells in response to brain-derived neurotrophic factor depends on their age. *Dev. Biol.* **136**, 296–303.

Rodriguez-Tébar, A., Dechant, G., and Barde, Y.-A. (1990). Brain-derived neurotrophic factor binds to the nerve growth factor receptor. *Neuron* **4**, 487–492.

Rosenthal, A., Goeddel, D. V., Nguyen, T., Lewis, M., Shih, A., Laramee, G. R., Nikolics, K, and Winslow, J. W. (1990). Primary structure and biological activity of a novel human neurotrophic factor. *Neuron* **4**, 767–773.

Schwartz, J. P., Pearson, J., and Johnson, E. M., Jr. (1982). Effect of exposure to anti-NGF on sensory neurons of adult rats and guinea pigs. *Brain Res.* **244**, 378–381.

Snider, W. D., and Johnson, E. M. (1989). Neurotrophic molecules. *Ann. Neurol.* **26**, 489–506.

Spina, M.-B., Hyman, C., Squinto, S., and Lindsay, R. M. (1992). Brain-derived neurotrophic factor (BDNF) protects dopaminergic cells from 6-hydroxydopamine toxicity. *Ann. N.Y. Acad. Sci.* (in press).

Spina, M. B., Squinto, S. P., Miller, F., Lindsay, R. M., and Hyman, C. (1992). BDNF protects dopamine neurons against 6-OHDA and MPP+ toxicity: Studies on the mechanism of protection. *J. Neurochem.* (in press).

Squinto, S. P., Stitt, T. S., Aldrich, T. H., Davis, S., Bianco, S. M., Radziejewski, C., Glass, D. J., Masiakowski, P., Furth, M. E., Valenzuela, D. M., DiStefano, P. S., and Yancouplos, G. D. (1991). trkB encodes a functional receptor for BDNF and NT-3 but not NGF. *Cell* **65**, 885–893.

Stöckel, K., Schwab, M., and Thoenen, H. (1975). Specificity of retrograde transport of nerve growth factor in sensory neurons. A biochemical and morphological study. *Brain Res.* **89**, 1–14.

Stöckli, K. A., Lottspeich, F., Sendtner, M., Masiakowski, P., Carroll, P., Götz, R., Lindholm, D., and

Thoenen, H. (1989). Molecular cloning, expression and regional distribution of rat ciliary neurotrophic factor. *Nature (London)* **342**, 920–923.

Sutter, A., Riopelle, R. J., Harris-Warrick, R. M., and Shooter, E. M. (1979). Nerve growth factor receptors. Characterization of two distinct classes of binding sites on chick embryo sensory ganglia cells. *J. Biol. Chem.* **254**, 5972–5982.

Tessler, A., Himes, B. T., Krieger, N. R., Murray, M., and Goldberger, M. (1985). Sciatic nerve transection produces death of dorsal root ganglion cells and a reversible loss of substance P in spinal cord. *Brain Res.* **332**, 209–219.

Thanos, S., Bahr, M., Barde, Y.-A., and Vaneslow, J. (1989). Survival and axonal elongation of adult rat retinal ganglion cells. In vitro effects of lesioned sciatic nerve and brain-derived neurotrophic factor. *Eur. J. Neurosci.* **1**, 19–26.

Thoenen, H. (1991). The changing scene of neurotrophic factors. *Trends Neurosci.* **14**, 165–170.

Thoenen, H., and Barde, Y.-A. (1980). Physiology of nerve growth factor. *Physiol. Rev.* **60**, 1284–1335.

Thoenen, H., Barde, Y.-A., and Edgar, D. (1982). Factors involved in the regulation of survival and differentiation of neurons. In "Repair and Regeneration of the Nervous System" (J. G. Nicholls, ed.), pp. 173–185. Dahlem Conference. Springer Verlag, Berlin.

Thoenen, H., Barde, Y.-A., and Edgar, D. (1984a). The role of nerve growth factor and related factors in the survival of peripheral neurons. *Adv. Biochem. Psychopharmacol.* **28**, 262–273.

Thoenen, H., Barde, Y.-A., and Edgar, D. (1984b). Macromolecular factors involved in the regulation of survival and differentiation of peripheral sensory and sympathetic neurons. In "Cellular and Molecular Biology of Neuronal Development" (I. B. Black, ed.), pp. 243–250. Plenum Press, New York.

Turner, J. E., Barde, Y.-A., Schwab, M. E., and Thoenen, H. (1983). Extracts from brain stimulate neurite outgrowth from fetal retinal explants. *Dev. Brain Res.* **6**, 77–83.

Tuszynski, M. H., U, H. S., Amaral, D. G., and Gage, F. H. (1990). Nerve growth factor infusion in the primate brain reduces lesion-induced cholinergic neuronal degeneration. *J. Neurosci.* **10**, 3604–3614.

Waldmann, T. A. (1991). The interleukin-2 receptor. *J. Biol. Chem.* **266**, 2681–2684.

Wetmore, C., Ernfors, P., Persson, H., and Olson, L. (1990). Localization of brain-derived neurotrophic factor messenger RNA to neurons on the brain by in situ-hybridization. *Exp. Neurol.* **109**, 141–152.

Whittemore, S. R., and Seiger, A. (1987). The expression, localization and functional significance of βnerve growth factor in the central nervous system. *Brain Res. Rev.* **12**, 439–464.

Williams, L. R., Varon, S., Peterson, G., Wictorin, K., Fischer, W., Bjorklund, A., and Gage, F. H. (1986). Continuous infusion of nerve growth factor prevents basal forebrain neuronal death after fimbria fornix transection. *Proc. Natl. Acad. Sci. U.S.A.* **83**, 9231–9235.

Winter, J., Forbes, C. A., Sternberg, J., and Lindsay, R. M. (1988). Nerve growth factor (NGF) regulates adult rat dorsal root ganglion (DRG) neuron responses to the excitotoxin capasicin. *Neuron* **1**, 973–981.

Wong, V., Arriaga, R., and Lindsay, R. M. (1990). Effects of ciliary neurotrophic factor (CNTF) on ventral spinal cord neurons in culture. *Soc. Neurosci. Abs.* **16**, 209.6.

Wong, V., Arriaga, R., Ip, N. Y., and Lindsay, R. M. (1992). BDNF and NT-3 but not NGF up-regulate cholinergic phenotypic properties of motor neurons in culture. (submitted)

Yamamoto, H., and Gurney, M. E. (1990). Human platelets contain brain-derived neurotrophic factor. *J. Neurosci.* **10**, 3469–3478.

Yan, Q., and Johnson, E. M. (1988). An immunohistochemical study of the nerve growth factor receptor in developing rats. *J. Neurosci.* **8**, 3481–3498.

Yancopoulos, G. D., Maisonpierre, P. C., Ip, N. Y., Aldrich, T. H. Belluscio, L., Boulton, T. G., Cobb, M. H., Squinto, S. P., and Furth, M. E. (1990). Neurotrophic factors, their receptors and the signal transduction pathways they activate. *Cold Spring Harbor Symp. Quant. Biol.* **55**, 371–379.

Yip, H. K., and Johnson, E. M. (1984). Developing dorsal root ganglion neurons require trophic support from their central processes: Evidence for a role of retrogradely transported nerve growth factor from the central nervous system to the periphery. *Proc. Natl. Acad. Sci. U.S.A.* **81**, 6245–6249.

Zafra, F., Hengerer, B., Leibrock, J., Thoenen, H., and Lindholm, D. (1990). Activity dependent regulation of BDNF and NGF messenger RNAs in the rat hippocampus is mediated by non-NMDA glutamate receptors. *EMBO J.* **9**, 3545–3550.

Zigmond, M. J., and Stricker, E. M. (1989). Animal models of Parkinsonism using selective neurotoxins: Clinical and basic implications. *Int. Rev. Neurobiol.* **31**, 1–79.

# 9 Biochemistry and Molecular Biology of Fibroblast Growth Factors

Kenneth A. Thomas

## I. INTRODUCTION

The mitosis, migration, and maintenance of cells in culture, and presumably in vivo, are controlled by intercellular signals mediated in large part by protein growth factors. Although growth factors can function systemically, they are thought to act principally in a local, or paracrine, fashion to coordinate cellular interactions in complex multicellular organisms. Numerous protein growth factors have been identified, belonging to the hematopoetic growth factor, epidermal growth factor (EGF), insulin-like growth factor (IGF), platelet-derived growth factor (PDGF), transforming growth factor β (TGFβ), nerve growth factor (NGF), and fibroblast growth factor (FGF) families.

Although first purified and structurally characterized only in the last decade, the fibroblast growth factors are recognized already to constitute a large family of homologous proteins of which seven members currently are known. The first two FGFs to be purified, acidic FGF (aFGF; Thomas et al., 1984) and basic FGF (bFGF; Lemmon and Bradshaw, 1983; Bohlen et al., 1984), have been the most thoroughly studied and are considered commonly as prototypic of the larger family. Five additional family members, INT-2 (Smith et al., 1988), HST/k-FGF (Delli-Bovi et al., 1987; Yoshida et al., 1987), FGF-5 (Zhan et al., 1988), FGF-6 (Marics et al., 1989), and keratinocyte growth factor (KGF; Finch et al., 1989; Rubin et al., 1989), were identified subsequently by oncogene assays, expression cloning, reduced stringency hybridization to other FGF cDNAs, or specific mitogenic assays. The FGFs are characterized functionally by their affinity for heparin and related molecules, which has led to the synonym "heparin-binding growth factors" (Lobb and Fett, 1984). Numerous other nomenclatures for aFGF and bFGF have been used previously, including endothelial cell growth factor, astrocyte growth factor, brain-derived growth factor, and prostatropin.

The members of the FGF family exhibit activities on many types of cells, including neurons, neuroblasts, astrocytes, and oligodendrocytes, through multiple plasma membrane receptors. The diversity of FGFs, receptors, target cells, and functions makes understanding the physiologic significance of these proteins both intriguing and challenging. The current status of knowledge pertaining to the genetic control, structures, and functions of FGFs and their receptors will be described. Previous reviews can be consulted for additional perspective (Thomas, 1987; Burgess and Maciag, 1989; Klagsbrun, 1989; Rifkin and Moscatelli, 1989; Ortega and Thomas, 1990).

## II. STRUCTURE AND EXPRESSION OF FGF GENES

The human chromosome locations for six of the seven known FGFs have been determined. Although these proteins clearly arose by duplication and divergence from a common ancestral gene, their single copy genes have been distributed over multiple chromosomes. Both bFGF (Mergia et al., 1986) and FGF-5 (Dionne et al., 1990b) are located on chromosome 4; aFGF is on chromosome 5 (Jaye et al., 1986); both *int*-2 (Casey et al., 1986) and *hst/k-fgf* (Adelaide et al., 1988) are within 35 kb of each other (Wada et al., 1988) on chromosome 11; and FGF-6 is on chromosome 12 (Marics et al., 1989). The proximity and common transcriptional orientation of *int*-2 and *hst/k-fgf* in both human and mouse might be a remnant of gene duplication (Wada et al., 1988), but also could reflect some elements of common control of expression. The genes appear to be coamplified in some tumors (Adelaide et al., 1988; Yoshida et al., 1988) and are under complementary transcriptional control during mouse F9 teratocarcinoma cell differentiation, in which the mRNA for *hst/k-fgf* decreases with a concomitant increase in *int*-2 transcription (Smith et al., 1988; Velcich et al., 1989).

All characterized FGF genes have two conserved introns that divide each of the protein coding regions into three similar exons. These two introns in aFGF are 13.6 and 5.3 kb (Chiu et al., 1990), whereas in bFGF they are at least 16 and 17.5 kb (Abraham et al., 1986b). In contrast, these two introns are only 1.7 and 1.8 kb in the mouse *int*-2 gene (Moore et al., 1986). The functional significance, if any, of the large aFGF and bFGF introns is not clear. The aFGF gene also encodes multiple 5' exons that are transcribed but, because of an in-frame translational stop signal two codons upstream of the AUG initiation site in the 155-residue translation product (Jaye et al., 1986), are not translated as part of the aFGF protein. The different 5'ends are presumably attached by alternative splicing to a site located 34 residues 5'to the translational start codon (Chiu et al., 1990; Crumley et al., 1990). Although the total number, functions, and sizes of the alternative 5' ends are unknown, the entire human aFGF genetic locus including introns is estimated currently to span at least 50 kb of genomic DNA (Chiu et al., 1990).

Various states of differentiation, tissue repair, and transformation have been identified that correlate with either enhancement or repression of mRNA transcription of the FGF genes, yet little is known about the details of transcriptional control. The array of mRNA sizes for each FGF might, in part, reflect alternative control of expression. In addition to aFGF, *int*-2 has multiple 5' ends. In the case of *int*-2, these ends are transcribed from two different promoters, the second of which is located in a region that is an intron of the first (Smith et al., 1988). A GC-rich promoter-like region and eight possible enhancer sequences have been located in the *hst/k-fgf* gene, four of which are 800–1600 bp upstream of the first transcribed nucleotide (Yoshida et al., 1987). In addition, a functional differentiation-specific enhancer has been mapped to an unusual position 3' of the coding region (Curatola and Basilico, 1990). Many of the FGFs use alternative 3' poly (A)+ sites, with the exception of *hst/k-fgf*, which appears to lack a poly (A)+ site (Yoshida et al., 1987). The effect, if any, of alternative 3' ends on expression is unknown.

Transcriptional efficiency of the bFGF gene also might be influenced by competition with transcription from an adjacent gene in the opposite orientation. In *Xenopus* frogs, this overlapping gene encodes a transcribed mRNA for a 217-amino-acid protein of unknown function. Although the coding regions of bFGF and this gene do not overlap, the complementary strand gene extends into exon 3 of bFGF with its poly(A)+ addition site in intron 2. A similar complementary gene appears to be conserved in human DNA (Volk et al., 1989).

Control of expression can be exerted not only at the level of mRNA synthesis but also at the level of stabilization. The mature mRNA for bFGF might be unstable based on the observation that unprocessed mRNA was cloned originally, implying that it represented a significant fraction of total bFGF mRNA (Abraham et al., 1986a). aFGF (Crumley et al., 1989), *hst/k-fgf* (Delli Bovi et al., 1987), and KGF (Finch et al., 1989) all contain ATTT(A) sequences that are associated with decreased stability, presumably by acting as recognition sites for specific ribonucleases.

In addition to transcriptional control, some FGFs might be under translational control. Both FGF-5 (Ferrara and Henzel, 1989) and *int*-2 have secondary short open reading frames that, in the case of *int*-2, decrease translational efficiency 6- to 8-fold (Dixon et al., 1989). *hst* mRNA also contains an open reading frame (Yoshida et al., 1987) at its 3' end, but the effect, if any, on translation is unknown. Also, both bFGF and *int*-2 mRNAs appear to be able to use multiple upstream CUG translation initiation sites (Florkiewicz and Summer, 1989; Iberg et al., 1989; Prats et al., 1989). These amino-terminally elongated products also appear to accumulate selectively in the nucleus, probably as a result of their added cationic sequences (Acland et al., 1990; Renko et al., 1990). The functional significance of nuclear targeting is unknown.

### III. PROTEIN STRUCTURES

Although the individual FGF family members have different amino- and carboxy-terminal lengths, as shown schematically in Fig. 1, they all share an approximately 125-amino-acid internal homologous region, as shown in Fig. 2. The percentage

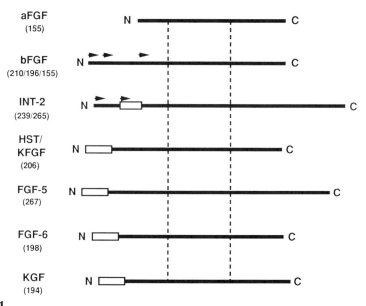

FIGURE 1

Relative lengths and alignments of FGFs. The full length translation products of the seven known FGFs are drawn to scale from the amino (N) to carboxy (C) terminus and positioned based on their homologous alignments. The numbers of amino acid residues in the primary translation products are given in parentheses below the protein names. All sequences are for human proteins except FGF-6, which is murine. The perpendicular dashed lines indicate the locations of the two Cys residues conserved among all known FGFs. The arrows above the bFGF and INT-2 polypeptides denote multiple translational initiation sites. Open rectangles at the amino termini indicate the presence of secretory leader sequences.

identity between any two members in this portion of the amino acid sequence ranges from 35% to nearly 80%. Based on their homology, the seven known FGFs all clearly are descendent from a common ancestral gene. The age and phylogenetic span of the FGFs is not known, but heparin-binding mitogens have been described in animals as primitive as dogfish (Lagente et al., 1986). The aFGF sequence also contains, within a region of maximum divergence among FGFs, a 10-residue tachykinin-like sequence, shown in Fig. 3. The decapeptide is flanked at both the amino- and the carboxy-terminal ends by basic dipeptides that are used frequently

FIGURE 2

Amino acid sequence homologies among FGFs. The regions common to all seven known FGFs are aligned. Identical amino acid residues are boxed. Residues that are identical among all seven FGFs are denoted by stippled backgrounds. The 12 β strands identified by X-ray crystal structure analysis are indicated by numbered double-headed arrows. The single letter amino acid code is used: A, alanine; C, cysteine; D, aspartic acid; E, glutamic acid; F, phenylalanine; G, glycine; H, histidine; I, isoleucine; K, lysine; L, leucine; M, methionine; N, asparagine; P, proline; Q, glutamine; R, arginine; S, serine; T, threonine; V, valine; W, tryptophan; Y, tyrosine.

```
aFGF       Ac-A-E-G-E-I-T-T-F-T-A-      L-T-E-           -K-F-N-L-P-P-G-N-T-K
bFGF       Ac-A-A-G-S-I-T-T-L-P-A-      L-P-E-D-G-G-S-G-A-F-P-P-G-H-F-K
INT-2      -E-P-G-W-P-A-A-G-P-G-A-R-L-R-R-D-A-G-G-R-G-G-V-Y-E-H-L-G-G-A-
HST/KFGF   -A-R-L-P-V-A-A-Q-P-K-    -E-A-A-V-Q-S-G-A-G-D-Y-L-L-G-I-K
FGF-5      -S-S-S-S-A-S-S-S-P-A-A-S-L-G-S-Q-    -G-S-G-L-E-Q-S-S-F-Q-W-S-
FGF-6                                                        G-I-K
KGF        -T-P-E-Q-M-A-T-N-V-N-C-S-S-P-E-R-H-T-R-S-Y-D-Y-M-E-G-G-D-I-
```

|←— 1 —→|←— 2 —→|←— 3 —→|

```
aFGF       -K-P-K-L-Y-C-S-N-G-    G-H-F-L-R-I-L-P-D-G-T-V-D-G-T-R
bFGF       -D-P-K-R-L-Y-C-K-N-G-  G-F-F-L-R-I-H-P-D-G-R-V-D-G-V-R
INT-2      R-R-R-K-L-Y-C-A-T-K-   -Y-H-L-Q-L-H-P-S-G-R-V-N-G-S-L
HST/KFGF   R-L-R-R-L-Y-C-N-V-G-I-G-F-H-L-Q-A-L-P-D-G-R-I-G-G-A-H
FGF-5      -L-G-A-R-T-G-S-L-Y-C-R-V-G-I-G-F-H-L-Q-I-Y-P-D-G-K-V-N-G-S-H-
FGF-6      R-Q-R-R-L-Y-C-N-V-G-I-G-F-H-L-Q-L-P-D-G-R-I-S-G-T-H-
KGF        R-V-R-R-L-F-C-R-T-     -Q-W-Y-L-R-I-D-K-R-G-K-V-K-G-T-Q-
```

|←— 4 —→|←— 5 —→|←—|

```
aFGF       -D-R-S-D-Q-H-I-Q-L-Q-L-S-A-E-S-V-G-E-V-Y-I-K-S-T-E-T-G-Q-Y-L-
bFGF       -E-K-S-D-P-H-I-K-L-Q-L-Q-A-E-E-R-G-V-V-S-I-K-G-V-C-A-N-R-Y-L-
INT-2      -E-N-S-A-Y-S-I-L-E-I-T-A-V-E-V-G-I-V-A-I-R-G-L-F-S-G-R-Y-L-
HST/KFGF   -A-D-T-R-D-S-L-L-E-L-S-P-V-E-R-G-V-V-S-I-F-G-V-A-S-R-F-F-V-
FGF-5      -E-A-N-N-L-S-V-L-E-I-F-A-V-S-Q-G-I-V-G-I-R-G-V-F-S-N-K-F-L-
FGF-6      -E-E-N-P-Y-S-L-L-E-I-S-T-V-E-R-G-V-V-S-L-F-G-V-R-S-A-L-F-V-
KGF        E-M-K-N-N-Y-N-I-M-E-I-R-T-V-A-V-G-I-V-A-I-K-G-V-E-S-E-F-Y-L-
```

|6 →|←— 7 —→|←— 8 —→|←|

```
aFGF       -A-M-D-T-D-G-L-L-Y-G-S-Q-T-P-N-E-E-C-L-F-L-E-R-L-E-E-N-H-Y-N-
bFGF       -A-M-K-E-D-G-R-L-L-A-S-K-C-V-T-D-E-C-F-F-F-E-R-L-E-S-N-N-Y-N-
INT-2      -A-M-N-K-R-G-R-L-Y-A-S-E-H-Y-S-A-E-C-E-F-V-E-R-I-H-E-L-G-Y-N-
HST/KFGF   -A-M-S-S-K-G-K-L-Y-G-S-P-F-F-T-D-E-C-T-F-K-E-I-L-L-P-N-N-Y-N-
FGF-5      -A-M-S-K-K-G-K-L-H-A-S-A-K-F-T-D-E-C-T-F-R-E-R-F-Q-E-N-S-Y-N-
FGF-6      -A-M-N-S-K-G-R-L-Y-A-T-P-S-F-Q-E-E-C-K-F-R-E-T-L-L-P-N-N-Y-N-
KGF        -A-M-N-K-E-G-K-L-Y-A-K-K-E-C-N-E-D-C-N-F-K-E-L-I-L-E-N-H-Y-N-
```

|— 9 →|←— 10 —|

```
aFGF       -T-Y-I-S-K-K-H-A-                       E-K-N-W-F-V-G-L-
bFGF       -T-Y-R-S-R-K-Y-T-                       -S-W-Y-V-A-L-
INT-2      -T-Y-A-S-R-L-Y-R-T-V-S-S-T-P-G-A-R-R-Q-P-S-A-E-R-L-W-Y-V-S-V-
HST/KFGF   -A-Y-E-S-Y-K-T-P-                       -G-M-F-I-A-L-
FGF-5      -T-Y-A-S-A-I-H-R-T-E-K-T-G-             R-E-W-Y-V-A-L-
FGF-6      -A-Y-E-S-D-L-Y-Q-                       G-T-Y-I-A-L-
KGF        -T-Y-A-S-A-K-W-T-H-N-G-G-               E-M-F-V-A-L-
```

|→|←— 11 —→|←— 12 —→|

```
aFGF       -K-K-N-G-S-C-K-R-G-    P-R-T-H-Y-G-Q-K-A-I-L-F-L-P-L-P-V-S-S-D
bFGF       -K-R-T-G-Q-Y-K-L-G-    S-K-T-G-P-G-Q-K-A-I-L-F-L-P-M-S-A-K-S
INT-2      N-G-K-G-R-P-R-R-G-     -F-K-T-R-R-T-Q-K-S-S-L-F-L-P-R-V-L-D-H-R-
HST/KFGF   S-K-N-G-K-T-K-K-G-     -N-R-V-S-P-T-M-K-V-T-H-F-L-P-R-L
FGF-5      N-K-R-G-R-K-R-G-C-S-P-R-V-K-P-Q-H-I-S-T-H-F-L-P-R-F-K-Q-S-E-
FGF-6      S-K-V-G-R-V-K-R-G-     S-K-V-S-P-I-M-T-V-T-H-F-L-P-R-I
KGF        N-Q-K-G-I-P-V-R-G-     -K-K-T-K-K-E-Q-K-T-A-H-F-L-P-M-A-I-T
```

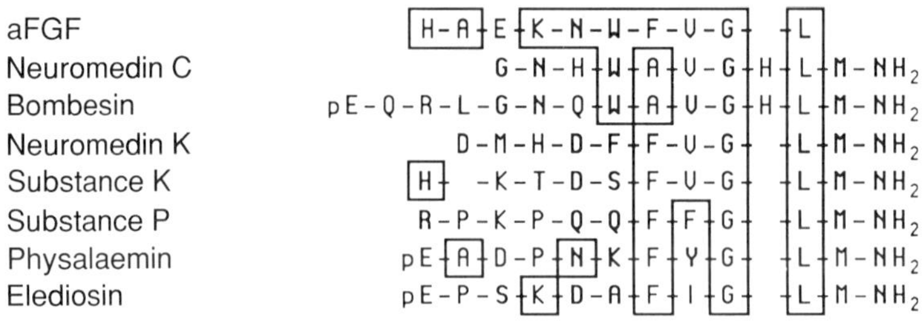

```
aFGF           H-A-E-K-N-W-F-U-G-    -L
Neuromedin C           G-N-H-W-A-U-G-H-L-M-NH₂
Bombesin       pE-Q-R-L-G-N-Q-W-A-U-G-H-L-M-NH₂
Neuromedin K           D-M-H-D-F-F-U-G-    -L-M-NH₂
Substance K            H-  -K-T-D-S-F-U-G-    -L-M-NH₂
Substance P            R-P-K-P-Q-Q-F-F-G-    -L-M-NH₂
Physalaemin            pE-A-D-P-N-K-F-Y-G-   -L-M-NH₂
Eledoisin              pE-P-S-K-D-A-F-I-G-   -L-M-NH₂
```

FIGURE 3

Sequence homologies among the aFGF neuropeptide-like region and known neuropeptides. The sequence of amino acid residues 102–111 of human aFGF (140-residue numbering system), from the end of β-strand 9 to the end of β-strand 10 (see Fig. 2), is aligned with sequences of known neuropeptides. Residues identical to aFGF are boxed.

as proteolytic cleavage sites to generate mature active polypeptides from larger precursors (Gimenez-Gallego et al., 1985).

Both aFGF and bFGF have been observed to have a distant amino acid sequence homology with another set of growth factors, the interleukin-1 (IL-1) family (Gimenez-Gallego et al., 1985; Thomas et al., 1985). The general conclusion that FGFs and IL-1s are homologous has been confirmed by the crystal structures of IL-1β, IL-1α, aFGF, and bFGF. All four proteins have the same overall tertiary folding pattern composed of 12 strands of β-pleated sheet, identified in Fig. 2, in a 4-lobed structure with a pseudo-3-fold rotation axis along the axis of a central 6-stranded β cylinder (Zhu et al., 1991), as shown in Fig. 4.

aFGF and bFGF also share with the IL-1s the absence of recognizable secretory signal sequences (Abraham et al., 1986a,b; Jaye et al., 1986). Virtually all bFGF protein remains sequestered in either normal or transformed cells; only 1–2% is released under normal tissue culture conditions (Sasada et al., 1988). In contrast, bFGF gene constructs containing either immunoglobulin (Rogeli et al., 1988) or growth hormone (Blam et al., 1988) secretory leader sequences appear to be secreted efficiently from transformed cells, as reflected by their transformed phenotype. The means by which leaderless proteins such as these, that presumably act on the external plasma membrane receptor binding sites, become excreted is not certain. Clearly, overt cell death would be expected to release paracrine stimulatory mitogens to support subsequent mitosis and migration, which would facilitate replacement of cells lost by normal turnover and promote repair of damaged tissue. Nonlethal plasma membrane wounding that temporarily allows leakage from cells could be an additional means of release (McNeil et al., 1989). Thus, aFGF and bFGF might act as mitogens that are released locally by damaged cells in a manner that complements plasma- and platelet-derived growth factors that become available after vascular injury.

The other five FGFs all appear to have functional leader sequences, shown schematically in Fig. 1, and either are presumed or, in some cases, have been

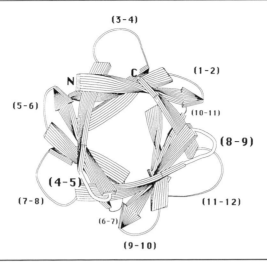

**FIGURE 4**

Schematic tertiary structure of FGF. The β strands of aFGF/bFGF are shown as broad arrows pointing in the amino-to-carboxy terminus direction. The loop nomenclature denotes the β strands, numbered from amino (N) to carboxy (C) terminus (see Fig. 2), that each loop connects. The view is down a 6-stranded antiparallel β cylinder.

demonstrated directly, to be secreted. One measure of secretion is glycosylation, which occurs along the secretory pathway. Neither aFGF nor bFGF appears to be glycosylated. In contrast, transfected cells produce glycosylated INT-2 (Dixon et al., 1989), HST/k-FGF (Delli Bovi et al., 1987), and, based on the presence of an asparagine consensus glycosylation site and larger than expected mass, KGF (Finch et al., 1989). Given the secretory leader and N-glycosylation sequences, both FGF-5 (Zhan et al., 1988) and FGF-6 (de Lapeyiere et al., 1990) also might be glycosylated.

Additional posttranslational modification has been detected, including phosphorylation of both aFGF and bFGF. In dark-isolated retinal rod outer segments, aFGF binds on illumination and is released by an ATP-dependent phosphorylation that is protein kinase C (PKC) but not protein kinase A (PKA) dependent (Plouet et al., 1988; Mascarelli et al., 1989). Phosphorylation of bFGF by cAMP-dependent kinase results in a 3- to 8-fold increase in receptor binding in vitro (Feige and Baird, 1989). PKC and PKA phosphorylate bFGF in vitro on $Ser_{64}$ and $Thr_{112}$, respectively. In the presence of heparin, however, PKC phosphorylation is inhibited and PKA no longer phosphorylates the threonine, but its rate of phosphorylation of $Ser_{64}$ is accelerated (Feige et al., 1989). The physiologic significance of these phosphorylation events is uncertain. The proteins could be phosphorylated prior to release from cells or afterward by ectokinases.

Proteolytic posttranslational modification also occurs. Truncated forms of both aFGF and bFGF have been purified that have up to 20 residues removed from the amino-terminal end (Thomas et al., 1985; Klagsbrun et al., 1987). The crystal structures of aFGF and bFGF indicate that the amino terminus is not required for formation of the compact tertiary structure consistent with the high specific mitogenic activities of these forms (Zhu et al., 1991). The previously described amino-terminally extended forms of bFGF accumulate selectively in the nucleus and appear to be modified, probably by methylation of arginine residues (Sommer et al., 1989).

Two cysteine residues are conserved among all seven known FGFs. Cysteine residues typically are conserved based on their participation in stabilizing disulfide bonds, yet in both aFGF and bFGF, conversion of all of the cysteines to serine residues has no adverse effect on mitogenic activity in culture (Arakawa et al., 1989; Ortega et al., 1991). In fact, disulfide bond formation between these residues in aFGF causes loss of greater than 99% of the mitogenic activity, which is restored completely by subsequent reduction (Linemeyer et al., 1990). One possibility, therefore, is that these cysteine residues are conserved to function in a biologically significant catabolic pathway in vivo.

## IV. RECEPTORS

Both low affinity and high affinity receptors have been observed for aFGF and bFGF. The low affinity receptor ($K_d = 2$ n$M$) appears to be a cell-surface heparan proteoglycan (Moscatelli, 1987). Expression cloning of the hamster FGF-binding heparan sulfate proteoglycan (FGF-HSRG) identified a gene containing a 309-amino-acid open reading frame that is 85% identical to murine syndecan (Kiefer et al., 1990). Syndecan is characterized by a large extracellular domain containing five Ser–Gly glycosaminoglycan attachment sites, a transmembrane spanning sequence, and a small 34-residue intracellular cytoskeletal-binding domain, and thus appears to couple extra- and intracellular interactions. Hamster FGF-HSPG contains six Ser–Gly sequences, five of which are surrounded by the acidic residues typically found in well-characterized glycosaminoglycan attachment sites. The 34-residue internal cellular domain is identical to that of murine syndecan. Despite the similarities between FGF-HSPG and syndecan, Southern blot analysis suggests two closely related FGF-HSPG-hybridizing genes. Therefore, FGF-HSPG and syndecan could be two homologous but separate rather than species-equivalent proteoglycans.

The function of the low affinity heparan proteoglycan binding sites in FGF cell physiology remains somewhat obscure. One possible function of cell-surface heparan proteoglycans could be to partition FGFs rapidly from solution onto cell surfaces, where they would be available at locally higher concentrations to redistribute onto the mitogenically functional high affinity plasma membrane receptors. If the low and high affinity sites are in equilibrium, then by increasing the local concentrations of FGFs in the vicinity of the high affinity sites, these low affinity sites could appear at the macroscopic level to enhance the affinity of the mitogenically functional receptor. Since the receptor and heparin binding sites are on different surface regions of bFGF (Kurokawa et al., 1989), formation of a tertiary complex containing FGF, FGF-HSPG, and an FGF high affinity receptor at the cellular plasma membrane might be possible. In addition, heparins and heparans not only bind to FGFs but have been shown to protect aFGF and bFGF from inactivation by heat, acid (Gospodarowicz and Cheng, 1986), and proteolytic degradation by trypsin, chymotrypsin, and plasmin (Rosengart et al., 1988; Saksela et al., 1988; Sommer and Rifkin, 1989), the last of which could be physiologically significant. The half-life of aFGF at physiologic temperature and pH in tissue culture medium also is increased significantly by heparin (Mueller et al., 1989; Ortega et al., 1991).

Although heparan proteoglycans can protect FGFs from degradation, FGF-HSPG itself contains an extracellular proteolytic cleavage sequence adjacent to the plasma membrane that might, as in the case of syndecan, serve as a processing site to yield a soluble form of the molecule which could act as an FGF carrier (Kiefer et al., 1990). Cleaved FGF-HSPG could diffuse to other cells to be recognized by a high affinity site, become incorporated into extracellular matrices and basement membranes, or be cleared through the circulatory and lymphatic systems. bFGF (Baird and Ling, 1987; Vlodavsky et al., 1987; Bashkin et al., 1989) and aFGF (Baird and Ling, 1987; Weiner and Swain, 1989) have been found associated with extracellular matrix in culture and bFGF has been localized within basement membranes in vivo (Folkman et al., 1988; Bashkin et al., 1989). Binding to these structures appears to be mediated through association of the mitogen with heparan proteoglycan, since it can be competed by heparin or abolished by prior treatment with either heparinase or heparitinase.

The mitogenic activities of FGFs are mediated through high affinity tyrosine kinase-coupled receptors with equilibrium dissociation constants in the range of 10–200 p$M$. Affinity cross-linking experiments typically have revealed multiple bands with masses in the range of 110–150 kDa. Based on partial amino acid sequence information of a purified chicken embryo bFGF receptor, a gene was cloned originally that correspond to a 90-kDa protein, shown in Fig. 5, containing a secretory leader sequence, three extracellular immunoglobulin-like domains, a transmembrane region, and an intracellular tyrosine kinase composed of two tandem domains separated by 14 amino acid residues. An unusual 8-residue polyacidic sequence occurs in the extracellular three-domain growth factor-binding region between the first and second domains. Glycosylation of either some or all of the nine N-glycosylation sites in the extracellular ligand region presumably contributes the extra mass found in the mature receptor (Lee et al., 1989).

This receptor was recognized to be highly homologous to human *flg* (*fms*-like gene), a gene encoding a receptor kinase-like structure located on human chromosome 8 (Dionne et al., 1990a). In fact, chicken bFGF, chicken Cek1 (chicken embryonic kinase; Pasquale et al., 1989), *Xenopus* embryonic FGF receptor (Musci et al., 1990), human bFGF receptor (Itoh et al., 1990), human FLG (Dionne et al., 1990a), and human N-*sam* (Hattori et al., 1990) all appear to be equivalent gene products and will be denoted "type I" receptor in the subsequent discussion. This FGF receptor is expressed in many embryonic tissues, including brain, intestine,

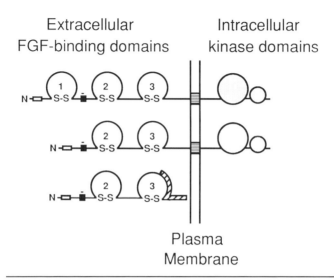

FIGURE 5

Domain structure of FGF receptors. The three- and two-immunoglobulin FGF-binding domain receptors are shown. The immunoglobulin domains contain the characteristic disulfide bond. A unique polyacidic stretch of amino acids (8 and 5 residues in the type I and II receptors, respectively) is present on the amino side of domain 2. Both receptor forms contain FGF-binding domains 2 and 3 but differ in the presence or absence of domain 1. The plasma membrane receptors contain an identical intracellular two-domain tyrosine kinase. A secreted type-I FGF receptor devoid of both a plasma membrane-spanning region and the intracellular tyrosine kinase is shown at the bottom. The carboxy terminal portion of domain 3 differs from that of the plasma membrane-anchored receptors.

gizzard, fibroblasts, lung, and heart, but is detected only in adult brain (Pasquale and Singer, 1989). The expressed receptor binds aFGF and bFGF with $K_d$s of 24 and 47 p$M$, respectively (Dionne et al., 1990a). k-FGF also binds this receptor but with lower affinity (Mansukhani et al., 1990).

Although a three-immunoglobulin-domain extracellular binding region is similar to that identified for the homologous IL-1 receptor, type I FGF receptors with only two extracellular domains also have been identified (Itoh et al., 1990; Johnson et al., 1990). These triple and double extracellular domain forms, shown in Fig. 5, might correspond to the previously observed 145- and 125-kDa affinity-labeled receptor species (Neufeld and Gospodarowicz, 1986). Both the three- and two-immunoglobulin-domain forms bind aFGF and bFGF, so the function of the first domain is not clear, although it could contribute to specificity among various other FGFs. An intron has been mapped between these first two domains, so the extracellular domain pattern presumably arises by alternative splicing of mRNA (Johnson et al., 1990).

A second, or type II, FGF receptor gene, denoted as murine *Bek* (bacterial expressed kinase; Dionne et al., 1990a), chicken Cek3 (Pasquale, 1990), human TK14 (Houssaint et al., 1990), and human K-*sam* (Hattori et al., 1990), has been cloned. This second receptor also recognizes both aFGF and bFGF with $K_d$s of approximately 70 and 115 p$M$, respectively (Dionne et al., 1990a). Mouse type II receptor is expressed in adult brain, liver, lung, and kidney (Kornbluth et al., 1988). Again, both a 90-kDa triple immunoglobulin extracellular domain, containing eight potential $N$-glycosylation sites, and a smaller double immunoglobulin domain form are found. Human type II FGF receptor is 71% identical in amino acid sequence to the corresponding human type I receptor. The tyrosine kinase domain is most similar (88% identity) whereas the second and third extracellular immunoglobulin domains are 74% identical to the type I receptor. In contrast, the dispensable first immunoglobulin domains of the type I and II FGF receptors are only 43% identical (Dionne et al., 1990a).

Finally, a third homologous receptor, Cek2, has been cloned and sequenced; it has 65% and 68% overall homology with the chicken type I and II receptors (Pasquale, 1990). The binding specificity of this receptor is not known, however, so its identity as an FGF receptor has not been proved formally. Nevertheless, a third type of FGF receptor has been defined functionally by its 20-fold tighter binding for KGF and aFGF than for bFGF (Bottaro et al., 1990).

The control of receptor expression is largely unknown. Functional FGF receptor binding is lost on acquisition of the postmitotic phenotype during skeletal muscle terminal differentiation in culture (Olwin and Hauschka, 1988). However, since the decline in receptors lags 8 hr behind the loss of FGF growth responsiveness, receptor loss does not appear to cause commitment to the postmitotic phenotype. In contrast, satellite cells, the residual 1–2% of muscle cells that remain mitotically responsive, still respond to FGF stimulation, clearly implying the persistence of receptor expression in these stem cells (Allen and Boxhorn, 1989). Finally, aFGF appears to up-regulate expression of the human type I receptor mRNA in vascular endothelial cells (Safran et al., 1990). Such induction of receptor expression by the corresponding ligand might be a cellular feedback control loop that promotes replacement of receptors lost by internalization and degradation after ligand binding.

In addition to the normal function of FGF receptors, they also appear to be a portal for the entry of herpes simplex virus type 1 (HSV-1). The virus can bind to one or more of the FGF receptors, inducing tyrosine phosphorylation (Kaner et al., 1990). Although HSV-1 encodes no proteins with significant homology to FGFs, it competes with bFGF for receptor binding. Surprisingly, Western blots of HSV-1 identify an 18-kDa band that cross-reacts with anti-bFGF antisera so bFGF appears to become associated with the virus, perhaps at the time of viral budding from infected cells (Baird et al., 1990).

## V. INHIBITORS

The previously described sequestration of FGFs by extracellular matrix and basement membrane heparan proteoglycans offers one potential means by which these factors can be stored temporarily in a fashion that could lead to later release. Thus, although the lifetime and fate of FGFs bound to these proteoglycans is unknown, they are, perhaps, storage reservoirs rather than sites of terminal inhibition.

In addition to the high affinity tyrosine kinase FGF receptors, two 70-kDa truncated forms of the type I receptor have been identified that contain the second and a modified third immunoglobulin binding domain, as shown in Fig. 5, but neither the plasma membrane spanning region nor the tyrosine kinase domain (Johnson et al., 1990). These truncated receptors could inhibit FGF activity by forming highly specific nonfunctional complexes. Therefore, selective induction of their expression might provide a biologically relevant means of modulating FGF activity in vivo. Normal expression of these receptor-based inhibitors, however, has not yet been documented, so their physiologic significance remains speculative.

Another clearance mechanism for aFGF, bFGF, and perhaps other FGF family members is through their binding to $\alpha_2$-macroglobulin, a large (~800 kDa) multisubunit circulating protein originally recognized for its ability to bind, inhibit, and clear proteases from plasma through a receptor-mediated uptake mechanism. Many proteases have been characterized to bind $\alpha_2$-macroglobulin covalently through activation of an unusual thioester bond. This mechanism apparently is not used to bind FGFs, however, since they complex with a form of the inhibitor that is cleared rapidly from circulation in which this thioester bond is already cleaved (Dennis et al., 1989). Further, neither plasmin nor trypsin competes with bFGF for binding to the inhibitor, so they must bind at different sites. In fact, $\alpha_2$-macroglobulin sequestration of FGF might be by at least two mechanisms, since only half of its binding to bFGF is sensitive to reduction. FGFs are not the only growth factors recognized to bind to $\alpha_2$-macroglobulin. Both PDGF and TGF$\beta$ are bound to this inhibitor, but only TGF$\beta$ competes with FGFs for binding. This plasma clearance mechanism presumably serves to remove circulating FGF, a mitogenic signal that may be intended to function principally in a local, or paracrine, fashion, and thus might account for the absence of detectable FGF mitogenic activity in normal plasma and serum.

## VI. SIGNAL TRANSDUCTION

As is currently the case with all growth factors, the mechanism by which FGFs stimulate mitogenic and nonmitogenic responses is understood only partially. One of the earliest events, if not the initial response to FGF binding to high affinity plasma membrane receptors, is the activation of intracellular receptor tyrosine kinase activity (Huang and Huang, 1986). The means by which the kinase is activated remains elusive. One possibility is that FGF induces clustering among receptor extracellular binding domain regions in the cell membrane, which induces association of intracellular tyrosine kinase domains on adjacent receptors to form a complete active site, much like those formed by subunit interactions in soluble enzymes. Whatever the mechanism, the result is that the receptor becomes either auto- or transphosphorylated and phosphorylates other proteins in a cascade of activation.

In Swiss 3T3 cells, one substrate for rapid FGF-induced tyrosine phosphorylation is a 90-kDa protein of unknown function. Phosphorylation of this protein is detectable within 30 sec of FGF binding and plateaus within 10 min (Coughlin et al., 1988). Another substrate in certain responsive cells is phospholipase C-γ, a 150-kDa enzyme involved in the breakdown of inositol, 1,4,5-triphosphate, with the subsequent release of $Ca^{2+}$ from intracellular stores and diacylglyerol, activating the PKC family of serine/threonine kinases (Burgess et al., 1990). Maximal tyrosine phosphorylation of phospholipase C-γ occurs within 10 min of the addition of aFGF to NIH 3T3 cells and follows the same dose–response curve as mitogenesis, including the high-concentration aFGF inhibition of maximal stimulation. This inositol phosphate pathway is implicated in signal transduction in both EGF- and PDGF-stimulated cells. Conflicting results have been reported, however, concerning accumulation of inositol phosphates in Swiss 3T3 cells in response to FGFs (Brown et al., 1989; Nanberg et al., 1990). Regardless of whether or not inositol phosphates participate in signal transduction in these cells, PKC does appear to be activated. Upon activation, PKC translocates from the cytoplasm to the inside of the plasma membrane where it actively phosphorylates specific protein substrates.

Based on either inhibition by $Ca^{2+}$ sequesterants or enzyme inhibition of PKC, $Ca^{2+}$ mobilization appears to be coupled to the induction of plasminogen activator expression in vascular endothelial cells, whereas PKC activation is correlated with initiation of mitogenesis (Presta et al., 1989). bFGF binding has been observed to result in the phosphorylation of an 80-kDa cellular protein that is a major substrate of PKC (Nanberg et al., 1990). Other serine/threonine kinases that are activated include the 74-kDa *raf*-1 (Morrison et al., 1988) and the 31-kDa S6 (Gavaret et al., 1989), a protein associated with the 40S ribosomal subunit that is suspected of modulating the efficiency of translation. *raf*-1 presumably is not activated directly by the FGF receptor, since phosphorylation is not primarily on tyrosine residues. Likewise, S6 is phosphorylated by a 95-kDa kinase rather than by the FGF receptor directly (Pelech et al., 1986).

The accumulation of *c-fos* and *c-jun*, two nuclear transcription factors associated with response to a variety of mitogenic stimuli, is stimulated by bFGF (Buchou and Mester, 1990). The induction was diminished by staurosporine, an inhibitor of PKC and other serine/threonine kinases, but not by the phorbol ester-mediated down-regulation of PKC. Although the interpretation of this result might be clouded by the recognition of multiple PKCs, it appears to indicate that accumulation of *c-fos* and *c-jun* is the result of a second PKC-independent mechanism. bFGF, like EGF and bombesin, also has been shown to induce the rapid serine phosphorylation of a 33-kDa nuclear nonhistone protein within 10 min that continues to increase in phosphorylation for at least 60 min (Mahadevan et al., 1988).

Additional activities are induced rapidly by FGF binding and are presumably either the direct or the indirect result of the phosphorylation cascade. Receptor binding by either aFGF or bFGF results in 80% inhibition of EGF binding to its receptor within 5 min by a PKC-independent mechanism (Brigstock et al., 1990). Neither the mechanism nor physiologic significance of this cross-modulation of EGF receptor function is clear. Within 5–10 min of FGF binding, the activity of the $Na^+$–$H^+$ exchanger promotes $Na^+$ influx and $H^+$ efflux, resulting in alkalinization of the cytoplasm. The elevated intracellular $Na^+$ is pumped out rapidly by the $Na^+$–$K^+$ pump, with activity that parallels the mitogenic dose–response curve of FGFs (Halperin and Lobb, 1987). The ion fluxes can be viewed as part of an orchestrated cascade of "pleiotypic" responses that prepares a cell for division. Included in this response are transport of nutrients including glucose. bFGF induces a 3- to 4-fold increase in glucose transporter mRNA in BALB/c and NIH 3T3 cells within 3–4 hr of stimulation (Hiraki et al., 1988).

Within 5 min of exposure of vascular endothelial cells to bFGF, the mitogen is detected in cytoplasmic vesicles, presumably "endosomes." Uptake of bFGF into the cytoplasm is continuous throughout the cell cycle, with a maximum in $G_2$. Late in $G_1$, however, up to 10% of the cytoplasmic bFGF is taken up into the nucleus and localized within the nucleolus, the nuclear compartment in which ribosomal gene transcription is centered (Baldin et al., 1990). The function of FGFs in the nucleus is not known, but 1 n$M$ bFGF enhances the activity of RNA polymerase 1 approximately 5-fold in isolated nuclei from sparse quiescent $G_0$ endothelial cells and increases transcription of ribosomal genes (Bouche et al., 1987). Preliminary results indicate that bFGF might act in the nucleolus by activating a cAMP-independent protein kinase, NII, whose specific substrate is the major nonhistone nucleolar protein nucleolin. The endoproteolytic cleavage of phosphorylated nucleolin is required to trigger ribosomal gene transcription.

Although the 18-kDa form of bFGF can enter the nucleus following receptor-mediated internalization, the longer 22- and 24-kDa forms that are initiated at the upstream CUG codons and contain very basic amino-terminal extensions appear to be able to enter the nucleus directly, without prior release and internalization (Renko et al., 1990). The INT-2 protein also is expressed from both AUG and CUG initiation codons. The AUG-initiated form, which contains a functional amino-terminal secretory leader sequence, is directed predominantly into the secretory pathway. In contrast, the CUG-initiated form, which contains a 29-residue amino-terminal extension rich in arginine and proline residues that is reminiscent of yeast nuclear targeting sequences, accumulates in the nucleus (Acland et al., 1990). No function, however, has as yet been observed to be associated with the direct nuclear localization of these elongated forms of INT-2.

The later response of cells to aFGF has been characterized by monitoring the relative abundance of approximately 600 [35]S-labeled polypeptides from cultured rat astroblasts on high resolution isoelectric focusing/SDS polyacrylamide two-dimensional gels (Loret et al., 1989). The intensities of 32 spots were increased, whereas 16 were decreased 18 hr after mitogenic stimulation by aFGF. Of these 48 modulated spots, 41 were similarly altered by treatment with EGF, so they could be part of a general response to mitogenic stimulation. However, 7 spots appeared selectively modulated by aFGF; the abundance of 3 was increased, whereas that of the remaining 4 was decreased. These uniquely aFGF responsive polypeptides might reflect a particular stimulatory pathway induced by this mitogen.

The half-lives of internalized aFGF and bFGF are unusually long compared with those of some other growth factors such as EGF and the homologous IL-1. After 2 hr, conversion of the parent 17-kDa aFGF molecule to 15- and 10-kDa fragments is detected in cultured cells. Within 12 hr, the 17-kDa parent molecule is depleted entirely whereas the 10-kDa form persists for at least 24 hr (Friesel and Maciag, 1988). Internalized bFGF is converted within 2 hr from 18 to 16 kDa, and subsequently degrades with a half-life of 8 hr (Moscatelli, 1988). Intracellular degradation of both aFGF and bFGF presumably occurs in lysosomes since, in the presence of chloroquine, an inhibitor of lysosomal function, proteolytic processing is prevented for at least 24 hr. Neither the reason for the unusual stability nor the mechanistic significance, if any, of the relatively slow intracellular degradation rates of aFGF and bFGF is clear. Resistance to rapid proteolytic degradation conceivably could arise from complexation with heparan proteoglycans that are either resident within cells or internalized with the FGFs.

## VII. BIOLOGIC FUNCTIONS

Growth factors apparently function to support cellular proliferation during embryonic development and subsequent postnatal growth. They also facilitate tissue repair and, presumably, mitosis associated with normal cellular replacement in adults. Perhaps less well recognized and understood is the chemotactic function of growth factors that could augment proliferation in organizing and coordinating cellular interactions in complex multicellular organisms.

Growth factors in general, and FGFs in particular, are relatively abundant during embryonic development. In *Xenopus* oocytes, bFGF mRNA has been detected, decreasing by 95% during maturation and returning to 60% of the early oocyte level at the midblastula stage (Kimelman and Kirschner, 1987). Following fertilization, the early embryo segregates into ectodermal animal and endodermal vegetal poles. Soon afterward, a morphogenic signal from the vegetal pole induces mesoderm formation in the ectoderm of the animal pole. This process can be blocked by heparin (Slack et al., 1987) and is mimicked, in part, by small amounts of bFGF, aFGF, HST/k-FGF or, to a lesser extent, INT-2 (Paterno et al., 1989). FGF-5 also might be involved in mesoderm induction, since the peak in expression of its mRNA either precedes or is coincident with the appearance of mesodermal cells in differentiating teratocarcinoma cells (Herbert et al., 1990).

In *Xenopus*, the type I FGF receptor mRNA can be detected in the oocyte. However, the oocyte is not responsive to FGF, so the mRNA might be under translational or posttranslational control (Musci et al., 1990). In ectodermal explants of *Xenopus* blastulas, the appearance of high affinity receptor binding sites, which peaks at the 1024-cell midblastula stage 6 hr after fertilization, parallels the development of competence to respond to mesoderm induction from the animal pole (Gillespie et al., 1989). At later times, FGF receptor binding, but not receptor mRNA, decreases. FGF receptors also are found in the vegetal zone, however, despite its lack of mesoderm formation.

Purified cloned *Xenopus* oocyte bFGF induces cardiac actin, a marker for mesoderm, but at only 10% or less of the normal levels (Kimelman and Kirschner, 1987). Expression is increased to normal levels by TGFβ1 or TGFβ2, which by themselves do not induce cardiac actin. The natural early embryonic TGFβ form is probably *Xenopus* tissue culture mesoderm-inducing factor (XTC-MIF). Expression of the *Xenopus* homeobox gene *xhox3*, which appears in a concentration gradient in mesoderm to control anteroposterior axis development, is in turn influenced by FGFs (Altaba and Melton, 1989). *Xenopus* bFGF and XTC-MIF have similar minimum active concentrations for induction in animal poles (0.1–1.0 ng/ml) but XTC-MIF is 40 times more active in inducing muscle. In vitro, isolated animal caps are induced to form anterior mesoderm structures such as neural tissue, notochord, and pronephrons by XTC-MIF or to form posterior structures by lower concentrations of XTC-MIF or by bFGF. These results are consistent with bFGF and XTC-MIF being ventroposterior and dorsoanterior inducers, respectively (Green et al., 1990).

Although INT-2 appears to be a weak mesoderm inducer, its mRNA is expressed in newly induced mouse mesoderm migrating through the primitive streak from approximately day 7 to day 9 postconceptus as observed by in situ hybridization (Wilkinson et al., 1988). Once these cells appear in the somites, they no longer express *int*-2 mRNA. The *int*-2 gene is expressed in later mouse fetal development; its mRNA is detected in cerebellum Purkinje cells, retina, sensory regions of the inner ear, and neural crest-derived mesenchyme of teeth (Wilkinson et al., 1989). The mRNA disappears, however, at later stages of growth. HST/k-FGF is expressed during early embryogenesis but not in normal adult tissue. *hst/k-fgf* mRNA is found in mouse embryos up until day 14 postconceptus, but is no longer detectable by day 17 (Yoshida et al., 1988). FGF-6 mRNA is present during mouse development appearing by at least 10.5 days after conception, peaking on day 15.5, and declining by a factor of approximately 3 by birth on day 21 (de Lapeyriere et al., 1990).

The prototypic aFGF and bFGF are observed later in development. Mouse aFGF mRNA is present at 13 days after fertilization in all tissues tested. At 16–17 days after conception, the mRNA levels appear 2–10 times more abundant in lung than in other tissues (Herbert et al., 1990). bFGF immunocrossreactivity has been reported to be distributed widely in 18-day fetal rats (Gonzalez et al., 1990). Intense staining is found in vascular endothelial cells from capillaries and large vessels, a location consistent with bFGF expression in these cells in tissue culture. Immunostaining is found in brain, skeletal and smooth muscle, mesenchyme underlying epithelial tissues, and multiple basement membranes. bFGF has been purified from human placenta, although its distribution and function in this transient tissue is unknown (Gospodarowicz et al., 1985a; Moscatelli et al., 1986).

Expression of FGF genes is not limited to embryonic development. Both aFGF

(Thomas et al., 1984; Gimenez-Gallego et al., 1986) and bFGF (Lemmon and Bradshaw, 1983; Gospodarowicz et al., 1984) have been purified from adult brain. In addition, aFGF has been purified from heart (Quinkler et al., 1989; Sasaki et al., 1989), kidney (Gautschi-Sova et al., 1987), omentum (Ohtaki et al., 1989), and uterus (Brigstock et al., 1990). Specific immunocrossreactivity or mRNA hybridization has been detected in CNS neurons including retinal neurons (Jacquemin et al., 1990), cardiac myocytes (Speir et al., 1988; Sasaki et al., 1989; Weiner and Swaim, 1989), vascular smooth muscle cells, dermal fibroblasts (Winkles et al., 1987), bone (Hauschka et al., 1986), and ovary (Koos and Seidel, 1989). bFGF has been purified from pituitary (Lemmon and Bradshaw, 1983; Bohlen et al., 1984; Gospodarowicz et al., 1984), cartilage (Sullivan and Klagsbrun, 1985), liver (Ueno et al., 1986), adrenal glands (Gospodarowicz et al., 1986), uterus (Brigstock et al., 1990), and testes (Ueno et al., 1987), and has been identified in heart (Quinkler et al., 1989), cardiac myocytes (Speir et al., 1988), capillary (Schweigerer et al., 1987) and aortic vascular endothelial cells (Casey et al., 1986), pituitary follicular cells (Ferrara et al., 1987), bone (Hauschka et al., 1986), retinal pigmented epithelial cells (Sommer et al., 1989), and corneal endothelial cells (Vlodavsky et al., 1987; Folkman et al., 1988). Expression of other FGF family members can be detected after birth. Human FGF-5 was cloned originally from a brain cDNA library generated from tissue of a 1-day-old neonate (Zhan et al., 1988). FGF-6 mRNA is found in adult mouse skeletal muscle, heart, and testis (de Lapeyriere et al., 1990). KGF mRNA has been detected in adults, specifically in kidney, gastrointestinal tract, and dermis (Finch et al., 1989). The latter source might reflect the ability of KGF secreted from dermal cells to drive proliferation of epidermal keratinocytes in culture.

The spectrum of target cells for FGFs in culture is remarkably broad. Most of the responsive cells have been identified with either aFGF or bFGF. Since these two FGFs act through at least some common receptors, it is not surprising that they stimulate many of the same cells. In general, these FGFs drive DNA synthesis and proliferation of cells of mesodermal and ectodermal origin. Specifically, one, or both, of these FGFs has been shown to stimulate dermal (Shipley et al., 1988) and continuous lines of fibroblasts (Thomas et al., 1984), epidermal keratinocytes (Shipley et al., 1988), vascular endothelial cells (Thomas et al., 1985; Gospodarowicz et al., 1985b), vascular smooth muscle cells (Gospodarowicz et al., 1985b, Winkles et al., 1987), muscle satellite cells (Allen and Boxhorn, 1989), chondrocytes (Kato and Gospodarowicz, 1985), osteoblasts (Rodan et al., 1987), corneal epithelial and stromal cells (Hecquet et al., 1990), corneal endothelial cells (Gospodarowicz et al., 1985b), lens epithelial cells (Moenner et al., 1986; Uhlrich et al., 1986), lung alveolar type II cells (Leslie et al., 1990), hepatocytes (Kan et al., 1989), adrenal cortical cells, granulosa cells (Gospodarowicz et al., 1985b), and prostatic epithelial cells (Crabb et al., 1986). FGFs clearly have potentially important roles in the central nervous system as described in detail in Chapter 10. In the nervous system, FGFs drive mitogenesis of astrocytes (Perraud et al., 1988), oligodendrocytes (Besnard et al., 1989), and neuroblasts (Gensburger et al., 1987; Wu et al., 1988). In addition, they have been shown to promote survival and neurite outgrowth of some but not all neurons (Morrison et al., 1986; Walicke et al., 1986; Lipton et al., 1988). This is a remarkably diverse set of responsive cells and is undoubtedly only a partial list. Control of FGF responsiveness in vivo might, therefore, be exerted in large part at the level of local expression. Multiple functions for a single type of activity could have contributed to the generation of a large family of homologous FGFs that have significant overlap of activity.

FGFs, like other growth factors, are also chemotactic for at least some of these same target cells. Specifically, both aFGF (Terranova et al., 1985) and bFGF (Moscatelli et al., 1986; Presta et al., 1986) are chemotactic in culture for vascular endothelial cells. In addition, aFGF has been demonstrated to promote migration of fibroblasts and astroglia (Senior et al., 1986). Although the list of chemotactically responsive cells is substantially shorter than that of those that respond mitogenically, no reports exist of cells that divide but do not migrate in response to FGFs. Therefore, many, if not all, mitogenically responsive cells also might be induced to migrate by these factors. Nevertheless, neither the identity of the specific receptors nor the mechanism of chemotaxis is known.

These mitogenic and chemotactic activities in culture have, in some cases, correlative functions in vivo. Both aFGF (Thomas et al., 1985; Lobb et al., 1985; Herbert et al., 1988; Thompson et al., 1989) and bFGF (Eppley et al., 1988; Herbert et al., 1988) have been observed to support blood vessel growth in animal models of angiogenesis. Also, both have been shown to accelerate dermal wound healing (Davidson et al., 1985; Mellin et al., 1988), a process that is probably dependent on not only angiogenesis but also mitogenesis of fibroblasts and epidermal keratinocytes. bFGF induces growth and/or migration of vascular endothelial cells in damaged arteries (Lindner et al., 1990) and aFGF is able to facilitate arterial repair in animal models (Dryjski et al., 1988). A mutant of bFGF has been described that promotes healing of chemically induced duodenal ulcers (Folkman et al., 1990). aFGF levels are increased dramatically in regenerating liver (Kan et al., 1989). Expression of both aFGF (Logan, 1988; Nieto-Sampedro et al., 1988), and bFGF (Finklestein et al., 1988) increases at sites of brain injury. Given the long lists of cellular sources and responsive cells, many other tissue reparative functions can be anticipated that might be of therapeutic use.

Inevitably, this same broad responsiveness of normal and injured tissues is reflected in pathologic activities. Most notably the FGFs, like virtually all other known growth factors, can if inappropriately and persistently expressed promote transformation, as reviewed by Ortega and Thomas (1990). Overexpression of both *hst/k-fgf* and *int*-2 genes, encoding secretable FGFs normally found only in relative abundance during development, has been identified in selected tumors. Integration of mouse mammary tumor virus (MMTV) into DNA sites near either of these two FGFs is correlated with their increased expression and the resulting induction of phenotypic transformation. Perhaps because of their very close proximity in the genome, *hst/k-fgf* and *int*-2 genes occasionally are coamplified in tumors. Further, transgenic mice that carry *int*-2 driven by the MMTV promoter selectively express *int-2* in enlarged mammary and prostate glands (Muller et al., 1990). The relevance of these observations to spontaneous human mammary carcinomas, prostatic cancers, or benign prostatic hyperplasia remains speculative.

The animal biology and physiology of fibroblast growth factors is clearly in a very early stage. All seven FGF family members have been identified at the molecular level in the past decade. No compelling reason exists to think that all the FGFs or all their receptors have been discovered. Moreover, elucidating the controls and functions of the members of this uniquely large family of protein growth factors will, in the years to come, advance our understanding of animal embryology and physiology substantially and offer insights into the therapeutic use of these proteins.

## REFERENCES

Abraham, J. A., Mergia, A., Whang, J. L., Tumolo, A., Friedman, J., Hjerrild, K. A., Gospodarowicz, D., and Fiddes, J. C. (1986a). Nucleotide sequence of a bovine clone encoding the angiogenic protein, basic fibroblast growth factor. *Science* **233**, 545–548.

Abraham, J. A., Whang, J. L., Tumolo, A., Mergia, A., Friedman, J., Gospodarowicz, D., and Fiddes, J. C. (1986b). Human basic fibroblast growth factor: Nucleotide sequence and genomic organization. *EMBO J.* **5**, 2523–2528.

Acland, P., Dixon, M., Peters, G., and Dickson, C. (1990). Subcellular fate of the int-2 oncoprotein is determined by choice of initiation codon. *Nature (London)* **343**, 662–665.

Adelaide, J., Mattei, M.-G., Marcis, I., Raybaud, F., Planche, J., Lapeyriere, O. D., and Birnbaum, D. (1988). Chromosomal localization of the *hst* oncogene and its co-amplification with the *int-2* oncogene in a human melanoma. *Oncogene* **2**, 413–416.

Albert, P., Boilly, B., Courty, J., and Barritault, D. (1987). Stimulation in cell culture of mesenchymal cells of newt limb blastemas by EDGF I or II (basic or acidic FGF). *Cell Diff.* **21**, 63–68.

Allen, R. E., and Boxhorn, L. K. (1989). Regulation of skeletal muscle satellite cell proliferation and differentiation by transforming growth factor-β, insulin-like growth factor I, and fibroblast growth factor. *J. Cell. Physiol.* **138**, 311–315.

Altaba, A. R. I., and Melton, D. A. (1989). Interaction between peptide growth factors and homeobox genes in the establishment of antero-posterior polarity in frog embryos. *Nature (London)* **341**, 33–38.

Arakawa, T., Hsu, Y.-R., and Schiffer, S. G. (1989). Characterization of a cysteine-free analog of recombinant human basic fibroblast growth factor. *Biochem. Biophys. Res. Commun.* **161**, 335–341.

Baird, A., and Ling, N. (1987). Fibroblast growth factors are present in the extracellular matrix produced by endothelial cells *in vitro*: Implications for a role of heparinase-like enzymes in the neovascular response. *Biochem. Biophys. Res. Commun.* **142**, 428–435.

Baird, A., Florkiewicz, R. Z., Maher, P. A., Kaner, R. J., and Hajjar, D. P. (1990). Mediation of virion penetration into vascular cells by association of basic fibroblast growth factor with herpes simplex virus type 1. *Nature (London)* **348**, 344–346.

Baldin, V., Roman, A.-M., Bosc-Bierne, I., Amalric, F., and Bouche, G. (1990). Translocation of bFGF to the nucleus is G1 phase cell cycle specific in bovine aortic endothelial cells. *EMBO J.* **9**, 1511–1517.

Bashkin, P., Doctrow, S., Klagsbrun, M., Svahn, C. M., Folkman, J., and Vlodavsky, I. (1989). Basic fibroblast growth factor binds to subendothelial extracellular matrix and is released by heparitinase and heparin-like molecules. *Biochemistry* **28**, 1737–1743.

Besnard, F., Perraud, F., Sensenbrenner, M., and Labourdette, G. (1989). Effects of acidic and basic fibroblast growth factors on proliferation and maturation of cultured rat oligodendrocytes. *Int. J. Dev. Neurosci.* **7**, 401–409.

Blam, S. B., Mitchell, R., Tischer, E., Rubin, J. S., Silva, M., Silver, S., Fiddes, J. C., Abraham, J. A., and Aaronson, S. A. (1988). Addition of growth hormone secretion signal to basic fibroblast growth factor results in cell transformation and secretion of aberrant forms of the protein. *Oncogene* **3**, 129–136.

Bohlen, P., Baird, A., Esch, F., Ling, N., and Gospodarowicz, D. (1984). Isolation and partial molecular characterization of pituitary fibroblast growth factor. *Proc. Natl. Acad. Sci. U.S.A.* **81**, 5364–5368.

Bottaro, D. P., Rubin, J. S., Ron, D., Finch, P. W., Florio, C., and Aaronson, S. A. (1990). Characterization of the receptor for keratinocyte growth factor. *J. Biol. Chem.* **265**, 12767–12770.

Bouche, G., Gas, I., Prats, H., Baldin, V., Tauber, J.-P., Teissie, J., and Amalric, F. (1987). Basic fibroblast growth factor enters the nucleolus and stimulates the transcription of ribosomal genes in ABAE cells undergoing $G_0$ to $G_1$ transition. *Proc. Natl. Acad. Sci. U.S.A.* **84**, 6770–6774.

Brigstock, D. R., Heap, R. B., Barker, P. J., and Brown, K. D. (1990). Purification and characterization of heparin-binding growth factors from porcine uterus. *Biochem. J.* **266**, 273–282.

Brockes, J. P. (1984). Mitogenic growth factors and nerve dependence of limb regeneration. *Science* **225**, 1280–1287.

Brown, K. D., Blakeley, D. M., and Brigstock, D. R. (1989). Stimulation of polyphosphoinositide hydrolysis in Swiss 3T3 cells by recombinant fibroblast growth factors. *FEBS Lett.* **247**, 227–231.

Buchou, T., and Mester, J. (1990). Fibroblast growth factor-dependent mitogenic signal transduction pathway in chemically transformed mouse fibroblasts is similar to but distinct from that initiated by phorbol esters. *J. Cell. Physiol.* **142**, 559–565.

Burgess, W. H., and Maciag, T. (1989). The heparin-binding (fibroblast) growth factor family of proteins. *Ann. Rev. Biochem.* **58**, 575–606.

Burgess, W. H., Dionne, C. A., Kaplow, J., Mudd, R., Friesel, R., Zilberstein, A., Schlessinger, J., and Jaye, M. (1990). Characterization and cDNA cloning of phospholipase C-γ, a major substrate for heparin-binding growth factor 1 (acidic fibroblast growth factor)-activated tyrosine kinase. *Mol. Cell. Biol.* **10**, 4770–4777.

Casey, G., Smith, R., McGillivray, D., Peters, G., and Dickson, C. (1986). Characterization and chromosome assignment of the human homolog *int*-2, a potential proto-oncogene. *Mol. Cell. Biol.* **6**, 502–510.

Chiu, I.-M., Wang, W.-P., and Lehtoma, K. (1990). Alternative splicing generates two forms of mRNA coding for human heparin-binding growth factor 1. *Oncogene* **5**, 755–762.

Coughlin, S. R., Barr, P. J., Cousens, L. S., Ferretto, L. J., and Williams, L. T. (1988). Acidic and basic fibroblast growth factors stimulate tyrosine kinase activity *in vivo*. *J. Biol. Chem.* **263**, 988–993.

Crabb, J. W., Armes, L. G., Carr, S. A., Johnson, C. M., Roberts, G. D., Bordoli, R. S., and McKeehan, W. L. (1986). Complete primary structure of prostatropin, a prostate epithelial cell growth factor. *Biochemistry* **25**, 4988–4993.

Crumley, G. R., Howk, R., Ravera, M. W., and Jaye, M. (1989). Multiple polyadenylation sites downstream from the human aFGF gene encoding acidic fibroblast growth factor. *Gene* **85**, 489–497.

Crumley, G., Dionne, C. A., and Jaye, M. (1990). The gene for human acidic fibroblast growth factor encodes two upstream exons alternatively spliced to the first coding exon. *Biochem. Biophys. Res. Commun.* **171**, 7–13.

Curatola, A. M., and Basilico, C. (1990). Expression of the k-fgf proto-oncogene is controlled by 3′regulatory elements which are specific for embryonal carcinoma cells. *Mol. Cell. Biol.* **10**, 2475–2484.

Davidson, J. M., Klagsbrun, M., Hill, K. E., Buckley, A., Sullivan, R., Brewer, P. S., and Woodward, S. C. (1985). Accelerated wound repair, cell proliferation, and collagen accumulation are produced by a cartilage-derived growth factor. *J. Cell. Biol.* **100**, 1219–1227.

Delli Bovi, P. D., Curatola, A. M., Kern, F. G., Greco, A., Ittmann, M., and Basilico, C. (1987). An oncogene isolated by transfection of Kaposi's sarcoma DNA encodes a growth factor that is a member of the FGF family. *Cell* **50**, 729–737.

Dennis, P. A., Saksela, O., Harpel, P., and Rifkin, D. B. (1989). α2-Macroglobulin is a binding protein for basic fibroblast growth factor. *J. Biol. Chem.* **264**, 7210–7216.

Dionne, C. A., Crumley, G., Bellot, F., Kaplow, J. M., Searfoss, G., Ruta, M., Burgess, W. H., Jaye, M., and Schlessinger, J. (1990a). Cloning and expression of two distinct high-affinity receptors cross-reacting with acidic and basic fibroblast growth factors. *EMBO J.* **9**, 2685–2692.

Dionne, C. A., Kaplan, R., Seuanez, H., O'Brien, S. J., and Jaye, M. (1990b). Chromosome assignment by polymerase chain reaction techniques: Assignment of the oncogene FGF-5 to human chromosome 4. *BioTechniques* **8**, 190–194.

Dixon, M., Deed, R., Acland, P., Moore, R., Whyte, A., Peters, G., and Dickson, C. (1989). Detection and characterization of the fibroblast growth factor-related oncoprotein INT-2. *Mol. Cell. Biol.* **9**, 4896–4902.

Dryjski, M., Bjornsson, T. D., Linemeyer, D., and Thomas, K. A. (1988). Recombinant human acidic fibroblast growth factor inhibits intimal hyperplasia in a vascular injury model. *Clin. Res.* **36**, A273.

Eppley, B. L., Doucet, M., Connolly, D. T., and Feder, J. (1988). Enhancement of angiogenesis by bFGF in mandibular bone graft healing in the rabbit. *J. Oral Maxil. Surg.* **46**, 391–398.

Esch, F., Baird, A., Ling, N., Ueno, N., Klepper, R., Gospodarowicz, D., Bohlen, P., and Guillemin, R. (1985). Primary structure of bovine basic fibroblast growth factor (FGF) and comparison with the amino-terminal sequence of bovine brain acidic FGF. *Proc. Natl. Acad. Sci. U.S.A.* **82**, 6507–6511.

Feige, J. J., and Baird, A. (1989). Basic fibroblast growth factor is a substrate for protein phosphorylation and is phosphorylated by capillary endothelial cells in culture. *Proc. Natl. Acad. Sci. U.S.A.* **86**, 3174–3178.

Feige, J.-J., Bradley, J. D., Fryburg, K., Farris, J., Cousens, L. C., Barr, P. J., and Baird, A. (1989). Differential effects of heparin, fibronectin, and laminin on the phosphorylation of basic fibroblast growth factor by protein kinase C and the catalytic subunit of protein kinase A. *J. Cell Biol.* **109**, 3105–3114.

Ferrara, N., and Henzel, W. J. (1989). Pituitary follicular cells secrete a novel heparin-binding growth factor specific for vascular endothelial cells. *Biochem. Biophys. Res. Commun.* **161**, 851–858.

Ferrara, N., Schweigerer, L., Neufeld, G., Mitchell, R., and Gospodarowicz, D. (1987). Pituitary follicular cells produce basic fibroblast growth factor. *Proc. Natl. Acad. Sci. U.S.A.* **84**, 5773–5777.

Finch, P. W., Rubin, J. S., Miki, T., Ron, D., and Aaronson, S. A. (1989). Human KGF is FGF-related with properties of a paracrine effector of epithelial cell growth. *Science* **245**, 752–755.

Finklestein, S. P., Apostolides, P. J., Caday, C. G., Prosser, J., Philips, M. F., and Klagsbrun, M. (1988). Increased basic fibroblast growth factor (bFGF) immunoreactivity at the site of focal brain wounds. *Brain Res.* **460**, 253–259.

Florkiewicz, R. F., and Sommer, A. (1989). Human basic fibroblast growth factor gene encodes four polypeptides: Three initiate translation from non-AUG codons. *Proc. Natl. Acad. Sci. U.S.A.* **86**, 3978–3981.

Folkman, J., Klagsbrun, M., Sasse, J., Wadzinski, M., Ingber, D., and Vlodavsky, I. (1988). A heparin-binding angiogenic protein—basic fibroblast growth factor—is stored within basement membrane. *Am. J. Pathol.* **130**, 393–400.

Folkman, J., Szabo, S., Vattay, P., Morales, R. E., Pinkus, G., and Kato, K. (1990). Effect of orally administered bFGF on healing of chronic duodenal ulcers, gastric secretion and acute mucosal lesions in rats. *Gastroenterology* **98**, A45.

Friesel, R., and Maciag, T. (1988). Internalization and degradation of heparin binding growth factor-I by endothelial cells. *Biochem. Biophys. Res. Commun.* **151**, 957–964.

Gautschi-Sova, P., Jiang, Z.-P., Frater-Schroder, M., and Bohlen, P. (1987). Acidic fibroblast growth factor is present in nonneural tissue: Isolation and chemical characterization from bovine kidney. *Biochemistry* **26**, 5844–5847.

Gavaret, J. M., Matricon, C., Pomerance, M., Jacquemin, C., Toru-Delbauffe, D., and Pierre, M. (1989). Activation of S6 kinase in astroglial cells by FGFa and FGFb. *Dev. Brain Res.* **45**, 77–82.

Gensburger, C., Labourdette, G. and Sensenbrenner, M. (1987). Brain basic fibroblast growth factor stimulates the proliferation of rat neuronal precursor cells in vitro. *FEBS Lett.* **217**, 1–5.

Gillespie, L. L., Paterno, G. D., and Slack, J. M. W. (1989). Analysis of competence: Receptors for fibroblast growth factor in early *Xenopus* embryos. *Development* **106**, 203–208.

Gimenez-Gallego, G., Rodkey, J., Bennett, C., Rios-Candelore, M., DiSalvo, J., and Thomas, K. A. (1985). Brain-derived acidic fibroblast growth factor: Complete amino acid sequence and homologies. *Science* **230**, 1385–1388.

Gimenez-Gallego, G., Conn, G., Hatcher, V. B., and Thomas, K. A. (1986). Human brain-derived acidic and basic fibroblast growth factors: Amino terminal sequences and specific mitogenic activities. *Biochem. Biophys. Res. Commun.* **135**, 541–548.

Gonzalez, A.-M., Buscaglia, M., Ong, M., and Baird, A. (1990). Distribution of basic fibroblast growth factor in the 18-day rat fetus: Localization in the basement membranes of diverse tissues. *J. Cell Biol.* **110**, 753–765.

Gospodarowicz, D., and Cheng, J. (1986). Heparin protects basic and acidic FGF from inactivation. *J. Cell Physiol.* **128**, 475–484.

Gospodarowicz, D., Cheng, J., Lui, G.-M., Baird, A., and Bohlen, P. (1984). Isolation of brain fibroblast growth factor by heparin–Sepharose affinity chromatography: Identity with pituitary fibroblast growth factor. *Proc. Natl. Acad. Sci. U.S.A.* **81**, 6963–6967.

Gospodarowicz, D., Cheng, J., Lui, G.-M., Fujii, D. K., Baird, A., and Bohlen, P. (1985a). Fibroblast growth factor in the human placenta. *Biochem. Biophys. Res. Commun.* **128**, 554–562.

Gospodarowicz, D., Massoglia, S., Cheng, J., Lui, G.-M., and Bohlen, P. (1985b). Isolation of bovine pituitary fibroblast growth factor by fast protein liquid chromatography (FPLC) partial chemical and biological characterization. *J. Cell. Physiol.* **122**, 323–332.

Gospodarowicz, D., Baird, A., Cheng, J., Lui, G. M., Esch, F., and Bohlen, P. (1986). Isolation of fibroblast growth factor from bovine adrenal gland: Physiochemical and biological characterization. *Endocrinology* **118**, 82–90.

Green, J. B. A., Howes, G., Symes, K., Cooke, J., and Smith, J. C. (1990). The biological effects of XTC-MIF: Quantitative comparison with *Xenopus* bFGF. *Development* **108**, 173–183.

Halperin, J. A., and Lobb, R. R. (1987). Effect of heparin-binding growth factors on monovalent cation transport in Balb/C 3T3 cells. *Biochem. Biophys. Res. Commun.* **144**, 115–122.

Hattori, Y., Odagiri, H., Nakatani, H., Miyagawa, H., Naito, K., Sakamoto, H., Katoh, O., Yoshida, T., Sugimura, T., and Terada, M. (1990). K-sam, an amplified gene in stomach cancer, is a member of the heparin-binding growth factor receptor genes. *Proc. Natl. Acad. Sci. U.S.A.* **87**, 5983–5987.

Hauschka, P. V., Mavrakos, A. E., Iafrati, M. D., Doleman, S. E., and Klagsbrun, M. (1986). Growth factors in bone matrix: Isolation of multiple types by affinity chromatography on heparin–Sepharose. *J. Biol. Chem.* **261**, 12665–12674.

Hecquet, C., Morisset, S., Lorans, G., Plouet, J., and Adolphe, M. (1990). Effects of acidic and basic fibroblast growth factors on the proliferation of rabbit corneal cells. *Curr. Eye Res.* **9**, 429–433.

Herbert, J. M., Laplace, M. C., and Maffrand, J. P. (1988). Effect of heparin on the angiogenic potency of basic and acidic fibroblast growth factors in the rabbit cornea assay. *Int. J. Tiss. Reac.* **10**, 133–139.

Herbert, J. M., Basilico, C., Goldfarb, M., Haub, O., and Martin, G. R. (1990). Isolation of cDNAs encoding four mouse FGF family members and characterization of their expression patterns during embryogenesis. *Dev. Biol.* **138**, 454–463.

Hiraki, Y., Rosen, O. M., and Birnbaum, M. J. (1988). Growth factors rapidly induce expression of the glucose transporter gene. *J. Biol. Chem.* **263**, 13655–13662.

Houssaint, E., Blanquet, P. R., Champion-Arnaud, P., Gesnel, M. C., Torriglia, A., Courtois, Y., and Breathnach, R. (1990). Related fibroblast growth factor receptor genes exist in the human genome. *Proc. Natl. Acad. Sci. U.S.A.* **87**, 8180–8184.

Huang, S. S., and Huang, J. S. (1986). Association of bovine brain-derived growth factor receptor with protein tyrosine kinase activity. *J. Biol. Chem.* **261**, 9568–9571.

Iberg, N., Rogelj, S., Fanning, P., and Klagsbrun, M. (1989). Purification of 18- and 22-kDa forms of basic fibroblast growth factor from rat cells transformed by the *ras* oncogene. *J. Biol. Chem.* **264**, 19951–19955.

Itoh, N., Terachi, T., Ohta, M., and Seo, M. K. (1990). The complete amino acid sequence of the shorter form of human basic fibroblast growth factor receptor deduced from its cDNA. *Biochem. Biophys. Res. Commun.* **169**, 680–685.

Jacquemin, E., Halley, C., Alterio, J., Laurent, M., Courtois, Y., and Jeanny, J. C. (1990). Localization of acidic fibroblast growth factor (aFGF) mRNA in mouse and bovine retina by *in situ* hybridization. *Neurosci. Lett.* **116**, 23–28.

Jaye, M., Howk, R., Burgess, W., Ricca, G. A., Chiu, I.-M., Ravera, M. W., O'Brien, S. J., Modi, W. S., Maciag, T., and Drohan, W. N. (1986). Human endothelial cell growth factor: Cloning, nucleotide sequence, and chromosome localization. *Science* **233**, 541–544.

Johnson, D. E., Lee, P. L., Lu, J., and Williams, L. T. (1990). Diverse forms of a receptor for acidic and basic fibroblast growth factors. *Mol. Cell. Biol.* **10**, 4728–4736.

Kan, M., Huang, J., Mansson, P.-E., Yasumitsu, H., Carr, B., and McKeehan, W. L. (1989). Heparin-binding growth factor type 1 (acidic fibroblast growth factor): A potential biphasic autocrine and paracrine regulator of hepatocyte regeneration. *Proc. Natl. Acad. Sci. U.S.A.* **86**, 7432–7436.

Kaner, R. J., Baird, A., Mansukhani, A., Basilico, C., Summers, B. D., Florkiewicz, R. Z., and Hajjar, D. P. (1990). Fibroblast growth factor receptor is a portal of cellular entry for herpes simplex virus type 1. *Science* **248**, 1410–1413.

Kato, Y., and Gospodarowicz, D. (1985). Sulfated proteoglycan synthesis by confluent cultures of rabbit costal chondrocytes grown in the presence of fibroblast growth factor. *J. Cell Biol.* **100**, 477–485.

Kiefer, M. C., Stephans, J. C., Crawford, K., Okino, K., and Barr, P. J. (1990). Ligand-affinity cloning and structure of a cell surface heparan sulfate proteoglycan that binds basic fibroblast growth factor. *Proc. Natl. Acad. Sci. U.S.A.* **87**, 6985–6989.

Kimelman, D., and Kirschner, M. (1987). Synergistic induction of mesoderm by FGF and TGF-β and the identification of an mRNA coding for FGF in the early *Xenopus* embryo. *Cell* **51**, 869–877.

Klagsbrun, M. (1989). The fibroblast growth factor family: Structural and biological properties. *Progr. Growth Factor Res.* **1**, 207–235.

Klagsbrun, M., Smith, S., Sullivan, R., Shing, Y., Davidson, S., Smith, J. A., and Sasse, J. (1987). Multiple forms of basic fibroblast growth factor: Amino-terminal cleavages by tumor cell- and brain cell-derived acid proteinases. *Proc. Natl. Acad. Sci. U.S.A.* **84**, 1839–1843.

Koos, R., and Seidel, R. (1989). Detection of acidic fibroblast growth factor mRNA in the rat ovary using reverse transcription–polymerase chain reaction amplification. *Biochem. Biophys. Res. Commun.* **165**, 82–88.

Kornbluth, S., Paulson, K. E., and Hanafusa, H. (1988). Novel tyrosine kinase identified by phosphotyrosine antibody screening of cDNA libraries. *Mol. Cell. Biol.* **8**, 5541–5544.

Kurokawa, M., Doctrow, S. R., and Klagsbrun, M. (1989). Neutralizing antibodies inhibit the binding of basic fibroblast growth factor to its receptor but not to heparin. *J. Biol. Chem.* **264**, 7686–7691.

Lagente, O., Diry, M., and Courtois, Y. (1986). Isolation of heparin-binding growth factors from dogfish (*Mustela canis*) brain and retina. *FEBS Lett.* **202**, 207–210.

de Lapeyriere, O., Rosnet, O., Benharroch, D., Raybaud, F., Marchetto, S., Planche, J., Galland, F., Mattei, M.-G., Copeland, N. G., Jenkins, N. A., Coulier, F., and Birnbaum, D. (1990). Structure, chromosome mapping and expression of the murine fgf-6 gene. *Oncogene* **5**, 823–831.

Lee, P., Johnson, D. E., Cousens, L. S., Fried, V. A., and Williams, L. T. (1989). Purification and complementary DNA cloning of a receptor for basic fibroblast growth factor. *Science* **245**, 57–60.

Lemmon, S. K., and Bradshaw, R. A. (1983). Purification and partial characterization of bovine pituitary fibroblast growth factor. *J. Cell. Biochem.* **21**, 195–208.

Leslie, C. C., McCormick-Shannon, K., and Mason, R. J. (1990). Heparin-binding growth factors stimulate DNA synthesis in rat alveolar type II cells. *Am. J. Respir. Cell Mol. Biol.* **2**, 99–106.

Lindner, V., Majack, R. A., and Reidy, M. A. (1990). Basic fibroblast growth factor stimulates endothelial regrowth and proliferation in denuded arteries. *J. Clin. Invest.* **85**, 2004–2008.

Linemeyer, D. L., Menke, J. G., Kelly, L. J., DiSalvo, J., Soderman, D., Schaeffer, M.-T., Ortega, S., Gimenez-Gallego, G., and Thomas, K. A. (1990). Disulfide bonds are neither required, present, nor compatible with full activity of human recombinant acidic fibroblast growth factor. *Growth Factors* **3**, 287–298.

Lipton, S. A., Wagner, J. A., Madison, R. D., and D'Amore, P. (1988). Acidic fibroblast growth factor enhances regeneration of processes by postnatal mammalian retinal ganglion cells in culture. *Proc. Natl. Acad. Sci. U.S.A.* **85**, 2388–2392.

Lobb, R. R., and Fett, J. W. (1984). Purification of two distinct growth factors from bovine neural tissue by heparin affinity chromatography. *Biochemistry* **23**, 6295–6299.

Lobb, R. R., Alderman, E. M., and Fett, J. W. (1985). Induction of angiogenesis by bovine class 1 heparin-binding growth factor. *Biochemistry* **24**, 4969–4973.

Logan, A. (1988). Elevation of acidic fibroblast growth factor mRNA in lesioned rat brain. *Mol. Cell. Endocrinol.* **58**, 275–278.

Loret, C., Sensenbrenner, M., and Labourdette, G. (1989). Differential phenotypic expression induced in cultured rat astroblasts by acidic fibroblast growth factor, epidermal growth factor, and thrombin. *J. Biol. Chem.* **264**, 8319–8327.

McNeil, P. L., Muthukrishman, L., Warder, E., and D'Amore, P. A. (1989). Growth factors are released by mechanically wounded endothelial cells. *J. Cell Biol.* **109**, 811–822.

Mahadevan, L. C., Heath, J. K., Leichtfried, F. E., Cumming, D. V. E., Hirst, E. M. A., and Foulkes, J. G. (1988). Rapid appearance of novel phosphoproteins in the nuclei of mitogen-stimulated fibroblasts. *Oncogene* **2**, 249–257.

Mansukhani, A., Moscatelli, D., Talarico, D., Levytska, V., and Basilico, C. (1990). A murine fibroblast growth factor (FGF) receptor expressed in CHO cells is activated by basic FGF and Kaposi FGF. *Proc. Natl. Acad. Sci. U.S.A.* **87**, 4378–4382.

Marics, I., Adelaide, J., Raybaud, F., Mattei, M.-G., Coulier, F., Planche, J., de Lapeyriere, O., and Birnbaum, D. (1989). Characteristic of the *hst*-related FGF-6 gene, a new member of the fibroblast growth factor gene family. *Oncogene* **4**, 335–340.

Mascarelli, F., Raulais, D., and Courtois, Y. (1989). Fibroblast growth phosphorylation and receptors in rod outer segments. *EMBO J.* **8**, 2265–2273.

Mellin, T. N., Busch, R. D., Capparella, J., DiSalvo, J., Linemeyer, D., Huber, C., Van Zwieten, M. J., Frank, J., and Thomas, K. A. (1988). Enhancement of wound closure and breaking strength in rodents with acidic fibroblast growth factor. *J. Invest. Dermatol.* **90**, 587.

Mergia, A., Eddy, R., Abraham, J. A., Fiddes, J. C., and Shows, T. B. (1986). The genes for basic and acidic fibroblast growth factors are on different human chromosomes. *Biochem. Biophys. Res. Commun.* **138**, 644–651.

Mergia, A., Tischer, E., Graves, D., Tumolo, A., Miller, J., Gospodarowicz, D., Abraham, J. A., Shipley, G. D., and Fiddes, J. C. (1989). Structural analysis of the gene for human acidic fibroblast growth factor. *Biochem. Biophys. Res. Commun.* **164**, 1121–1129.

Moenner, M., Chevallier, B., Badet, J., and Barritault, D. (1986). Evidence and characterization of the receptor to eye-derived growth factor I, the retinal form of basic fibroblast growth factor, on bovine epithelial lens cells. *Proc. Natl. Acad. Sci. U.S.A.* **83**, 5024–5028.

Moore, R., Casey, G., Brooks, S., Dixon, M., Peters, G., and Dickson, C. (1986). Sequence, topography and protein coding potential of mouse int-2, a putative oncogene activated by mouse mammary tumour virus. *EMBO J.* **5**, 919–924.

Morrison, D. K., Kaplan, D. R., Rapp, U., and Roberts, T. M. (1988). Signal transduction from membrane to cytoplasm: Growth factors and membrane-bound oncogene products increase *raf*-1 phosphorylation and associated protein kinase activity. *Proc. Natl. Acad. Sci. U.S.A.* **85**, 8855–8859.

Morrison, R. S., Sharma, A., DeVellis, J., and Bradshaw, R. A. (1986). Basic fibroblast growth factor supports the survival of cerebral cortical neurons in primary culture. *Proc. Natl. Acad. Sci. U.S.A.* **83**, 7537–7541.

Moscatelli, D. (1987). High and low affinity binding sites for basic fibroblast growth factor on cultured

cells: Absence of a role for low affinity binding in the stimulation of plasminogen activator production by bovine capillary endothelial cells. *J. Cell. Physiol.* **131**, 123–130.
Moscatelli, D. (1988). Metabolism of receptor-bound and matrix-bound basic fibroblast growth factor by bovine capillary endothelial cells. *J. Cell Biol.* **107**, 753–759.
Moscatelli, D., Presta, M., and Rifkin, D. B. (1986). Purification of a factor from human placenta that stimulates capillary endothelial cell protease production. DNA synthesis, and migration. *Proc. Natl. Acad. Sci. U.S.A.* **83**, 2091–2095.
Mueller, S. N., Thomas, K. A., DiSalvo, J., and Levine, E. M. (1989). Stabilization by heparin of acidic fibroblast growth factor mitogenicity for human endothelial cells *in vitro*. *J. Cell. Physiol.* **140**, 439–448.
Muller, W. J., Lee, F. S., Dickson, C., Peters, G., Pattengale, P., and Leder, P. (1990). The *int-2* gene product acts as an epithelial growth factor in transgenic mice. *EMBO J.* **9**, 907–913.
Musci, T. J., Amaya, E., and Kirschner, M. W. (1990). Regulation of the fibroblast growth factor receptor in early *Xenopus* embryos. *Proc. Natl. Acad. Sci. U.S.A.* **87**, 8365–8369.
Nanberg, E., Morris, C., Higgins, T., Vara, F., and Rozengurt, E. (1990). Fibroblast growth factor stimulates protein kinase C in quiescent 3T3 cells without $Ca^{2+}$ mobilization or inositol phosphate accumulation. *J. Cell. Physiol.* **143**, 232–242.
Neufeld, G., and Gospodarowicz, D. (1986). Basic and acidic fibroblast growth factors interact with the same cell surface receptors. *J. Biol. Chem.* **261**, 5631–5637.
Nieto-Sampedro, M., Lim, R., Hicklin, D. J., and Cotman, C. W. (1988). Early release of glia maturation factor and acidic fibroblast growth factor after rat brain injury. *Neurosci. Lett.* **86**, 361–365.
Ohtaki, T., Wakamatsu, K., Mori, M., Ishibashi, Y., and Yasuhara, T. (1989). Purification of acidic fibroblast growth factor from bovine omentum. *Biochem. Biophys. Res. Commun.* **161**, 169–175.
Olwin, B. B., and Hauschka, S. D. (1988). Cell surface fibroblast growth factor and epidermal growth factor receptors are permanently lost during skeletal muscle terminal differentiation in culture. *J. Cell Biol.* **107**, 761–769.
Ortega, S., and Thomas, K. A. (1990). The oncogenic potential of fibroblast growth factor genes. *In* "Molecular Biology of Cancer Genes" (M. Sluyser, ed.), pp. 134–149. Ellis Horwood, Chichester, England.
Ortega, S., Schaeffer, M.-T., Soderman, D., DiSalvo, J., Linemeyer, D., Gimenez-Gallego, G., and Thomas, K. A. (1991). Conversion of cysteine to serine residues alters the activity, stability and heparin dependence of acidic fibroblast growth factor. *J. Biol. Chem.* **266**, 5842–5846.
Pasquale, E. B. (1990). A distinctive family of embryonic protein-tyrosine kinase receptors. *Proc. Natl. Acad. Sci. U.S.A.* **87**, 5812–5816.
Pasquale, E. B., and Singer, S. J. (1989). Identification of a developmentally regulated protein-tyrosine kinase by using anti-phosphotyrosine antibodies to screen a cDNA expression library. *Proc. Natl. Acad. Sci. U.S.A.* **86**, 5449–5453.
Paterno, G. D., Gillespie, L. L., Dixon, M. S., Slack, J. M. W., and Heath, J. K. (1989). Mesoderm-inducing properties of INT-2 and kFGF: Two oncogene-encoded growth factors related to FGF. *Development* **106**, 79–83.
Pelech, S. L., Olwin, B. B., and Krebs, E. G. (1986). Fibroblast growth factor treatment of Swiss 3T3 cells activates a subunit S6 kinase that phosphorylates a synthetic peptide substrate. *Proc. Natl. Acad. Sci. U.S.A.* **83**, 5968–5972.
Perraud, F., Besnard, F., Pettmann, B., Sensenbrenner, M., and Labourdette, G. (1988). Effects of acidic and basic fibroblast growth factors (aFGF and bFGF) on the proliferation and the glutamine synthetase expression of rat astroblasts in culture. *Glia* **1**, 124–131.
Plouet, J., Mascarelli, F., Loret, M. D., Faure, J. P., and Courtois, Y. (1988). Regulation of eye-derived growth factor binding to membranes by light, ATP or GTP in photoreceptor outer segments. *EMBO J.* **7**, 373–376.
Prats, H., Kaghad, M., Prat, A. C., Klagsbrun, M., Lelias, J. M., Liauzun, P., Chalon, P., Tauber, J. P., Amalric, F., Smith, J. A., and Caput, D. (1989). High molecular mass forms of basic fibroblast

growth factor are initiated by alternative CUG codons. *Proc. Natl. Acad. Sci. U.S.A.* **86**, 1836–1840.

Presta, M., Moscatelli, D., Joseph-Silversteiner, J., and Rifkin, D. B. (1986). Purification from a human hepatoma cell line of a basic fibroblast growth factor-like molecule that stimulates capillary endothelial cell plasminogen activator production, DNA synthesis, and migration. *Mol. Cell. Biol.* **6**, 4060–4066.

Presta, M., Maier, J. A. M., and Ragnotti, G. (1989). The mitogenic signaling pathway but not the plasminogen activator-inducing pathway of basic fibroblast growth factor is mediated through protein kinase C in fetal bovine aortic endothelial cells. *J. Cell Biol.* **109**, 1877–1884.

Quinkler, W., Maasberg, M., Bernotat-Danielowski, S., Luthe, N., Sharma, H. S., and Schaper, W. (1989). Isolation of heparin-binding growth factors from bovine, porcine and canine hearts. *Eur. J. Biochem.* **181**, 67–73.

Renko, M., Quarto, N., Morimoto, T., and Rifkin, D. B. (1990). Nuclear and cytoplasmic localization of different basic fibroblast growth factor species. *J. Cell. Physiol.* **144**, 108–114.

Rifkin, D. B., and Moscatelli, D. (1989). Recent developments in the cell biology of basic fibroblast growth factor. *J. Cell Biol.* **109**, 1–6.

Rodan, S. B., Wesolowski, G., Thomas, K., and Rodan, G. A. (1987). Growth stimulation of rat calvaria osteoblastic cells by acidic fibroblast growth factor. *Endocrinology* **121**, 1917–1923.

Rogelj, S., Weinberg, R. A., Fanning, P., and Klagsbrun, M. (1988). Basic fibroblast growth factor fused to a signal peptide transforms cells. *Nature (London)* **331**, 173–175.

Rosengart, T., Johnson, W., Friesel, R., Clark, R., and Maciag, T. (1988). Heparin protects heparin-binding growth factor-I from proteolytic inactivation *in vitro*. *Biochem. Biophys. Res. Commun.* **152**, 432–440.

Rubin, J. S., Osada, H., Finch, P. W., Taylor, W. G., Rudikoff, S., and Aaronson, S. A. (1989). Purification and characterization of a newly identified growth factor specific for epithelial cells. *Proc. Natl. Acad. Sci. U.S.A.* **86**, 802–806.

Safran, A., Avivi, A., Orr-Urtereger, A., Neufeld, G., Lonai, P., Givol, D., and Yarden, Y. (1990). The murine flg gene encodes a receptor for fibroblast growth factor. *Oncogene* **5**, 635–643.

Saksela, O., Moscatelli, D., Sommer, A., and Rifkin, D. B. (1988). Endothelial cell-derived heparin sulfate binds basic fibroblast growth factor and protects it from proteolytic degradation. *J. Cell Biol.* **107**, 743–751.

Sasada, R., Kurokawa, T., Iwane, M., and Igarashi, K. (1988). Transformation of mouse BALB/c 3T3 cells with human basic fibroblast growth factor cDNA. *Mol. Cell. Biol.* **8**, 588–594.

Sasaki, H., Hoshi, H., Hong, Y.-M., Suzuki, T., Kato, T., Sasaki, H., Saito, M., Youki, H., Kumiko, K., Konno, S., Ondera, M., Saito, T., and Aoyagi, S. (1989). Purification of acidic fibroblast growth factor from bovine heart and its localization in the cardiac myocytes. *J. Biol. Chem.* **264**, 17606–17612.

Schweigerer, L., Neufeld, G., Friedman, J., Abraham, J. A., Fiddes, J. C., and Gospodarowicz, D. (1987). Capillary endothelial cells express basic fibroblast growth factor, a mitogen that promotes their own growth. *Nature (London)* **325**, 257–259.

Senior, R. M., Huang, S. S., Griffin, G. L., and Huang, J. S. (1986). Brain-derived growth factor is a chemoattractant for fibroblasts and astroglial cells. *Biochem. Biophys. Res. Commun.* **141**, 67–72.

Shipley, G. D., Keeble, W. W., Hendrickson, J. E., Coffey, J. R. J., and Pittelkow, M. R. (1988). Growth of normal human keratinocytes and fibroblasts in serum-free medium is stimulated by acidic and basic fibroblast growth factor. *J. Cell. Physiol.* **138**, 511–518.

Slack, J. M. W., Darlington, B. G., Health, J. K., and Godsave, S. F. (1987). Mesoderm induction in early *Xenopus* embryos by heparin-binding growth factors. *Nature (London)* **326**, 197–200.

Smith, R., Peters, G., and Dickson, C. (1988). Multiple RNAs expressed from the *int-2* gene in mouse embryonal carcinoma cell lines encode a protein with homology to fibroblast growth factors. *EMBO J.* **7**, 1013–1021.

Sommer, A., and Rifkin, D. B. (1989). Interaction of heparin with human basic fibroblast growth factor: Protection of the angiogenic protein from proteolytic degradation by a glycosaminoglycan. *J. Cell. Physiol.* **138**, 215–220.

Sommer, A., Moscatelli, D., and Rifkin, D. B. (1989). An amino-terminally extended and post-translationally modified form of a 25-kD basic fibroblast growth factor. *Biochem. Biophys. Res. Commun.* **160**, 1267–1274.
Speir, E., Yi-Fu, Z., and Lee, M. (1988). Fibroblast growth factors are present in adult cardiac myocytes, *in vivo. Biochem. Biophys. Res. Commun.* **157**, 1336–1340.
Sullivan, R., and Klagsbrun, M. (1985). Purification of cartilage-derived growth factor by heparin affinity chromatography. *J. Biol. Chem.* **260**, 2399–2403.
Terranova, V. P., DiFlorio, R., Lyall, R. M., Hic, S., Friesel, R., and Maciag, T. (1985). Human endothelial cells are chemotactic to endothelial cell growth factor and heparin. *J. Cell Biol.* **101**, 2330–2334.
Thomas, K. A. (1987). Fibroblast growth factors. *FASEB J.* **1**, 434–440.
Thomas, K. A., Rios-Candelore, M., and Fitzpatrick, S. (1984). Purification and characterization of acidic fibroblast growth factor from bovine brain. *Proc. Natl. Acad. Sci. U.S.A.* **81**, 357–361.
Thomas, K. A., Candelore, M.-R., Gimenez-Gallego, G., DiSalvo, J., Bennett, C., Rodkey, J., and Fitzpatrick, S. (1985). Pure brain-derived acidic fibroblast growth factor is a potent angiogenic vascular endothelial cell mitogen with sequence homology to interleukin 1. *Proc. Natl. Acad. Sci. U.S.A.* **82**, 6409–6413.
Thompson, J. A., Haudenschild, C. C., Anderson, K. D., DiPietro, J. M., Anderson, W. F., and Maciag, T. (1989). Heparin-binding growth factor 1 induces the formation of organoid neovascular structures *in vivo. Proc. Natl. Acad. Sci. U.S.A.* **86**, 7928–7932.
Ueno, N., Baird, A., Esch, F., Shimasaki, S., Ling, N., and Guillemin, R. (1986). Purification and partial characterization of a mitogenic factor from bovine liver: Structural homology with basic fibroblast growth factor. *Regulatory Peptides* **16**, 135–145.
Ueno, N., Baird, A., Esch, F., Ling, N., and Guillemin, R. (1987). Isolation and partial characterization of basic fibroblast growth factor from bovine testis. *Mol. Cell. Endocrinol.* **49**, 189–194.
Uhlrich, S., Lagente, O., Lenfant, M., and Courtois, Y. (1986). Effect of heparin on the stimulation of non-vascular cells by human acidic and basic FGF. *Biochem. Biophys. Res. Commun.* **137**, 1205–1213.
Velcich, A., Delli-Bovi, P., Mansukhani, A., Ziff, E. B., and Basilico, C. (1989). Expression of the k-fgf protooncogene is repressed during differentiation of F9 cells. *Oncogene Res.* **5**, 31–37.
Vlodavsky, I., Folkman, J., Sullivan, R., Fridman, R., Ishai-Michaeli, R., Sasse, J., and Klagsbrun, M. (1987). Endothelial cell-derived basic fibroblast growth factor: synthesis and deposition into subendothelial extracellular matrix. *Proc. Natl. Acad. Sci. U.S.A.* **84**, 2292–2296.
Volk, R., Koster, M., Poting, A., Hartmann, L., and Knochel, W. (1989). An antisense transcript from the *Xenopus laevis* bFGF gene coding for an evolutionarily conserved 24-kD protein. *EMBO J.* **8**, 2983–2988.
Wada, A., Sakamoto, H., Katoh, O., Yoshida, T., Yokota, J., Little, P. F. R., Sugimura, T., and Terada, M. (1988). Two homologous oncogenes, hst-1 and int-2, are closely located in human genome. *Biochem. Biophys. Res. Commun.* **157**, 828–835.
Walicke, P., Cowan, W. M., Ueno, N., Baird, A., and Guillemin, R. (1986). Fibroblast growth factor promotes survival of dissociated hippocampal neurons and enhances neurite extension. *Proc. Natl. Acad. Sci. U.S.A.* **83**, 3012–3016.
Weiner, H. J., and Swain, J. L. (1989). Acidic fibroblast growth factor mRNA is expressed by cardiac myocytes in culture and the protein is localized to the extracellular matrix. *Proc. Natl. Acad. Sci. U.S.A.* **86**, 2683–2687.
Wilkinson, D. G., Peters, G., Dickson, C., and McMahon, A. P. (1988). Expression of the FGF-related proto-oncogene *int-2* during gastrulation and neurolation in the mouse. *EMBO J.* **7**, 691–695.
Wilkinson, D. G., Bhatt, S., and McMahon, A. P. (1989). Expression pattern of the FGF-related proto-oncogene *int-2* suggests multiple roles in fetal development. *Development* **105**, 131–136.
Winkles, J. A., Friesel, R., Burgess, W. H., Howk, R., Mehlman, T., Weinstein, R., and Maciag, T. (1987). Human vascular smooth muscle cells both express and respond to heparin-binding growth factor I (endothelial cell growth factor). *Proc. Natl. Acad. Sci. U.S.A.* **84**, 7124–7128.

Wu, D. K., Maciag, T., and Vellis, J. D. (1988). Regulation of neuroblast proliferation by hormones and growth factors in chemically defined medium. *J. Cell. Physiol.* **136**, 367–372.

Yoshida, M. C., Wada, M., Satoh, H., Yoshida, T., Sakamoto, H., Miyagawa, K., Yokota, J., and Terada, M. (1988). Human HST1 (HSTF1) gene maps to chromosome band 11q13 and coamplifies with the INT2 gene in human cancer. *Proc. Natl. Acad. Sci. U.S.A.* **85**, 4861–4864.

Yoshida, T., Miyagawa, K., Odagiri, H., Sakamoto, H., Little, P. F. R., Terada, M., and Sugimura, T. (1987). Genomic sequence of *hst*, a transforming gene encoding a protein homologous to fibroblast growth factors and the *int*-2-encoded protein. *Proc. Natl. Acad. Sci. U.S.A.* **84**, 7305–7309.

Yoshida, T., Tsutsumi, M., Sakamoto, H., Miyagawa, K., Teshima, S., Sugimura, T., and Terada, M. (1988). Expression of the HST1 oncogene in human germ cell tumors. *Biochem. Biophys. Res. Commun.* **155**, 1324–1329.

Yue, B. Y. J. T., Sugar, J., Gilboy, J. E., and Elvart, J. L. (1989). Growth of human corneal endothelial cells in culture. *Invest. Opthalmol. Vision Sci.* **30**, 248–253.

Zhan, X., Bates, B., Hu, X., and Goldfarb, M. (1988). The human FGF-5 oncogene encodes a novel protein related to fibroblast growth factor. *Mol. Cell. Biol.* **8**, 3487–3495.

Zhu, X., Komiya, H., Chirino, A., Faham, S., Hsu, B. T., and Rees, D. C. (1991). Three-dimensional structures of acidic and basic fibroblast growth factors. *Science* **251**, 90–93.

# 10 Fibroblast Growth Factors: Their Roles in the Central and Peripheral Nervous System

Klaus Unsicker, Claudia Grothe, Gerson Lüdecke, Dörte Otto, and Reiner Westermann

## I. INTRODUCTION: THE FGF GENE FAMILY

Although fibroblast growth factors (FGFs) are ubiquitous in nature, their discovery is linked essentially to their localization in the brain and pituitary and dates back to experiments showing that extracts from these tissues promote growth of fibroblasts (Trowell et al., 1939; Hoffman, 1940). Mitogenicity is no longer the sole, and perhaps not even the most important, function of FGFs. The spectrum of biologic actions of the FGF gene family spans from cell proliferation and differentiation to angiogenesis, cell maintenance, chemotaxis, and repair (for reviews, see Gospodarowicz et al., 1986; Lobb, 1988; Burgess and Maciag, 1989; Rifkin and Moscatelli, 1989; Baird and Böhlen, 1990). Heterogeneity of biologic activities is paralleled by a considerable molecular diversity of the FGF family, raising the question of why there may be a need for so many, often very closely related, molecular forms.

Acidic and basic FGF are the longest established members of the FGF family, to which INT-2, HST/FGF, FGF-5, FGF-6, and KGF have been added, based on their sequence homologies to acidic and basic FGF (Delli Bovi et al., 1987; Dickson and Peters, 1987; Taira et al., 1987; Marics et al., 1989; see also Chapter 9 and Baird and Klagsbrun, 1991, for reviews). Some of these FGFs have molecular and biologic properties that differ from those of acidic and basic FGF. These include the presence of a signal peptide, glycosylation, evidence for secretion, restricted expression during development, specific target cells, and encoding by oncogenes (Delli Bovi et al., 1987; Wilkinson et al., 1988).

Acidic and basic FGF exist in multiple molecular extended and truncated forms, but all forms lack signal sequences (Moscatelli et al., 1987; Florkiewicz and Sommer, 1989; Prats et al., 1989). The question of whether the presence of multiple molecular weight forms of a single FGF in a tissue such as the brain has a physiologic meaning is compelling, but unresolved. With several other growth factors, FGFs (except INT-2) share a high affinity for heparin, which can modulate FGF activity and may be involved in storage and liberation from extracellular matrix constituents (Lobb et al., 1986; Lobb, 1988).

FGF receptors seem to exist under a variety of molecular forms (Neufeld and Gospodarowicz, 1985; Gospodarowicz et al., 1986; Olwin and Hauschka, 1986,1990; Ruta et al., 1989; Reid et al., 1990; Williams, 1991). Chick, mouse, and human receptors display high sequence homology, incorporating three immunoglobulin-like domains, an acid region, a single transmembrane region, and an intracellular tyrosine kinase domain (Kornbluth et al., 1988; Lee et al., 1989; Reid et al., 1990). Receptors with two immunoglobulin-like domains and a soluble form also exist (Williams, 1991). A notable increase in the number of cDNA sequences encoding putative FGF receptors and binding sites, adding to the expansion of the FGF family, raises new questions about the physiologic significance of these diversities (Chapter 9).

Several comprehensive reviews covering structure, distribution, regulation of expression, and functions of FGFs in various normal and malignant cells and tissues are available (Gospodarowicz et al., 1986; Lobb, 1988; Baird and Böhlen, 1990; Baird and Klagsbrun, 1991). Presence and functions of FGFs in the nervous system also have been reviewed, but mainly with a focus on data provided by our laboratories and those of Walicke (Walicke, 1989,1990; Otto and Unsicker, 1990b; Unsicker et al., 1990; Otto et al., 1991; Westermann et al., 1991). The purpose of this chapter is to provide a more complete and critical overview of recent discoveries with regard to the distributions and putative roles of members of the FGF gene family and their receptors in the developing, mature, and lesioned nervous system.

## II. LOCALIZATION OF THE FGF GENE FAMILY IN NEURAL CELLS AND TISSUES

Both acidic and basic FGF have been isolated from the embryonic and adult mammalian and chick central nervous system (CNS) (Gospodarowicz et al., 1984,1986; Gimenez-Gallego et al., 1985,1986; Logan et al., 1985; Pettmann et al., 1985; Moenner et al., 1986; Alterio et al., 1988; Risau et al., 1988; Caruelle et al., 1989). Acidic FGF seems to be the predominant form. Molecular mass forms of basic FGF greater than 18 kDa also have been purified from the brain (Moscatelli et al., 1987; Presta et al., 1988a). INT-2, another member of the FGF family, is found in the early developing hindbrain, cerebellum, retina, and inner ear of mouse embryos (Wilkinson et al., 1988,1989). FGF-5 mRNA occurs in the embryonic acoustic ganglion and in the CA3 region of the adult mouse hippocampus (Goldfarb, 1991).

It is possible that basic and acidic FGF are expressed in temporally and spatially distinct patterns during nervous system development. Thus, Kalcheim and Neufeld

(1990), using a radioimmunoassay for basic FGF, detected basic FGF at low concentrations (1.4 ng/mg protein) as early as embryonic day (ED) 3 in quail spinal cord extracts. The concentration of basic FGF-immunoreactive material increased with age, reached a peak at ED 10 (18 ng/mg protein) and subsequently decreased to the initial level by ED 14. Ishikawa et al. (1991a,b) developed a sensitive (0.2 ng/ml) two-site enzyme immunoassay for acidic FGF and reported low levels (less than 10 ng/g tissue) in all regions of the neonatal rat brain. In young adult rats, concentrations of acidic FGF varied widely in different brain areas. Highest concentrations were found in the pons-medulla (70–90 ng/g tissue), intermediate levels in the diencephalon and mesencephalon, and comparably low levels (5–15 ng/g tissue) in other regions, such as the frontal cortex, piriform cortex, hippocampus, olfactory bulb, cerebellum, and striatum.

Pettmann et al. (1986) were the first to immunocytochemically localize FGF in the nervous system. Using an antibody that recognized both basic and acidic FGF, these investigators found that virtually all neurons in the brain of 18–20-day-old rats immunostained, but glial cells were consistently devoid of immunoreactivity. At an ultrastructural level (Janet et al., 1987), the reaction product was localized in the cytosol of neuronal perikarya and large processes; most small diameter processes were unstained. No immunostaining was observed in the endoplasmic reticulum and Golgi apparatus, a finding that is consistent with the lack of a signal peptide in basic and acidic FGF. This study also reported a decrease in the proportion of immunostained neurons in the adult compared with the 15- and 20-day-old brain. Again, glial, ependymal, and brain endothelial cells were unstained. This result is particularly noteworthy, since astrocytes and brain capillary endothelial cells in culture express basic FGF (Gospodarowicz et al., 1986; Ferrara et al., 1988; Hatten et al., 1988), possibly as a consequence of their isolation and lesioning.

More recent studies on the cellular localization of FGFs in the nervous system have confirmed and extended the initial observations, but also have produced several conflicting results. In a study with no particular focus on the nervous system, Gonzalez et al., (1990) reported widely distributed staining for basic FGF in the ED 18 rat brain. In some areas, such as diencephalon, mesencephalon, and spinal cord, staining seemed to be associated with fibers whereas, in the telencephalon, neuronal cell bodies showed intense immunoreactivity. In peripheral ganglia, nerve fibers reacted more strongly than did perikarya. Somehow in conflict with these observations is a statement made by the authors in the discussion of the same paper that basic FGF is "clearly more concentrated in endothelial cells," both in the brain and in the spinal cord. Studies conducted in our laboratory using the same antibody, which was raised against an amino-terminal 1–24 synthetic peptide of bovine basic FGF, have failed to reveal any staining of endothelial or glial cells in the central or peripheral nervous system (Grothe and Unsicker, 1990; Grothe, 1991; Grothe et al., 1991; C. Grothe, unpublished observations). In contrast, Cotman et al. (1991) have localized bFGF-like immunoreactivity in astrocytes. There is no doubt, however, that the staining pattern may vary depending on the type of fixative used, a fact known by many who work on the immunolocalization of growth factors. It is therefore important to interpret the available data on the localization of bFGF with caution and, whenever possible, correlate them with results obtained by an independent method, for example, in situ hybridization (see subsequent text).

An immunocytochemical mapping of basic FGF in the developing and adult rat brain stem (Grothe et al., 1991) has provided evidence for its localization in specific neuronal subsets and for a distinct regulatory pattern of basic FGF expression during development. For example, although basic FGF-like immunoreactivity was absent from most neuronal cell bodies in the red nucleus, many but not all perikarya in motor nuclei were stained. The appearance of basic FGF in neurons of the hypoglossal nerve preceded its initial expression in the dorsal nucleus of the vagus nerve. Many areas contained a higher proportion of immunoreactive cell bodies and fibers in 1–2-week-old rats than in adult animals. A clearly transient expression of basic FGF was obvious in two subnuclei of the inferior colliculus, in the medial geniculate nucleus, and in the gracile and cuneate nuclei during the early postnatal period. The reverse was found in the substantia nigra and interpeduncular nucleus, where immunoreactivity was not detectable before adulthood. Although straightforward conclusions concerning functional roles of basic FGF in the developing and adult brain cannot be drawn from these data, they permit the postulation of its involvement in developmental processes (transient expression) and events associated with a differentiated neuronal status (high levels during adulthood).

An important role of basic FGF during neural development can be deduced from its localization in ED 6 to ED 10 avian spinal cord and ED 7 dorsal root ganglion neurons (Kalcheim and Neufeld, 1990). These investigators also corroborated data reported in a previous study that had shown FGF immunoreactivity in cultured neurons (Janet et al., 1988). Interestingly, basic FGF immunoreactivity also was localized to the cytoplasm of neuroepithelial cells from ED 2 quail neural tubes and a subpopulation of nonneuronal cells from sensory ganglia in culture (Kalcheim and Neufeld, 1990).

The observation that the central nervous system is the only normal adult tissue that has detectable levels of basic FGF mRNA (Abraham et al., 1986a,b; Shimasaki et al., 1988) led Emoto and collaborators (1989) to investigate the distribution of basic FGF mRNA in different areas of the adult rat brain and in CNS cultures derived from ED 18 rat telencephalon. Northern blots revealed a wide distribution in the cortex, hippocampus, hypothalamus, and pons. In contrast, only discrete neuronal populations, such as the CA2 region of the hippocampus, gave a strong signal in the in situ hybridizations. Only scattered positive cells were seen in the hypothalamus and pons. RNA from both glial and neuronal cell cultures gave 6-kb hybridizing bands indicative of basic FGF mRNA signals; these were considerably weaker in preparations of neuronal than of astroglial RNA. These data were considered to support the notion that glial cells, in contrast to most neurons, make low levels of basic FGF mRNA, disregarding the fact that glial cells in situ have not been unequivocally shown to contain the mRNA or the protein and, further, that detection of low abundance messages by in situ hybridization, well exemplified by the mRNA for nerve growth factor, may be extremely difficult (cf. Bandtlow et al., 1988; Korsching et al., 1988).

The retina, an outpost of the diencephalon, long has been known to contain FGFs (Baird et al., 1985; Courty et al., 1985). Immunohistologic staining has revealed acidic FGF, primarily in the nerve fiber layer and the inner and outer segments of photoreceptors (Caruelle et al., 1989). Rod outer segments have been shown to release acidic FGF and possess specific binding sites for this growth factor, suggesting a role in phototransduction (Plouet et al., 1988).

Basic FGF has been found in the peripheral nervous system (PNS). Developing sympathetic neurons (Kalcheim and Neufeld, 1990), adrenal chromaffin cells (Grothe and Unsicker, 1990), and a subpopulation of dorsal root ganglion neurons (comprising most somatostatin-containing neurons; C. Grothe, unpublished observations) contain basic FGF-like immunoreactivity. The localization of basic FGF in the adrenal medulla has been most thoroughly studied (Blottner et al., 1989; Grothe and Unsicker, 1990; Grothe et al., 1990; Westermann et al., 1990). Basic FGF is localized in a subpopulation of chromaffin cells that synthesize noradrenalin. Western blots of materials (adrenal medulla, purified chromaffin granules) enriched for basic FGF by antibody affinity chromatography reveal immunoreactive bands at apparent molecular masses of approximately 18, 24, 30, and 46 kDa (Grothe et al., 1990; Westermann et al., 1990). An identification by sequencing, of the higher molecular weight forms as members of the FGF family remains to be established. Most interestingly, basic FGF-like immunoreactivity has been localized inside chromaffin granules by immunoelectron microscopy (Westermann et al., 1990), a finding that is consistent with the presence of multiple immunoreactive forms of basic FGF in extracts of chromaffin vesicles. Presta et al. (1991) have confirmed that basic FGF is a constituent of soluble chromaffin vesicle proteins. The most important aspects of this observation relate to the uptake mechanism of basic FGF into the granule and its possible release by exocytosis. Putative functions of the adrenomedullary basic FGF will be discussed subsequently.

## III. FGF RECEPTORS ON NEURAL CELLS

High affinity receptors for FGF in the range of 110–150 kDa have been identified on a variety of cell types by chemical cross-linking (Friesel et al., 1986; Neufeld and Gospodarowicz, 1986; Olwin and Hauschka, 1986; Libermann et al., 1987; Feige and Baird, 1988; Imamura et al., 1988). Low affinity binding sites also seem to exist and appear to be glycosaminoglycans (Saksela et al., 1988). Various types of receptors for basic and acidic FGF have been cloned and characterized (Lee et al., 1989; Ruta et al., 1989; Williams, 1991).

FGF receptors on neural cells were characterized first in cultures of fetal hippocampal neurons (Walicke et al., 1989). These neurons bind FGF specifically, as visualized by autoradiography. Subsequently bound FGF is internalized and degraded. Biochemically, two binding sites can be distinguished: (1) a low affinity system likely to be related to heparan sulfate and (2) a saturable high affinity membrane receptor. The $K_d$ of the receptor is approximately 0.2 n$M$.

Affinity labeling suggests the presence of two glycoprotein receptors of about 135 and 90 kDa. PC12 pheochromocytoma cells bear a receptor that binds both basic and acidic FGF. Binding characteristics and molecular weight are identical to those of the mesenchymal receptor (Neufeld and Gospodarowicz, 1986; Wagner and D'Amore, 1986; Schubert et al., 1987).

The present status of FGF receptors in the brain is that FGF receptors type 1 (highly homologous to human *flg,* chicken C-ek1, *Xenopus* embryonic FLG, and human N-*sam*; see Chapter 9), type 2 (also denoted as murine Bek, chicken Cek3, human TK14, and human K-*sam*), and type 3 (Cek2) are all expressed in brain, type 1 on neurons and type 2 on glial cells (Williams, 1991).

Two forms of basic FGF receptor mRNAs are expressed in the neuroepithelium of the mouse embryo (Reid et al., 1990). A shorter form of the murine basic FGF receptor mRNA, containing 75% of the basic FGF receptor cDNA, appears to be expressed at higher levels in neuronal cells in early developmental stages. A recent in situ hybridization study localizing transcripts of the (basic) FGF receptor in the developing chick nervous system (Heuer et al., 1990) has detected abundant expression in the germinal neuroepithelial layer. FGF receptor mRNA was absent from migrating cells, but returned during later stages of neuronal differentiation when several neuron populations, including cholinergic forebrain, brainstem reticular and motor, and cerebellar Purkinje and granule neurons, gave a clear signal. In the adult rat CNS (Wanaka et al, 1990), FGF receptor mRNA is distributed widely. Most intense signals were seen in the hippocampus and in pontine cholinergic nuclei. Moreover, several limbic areas such as hypothalamus and amygdala contained groups of neurons expressing FGF receptor mRNA. Motor nuclei in the rhombencephalon and spinal cord were labeled intensely, but other neurons in laminae 2–8 in the spinal cord also reacted. No significant signals were found in the caudateputamen, septal nuclei, and mesencephalic trigeminal nucleus. These findings have important implications for functional considerations to be discussed later. In several brain areas, for example, the cerebral cortex, it was not possible to distinguish between neuronal, astroglial, or endothelial localization of the label. As a whole, these results suggest a broad distribution of putative target cells for FGFs or related molecules in the embryonic and adult CNS, with no, or very little, expression of the receptor message by glial cells. Determining whether this means that glial cell responses to FGFs seen in vitro are the result of a de novo induction of receptor message and protein will require more detailed investigation.

Information on the transduction pathways used by the FGFs in neurons and glial cells subsequent to their binding to the receptor is scarce. Morrison and co-workers (1988) have suggested that down-regulation of protein kinase is required for the action of basic FGF on neurons. Activation of protein kinase apparently is required for acidic and basic FGF to induce ornithine decarboxylase, transcription of early response genes, and neurite growth on PC12 cells maximally (Damon et al., 1990).

## IV. FGFs AND THEIR EFFECTS ON NEURONS

### A. In Vitro Effects

Basic and acidic FGF exert a variety of neurotrophic actions on neurons cultured from fetal and postnatal CNS and PNS, inducing increased survival, neurotransmitter synthesis, and neurite growth (Morrison et al., 1986; Walicke et al., 1986; Morrison, 1987; Schubert et al., 1987; Unsicker et al., 1987; Hatten et al., 1988; Walicke, 1988; Ferrari et al., 1989; Grothe et al., 1989; Knusel et al., 1990; Matsuda et al., 1990). Neuron populations from a variety of CNS regions (neocortex, hippocampus, septum, striatum, thalamus, mesencephalon, cerebellum, spinal cord) and one peripheral ganglion (ciliary), belonging to several identified (cholinergic, GABAergic, dopaminergic) and unidentified transmitter phenotypes, respond to FGFs. FGF even promotes survival and transmitter development of neurons from those areas that have been shown to lack FGF receptor mRNA in the adult rat brain, for example, septum and striatum (Wanaka et al., 1990). Whether this result argues in favor of a glial cell-mediated effect or loss of neuronal receptors in the adult brain or in favor of de novo induction of FGF receptors on cultured neurons is uncertain.

The neurotrophic effects of FGF initially were met with great and partly justified skepticism. Major reasons included abundance and wide distribution in many tissues (in apparent contrast to nerve growth factor, NGF) and the fact that FGFs were mitogens for glial and other nonneuronal cells, which could contribute potentially to the trophic support. Although a cotrophic effect and corequirement of unknown agents (provided, for example, by dying neurons in vitro) can never be ruled out, not even in the case of the best-established neurotrophic factor, several observations make it unlikely that FGFs consistently would require the presence of nonneuronal cells to mediate their functions. (1) Several culture systems employed have less than 5% nonneuronal cells, and this proportion decreases rather than increases during the experimental period (Walicke and Baird, 1988), and (2) FGFs support a single ciliary ganglionic neuron in one-neuron-per-well cultures (Unsicker et al., 1991). Although FGFs support a broader spectrum of neurons than does NGF, there are clearly neuron populations that cannot be maintained in vitro with FGF, for example, sympathetic and sensory dorsal root ganglion neurons (Unsicker et al., 1987) and purified embryonic chick retinal ganglion cells (Lehwalder et al., 1989). On most neurons, basic FGF is about 200-fold more potent than acidic FGF, whose potency in vitro can be increased dramatically by heparin (Unsicker et al., 1987), a phenomenon also known by its actions on mesenchymal cells (Gospodarowicz et al., 1986; Baird and Böhlen, 1990). Retinal explants, which extend neurites after exposure to FGF, seem to be the only system in which acidic FGF may be more potent than basic FGF (Lipton et al., 1988). Half-maximally effective concentrations for the survival-promoting actions of basic FGF range from 1 p$M$ (Walicke, 1988) to about 50 p$M$ (Morrison et al., 1986; Unsicker et al., 1987; Hatten et al., 1988). The factor is also active when bound to a cell culture substratum (Unsicker et al., 1987), an observation that has raised speculations with respect to a presentation of FGF in vivo by extracellular matrix constituents. Stability of FGF in culture depends on cell density and types of cells present. Certain batches of serum, particularly horse serum, can inactivate or destroy basic FGF rapidly. (M. Sendtner, personal communication; H. Reichert-Preibsch and K. Unsicker, unpublished observations).

FGFs share several target cell populations with other (nerve) growth factors, for example, NGF and ciliary neurotrophic factor (CNTF). Careful analysis, however, often reveals that their effects differ somewhat in detail. For example, both NGF and basic FGF enhance survival of fetal rat septal neurons and stimulate choline acetyltransferase activity (Grothe et al., 1989; Knusel et al., 1990). Yet, with respect to enzyme activity, FGF is most effective when given early in culture, and NGF when given later (Knusel et al., 1990). Both FGF and CNTF increase survival, neurite growth, and choline acetyltransferase activity in cultured chick ciliary ganglion neurons (Schubert et al., 1987; Unsicker et al., 1987), but basic FGF is more potent with respect to enzyme induction (Fuller et al., 1990). Such subtle differences may be understood better once our knowledge of the temporal and spatial pattern of local distribution of these factors in vivo becomes more comprehensive.

Related to the results just discussed is the finding that FGF may act more potently on one than on another neuron system. Thus, basic FGF stimulates uptake of dopamine in a more pronounced fashion than uptake of GABA in cultures of fetal mesencephalic neurons (Ferrari et al., 1989). Finally, FGF may influence one parameter of neuronal function directly (e.g., survival), but another one through the intermediate effect on glial cells (e.g., dopamine uptake; cf. Knusel et al., 1990). Such aspects of the multifunctionality of FGFs will be further discussed in later sections.

FGF also may have neuromodulatory effects, as demonstrated by the interference of exogenous FGF with long-term potentiation in hippocampal slices (Terlau and Seifert, 1990) and its ability to augment sodium-channel density in PC12 cells (Pollock et al., 1990).

Whether members of the FGF gene family other than acidic and basic FGF exert similar effects on neurons is not known, but it is conceivable, given the established similarities of the molecular structure of FGF receptors and the biologic actions of several FGFs (Baird and Böhlen, 1990; Merlo et al., 1990).

## B. In Vivo Effects

Data providing an idea of how the FGFs may act in the normal unlesioned nervous system are still scarce. In attempts to test the neurotrophic hypothesis, at least three groups of investigators have studied whether FGF can prevent or reduce ontogenetic neuron death, a phenomenon known to be influenced by neurotrophic factors, most notably NGF (Barde, 1989; Oppenheim, 1989). Dreyer et al. (1989) reported that basic FGF (recombinant) administered to the chorioallantoic membrane of chick embryos during the period of developmental neuron death in the ciliary ganglion, as judged by neuron cell counts, completely prevented death. In a follow-up study, these authors did not find any evidence for an increase in number of intraganglionic nonneuronal cells as determined by [$^3$H]thymidine incorporation (C. Grothe, unpublished observations), suggesting that FGF was not acting through a stimulation of glial cell proliferation. Hendry et al. (1990), in abstract form, communicated that both basic and acidic FGF reversed naturally occurring cell death in the chick ciliary ganglion in vivo and that antibodies to acidic FGF prevented the normal development of parasympathetic nerves supplying the iris in mice. The mode of action of FGF in this in vivo model is obscure as yet, since no specific retrograde transport of $^{125}$I-labeled basic or acidic FGF from the anterior chamber of the eye to the ciliary ganglion was found (Hendry and Belford, 1990). In contrast to these reports, Oppenheim and co-workers (1990) could not find any alterations in the course or magnitude of neuron losses in the development of chick spinal motoneurons or ciliary, sympathetic, and dorsal root ganglion neurons treated with basic FGF in vivo, although basic FGF was mitogenic for nonneuronal cells in the CNS. Reasons for these conflicting data regarding ontogenetic neuron death in the ciliary ganglion are not readily apparent. There is consistency, however, concerning a lack of effect of basic FGF on ontogenetic cell death of spinal motoneurons (cf. McManaman et al., 1990; K. Wewetzer and C. Grothe, unpublished observations). Nonetheless, basic FGF induces choline acetyltransferase activity in embryonic chick spinal cord neurons in vivo (C. Grothe and K. Wewetzer, unpublished observations). Since basic FGF is present in embryonic and adult muscle (Vaca et al., 1989), it is conceivable that it may act as a regulator of differentiation for cholinergic neurons. Interestingly, FGF also has been shown to increase aggregation of acetylcholine receptors of aneurally cultured muscle (Askanas et al., 1985).

Whether antibodies to basic FGF interfere with normal ontogenetic neuron death, causing a "paraimmunosympathectomy" or related numerical deficits in other neuron populations, would be important to know.

To our knowledge, FGF has not been applied to the adult unlesioned nervous system. However, basic FGF has been administered in gel foam locally to the striatum. Under these circumstances, basic FGF paradoxically decreases striatal levels of dopamine and tyrosine hydroxylase activity significantly (D. Otto, unpublished results).

## V. FGFs AND THEIR EFFECTS ON GLIAL CELLS

It has been known for some time that brain soluble extracts elicit morphologic changes, proliferation, and increased levels of S-100 protein and glutamine synthetase in cultured astroblasts (Sensenbrenner et al., 1980; Pettmann et al., 1982). Purification of the activities responsible for these effects led to the identification of two astroglial growth factors (Pettmann et al., 1985) that later were found to be acidic and basic FGF. Several other in vitro functions of FGFs for astroglial cells were described subsequently, including stimulation of release of plasminogen activator (Rogister et al., 1988), cell migration (Senior et al., 1986), expression of glial fibrillary acidic protein and tubulin (Morrison et al., 1985; Weibel et al., 1985,1987), and changes in membrane structure that possibly are related to altered expression of ion channels (Wolburg et al., 1986). By two-dimensional gel electrophoresis, Loret et al. (1988) have found FGF-dependent alterations of several proteins in cultured astroblasts. NGF and transforming growth factor βs (TGFβs) are among a potentially larger set of growth factors synthesized by astroglial cells that are regulated by FGF (Lindholm et al., 1990; Yoshida and Gage, 1990; K. C. Flanders, personal communication).

Proliferation of oligodendrocytes and Schwann cells also has been shown to be regulated by FGFs in vitro (Pruss et al, 1981; Eccleston and Silberberg, 1985; Saneto and deVellis, 1985; Besnard et al., 1989; Davis and Stroobant, 1990). In addition, FGF modulates development and induces carbonic anhydrase in oligodendrocytes (Delaunoy et al., 1988; McKinnon et al., 1990). These data from cell culture studies do not permit us to conclude that glial cell performances in the normal unlesioned nervous system are regulated by FGFs. Reports by Barotte et al. (1989) and Eclancher et al. (1990), suggesting that basic FGF injected into the brain (ventricle or lesion locus) increases the apparent number of reactive astrocytes, were based on experiments in rats with partial fimbria–fornix and stab or electrolytic lesions, respectively. However, basic FGF applied in gel foam to the striatum of parkinsonian mice failed to elicit a gliosis reaction overtly different from that induced by a nontrophic control protein, cytochrome C (D. Otto, unpublished observations). Studies to evaluate several astroglial parameters quantitatively, for example, GFAP protein and mRNA, in this experimental paradigm are in progress.

## VI. FGFs IN THE LESIONED NERVOUS SYSTEM: REGULATION AND EFFECTS

Several studies have reported alterations of FGF levels in the brain after mechanical and chemical trauma. Nieto-Sampedro et al. (1988) found a striking 13-fold increase of immunoreactive acidic FGF in a gel foam within 1 hr of making a cortical aspiration lesion, compared with the tissues immediately adjacent to the wound. The gel foam content of acidic FGF sharply decreased on day 3 after surgery, but there were no significant changes in brain tissue levels during the length of the experiment (20 days). A different time course for acidic FGF in the rat brain after cortical cavity lesioning was reported by Ishikawa et al. (1991a). These authors were unable to detect acidic FGF in the cavity fluid until 10 days after lesioning, in contrast to NGF, which appeared already after 16 hr and declined after 6 days. Levels of acidic FGF in the wound fluid increased gradually until day 30, whereas levels of both NGF and acidic FGF in the adjacent tissues were not changed significantly during the 30-day experimental period.

Finklestein and collaborators (1988), studying basic FGF-like immunoreactivity in sections of rat brain, found a marked increase in staining at the borders of the lesion to be localized in cells that morphologically resembled reactive astrocytes. This pattern of staining did not appear until 1 week after injury; normal astrocytes did not exhibit FGF-like immunoreactivity. In an extension of this work, Finklestein (1990), using heparin affinity chromatography coupled with a 3T3 cell mitogenic assay, found a 6-fold increase of basic FGF biologic activity in tissue surrounding the injury site after 1 week. This increase in basic FGF biologic activity was paralleled by increased levels of immunoreactive basic FGF (Western blots) and an 80-fold increase in basic FGF mRNA levels. In confirmation of results reviewed earlier, Finklestein and co-workers found no increases in tissue levels of acidic FGF bioactivity, immunoreactivity, and mRNA, supporting the notion that basic rather than acidic FGF is regulated specifically in the lesioned brain and possibly is involved in cellular events associated with brain tissue repair. In the retina of newborn mice, hyperoxia induces increased levels of immunoreactive basic FGF (Nyberg et al., 1990), which may be responsible for vasoproliferation following hyperoxia. Focal cerebral ischemia also leads to sustained enhanced FGF levels (bioactivity, immunoreactivity) that however, do not exceed 1.5- to 2-fold normal levels (Finklestein et al., 1989).

Although studies investigating temporal and spatial changes in FGF levels in neurological diseases such as Parkinson's disease and Alzheimer's disease are only beginning, but are of utmost importance to improving our understanding of the roles of FGF in brain pathology, several groups have set out to study FGFs as pharmacologic agents in a variety of neurologic disorders.

A deleterious consequence of axotomy, which aborts potential regeneration immediately, is neuron death. NGF has been shown to improve neuron survival dramatically in several lesion paradigms, for example, after sciatic nerve, fimbria–fornix, and optic nerve transections (Williams et al., 1986; Maffel et al., 1987; Otto et al., 1987,1989). FGFs also show such a neurotrophic capacity in vivo. Basic FGF administered to the transection site of an adult rat sciatic nerve maintains almost the same number of dorsal root ganglion neurons in spinal ganglia L4–6 as does NGF (Otto et al., 1987). Likewise, basic FGF chronically infused into the lateral ventricle or applied locally in gel foam to a transected fimbria–fornix rescues most cholinergic neurons in the ipsilateral medial septal nuclei that would otherwise degenerate (Anderson et al., 1988; Otto et al., 1989). In the adult rat retina, basic FGF partially prevents the massive degeneration of ganglion cells following an optic nerve cut (Sievers et al., 1987). Although these effects may be of potential clinical relevance, they also raise important questions regarding their underlying mechanisms. Conceptually, the beneficial effect of NGF on the maintenance of axotomized neuronal cell bodies can be attributed to a substitution for target organ-derived factor, whose retrograde axonal transport has been interrupted by the axonal lesion. Retrograde transport requires initial binding to axonal surface receptors and subsequent internalization. Although FGF has been shown to be transported axonally in several systems in the CNS (Ferguson et al., 1988b) and in the hypoglossal nerve (C. Grothe, unpublished observations), a specific retrograde transport in the sciatic nerve and in the septo-hippocampal system does not seem to occur (Ferguson et al., 1990a,b). Accordingly, maintenance of axotomized sensory and septal neurons in vivo by FGF must either be indirect and mediated by glial, vascular, or inflammatory cells or require an initial induction of axonal FGF receptors after lesioning in those systems that lack FGF receptors. Retinal ganglion cells have specific receptors for FGF, but transport the protein anterogradely from the retina to the lateral geniculate body and superior colliculus (Ferguson et al., 1990a). It seems, therefore, that FGF applied to the transected optic nerve must employ indirect mechanisms to rescue the retinal ganglion cells (Sievers et al., 1987).

The dopaminergic nigro-striatal projection is among those in the CNS for which specific retrograde transport of basic FGF and receptors has been documented (Ferguson et al., 1990b). Nigro-striatal neurons are affected severely in Parkinson's disease. Trophic effects of basic FGF on cultured mesencephalic neurons (Ferrari et al., 1989), isolation of a growth factor possibly related to FGF from the bovine striatum (Dal Toso et al., 1988), presence of FGF receptors on nigral neurons, and transport of FGF from the striatum to the substantia nigra provided rationales for administering basic FGF in animal models of Parkinson's disease (Otto and Unsicker, 1990; D. Otto, unpublished observations). Local unilateral application of basic FGF in gel foam to the striatum of mice treated simultaneously with or 1 week preceding FGF administration with the neurotoxin MPTP caused reappearance of tyrosine hydroxylase-immunoreactive nerve fibers in the ipsilateral striatum after 2 weeks. A particularly dense fiber plexus was found in close proximity to the implant. In contrast to the pronounced side differences in restoration of dopaminergic nerve terminals, several parameters of dopamine metabolism, including striatal levels of dopamine and tyrosine hydroxylase activity, were affected by FGF on either side. Discordant effects included higher levels of tyrosine hydroxylase protein on the ipsilateral than on the contralateral side, and a concomitant higher dopamine turnover on the contralateral side, as indicated by elevated levels of dihydroxyphenylacetic acid. Basic FGF was no longer detectable in eluates from the gel foams after 2 weeks, and was replaced by a neurotrophic activity distinct from basic FGF that supported embryonic chick ciliary ganglion neurons in culture. Gel foams soaked with cytochrome C also contained such a trophic material. Since cytochrome C-containing gel foams failed to reverse any of the chemical and morphologic deficits, it is obvious that neurotrophic activities accumulating in FGF- as opposed to cytochrome c-containing implants must differ, although either one promoted ciliary ganglion neurons. Determining whether the activity present in the FGF gel foam plays a role in the restoration of dopaminergic functions or whether FGF directly acts on dopaminergic nigro-striatal neurons in vivo requires further studies.

The capacity of FGF to antagonize a neurotoxic effect was underscored in a study using an excitatory amino acid transmitter, glutamate, on cultured hippocampal neurons (Mattson et al., 1989). This study, however, also pointed at another important putative role of FGF: regulating in concert with neurotransmitters, neuritic and dendritic pattern formation during development, that is, determining the degree of complexity of neuronal circuits.

A role for FGF in vivo to stimulate axonal regrowth after a peripheral nerve lesion is apparent from several studies. When applied in nerve guidance channels, both acidic and basic FGF enhance nerve fiber growth and axonal myelination (Cuevas et al., 1988; Danielsen et al., 1988; Aebischer et al., 1989; Cordeiro et al., 1989).

A central issue regarding the trophic and regenerative effects of FGFs in vivo concerns their mechanisms of action. Studies employing the adrenal medulla and its innervating neuron population, which is located in the intermediolateral column of the spinal cord, as a model have provided results compatible with the idea of several indirect modes of action in addition to possible direct modes (Blottner et al., 1989; Blottner and Unsicker, 1990). FGF is contained in the adrenal medulla and its chromaffin cells (Blottner et al., 1989; Grothe and Unsicker, 1990; Westermann et al., 1990). Selective destruction of the medulla causes substantial cell losses among the spinal cord neurons innervating the adrenal medulla; substitution of the medulla with basic FGF-soaked gel foam prevents these neuron losses. The protective effect of FGF is abolished when axons connecting spinal cord and adrenal gland are disrupted by splanchnicotomy, suggesting that a retrograde messenger, that may or may not be FGF, is essential for mediating the maintenance effect. Retrograde axonal transport of FGF in this system has not been shown to date. However, ultrastructurally, FGF adrenal implants, in contrast to the cytochrome C-containing ones, have numerous regenerated axons, well-organized vascular structures, and fibroblast-like cells. These components possibly provide a scenario for a cascade of events initiated, but not necessarily finalized, by FGF. FGF may trigger or enhance local production of unknown neurotrophic factors that may then find access to an axonal terminal field enlarged by the neurite-promoting factor FGF. If such a sequence of events accounts for the beneficial effects of the factor in the context of neurotrauma, degeneration, and toxicity, a brief period of application of the factor would perhaps suffice to generate the full set of desired effects. This hypothesis, if experimentally verified and correct, would also have implications for applications of FGF in the human brain (cf. Hefti and Knusel, 1988), alone or in combination with neuronal cell grafts that show improved survival when exposed to FGF (Giacobini et al., 1990).

## VII. FGFs IN NEURAL TUMORS

FGFs occur in neuroblastoma cells (Heymann et al., 1989; K. Wewetzer and G. Lüdecke, unpublished observations), gliomas, and glioma cell lines (see Murphy et al., 1989; Paulus et al., 1990; Westermann et al., 1991, for references). Although FGFs are present in many different types of tumor cells (Lobb, 1988; Baird and Böhlen, 1990), they are undetectable in others (Murphy et al., 1989), raising the question of whether they are essential for tumorigenesis and tumor growth.

Pheochromocytoma cells of the rat PC12 cell line have been very useful for characterizing responses of neuronal cells to FGFs at a molecular level and for comparing them with those elicited by NGF (Togari et al., 1985; Wagner and D'Amore, 1986; Neufeld et al., 1987; Rydel and Greene, 1987; Schubert et al., 1987; Sigmund et al., 1990). These studies, reviewed more extensively by Westermann et al. (1990), have demonstrated that NGF and FGF act through partially distinct mechanisms to elicit effects that differ, for example, with respect to neurite stability, induction of the β-amyloid precursor protein, and additivity in the induction of ornithine decarboxylase. These studies on PC12 cells were also important, since these cells are closely related to the chromaffin precursor cells that provided the first evidence that FGFs may be instrumental in cell lineage decisions during development.

## VIII. FGFs IN CELL LINEAGE DECISIONS

Chromaffin cells, the endocrine cells of the adrenal medulla, and sympathetic neurons share a common precursor cell that, in response to specific microenvironmental signals, develops endocrine or neuronal traits (Landis and Patterson, 1980; Unsicker et al., 1989). Stemple and collaborators (1988) have established that basic FGF induces a neuronal phenotype in chromaffin precursor cells or their immortalized counterparts, but also drives the cells into NGF dependence typical of sympathetic neurons. However, FGF is not only a differentiation signal, but also a mitogen for chromaffin cells. This ambivalence is reminiscent of the diverse effects that FGF induces on neuroblastoma cells, including mitogenicity, neuronal differentiation, and, in a few cell lines, arrest of cell division (Lüdecke and Unsicker, 1990a,b). The molecular bases of these phenomena remain to be investigated.

With respect to FGF and chromaffin precursor decisions, there is no evidence from FGF localization and in situ hybridization studies that FGF actually is present in the sympathetic ganglion primordia, a place one would expect to find it if its function is to channel a precursor cell in a neuronal direction. FGF seems to be absent from the embryonic adrenal medulla (Grothe and Unsicker, 1990), which could permit endocrine chromaffin cells to develop.

The possibility that FGF has important roles in other cell lineage decisions, as documented by in vitro studies, is questioned occasionally because of a lack of evidence for FGF localization in the right place at the right time. Thus, FGF stimulates differentiation of mouse neuroepithelial cells (Bartlett et al., 1988; Murphy et al., 1990) and proliferation of rat neuronal precursor cells (Gensburger et al., 1987), and serves as a decisive signal permitting NGF to act as a mitogen for neuronal precursor cells (Cattaneo and McKay, 1990). Whether, in all these instances, FGF is the correct factor, mimics the effect of another FGF family member, or serves as a tool to elucidate mechanisms essential for a particular development step, is still an open question.

## IX. FGFs AND THEIR INTERACTIONS WITH OTHER GROWTH FACTORS

One of the conceptually and experimentally most rewarding breakthroughs in growth factor research during the past 5 years has been the understanding of context dependency in the actions of growth factors (Sporn and Roberts, 1990). FGFs are no exception in this regard; from their multiple interactions with growth factors such as platelet-derived growth factor (PDGF) and TGFβs in eliciting specific effects on nonneural cells (Presta et al., 1988b; Noda and Vogel, 1989; Plouet and Gospodarowicz, 1989; Baird and Böhlen, 1990; Sporn and Roberts, 1990), such actions also may be inferred to occur in the nervous system. Unfortunately, interactions of FGFs with other factors on neurons and glial cells are only beginning to be explored.

We have studied putative interactions of basic FGF and TGFβs 1, 2, and 3 for the in vitro survival of several purified and enriched neuron populations, including chick ciliary, dorsal root, and spinal cord motoneurons, without noting significant changes in FGF-plus-TGFβ-treated cultures compared with cultures that received FGF only (Flanders et al., 1991; K. Unsicker and H. Reichert-Preibsch, unpublished observations). Walicke (1990) found that TGFβ1, when applied in combination with basic FGF, increased survival of cultured hippocampal neurons and decreased glial cell proliferation. Since receptors for TGFβ visualized by affinity labeling were detectable on astrocytes, but not on neurons, she concluded that the stimulatory effect of TGFβ on neurons was mediated through glial cells.

Modulation of the effects of basic FGF on proliferation and phenotypic expression of astroblasts in vitro was reported by Labourdette and co-workers (1990). A suppressive effect of TGFβs on proliferation that varied quantitatively with the respective TGFβ isoforms employed also was noted in studies on cultured astroglial cells (G. Lüdecke, unpublished observations). Although still largely descriptive, these findings are likely to have a considerable impact on formulating a conceptual framework for the multifunctional neural growth and differentiation factor family of the FGFs.

## X. CONCLUSIONS

### FGFs and the Need for a New Concept of Neurotrophic and Multifunctional Growth Factors in the Nervous System

For more than four decades, NGF has been the neurotrophic protein that was instrumental in shaping the theoretical model of a neurotrophic factor. Limiting amounts in target tissues, peaks of expression in target tissues during the period of physiologic cell death of the respective neuron populations, exclusive action on neurons, retrograde axonal transport, and neuron numbers in relation to ontogenetic neuron death affected by administration of the protein and antibodies in vivo were criteria for defining a neurotrophic factor. Based on the available information concerning brain-derived neurotrophic factor (BDNF) and neurotrophin-3 (NT-3), two additional members of the NGF gene family, it is already likely that more liberal definitions for a neurotrophic protein will replace the very stringent ones just outlined. For example, BDNF mRNA levels are higher in the adult than in the newborn brain, suggesting important actions on mature neurons or glial cells (Maisonpierre et al., 1990). Both factors and NGF occur in several nonneural tissues, for example, submandibular gland, seminal vesicle, testis, and ovary, indicating functions different from those in the nervous system (Ayer-Le Lièvre et al., 1988). BDNF mRNA has been found by in situ hybridization in sensory neurons (H. Persson, personal communication), suggesting autocrine functions or actions on satellite glial cells.

The dependence of a neuron on a sole neurotrophic factor, as suggested by the "immunosympathectomy" resulting from administration of NGF antibodies, also may be a simplification that fails to described correctly the putative dependence of a neuron on multiple trophic factors provided by multiple targets, target cells and glia, or retro- and anterogradely acting trophic molecules. Spinal cord motoneuron death, which in vivo can be affected by at least two different molecules [CAT development factor (CDF; McManaman et al., 1990) and CNTF; Wewetzer et al., 1990)], may become the first example of the convergent actions of two or more trophic factors determining neuronal survival.

FGFs, with their plethora of in vitro activities addressing neurons, glia, and other cells, definitely cannot be accommodated by a theorem based on NGF. Unfortunately, construction of a new and broader model, setting the stage for a comprehensive understanding of multifunctional growth factor actions in the nervous system, still is hampered by a lack of knowledge about such basic questions as whether FGFs in vivo regulate glial cell functions, determine cell fate, or rescue neurons from ontogenetic cell death. Most importantly, how FGFs are released from cells must be determined.

Certainly the reductionistic bias provided by a growth factor with the particular features and, perhaps, restricted functional repertoire of NGF cannot be employed to argue against the possibility that proteins exist that, unlike NGF, represent parts of rather than complete intercellular messages and, hence, require the presence of additional growth factors to complete their actions (cf. Sporn and Roberts, 1990), or represent signals for both neurons and nonneuronal cells with different resultant actions. Competition of neurons and nonneurons for the same factor, using receptors with different affinities and different degrading mechanisms could create situations in which the factor is available in limited amounts for one cell type but not for the other one. Restricted mobilization of a factor may be another means to ensure its limited availability, an important criterion for a neurotrophic factor. Factors with neurotrophic and glia-addressing properties could serve to reinforce the trophic response of the neuron by inducing secretion of another trophic protein from glial cells or to inhibit the trophic effect by inducing secretion of an antagonizing factor. These few examples may help illustrate that new models based on the idea of context for growth factor actions and cooperativity of different cell types, although more complex than a model based on NGF, may be closer to the complex reality of the nervous system and, eventually, may be more useful for guiding research on FGFs and other growth factors in neurobiology.

## ACKNOWLEDGMENTS

Work from our laboratory described in this article was supported by grants from the Deutsche Forschungsgemeinschaft, Deutsche Krebshilfe—Mildred Scheel Stiftung, and the Bundesministerium für Forschung und Technologie. The text of this chapter includes materials published through February, 1991, at which time it was submitted to the editors. For an update, see Unsicker et al., Cytokines in neural regeneration. *In* "Current Opinion in Neurobiology," Vol. 2, No. 5.

## REFERENCES

Abraham, J. A., Whang, J. L., Tumulo, A., Mergia, A., Friedman, J., Gospodarowicz, D., and Fiddes, J. C. (1986a). Human basic fibroblast growth factor: Nucleotide sequence and genomic organization. *EMBO J.* **5,** 2523–2528.

Abraham, J. A., Mergia, A., Whang, J. L., Tumolo, A., Friedman, J., Hjerrild, K. A., Gospodarowicz, D., and Fiddes, J. C. (1986b). Nucleotide sequence of a bovine clone encoding the angiogenic protein, basic fibroblast growth factor. *Science* **233,** 545–548.

Aebischer, P., Salessiotis, A. N., and Winn, S. R. (1989). Basic fibroblast growth factor released from synthetic guidance channels facilitates peripheral nerve regeneration across long nerve gaps. *J. Neurosci. Res.* **23,** 282–289.

Alterio, J., Halley, C., Brou, C., Soussi, T., Courtois, Y., and Laurent, M. (1988). Characterization of a bovine acidic FGF cDNA clone and its expression in brain and retina. *FEBS Lett.* **242,** 41–46.

Anderson, K. J., Dam, D., Lee, S., and Cotman, C. W. (1988). Basic fibroblast growth factor prevents death of lesioned cholinergic neurons in vivo. *Nature (London)* **332,** 360–362.

Askanas, V., Cave, S., Gallez-Hawkins, G., and King Engel, W. (1985). Fibroblast growth factor, epidermal growth factor and insulin exert a neuronal-like influence on acetylcholine receptors in aneurally cultured human muscle. *Neurosci. Lett.* **61,** 213–219.

Ayer-Le Lièvre, C., Olson, L., Ebendal, T., Hallböck, F., and Persson, H. (1988). Nerve growth factor mRNA and protein in the testis and epididymis of mouse and rat. *Proc. Natl. Acad. Sci. U.S.A.* **85,** 2628–2632.

Bandtlow, C. E., Heumann, R., Schwab, M. E., and Thoenen, H. (1988). Cellular localization of nerve growth factor synthesis in various organs of the peripheral nervous system. *In* "Neuronal Plasticity and Trophic Factors" (G. Biggio, P. F. Spano, G. Toffano, S. H. Appel, and G. L. Gesa, eds.). Liviana Press, Padova, Italy.

Baird, A., and Böhlen, P. (1990). Fibroblast growth factors. *In* "Peptide Growth Factors and Their Receptors I" (M. B. Sporn and A. B. Roberts, eds.), Vol. 95/1. Springer-Verlag, Berlin.

Baird, A., and Klagsbrun, M. (1991). The fibroblast growth factor family. *Ann. N.Y. Acad. Sci.* **638**.

Baird, A., Esch, F., Gospodarowicz, D., and Guillemin, R. (1985). Retina and eye-derived endothelial cell growth factors: Partial molecular characterization and identity with acidic and basic fibroblast growth factors. *Biochemistry* **24,** 7855–7860.

Barde, Y. A. (1989). Trophic factors and neuronal survival. *Neuron* **2,** 1525–1534.

Barotte, C., Eclancher, F., Ebel, A., Labourdette, G., Sensenbrenner, M., and Will, B. (1989). Effects of basic fibroblast growth factor (bFGF) on choline acetyltransferase activity and astroglial reaction in adult rats after partial fimbria transection. *Neurosci. Lett.* **101,** 197–202.

Bartlett, P. F., Reid, H. H., Bailey, K. A., and Bernard, O. (1988). Immortalization of mouse neural precursor cells by the c-*myc* oncogene. *Proc. Natl. Acad. Sci. U.S.A.* **85,** 3255–3259.

Besnard, F., Perraud, F., Sensenbrenner, M., and Labourdette, G. (1989). Effects of acidic and basic fibroblast growth factors on proliferation and maturation of cultured rat oligodendrocytes. *Int. J. Dev. Neurosci.* **7,** 401–409.

Birren, S. J., and Anderson, D. J. (1990). A v-*myc*-immortalized sympathoadrenal progenitor cell line in which neuronal differentiation in initiated by FGF but not NGF. *Neuron* **4,** 189–201.

Blottner, D., and Unsicker, K. (1990). Maintenance of intermediolateral spinal cord neurons by fibroblast growth factor administered to the medullectomized rat adrenal gland: Dependence on intact adrenal innervation and cellular organization of implants. *Eur. J. Neurosci.* **2,** 378–382.

Blottner, D., Westermann, R., Grothe, C., Böhlen, P., and Unsicker, K. (1989). Basic fibroblast growth factor in the adrenal gland. *Eur. J. Neurosci.* **1,** 471–478.

Burgess, W. H., and Maciag, T. (1989). Heparin binding (fibroblast) growth factor family of proteins. *Ann. Rev. Biochem.* **58,** 575–606.

Carmignoto, G., Maffei, L., Candeo, P., Canella, R., and Comelli, C. (1988). Effect of NGF on the survival of rat retinal ganglion cells following optic nerve transection. *J. Neurosci.* **9,** 1263.

Caruelle, D., Groux-Muscatelli, B., Gaudric, A., Sestier, C., Coscas, G., Caruelle, J. P., and Barritault, D. (1989). Immunological study of acidic fibroblast growth factor (aFGF) distribution in the eye. *J. Cell. Biochem.* **39,** 117–128.

Cattaneo, E., and McKay, R. (1990). Proliferation and differentiation of neuronal stem cells regulated by nerve growth factor. *Nature (London)* **347,** 762–765.

Cordeiro, P. G., Seckel, B. R., Lipton, S. A., D'Amore, P. A., Wagner, J., and Madison, R. (1989). Acidic fibroblast growth factor enhances peripheral nerve regeneration in vivo. *Plastic Reconstr. Surg.* **83,** 1013–1019.

Cotman, C. W. and Gómez-Pinilla (1991). Basic fibroblast growth factor in the mature brain. *Ann. N.Y. Acad. Sci.* **638,** 221–231.

Courty, J., Loret, C., Moenner, M., Chevallier, B., Lagente, O., Courtois, Y., and Barritault, D. (1985). Bovine retina contains three growth factor activities with different affinity to heparin: Eye derived growth factor I, II, III. *Biochimie* **67,** 265–269.

Courty, J., Loret, C., Chevallier, B., Moenner, M., and Barritault, D. (1987). Biochemical comparative studies between eye- and brain-derived growth factors. *Biochimie* **69,** 511–516.

Cuevas, P., Carceller, F., Esteban, A., Baird, A., and Guillemin, R. (1988). Basic fibroblast growth factor (bFGF) enhances retinal ganglion cells survival and promotes axonal growth of rat transected optic nerve. *3rd Joint Meeting on Neurochemical Approaches to the Understanding of Cerebral Disorders.* Copenhagen, June 9–12 (Abstr.).

Dal Toso, R., Giorgi, O., Soranzo, C., Kirschner, G., Ferrari, G., Favaron, M., Benvegnu, D., Presti, D., Vicini, S., Toffano, G., Azzone, G. F., and Leon, A. (1988). Development and survival of neurons in dissociated fetal mesencephalic serum-free cell cultures. I. Effects of cell density and of an adult mammalian striatal-derived neuronotrophic factor (SDNF). *J. Neurosci.* **8,** 733–745.

Damon, D. H., D'Amore, P. A., and Wagner, J. A. (1988). Sulfated glycosaminoglycans modify growth factor-induced neurite outgrowth in PC12 cells. *J. Cell. Physiol.* **135,** 293–300.

Damon, D. H., D'Amore, P. A., and Wagner, J. A. (1990). Nerve growth factor and fibroblast growth factor regulate neurite outgrowth and gene expression in PC12 cells via protein kinase- and cAMP-independent mechanisms. *Cell Biol.* **110,** 1333–1339.

Danielsen, N., Pettmann, B., Valhising, H. L., Manthorpe, M., and Varon, S. (1988). Fibroblast growth factor effects on peripheral nerve regeneration in a silicone chamber model. *J. Neurosci. Res.* **20,** 320–330.

Davis, J. B., and Stroobant, P. (1990). Platelet-derived growth factors and fibroblast growth factors are mitogens for rat Schwann cells. *J. Cell Biol.* **110,** 1353–1360.

Delaunoy, J. P., Langui, D., Ghandour, G., Labourdette, G., and Sensenbrenner, M. (1988). Influence of basic fibroblast growth factor on carbonic anhydrase expression by rat glial cells in primary culture. *Int. J. Dev. Neurosci.* **6,** 129–136.

Delli-Bovi, P., Curatola, A. M., Kern, F. G., et al. (1987). An oncogene isolated by transfection of Kaposi's sarcoma DNA encodes a growth factor that is a member of the FGF family. *Cell* **50**, 729–737.
Dickson, C., and Peters, G. (1987). Potential oncogene product related to growth factors. *Nature (London)* **326**, 833.
Dreyer, D., Lagrange, A., Grothe, C., and Unsicker, K. (1989). Basic fibroblast growth factor prevents ontogenetic neuron death in vivo. *Neurosci. Lett.* **99**, 35–38.
Eccleston, P. A., and Silberberg, D. H. (1985). Fibroblast growth factor is a mitogen for oligodendrocytes in vitro. *Dev. Brain Res.* **21**, 315–318.
Eclancher, F., Perraud, F., Faltin, J., Labourdette, G., and Sensenbrenner, M. (1990). Reactive astrogliosis after basic fibroblast growth factor (bFGF) injection in injured neonatal rat brain. *Glia* **3**, 502–509.
Emoto, N., Gonzalez, A. M., Walicke, P. A., Wada, E., Simmons, D. M., Shimasaki, S., and Baird, A. (1989). Basic fibroblast growth factor (FGF) in the central nervous system: Identification of specific loci of basic FGF expression in the rat brain. *Growth Factors* **2**, 21–29.
Feige, J. J., and Baird, A. (1988). Phosphorylation of basic FGF: A new substrate for protein kinase C (Abstract). *Endocrinology* **122**, 1237 A.
Ferguson, I. A., Schweitzer, J. B., and Johnson, E. M., Jr. (1990a). Basic fibroblast growth factor: receptor-mediated internalization, metabolism, and anterograde axonal transport in retinal ganglion cells. *J. Neurosci.* **10**, 2176–2189.
Ferguson, I. A., Wanaka, A., and Johnson, E. M., Jr. (1990b). bFGF undergoes receptor-mediated retrograde transport in CNS neurons. *Soc. Neurosci. Abstr.* **824**.
Ferrara, N., Ousley, F., and Gospodarowicz, D. (1988). Bovine brain astrocytes express basic fibroblast growth factor, a neurotrophic and angiogenic mitogen. *Brain Res.* **462**, 223–323.
Ferrari, G., Minozzi, M.-C., Toffano, G., Leon, A., and Skaper, S. D. (1989). Basic fibroblast growth factor promotes the survival and development of mesencephalic neurons in culture. *Dev. Biol.* **133**, 140–147.
Finklestein, S. P., Apostolides, P. J., Caday, C. G., Prosser, J., Philips, M. F., and Klagsbrun, M. (1988). Increased basic fibroblast growth factor (bFGF) immunoreactivity at the site of focal brain wounds. *Brain Res.* **460**, 253–259.
Finklestein, S. P., Flanning, P. J., Caday, C. G., Powell, P. P., and Klagsbrun, M. (1989). Preferential induction of basic fibroblast growth factor following focal brain injury. *J. Neurosci.* **9**, 123.
Flanders, K. C., Lüdecke, G., Engels, S., Cissel, D. S., Roberts, A. B., Kondaiah, P., Lafyatis, R., Sporn, M. B., and Unsicker, K. (1991). Localization and actions of transforming growth factors-βs in the embryonic nervous system. *Development* **113**, 183–191.
Florkiewicz, R. Z., and Sommer, A. (1989). Human basic fibroblast growth factor gene encodes four polypeptides: Three initiate translation from non-AUG codons. *Proc. Natl. Acad. Sci. U.S.A.* **86**, 3978–3981.
Friesel, R., Burgess, W. H., Mehlman, T., and Maciag, T. (1986). The characterization of the receptor for endothelial cell growth factor by covalent ligand attachment. *J. Biol. Chem.* **261**, 7581–7584.
Fuller, F., Lam, A., Kloss, J., Cordell, B., Varon, S., and Manthorpe, M. (1990). Depolarization modulates ciliary ganglion neuronal responses to neurokines. *Soc. Neurosci. Abstr.* **16**, 467.1.
Gensburger, C., Labourdette, G., and Sensenbrenner, M. (1987). Brain basic fibroblast growth factor stimulates the proliferation of rat neuronal precursor cells in vivo. *FEBS Lett.* **217**, 1–5.
Giacobini, M. B. M. J., Hoffer, B. J., and Olson, L. (1990). Trophic influences of fibroblast growth factors on fetal brain tissue grafts. *Soc. Neurosci. Abstr.* **16**, 485.
Gimenez-Gallego, G., Rodkey, J., Bennett, C., Rios-Candelore, M., DiSalvo, J., and Thomas, K. (1985). Brain-derived acidic fibroblast growth factor: Complete amino acid sequence and homologies. *Science* **230**, 1385–1388.
Gimenez-Gallego, G., Conn, G., Hatcher, V. B., and Thomas, K. A. (1986). Human brain-derived acidic and basic fibroblast growth factors: Amino terminal sequences and specific mitogenic activities. *Biochem. Biophys. Res. Commun.* **135**, 541–548.

Goldfarb, M., Bates, B., Drucker, B., Hardin, J., and Haub, O. (1991). Expression and possible functions of the FGF-5 gene. *Ann. N.Y. Acad. Sci.* **638**, 38–52.

Gonzalez, A.-M., Buscaglia, M., Ong, M., and Baird, A. (1990). Distribution of basic fibroblast growth factor in the 18-day rat fetus: Localization in the basement membranes of diverse tissues. *J. Cell Biol.* **110**, 753–765.

Gospodarowicz, D., Cheng, J., Lui, G. M., Baird, A., and Boehlen, P. (1984). Isolation of brain fibroblast growth factor by heparin-sepharose affinity chromatography: Identity with pituitary fibroblast growth factor. *Proc. Natl. Acad. Sci. U.S.A.* **81**, 6963–6967.

Gospodarowicz, D., Neufeld, G., and Schweigerer, L. (1986). Molecular and biological characterization of fibroblast growth factor, an angiogenic factor which also controls the proliferation and differentiation of mesoderm and neuroectoderm derived cells. *Cell Diff.* **19**, 1–17.

Grothe, C. (1991). Untersuchungen zur funktionellen Bedeutung des basischen Fibroblastenwachstumsfaktors im Nervensystem. Habilitationsschrift, Universität Marburg.

Grothe, C., and Unsicker, K. (1990). Immunocytochemical mapping of basic fibroblast growth factor in the developing and adult rat adrenal gland. *Histochemistry* **94**, 141–147.

Grothe, C., Otto, D., and Unsicker, K. (1989). Basic fibroblast growth factor promotes in vitro survival and cholinergic development of rat septal neurons: Comparison with the effects of nerve growth factor. *Neurosci.* **31**, 649–661.

Grothe, C., Zachman, K., Unsicker, K., and Westermann, R. (1990). High molecular weight forms of basic fibroblast growth factor recognized by a new anti-bFGF antibody. *FEBS Lett.* **260**, 35–39.

Grothe, C., Zachmann, K., and Unsicker, K. (1991). Basic FGF-like immunoreactivity in the developing and adult rat brain stem. *J. Comp. Neurol.* **305**, 1–9.

Hatten, M. E., Lynch, M., Rydel, R. E., Sanchez, J., Joseph-Silverstein, J., Moscatelli, D. and Rifkin, D. B. (1988). In vitro neurite extension by granule neurons is dependent upon astroglial-derived fibroblast growth-factor. *Dev. Biol.* **125**, 280–289.

Hefti, F., and Knusel, B. (1988). Chronic administration of nerve growth factor and other neurotrophic factors to the brain. *Neurobiol. Aging* **9**, 689–690.

Hendry, I. A., and Belford, D. A. (1990). Lack of retrograde axonal transport of fibroblast growth factors in peripheral neurons. *In* "Program and Abstracts of the 8th Biennial Meeting of the International Society for Developmental Neuroscience, June 16–22, Bal Harbour, Florida." *Int. J. Dev. Neurosci. Supp.* Vols. 1–8 Pergamon Press.

Hendry, I. A., Belford, D. A., Crouch, M. F., and Hill, C. E. (1990). Parasympathetic neurotrophic factors. *In* "Program and Abstracts of the 8th Biennial Meeting of the International Society for Developmental Neuroscience, June 16–22, Bal Harbour, Florida." *Int. J. Dev. Neurosci Supp.* Vols. 1–8 Pergamon Press.

Heuer, J. G., von Bartheld, C. S., Kinoshita, Y., Evers, P. C., and Bothwell, M. (1990). Alternating phases of FGF receptor and NGF receptor expression in the developing chicken nervous system. *Neuron* **5**, 283–296.

Heymann, D., Böhlen, P., Gautschi, P., Grothe, C., Lüdecke, G., and Unsicker, K. (1989). Expression of basic fibroblast growth factor and its effect on neuroblastoma cells. *J. Neurochem. (Suppl.)* **52**, S193.

Hoffman, R. S. (1940). The growth-activating effect of extracts of adult and embryonic tissues of the rat on fibroblast colonies in culture. *Growth* **4**, 361–376.

Imamura, T., Tokita, Y., and Mitsui, Y. (1988). Purification of basic FGF receptors from rat brain. *Biochem. Biophys. Res. Commun.* **155**, 583–590.

Ishikawa, R., Nishikori, K., and Furukawa, S. (1991a). Appearance of nerve growth factor and acidic fibroblast growth factor with different time courses in the cavity-lesioned cortex of the rat brain. *Neurosci. Lett.* **127**, 70–72.

Ishikawa, R., Nishikori, K., and Furukawa, S. (1991b). Developmental changes in distribution of acidic fibroblast growth factor in rat brain evaluated by a sensitive two-site enzyme immunoassay. *J. Neurochem.* **56**, 836–841.

Janet, T., Miehe, M., Pettmann, B., Labourdette, G., and Sensenbrenner, M. (1987). Ultrastructural localization of fibroblast growth factor in neurons of rat brain. *Neurosci. Lett.* **80**, 153–157.

Janet, T., Grothe, C., Pettmann, B., Unsicker, K., and Sensenbrenner, M. (1988). Immunocytochemical demonstration of fibroblast growth factor in cultured chick and rat neurons. *J. Neurosci. Res.* **19**, 195–201.

Kalcheim, C. (1989). Basic fibroblast growth factor stimulates survival of nonneuronal cells developing from trunk neural crest. *Dev. Biol.* **134**, 1–10.

Kalcheim, C., and Neufeld, G. (1990). Expression of basic fibroblast growth factor in the nervous system of early avian embryos. *Development* **109**, 203–215.

Knusel, B., Michel, P. P., Schwaber, J. S., and Hefti, F. (1990). Selective and nonselective stimulation of central cholinergic and dopaminergic development in vitro by nerve growth factor, basic fibroblast growth factor, epidermal growth factor, insulin and the insulin-like growth factors I and II. *J. Neurosci.* **10**, 558–570.

Kornbluth, S., Paulson, K. E., and Hanafusa, H. (1988). Novel tyrosine kinase identified by phosphotyrosine antibody screening of cDNA libraries. *Mol. Cell. Biol.* **8**, 5541–5544.

Korsching, S., Heumann, R., Davies, A. M., and Thoenen, H. (1988). Levels of nerve growth factor and its mRNA during development and regeneration of the peripheral nervous system. *In* "Neuronal Plasticity and Trophic Factors" (G. Biggio, P. F. Spano, G. Toffano, S. H. Appel, and G. L. Gesa, eds.). Liviana Press, Padova, Italy.

Labourdette, G., Janet, T., Laeng, P., Perraud, F., Lawrence, D., and Pettmann, B. (1990). Transforming growth factor type $\beta 1$ modulates the effects of basic fibroblast growth factor on growth and phenotypic expression of rat astroblasts in vitro. *J. Cell. Physiol.* **144**.

Landis, S. C., and Patterson, P. H. (1981). Neural crest cell lineages. *Trends Neurosci.* **4**, 172–175.

Lee, P. L., Johnson, D. E., Cousens, L. S., Fried, V. A., and Williams, L. T. (1989). Purification and complementary DNA cloning of a receptor for basic fibroblast growth factor. *Science* **245**, 57–60.

Lehwalder, D., Jeffrey, P. L., and Unsicker, K. (1989). Survival of purified embryonic chick retinal ganglion cells in the presence of neurotrophic factors. *J. Neurosci. Res.* **24**, 329–337.

Libermann, T. A., Friesel, R., Jaye, M., Lyall, R. M., Westermark, B., Drohan, W., Schmidt, A., Maciag, T., and Schlessinger, J. (1987). An angiogenic growth factor is expressed in human glioma cells. *EMBO J.* **6**, 1627–1632.

Lindholm, D., Hengerer, B., Zafra, F., and Thoenen, H. (1990). Transforming growth factor-$\beta 1$ stimulates expression of nerve growth factor in the rat CNS. *NeuroReport* **1**, 9–12.

Lipton, S. A., Wagner, J. A., Madison, R. D., and D'Amore, P. A. (1988). Acidic fibroblast growth factor enhances regeneration of processes by postnatal mammalian retinal ganglion cells in culture. *Proc. Natl. Acad. Sci. U.S.A.* **85**, 2388–2392.

Lobb, R. R. (1988). Clinical applications of heparin-binding growth factors. *Eur. J. Clin. Invest.* **18**, 321–336.

Lobb, R. R., Harper, J. W., and Fett, J. W. (1986). Purification of heparin-binding growth factors. *Anal. Biochem.* **154**, 1–14.

Logan, A., and Logan, S. D. (1986). Distribution of fibroblast growth factor in the central and peripheral nervous system of various mammals. *Neurosci. Lett.* **69**, 162–165.

Logan, R. R., Berry, M., Thomas, G. H., Gregson, N. A., and Logan, S. D. (1985). Identification and partial purification of fibroblast growth factor from the brains of developing rats and leucodystrophic mutant mice. *Neurosci.* **15**, 1239–1246.

Loret, C., Sensenbrenner, M., and Labourdette, G. (1988). Maturation-related gene expression of rat astroblasts in vitro studied by two-dimensional polyacrylamide gel electrophoresis. *Cell Diff. Dev.* **25**, 37–46.

Lüdecke, G., and Unsicker, K. (1990a). Basic FGF and NGF as mitogens for human neuroblastoma cells. *Clin. Chem. Enzym. Comm.* **2**, 293–298.

Lüdecke, G., and Unsicker, K. (1990b). Mitogenic effect of neurotrophic factors on human IMR 32 neuroblastoma cells. *Cancer* **65**, 2270–2278.

McKinnon, R. D., Matsui, T., Dubois-Dalcq, M., and Aaronson, S. A. (1990). FGF modulates the PDGF-driven pathway of oligodendrocyte development. *Neuron* **5**, 603–614.

McManaman, J. L., Oppenheim, R. W., Prevette, D., and Marchetti, D. (1990). Rescue of motoneurons from cell death by a purified skeletal muscle polypeptide: Effects of the chat development factor, CDF. *Neuron* **4,** 891–898.

Maffel, L., Carmignoto, G., Perry, V. H., Candeo, P., and Ferrari, G. (1987). Schwann cells promote the survival of rat ganglion cells after optic nerve transection. *Proc. Natl. Acad. Sci. U.S.A.* **84,** 1855–1859.

Maisonpierre, P. C., Belluscio, L., Squinto, S., Yo, N. Y., Furth, M. E., Lindsay, R. M., and Yancopoulos, G. D. (1990). Neurotrophin-3: A neurotrophic factor related to NGF and BDNF. *Science* **247,** 1446–1451.

Marics, I., Adelaide, J., Raybaud, F., Mattei, M. G., Coulier, F., Planche, J., de Lapeyriere, O., and Birnbaum, D. (1989). Characterization of the HST-related FGF-6 gene, a new member of the fibroblast growth factor gene family. *Oncogene* **4,** 335–340.

Matsuda, S., Saito, H., and Nishiyama, N. (1990). Effect of basic fibroblast growth factor on neurons cultured from various regions of postnatal rat brain. *Brain Res.* **520,** 310–316.

Mattson, M. P., Murrain, M., Guthrie, P. B., and Kater, S. B. (1989). Fibroblast growth factor and glutamate: Opposing roles in the generation and degeneration of hippocampal neuroarchitecture. *J. Neurosci.* **9,** 3728–3740.

Merlo, G. R., Blondel, B. J., Dreed, R., MacAllan, D., Peters, G., Dickson, C., Liscia, D. S., Ciardiello, F., Valverius, E. M., Salomon, D. S., and Callahan, R. (1990). The mouse int-2 gene exhibits basic fibroblast growth factor activity in a basic fibroblast growth factor responsive cell line. *Cell Growth Diff.* **1,** 463–472.

Moenner, M., Chevallier, B., Badet, J., and Barritault, D. (1986). Evidence and characterization of the receptor to eye-derived growth factor I, the retinal form of basic fibroblast growth factor, on bovine epithelial lens cells. *Proc. Natl. Acad. Sci. U.S.A.* **83,** 5024–5028.

Morrison, R. S. (1987). Fibroblast growth factors: Potential neurotrophic agents in the central nervous system. *J. Neurosci. Res.* **17,** 99–101.

Morrison, R. S., De Vellis, J., Lee, Y. L., Bradshaw, R., and Eng, L. F. (1985). Hormones and growth factors induce the synthesis of glial fibrillary acidic protein in rat brain astrocytes. *J. Neurosci. Res.* **14,** 167–176.

Morrison, R. S., Sharma, A., de Vellis, J., and Bradshaw, R. A. (1986). Basic fibroblast growth factor supports the survival of cerebral cortical neurons in primary culture. *Proc. Natl. Acad. Sci. U.S.A.* **83,** 7537–7541.

Morrison, R. S., Gross, J. L., and Moskal, J. R. (1988). Inhibition of protein kinase C activity promotes the neurotrophic action of epidermal and basic fibroblast growth factors. *Brain Res.* **473,** 141–146.

Moscatelli, D., Joseph-Silverstein, J., Manejjas, R., and Rifkin, D. B. (1987). $M_r$ 25,000 heparin-binding protein from guinea pig brain is a high molecular weight form of basic fibroblast growth factor. *Proc. Natl. Acad. Sci. U.S.A.* **84,** 5778–5782.

Murphy, M., Drago, J., and Bartlett, P. F. (1990). Fibroblast growth factor stimulates the proliferation and differentiation of neural precursor cells in vivo. *J. Neurosci. Res.* **25,** 463–475.

Murphy, P. R., Myal, Y., Sato, Y., Sato, R., West, M., and Friesen, H. G. (1989). Elevated expression of basic fibroblast growth factor messenger ribonucleic acid in acoustic neuromas. *Mol. Endocrinol.* **3,** 225–231.

Neufeld, G., and Gospodarowicz, D. (1985). The identification and partial characterization of the fibroblast growth factor receptor of baby hamster kidney cells. *J. Biol. Chem.* **260,** 13860–13868.

Neufeld, G., and Gospodarowicz, D. (1986). Basic and acidic fibroblast growth factors interact with the same cell surface receptors. *J. Biol. Chem.* **261,** 5631–5637.

Neufeld, G., Gospodarowicz, D., Dodge, L., and Fuji, D. K. (1987). Heparin modulation of the neurotrophic effects of acidic and basic fibroblast growth factors and nerve growth factor on PC12 cells. *J. Cell. Physiol.* **131,** 131–140.

Nieto-Sampedro, M., Lim, R., Hicklin, D. J., and Cotman, C. W. (1988). Early release of glia maturation factor and acidic fibroblast growth factor after rat brain injury. *Neurosci. Lett.* **86,** 361–365.

Noda, M., and Vogel, R. (1989). Fibroblast growth factor enhances type β1 transforming growth factor gene expression in osteoblast-like cells. *J. Cell Biol.* **109,** 2529–2535.

Nyberg, F., Hahnenberger, R., Jakobson, A. M., and Terenius, L. (1990). Enhancement of FGF-like polypeptides in the retinae of newborn mice exposed to hyperoxia. *FEBS Lett.* **267**, 75–77.

Olwin, B. B., and Hauschka, S. D. (1986). Identification of the fibroblast growth factor receptor of Swiss 3T3 cells and mouse skeletal muscle myoblasts. *Biochemistry* **25**, 3487–3492.

Olwin, B. B., and Hauschka, S. D. (1990). Fibroblast growth factor receptor levels decrease during chick embryogenesis. *J. Cell Biol.* **110**, 503–509.

Oppenheim, R. W. (1989). The neurotrophic theory and naturally occurring montoneuron death. *Trends Neurosci.* **12**, 252–255.

Oppenheim, R. W., Prevette, D., and Fuller, F. H. (1990). In vivo treatment with basic fibroblast growth factor during development does not alter naturally occurring neuronal death. *Soc. Neurosci. Abstr.* **16**, 1135.

Otto, D., and Unsicker, K. (1990a). Basic FGF reverses chemical and morphological deficits in the nigrostriatal system of MPTP-treated mice. *J. Neurosci.* **10**, 1912–1921.

Otto, D., and Unsicker, K. (1990b). Basic FGF, lesion-induced neuron death and neural regeneration. *Adv. Neural Regen.* 115–124.

Otto, D., Unsicker, K., and Grothe, G. (1987). Pharmacological effects of nerve growth factor and fibroblast factor applied to the transectioned sicatic nerve on neuron death in adult rat dorsal root ganglia. *Neurosci. Lett.* **83**, 156–160.

Otto, D., Frotscher, M., and Unsicker, K. (1989). Basic fibroblast growth factor and nerve growth factor administered in gel foam rescue medial septal neurons after fimbria fornix transection. *J. Neurosci. Res.* **22**, 83–91.

Otto, D., Grothe, C., Westermann, R., and Unsicker, K. (1991). Basic FGF and its actions on neurons: A group account with special emphasis on the Parkinsonian brain. *In* "Plasticity and Regeneration of the Nervous System" (P. S. Timiras, ed.). Plenum Press, New York.

Paulus, W., Grothe, C., Sensenbrenner, M., Janet, T., Baur, I., Graf, M., and Roggendorf, W. (1990). Localization of basic fibroblast growth factor, a mitogen and angiogenic factor, in human brain tumors. *Acta Neuropathol.* **79**, 418–423.

Perraud, F., Labourdette, G., Miche, M., Loret, C., and Sensenbrenner, M. (1988a). Comparison of the morphological effects of acidic and basic fibroblast growth factors on rat astroblasts in culture. *J. Neurosci. Res.* **20**, 1–11.

Perraud, F., Besnard, F., Pettmann, B., Sensenbrenner, M., and Labourdette, G. (1988b). Effects of acidic and basic fibroblast growth factors (aFGF and bFGF) on the proliferation and the glutamine synthetase expression of rat astroblasts in culture. *Glia* **1**, 124–131.

Pettmann, B., Weibel, M., Daune, G., Sensenbrenner, M., and Labourdette, G. (1982). Stimulation of proliferation and maturation of rat astroblasts in serum-free culture by an astroglial growth factor. *J. Neurosci. Res.* **8**, 463–476.

Pettmann, B., Weibel, M., Sensenbrenner, M., and Labourdette, G. (1985). Purification of two astroglial growth factors from bovine brain. *FEBS Lett.* **189**, 102–108.

Pettmann, B., Labourdette, G., Weibel, M., and Sensenbrenner, M. (1986). The brain fibroblast growth factor (FGF) is localized in neurons. *Neurosci. Lett.* **68**, 175–180.

Plouet, J., and Gospodarowicz, D. (1989). Transforming growth factor β-1 positively modulates the bioactivity of fibroblast growth factor on corneal endothelial cells. *J. Cell. Physiol.* **141**, 392–399.

Plouet, J. Mascarelli, F., Loret, M. D., Faure, J. P., and Courtois, Y. (1988). Regulation of eye derived growth factor binding to membranes by light, ATP or GTP in photoreceptor outer segments. *EMBO J.* **7**, 373–376.

Pollock, J. D., Krempin, M., and Rudy, B. (1990). Differential effects of NGF, FGF, EGF, cAMP and dexamethasone on neurite outgrowth and sodium channel expression in PC12 cells. *J. Neurosci.* **10**, 2626–2637.

Prats, H., Kaghad, M., Prats, A. C., Klagsbrun, M., Lelias, J. M., Liauzun, P., Chalon, P., Tauber, J. P., Amalric, F., Smith, J. A., and Caput, D. (1989). High molecular mass forms of basic fibroblast growth factor are initiated by alternative CUG codons. *Proc. Natl. Acad. Sci. U.S.A.* **86**, 1836–1840.

Presta, M., Rusnati, M., Maier, J. A. M., and Ragnotti, G. (1988a). Purification of basic fibroblast growth factor from rat brain: Identification of a $M_r$ 22,000 immunoreactive form. *Biochem. Biophys. Res. Commun.* **155**, 1161–1172.

Presta, M., Maier, J. A. M., Rusnati, M., Moscatelli, D., and Ragnotti, G. (1988b). Modulation of plasminogen activator activity in human endometrial adenocarcinoma cells by fms-like gene (FLG). *Proc. Natl. Acad. Sci. U.S.A.* **86**, 8722–8726.

Rydel, R. E., and Greene, L. A. (1987). Acidic and basic fibroblast growth factors promote stable neurite outgrowth and neuronal differentiation in cultures of PC12 cells. *J. Neurosci.* **7**, 3639–3653.

Saksela, O., Moscatelli, D., and Rifkin, D. B. (1988). Endothelial cell-derived heparin sulfate bind basic fibroblast growth factor and protects it from proteolytic degradation. *J. Cell Biol.* **107**, 743–751.

Saneto, R. P., and de Vellis, J. (1985). Characterization of cultured rat oligodendrocytes proliferating in a serum-free, chemically defined medium. *Proc. Natl. Acad. Sci. U.S.A.* **82**, 3509–3513.

Schubert, D., Ling, N., and Baird, A. (1987). Multiple influences of a heparin-binding growth factor on neuronal development. *J. Cell Biol.* **104**, 635–643.

Senior, R. M., Huang, S. S., Griffin, G. L., and Huang, J. S. (1986). Brain-derived growth factor is a chemoattractant for fibroblasts and astroglial cells. *Biochem. Biophys. Res. Commun.* **141**, 67–72.

Sensenbrenner, M., Devilliers, G., Brock, E., and Porte, A. (1980). Biochemical and ultrastructural studies of cultured rat astroglial cells. Effect of brain extract and dibutyryl cyclic AMP on glial fibrillary acidic protein and glial filaments. *Differentiation* **17**, 51–61.

Shimasaki, S., Emoto, N., Koba, A., Mercado, M., Shibata, F., Cooksey, K., Baird, A., and Ling, N. (1988). Complementary DNA cloning and sequencing of rat ovarian basic fibroblast basic fibroblast growth factor and transforming growth factor β1. *Cancer Res.* **48**, 6384–6389.

Presta, M., et al. (1991). Letter to the Editor. *J. Neurochem.* (in press).

Pruss, R. M., Bartlett, P. F., Gavrilovic, J., Lisak, R. P., and Rattray, S. (1981). Mitogens for glial cells: A comparison of the response of cultured astrocytes, oligodendrocytes and Schwann cells. *Brain Res.* **254**, 19–35.

Reid, H. H., Wilks, A. F., and Bernard, O. (1990). Two forms of the basic fibroblast growth factor receptor-like mRNA are expressed in the developing mouse brain. *Proc. Natl. Acad. Sci. U.S.A.* **87**, 1596–1600.

Rifkin, D. B., and Moscatelli, D. (1989). Recent developments in the cell biology of basic fibroblast growth factor. *J. Cell Biol.* **109**, 1–6.

Risau, W., Gautschi-Sova, P., and Böhlen, P. (1988). Endothelial cell growth factors in embryonic and adult chick brain are related to human acidic fibroblast growth factor. *EMBO J.* **7**, 959–962.

Rogister, B., Leprince, P., Pettmann, B., Labourdette, G., Sensenbrenner, M., and Moonen, G. (1988). Brain basic fibroblast growth factor stimulates the release of plasminogen activators by newborn rat cultured astroglial cells. *Neurosci. Lett.* **91**, 321–326.

Ruta, M., Burgess, W., Givol, D., Epstein, J., Neiger, N., Kaplow, J., Crumley, G., Dionne, C., Jaye, M., and Schlessinger, J. (1989). Receptor for acidic fibroblast growth factor is related to tyrosine kinase encoded by the growth factor and tissue distribution study of its mRNA. *Biochem. Biophys. Res. Commun.* **157**, 256–263.

Sievers, J., Hausmann, B., Unsicker, K., and Berry, M. (1987). Fibroblast growth factors promote the survival of adult rat retinal ganglion cells after transection of the optic nerve. *Neurosci. Lett.* **76**, 157–162.

Sigmund, O., Naort, Z., Anderson, D. J., and Stein, R. (1990). Effect of nerve growth factor and fibroblast growth factor on SCG10 and c-fos expression and neurite outgrowth in protein kinase C-depleted PC12 cells. *J. Biol. Chem.* **265**, 2257–2261.

Sporn, M. B., and Roberts, A. B. (1990). The multifunctional nature of peptide growth factors. *Handbook Exp. Pharmacol.* **195/I**, 3–15.

Stemple, D. L., Mahanthappa, N. K., and Anderson, D. J. (1988). Basic FGF induces neuronal differentiation, cell division, and NGF dependence in chromaffin cells: A sequence of events in sympathetic development. *Neuron* **1**, 517–525.

Taira, M., Yoshida, T., Miyagawa, K., et al. (1987). cDNA sequence of human transforming gene hst and identification of the coding sequence required for transforming activity. *Proc. Natl. Acad. Sci. U.S.A.* **84**, 2980–2984.

Terlau, H., and Seifert, W. (1990). Fibroblast growth factor enhances long-term potentiation in the hippocampal slice. *Eur. J. Neurosci.* **2**, 973–977.
Togari, A., Dickens, G., Kuzuya, H., and Guroff, G. (1985). The effect of fibroblast growth factor on PC12 cells. *J. Neurosci.* **5**, 307–316.
Trowell, O. A., Chir, B., and Willmer, E. N. (1939). Growth of tissues in vitro. IV. The effects of some tissue extracts on the growth of periosteal fibroblasts. *J. Exp. Biol.* **16**, 60–70.
Unsicker, K., Reichert-Preibsch, H., Schmidt, R., Pettmann, B., Labourdette, G., and Sensenbrenner, M. (1987). Astroglial and fibroblast growth factors have neurotrophic functions for cultured peripheral and central nervous system neurons. *Proc. Natl. Acad. Sci. U.S.A.* **84**, 5459–5463.
Unsicker, K., Seidl, K., and Hofmann, H. D. (1989). The neuroendocrine ambiguity of sympathoadrenal cells. *Int. J. Dev. Neurosci.* **7**, 413–417.
Unsicker, K., Reichert-Preibsch, H. and Wewetzer, K. (1991). FGF and CNTF: Evidence for a direct effect on neurons from low density and single cell culture.
Vaca, K., Stewart, S. S., and Appel, S. H. (1989). Identification of basic fibroblast growth factor as a cholinergic growth factor from human muscle. *J. Neurosci. Res.* **23**, 55–63.
Wagner, J. A., and D'Amore, P. (1986). Neurite outgrowth induced by an endothelial cell mitogen isolated from retina. J. Cell Biol. **103**, 1363–1367.
Walicke, P. (1988). Basic and acidic fibroblast growth factors have trophic effects on neurons from multiple CNS regions. *J. Neurosci.* **8**, 2618–2627.
Walicke, P. (1989). Novel neurotrophic factors, receptors, and oncogenes. *Ann. Rev. Neurosci.* **12**, 103–126.
Walicke, P. (1990). Fibroblast growth factor (FGF): A multifunctional growth factor in the CNS. *Adv. Neural Regen. Res.* 103–114.
Walicke, P., and Baird, A. (1988). Neurotrophic effects of basic and acidic fibroblast growth factors are not mediated through glial cells. *Dev. Brain Res.* **40**, 71–79.
Walicke, P., Cowan, W. M., Ueno, N., Baird, A., and Guillemin, R. (1986). Fibroblast growth factor promotes survival of dissociated hippocampal neurons and enhances neurite extension. *Proc. Natl. Acad. Sci. U.S.A.* **83**, 3012–3016.
Walicke, P., Feige, J. J., and Baird, A. (1989). Characterization of the neuronal receptor for basic fibroblast growth factor (bFGF): Comparison to mesenchymal cell receptors. *J. Biol. Chem.* **264**, 4120–4126.
Wanaka, A., Johnson, M., Jr., and Milbrandt, J. (1990). Localization of FGF receptor mRNA in the adult rat central nervous system by in situ hybridization. *Neuron* **5**, 267–281.
Weibel, M., Pettmann, M., Labourdette, G., Miehe, M., Bock, E., and Sensenbrenner, M. (1985). Morphological and biochemical maturation of rat astroglial cells grown in a chemically defined medium: Influence of an astroglial growth factor. *Int. J. Dev. Neurosci.* **3**, 617–630.
Weibel, M., Fages, C., Belakebi, M., Tardy, M., and Nuñez, J. (1987). Astroglial growth factor-2 (AGF2) increases tubulin in astroglial cells cultured in a defined medium. *Neurochem. Int.* **11**, 223–228.
Westermann, R., Johannsen, M., Unsicker, K., and Grothe, C. (1990). Basic fibroblast growth factor (bFGF) immunoreactivity is present in chromaffin granules. *J. Neurochem.* **55**, 285–292.
Westermann, R., Grothe, C., and Unsicker, K. (1991). Basic fibroblast growth factor (bFGF), a multifunctional growth factor for neuroectodermal cells. *J. Cell Sci. (Suppl.)* **13**, 97–117.
Wewetzer, K., MacDonald, J. R., Collins, F., and Unsicker, K. (1990). CNTF rescues motoneurons from ontogenetic cell death in-vivo, but not in-vitro. *NeuroReport* **1**, 203–206.
Williams, L. R., Varon, S., Peterson, G. M., Wictorin, K., Fischer, W., Björklund, A., and Gage, F. H. (1986). Continuous infusion of nerve growth factor prevents basal forebrain neuronal death after fimbria fornix transection. *Proc. Natl. Acad. Sci. U.S.A.* **83**, 9231–9235.
Williams, L. T. (1991). Expression and function of multiple forms of FGF receptors. *Ann. N.Y. Acad. Sci.* (in press).
Wilkinson, D. G., Peters, G., Dickson, C., and McMahon, A. P. (1988). Expression of the FGF-related proto-oncogene int-2 during gastrulation and neurulation in the mouse. *EMBO J.* **7**, 691–695.
Wilkinson, D. G., Bhatt, S., and McMahon, A. P. (1989). Expression pattern of the FGF-related protooncogene int-2 suggests multiple roles in fetal development. *Development* **105**, 131–136.

Wolburg, H., Neuhaus, J., Pettmann, B., Labourdette, G., and Sensenbrenner, M. (1986). Decrease in the density of orthogonal arrays of particle in membranes of cultured rat astroglial cells by the brain fibroblast growth factor. *Neurosci. Lett.* **72,** 25–30.

Yoshida, K., and Gage, F. H. (1990). Regulation of nerve growth factor synthesis by fibroblast growth factor in astrocytes. *Soc. Neurosci. Abstr.* **16,** 992.

# 11  Epidermal Growth Factor: Structure, Expression, and Functions in the Central Nervous System

Richard Morrison

## I. IDENTIFICATION

Epidermal growth factor (EGF) is a small mitogenic polypeptide that was isolated originally from the submaxillary glands of male mice (Cohen, 1962; Savage and Cohen, 1972). Its activity was discovered unexpectedly in a bioassay of crude fractions of nerve growth factor (NGF) that resulted in accelerated incisor eruption and precocious opening of the eyelids after injection into perinatal mammals. EGF subsequently was isolated from the extracts and shown to be a low molecular weight, heat stable, nondialyzable protein. Moreover, EGF accounted for approximately 0.5% of the protein content of the submaxillary gland. These studies indicated that EGF might have important physiologic functions in vivo. Although EGF has a variety of effects on cells, it has been studied primarily as a stimulator of cellular proliferation. Because of its abundance in the submaxillary gland and its ease of purification, it has been well characterized. The human equivalent is urogastrone, which shares extensive sequence homology with EGF and is a gut peptide bearing antigastric secretory activity.

## II. MOLECULAR STRUCTURE AND HOMOLOGIES

The structure of the mRNA encoding the precursor to mouse submaxillary EGF has been determined from a set of overlapping complementary DNAs (Gray et al., 1983; Scott et al., 1983). The EGF moiety consists of 53 amino acid residues and is synthesized as a portion of a 1217-amino-acid precursor molecule (Gray et al., 1983; Scott et al., 1983). The EGF precursor is predicted to be a transmembrane protein with an extracellular domain that contains seven peptides with sequences that are similar but not identical to EGF. In addition, the precursor contains a 165-amino-acid cytoplasmic domain. The boundaries of each EGF-like peptide can be defined by a basic amino acid, indicating that each could be released by trypsin-like enzymes (Scott et al., 1983). Thus, the preproEGF molecule may be processed to yield a number of different peptides. Tissue- or temporal-specific processing of the precursor molecule could generate a spectrum of structurally similar but functionally different proteins with a variety of effects on development. It is not presently known whether the EGF-like sequences are processed. There is certainly precedence for such processing, for example, processing of the glucagon and pro-opiomelanocortin precursors (Seger and Bennet, 1986; Mojsov et al., 1987). Further, the homeotic proteins NOTCH and lin-12 also contain multiple EGF-like sequences (Bender, 1985; Wharton et al., 1985).

Perhaps even more surprising was the observation that the precursor molecule also contains a hydrophobic domain adjacent to the EGF moiety. This segment could function as a membrane spanning region. If so, the EGF precursor could, in a less processed form, act as a transmembrane receptor. Evidence for this hypothesis was obtained when a cDNA was used to probe for the presence of preproEGF mRNA in mouse kidney (Rall et al., 1985). A substantially greater amount of this message is present in mouse kidney than is necessary to express the levels of EGF found in this tissue. Further, a larger antigen precipitated by antiserum to preproEGF was observed in mouse kidney, indicating an unprocessed EGF molecule. Recent analysis of the purified precursor shows it to be a glycosylated integral membrane protein with an apparent molecular mass of 140–150 kDa (Breyer and Cohen, 1990). The solubilized precursor protein exhibits biologic activity, as demonstrated by its ability to compete with $^{125}$I-labeled EGF for binding to the EGF receptor in intact fibroblasts and by its ability to stimulate the growth of mouse keratinocytes. Finally, extensive enzymatic digestion of the EGF precursor with pepsin yields a biologically and immunologically active protein similar in size to mature EGF. Thus, it is conceivable that, in the absence of specific proteases, the precursor may have localized effects. In support of this concept, the transforming growth factor α (TGFα) precursor, when expressed on the cell surface, binds to the EGF receptor on adjacent cells, resulting in signal transduction (Brachmann et al., 1989; Wong et al., 1989). These effects may be brought about by both homotypic and heterotypic cellular interactions between neurons and glia. For example, contact between neurons bearing the precursor and those bearing the receptor (Fig. 1) presumably could influence such processes as neuronal survival, neurite outgrowth, or neurotransmitter expression in restricted cell populations. Conversely, astrocytes bearing the EGF precursor could influence neurons bearing the EGF receptor. In the presence of proteases, the mature growth factor could be released for more distant effects.

The mature form of EGF is a single polypeptide chain containing 53 amino acids with an isoelectric point at pH 4.6. EGF contains six cysteine residues that are involved in the formation of three intramolecular disulfide bonds, as seen in Fig. 2. The disulfide bonds are essential for biologic activity (Taylor et al., 1972).

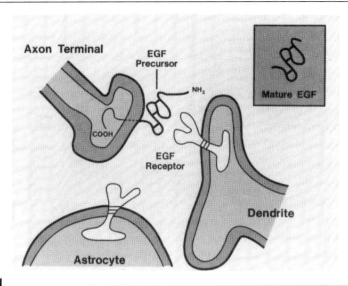

FIGURE 1

EGF precursor functions as a transmembrane receptor. Because of the transmembrane domain, the EGF precursor is pictured as an integral membrane protein with the active EGF moiety found in the extracellular domain. The EGF moiety of the precursor is visualized as able to interact with an EGF receptor on neighboring cells. These interactions may occur between similar cells (homotypic interactions, e.g., neuron–neuron or astrocyte–astrocyte) or dissimilar cells (heterotypic interactions, e.g., neuron–astrocyte). The interaction between an EGF precursor molecule and a neighboring EGF receptor is thought to result in activation of the EGF receptor, with ensuing physiological changes.

Reduction of the disulfide bonds in the presence of mercaptoethanol and urea inactivates biologic activity. The precise spacing of the six cysteines involved in disulfide bond formation is very characteristic of EGF and helps to define the secondary structure of EGF. A similar pattern of cysteine repeats is found in the other EGF-like sequences contained in the precursor also.

The pattern of cysteine repeats identified in EGF has been observed in other structurally related molecules, implying the presence of an EGF family of proteins. There are presently five members in this family, including EGF, TGFα, amphiregulin (AR), vaccinia growth factor (VGF), and schwannoma-derived growth factor (SDGF). SDGF has just been cloned and sequenced and is not very well characterized (Kimura et al., 1990). At least three members, TGFα (Marquardt et al., 1984), AR (Shoyab et al., 1989), and VGF (Brown et al., 1985), display a secondary structure similar to that of EGF, as seen in Fig. 2. Each member of the family adapts the EGF conformation because of the three disulfide bridges. The sequence homology (Table 1) and the similarities in secondary structure provide a molecular explanation for the interaction of these proteins with the same cellular receptor (Massague, 1983). EGF, TGFα, AR, and VGF exhibit similar biologic activity, although differences in function have been documented (Schreiber et al.,

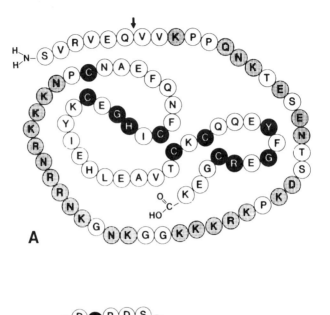

A

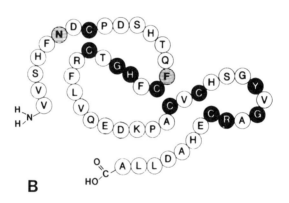

B

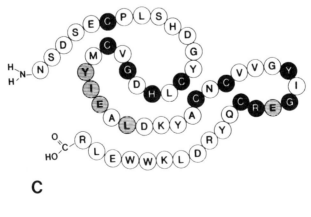

C

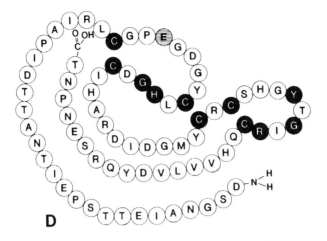

FIGURE 2

Protein sequence homologies among the EGF-like growth modulations, using predicted secondary structure of human amphiregulin (A), EGF (C), and TGFα (B), and vaccinia virus growth factor (VGF) (D). Residues that are completely conserved among all four proteins are black. Additional residues that EGF, TGFα, or VGF share with amphiregulin are shaded in those sequences. Hydrophilic residues in the N-terminal portion of amphiregulin are shaded in that sequence. An arrow marks the alternate cleavage site for the 78-aa form of mature amphiregulin. Reprinted from Plowman et al. (1990).

1986). Differences in activity could reflect the presence of multiple receptor subtypes (Winkler et al., 1989).

TGFα is a secreted polypeptide originally identified in the medium of retrovirus-transformed cells (De Larco and Todaro, 1978; see also Chapter 12). This activity was named based on the observation that preparations of the protein reversibly transformed normal rat kidney (NRK) fibroblasts (Todaro et al., 1980). However, on subsequent examination the preparation was found to contain a second factor, TGFβ, that was needed also for efficient induction of transformation in NRK cells. These two factors are structurally unrelated and express different biologic activities in most instances. TGFα is expressed in a variety of tumors, but also is expressed in nontransformed tissues. TGFα has been detected in a variety of normal tissues during embryonic development including placenta, kidney, and otic vesicle (Lee et al., 1985). TGFα also has been demonstrated in certain adult tissues including pituitary cells, skin keratinocytes, activated macrophages (Derynck, 1988), and brain (Wilcox and Derynck, 1988). With its identification in normal tissues, it becomes clear that TGFα may play a role in the development and maintenance of normal tissues, including brain.

AR is a secreted glycoprotein isolated from the conditioned medium of human breast adenocarcinoma cells following treatment with 12-O-tetradecanoyl phorbol-13-acetate (TPA) (Shoyab et al., 1989). AR derived its name from the observation that it was capable of inhibiting the growth of many tumor cells as well as stimulat-

TABLE 1

Properties of the EGF family of proteins

| Factor | Size | | Disulfide bonds | Homology with EGF[a] (%) | CNS localization |
|---|---|---|---|---|---|
| | kDa | AA | | | |
| EGF | 6045 | 53 | 3 | — | Yes |
| TGFα | 6000[b] | 50 | 3 | 49 | Yes |
| AR | 9700 | 84 | 3 | 43 | ND[c] |
| VGF | 19000 | 140 | 3 | 46 | No |

[a]Sequence comparisons were made between the 37 positions contained in the regions spanning the three disulfide loops (Savage et al.).
[b]Multiple species of TGFα have been identified ranging in molecular size from 5 to 20 kDa (Gentry et al., 1987; Derynck, 1988; Teixido and Massague, 1988). The intermediate forms correspond to partially processed and glycosylated forms of TGFα.
[c]ND, Not detected in human cerebral cortex (Plowman et al., 1990).

ing the proliferation of normal fibroblasts and keratinocytes. AR exhibits structural and functional homology with EGF and TGFα. AR partially competes with EGF for binding to the EGF receptor and supercedes the requirement for EGF or TGFα to maintain keratinocytes in culture. AR also is observed in many of the same normal tissues and tumor cells as TGFα (Plowman et al., 1990). In contrast to EGF and TGFα, AR initially was not detected in normal human brain (cerebral cortex). However, it is conceivable that AR is expressed only in specific subregions of brain. In addition, AR may be present in brain only at select times during development. Further studies are required to clarify this issue.

VGF is a secreted protein initially found in monkey cells infected with vaccinia virus (Twardzik et al., 1985). Vaccinia virus is a double-stranded DNA virus. The 19-kDa VGF is encoded by the viral genome and is transcribed early in infection. VGF shares biologic activity with EGF and TGFα and competes with EGF for binding to EGF membrane receptors (Twardzik et al., 1987). A segment of the 140-amino-acid VGF exhibits a similar alignment of cysteine residues as that found in three other family members, predicting a common secondary structure and presumably accounting for similar activities. Although VGF is not found in normal tissues, its presence in a DNA virus may be responsible for hyperplastic responses associated with some pox virus infections, and underlies the potential extent and use of EGF-like domains in biologic systems.

The final member of this family was purified and cloned recently (Kimura et al., 1990). The protein, termed SDGF was isolated from the conditioned medium of a schwannoma cell line. SDGF bears 76% homology to AR. Further, SDGF also appears to be synthesized as a larger precursor, in a manner similar to the other members of the family. The protein is mitogenic for rat cortical astrocytes but neuronotrophic activity has not been reported. Northern analysis demonstrates a weak reaction with fetal brain. No detectable hybridization is observed in adult brain. The weak reaction with fetal brain suggests that SDGF may exist in select brain subregions, as suggested for AR. The observation that SDGF mRNA is present in fetal brain but not adult brain contrasts with evidence for TGFα, which clearly is expressed in adult brain. These preliminary results indicate that EGF-family members may have distinct patterns of expression during the development of the central nervous system.

## III. EGF LOCALIZATION IN BRAIN

Any unifying hypothesis pertaining to EGF action in brain must include an identification of the ligand or related ligands in neural tissue. Identifying EGF in brain has not been straightforward and is still a controversial issue for many investigators working in this field. Evidence for localization of the *protein* in brain is still limited. Evidence for expression of the *mRNA* in brain is now forthcoming.

EGF-immunoreactive material (EGF-IRM) was first identified in brain using immunofluorescent techniques (Fallon et al., 1984). EGF-IRM was identified in forebrain and midbrain structures of the rat, but the cellular source of the EGF-IRM was not determined. However, the morphology, location, and length of the fibers in which EGF-IRM was observed suggested that the localization was neuronal rather than glial. In a separate study, EGF could not be detected in developing or adult rodent brain using a sensitive double-site enzyme immunoassay (Probstmeier and Schachner, 1986). In contrast, Lakshmanan and co-workers were able to detect EGF-IRM in synaptosomal fractions prepared from mouse cerebral cortex (Lakshmanan et al., 1986). The reason for the discrepancy between these studies is not clear. Synaptosomal preparations may have enriched EGF-IRM to detectable levels. In addition, it is not clear that the sensitivities of the assays were equivalent. In a more recent study, EGF has been detected in rat brain using a homologous radioimmunoassay; the immunoreactive proteins were characterized partially (Schaudies et al., 1989). The competition curves generated in this assay with rat brain did not parallel those generated with standard rat EGF, suggesting that the immunoreactive material in rat brain is structurally distinct from standard rat EGF. Extracts prepared from rat submandibular gland and blood did produce parallel competition curves. The immunoreactive species were characterized further by separation on nondenaturing polyacrylamide gel electrophoresis. Electrophoretic analysis indicated the presence of multiple immunoreactive bands. These bands did not migrate to the same position as standard rat EGF, suggesting that the proteins are structurally dissimilar. As a whole, these data indicate that EGF-IRM is present in rat brain although a precise structural determination has not been made. Further structural characterization will be required to determine if these proteins are derived from processing of the EGF precursor or if they are distinct gene products.

Evidence for EGF gene expression in brain was originally provided by Northern analysis of poly(A)+ RNA from 8-week-old mouse brain (Rall et al., 1985). When a cDNA for preproEGF was used to measure the expression of EGF mRNA in brain, mRNA was present, but the relative abundance of the message was lower than in many other tissues. The predicted frequency of proproEGF mRNA in poly(A)+ RNA isolated from 8-week-old mouse brain was only 1:200,000. Much more convincing data have been obtained by employing a sensitive solution hybridization assay. Using this assay, Blum and co-workers have been able to detect and quantitate EGF mRNA in several distinct regions of mouse brain (Lazar and Blum, 1992). The greatest levels of EGF message were contained in the cerebellum, hypothalamus, and olfactory bulb. EGF mRNA also was present in other brain regions, including cortex, striatum, and brain stem.

RNA for two other members of the EGF family also is expressed in brain. TGFα mRNA expression has been detected in brain by in situ hybridization (Wilcox and Derynck, 1988) and by a solution hybridization assay (Lazar and M. Blum, 1992). TGFα mRNA appears to be expressed in many of the same brain regions as EGF. TGFα mRNA also has been localized by in situ hybridization to cell bodies of the caudate nucleus, dentate gyrus, anterior olfactory nuclei, and the olfactory bulb. However, in contrast to EGF, TGFα mRNA is expressed at much higher levels in all brain regions examined. Despite the presence of TGFα mRNA, evidence for expression of the protein in brain is not well characterized. TGFα immunoreactivity has been detected in neurons (Code et al., 1987; Kudlow et al., 1989) and in astrocytes (Fallon et al., 1988,1990; Loughlin et al., 1989). The discrepancy between these reports has not been resolved. Further, it remains to be determined if brain expresses TGFα protein predominantly in the membrane precursor form or as a cleaved soluble product. mRNA for SDGF also is expressed in fetal brain, as previously discussed. The expression of these proteins in brain suggests that each factor may subserve separate functions. Alternatively, the factors may share similar functions, but may affect different target populations. Initially, there was no demonstration of AR expression in brain, but this issue apparently was not examined rigorously (Plowman et al., 1990). It has been suggested now that AR mRNA is expressed in brain (G.D. Plowman, personal communication). Thus, there is evidence using Northern analysis, solution hybridization assays, or in situ hybridization histochemistry to support the presence of EGF, TGFα, SDGF, and AR in the central nervous system.

## IV. EGF RECEPTOR

EGF, like other protein growth factors, initiates cellular responses by first binding to specific receptors on the surface of target cells. The binding of EGF to its cell-surface receptor induces a cascade of intracellular events, including induction of a tyrosine kinase activity intrinsic to the EGF receptor (Ushiro and Cohen, 1980). Shortly after binding, EGF–receptor complexes are localized in clathrin-coated regions of the plasma membrane and internalized into the cell. The cellular receptor for EGF is a 170-kDa membrane glycoprotein. The sequence of the EGF receptor has been determined from a full-length cDNA (Ullrich et al., 1984). The receptor is a 1210-amino-acid protein consisting of a 621-amino-acid extracellular component, a single transmembrane domain, and a 542-amino-acid cytoplasmic region. Sequence analysis has demonstrated a close similarity between the EGF receptor and the avian erythroblastosis virus transforming protein, v-erbB. This viral transforming protein contains the transmembrane and internal tyrosine kinase domains of the avian EGF receptor. Transformation of cells by this virus may lead to the expression of a truncated EGF receptor that lacks the extracellular control domain. Such aberrant control mechanisms may have implications for the development and propagation of certain glial tumors, many of which overexpress the EGF receptor.

EGF activates a number of biochemical events after binding to its receptor, including alterations in intracellular calcium, pH, and transcription of several responsive genes. The EGF receptor also possesses intrinsic protein tyrosine kinase activity that is activated by EGF binding (Ushiro and Cohen, 1980). Activated tyrosine kinase activity leads to increased intracellular substrate phosphorylation and self-phosphorylation. The EGF receptor normally exists as a phosphoprotein in vivo; the phosphates are distributed primarily on serine and threonine residues (Cochet et al., 1984). EGF activation results in receptor autophosphorylation on tyrosines contained in the extreme carboxyterminal region of the molecule (Downward et al., 1984). The tyrosine kinase activity associated with the EGF receptor is essential for many of its diverse actions (Chen et al., 1987). Eliminating tyrosine kinase activity from the EGF receptor with a single amino acid mutation abolishes the mitogenic action of EGF and EGF stimulation of c-*fos* expression (Chen et al., 1987). Interestingly, this particular mutation does not appear to affect synthesis or insertion of mature EGF-binding receptors into the cell. Intact tyrosine kinase activity also is required for normal cellular routing of the EGF–receptor complex after EGF binding (Honegger et al., 1987). Using the single amino acid mutation just described, it was demonstrated that the mutated receptors were not down-regulated appropriately. The mutant receptors were internalized (Livneh et al., 1987) but, instead of being degraded, they were recycled from a cytoplasmic compartment back to the cell surface. Thus, the EGF receptor-associated protein tyrosine kinase activity is necessary for EGF-induced changes in cellular phenotype.

## V. EGF RECEPTOR EXPRESSION IN BRAIN

EGF receptor expression has been demonstrated in brain using receptor binding assays (Adamson and Meek, 1984), immunocytochemistry (Loy et al., 1987; Gomez-Pinilla et al., 1988; Werner et al., 1988), and $^{125}$I-labeled EGF receptor autoradiography (Quirion et al., 1988; Wiedermann et al., 1988). EGF receptor binding sites were identified in brain by examining $^{125}$I-labeled EGF binding to membrane preparations obtained from embryonic mouse brain (Adamson and Meek, 1984). EGF receptor binding sites were first detected in mouse brain membranes at 15 days gestation. The highest level of receptor sites was observed at 19 days gestation, but no information was available for older brain tissues. These data imply that EGF receptors are present in developing brain at a time corresponding to the maturation of neurons and the genesis of glia.

EGF receptor immunoreactivity (EGFR-IR) has been demonstrated in brain by several different groups of investigators. EGFR-IR was observed in cerebellar Purkinje cells as early as day 11 postnatal and was maintained into adulthood (Gomez-Pinilla et al., 1988). In adult and aged animals, prominent EGFR-IR was observed in cerebral cortex neurons (layers IV and V) that had the morphology of basket cells. Immunoreactive neurons were also abundant in the cingulate, frontal, frontoparietal, and striate cortices. Glial immunoreactivity also was observed, but the staining was transient compared with that seen in neurons. Astroglial receptor immunoreactivity was present beginning at postnatal day 16. However, in contrast to neuronal immunoreactivity, astrocyte EGFR-IR reached maximal intensity at day 19 postnatal and declined at approximately 1 month. At 2 months after birth, glial staining was weak or absent and neuronal staining was established fully. Developmental time points preceding postnatal day 11 were not examined, so it was not possible to correlate the immunocytochemical results with the receptor binding data.

In a separate study, EGFR-IR was demonstrated in rat brain neurons as early as embryonic day 14 (Loy et al., 1987). In the adult, EGF receptor immunoreactive cell bodies were identified in various basal forebrain structures, the caudate putamen, lateral septal nucleus, olfactory tubercle, neocortex, and hippocampal formation. EGFR-IR also has been demonstrated in the human nervous system (Werner et al., 1988). A wide variety of nerve cells in human brain contained EGFR-IR. Immunoreactive neurons were observed in human brain samples ranging in age from 28 weeks gestation to 73 years of age. Glial immunoreactivity was not detected in normal brain, although prominent intracytoplasmic EGFR-IR was reported for specimens obtained from an anaplastic astrocytoma and two recent cerebral infarctions. This finding is consistent with previous reports demonstrating enhanced EGF receptor expression in neoplastic (Libermann et al., 1984; Wong et al., 1987) and reactive astrocytes (Nieto-Sampedro, 1988). Vascular smooth muscle cells in human brain also exhibited EGFR-IR. EGF receptor expression also has been documented in demented human brain (Styren et al., 1990). EGFR-IR was localized to the luminal surfaces of endothelial cells in these patients, whereas staining was absent in most nondemented elderly patients. EGFR-IR was not detected in any other brain cell populations, as described in previous studies. However, differences between this study and others may reflect the use of different antibodies. It is interesting that only endothelial cells in the brains of demented patients exhibited luminal immunoreactivity. It was suggested by the authors that the antigen could be serum derived. The cellular origin and actual identity of this antigen and its relationship to dementia still must be elucidated.

Growth factor receptors related to the EGF receptor may also exist in brain, consistent with the identification of EGF-related proteins in brain. Thus, another transmembrane glycoprotein with tyrosine kinase activity also has been identified in developing rat brain. The glycoprotein is termed p185 (185 kDa) and is encoded by the *neu* oncogene (also referred to as c-*erbB2* and *HER2*). The *neu* oncogene first was identified as a transforming gene in rat neuroblastomas (Schecter et al., 1984). p185 is structurally related to, but clearly distinct from, the EGF receptor. Expression of the *neu* gene has been detected in midgestation embryos in a variety of tissues, including the nervous system, connective tissue, and secretory epithelium (Kokai et al., 1987). The protein product was detected by immunocytochemical staining in neural cell bodies in the region of the developing dorsal root ganglia in embryonic day (ED) 14 embryos. Neurofilament positive cell bodies in the external portions of the neural tube also expressed p185 in their cell processes. During late gestation (ED 18) and at later times, p185 immunoreactivity no longer could be detected. Although the cognate ligand for the receptor-like p185 protein has not been identified, it is conceivable that EGF (or a related protein) is the activating ligand. The striking differences in developmental expression between the EGF receptor and the closely related p185 protein in brain imply that EGF and related family members may subserve a variety of roles during the development and maintenance of the central nervous system.

## VI. EGF AND NEUROGLIA

In 1976, EGF was first demonstrated to stimulate the division of a human diploid "glia-like" cell line in serum-free medium (Westermark, 1976). These observations were extended using isolated homotypic populations of central nervous system glia derived from neonatal rat brain. EGF has been shown to stimulate cell division in astrocytes isolated from early postnatal cerebral cortex and cerebellum (Leutz and Shachner, 1981; Simpson et al., 1982). EGF also can have nonmitogenic effects on cultured astrocytes. For example, EGF can inhibit intracellular elevations of AMP in astrocytes in response to β-adrenergic stimulation (Wu et al., 1985).

Purified populations of astrocytes contain receptors for EGF. Astrocytes reportedly contain between $4 \times 10^4$ and $1 \times 10^5$ EGF receptors per cell. EGFR-IR also has been demonstrated on astrocytes in frozen sections of rat brain, using antibodies generated against the human EGF receptor. Reactive astrocytes adjacent to a site of injury have been reported to show intensely positive immunoreactivity for the expression of EGF receptor (Nieto-Sampedro, 1988). Oligodendrocytes have been reported to express receptors for EGF (Simpson et al., 1982). Oligodendrocytes contain between 6,000 and 10,000 receptors per cell, but EGF has not been observed to stimulate the proliferation of these cells. However, EGF has been shown to stimulate the activity of two oligodendrocyte markers in brain cell aggregates cultured in a serum-free chemically defined medium (Honneger and Guentert-Lauber, 1983; Almazan et al., 1985). EGF concentrations as low as 10 ng/ml were effective in increasing the content of 2',3'-cyclic nucleotide 3'-phosphodiesterase and myelin basic protein. Thus, EGF may exert nonmitogenic effects on oligodendrocytes. However, these results contrast with a report by Sheng et al.

(1989) demonstrating that EGF inhibits the expression of myelin basic protein in cultured brain cells isolated from 4-day-old mice. The reason for the discrepancy is not clear, but it may be related to differences between monolayer cultures and aggregate cultures.

The EGF receptor also has been associated with pathology of the nervous system. As previously discussed, enhanced EGFR-IR has been associated with astrocytes adjacent to a central nervous system lesion site. Enhanced levels of EGF receptor expression have been detected in malignant gliomas (Libermann et al., 1984; Wong et al., 1987). Approximately 40% of malignant gliomas in one study were found to have amplified copies of the EGF receptor gene (Wong et al., 1987). Increased expression of the EGF receptor was found only in gliomas expressing alterations in the structure of the EGF receptor gene (e.g., amplification). The exact contribution made by EGF receptors toward the regulation of normal and abnormal astrocyte proliferation is not understood. The observation that the EGF receptor locus is not overexpressed in all gliomas suggests that changes in this particular locus may not be a mechanism for the formation of all malignant gliomas. However, the differential expression of EGF receptor in normal and neoplastic tissue, coupled with the responses of these tissues to EGF in vitro, suggests that the receptor plays a critical role in regulating the growth of these cells in vivo.

## VII. EGF AND NEURONAL FUNCTION

EGF has been studied primarily in the context of being able to stimulate cellular proliferation. The first report of a neuron-specific response for EGF was made using a rat pheochromocytoma cell line, PC12. When these tumor cells are treated with NGF, they cease dividing and elaborate an extensive system of neurites much like sympathetic neurons do in culture. A subclone of PC12, PCG2, was found to respond to EGF with a marked elevation in the activity of the neurotransmitter-synthesizing enzyme tyrosine hydroxylase (TH) (Goodman et al., 1980). Moreover, the concentration of EGF required for maximal TH induction was three orders of magnitude below that required for equivalent induction with NGF. EGF subsequently was shown to elicit other responses from PC12 cells, but these would not be considered uniquely neuronal responses. The effects of EGF next were evaluated using organ cultures of superior cervical ganglia from the rat. The level of TH in the EGF-treated organ cultures was elevated significantly over the levels in control cultures (Herschman et al., 1983). However, whether there was an EGF-induced increase in TH was not clear, since control cultures simply may have degenerated during the experiment. If that were the case, EGF could have raised the levels of TH relative to controls by acting simply as a survival or maintenance factor for the ganglionic neurons. Further, although the results with sympathetic ganglia suggest that EGF had an effect on this target tissue, they do not distinguish between a direct effect on the neurons and an indirect effect mediated by nonneuronal cells in these cultures. Although EGF does influence one neuron-specific characteristic in a

PC12-derived clone, it does not stimulate neurite outgrowth from these cells or from sympathetic neurons in culture (Chalazonitis, et al., 1992). Therefore, its role in regulating the development of peripheral neurons remains equivocal.

EGF has been tested on a variety of other peripheral nervous system neurons in culture without effect (Chalazonitis, et al., 1992). EGF failed to support the survival of trigeminal, nodose, and dorsal root ganglion (DRG) neurons in vitro. Concentrations up to 500 ng/ml had no effect on survival or neurite outgrowth. Normally, concentrations as low as 1 ng/ml significantly enhance both neuronal survival and neurite outgrowth from central nervous system neurons (Morrison et al., 1987, 1988; Kornblum et al., 1990). TGFα was found ineffective on peripheral neurons with one surprising exception. TGFα was found to enhance survival of cultured neonatal rat DRG neurons in vitro. This effect was dose dependent; TGFα concentrations as low as 500 pg/ml proved effective. As previously stated, EGF at concentrations up to 500 ng/ml was ineffective. The actions of TGFα on DRG neurons did not appear to be mediated through glial cells, since its trophic effects were not diminished in the presence of mitotic inhibitors. However, it is still premature to conclude that the actions were mediated directly through neurons in the absence of data demonstrating neuronal expression of a TGFα/EGF receptor. It is interesting to note that the EGF receptor and the EGF receptor-related *neu* oncogene product both have been identified on DRG neurons using immunocytochemical techniques (Kokai et al., 1987; Werner et al., 1988). Further, TGFα is expressed by skin keratinocytes, providing a potential source of the factor for developing and mature DRG neurons (Coffey et al., 1987). Thus TGFα, in contrast to EGF, may act as a survival or maintenance factor for a subset of rat sensory neurons, possibly via a novel form of receptor or EGF-receptor companion molecule that modifies receptor function.

Establishing a role for EGF as a neurotrophic factor in the central nervous system has been difficult also. Until recently, very few good in vitro central nervous system models were available. In contrast to results observed with peripheral neurons, the addition of EGF to primary central nervous system neurons significantly enhanced their survival (from days to weeks) and neurite outgrowth (Morrison et al., 1987,1988; Kinoshita et al., 1990; Kornblum et al., 1990). This response has been observed in cultures established from different regions of the rat central nervous system. Neurons from multiple regions of the brain respond to EGF in a similar dose-responsive manner and with the same time course (R. Morrison, unpublished observations). These effects have been observed with EGF concentrations as low as 100 pg/ml (16 p$M$). EGF maximally stimulates neuronal survival and process outgrowth at approximately 10 ng/ml (1.6 n$M$).

EGF acts similarly to another growth factor with recently described neuronotrophic activity, basic fibroblast growth factor (bFGF). Recent studies suggest that EGF and bFGF affect overlapping populations of neurons in certain brain regions such as the cerebral cortex (Kornblum et al., 1990). However, differential trophic effects also have been observed. Whereas EGF has been shown to support the survival of neonatal rat cerebellar neurons, bFGF was inactive on this neuronal population (Morrison et al., 1988). Granule cells, a neuronal cell type in the cerebellum that develops after birth, eventually exhibit a requirement for bFGF (Hatten et al., 1988). These responses to EGF and bFGF appear to correlate with EGF and bFGF receptor localization in situ (Gomez-Pinilla et al., 1988; Wanaka et al., 1990). In contrast to the trophic effects observed with EGF and bFGF, other growth factors such as platelet derived growth factor, NGF, interleukins 1 and 2, thrombin, and transforming growth factor beta had no survival-promoting effect when tested on cortical (Morrison et al., 1986; R. Morrison, unpublished observations) or subneocortical neurons (Morrison et al., 1987). However, the EGF-related protein TGFα appears to support the same central nervous system neuronal populations as EGF (R. Morrison, unpublished observations). Thus, the neuronotrophic actions of EGF and bFGF appear to be specific and selective.

EGF and TGFα also enhance expression of neurotransmitter-synthesizing enzymes in addition to promoting both neuron survival and neurite outgrowth. EGF elevates the activity of the acetylcholine synthetic enzyme, choline acetyltransferase (ChAT), in cultured basal forebrain neurons (Yokoyama et al., 1990). EGF also elevates the activity of glutamic acid decarboxylase in cultured basal forebrain neurons as well as dopamine uptake in cultured mesencephalic neurons (Casper et al., 1990; Knusel et al., 1990). Although much of this work has been done in vitro, EGF and TGFα have been shown to elevate hippocampal ChAT activity in vivo following application of the neurotoxin AF64A (Potter and R. Morrison, 1991). Infusion of AF64A into the ventricles of adult rats significantly depresses hippocampal ChAT activity (Hortnagl et al., 1987). Injection of EGF, TGFα, or bFGF into the ventricles immediately after AF64A administration dramatically elevated hippocampal ChAT activity. Among the various growth factors tested, bFGF was the most potent at restoring ChAT activity the lesion, but the activities of EGF and TGFα closely parallel that of NGF. These results suggest that EGF and related family members may exert important trophic influences on developing and mature central nervous system neurons. Further, the actions of EGF do not appear to be restricted to a single neuronal phenotype.

The mechanism by which EGF and related proteins enhance neuronal survival and elevate the expression of neurotransmitter-synthesizing enzymes is still controversial. Several groups have shown independently that the neurotrophic actions of EGF are significantly attenuated in the presence of mitotic inhibitors (Knusel et al., 1990; Yokoyama et al., 1990). These results have been interpreted to mean that EGF initiates its trophic actions through an intermediate cell type. The most logical targets would be glial cells, since EGF has demonstrated effects on both astrocytes and oligodendrocytes. These results contrast with earlier work with cortical and subneocortical neurons (Morrison et al., 1986,1987; Kornblum et al., 1990). The results of these earlier studies demonstrated that EGF enhanced neuronal survival in the presence of mitotic inhibitors. The differences among these studies may reflect differences in the origin of the cell types. Neurons from the septum and mesencephalon may not express EGF receptors. These neurons conceivably could be influenced by additional factors secreted by glia in response to EGF. However, immunocytochemical localization of the EGF receptor clearly has been made to neurons in multiple regions of the central nervous system. In addition, Kinoshita and co-workers have detected the EGF receptor on cerebral cortical neurons in culture using anti-EGF receptor antibodies (Kinoshita et al., 1990). Approximately 70% of cultured neurons expressed EGFR-IR after 3 days in culture. EGFR-IR was localized preferentially to the neuronal cell body. EGF receptor expression also was demonstrated using a receptor binding assay. Thus it seems clear that, in certain instances, EGF may mediate changes in neuronal survival and phenotype by interacting directly with neuronal EGF receptors. In some circumstances, as in septal or mesencephalic neurons, glial cells may be required for inducing EGF responsivity in neurons. Neuronal–glial interactions may be required for expression of the neuronal EGF receptor. Alternatively, neuronal–glial interactions may facilitate coupling between activation of neuronal EGF receptors and transduction of the signal. Clearly more work is required to distinguish between these various alternatives.

In summary, at least five different proteins constitute the EGF family of mitogens. Three, and perhaps as many as four, of these proteins are expressed in brain at different times during development. All five family members express mitogenic activity; at least two support neuronal survival based on tissue culture assays and brain lesion paradigms. The EGF family of proteins is similar to the NGF family since NGF also has been shown to contain three family members. Fibroblast growth factor (see Chapters 9 and 10) also contains seven family members; several of these FGF-related proteins are expressed in brain. It is not yet clear how these growth factors interact with one another in neural tissues. In the case of EGF, it is not clear whether separate receptors exist for each family member. However, the EGF-related proteins exhibit similar structural conformations and, in many cases, interact with a common receptor. This receptor has been identified in the central nervous system. Thus, members of the EGF family may be engaged in a variety of processes critical to brain development, maintenance, and pathology.

## REFERENCES

Adamson, E. D., and Meek, J. (1984). The ontogeny of epidermal growth factor receptors during mouse development. *Dev. Biol.* **103**, 62–70.
Almazan, G., Honegger, P., Matthieu, J.-M., and Guentert-Lauber, B. (1985). Epidermal growth factor and bovine growth hormone stimulate differentiation and myelination of brain cell aggregates in culture. *Dev. Brain Res.* **21**, 257–264.
Bender, W. (1985). Homeotic gene products as growth factors. *Cell* **43**, 559–560.
Brachmann, R., Lindquist, P. B., Nagashima, M., Kohr, W., Lipari, T., Napier, M., and Derynck, R. (1989). Transmembrane TGFα precursors activate EGF/TGFα receptors. *Cell* **56**, 691–700.
Breyer, J. A., and Cohen, S. (1990). The epidermal growth factor precursor isolated from murine kidney membranes. Chemical characterization and biological properties. *J. Biol. Chem.* **265**, 16564–16570.
Brown, J. P., Twardzik, D. K., Marquardt, H., and Todaro, G. J. (1985). Vaccinia virus encodes a

polypeptide homologous to epidermal growth factor and transforming growth factor. *Nature (London)* **313**, 491–492.
Casper, D., Mytilineou, C., and Blum, M. (1991). EGF enhances the survival of Dopamine Neurons in rat embryonic mesencephalon primary cell culture. *J. Neurosci. Res.* **30**, 372–381.
Chalazonitis, A., Kessler, J. A., Twardzik, D. R. and Morrison, R. S. (1992). Transforming growth factor α, but not epidermal growth factor, promotes the survival of sensory neurons in vitro. *J. Neurosci.* **12**:583–594.
Chen, W. S., Lazar, C. S., Poenie, M., Tsien, R. Y., Gill, G. N., and Rosenfeld, M. G. (1987). Requirement for intrinsic protein tyrosine kinase in the immediate and late actions of the EGF receptor. *Nature (London)* **328**, 820–823.
Cochet, C., Gill, G. N., Meisenhelder, J., Cooper, J. A., and Hunter, T. (1984). C-Kinase phosphorylates the epidermal tyrosine protein kinase activity. *J. Biol. Chem.* **259**, 2553–2558.
Code, R. A., Seroogy, K. B., and Fallon, J. H. (1987). Connections of some transforming growth factor containing neurons in the CNS. *Brain Res.* **421**, 401–405.
Coffey, R. J., Derynck, R., Wilcox, J. N., Bringman, T. S., Goustin, A. S., Moses, H. L., and Pittelkow, M. R. (1987). Production and auto-induction of transforming growth factor-α in human keratinocytes. *Nature (London)* **328**, 817–820.
Cohen, S. (1962). Isolation of a mouse submaxillary protein accelerating incisor eruption and eyelid opening in the newborn animal. *J. Biol. Chem.* **237**, 1555–1562.
DeLarco, J. E., and Todaro, G. J. (1978). Growth factors from murine sarcoma virus-transformed cells. *Proc. Natl. Acad. Sci. U.S.A.* **75**, 4001–4005.
Derynck, R. (1988). Transforming growth factor alpha. *Cell* **54**, 593–595.
Downward, J., Parker, P., and Waterfield, M. D. (1984). Autophosphorylation sites on the epidermal growth factor receptor. *Nature (London)* **311**, 483–485.
Fallon, J. H., Seroogy, K. B., Loughlin, S. E., Morrison, R. S., Bradshaw, R. A., Knauer, D. J., and Cunningham, D. D. (1984). Epidermal growth factor immunoreactive material in the central nervous system: Location and development. *Science* **224**, 1107–1109.
Fallon, J. H., Loughlin, S. E., Kornblum, H. I., and Leslie, F. M. (1988). Growth factors and their receptors in opiod-rich brain regions. *INRC Abstr.* **122**.
Fallon, J. H., Annis, C. M., Gentry, L. E., Twardzik, D. R., and Loughlin, S. E. (1990). Localization of cells containing transforming growth factor-α precursor immunoreactivity in the basal ganglia of the adult rat brain. *Growth Factors* **2**, 241–250.
Gentry, L. E., Twardzik, D. R., Lim, G. J., Ranchalis, J. E., and Lee, D. C. (1987). Expression and characterization of transforming growth factor α precursor protein in transfected mammalian cells. *Mol. Cell. Biol.* **7**, 1585–1591.
Gomez-Pinilla, F., Knauer, D. J., and Nieto-Sampedro, M. (1988). Epidermal growth factor receptor immunoreactivity in rat brain. Development and cellular localization. *Brain Res.* **438**, 385–390.
Goodman, R., Slater, E., and Herschman, H. R. (1980). Epidermal growth factor induces tyrosine hydroxylase in a clonal pheochromocytoma cell line, PC-G2. *J. Cell Biol.* **84**, 495–500.
Gray, A., Dull, T. J., and Ullrich, A. (1983). Nucleotide sequence of epidermal growth factor cDNA predicts a 128,000-molecular weight protein precursor. *Nature (London)* **303**, 722–725.
Hatten, M. E., Lynch, M., Rydel, R. E., Sanchez, J., Joseph-Silverstein, J., Moscatelli, D., and Rifkin, D. B. (1988). In vitro neurite extension by granule neurons is dependent upon astrological-derived fibroblast growth factor. *Dev. Biol.* **125**, 280–289.
Herschman, H. R., Goodman, R., Chandler, C., Simpson, D., Cawley, D., Cole, R., and de Vellis, J. (1983). Is epidermal growth factor a modulator of nervous system function? *In* "Nervous System Regeneration" (B. Haber and J. R. Perez-Polo, eds.), pp. 79–94. Liss, New York.
Honegger, A. M., Dull, T. J., Felder, S., Van Obberghen, E., Bellot, F., Szapary, D., Schmidt, A., Ullrich, A., and Schlessinger, J. (1987). Point mutation at the ATP binding site of EGF receptor abolishes protein-tyrosine kinase activity and alters cellular routing. *Cell* **51**, 199–209.
Honegger, P., and Guentert-Lauber, B. (1983). Epidermal growth factor (EGF) stimulation of cultured brain cells. I. Enhancement of the developmental increase in glial enzymatic activity. *Dev. Brain Res.* **11**, 245–251.

Hortnagl, H., Potter, P. E., and Hanin, I. (1987). Effect of cholinergic deficit induced by ethylcholine aziridinium (AF64A) on noradrenergic and dopaminergic parameters in rat brain. *Brain Res.* **421**, 75–84.

Kimura, H., Fischer, W. H., and Schubert, D. (1990). Structure, expression and function of a schwannoma-derived growth factor. *Nature (London)* **348**, 257–260.

Kinoshita, A., Yamada, K., Hayakawa, T., Kataoka, K., Mushiroi, T., Kohmura, E., and Mogami, H. (1990). Modification of anoxic neuronal injury by human recombinant epidermal growth factor and its possible mechanism. *J. Neurosci. Res.* **25**, 324–330.

Knusel, B., Michel, P. P., Schwaber, J. S., and Hefti, F. (1990). Selective and nonselective stimulation of central cholinergic and dopaminergic development in vitro by nerve growth factor, basic fibroblast growth factor, epidermal growth factor, insulin, and the insulin-like growth factors I and II. *J. Neurosci.* **10**, 558–570.

Kokai, Y., Cohen, J. A., Drebin, J. A., and Greene, M. I. (1987). Stage- and tissue-specific expression of the *neu* oncogene in rat development. *Proc. Natl. Acad. Sci. U.S.A.* **84**, 8498–8501.

Kornblum, H. I., Raymon, H. K., Morrison, R. S., Cavanaugh, K. P., Bradshaw, R. A., and Leslie, F. M. (1990). Epidermal growth factor and basic fibroblast growth factor: Effects on an overlapping population of neocortical neurons in vitro. *Brain Res.* **535**, 255–263.

Kudlow, J. E., Leung, A. W. C., Kobrin, M. S., Paterson, A. J., and Asa, S. L. (1989). Transforming growth factor-α in the mammalian brain: Immunohistochemical detection in neurons and characterization of its mRNA. *J. Biol. Chem.* **264**, 3880–3883.

Lakshmanan, J., Weichsel, M. E., Jr., and Fisher, D. A. (1986). Epidermal growth factor in synaptosomal fractions of mouse cerebral cortex. *J. Neurochem.* **46(4)**, 1081–1085.

Lazar, L. M. and Blum, M. (1992). Regional distribution and developmental expression of epidermal growth factor and transforming growth factor alpha mRNA in mouse brain by a quantitative nuclease protection assay. *J. Neurosci.* **12**:1688–1697.

Lee, D. C., Rochford, R., Todaro, G. J., and Villarreal, L. P. (1985). Developmental expression of rat transforming growth factor-alpha mRNA. *Mol. Cell. Biol.* **5(12)**, 3644–3646.

Leutz, A., Schachner, M. (1981). Epidermal growth factor stimulates DNA synthesis of astrocytes in primary cerebellar cultures. *Cell Tissue Res.* **220**:393–404.

Libermann, T. A., Razon, N., Bartal, A. D., Yarden, Y., Schlessinger, J., and Soreq, H. (1984). Expression of epidermal growth factor receptors in human brain tumors. *Cancer Res.* **44**, 753–760.

Livneh, E., Reiss, N., Berent, E., Ullrich, A., and Schlessinger, J. (1987). An insertional mutant of epidermal growth factor receptor allows dissection of diverse receptor functions. *EMBO J.* **6(9)**, 2669–2676.

Loughlin, S. E., Annis, C. M., Twardzik, D. R., Lee, D. C., Gentry, L. E., and Fallon, J. H. (1989). Growth factors in opioid rich brain regions: Distribution and response to intrastraital transplants. *Adv. Biosci.* **75**, 403–406.

Loy, R., Springer, J. E., and Koh, S. (1987). Localization of cells immunoreactive for epidermal growth factor receptor in the adult and developing rat forebrain. *Soc. Neurosci. Abstr.* **13(1)**, 575–575.

Marquardt, H., Hunkapiller, M. W., Hood, L. E., and Todaro, G. J. (1984). Rat transforming growth factor type 1: Structure and relation to epidermal growth factor. *Science* **223**, 1079–1082.

Massague, J. (1983). Epidermal growth factor-like transforming growth factor. II. Interaction with epidermal growth factor receptors in human placenta membranes and A431 cells. *J. Biol. Chem.* **258**, 13614–13620.

Mojsov, S., Weir, G. C., and Habener, J. F. (1987). Insulinotropin: Glucagon-like peptide I (7-37) co-encoded in the glucagon gene is a potent stimulator of insulin release in the perfused rat pancreas. *J. Clin. Invest.* **79**, 616–619.

Morrison, R. S., Sharma, A., DeVellis, J., and Bradshaw, R. (1986). Basic fibroblast growth factor supports the survival of cerebral cortical neurons in primary culture. *Proc. Natl. Acad. Sci. U.S.A.* **83**, 7537–7541.

Morrison, R. S., Kornblum, H. I., Leslie, F. M., and Bradshaw, R. A. (1987). Trophic stimulation of cultured neurons from neonatal rat brain by epidermal growth factor. *Science* **238**, 72–75.

Morrison, R. S., Keating, R. F., and Moskal, J. R. (1988). Basic fibroblast growth factor and epidermal growth factor exert differential trophic effects on CNS neurons. *J. Neurosci. Res.* **21**, 71–79.

Nieto-Sampedro, M. (1988). Astrocyte mitogen inhibitor related to epidermal growth factor receptor. *Science* **240**, 1784–1786.

Plowman, G. D., Green, J. M., McDonald, V. L., Neubauer, M. G., Disteche, C. M., Todaro, G. J., and Shoyab, M. (1990). The amphiregulin gene encodes a novel epidermal growth factor-related protein with tumor-inhibitory activity. *Mol. Cell. Biol.* **10**, 1969–1981.

Potter, P. E. and Morrison, R. S. (1991). Basic fibroblast growth factor protects septal-hippocampal cholinergic neurons against lesions induced by AF64A. Chapter 82: 639–642. *In* "Alzheimer's Disease: Basic Mechanisms, Diagnosis and Therapeutic Strategies." (K. Iqbal, D. R. C. McLachlan, B. Winblad and H. M. Wisniewski, eds.). John Wiley and Sons Ltd.

Probstmeier, R., and Schachner, M. (1986). Epidermal growth factor is not detectable in developing and adult rodent brain by a sensitive double-site enzyme immunoassay. *Neurosci. Lett.* **63**, 290–294.

Quirion, R., Araujo, D., Nair, N. P. V., and Chabot, J.-G. (1988). Visualization of growth factor receptor sites in rat forebrain. *Synapse* **2**, 212–218.

Rall, L. B., Scott, J., Bell, G. I., Crawford, R. J., Penschow, J. D., Niall, H. D., and Coghlan, J. P. (1985). Mouse prepro-epidermal growth factor synthesis by the kidney and other tissues. *Nature (London)* **313**, 228–231.

Savage, C. R., Jr., and Cohen, S. (1972). Epidermal growth factor and a new derivative. *J. Biol. Chem.* **247**, 7609–7611.

Schaudies, R. P., Christian, E. L., and Savage, C. R. Jr. (1989). Epidermal growth factor immunoreactive material in the rat brain. *J. Biol. Chem.* **264(18)**, 10447–10450.

Schechter, A. L., Stern, D. F., Vaidyanathan, L., Decker, S. J., Drebin, J. A., Greene, M. I., and Weinberg, R. A. (1984). The *neu* oncogene: An *erb*-B-related gene encoding a 185,000 $M_r$ tumour antigen. *Nature (London)* **312**, 513–516.

Schreiber, A. B., Wendler, M. E., and Derynck, R. (1986). Transforming growth factor-α is a more potent angiogenic mediator than epidermal growth factor. *Science* **232**, 1250–1253.

Scott, J., Urdea, M., Quiroga, M., Sanchez-Pescador, R., Fong, N., Selby, M., Rutter, W. J., and Bell, G. L. (1983). Structure of a mouse submaxillary messenger RNA encoding epidermal growth factor and seven related proteins. *Science* **221**, 236–240.

Seger, M. A., and Bennett, H. P. J. (1986). Structure and bioactivity of the amino-terminal fragment of pro-opiomelanocortin. *J. Steroid Biochem.* **25**, 703–710.

Sheng, H. Z., Turnley, A., Murphy, M., Bernard, C. C. A., and Bartlett, P. F. (1989). Epidermal growth factor inhibits the expression of myelin basic protein in oligodendrocytes. *J. Neurosci. Res.* **23**, 425–432.

Shoyab, M., Plowman, G. D., McDonald, V. L., Bradley, J. G., and Todaro, G. J. (1989). Structure and function of human amphiregulin: A member of the epidermal growth factor family. *Science* **243**, 1074–1076.

Simpson, D. L., Morrison, R. S., de Vellis, J., and Herschman, H. R. (1982). Epidermal growth factor binding and mitogenic activity on purified populations of cells from the central nervous system. *J. Neurosci. Res.* **8**, 453–462.

Styren, S. D., Mufson, E. J., Styren, G. C., Civin, W. H., and Rogers, J. (1990). Epidermal growth factor receptor expression in demented and aged human brain. *Brain Res.* **347**, 347–352.

Taylor, J. M., Mitchell, W. M., and Cohen, S. (1972). Epidermal growth factor. Physical and chemical properties. *J. Biol. Chem.* **247**, 5928–5934.

Teixido, J., and Massague, J. (1988). Structural properties of a soluble bioactive precursor for transforming growth factor-α. *J. Biol. Chem.* **263**, 3924–3929.

Todaro, G. J., Fryling, C., and DeLarco, J. E. (1980). Transforming growth factors produced by certain human tumor cells: Polypeptides that interact with human EGF receptors. *Proc. Natl. Acad. Sci. U.S.A.* **77**, 5258–5262.

Twardzik, D. R., Brown, J. P., Ranchalis, J. E., Todaro, G. J., and Moss, B. (1985). Vaccinia virus infected cells release a novel polypeptide functionally related to transforming and epidermal growth factors. *Proc. Natl. Acad. Sci. U.S.A.* **82**, 5300–5304.

Twardzik, D. R., Ranchalis, J. E., Moss, B., and Todaro, G. J. (1987). Vaccinia growth factor: Newest member of the family of growth modulators which utilize the membrane receptor for EGF. *Acta Neurochirurgica* **41**, 104–109.

Ullrich, A., Coussens, L., Hayflick, J. S., Dull, T. J., Gray, A., Tam, A. W., Lee, J., Yarden, Y., Libermann, T. A., Schlessinger, J., Downward, J., Mayes, E. L. V., Whittle, N., Waterfield, M. D., and Seeburg, P. H. (1984). Human epidermal growth factor receptor cDNA sequence and aberrant expression of the amplified gene in A431 epidermoid carcinoma cells. *Nature (London)* **309**, 418–425.

Ushiro, H., and Cohen, S. (1980). Identification of phosphotyrosine as a product of epidermal growth factor-activated protein kinase in A-431 cell membrane. *J. Biol. Chem.* **255**, 8363–8365.

Wanaka, A., Johnson, E. M., Jr., and Milbrandt, J. (1990). Localization of FGF receptor mRNA in the adult rat central nervous system by in situ hybridization. *Neuron* **5**, 267–281.

Werner, M. H., Nanney, L. B., Stoscheck, C. M., and King, L. E. (1988). Localization of immunoreactive epidermal growth factor receptors in human nervous system. *J. Histochem. Cytochem.* **36(1)**, 81–86.

Westermark, B. (1976). Density dependent proliferation of human glia cells stimulated by epidermal growth factor. *Biochem. Biophys. Res. Commun.* **69**, 304–310.

Wharton, K. A., Johansen, K. M., Xu, T., and Artavanis-Tsakonas, S. (1985). Nucleotide sequence from the neurogenic locus notch implies a gene product that shares homology with proteins containing EGF-like repeats. *Cell* **43**, 567–581.

Wiedermann, C. J., Jelesof, N. J., Pert, C. B., Hill, J. M., and Braunsteiner, H. (1988). Neuromodulation by polypeptide growth factors: Preliminary results on the distribution of epidermal growth factor receptors in adult brain. *Wiener Klin. Wochenschr.* **100(23)**, 760–763.

Wilcox, J. N., and Derynck, R. (1988). Localization of cells synthesizing transforming growth factor-alpha mRNA in the mouse brain. *J. Neurosci.* **8(6)**, 1901–1904.

Winkler, M. E., O'Conner, L., Winget, M., and Fendly, B. (1989). Epidermal growth factor and transforming growth factor alpha bind differently to the epidermal growth factor receptor. *Biochemistry* **28**, 6373–6378.

Wong, A. J., Bigner, S. H., Bigner, D. D., Kinzler, K. W., Hamilton, S. R., and Vogelstein, B. (1987). Increased expression of the epidermal growth factor receptor gene in malignant gliomas is invariably associated with gene amplifications. *Proc. Natl. Acad. Sci. U.S.A.* **84**, 6899–6903.

Wong, S. T., Winchell, L. F., McCune, B. K., Earp, H. S., Teixido, J., Massague, J., Herman, B., and Lee, D. C. (1989). The TGFα precursor expressed on the cell surface binds to the EGF receptor on adjacent cells, leading to signal transduction. *Cell* **56**, 495–506.

Wu, D. K., Morrison, R. S., and de Vellis, J. (1985). Modulation of beta-adrenergic response in rat brain astrocytes by serum and hormones. *J. Cell. Physiol.* **122**, 73–80.

Yokoyama, M., Morrison, R. S., Black, I. B., and Dreyfus, C. F. (1990). Multiple factors regulate cultured basal forebrain (BF) neurons through different mechanisms. *Soc. Neurosci. Abstr.* **16(1)**, 819.

# 12. Transforming Growth Factors Alpha and Beta

Pauli Puolakkainen and Daniel R. Twardzik

## I. INTRODUCTION

In the past decade, growth factors and their actions in the regulation of cell growth have become an important research area in modern cell biology. These growth regulatory polypeptides affect many different aspects of cellular behavior, including cell growth, differentiation, and such diverse activities as migration and receptor modulation. Several different growth factors have been discovered and, according to their molecular structure and functions, they can be classified into different groups. In this chapter, we are going to concentrate on transforming growth factors alpha and beta (TGFα and β). Their discovery, structure, function, and cellular and molecular biology, as well as in vivo effects and their possible clinical implications will be discussed. Neurotrophic effects will be reviewed also, albeit only briefly because these effects are just beginning to be explored.

## II. HISTORICAL BACKGROUND

Epidermal growth factor (EGF) was discovered almost 30 years ago as a contaminant of nerve growth factor (Cohen, 1962). Thereafter, the research in the growth factor area expanded remarkably. The initial observation that led to the identification of TGFα was that several retrovirus-transformed rodent cells no longer contained cell-surface receptors capable of binding EGF (Todaro and De Larco, 1976). In contrast, cells transformed with DNA viruses or with chemicals displayed normal levels of EGF receptor. It was suggested that retrovirus-transformed cells release a factor that is able to bind to the EGF receptors and compete with the externally added ligand for binding to the receptor. Such a factor was identified subsequently in the medium of Moloney sarcoma virus-transformed 3T3 cells. Abelson murine leukemia-transformed rat embryo cells (Twardzik et al., 1982b), and a human melanoma culture (Marquardt et al., 1983). Initially called sarcoma growth factor (SGF) (De Larco and Todaro, 1978), the molecule was able to induce a transformed phenotype in NRK49F fibroblasts, consisting of loss of contact inhibition in liquid culture, morphologic changes, and the ability to form colonies in soft agar (Fig. 1). These effects were shown to be reversible; cells behaved normally after removal of medium containing SGF. The SGF activity later was shown to consist of two distinct proteins, an EGF-like molecule termed TGFα (Todaro et al., 1980) and a distinct unrelated polypeptide designated TGFβ (Roberts et al., 1981; Massagué, 1984; Moses et al., 1985). The name transforming growth factor was based on the fact that preparations of this factor were able to convert normal rat kidney (NRK) cells into phenotypically transformed cells, as just described, and on the fact that this factor was synthesized by many different transformed cells.

TGFα and TGFβ are distinct factors; they bind to different cell-surface receptors and are unrelated both at the structural level and in most of their biologic activities and properties. Therefore, this chapter is divided into two parts. The current knowledge about TGFα is reviewed first, followed by that about TGFβ.

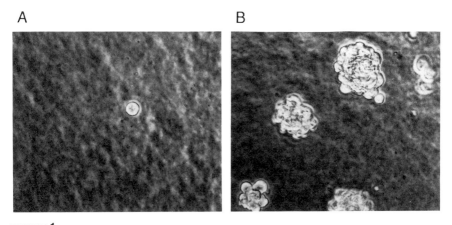

FIGURE 1

Transforming growth factor alpha (TGFα) reversibly induces transformed phenotype in fibroblasts as measured by its ability to stimulate NRK fibroblasts (A) to form progressively growing colonies in soft agar (B).

## III. ASSAYS FOR TGFα

Determination of the presence of TGFα can be accomplished using several different methods. Mitogenic, radioreceptor, eyelid-opening, and angiogenesis assay methods are discussed.

### A. Mitogenic Assay of TGFα

The mitogenic assay is based on the ability of TGFα to stimulate DNA synthesis in various cell lines (Schreiber et al., 1986). In this assay, cells grown to confluence in 48-well dishes (Costar) are brought to quiescence by incubation at 37°C for 48 hr with Dulbecco's modified Eagle's medium (DMEM) and 0.5% Fetal bovine serum (FBS). The cells are then incubated for 18 hr with the sample containing TGFα before being exposed for 4 hr to [methyl-$^3$H]thymidine (New England Nuclear) (1 µCi per well). The stimulation of DNA synthesis and the amount of growth factor in the sample can be determined by measuring the trichloroacetic acid precipitable radioactivity. On a molar basis, TGFα and EGF are nearly equivalent in their ability to stimulate DNA synthesis in various cell lines. In a related assay, the biologic activity of TGFα is measured by determining its ability to induce the anchorage-independent growth of nontransformed cells such as NRK cells (De Larco and Todaro, 1978; Twardzik et al., 1983).

### B. Radioreceptor Assay

By far the most used and the most rapid of all assays for TGFα levels is the receptor binding assay (radioreceptor assay), which is based on competition of TGFα with $^{125}$I-labeled EGF for the same cell-surface receptor (Todaro et al., 1980; Massagué, 1983a,b). Human epidermoid carcinoma cells A431 and mink lung cells (CCL64) often are used as sources of receptor sites in this assay. Results of the radioreceptor assay are expressed as EGF ng equivalents. One EGF ng equivalent is the amount of TGFα that is equipotent to 1 ng of mouse EGF in the standard radioreceptor assay (Massagué, 1983a,b).

### C. Other Assays for TGFα

Other assays for TGFα include the angiogenesis assay (Schreiber et al., 1986) and the eyelid-opening assay (Smith et al., 1985). Recently, an enzyme-linked immunosorbent assay (ELISA) has been developed for the determination of TGFα in human urine and plasma (Katoh et al., 1990). Also, immunohistochemical detection of the growth factor has been used (Fallon et al., 1990; Teerds et al., 1990).

## IV. STRUCTURE AND PROPERTIES OF TGFα

### A. Structure of TGFα Precursor

The TGFα precursor is shown to be a protein consisting of 160 (human; Derynck et al., 1984) or 159 (rat; Lee et al., 1985a) amino acids. This transmembrane protein is anchored to the cell surface via its C terminus (Fig. 2). The precursor contains an extracellular domain of about 100 amino acids that includes the N-terminal signal sequence and the mature 50-amino-acid TGFα, a hydrophobic transmembrane domain, and a 35-residue cytoplasmic domain (Bringman et al., 1987; Gentry et al., 1987; Teixido et al., 1987). The portion of the TGFα sequence that is anchored in the cell membrane contains seven cysteine residues and is palmitoylated (presumably at cysteine residues). It has been suggested that this palmitoylation could slow the passage of the TGFα precursor through the cell surface, allowing more efficient cleavage of the precursor and, thus, TGFα production (Burgess, 1989).

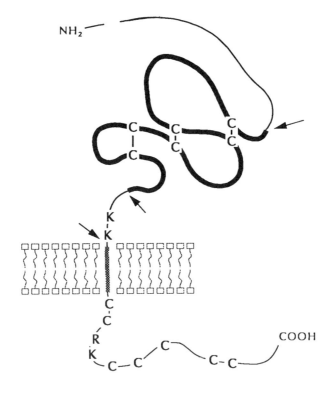

FIGURE 2

Schematic structure of the transmembrane TGFα precursor. The 50-amino acid TGFα, with its three disulfide bonds, is shown as a bold line. The arrows indicate the potential proteolytic cleavage sites.

## B. Structure of TGFα

The active form of TGFα (human and rat) is derived from the precursor and contains 50 amino acids (Fig. 3) (Marquardt et al., 1983, 1984; Derynck et al., 1984). The secreted TGFα exists as multiple species, ranging in apparent size from 5–34 kDa (Marquardt et al., 1983; Sherwin et al., 1983; Derynck, 1988). Experimental evidence indicates that this size heterogeneity can be explained by differential processing of the extracellular protein and by glycosylation of the larger forms. The smallest form of TGFα contains 50 amino acid residues. It shares about 30% structural similarity with the 53-amino-acid form of EGF, including the conservation of all six cysteines (Marquardt et al., 1984). This sequence is extremely conserved across species and has strong homology with the vaccinia virus protein and Shope fibroma virus protein (Twardzik et al., 1985). Human TGFα contains four amino acid substitutions compared with the rat protein, indicating a sequence conservation greater than 90%. This compares with a 70% conservation of sequence between rodent and human EGFs (Lee et al., 1987) (Fig. 4). The similarity in the sequence and the presumed formation of three similar disulfide bonds in TGFα and EGF provide a molecular explanation for the interaction of both TGFα and EGF with the same receptor (Massagué, 1983). The data on the structure of TGFα have been confirmed and extended by the analysis of TGFα cDNA clones.

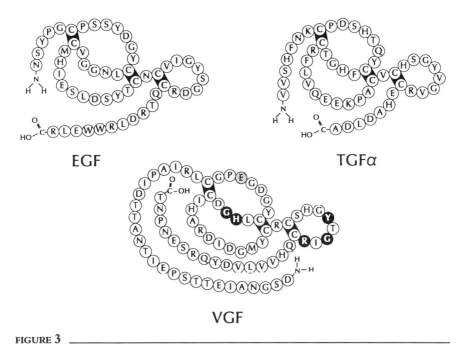

FIGURE 3

The related amino acid sequences of epidermal growth factor (EGF), transforming growth factor alpha (TGFα) and vaccinia virus growth factor (VGF).

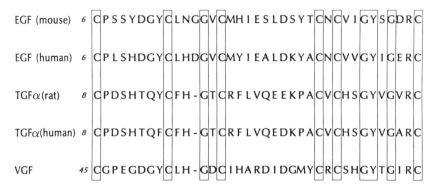

FIGURE 4

The amino acid alignment of EGF-related proteins. Only partial sequences are shown; the position of the first amino acid in the entire sequence is indicated. For EGF and TGFα, sequences are numbered relative to the amino termini of the mature growth factors. For VGF, the sequence is numbered relative to the amino terminus of the membrane-bound form. Residues conserved in all five sequences are boxed.

### C. Synthesis, Release, and Activation of TGFα

Experimental evidence indicates that the TGFα precursor is synthesized as a transmembrane molecule (Fig. 2). To generate the 50-amino-acid TGFα, proteolytic cleavage of the precursor must occur at both the N and the C terminus between an alanine residue and a valine dipeptide. This Ala–Val–Val trimer is located in the sequence Val–Ala–Ala–Ala–Val–Val at the N terminus and in the similar sequence Ala–Val–Val–Ala–Ala– at the C terminus (Derynck et al., 1984; Lee et al., 1985b). Proteolytic processing of a precursor protein by a protease with such specificity has not been described for any other polypeptide, although there is some similarity with the processing of elastase. According to the model, cleavage of the TGFα precursor by this novel protease would occur either outside the cell or in intracellular vesicles (Lee et al., 1987). The sites of TGFα synthesis are discussed in detail in Section V,A. As mentioned, the release and activation of mature TGFα occurs by the action of a specific membrane-associated protease (Bringman et al., 1987; Teixido et al., 1987). The size heterogeneity of the mature TGFα among species can be explained by differences in this proteolytic cleavage of the external precursor domain and by N- or O-glycosylation of the larger forms (Derynck, 1988).

### D. Mode and Mechanism of Action of TGFα and Its Regulation

Current evidence suggests that TGFα can act by autocrine, paracrine, and endocrine mechanisms (Burgess, 1989). Also, it has been shown that pro-TGFα and the EGF receptor can function as mediators of intercellular adhesion and, by this interaction, promote a mitogenic response. This form of stimulation between adjacent cells has been termed a juxtacrine mechanism of action (Anklesaria et al., 1990). The actions of TGFα are modulated by several other growth factors. Also, molecules other than growth factors are known to regulate the various effects of TGFα. An important observation is that TGFα itself can induce TGFα expression in keratinocytes (Coffey et al., 1987). Such autostimulation has been seen with other growth factors and may be a general amplification mechanism among growth modulators (Derynck, 1988). Treatment of normal human keratinocytes in culture with TPA (12-O-tetradecanoyl phorbol-13-acetate), a very potent tumor promoter, has been shown to induce TGFα mRNA expression and secretion of TGFα. This enhancement of TGFα production is suggested to be mediated through activation of protein kinase C (Pittelkow et al., 1989). The actions of TGFα, on the other hand, are mediated through its interactions with the EGF receptor, resulting in the activation of the receptor-associated tyrosine kinase (Reynolds et al., 1981). It has been shown that TGFα, like EGF, also stimulates a phosphoinositol pathway (Kato et al., 1988).

### E. Receptor Binding of TGFα

The actions of TGFα are thought to be mediated through its binding and activation of EGF receptor. This common 180-kDa membrane receptor is part of the tyrosine kinase receptor family and has homology with the oncogenic protein of viral erythroblastosis (Carpenter and Cohen, 1979; Reynolds et al., 1981). These receptors of the same family are phosphorylated through interaction with their ligands, thus inducing pleiotropic metabolic responses, including cell division, differentiation, and development (Adamson and Rees, 1981; Xu et al., 1984). Although available comparative data on receptor binding properties are scarce, TGFα and EGF may interact with their common receptor in nonidentical ways or the ligand–receptor complexes may be processed differently in the cell. Either hypothesis might explain the differences between the actions of TGFα and EGF. Moreover, existence of additional separate TGFα receptors in some cells cannot be excluded.

### V. MOLECULAR BIOLOGY AND GENETICS OF TGFα

The human gene encoding TGFα is 70–100 kb and is located on chromosome 2 (Derynck et al., 1985a). This DNA sequence is spread over six exons in the chromosomal gene (Derynck et al., 1985a). The TGFα peptide precursor (160 amino acids) is encoded by a transcript of 4.5–4.8 kb (Gentry et al., 1987a). Blasband et al. (1990) reported that they had determined the complete nucleotide sequence of rat TGFα mRNA and characterized the six exons that encode this transcript.

## A. TGFα Expression, Tissue Distribution, and Role in Transformation

The production of TGFα by many tumor cells suggests that it plays an important role in malignant transformation. Transformation of rodent cells with retroviruses (Twardzik et al., 1982b), polyoma virus (Kaplan and Ozanne, 1980), or SV40 (Kaplan et al., 1981) has been shown to result in secretion of TGFα. Further, TGFα has been detected in the urine of patients with advanced cancer (Twardzik et al., 1982a; Sherwin et al., 1983). Also, transformation by the *ras* oncogene leads to induction of TGFα gene expression in both immortalized fibroblasts and epithelial cells (Ozanne et al., 1980; Salomon et al., 1987; Cutry et al., 1988). Expression of the TGFα gene has been demonstrated in a variety of tumors, mostly carcinomas and sarcomas, but not in hematopoietic tumor cell lines (Derynck, 1988). TGFα gene expression is shown to be consistent in squamous cell carcinomas and renal carcinomas (Derynck et al., 1987a). Also, breast, gastric, colon, liver, ovarial, and lung carcinomas, melanomas, and tumors of neuronal origin such as gliomas synthesize TGFα (Derynck, 1989; Murthy et al., 1989; Yeh and Yeh, 1989; Yoshida et al., 1990). Some studies suggest that TGFα can act as an oncogene and can predispose mammary epithelium to neoplasia and carcinoma (Matsui et al., 1990). The occurrence of TGFα mRNA in such a large variety of solid tumors suggests that the synthesis of TGFα may play a biologic role in malignant transformation and tumor development in vivo. Thus, the autocrine phenomena proposed by Todaro and Sporn may be invoked (Sporn and Todaro, 1980). The synthesis of TGFα by these tumors could also explain the presence of a high molecular weight EGF receptor-binding factor that reacts with anti-TGFα antibodies in the urine of some cancer patients and its absence in the urine of normal controls (Sherwin et al., 1983; Twardzik and Sherwin, 1985). Further, liver and mammary neoplasias have been shown to be induced in transgenic mice by TGFα overexpression (Jhappan et al., 1990; Sandgren et al., 1990). In spite of the accumulating evidence, the true role of TGFα in the development of human tumors remains to be evaluated.

Initially it was thought that TGF was uniquely associated with transformed cells. However, subsequently it was realized that both TGFα and TGFβ also can be extracted from normal cells (Roberts et al., 1981; Sporn et al., 1986). The levels of TGFα in many tissues were thought to be quite low. Observations on self-renewing epithelial tissues (e.g., skin and gastrointestinal tissue), however, have shown significant concentrations of TGFα and TGFα mRNA (Derynck et al., 1984). TGFα biosynthesis is shown to occur in keratinocytes throughout the stratified epidermis (Derynck et al., 1984); elevated expression has been observed in psoriatic epidermis (Turbitt et al., 1990). Also, activated macrophages synthesize TGFα (Madtes et al., 1988). Likewise, human eosinophils have been shown to express TGFα (Wong et al., 1990). TGF-like activity was first detected during fetal development (Twardzik et al., 1982) and, subsequently, TGFα was shown to be expressed transiently during embryonic development in the kidney, the placenta, the nasopharyngeal pouch, and the otic vesicle (Lee et al., 1985b; Wilcox and Derynck, 1988b). It is present in pituitary cells and in several areas of brain (Kobrin et al., 1986; Wilcox and Derynck, 1988a; Fallon et al., 1990). Other locations at which TGFα is expressed include smooth muscle cells in the arterioles (Mueller et al., 1990), normal adult human kidney (Gomella et al., 1990), ovary (Adashi et al., 1989), and normal adult plasma (Katoh et al., 1990). Most studies of TGFα typically have measured mRNA levels or receptor binding activity. Therefore, the exact form and location (membrane bound or secreted) of the TGFα gene product are not known in most cases (Derynck, 1988).

## VI. EFFECTS OF TGFα IN VIVO AND IN VITRO

TGFα is a potent modulator of cell growth and differentiation. Initial observations centered around the proliferative effects on fibroblasts, keratinocytes, and epithelial cells; observations of other effects have been made more recently. On the whole, many effects are dependent on the mode of delivery and the status of the target cells as well as on the amount of TGFα and the presence of other growth factors and other substances in the cellular milieu. TGFα is, for example, shown to inhibit the growth of an endometrial carcinoma cell line when the cells are seeded at low densities and to stimulate it when the cells are seeded at higher densities (Korc et al., 1987). The possible effects of TGFα on malignant transformation have been discussed earlier.

TGFα has demonstrable effects on skin thickening, retardation of hair production by follicles, cell production and function in the gastrointestinal tract, craniofacial morphogenesis, blood vessel development, neonatal growth, and the immune and endocrine systems (Burgess, 1989). Wound healing and subsequent reepithelialization requires both the proliferation and migration of resident keratinocytes as well as deposition of granulation material (angiogenesis). TGFα has been shown to stimulate all these processes (Schultz et al., 1987).

TGFα also is able to stimulate DNA synthesis in various cell lines, to induce anchorage independence, to induce eyelid opening in newborn mice, to increase arterial blood flow, and to enhance bone resorption in vitro (Anzano et al., 1983; Ibbotson et al., 1983; Smith et al., 1985; Schreiber et al., 1986; Gan et al., 1987). In addition, TGFα has been suggested to play a role in lung maturation and postinjury repair, liver repair, and regeneration, as well as neuronal cell growth (Brown et al., 1990).

In spite of all this knowledge, the exact physiologic role of TGFα remains unknown, although one can speculate that many of its pleiotropic effects are regulated stringently in vivo and modulated via site-specific physiologically induced signals.

## VII. CLINICAL IMPLICATIONS

Two major clinical implications of TGFα are neoplasia and wound healing. TGFα may have some use as a diagnostic marker for cancer, since it is expressed in a variety of different tumor cells (Table 1) (Yeh and Yeh, 1989). Even in the early experiments, TGFα was found to be able to induce the growth of normal fibroblasts in soft agar—a property usually associated with the malignant phenotype. However, the role of TGFα in the initiation or maintenance of naturally occurring tumors is still far from clear (Burgess, 1989). However, antagonists of TGFα and its angiogenic effects could decrease the vascularization of tumor mass. Likewise, autocrine stimulation of tumor cells also may be susceptible to intervention therapies.

As mentioned earlier, TGFα promotes the reepithelialization and neovascularization of the woundbed (Schultz et al., 1987). Therefore, it would seem reasonable to use TGFα clinically to improve the impaired healing of complex wounds (both rate and quality of repair are a consideration). In fact, in experimental burns in a porcine model, TGFα was shown to accelerate the healing rate of the full thickness wounds (Schultz et al., 1987). Clinically, diabetics with slowly healing chronic wounds and patients with peripheral vascular disease would constitute an appropriate patient population that might response to a growth factor regimen of which TGFα is a component.

Use of EGF for the treatment of gastric ulcers has been considered since its initial discovery. However, the short half-life of EGF in vivo, as well as its former limited availability, has restricted the possible exploration of its usefulness in this area. However, the use of TGFα in the treatment of duodenal and gastric ulcers should be considered since TGFα has been shown to inhibit parietal cell $H^+$ secretion (Lewis et al., 1990).

TGFα is shown to be strongly overexpressed in the skin of psoriatic patients; thus, it could be assumed that TGFα is a mediator of the epidermal hyperproliferative response in psoriasis (Elder et al., 1989) since keratinocytes both respond to TGFα mitogenicity and are superinducible for TGFα itself.

TGFα has been shown to be present in embryos and in normal tissues.

TABLE 1
Increased TGFα in human cancers

| Human cancer | Locations of elevated TGFα | Reference |
|---|---|---|
| Brain | Malignant forms of gliomas | Samuels et al., 1989 |
| Skin | Urine of melanoma patients | Kim et al., 1985 |
| Breast | Urine of breast cancer patients | Stromberg et al., 1987 |
| | Effusions of patients with breast cancer | Arteaga et al., 1988 |
| Lung | Effusions of patients with lung cancer | Arteaga et al., 1988 |
| | Urine of lung cancer patients | Sherwin et al., 1983 |
| Liver | Urine of hepatoma patients | Yeh et al., 1987 |
| Abdomen | Effusions of patients with colon, stomach, and pancreas cancers | Arteaga et al., 1988 |
| | Urine of colon cancer patients | Sherwin et al., 1983 |
| Ovary | Urine of patients with ovarian cancer | Sherwin et al., 1983 |
| | Effusions of patients with ovarian cancer | Arteaga et al., 1988 |

Therefore it seems reasonable to assume that this molecule has no adverse effects on humans. Further, no such effects have been found in studies done so far. Thus, TGFα might be of some clinical use in the future.

## VIII. TGFα AND NEUROBIOLOGY

TGFα mRNA, as well as TGFα precursor immunoreactivity, has been found in pituitary cells and in several areas of the adult brain (Kobrin et al., 1986; Wilcox and Derynck, 1988a; Fallon et al., 1990). In adult brain, precursor immunoreactivity is localized in astrocytes (Fallon et al., 1990). This immunoreactivity is increased by disruption of the tissue, especially by fetal cell transplants (Loughlin et al., 1989). TGFα also is known to be expressed in gliomas and many other tumors of neural origin. There is convergent evidence that TGFα is present in neurons that contain enkephalin or interact with enkephalinergic neurons, and that TGFα is synthesized in the same brain regions, although it is not clear whether neurons or glia actually contain mRNA for TGFα. It has been suggested that the role of a peptide fragment in the TGFα precursor outside the mature TGFα peptide might be important (Fallon et al., 1990). That is, peptide fragments in the precursor may have an important neuromodulatory role. In the future, it will be important to determine the generation and fate of the precursor sequences in brain cells, whether they are neurons or glia. Also it will be important to determine which cell types synthesize the precursor in order to understand the function of TGFα in the central nervous system. At the moment, the role of TGFα in brain physiology is not known, although its neurotrophic expression has been documented.

## IX. ASSAYS FOR TGFβ

### A. Assay for the Stimulation of Anchorage-Independent Growth by TGFβ

TGFβ originally was identified in an assay that measured its ability to enhance the growth of NRK fibroblasts in soft agar (Roberts et al., 1981). This assay system was described originally by De Larco and Todaro (1978) to measure the activity of SGF, which later was found to be a mixture of TGFα and TGFβ.

### B. Growth Inhibition and Differentiation Assays

TGFβ has been purified using assays measuring the differentiation of primitive mesenchymal cells into cells expressing a cartilaginous phenotype (Seyedin et al., 1985, 1986). In addition, TGFβ also has been purified using assays that measure inhibition of growth of mink lung epithelial cells (CCL64) and inhibition of C3H/HeJ mouse thymocyte mitogenesis (Ikeda et al., 1987; Wrann et al., 1987).

Assays measuring inhibition of growth of BSC-1 monkey kidney cells and inhibition of myoblast differentiation also have been used (Holley et al., 1980; Tucker et al., 1984b; Florini et al., 1986). Growth inhibition assays are likely to be less specific than assays for colony growth in soft agar since peptides other than TGFβ also can inhibit the growth of the target cells (Roberts and Sporn, 1988).

### C. Immunological Assays for TGFβ

Various antisera against different regions of TGFβ have been raised. They have been used to detect the native peptide by radioimmunoassays, immunoblots, and receptor binding assays (radioreceptor assay) (Heldin et al., 1981; Frolik and De Larco, 1987). One of the most important applications of these antibodies, however, has been in immunohistochemical studies. Most often used have been two antisera against a peptide representing the N-terminal 30 amino acids of TGFβ (Ellingsworth et al., 1986; Thompson et al., 1989).

## X. STRUCTURE AND PROPERTIES OF TGFβ

### A. TGFβ Gene Family

Five different forms of TGFβ have been found so far (TGFβ1–5). TGFβ1 and TGFβ2 have been isolated from natural sources, including human and porcine platelets, bovine bone, and human glioblastoma cells (Assoian et al., 1983; Seyedin et al., 1985; Cheifetz et al., 1987; Wrann et al., 1987). The three new forms TGFβ(3–5) have been identified in the past few years by screening of cDNA libraries. None of these peptides has been isolated from natural sources (Roberts and Sporn, 1990).

TGFβ1–5 and mammalian inhibins (Mason et al., 1985), activins (Ling et al., 1986), and Mullerian inhibitory substances (MIS; Cate et al., 1986), as well as the predicted products of both a pattern gene in *Drosophila* (*DPP-C;* Padgett et al., 1987) and an amphibian gene expressed in frog oocytes (*Vg1;* Weeks and Melton, 1987), form the larger gene family of structurally related (conservation of seven of the nine cysteine residues of TGFβ) but functionally partially distinct proteins. More recently, three new proteins called bone morphogenetic proteins (BMPs) (Wozney et al., 1988), as well as the product of the *vgr-1* gene (Lyons et al., 1989), have been added to the family. The unifying properties of these peptides are their similarities in structure and their ability to regulate development and cell differentiation. This regulation is mediated by receptors that, presumably, are distinct for every peptide (Roberts and Sporn, 1990).

### B. TGFβ Precursor

The protein structure of TGFβ is far more complicated than that of TGFα. The structure of the latent TGFβ1 complex of platelets consists of three components:

a 125–160-kDa TGFβ modulator protein, a latency protein consisting of the remainder of the precursor (the precursor without the N-terminal 29-amino-acid signal peptide sequence and the C-terminal 112 amino acids of the mature TGFβ1), and the mature dimeric TGFβ1 (Miyazono et al., 1988; Wakefield et al., 1988). Thus, the precursor complex for TGFβ1 contains a typical hydrophobic signal peptide and three potential N-linked glycosylation sites (Gentry et al., 1988). The latency protein is glycosylated and forms a disulfide-bonded dimer of approximately 75 kDa (Miyazono et al., 1988). TGFβ1 is encoded as a 390-amino-acid precursor (Derynck et al., 1985b) and TGFβ2 as a 412-amino-acid precursor (De Martin et al., 1987; Madisen et al., 1988). Each has a signal peptide of 20–23 amino acids at the N terminus. The numbers of amino acids in the precursors for TGFβ3, 4, and 5 are 412, 304, 382, respectively. The processed 112-amino-acid chains of TGFβ1 and 2 are 72% identical, with the conservation of all nine cysteine residues. In addition to TGFβ of platelet origin, biologically active recombinant TGFβ1 now can be expressed at high levels in Chinese hamster ovary cells (CHO) using gene amplification with dihydrofolate reductase (Gentry et al., 1987b). The expressed gene encodes the entire precursor form of the peptide. This precursor is glycosylated and phosphorylated (Brunner et al., 1988).

It has been shown that the latent form of TGFβ is unable to bind to TGFβ receptors and is thus inactive (Wakefield et al., 1987). Further, TGFβ in serum is known to be its latent form and is thought to be carried as a complex with $\alpha_2$-macroglobulin (O'Connor-McCourt and Wakefield, 1987).

## C. Mature TGFβ

TGFβ1 is an acid and heat stable, disulfide-linked, homodimeric 24-kDa protein that consists of two identical 112-amino-acid subunits (Fig. 5). Each subunit contains nine cysteines. On reduction it yields two identical peptides of approximately 12 kDa. The nonreduced dimeric form is known to be the biologically active form of the peptide. As mentioned earlier, there are five different TGFβs (TGFβ1–5). They are, however, structurally and functionally closely related and are therefore discussed together in this context. TGFβ first was isolated from human platelets and later was cloned from a human DNA library (recombinant TGFβ) (Assoian et al., 1983; Sharples et al., 1987; Miyazono et al., 1988). The universality of its action is emphasized by the identical amino acid sequence in human, monkey, cow, pig, and chicken (Sporn and Roberts, 1989). Mature TGFβ1 from human platelets is a disulfide-linked homodimer of about 25 kDa that is associated with a disulfide-linked dimer of the N terminus of its precursor (Miyazono et al., 1988). The dimeric N-terminal portion of the precursor is noncovalently associated with the mature TGFβ and is associated also through disulfide bonds with TGFβ binding protein. This glycosylated protein (125–160 kDa) is a dimer of two 70-kDa polypeptides and is not related to TGFβ, either immunologically or in sequence (Miyazono et al., 1988; Wakefield et al., 1988). The carbohydrate on the TGFβ precursor is necessary for the interaction of TGFβ with the complex (Miyazono and Heldin,

1989). This TGFβ binding protein present in platelet TGFβ has not been found in recombinant TGFβ and is thus the main structural difference between these forms.

## D. Synthesis, Activation, and Release of TGFβ and Its Regulation

Almost all cells can synthesize TGFβ in one or more of its molecular forms. Platelets are the most concentrated source of TGFβ in the human body (Sporn and Roberts, 1989). The sites of synthesis are further discussed in Section XI,A.

Like TGFα, TGFβ is synthesized as a larger biologically inactive precursor protein that is unable to bind to cellular receptors (Pircher et al., 1986; Miyazono et al., 1988). Mature active TGFβ is produced only after cleavage. The activation of TGFβ can be accomplished in vitro by transient acid treatment (pH<4), by alkalinization (pH>9), by proteases such as plasmin and cathepsin D, by exposure to denaturing agents, and by γ-interferon (Lawrence et al., 1985; Keski-Oja et al., 1987a; Lyons et al., 1988; Twardzik et al., 1990). The physiologic mechanisms of activation of the different latent forms may be specific for different sites and types of TGFβ and are currently under investigation. The N-linked glycosylation sites of the latency protein of TGFβ1 have been shown to play a critical role in latency; treatment of latent TGFβ1 with endoglycosidase F or sialidase and addition of sialic acid or mannose 6-phosphate results in dose-dependent activation (Miyazono and Heldin, 1989). In healing wounds, the acid microenvironment and the production of sialidase and proteases by activated macrophages could contribute to activation of latent TGFβ. The exact sequence of events in the processing of TGFβ1 is uncertain but appears to include cleavage of a 29-amino-acid signal sequence, glycosylation and mannose 6-phosphorylation of the precursor, cleavage of the

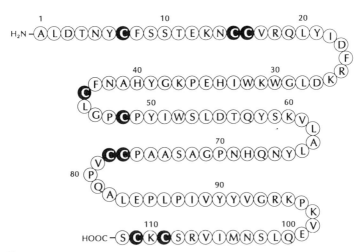

**FIGURE 5**

The deduced amino acid sequence of the processed 112-amino acid monomeric unit of human TGFβ.

C-terminal 112-amino-acid (12-kDa) monomer from the 390-amino-acid precursor, and disulfide isomerization (Fig. 6) (Gentry et al., 1987b, 1988; Purchio et al., 1988).

As mentioned, biologically active recombinant TGFβ now can be expressed at high levels in CHO cells (Gentry et al., 1987b). In CHO cells, the signal peptide is cleaved between $Gly_{29}$ and $Leu_{30}$, whereas the precursor is cleaved from the mature form at the basic site preceding $Ala_{279}$ (Gentry et al., 1988).

Synthesis of TGFβ is regulated by various mechanisms. For example, the loss of TGFβ receptors might be a mechanism by which preneoplastic cells progress to tumor cells that lack negative growth control (Sporn and Roberts, 1985). Tyrosine kinase activity as a mechanism of receptor signal transduction has been suggested. However, no such activity has been found (Libby et al., 1986). Thus, signals transduced from the TGFβ receptor seem to involve some novel pathways yet to be discovered. There is evidence that a G protein, perhaps $G_i$, is involved in transducing the TGFβ signal (Murthy et al., 1988). Another possibility is that the mechanism of action of TGFβ is distinct from those involving tyrosine kinases, G proteins, ion channels, or phosphatidylinositol phospholipases and C kinases and involves several different signal transduction pathways. This kind of transduction could explain the very varied, pleiotropic, and sometimes even contradictory cellular responses to TGFβ.

### E. Mode and Mechanism of Action of TGFβ

The activity of TGFβ is regulated in many ways, as indicated earlier. It depends, for example, on the cell type, the culture conditions, and the presence of other growth

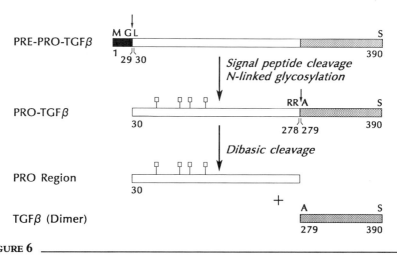

FIGURE 6

Processing events of TGFβ precursor protein in transfected CHO cells. Proteolytic processing sites are indicated.

factors (Derynck, 1989). TGFβ has a mitogenic effect on fibroblast cultures. This may, in part, be caused by an indirect stimulation, since TGFβ can induce the expression of c-*sis* mRNA in these cells, which encodes the B chain of platelet-derived growth factor (PDGF; Leof et al., 1986). TGFβ synthesis in vivo has been shown to be stimulated by tamoxifen (Knabbe et al., 1987). Also, malignant transformation of cells is shown to increase TGFβ synthesis (Anzano et al., 1985). TGFβ exerts its actions on the targets by both paracrine and autocrine mechanisms (Van Obberghen-Schilling et al., 1988).

### F. Receptor Binding of TGFβ

The action of TGFβ is mediated through binding to specific cell-membrane receptors. Almost all normal cells are shown to have receptors for TGFβ (Sporn and Roberts, 1989; Nilsen-Hamilton, 1990). Typically, there are between 10,000 and 50,000 receptor sites per cell (Tucker et al., 1984a); the binding of TGFβ to its receptor is saturable with an affinity of $3-9 \times 10^{-11}$ M (Frolik et al., 1984; Tucker et al., 1984a). Currently, three different classes of integral cell-membrane components that bind TGFβ specifically with high affinity have been described (Sporn and Roberts, 1989). Class I components are 65-kDa glycoproteins in all species, whereas class II components are glycoproteins that range from 85 to 110 kDa, depending on the species. Class III receptors are considered to be dimeric and to be composed of proteoglycan subunits of 250–350 kDa (Massagué, 1985; Cheifetz et al., 1988b; Segarini and Seyedin, 1988). Most frequently, all three classes of receptors coexist on cells. Recently, a class IV receptor (60 kDa) has been found in GH3 pituitary cells (Cheifetz et al., 1988c). Some controversy exists concerning the roles of the different receptors (Massagué et al., 1990). It has been proposed that class III receptors mediate all functions of TGFβ when TGFβs 1 and 2 are equipotent (Cheifetz et al., 1987, 1988a). Moreover, they suggest that biologic activities specific to TGFβ1 are mediated through class I and II receptors. However, currently, specific cellular responses cannot be assigned confidently to any of the three receptor types (Barnard et al., 1990). Unlike normal cells, several neoplastic cells appear to lack TGFβ receptors (Keller et al., 1989). This lack correlates with the resistance of the cells to the inhibitory effects of TGFβ; it has been suggested that the loss of TGFβ receptors might be a mechanism by which preneoplastic cells could progress to tumor cells by escaping negative growth control.

### XI. MOLECULAR BIOLOGY AND GENETICS OF TGFβ

The human genes for TGFβs have been localized as follows: 19 q13.1–q13.3 (TGFβ1), 1 q41 (TGFβ2), and 14 q23–q24 (TGFβ3) (Fujii et al., 1986; Barton et al., 1988; ten Dijke et al., 1988). The locations for TGFβ4 and 5 are not known currently. The human TGFβ1 precursor is encoded by seven exons; the size of the introns is unknown, but the gene is estimated to be greater than 100 kb (Derynck et al., 1987b). The sizes of the human mRNAs are 2.5 kb for TGFβ1, 4.1, 5.1, and 6.5 kb for TGFβ2, and 3.0 kb for TGFβ3 (Roberts and Sporn, 1990). Two major transcriptional start sites of human TGFβ1 mRNAs, 271 nucleotides apart, have been found (Kim et al., 1989). The identification and cloning of the genes for TGFβs will allow large scale production of their respective peptides by recombinant DNA technology (Sporn and Roberts 1989).

### A. TGFβ Expression, Tissue Distribution, and Role in Transformation

As mentioned earlier, TGFβ is expressed in almost all cells, both normal and neoplastic; thus, listing all these different types of cells is beyond the scope of this review. Of special importance, however, is its location in mammalian embryo, in platelets, in extracellular matrix, in bone and cartilage tissue, in T and B lymphocytes, and in cardiac myocytes.

TGFβ is known to be a potent antiproliferative agent for most epithelial cells, the cells from which most human cancers arise (Barnard et al., 1988). During transformation, the epithelial cells, whose proliferation is normally suppressed by TGFβ, may escape control and become autonomous. The proposed mechanisms of this process of carcinogenesis include loss of TGFβ itself, loss of the TGFβ receptor, or a failure in the intracellular pathway mediated by TGFβ (Sporn and Roberts, 1985, 1989). Further, TGFβ might have its effects on cell transformation by regulating the expression of several oncogenes such as c-*sis* (Leof et al., 1986).

## XII. EFFECTS OF TGFβ IN VITRO AND IN VIVO

The biologic activities of TGFβ are very variable and depend on the cell type, on culture conditions, and on the presence of other substances such as other growth factors. Its main effects are its bifunctional mitogenic properties on the regulation of cell growth (activation or inhibition) of cell behavior (migration, adhesion), of cell differentiation, and of the status of the extracellular matrix.

### A. Growth Stimulation

TGFβ is known to stimulate the growth of some fibroblasts as well as certain mesenchymal, endothelial, and transformed cells. When TGFβ is added to AKR-2B cells, it introduces the expression of c-*sis* mRNA and the production of PDGF-like substance (Leof et al., 1986). The growth-promoting activity that occurs on this cell type suggests the autocrine stimulation of cellular growth by the PDGF-like substance.

## B. Growth Inhibition

TGFβ inhibits the growth of most normal cells including epithelial cells, endothelial cells (Heimark et al., 1986), blood lymphocytes (Kehrl et al., 1986), hepatocytes (Carr et al., 1986), keratinocytes (Shipley et al., 1986), and hematopoietic progenitor cells (Keski-Oja et al., 1987b). Likewise, it inhibits the growth of different tumor cells (Tucker et al., 1984b; Moses et al., 1985). For example, TGFβ inhibits in a dose-dependent manner, the growth of human lung adenocarcinoma A549 in male athymic BALB/c mice (Twardzik et al., 1989).

## C. Cell Behavior

TGFβ is shown to stimulate anchorage-independent growth of nonmalignant fibroblast cells (Shipley et al., 1985) as well as have an effect on cell migration. It acts also as a potent immunosuppressive agent. In addition to inhibiting the growth of lymphocytes, it has been shown to inhibit the function of natural killer cells (Sporn et al., 1987). TGFβ is also a potent chemotactic factor for monocytes, macrophages, and fibroblasts (Postlethwaite et al., 1987; Wahl et al., 1987).

## D. Cell Differentiation

TGFβ modulates the differentiation of various kinds of cells. It is shown to stimulate differentiation of human bronchial epithelial cells (Masui et al., 1986) and intestinal crypt epithelial cells (Kurokawa et al., 1987). In contrast, it inhibits the differentiation of adipocytes (Ignotz and Massagué, 1985) and keratinocytes (Reiss and Sartorelli, 1987) as well as B lymphocytes (Lee et al., 1987). TGFβ regulates the differentiation of several steroidogenic cell types such as the ovarian granulosa cell and the adrenocortical cell. In neonatal mice, TGFβ has been shown to be a very potent inhibitor of mammary duct development and differentiation (Silberstein and Daniel, 1987).

So far, most studies of the effects of TGFβ on cell differentiation have been done with normal cells. Studies showing that TGFβ induces tumor cell differentiation are few. Pfeilschifter et al. (1987) showed that TGFβ stimulates osteoblastic differentiation of osteosarcoma cells. In experiments by Block et al. (Block, personal communication) and Kamijo et al. (1990), TGFβ induced cell differentiation of leukemic cells. Zugmaier and Lippman (1990) examined the effects of TGFβ on differentiation in oncogene-transformed human mammary epithelial cells. They found that the expression of human milk fat globule antigen as a differentiation marker was increased 2- to 3-fold in transformed cells after TGFβ treatment (Zugmaier and Lippman, 1990). Moreover, TGFβ was able to enhance differentiation of colon carcinoma cells by increasing the expression of an integrin RGDT collagen receptor (Pignatelli and Bodmer, 1990). Choi and Fuchs (1990) reported the effects of TGFβ on the differentiation of human skin squamous cell carcinoma (SCC-13). Further evidence of the ability of TGFβ to induce tumor differentiation is shown by Twardzik et al. (1989).

Based on current knowledge, it seems most plausible that TGFβ functions in concert with a variety of other factors, including extracellular matrix proteins and other growth factors, to affect cellular differentiation (Barnard et al., 1990). In support of this concept, the cellular distribution of TGFβ expression in situ and in vivo frequently is associated with the differentiated phenotype (Barnard et al., 1989).

### E. Effects on Extracellular Matrix

TGFβ is shown to enhance the accumulation of extracellular matrix components by fibroblasts (Table 2). It stimulates the production of collagens, fibronectins, thrombospondin, and tenascin; inhibits the effects of proteolytic enzymes such as plasminogen activator on matrix proteins; and increases the synthesis of receptors for cell attachment factors (up-regulation of fibronectin receptors), matrix proteoglycans, and glycosaminoglycans (Ignotz and Massagué, 1987; Saksela et al., 1987; Penttinen et al., 1988; Laiho and Keski-Oja, 1989). Further, TGFβ is shown to be an angiogenic factor promoting the development of new blood vessels at the site of tissue injury. As a net result of these effects, TGFβ may induce an increased deposition and decreased degradation of the extracellular matrix and form new granulation tissue.

### F. In Vivo Effects of TGFβ

TGFβ has been implicated to play a central role in many in vivo processes, including embryogenesis, immunoregulation, bone remodeling, and wound healing.

TGFβ has morphogenic activity during embryogenesis. The development of mesenchymal cells in mammals appears to require TGFβ. It has been shown by immunohistochemical methods that areas of mesenchymal differentiation in palate, larynx, teeth, cardiac valves, hair follicles, bones, and cartilage correlate with high TGFβ1 expression in the tissues (Heine et al., 1987).

As mentioned, TGFβ is a very potent suppressor of lymphocyte proliferation and function. It inhibits the proliferation of T cells stimulated by interleukin 1 or

TABLE 2
**TGFβ Enhances the formation of extracellular matrix**

| TGFβ stimulates matrix synthesis | TGFβ inhibits matrix degradation |
|---|---|
| Activation of gene transcription for collagen, fibronectin, and other matrix proteins<br>Increased synthesis and secretion of matrix proteins<br>Increased synthesis of protease inhibitors<br>Increased transcription, translation, and processing of cellular receptors for matrix proteins | Decreased synthesis of proteolytic enzymes |

2, inhibits proliferation and antibody production in B cells stimulated by any activating factor, depresses the cytolytic activity of natural killer cells, and inhibits the generation of cytotoxic T cells and lymphokine-activated killer cells (Roberts and Sporn, 1990). TGFβ seems to act like the brake of the immune system to arrest its excessive proliferation when stimulated by antigens. Thus, the clinical use of TGFβ as an immunosuppressive agent in patients with organ transplants is being studied currently (Sporn, 1989).

Large amounts of TGFβ have been found in cardiac myocytes but its physiologic role in normal cardiac function is not known. Recently, Lefer et al. reported the cardioprotective properties of TGFβ in ischemic cardiac injury (Lefer et al., 1990).

## XIII. TGFβ AND OTHER GROWTH FACTORS

TGFβ is known to have interactions with various other growth factors, such as PDGF, EGF, fibroblast growth factor (FGF), and insulin-like growth factor. In fact, these interactions are so common that it has been suggested that one of the main mechanisms of action of TGFβ is to modify and regulate the effects of other growth factors. As well as regulating the expression of genes that encode growth factors, TGFβ alters and modifies cellular responses to some growth factors. The interactions between TGFβ and other growth factors have been reviewed (Nilsen-Hamilton, 1990).

## XIV. CLINICAL IMPLICATIONS OF TGFβ

### A. Bone Remodeling

TGFβ acts as a mitogen for osteoblasts and as a stimulator of their production of extracellular matrix. Enhancing bone formation, it thus offers a potential treatment of delayed union by accelerating fracture healing (Joyce et al., 1990). Studies on the potential role of TGFβ in the treatment of osteoporosis are also in progress. The role of TGFβ in bone remodelling has been reviewed extensively by Mundy and Bonewald (1990).

### B. Wound Healing

TGFβ is known to enhance wound healing. As discussed earlier, TGFβ is strong chemotactic for macrophages and fibroblasts (Wahl et al., 1987; Moses et al., 1990). It stimulates the production of connective tissue components, prevents reepithelialization from occurring too early in the healing process by inhibiting the epithelial cells (Shipley et al., 1986; Choi and Fuchs, 1990), and stimulates angiogenesis (Madri et al., 1988) in the wound area by stimulating FGF. Further, TGFβ suppresses the release of $H_2O_2$ from macrophages (Tsunawaki et al., 1988). The net result of these actions is the formation of new granulation tissue at the site of TGFβ action. This can be seen in experimental animals injected subcutaneously with TGFβ; highly vascularized granulation tissue forms rapidly at the injection site. These effects of TGFβ on wound healing might have significant clinical implications in the treatment of wounds with impaired healing, for example, in patients with large burns, diabetes mellitus, or peripheral vascular disease, or in patients undergoing chemotherapy. Studies have shown that TGFβ can enhance wound healing in animals (Mustoe et al., 1987; Hebda, 1988; Amman et al., 1990; Beck et al., 1990; Puolakkainen et al., 1992), but human studies are few (Roberts et al., 1988).

In addition, TGFβ might have clinical implications in the treatment of retinal detachment of the eye (Smiddy et al., 1989). TGFβ has been suggested to be involved in the pathogenesis of diseases such as rheumatoid arthritis, pulmonary fibrosis, hepatic cirrhosis, and keloids (Sporn and Roberts, 1989).

## XV. TGFβ AND NEUROBIOLOGY

Various brain tumors are known to contain TGFβ. Indeed, TGFβ2 was isolated first from a glioblastoma cell line (Wrann et al., 1987). In addition, TGFβ affects normal pituitary hormone synthesis and stimulates follicle stimulating hormone secretion by cultured pituitary cells (Ying et al., 1986). Whether TGFβ also has effects on synthesis of other pituitary hormones or of hypothalamic releasing factors remains unknown. TGFβ also is expressed in embryonic brain (Flanders et al., 1990). By immunohistochemical methods, TGFβ was found in neurons in both the brain and the spinal cord (Flanders et al., 1990). Whether the potential neurotrophic action of TGFβ can be used to repair ischemic or other injury to the brain or spinal cord remains to be determined (Sporn and Roberts, 1990). Further, it has been shown that TGFβ has a strong mitogenic effect on Schwann cells of the peripheral nervous system, suggesting that TGFβ might act as an intrinsic mediator of repair of peripheral neuronal damage (Roberts and Sporn, 1988). Whether TGFβ could be used in the treatment of injury in peripheral neurons is a question to be answered in the future.

## XVI. CONCLUSIONS AND FUTURE ASPECTS

It is difficult to predict the future of TGF research. A continuous series of new findings has indicated that TGFβ plays an important role in almost every tissue. It would seem likely that TGFβ will have many clinical implications, including the improvement of wound healing in debilitated patients as well as the treatment of bone fractures or the induction of repair in damaged cartilage. TGFβ is also potentially of use in the repair of cardiac injury after myocardial infarction. Its role in the nervous system needs further investigation. Before clinical applications reach fruition, much must be learned about basic mechanisms, especially about the signal transduction pathways by which the effects of these potent cytokines are mediated. On the whole, the future of TGF research will involve a very wide range of both basic and applied studies. This field holds important promise for discovering and developing novel modes of treatment for a variety of different diseases. Many significant and interesting findings are yet to be discovered.

## REFERENCES

Adamson, E. D., and Rees, A. R. (1981). Epidermal growth factor receptors. *Mol. Cell. Biochem.* **34**, 129–152.

Adashi, E. Y., Resnick, C. E., Hernandez, E. R., May, J. V., Purchio, A. F., and Twardzik, D. R. (1989). Ovarian transforming growth factor-beta (TGF-beta): Cellular site(s), and mechanism(s) of action. *Mol. Cell. Endocrinol.* **61**, 247–256.

Amman, A. J., Beck, L. S., DeGuzman, L., Hirabayashi, S. E., Lee, W. P., McFatridge, L., Nguyen, T., Xu, Y., and Mustoe, T. A. (1990). Transforming growth factor beta. Effect on soft tissue repair, *Ann. N.Y. Acad. Sci.* **593**, 124–134.

Anklesaria, P., Teixido, J., Laiho, M., Pierce, J. H., Greenberger, J. S., and Massague, J. (1990). Cell–cell adhesion mediated by binding of membrane-anchored transforming growth factor alpha to epidermal growth factor receptors promotes cell proliferation. *Proc. Natl. Acad. Sci. U.S.A.* **87**, 3289–3293.

Anzano, M. A., Roberts, A. B., Smith, J. M., Sporn, M. B., and De Larco, J. E. (1983). Sarcoma growth factor from conditioned medium of virally transformed cells is composed of both type alpha and type beta transforming growth factors. *Proc. Natl. Acad. Sci. U.S.A.* **80**, 6264–6268.

Anzano, M. A., Roberts, A. B., De Larco, J. E., Wakefield, L. M., Assoian, R. K., Roche, N. S., Smith, J. M., Lazarus, J. E., and Sporn, M. B. (1985). Increased secretion of type beta transforming growth factor accompanies viral transformation of cells. *Mol. Cell. Biol.* **5**, 242–247.

Arteaga, C. L., Hanauske, A. R., Clark, G. M., Osborne, C. K., Hazarika, P., Pardue, R. L., Tio, F., and Von Hoff, D. D. (1988). Immunoreactive alpha transforming growth factor activity in effusions from cancer patients as a marker of tumor burden and patient prognosis. *Cancer Res.* **48**, 5023–5028.

Assoian, R. K., Komoriya, A., Meyers, C. A., Miller, D. M., and Sporn, M. B. (1983). Transforming growth factor beta in human platelets. Identification of a major storage site, purification and characterization. *J. Biol. Chem.* **258**, 7155–7160.

Barnard, J. A., Bascom, C. C., Lyons, R. M., Sipes, N. J., and Moses, H. L. (1988). Transforming growth factor beta in the control of epidermal proliferation. *Ann. J. Med. Sci.* **296**, 159–163.

Barnard, J. A., Beauchamp, R. D., Coffey, R. J., and Moses, H. L. (1989). Regulation of intestinal epithelial cell growth by transforming growth factor type beta. *Proc. Natl. Acad. Sci. U.S.A.* **86**, 1578–1582.

Barnard, J. A., Lyons, R. M., and Moses, H. L. (1990). The cell biology of transforming growth factor beta. *Biochim. Biophys. Acta* **1032**, 79–87.

Barton, D. E., Foellmer, B. E., Du, J., Tamm, J., Derynck, R., and Francke, U. (1988). Chromosomal mapping of genes for transforming growth factors beta 2 and 3 in man and mouse: Dispersion of TGFβ gene family. *Oncogene Res.* **3**, 323–331.

Beck, S. L., Chen, T. L., Hirabayashi, S. E., Deguzman, L., Lee, W. P., McFatridge, L. L., Xu, Y., Bates, R. L., and Ammann, A. J. (1990). Accelerated healing of ulcer wounds in the rabbit ear by recombinant human transforming growth factor beta 1. *Growth Factors* **2**, 273–282.

Blasband, A. J., Rogers, K. T., Chen, X. R., Azizkhan, J. C., and Lee, D. C. (1990). Characterization of the rat transforming growth factor alpha gene and identification of promoter sequences. *Mol. Cell. Biol.* **10,** 2111–2121.

Bringman, T. S., Lindquist, P. B., and Derynck, R. (1987). Different transforming growth factor alpha species are derived from a glycosylated and palmitoylated transmembrane precursor. *Cell* **48,** 429–440.

Brown, P. I., Lam, R., Lakshmanan, J., and Fisher, D. A. (1990). Transforming growth factor alpha in developing rats. *Am. J. Physiol.* **259,** E256–260.

Brunner, A. M., Gentry, L. E., Cooper, J. A., and Purchio, A. F. (1988). Recombinant type 1 transforming growth factor beta precursor produced in Chinese hamster ovary cells is glycosylated and phosphorylated. *Mol. Cell. Biol.* **8,** 2229–2232.

Burgess, A. W. (1989). Epidermal growth factor and transforming growth factor alpha. *Br. Med. Bull.* **45,** 401–424.

Carpenter, G., and Cohen, S. (1979). Epidermal growth factor. *Ann. Rev. Biochem.* **48,** 193–216.

Carr, B. L., Hayashi, I., Branum, E. L., and Moses, H. L. (1986). Inhibition of DNA synthesis in rat hepatocytes by platelet-derived type beta transforming growth factor. *Cancer Res.* **46,** 2330–2334.

Cate, R. L., Mattaliano, R. J., Hession, C., Tizard, R., Farber, N. M., Cheung, A., Ninfa, E. G., Frey, A. Z., Gash, D. J., and Chow, E. P. (1986). Isolation of the bovine and human genes for Mullerian inhibiting substance and expression of the human gene in animal cells. *Cell* **45,** 685–698.

Cheifetz, S., Weatherbee, J. A., Tsang, M. L., Anderson, J. K., Mole, J. E., Lucas, R., and Massague, J. (1987). The transforming growth factor beta system, a complex pattern of cross-reactive ligands and receptors. *Cell* **48,** 409–415.

Cheifetz, S., Bassols, A., Stanley, K., Ohta, M., Greenberger, J., and Massague, J. (1988a). Heterodimeric transforming growth factor beta. Biological properties and interaction with three types of cell surface receptors. *J. Biol. Chem.* **263,** 10783–10789.

Cheifetz, S., Andres, J. L., and Massagué, J. (1988b). The transforming growth factor-beta receptor type III is a membrane proteoglycan. Domain structure of the receptor. *J. Biol. Chem.* **263,** 16984–16991.

Cheifetz, S., Ling, N., Guillemin, R., and Massagué, J. (1988c). A surface component on GH3 pituitary cells that recognizes transforming growth factor-beta, activin and inhibin. *J. Biol. Chem.* **263,** 17225–17228.

Choi, Y., and Fuchs, E. (1990). TGF-β and retinoic acid: Regulators of growth and modifiers of differentiation in human epidermal cells. *Cell Regulation* **1,** 791–809.

Coffey, R. J., Jr., Derynck, R., Wilcox, J. N., Bringman, T. S., Goustin, A. S., Moses, H. L., and Pittelkow, M. R. (1987). Production and autoinduction of transforming growth factor-alpha in human keratinocytes. *Nature (London)* **328,** 817–820.

Cohen, S. (1962). Isolation of a mouse submaxillary gland protein accelerating incisor eruption and eyelid opening in the newborn animal. *J. Biol. Chem.* **237,** 1555–1562.

Cutry, A. F., Kinninburg, A. J., Twardzik, D. R., and Wenner, C. E. (1988). Transforming growth factor alpha (TGF alpha) induction of c-*fos* and c-*myc* expression in C3H 10T1/2 cells. *Biochem. Biophys. Res. Commun.* **152,** 216–222.

De Larco, J. E., and Todaro, G. J. (1978). Growth factors from murine sarcoma virus-transformed cells. *Proc. Natl. Acad. Sci. U.S.A.* **75,** 4001–4005.

De Martin, R., Haendler, B., Hofer-Warbinek, R., Gaugitsch, H., Wrann, M., Schlüsener, H., Seifert, J. M., Bodmer, S., Fontana, A., and Hofer, E. (1987). Complementary DNA for human glioblastoma-derived T-cell suppressor factor, a novel member of the transforming growth factor-beta gene family. *EMBO J.* **6,** 3673–3677.

Derynck, R. (1988). Transforming growth factor alpha. *Cell* **54,** 593–595.

Derynck, R. (1989). Transforming growth factors-alpha and -beta and their potential roles in neoplastic transformation. *Cancer Treat. Res.* **47,** 177–195.

Derynck, R., Roberts, A. B., Winkler, M. E., Chen, E. Y., and Goeddel, D. V. (1984). Human transforming growth factor alpha: Precursor structure and expression in *E. coli. Cell* **38,** 287–297.

Derynck, R., Robert, A. B., Eaton, D. H., Winkler, M. E., and Goeddel, D. V. (1985a). Human transforming growth factor-α: Precursor sequence, gene structure, and heterologous expression. *Cancer Cells* **3**, 79–86.

Derynck, R., Jarrett, J. A., Chen, E. Y., Eaton, D. H., Bell, J. R., Assoian, R. K., Roberts, A. B., Sporn, M. B., and Goeddel D. V. (1985b). Human transforming growth factor-beta: Complementary DNA sequence and expression in normal and transformed cells. *Nature (London)* **316**, 701–705.

Derynck, R., Goeddel, D. V., Ullrich, A., Gutterman, J. U., Williams, R. D., Bringman, T. S., and Berger, W. H. (1987a). Synthesis of messenger RNAs for transforming growth factors alpha and beta and the epidermal growth factor receptor by human tumors. *Cancer Res.* **47**, 707–712.

Derynck, R., Rhee, L., and Chen, E. Y. (1987b). Intron–exon structure of the human transforming growth factor beta precursor gene. *Nucleic Acids Res.* **15**, 3188–3189.

Elder, J. T., Fisher, G. J., Lindquist, P. B., Bennett, G. L., Pittelkow, M. R., Coffey, R. J., Jr., Ellingsworth, L., Derynck, R., and Voorhees, J. J. (1989). Overexpression of transforming growth factor alpha in psoriatic epidermis. *Science* **243**, 811–814.

Ellingsworth, L. R., Brennan, J. E., Fok, K., Rosen, D. M., Bentz, H., Piez, K. A., and Seyedin, S. M. (1986). Antibodies to the N-terminal portion of cartilage-inducing factor A and transforming growth factor beta. Immunohistochemical localization and association with differentiating cells. *J. Biol. Chem.* **261**, 12362–12367.

Fallon, J. H., Annis, C. M., Gentry, L. E., Twardzik, D. R., and Loughlin, S. E. (1990). Localization of cells containing transforming growth factor-alpha precursor immunoreactivity in the basal ganglia of the adult rat brain. *Growth Factors* **2**, 241–250.

Flanders, K. C., Cissel, D. S., Jakowlew, S. B., Roberts, A. B., Danielpour, D., and Sporn, M. B. (1990). Immunohistochemical localization of TGFβ2 and β3 in the nervous system. *Ann. N.Y. Acad. Sci.* **593**, 338–339.

Florini, J. R., Roberts, A. B., Ewton, D. Z., Falen, S. L., Flanders, K. C., and Sporn, M. B. (1986). Transforming growth factor-beta. A very potent inhibitor of myoblast differentiation, identical to the differentiation inhibitor secreted by Buffalo rat liver cells. *J. Biol. Chem.* **261**, 16509–16513.

Frolik, C. A., and De Larco, J. E. (1987). Radioreceptor assays for transforming growth factors. *Methods Enzymol.* **146**, 95–102.

Frolik, C. A., Wakefield, L. M., Smith, D. M., and Sporn, M. B. (1984). Characterization of a membrane receptor for transforming growth factor-beta in normal rat kidney fibroblasts. *J. Biol. Chem.* **259**, 10995–11000.

Fujii, D., Brissenden, J. E., Derynck, R., and Francke, U. (1986). Transforming growth factor beta gene maps to human chromosome 19 long arm and to mouse chromosome 7. *Somatic Cell Mol. Genet,* **12**, 281–288.

Gan, B. S., Hollenberg, M. D., MacCannell, K. L., Lederis, K., Winkler, M. E., and Derynck, R. (1987). Distinct vascular actions of epidermal growth factor-urogastrone and transforming growth factor-alpha. *J. Pharmacol. Ext. Ther.* **242**, 331–337.

Gentry, L. E., Twardzik, D R., Lim, G. J., Ranchalis, J. E., and Lee, D. C. (1987a). Expression and characterization of transforming growth factor alpha precursor protein in transfected mammalian cells. *Mol. Cell. Biol.* **7**, 1585–1591.

Gentry, L. E., Webb, N. R., Lim, G. J., Brunner, A. M., Ranchalis, J. E., Twardzik, D. R., Lioubin, M. N., Marquardt, H., and Purchio, T. (1987b). Type 1 transforming growth factor beta: Amplified expression and secretion of mature and precursor polypeptides in Chinese hamster ovary cells. *Mol. Cell. Biol.* **7**, 3418–3427.

Gentry, L. E., Lioubin, M. N., Purchio, A. F., and Marquardt, H. (1988). Molecular events in the processing of recombinant type 1 pre-pro-transforming growth factor beta to the mature polypeptide. *Mol. Cell. Biol.* **8**, 4162–4168.

Gomella, L. G., Sargent, E. R., Wade, T. P., Anglard, P., Linehan, W. M., and Kasid, A. (1989). Expression of transforming growth factor alpha in normal human adult kidney and enhanced expression of transforming growth factors alpha and beta 1 in renal cell carcinoma. *Cancer Res.* **49**, 6972–6975.

Hebda, P. A. (1988). Stimulatory effects of transforming growth factor-beta and epidermal growth factor on epidermal cell outgrowth from porcine skin explant cultures. *J. Invest. Dermatol.* **9**, 440–445.

Heimark, R. L., Twardzik, D. R., and Schwartz, S. (1986). Inhibition of endothelial regeneration by type-beta transforming growth factor from platelets. *Science* **233**, 1078–1080.

Heine, U. I., Munoz, E. F., Flanders, K. C. Ellingsworth, L. R., Lam, H. Y., Thompson, N. L., Roberts, A. B., and Sporn, M. B. (1987). Role of transforming growth factor beta in the development of the mouse embryo. *J. Cell Biol.* **105**, 2861–2876.

Heldin, C. H., Westermark, B., and Wasteson, A. (1981). Specific receptors for platelet-derived growth factor on cells derived from connective tissue and ganglia. *Proc. Natl. Acad. Sci. U.S.A.* **78**, 3664–3668.

Holley, R. W., Böhlen, P., Fava, R., Baldwin, J. H., Kleeman, G., and Armour, R. (1980). Purification of kidney epithelial cell growth inhibitors. *Proc. Natl. Acad. Sci. U.S.A.* **77**, 5989–5992.

Ibbotson, K. J., D'Souza, S. M., Ng, K. W., Osborne, C. K., Nial, M., Martin, T. J., and Mundy, G. R. (1983). Tumor-derived growth factor increases bone resorption in a tumor associated with humoral hypercalcemia of malignancy. *Science* **221**, 1292–1294.

Ignotz, R. A., and Massagué, J. (1985). Type beta transforming growth factor controls the adipogenic differentiation of 3T3 fibroblasts. *Proc. Natl. Acad. Sci. U.S.A.* **82**, 8530–8534.

Ignotz, R. A., and Massagué, J. (1987). Cell adhesion protein receptors as targets for transforming growth factor-beta action. *Cell* **51**, 189–197.

Ikeda, T., Lioubin, M. N., and Marquardt, H. (1987). Human transforming growth factor type beta 2: Production by a prostatic adenocarcinoma cell line, purification, and initial characterization. *Biochemistry* **26**, 2406–2410.

Jhappan, C., Stahle, C., Harkins, R. N., Fausto, N., Smith, G. H., and Merlino, G. T. (1990). TGF alpha overexpression in transgenic mice induces liver neoplasia and abnormal development of the mammary gland and pancreas. *Cell* **61**, 1137–1146.

Joyce, M. E., Terek, R. M., Jingushi, S., and Bolander, M. E. (1990). Role of transforming growth factor-beta in fracture repair. *Ann. N.Y. Acad. Sci.* **593**, 107–123.

Kamijo, R., Takeda, K., Nagumo, M., and Konno, K. (1990). Effects of combinations of transforming growth factor-beta 1 and tumor necrosis factor on induction of differentiation of human myelogenous leukemic cell lines. *J. Immunol.* **144**, 1311–1316.

Kaplan, P. L., Topp, W. C., and Ozanne, B. (1981). Simian virus 40 induces the production of polypeptide transforming growth factor(s). *Virology* **108**, 484–490.

Kaplan, P. L., and Ozanne, B. (1982). Polyoma-virus transformed cells produce transforming growth factor(s) and grow in serum free medium. *Virology* **123**, 372–380.

Kato, M., Takenawa, T., and Twardzik, D. R. (1988). Effect of transforming growth factor-alpha on inositol phospholipid metabolism in human epidermoid carcinoma cells. *J. Cell Biochem.* **37**, 339–345.

Katoh, M., Inagaki, H., Kurosawa-Ohsawa, K., Katsuura, M., and Tanaka, S. (1990). Detection of transforming growth factor alpha in human urine and plasma. *Biochem. Biophys. Res. Commun.* **167**, 1065–1072.

Kehrl, J. H., Wakefield, L. M., Roberts, A. B., Jakowlew, S., Alvarez-Mon, M., Derynck, R., Sporn, M. B., and Fauci, A. S. (1986). Production of transforming growth factor beta by human T lymphocytes and its potential role in the regulation of T cell growth. *J. Exp. Med.* **163**, 1037–1050.

Keller, J. R., Sing, G. K., Ellinsworth, L. R., and Ruscetti, F. W. (1989). Transforming growth factor beta: Possible roles in the regulation of normal and leukemic hematopoietic cell growth. *J. Cell Biochem.* **39**, 175–184.

Keski-Oja, J., Lyons, R. M., and Moses, H. L. (1987a). Inactive secreted form(s) of transforming growth factor-β (TGFβ): Activation by proteolysis. *J. Cell Biochem. (Suppl.)* **11A**, 60.

Keski-Oja, J., Leof, E. B., Lyons, R. M., Coffey, R. J., Jr., and Moses, H. L. (1987b). Transforming growth factors and control of neoplastic cell growth. *J. Cell Biochem.* **33**, 95–107.

Kim, M. K., Warren, T. C., and Kimball, E. S. (1985). Purification and characterization of a low molecular weight transforming growth factor from the urine of melanoma patients. *J. Biol. Chem.* **260**, 9237–9243.

Kim, S. J., Glick, A., Sporn, M. B., and Roberts, A. B. (1989). Characterization of the promoter region of the human transforming growth factor-beta 1 gene. *J. Biol. Chem.* **264,** 402–408.

Knabbe, C., Lippman, M. E., Wakefield, L. M., Flanders, K. C., Kasid, A., Derynck, R., and Dickson, R. B. (1987). Evidence that transforming growth factor beta is a hormonally regulated negative growth factor in human breast cancer cells. *Cell* **48,** 417–428.

Kobrin, M. S., Samsoondar, J., and Kudlow, J. E. (1986). Alpha-transforming growth factor secreted by untransformed bovine anterior pituitary cells in culture. II. Identification using a sequence-specific monoclonal antibody. *J. Biol. Chem.* **261,** 14414–14419.

Korc, M., Haussler, C. A., and Trookman, N. S. (1987). Divergent effects of epidermal growth factor and transforming growth factors on a human endometrial carcinoma cell line. *Cancer Res.* **47,** 4909–4914.

Kurokawa, M., Lynch, K., and Podolsky, D. K. (1987). Effects of growth factors on an intestinal epithelial cell line: Transforming growth factor beta inhibits proliferation and stimulates differentiation. *Biochem. Biophys. Res. Commun.* **142,** 775–782.

Laiho, M., and Keski-Oja, J. (1989). Growth factors in the regulation of pericellular proteolysis: A review. *Cancer Res.* **49,** 2533–2553.

Lawrence, D. A., Pircher, R., and Jullien, P. (1985). Conversion of a high molecular weight latent beta-TGF from chicken embryo fibroblasts into a low molecular weight active beta-TGF under acidic conditions. *Biochem. Biophys. Res. Commun.* **133,** 1026–1034.

Lee, D. C., Rose, T. M., Webb, N. R., and Todaro, G. J. (1985a). Cloning and sequence analysis of a cDNA for rat transforming growth factor-alpha. *Nature (London)* **313,** 489–491.

Lee, D. C., Rochford, R., Todaro, G. J., and Villarreal, L. P. (1985b). Developmental expression of rat transforming growth factor-alpha mRNA. *Mol. Cell. Biol.* **5,** 3644–3646.

Lee, D. C., Rose, T. M., Webb, N. R., Twardzik, D. R., Ranchalis, J. E., Marquardt, H., and Todaro, G. (1987). Rat transforming growth factor-alpha is apparently cleaved from a transmembrane precursor by an unusual protease. *Roy. Soc. Med. Serv. Int. Congr. Symp.* **121,** 187–197.

Lefer, A. M., Tsao, P., Aoki, N., and Palladino, M. A., Jr. (1990). Mediation of cardioprotection by transforming growth factor-beta. *Science* **249,** 61–64.

Leof, E. B., Proper, J. A., Goustin, A. S., Shipley, G. D., DiCorleto, P. E., and Moses, H. L. (1986). Induction of c-*sis* mRNA and activity similar to platelet derived growth factor by transforming growth factor-beta: A proposed model for indirect mitogenesis involving autocrine activity. *Proc. Natl. Acad. Sci. U.S.A.* **83,** 2453–2457.

Lewis, J. J., Goldenring, J. R., Modlin, I. M., and Coffey, R. J. (1990). Inhibition of parietal cell $H^+$ secretion by transforming growth factor alpha: A possible autocrine regulatory mechanism. *Surgery* **108,** 220–227.

Libby, J., Martinez, R., and Weber, M. J. (1986). Tyrosine phosphorylation in cells treated with transforming growth factor-beta. *J. Cell. Physiol.* **129,** 159–166.

Ling, N., Ying, S. Y., Ueno, N., Shimasaki, S., Esch, F., Hotta, M., and Guillemin, R. (1986). Pituitary FSH is released by a heterodimer of the beta-subunits from the two forms of inhibin. *Nature (London)* **321,** 779–782.

Loughlin, S. E., Annis, C. M., Twardzik, D. R., Lee, D. C., Gentry, L., and Fallon, J. H. (1989). Growth factors in opioid rich brain regions: Distribution and response to intrastriatal transplants. *Adv. Biosci.* **75,** 403–406.

Lyons, R. M., Keski-Oja, J., and Moses, H. L. (1988). Proteolytic activation of latent transforming growth factor-beta from fibroblast-conditioned medium. *J. Cell Biol.* **106,** 1659–1665.

Lyons, K., Graycar, J. L., Lee, A., Hashmi, S., Lindquist, P. B., Chen, E. Y., Hogan, B. L., and Derynck, R. (1989). *Vgr-1*, a mammalian gene related to *Xenopus Vg-1*, is a member of the transforming growth factor beta gene superfamily. *Proc. Natl. Acad. Sci. U.S.A.* **86,** 4554–4558.

Madisen, L., Webb, N. R., Rose, T. M., Marquardt, H., Ikeda, T., Twardzik, D., Seyedin, S., and Purchio, A. F. (1988). Transforming growth factor-beta 2: cDNA cloning and sequence analysis. *DNA* **7,** 1–8.

Madri, J. A., Pratt, B. M., and Tucker, A. M. (1988). Phenotypic modulation of endothelial cells by

transforming growth factor-beta depends upon the composition and organization of the extracellular matrix. *J. Cell Biol.* **106**, 1375–1384.
Madtes, D. K., Raines, E. W., Sakariassen, K. S., Assoian, R. K., Sporn, M. B., Bell, G. I., and Ross, R. (1988). Induction of transforming growth factor-alpha in activated human alveolar macrophages. *Cell* **53**, 285–293.
Marquardt, H., Hunkapiller, M. W., Hood, L. E., Twardzik, D. R., DeLarco, J. E., Stephenson, J. R., and Todaro, G. J. (1983). Transforming growth factors produced by retrovirus-transformed rodent fibroblasts and human melanoma cells: Amino acid sequence homology with epidermal growth factor. *Proc. Natl. Acad. Sci. U.S.A.* **80**, 4684–4688.
Marquardt, H., Hunkapiller, M. W., Hood, L. E., and Todaro, G. J. (1984). Rat transforming growth factor type 1: Structure and relation to epidermal growth factor. *Science* **223**, 1073–1082.
Mason, A. J., Hayflick, J. S., Ling, N., Esch, F., Ueno, N., Ying, S. Y., Guillemin, R., Niall, H., and Seeburg, P. H. (1985). Complementary DNA sequences of ovarian follicular fluid inhibin show precursor structure and homology with transforming growth factor-beta. *Nature (London)* **318**, 659–663.
Massagué, J. (1983a). Epidermal growth factor like transforming growth factor. II. Interaction with epidermal growth factor receptors in human placenta membranes and A431 cells. *J. Biol. Chem.* **258**, 13606–13613.
Massagué, J. (1983b). Epidermal growth factor-like transforming growth factor. I. Isolation, chemical characterization, and potentiation by other transforming growth factors from feline sarcoma virus-transformed rat cells. *J. Biol. Chem.* **258**, 13614–13620.
Massagué, J. (1984). Type beta transforming growth factor from feline sarcoma virus-transformed rat cells. Isolation and biological properties. *J. Biol. Chem.* **259**, 9756–9761.
Massagué, J. (1985). Subunit structure of a high-affinity receptor for type beta-transforming growth factor. Evidence for a disulfide-linked glycosylated receptor complex. *J. Biol. Chem.* **260**, 7059–7066.
Massagué, J., Cheifetz, S., Boyd, F. T., and Andres, J. L. (1990). TGF-beta receptors and TGF-beta binding proteoglycans: Recent progress in identifying their functional properties. *Ann. N.Y. Acad. Sci.* **593**, 59–72.
Masui, T., Wakefield, L. M., Lechner, J. F., LaVeck, M. A., Sporn, M. B., and Harris, C. C. (1986). Type beta transforming growth factor is the primary differentiation-inducing serum factor for normal human bronchial epithelial cells. *Proc. Natl. Acad. Sci. U.S.A.* **83**, 2438–2442.
Matsui, Y., Halter, S. A., Holt, J. T., Hogan, B. L., and Coffey, R. J. (1990). Development of mammary hyperplasia and neoplasia in MMTV-TGF alpha transgenic mice. *Cell* **61**, 1147–1155.
Miyazono, K., Hellman, U., Wernstedt, C., and Heldin, C. H. (1988). Latent high molecular weight complex of transforming growth factor beta 1. Purification from human platelets and structural characterization. *J. Biol. Chem.* **263**, 6407–6415.
Miyazono, K., and Heldin, C.-H. (1989). Role of carbohydrate structures in TGF-beta 1 latency. *Nature (London)* **338**, 158–160.
Moses, H. L., Tucker, R. F., Leof, E. B., Coffey, R. J., Halper, J., and Shipley, G. D. (1985). Type-β transforming growth factor is a growth stimulator and a growth inhibitor. *Cancer Cells* **3**, 65–71.
Moses, H. L., Yang, E. Y., Pietenpol, J. A. (1990). TGF-β stimulation and inhibition of cell proliferation: New mechanistic insights. *Cell* **63**, 245–247.
Mueller, S. G., Paterson, A. J., and Kudlow, J. E. (1990). Transforming growth factor alpha in arterioles: Cell surface processing of its precursor by elastases. *Mol. Cell. Biol.* **10**, 4596–4602.
Mundy, G. R., and Bonewald, L. F. (1990). Role of TGF-beta in bone remodelling. *Ann. N.Y. Acad. Sci.* **593**, 91–97.
Murthy, U. S., Anzano, M. A., Stadel, J. M., and Greig, R. (1988). Coupling of TGF-beta-induced mitogenesis to G-protein activation in AKR-2B cells. *Biochem. Biophys. Res. Commun.* **152**, 1228–1235.
Murthy, U., Anzano, M. A., and Greig, R. G. (1989). Expression of TGF-alpha/EGF and TGF-beta receptors in human colon carcinoma cell lines. *Int. J. Cancer* **44**, 110–115.
Mustoe, T. A., Pierce, G. F., Thomason, A., Gramates, P., Sporn, M. B., and Deuel, T. F. (1987).

Accelerated healing of incisional wounds in rats induced by transforming growth factor-beta. *Science* **237**, 1333–1336.
Nilsen-Hamilton, M. (1990). Transforming growth factor beta and its actions on cellular growth and differentiation. *Curr. Top. Dev. Biol.* **24**, 95–136.
O'Connor-McCourt, M. D., and Wakefield, L. M. (1987). Latent transforming growth factor-beta in serum. A specific complex with alpha 2-macroglobulin. *J. Biol. Chem.* **262**, 14090–14099.
Ozanne, B., Fulton, R. J., and Kaplan, P. L. (1980). Kirsten murine sarcoma virus transformed cell lines and a spontaneously transformed rat cell-line produce transforming growth factors. *J. Cell Physiol.* **105**, 163–180.
Padgett, R. W., St. Johnston, R. D., and Gelbart, W. M. (1987). A transcript from a *Drosophila* pattern gene predicts a protein homologous to the transforming growth factor-β family. *Nature (London)* **325**, 81–84.
Penttinen, R. P., Kobayashi, S., and Bornstein, P. (1988). Transforming growth factor beta increases mRNA for matrix proteins both in the presence and in the absence of changes in mRNA stability. *Proc. Natl. Acad. Sci. U.S.A.* **85**, 1105–1108.
Pfeilschifter, J., and Mundy, G. R. (1987). Modulation of type beta transforming growth factor activity in bone cultures by osteotropic hormones. *Proc. Natl. Acad. Sci. U.S.A.* **84**, 2024–2028.
Pignatelli, M., and Bodmer, W. F. (1990). The role of TGFβs in controlling cell adhesion and differentiation of colon carcinoma cells. *Ann. N.Y. Acad. Sci.* **593**, 360–362.
Pircher, R., Jullien, P., and Lawrence, D. A. (1986). Beta-transforming growth factor is stored in human blood platelets as a latent high molecular weight complex. *Biochem. Biophys. Res. Commun.* **136**, 30–37.
Pittelkow, M. R., Lindquist, P. B., Abraham, R. T., Graves-Deal, R., Derynck, R., and Coffey, R. J., Jr. (1989). Induction of transforming growth factor alpha expression in human keratinocytes by phorbol esters. *J. Biol. Chem.* **264**, 5164–5171.
Postlethwaite, A. E., Keski-Oja, J., Moses, H. L., and Kang, A. H. (1987). Stimulation of the chemotactic migration of human fibroblasts by transforming growth factor beta. *J. Exp. Med.* **165**, 251–256.
Puolakkainen, P., Ranchalis, J. E., Reed, M., Pankey, S., Gombotz, W. (1992). TGF-β1's wound healing effect depends on the carrier. Second Annual Meeting of the Wound Healing Society, Richmond, Virginia, April 23–26, 1992 (abstract).
Purchio, A. F., Cooper, J. A., Brunner, A. M., Lioubin, M. N., Gentry, L. E., Kovacina, K. S., Roth, R. A., and Marquardt, H. (1988). Identification of mannose 6-phosphate in two asparagine-linked sugar chains of recombinant transforming growth factor beta 1 precursor. *J. Biol. Chem.* **263**, 14211–14215.
Reiss, M., and Sartorelli, A. C. (1987). Regulation of growth and differentiation of human keratinocytes by type beta transforming growth factor and epidermal growth factor. *Cancer Res.* **47**, 6705–6709.
Reynolds, F. H., Jr., Todaro, G. J., Fryling, C., and Stephenson, J. R. (1981). Humans transforming growth factors induce tyrosine phosphorylation of EGF receptors. *Nature (London)* **292**, 259–262.
Roberts, A. B., Anzano, M. A., Lamb, L. C., Smith, J. M., and Sporn, B. (1981). New class of transforming growth factors potentiated by epidermal growth factor: Isolation from nonneoplastic tissues. *Proc. Natl. Acad. Sci. U.S.A.* **78**, 5339–5343.
Roberts, A. B., Flanders, K. C., Kondaiah, Thompson, N. L., Van Obberghen-Schilling, E. Wakefield, L., Rossi, P., de Crombrugghe, B., Heine, U., and Sporn, M. B. (1988). Transforming growth factor beta: Biochemistry and roles in embryogenesis, tissue repair and remodeling, and carcinogenesis. *Recent Prog. Horm. Res.* **44**, 157–197.
Roberts, A. B., and Sporn, M. B. (1988). Transforming growth factor beta. *Adv. Cancer Res.* **51**, 107–145.
Roberts, A. B., and Sporn, M. B. (1990). Transforming growth factor-βs. *In* "Handbook of Experimental Pharmacology" Vol. 95/I, pp. 419–472. Springer-Verlag, Berlin.
Saksela, O., Moscatelli, D., and Rifkin, D. B. (1987). The opposite effects of basic fibroblast growth factor and transforming growth factor beta on the regulation of plasminogen activator activity in capillary endothelial cells. *J. Cell Biol.* **105**, 957–963.
Salomon, D. S., Perroteau, I., Kidwell, W. R., Tam, J., and Derynck, R. (1987). Loss of growth

responsiveness to epidermal growth factor and enhanced production of alpha-transforming growth factors in *ras*-transformed mouse mammary epithelial cells. *J. Cell. Physiol.* **130**, 297–309.
Samuels, V., Barrett, J. M., Bockman, S., Pantazis, C. G., and Allen, M. B., Jr. (1989). Immunocytochemical study of transforming growth factor expression in benign and malignant gliomas. *Am. J. Pathol.* **134**, 894–902.
Sandgren, E. P., Luetteke, N. C., Palmiter, R. D., Brinster, R. L., and Lee, D. C. (1990). Overexpression of TGF alpha in transgenic mice: Induction of epithelial hyperplasia, pancreatic metaplasia, and carcinoma of the breast. *Cell* **61**, 1121–1135.
Schreiber, A. B., Winkler, M. E., and Derynck, R. (1986). Transforming growth factor-alpha: A more potent angiogenic mediator than epidermal growth factor. *Science* **232**, 1250–1253.
Schultz, G. S., White, M., Mitchell, R., Brown, G., Lynch, J., Twardzik, D. R., and Todaro, G. J. (1987). Epithelial wound healing enhanced by transforming growth factor-alpha and vaccinia growth factor. *Science* **235**, 350–352.
Segarini, P. R., and Seyedin, S. M. (1988). The high molecular weight receptor to transforming growth factor-beta contains glycosaminoglycan chains. *J. Biol. Chem.* **263**, 8366–8370.
Seyedin, S. M., Thomas, T. C., Thompson, A. Y., Rosen, D. M., and Piez, K. A. (1985). Purification and characterization of two cartilage-inducing factors from bovine demineralized bone. *Proc. Natl. Acad. Sci. U.S.A.* **82**, 2267–2271.
Seyedin, S. M., Thompson, A. Y., Bentz, H., Rosen, D. M., McPherson, J. M., Conti, A., Siegel, N. R., Galluppi, G. R., and Piez, K. A. (1986). Cartilage-inducing factor-A. Apparent identity to transforming growth factor-beta. *J. Biol. Chem.* **261**, 5693–5695.
Sharples, K., Plowman, G. D., Rose, T. M., Twardzik, D. R., and Purchio, A. F. (1987). Cloning and sequence analysis of simian transforming growth factor-beta cDNA. *DNA* **6**, 239–244.
Sherwin, S. A., Twardzik, D. R., Bohn, W. H., Cockley, K. D., and Todaro, G. J. (1983). High-molecular-weight transforming growth factor activity in the urine of patients with disseminated cancer. *Cancer Res.* **43**, 403–407.
Shipley, G. D., Tucker, R. F., and Moses, H. L. (1985). Type beta transforming growth factor/growth inhibitor stimulates entry of monolayer cultures of AKR-2B cells into S phase after a prolonged prereplicative interval. *Proc. Natl. Acad. Sci. U.S.A.* **82**, 4147–4151.
Shipley, G. D., Pittelkow, M. R., Willie, J. J., Jr., Scott, R. E., and Moses, H. L. (1986). Reversible inhibition of normal human prokeratinocyte proliferation by type beta transforming growth factor-growth inhibitor in serum-free medium. *Cancer Res.* **46**, 2068–2071.
Silberstein, G. B., and Daniel, C. W. (1987). Reversible inhibition of mammary gland growth by transforming growth factor-beta. *Science* **237**, 291–293.
Smiddy, W. E., Glaser, B. M., Green, R., Connor, T. B., Jr., Roberts, A. B., Lucas, R., and Sporn, M. B. (1989). Transforming growth factor beta. A biologic chorioretinal glue. *Arch. Ophtalmol.* **107**, 577–580.
Smith, J. M., Sporn, M. B., Roberts, M. B., Derynck, R., Winkler, M. E., and Gregory, H. (1985). Human transforming growth factor-alpha causes precocious eyelid opening in newborn mice. *Nature (London)* **315**, 515–516.
Sporn, M. B., and Todaro, G. J. (1980). Autocrine secretion and malignant transformation of cells. *N. Engl. J. Med.* **303**, 878–882.
Sporn, M. B., and Roberts, A. B. (1985). Autocrine growth factors and cancer. *Nature (London)* **313**, 745–747.
Sporn, M. B., and Roberts, A. B. (1989). Transforming growth factor-beta. Multiple actions and potential clinical applications. *JAMA* **262**, 938–941.
Sporn, M. B., and Roberts, A. B. (1990). The transforming growth factor betas: Past, present, and future. *Ann. N.Y. Acad. Sci.* **593**, 1–6.
Sporn, M. B., Roberts, A. B., Wakefield, L. M. (1986). Transforming growth factor β: Biological function and chemical structure. *Science* **233**, 532–534.
Sporn, M. B., Roberts, A. B., Wakefield, L. M., and Crombrugghe, B. (1987). Some recent advances in the chemistry and biology of transforming growth factor-beta. *J. Cell Biol.* **105**, 1039–1045.
Stromberg, K., Hudgins, W. R., and Orth, D. N. (1987). Urinary TGFs1 in neoplasia: Immunoreactive

TGF-alpha in the urine of patients with disseminated breast carcinoma. *Biochem. Biophys. Res. Commun.* **144,** 1059-1068.

Teerds, K. J., Rommerts, F. F., and Dorrington, J. H. (1990). Immunohistochemical detection of transforming growth factor-alpha in Leydig cells during the development of the rat testis. *Mol. Cell. Endocrinol.* **69,** R1-6.

Teixido, J., Gilmore, R., Lee, D. C., and Massaque, J. (1987). Integral membrane glycoprotein properties of the prohormone pro-transforming growth factor α. *Nature (London)* **326,** 883-885.

ten Dijke, P., Hansen, P., Iwata, K. K., Pielr, C., and Foulkes, J. G. (1988). Identification of another member of the transforming growth factor type beta gene family. *Proc. Natl. Acad. Sci. U.S.A.* **85,** 4715-4719.

Thompson, N. L., Flanders, K. C., Smith, M., Ellingsworth, L. R., Roberts, A. B., and Sporn, M. B. (1989). Expression of transforming growth factor beta 1 in specific cells and tissues of adult and neonatal mice. *J. Cell Biol.* **108,** 661-669.

Todaro, G. J., and De Larco, J. E. (1976). Transformation by murine and feline sarcoma viruses specifically blocks binding of epidermal growth factor to cells. *Nature (London)* **264,** 26-31.

Todaro, G. J., Fryling, C., and De Larco, J. E. (1980). Transforming growth factors produced by certain human tumor cells: Polypeptides that interact with epidermal growth factor receptors. *Proc. Natl. Acad. Sci. U.S.A.* **77,** 5258-5262.

Tsunawaki, S., Sporn, M., Ding, A., and Nathan, C. (1988). Deactivation of macrophages by transforming growth factor-beta. *Nature (London)* **334,** 260-262.

Tucker, R. F., Branum, E. L., Shipley, G. D., Ryan, R. J., and Moses, H. L. (1984a). Specific binding to cultured cells of $^{125}$I-labeled type beta transforming growth factor from human platelets. *Proc. Natl. Acad. Sci. U.S.A.* **81,** 6757-6761.

Tucker, R. F., Shipley, G. D., Moses, H. L., and Holley, R. W. (1984b). Growth inhibitor from BSC-1 cells closely related to platelet type beta transforming growth factor. *Science* **226,** 705-707.

Turbitt, M. L., Akhurst, R. J., White, S. I., and MacKie, R. M. (1990). Localization of elevated transforming growth factor-alpha in psoriatic epidermis. *J. Invest. Dermatol.* **95,** 229-232.

Twardzik, D. R., and Sherwin, S. (1985). Transforming growth factor activity (TGF) in human urine: Synergism between TGF-beta and urogastrone. *J. Cell. Biochem.* **28,** 289-297.

Twardzik, D. R., and Ranchalis, J. E. (1987). Growth modulating peptide that utilizes the receptor for epidermal growth factor. "UCLA Symposium on Development and Diseases of Cartilage and Bone Matrix," pp. 421-437. Liss, New York.

Twardzik, D. R., Sherwin, S. A., Ranchalis, J. E., and Todaro, G. J. (1982a). Transforming growth factors in the urine of normal, pregnant and tumorbearing humans. *J. Natl. Cancer Inst.* **6,** 9793-9798.

Twardzik, D. R., Todaro, G. J., Reynolds, F. H., and Stephenson, J. R. (1982b). Abelson MuLV induced transformation involves production of a polypeptide growth factor. *Science* **216,** 894-897.

Twardzik, D. R., Todaro, G. J., Reynolds, F. H., Jr., and Stephenson, J. R. (1983). Similar transforming growth factors (TGFs) produced by cells transformed by different isolates from feline sarcoma virus. *Virology* **124,** 201-207.

Twardzik, D. R., Brown, J. P., Ranchalis, J. E., Todaro, G. J., and Moss, B. (1985). Vaccinia virus infected cells release high levels of a growth factor functionally related to EGF. *Proc. Natl. Acad. Sci. U.S.A.* **82,** 5300-5304.

Twardzik, D. R., Ranchalis, J. E., McPherson, J., Ogawa, I., Gentry, L., Purchio, A. F., and Todaro, G. J. (1989). Natural and recombinant transforming growth factor-βs inhibit tumor growth and promote differentiation of a human lung carcinoma in athymic mice. *J. Natl. Cancer Inst.* **81,** 1182-1185.

Twardzik, D. R., Mikovits, J. A., Ranchalis, J. E., Purchio, A. F., Ellingsworth, L., and Ruscetti, F. W. (1990). Gamma-interferon-induced activation of latent transforming growth factor-β by human monocytes. *Ann. N.Y. Acad. Sci.* **593,** 276-284.

Van Obberghen-Schilling, E., Roche, N. S., Flanders, K. C., Sporn, M. B., and Roberts, A. B. (1988).

Transforming growth factor beta 1 positively regulates its own expression in normal and transformed cells. *J. Biol. Chem.* **263**, 7741–7746.
Wahl, S. M., Hunt, D. A., Wakefield, L. M., McCartney-Francis, N., Wahl, L. M., Roberts, A. B., and Sporn, M. B. (1987). Transforming growth factor type beta induces monocyte chemotaxis and growth factor production. *Proc. Natl. Acad. Sci. U.S.A.* **84**, 5788–5792.
Wakefield, L. M., Smith, D. M., Masui, T., Harris, C. C., and Sporn, M. B. (1987). Distribution and modulation of the cellular receptor for transforming growth factor-beta. *J. Cell Biol.* **105**, 965–975.
Wakefield, L. M., Smith, D. M., Flanders, K. C., and Sporn, M. B. (1988). Latent transforming growth factor-beta from human platelets. A high molecular weight complex containing precursor sequences. *J. Biol. Chem.* **263**, 7646–7654.
Weeks, D. L., and Melton, D. A. (1987). A maternal mRNA localized to the vegetal hemisphere in *Xenopus* eggs codes for a growth factor related to TGF-beta. *Cell* **51**, 861–867.
Wilcox, J. N., and Derynck, R. (1988a). Localization of cells synthesizing transforming growth factor-alpha mRNA in the mouse brain. *J. Neurosci.* **8**, 1901–1904.
Wilcox, J. N., and Derynck, R. (1988b). Developmental expression of transforming growth factors alpha and beta in mouse fetus. *Mol. Cell. Biol.* **8**, 3415–3422.
Wong, D. T., Weller, P. F., Galli, S. J., Elovic, A., Rand, T. H., Gallagher, G. T., Chiang, T., Chou, M. Y., Matossian, K., and McBride, J. (1990). Human eosinophils express transforming growth factor alpha. *J. Exp. Med.* **172**, 673–681.
Wozney, J. M., Rosen, V., Celeste, A. J., Mitsock, L. M., Whitters, M. J., Kriz, R. W., Hewick, R. M., and Wang, E. A. (1988). Novel regulators of bone formation: Molecular clones and activities. *Science* **242**, 1528–1534.
Wrann, M., Bodmer, S., de Martin, R., Siepl, C., Hofer-Warbinek, R., Frei, K., Hofer, E., and Fontana, A. (1987). T cell suppressor factor from human glioblastoma cells is a 12.5-kD protein closely related to transforming growth factor-beta. *EMBO J.* **6**, 1633–1636.
Xu, Y.-H., Richert, N., Ito, S., Merlino, G. T., and Pastan, I. (1984). Characterization of epidermal growth factor receptor gene expression in malignant and normal human cell lines. *Proc. Natl. Acad. Sci. U.S.A.* **81**, 7308–7312.
Yeh, Y., and Yeh, Y. C. (1989). Transforming growth factor alpha and human cancer. *Biomed. Pharmacother.* **43**, 651–659.
Yeh, Y. C., Tsai, J. F., Chuang, L. Y., Yeh, H. W., Tsai, J. H., Florine, D. L., and Tam, J. P. (1987). Elevation of transforming growth factor alpha and its relationship to the epidermal growth factor and alpha-fetoprotein levels in patients with hepatocellular carcinoma. *Cancer Res.* **47**, 896–901.
Ying, S.-Y., Becker, A., Baird, A., Ling, N., and Ueno, N. (1986). Type beta transforming growth factor (TGF-β) is a potent stimulator of the basal secretion of follicle stimulating hormone (FSH) in pituitary monolayer system. *Biochem. Biophys. Res. Commun.* **135**, 950–956.
Yoshida, K., Kyo, E., Tsujino, T., Sano, T., Niimoto, M., and Tahara, E. (1990). Expression of epidermal growth factor, transforming growth factor-alpha and their receptor genes in human gastric carcinomas; implication for autocrine growth. *Jpn. J. Cancer Res.* **81**, 43–51.
Zugmaier, G., and Lippman, M. E. (1990). Effects of TGFβ on normal and malignant mammary epithelium. *Ann. N.Y. Acad. Sci.* **593**, 272–275.

# 13 Insulin-Like Growth Factors in the Brain

Derek LeRoith, Charles T. Roberts Jr., Haim Werner,
Carolyn Bondy, Mohan Raizada, and Martin L. Adamo

## I. INTRODUCTION

The insulin gene family comprises the genes for insulin and the insulin-like growth factors I and II (IGF-I and IGF-II). Traditionally these peptides have been considered to act in an endocrine manner. Insulin is released from the pancreas to affect metabolism and the IGFs (primarily IGF-I) are secreted from the liver to affect the growth and development of the organism. Recent evidence strongly suggests that these growth factors, especially the IGFs, are synthesized by most tissues of the body and function as paracrine or autocrine agents in a tissue-specific manner. These paracrine or autocrine actions influence not only the growth and development of a given tissue, but also certain differentiated functions. Of particular interest is the involvement of this family of peptides in the nervous system. Insulin, IGF-I, and IGF-II, as well as the IGF-binding proteins, are present in various regions of the brain. Detection of the specific mRNAs demonstrates that the IGFs and their binding proteins are synthesized locally in the nervous system. Similarly, the presence of specific receptors for these ligands throughout the nervous system suggests that insulin and the IGFs play an important role in nervous system physiology. Indeed, evidence exists for biologic actions of insulin and IGFs in different neural tissues (see Chapter 14).

In this chapter we will discuss aspects of the various components of the IGF system, including the ligands, their receptors, and the binding proteins as they relate to the nervous system in general and to the brain in particular. We will stress the region-specific production and localization of each component as well as the regulation of this production during development and in adulthood by hormones and other factors. In addition, we will describe certain aspects of these ligands and receptors using examples, where appropriate, from primary cell cultures as well as from transformed cell lines of neural origin.

Our current understanding of the insulin-like growth factor family of peptides, binding proteins, and receptors and the interaction of each component in this system is depicted schematically in Fig. 1 and reviewed more extensively elsewhere (Rechler and Nissley, 1985; Roberts and LeRoith, 1988; Daughaday and Rotwein, 1989).

## II. INSULIN-LIKE GROWTH FACTORS IN THE BRAIN

### A. Insulin

Insulin-like material isolated from the brain is both immunologically and biologically similar to pancreatic insulin. It is found in many brain regions and peripheral nerves, with highest concentrations in the olfactory bulb and hypothalamus (Havrankova et al., 1978). The source of brain insulin remains controversial. Some studies have indicated that the major contribution to brain insulin comes from the plasma (Baskin et al., 1983). Circulating plasma insulin may reach the brain via the circumventricular organs, which include the subfornical organ, the organum vasculosum of the lamina terminalis, and the median eminence, all of which lack a blood–brain barrier, or by transfer of plasma insulin to the brain across the blood–brain barrier into the cerebrospinal fluid (CSF). This latter mechanism can occur by transcytosis of plasma insulin across endothelial cells, that is, endothelial cell insulin receptors bind, internalize, and subsequently transport the ligand in an intact form (King and Johnson, 1985; Duffy and Pardridge, 1987). Alternatively, brain insulin could be synthesized locally (Havrankova et al., 1978) although the evidence for the local production of insulin is still fragmentary and inconclusive. Neuronal cells in primary culture contain proinsulin and insulin immunoreactive material and incorporate [$^3$H]leucine into insulin-like material (Birch et al., 1984a,b; Clarke et al., 1986), suggesting that brain-derived cells have the capacity to synthesize insulin. More convincing evidence for insulin synthesis in brain includes the presence of insulin-specific mRNA in primary neuronal cells and in the hypothalamic periventricular nuclei (Young, 1986; Schechter et al., 1988).

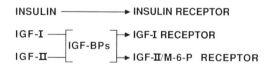

FIGURE 1

Interactions of insulin and the insulin-like growth factors and their homologous receptors. Both IGF-I and IGF-II circulate bound to IGF-binding proteins (IGF-BPs). These binding proteins also modulate the interactions of IGFs with their respective receptors.

Interestingly, cultured pituitary cells also contain insulin-like mRNA by in situ hybridization (Budd et al., 1986). Finally, recent studies using in situ hybridization and polymerase chain reaction (PCR) technology suggest that insulin synthesis also may occur in the microvasculature of the brain (J. Roth, personal communication).

## B. Insulin-Like Growth Factor I

IGF-I mRNA is present in rat central nervous system and is most abundant during development (Rotwein et al., 1988). Recent in situ hybridization studies employing highly sensitive and specific cRNA probes have shown that IGF-I mRNA is localized focally in peripheral target zones of trigeminal and sympathetic nerves during the period of their innervation (Fig. 2A; Bondy et al., 1990; Bondy and Chin, 1991). IGF-I mRNA also is highly abundant in a very selective pattern in the developing rat brain. It is localized in large projection neurons of maturing sensory and cerebellar relay systems and in nonpyramidal cells of the hippocampus and cerebral cortex (Bondy, 1991). Immunocytochemical evidence supports the local synthesis of IGF-I by Purkinje cells in the developing cerebellar cortex (Andersson et al., 1988).

In many brain regions, such as the olfactory bulb and cerebral cortex, IGF-I mRNA levels decline dramatically with maturity (Bach et al., 1991). Despite this decrease in the olfactory bulb (Rotwein et al., 1988), mitral and tufted neurons contain significant levels of IGF-I mRNA in adulthood (Bondy, 1991). In other regions, such as the hypothalamus, brainstem, and cerebellum, mRNA levels generally are unchanged during development (Bach et al., 1991). Persistence of IGF-I gene expression in these cells may be associated with the lifelong process of new synapse formation that characterizes the olfactory bulb. The temporal and topographic pattern of IGF-I gene expression in the rat nervous system suggests a role for IGF-I in a phase of neural differentiation involving dendritic maturation, synaptogenesis, or myelinization. Support for a potential role for IGF-I in these aspects of development comes from in vitro studies that have demonstrated that insulin-like growth factors promote embryonic neuronal survival and neurite outgrowth in cultured sensory, sympathetic, cortical, and motor neurons (Recio-Pinto et al., 1986; Aizenman and de Vellis, 1987; Caroni and Grandes, 1990) and induce oligodendrocyte differentiation and myelin synthesis (McMorris and Dubois-Dalq, 1988; Saneto et al., 1988).

IGF-I immunoreactivity also is found in human (Sara et al., 1982) and mouse (Noguchi et al., 1987) brain.

A variant IGF-I peptide lacking the first three amino acids has been detected in both fetal and adult human brain (Carlsson-Skwirut et al., 1986; Sara et al., 1986). This N-terminally truncated form is the only species found in brain extracts and arises by posttranslational modification of the molecule. It has been postulated that this form represents the paracrine form of the IGF-I molecule in brain and other extrahepatic tissues (Sara and Hall, 1990). Interestingly, since the N terminus of the IGF-I peptide is partially responsible for the interaction of IGF-I with

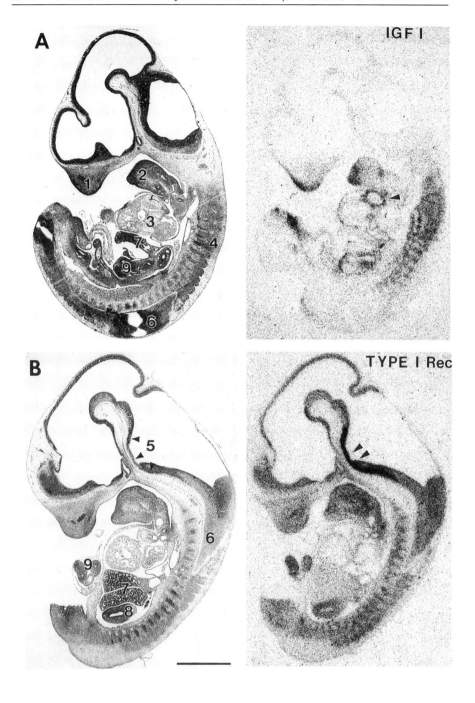

FIGURE 2

IGF-I mRNA and IGF-I receptor mRNA localization in embryonic day 14 rat as shown by in situ hybridization. Hematoxylin- and eosin-stained sections are shown on the left; autoradiographs are shown on the right. Section A was hybridized to a cRNA probe for the rat IGF-I. Section B was hybridized to a cRNA probe for the rat type-I IGF receptor. Details of the in situ hybridization protocol and probe synthesis are given in Bondy et al. (1990). 1, maxillary process; 2, mandibular process (and tongue). These two regions show graded IGF-I gene expression with levels peaking at their tips. These are the target zones of the maxillary and mandibular branches of the trigeminal nerve, which is innervating the face at this time in development. 3, heart. A rim of IGF-I mRNA-containing cells encircles the developing atrial chamber. 4, vertebral column; 5, vental floorplate of the hindbrain. Type-I receptor mRNA is concentrated heavily in this structure. 6, spinal cord. IGF-I receptor mRNA but not IGF-I mRNA is abundant in the developing brain and spinal cord at this midgestational stage of development. 7, liver; 8, stomach; 9, loops of bowel. Bar, 1mm.

IGF-binding proteins (IGF-BPs), the brain IGF-I molecule is bound less by IGF-BPs than intact circulating IGFs are. As a result, this N-terminally truncated IGF-I molecule binds the IGF-I receptor with a slightly higher affinity than intact IGF-I and is therefore more biologically potent, in terms of the neurotrophic and proliferative effects of IGF-I (Carlsson-Skwirut et al., 1989; Giacobini et al., 1990).

The tripeptide that supposedly is cleaved from the N terminus (Gly–Pro–Glu) may have biologic activity that differs from that of the N-terminally truncated IGF-I molecule itself. The synthetic version of this tripeptide can potentiate potassium-induced release of acetylcholine from rat cortical slices and can act through $N$-methyl-D-aspartate (NMDA) receptors to increase dopamine release from striatum (Sara et al., 1989).

## C. Insulin-Like Growth Factor II

IGF-II and a higher molecular weight form termed "big IGF-II" have been isolated from brain (Haselbacher et al., 1985). The higher $M_r$ form (13 kDa) probably represents a partially processed product of the prohormone that contains a C-terminal extension. Both forms are present in fetal and adult brain, although fetal IGF-II levels are higher than adult levels. Significant levels of IGF-II immunoreactivity are found in the anterior pituitary as well as in the dorsomedial hypothalamus and supraoptic nucleus. IGF-II levels in the CSF are particularly high (Haselbacher and Humbel, 1982), because of local synthesis in the choroid plexus (as has been shown in humans) and because of transport from the circulation via an IGF receptor present in the blood–brain barrier that has higher affinity for IGF-II (Duffy et al., 1988).

IGF-II mRNA, whose presence constitutes presumptive evidence of local synthesis, is found at higher levels in fetal than adult brain. In adult rats, however, the brain still appears to express the highest levels of IGF-II mRNA. Although the peptide is found in different regions in adult brain, the mesenchymal cells of the choroid plexus and the leptomeninges appear to be the major site of IGF-II synthesis (Hynes et al., 1988; Ichimiya et al., 1988; Stylianopoulou et al., 1988; Wood et al., 1990). Interestingly, neuronal and glial cell cultures both express IGF-I mRNA, whereas only glial cells express IGF-II mRNA (Adamo et al., 1988; Rotwein et al., 1988).

## D. Physiologic Regulation of the Expression of Insulin-Like Peptides in Brain

The regulation of production of the IGFs in the central nervous system (CNS) has been studied most effectively by examining the steady-state levels of the pertinent mRNAs. The developmental aspect of regulation of IGFs has been discussed already. Some models in which the physiologic regulation of these peptides was studied are presented here. Their possible physiologic significance, including potential autocrine or paracrine roles for these growth factors, is discussed. These examples also illustrate that regulation of central and peripheral IGF mRNA levels is sometimes divergent.

### 1. Growth hormone

Growth hormone (GH) is the principal regulator of hepatic IGF-I production, operating mainly at the transcriptional level (Mathews et al., 1986). GH also regulates IGF-I mRNA levels in extrahepatic tissues, including brain. For example, the congenitally GH-deficient lit/lit mouse has reduced brain IGF-I mRNA levels (Mathews et al., 1986) and hypophysectomy of normal rats reduces brain IGF-I mRNA levels (Hynes et al., 1987). These reduced levels could be restored almost to normal 4 hr after intracerebroventricular (ICV) GH administration. Although these results suggest that GH is capable of regulating brain IGF-I mRNA levels, they raise questions concerning the physiologic significance of the GH effect, that is, the ability of peripheral GH to regulate brain IGF-I gene expression. In the pituitary cell line $GH_3$, GH also was capable of increasing IGF-I mRNA levels (Fagin et al., 1989a).

Unlike IGF-II mRNA levels in peripheral tissue, brain IGF-II mRNA levels may be GH responsive. This possibility is based on the observation that hypophysectomy of rats was associated with decreased brain IGF-II mRNA levels (Hynes et al., 1987). Peripheral GH replacement caused slight increases in brain IGF-II mRNA, whereas ICV-administered GH partially restored IGF-II mRNA levels (Hynes et al., 1987). Collectively, these studies suggest that GH may regulate CNS IGF-I and IGF-II mRNA levels, but their strong developmental patterns of expression suggest that other factors are of greater importance in their regulation.

### 2. Glucocorticoids

Glucocorticoid excess may impede growth or development of the organism by mechanisms that include suppressive effects on the brain (Devenport and Devenport, 1985). Dexamethasone has been shown to reduce IGF-I mRNA levels in liver and extrahepatic tissues (Luo and Murphy, 1989), as well as in neuronal and glial cells in primary culture (Adamo et al., 1988). These results suggest that the inhibitory effects of glucocorticoids in brain may be mediated by a reduction in the steady-state levels of IGF-I mRNA. In contrast, glucocorticoids appear to cause decreased IGF-II mRNA levels in neonatal rat liver but not in choroid plexus (Beck et al., 1988; Levinovitz and Norstedt, 1989; Orlowski et al., 1990). This result suggests that glucocorticoids may, in part, mediate postnatal reduction in hepatic IGF-II mRNA, and that the effect may be tissue specific (Beck et al., 1988).

### 3. Insulin

Insulin has been shown to regulate IGF-I mRNA levels in liver and extrahepatic tissues (Fagin et al., 1989b). To date, no such effect on brain has been reported, although it cannot be ruled out. IGF-II mRNA levels in brain have been shown to respond to peripheral insulin. Chronic insulin injection decreased ventral hypothalamic IGF-II mRNA levels and increased lateral and dorsal hypothalamic and cerebral cortex IGF-II mRNA levels (Lauterio et al., 1990). The investigators performing these studies suggested that such regulation in the hypothalamus may point to a role for IGF-II (or insulin acting via local production of IGF-II) in the regulation of food intake or the regulation of GH release by growth hormone releasing factor (GRF).

### 4. Altered nutritional states

Previous studies have shown that rats fasted for 48 hr exhibit a large reduction in hepatic IGF-I mRNA levels, compared with marginal reduction of brain IGF-I mRNA levels (Lowe et al., 1989). In another study, dietary protein restriction reduced both hepatic IGF-I mRNA levels and brain IGF-II mRNA levels (Straus and Takemoto, 1990). The mechanism by which the nutritional status of the organism regulates the levels of IGF-II mRNA in brain is unclear.

## III. IGF-BINDING PROTEINS IN THE BRAIN

Several IGF-BPs secreted by numerous different tissues have been detected in blood (Martin and Baxter, 1986) (Table 1). At least four of these have been characterized by sequence analysis and found to exhibit many similarities, including a highly conserved (60–70%) cysteine-rich region with 18 invariant cysteine residues (Brewer et al., 1988; Lee et al., 1988; Wood et al., 1988). IGF-BP1 and IGF-BP2 also contain an Arg–Gly–Asp (RGD) sequence near their C terminus; this sequence has been demonstrated previously to be involved in binding of matrix proteins to the cell surface. In contrast, IGF-BP3 does not contain an RGD sequence and is the only IGF-BP that is significantly $N$-glycosylated. Further, IGF-BP3 is the major circulating IGF-BP postnatally, and is associated with an acid-labile component, as well as with IGF-I or II, to form a 150-kDa complex. Other less characterized, although immunologically and structurally distinct, IGF-BPs include one found in human CSF (IGF-BP5) that has unusually greater affinity for IGF-II than for IGF-I (Roghani et al., 1989).

TABLE 1

Characterization of IGF binding proteins

| IGF binding protein | Origin | Approximate mass (kDa) | Features |
|---|---|---|---|
| 1 | Hepatocytes; uterus; fibroblasts | 25 | R-G-D sequences |
| 2 | Hepatocytes; decidua; smooth muscle | 32 | R-G-D sequence |
| 3 | Hepatocytes; fibroblasts; endothelium | 54 | glycosylated; associates with labile component and ligand to form 150-kDa complex GH-stimulated |
| 4 | Hepatocytes; many other tissues | 20 | No R-G-D sequence |

The circulating IGF-BPs initially were believed to transport IGFs to their target tissue, to prevent their degradation, and thereby to prolong their half-life. Recent studies have suggested additional roles for the IGF-BPs, namely, the modulation of IGF action either by inhibiting or possibly enhancing IGF action under certain conditions in certain cell types (McCusker et al., 1990). In both human and rat CSF, multiple IGF-BPs have been detected with apparent $M_r$ ranging from 24 to 42 kDa (Binoux et al., 1982; Hossenlopp et al., 1986; Roghani et al., 1989; Romanus et al., 1989; Ocrant et al., 1990). The predominant IGF-BP found in both human and rat central nervous tissue and CSF is IGF-BP2, with three N-terminal amino acids missing. During early embryonic development in the rat, IGF-BP2 mRNA is localized in the ventral floorplate of the spinal cord and hind brain and in the infundibulum. These are specialized areas of neuroepithelium thought to be involved in axonal targeting (Wood et al., 1990). Later in embryonic development, IGF-BP2 gene expression is localized in the choroid plexus; this pattern persists into adulthood (Tseng et al., 1989). This binding protein is secreted from cultured choroidal epithelium and is found in CSF (Tseng et al., 1989). Although IGF-BP2 predominates in CSF, two additional IGF-BPs are also detectable, with apparent $M_r$s of 30–32 and 24 kDa. A minor 30–31-kDa CSF binding protein has an N terminus similar to that of IGF-BP3 and may represent a C-terminally truncated form, possibly secreted by astrocytes or neuronal cells. Another CSF binding protein of 24 kDa has not been characterized fully. Further, primate and bovine retinal pigment epithelial cells, functional and embryological equivalents of the choroid plexus, secrete IGF-BPs, in the former case shown to be IGF-BP2 (Ocrant et al., 1991; Waldbillig et al., 1992).

In culture, astroglial and neuronal cells synthesize multiple binding proteins, including IGF-BP2, suggesting that cells in addition to those in the choroid plexus express IGF-BPs (Lamson et al., 1989; Ocrant et al., 1989, 1990). The role of

IGF-BP2 (and other IGF-BPs) in CNS development and in the choroid plexus is under active investigation at present.

## IV. INSULIN AND IGF RECEPTORS

Insulin and IGFs elicit their biologic actions by binding to specific cell-surface receptors. Activation of these receptors results in the transduction of a signal across the membrane (Fig. 3). Each receptor has been well characterized at both the protein and nucleic acid levels, the latter including the cloning and sequencing of the appropriate cDNAs as well as, in the case of the insulin receptor, determination of the gene structure (Ebina et al., 1985; Ullrich et al., 1985, 1986; Seino et al., 1989).

The insulin and IGF-I receptors are very similar in structure. Each receptor is a heterotetrameric integral membrane glycoprotein, consisting of two α and two β subunits that are joined by disulfide bridges. The two α subunits lie entirely outside the cell and bind the ligand at or near a cysteine-rich domain, whereas the two β subunits contain a short extracellular domain, a transmembrane segment, and a

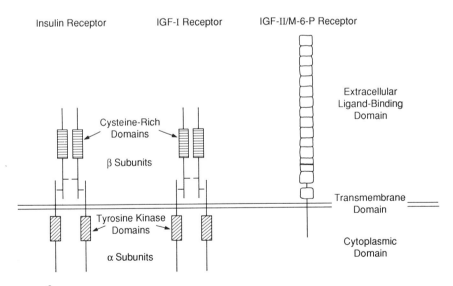

**FIGURE 3**

Schematic representation of the insulin and insulin-like growth factor receptors. The insulin and IGF-I receptors are both heterotetrameric glycoproteins, with strong homology to each other and similar tyrosine kinase domains in the cytoplasmic portion of the β subunits. The IGF-II/M-6-P receptor is completely different, containing 15 extracellular cysteine-rich repeat regions. The IGF-II/M-6-P receptor contains no tyrosine kinase domain.

cytoplasmic domain. The cytoplasmic portion of the β subunit contains an ATP binding site and a tyrosine kinase domain that is typically seen in the family of growth factor receptors related to the *src* family of oncogenic proteins. Activation of the receptor by ligand binding to the α subunit causes a conformational change in the β subunits, leading to autophosphorylation and increased receptor tyrosine kinase activity. These changes are associated with phosphorylation of certain endogenous substrates closely associated with the receptor and activation of a cascade of events culminating in the final biologic effect (Rosen, 1987).

The structure of the IGF-II receptor is completely different. The IGF-II (or cation-independent mannose 6-phosphate) receptor (IGF-II/M-6-P) is a single polypeptide chain with a large extracellular domain, composed of 15 repeats of a cysteine-rich region, and a short cytoplasmic domain. Until recently, the role of this receptor in signal transduction was unclear, since the very short cytoplasmic domain bears no resemblance to the tyrosine kinase domains of the insulin and IGF-I receptors. Recent evidence suggests that signaling by the IGF-II/M-6-P receptor may involve G protein activation. Indeed, there are sequences in the cytoplasmic domain that can interact with and activate G proteins in a fashion similar to that of the classical G protein-related receptors (Okamoto et al., 1990, 1991).

## A. Insulin and IGF Receptors in Brain

Receptors for the insulin-like growth factors are expressed throughout the brain. Characterization of the receptors by classical competitive ligand binding studies, photoaffinity and chemical affinity cross-linking, and gel electrophoresis has shown that, in general, these receptors resemble their counterparts in nonneural tissues (Gammeltoft et al., 1985). However, there are some differences in the structures of the brain receptors. In particular, the apparent $M_r$s of the insulin, IGF-I, and IGF-II receptors in brain are significantly smaller than those of their nonneural peripheral counterparts (Heidenreich et al., 1983, 1986; Hendricks et al., 1984; Heidenreich and Brandenburg, 1986; Lowe et al., 1986; McElduff et al., 1987; McElduff et al., 1988). This difference in $M_r$ is primarily due to alterations in glycosylation and is specific to this family of receptors, since neuronal basic fibroblast growth factor receptors do not show similar differences (Walicke et al., 1989). Most of the glycosylation is normally *N*-linked and contains high mannose and complex carbohydrate residues with terminal sialic acid residues. Studies using enzymes such as neuraminidase and exo- and endoglycosidases suggest that the brain receptor may contain polysialic acid residues and less *N*-linked glycosylation (Ota et al., 1988a,b). In addition, brain insulin receptors, in contrast to their nonneural peripheral counterparts, fail to demonstrate "negative cooperativity" with the ligand and show increased affinity for certain insulin analogs (Gammeltoft et al., 1984).

Interestingly, these structural differences also are present in receptors from primary cultures of neuronal cells, but not of glial cells, suggesting that the brain-type receptor is neuron-specific (Lowe et al., 1986; Burgess et al., 1987; Shemer et

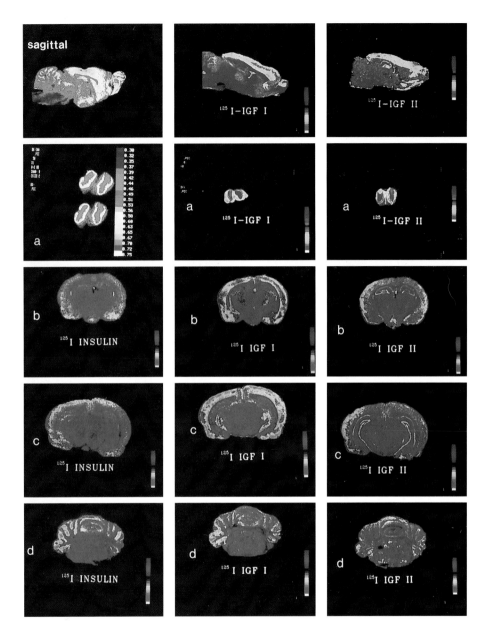

**FIGURE 4**

Autoradiographic localization of insulin and IGF receptors in rat brain. Specific receptors for insulin (*left*), IGF-I (*middle*), and IGF-II (*right*) were identified with the use of [125]I-labeled ligands. High concentrations of receptors were found in olfactory bulbs, hypothalamus, cerebellum, and choroid plexus. Within the same region each receptor was identified in different cell layers. Thus, in the olfactory bulbs, insulin receptors are concentrated in the external plexiform layer, whereas IGF-I receptors are concentrated in the glomerular and nerve fiber layers. In the hypothalamus, insulin binding sites are concentrated in the arcuate and dorsomedial nuclei, whereas IGF-I binding sites are concentrated in the median eminence.

al., 1987b). These structural features have been evaluated in the neuroblastoma cell line SK-N-SH in which neuronal-type insulin and IGF-I receptors are expressed (Ota et al., 1988a,b).

## B. Developmental Expression of Brain Receptors

The concentration of insulin receptors in brain peaks during the perinatal period (Kappy et al., 1984; Devaskar et al., 1986; Lowe et al., 1986), whereas the concentrations of IGF-I receptors and IGF-II/M-6-P receptors appear to peak earlier in fetal development (Sara et al., 1983; Valentino et al., 1990). These developmental trends, initially characterized by ligand binding, subsequently have been shown to be associated with similar trends in the levels of the respective receptor mRNAs (Werner et al., 1989c). Whereas IGF-I receptor mRNA is abundant throughout the nervous system during early stages of development, it is most heavily concentrated in the ventral floorplate (Fig. 2B; Bondy et al., 1990) in a pattern similar to that of IGF-BP2 (Wood et al., 1990).

## C. Regional Localization of Brain Receptors

The receptors for the insulin-like growth factors are expressed in numerous regions of the brain. Autoradiographic and in situ hybridization studies have enabled investigators to localize receptor distribution more precisely to particular nuclear regions and even to cell types. This information can be very useful in furthering our understanding of the possible role these growth factors play in controlling nervous system function. Quantitative autoradiographic assays using $^{125}$I-labeled ligand binding have demonstrated high concentrations of both insulin and IGF receptors in olfactory bulb, choroid plexus, cerebellum, and other sites (Baskin et al., 1986, 1990; Bohannon et al., 1986, 1988; Hill et al., 1986; Lesniak et al., 1988; also see Fig. 4). During postnatal development and in the mature brain, IGF-I receptor mRNA is concentrated in specific neuronal groups, including sensory and cerebellar relay systems in pyramidal cells of the piriform cortex, lamina VI of the frontal cortex, and Ammon's horn (Bondy et al., 1992). IGF-I receptor mRNA is also abundant in the nuclei of the amygdaloid complex and the suprachiasmatic nuclei (Fig. 5). These findings indicate that the CNS IGF-I receptor is synthesized close to IGF-I binding sites, as demonstrated by autoradiography (Bohannon et al., 1988; Lesniak et al., 1988), and close to IGF-I synthesis sites (Andersson et al., 1988; Bondy 1991b), supporting the concept of a local field of action for CNS IGF-I. Under certain circumstances, confirmation of these results has been obtained using antibodies specifically directed against particular receptors. Interestingly, using an antiphosphotyrosine antibody, Moss et al. (1990) have demonstrated that phosphotyrosine-containing proteins in brain have a regional distribution very similar to that of the insulin and IGF-I receptors. Indeed, the distribution is more specific for the

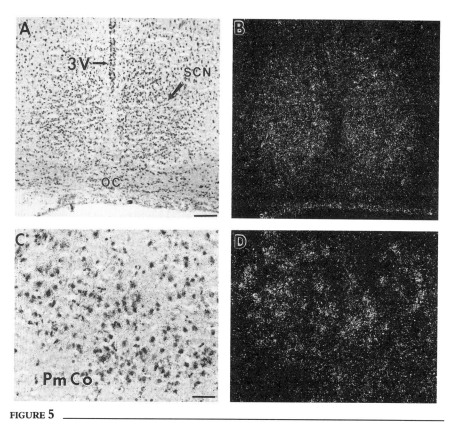

FIGURE 5

IGF-I receptor mRNA localization in the adult rat brain shown by in situ hybridization. (A, B) Bright and dark field views of the suprachiasmatic nucleus (SCN) in which IGF-I receptor mRNA is moderately abundant. 3V, Third ventricle; OC, optic chiasm. Bar, 100 µm. (C, D) Bright and dark field views of the posteromedial cortical amygdaloid nucleus (PmCo). IGF-I receptor mRNA is abundant in large projection neurons and less abundant in local interneurons in this and other nuclei of the amygdala. Bar, 25 µm.

insulin receptor since the arcuate nucleus, which has no IGF-I receptors, is phosphotyrosine positive, whereas cortical layer VI, which has no insulin receptors but is rich in IGF-I receptors, is phosphotyrosine negative (Bohannon et al., 1988; Lesniak et al., 1988). Insulin and IGF-I receptors also have been demonstrated in the neural retina. Higher levels of IGF-I receptors are found in all layers of the retina, nonphotoreceptor as well as photoreceptor layers (Waldbillig et al., 1987a,b; Rosenzweig et al., 1990). Since IGF-I and IGF-II are expressed by retinal tissue (Danias and Stylianopoulou, 1990), the presence of these receptors suggests a local autocrine or paracrine role for the IGFs in retina.

## V. POSTRECEPTOR MECHANISMS

Postreceptor mechanisms that have been studied in brain and neural-derived cells include the phosphorylation of endogenous and exogenous substrates by the tyrosine kinase activity of the insulin and IGF-I receptors. Both ligands bind with high affinity to their homologous receptors and with lower affinity to each others' receptors (reviewed in Adamo et al., 1989). As mentioned earlier, this binding is associated with autophosphorylation of the β subunit of the receptor as well as with induction of tyrosine kinase activity, exemplified by increased tyrosine phosphorylation of both endogenous and exogenous substrates. One such endogenous substrate, pp185, is present in neural-derived cell lines as well as in other cell lines, and is phosphorylated rapidly by both insulin and IGF-I receptors (Shemer et al., 1987a). Interestingly, in addition to expressing insulin and IGF-I receptors, these cell lines appear to express significant amounts of "hybrid" receptors. These "hybrid" receptors were postulated to be composed of one αβ insulin hemireceptor and one αβ IGF-I hemireceptor (Treadway et al., 1989). Indeed, gel electrophoresis of neural-derived proteins following phosphorylation experiments showed the presence of doublets of β subunits after gel electrophoresis, the larger subunit ($M_r$, 105 kDa) derived from the IGF-I receptor and the smaller subunit ($M_r$, 95 kDa) derived from the insulin receptor (Shemer et al., 1987a, 1989). Although the functional significance of these hybrids in neural and other tissues is undefined, they potentially contribute complexity to the role of insulin and IGF-I in the nervous system.

## VI. BIOLOGIC ACTION OF INSULIN AND IGFs

Most known growth factors act via paracrine or autocrine mechanisms; that is, they are synthesized, are secreted, and act locally. The fact that the insulin-like growth factors and their receptors and binding proteins are produced in the CNS strongly suggests an important role for this growth factor family in nerve function. The developmental regulation and regional specificity also suggest a specific physiologic role. Whether the IGFs act as neurotrophic factors at early stages of development and differentiation of the nervous system and are involved in neuromodulation of neurotransmission at later stages must be defined fully. We have reviewed the biologic actions of insulin and the IGFs in the nervous system (Adamo et al., 1989). Table 2 presents a summary of these effects with more recent references. At this early stage, it is difficult to form preliminary data into a single working hypothesis. However, the fact that certain biologic actions are strongly demonstrable supports the hypothesis that this family of growth factors has an important role in normal brain function. Since IGF-I, like most growth factors, affects the growth of cells, we will outline the recent studies of IGF-I effects on cellular function in nervous tissue with respect to metabolism and growth.

TABLE 2

**Potential biological actions of insulin and IGFs in the nervous system**

| | |
|---|---|
| Growth and development | |
| Cellular proliferation in vitro | McMorris and Dubois-Dalcq, 1988; DiCicco-Bloom and Blanc, 1989; Grant et al., 1990; Torres-Aleman et al., 1990 |
| Development and differentiation in vitro (e.g., specific neurotransmitter function or myelination) | McMorris and Dubois-Dalcq, 1988; Saneto et al., 1988; Knusel et al., 1990 |
| Synaptogenesis; nerve regeneration; neurite outgrowth | Bondy et al., 1990; Ishii, 1989; Kanje et al., 1989; Sjoberg and Kanje, 1989; Caroni and Grandes, 1990 |
| Tropic in vivo | Behringer et al., 1990; Giacobini et al., 1990 |
| Neuromodulatory | |
| Neurotransmitter release | Dahmer and Perlman, 1988; Nilsson et al., 1988 |
| Neurotransmitter synthesis (adults?) | Knusel et al., 1990 |
| Metabolic | |
| Glucose transport | Werner et al., 1989 |
| Protein synthesis and S6 kinase activity (growth related?) | Heidenreich and Toledo, 1989a,b |
| Protein kinase C activity (growth related?) | Heidenreich et al., 1990 |

## A. Brain Glucose Metabolism

Perhaps the most central aspect of cerebral metabolism, and a potential focal point for metabolic control by the IGF system, is glucose transport and subsequent use. Cellular glucose uptake, in the brain as well as in all other tissues, is dependent on glucose transporter proteins (reviewed by Simpson and Cushman, 1986). These membrane-spanning proteins are divided into two major classes (Table 3). The facilitated transport of glucose is achieved by a family of glycoproteins with multiple membrane spanning domains, of which five distinct gene products have been characterized (Glut1–5). (For more detail, see the review by Bell et al., 1990.) Of particular interest are Glut1, Glut3, and Glut4. Glut1 is expressed in most tissues, especially in brain and erythrocytes, the two tissues in which it was initially characterized. Glut3 seems to be found primarily in brain; however, only limited studies have been published on this glucose transport protein (Sadiq et al., 1990). Glut4, on the other hand, is the insulin-responsive glucose transporter expressed primarily in adipose tissue and cardiac and skeletal muscle. Glut1 has been considered responsible for basal insulin-independent glucose uptake by cells.

Glut1 has been detected at the level of both mRNA and protein in fetal, neonatal, and adult brain in brain microvasculature and neuronal and glial cells. The level of expression is regulated developmentally. Of those tissues examined in adult rats and rabbits, the brain contains the highest levels of Glut1 (Werner et al., 1989a; Sadiq et al., 1990). Using in situ hybridization, Pardridge et al. (1990) demonstrated that the major site of Glut1 expression in adult bovine brain is the microvasculature.

**TABLE 3**
**Glucose transporters**

| Designation | Major sites of expression |
|---|---|
| Facilitated diffusion | |
| Glut 1 (erythrocyte) | Brain, kidney, placenta |
| Glut 2 (liver) | Liver, β-cell (pancreas), kidney |
| Glut 3 (brain) | Brain, kidney, placenta |
| Glut 4 (muscle/fat) | Skeletal muscle, heart, adipocytes |
| Glut 5 (small intestine) | Jejunum |
| Active transport | |
| $Na^+$/glucose co-transporter | Small intestine, renal tubules |

On the other hand, neonatal rat neuronal and glial cells express Glut1 in culture (Walker et al., 1988; Werner et al., 1989b). Glut1 gene expression in both neuronal and glial cells is enhanced by both insulin and IGF-I at physiologic concentrations. However, insulin increases glucose uptake in glial but not neuronal cells (Werner et al., 1989b). These studies suggest, therefore, that Glut1 expression is regulated by IGF-I at the level of transcription, but that increased glucose transport in response to IGF-I occurs in glial cells, not in neuronal cells. Further, although the blood–brain barrier microvasculature probably regulates a large degree of the glucose transport from the serum, brain cells also demonstrate a degree of regulated glucose uptake. This is further exemplified by the finding that reductions in extracellular glucose resulted in increased glucose transporter mRNA and glucose transport in glial cells (Walker et al., 1988). It is possible that Glut1 is the major transporter in neonatal brain and Glut3 is the major form in adult brain, since Glut3 is highly expressed in adult brain (G.I. Bell, personal communication).

## VII. HYPOTHESIS

As outlined earlier, insulin and IGF-I have important effects on neuronal and glial cell metabolism. We propose that IGF-I, its receptors, and IGF-BPs constitute one set of chemical signals that control trophic and neuromodulatory activities in the brain (Fig. 6). We and others have developed evidence in support of such an hypothesis with the use of different cell culture techniques. Neurons and both types of macroglial cells in the brain express IGF-I receptors and exhibit IGF-I-induced physiologic changes in vitro. IGF-I not only stimulates proliferation of glial progenitor cells but also induces these progenitors to undergo differentiation into mature oligodendrocytes. This proliferative and differentiative effect is associated with induction of enzymes unique to oligodendrocytes (Lenoir and Honegger, 1983; Debbage, 1986; McMorris et al., 1986; Saneto et al., 1988; Van der Pal et al., 1988). In astroglial cells, IGF-I is a potent stimulator of proliferation that is associated with effects on Glut1 gene expression and DNA synthesis (Shemer et al., 1976; Werner et al., 1989b). In neuronal cells, IGF-I has many trophic and neuromodulatory effects. Stimulation of neurite outgrowth, proliferation and maintenance of neurons (particularly cholinergic and dopaminergic cells), and alterations in the expression of proteins associated with neurofilaments (Ishii and Recio-Pinto, 1987; Nilsson et al., 1988; Knusel et al., 1990) are some of the effects of IGF-I on neurons.

This review of our data and those of other investigators leads us to propose the following model for a potential role of the IGF-I system in physiologic signaling among neuronal and glial cells. IGF-I produced locally in neurons and astrocytes is released into the extracellular space and acts in an autocrine or paracrine fashion to induce unique effects on neurons, astrocytes, and oligodendrocytes, as shown in Fig. 6. These IGF-I effects are regulated not only by IGF-I synthesis and release and by changes in the levels of IGF-I receptors, but also by the expression of IGF-BPs, which predominantly are synthesized by and secreted from astrocytes (Han et al., 1988; Ocrant et al., 1989). Thus, physiologic and pathophysiogic situations may influence one or more of these components in the IGF-I cascade to alter brain cell activity. For example, hypoglycemia, which stimulates sympathetic activity and changes in the brain catecholamine levels, could be one such situation. Activation of sympathetic-stimulated release of norepinephrine can activate β-adrenergic re-

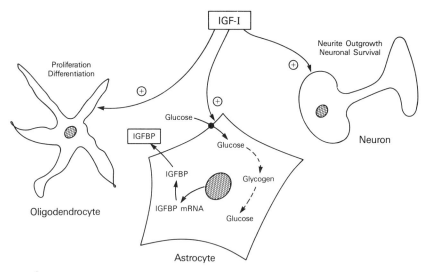

FIGURE 6

Hypothetical effects of IGF-I on brain cells. In astrocytes, IGF-I receptor activation stimulates *Glut1* gene expression and glucose transport. This leads to accumulation of glycogen. In addition, expression of the IGF-BP gene and release of BPs may regulate the actions of IGF-I on astrocytes. In neurons, IGF-I stimulates neurite outgrowth and neuronal survival. In oligodendrocytes, IGF-I stimulates proliferation and differentiation of glial progenitor cells.

ceptors on astrocytes (Baker et al., 1986), which in turn can regulate IGF-I-induced changes in Glut1 gene expression and metabolism of the glycogen predominantly stored in astrocytes (Saneto and deVellis, 1987). Such a mechanism may be of great significance in maintaining normal glucose levels in the brain by providing a constant although limited source of glucose.

Elucidation of interactions of the components of the IGF system among various cell types at the cellular level will have major implications for an understanding of why degeneration of specific neurons occurs in certain brain diseases and what role IGFs may play in neuronal regeneration. In addition, approaches such as these may form the basis for investigating the mechanisms of action of growth factors in the brain.

## ACKNOWLEDGMENTS

We wish to thank Violet Katz for her excellent secretarial assistance.

Martin L. Adamo is the recipient of a postdoctoral fellowship from the Juvenile Diabetes Foundation. This work was supported, in part, by a grant from the Washington, D.C., affiliate of the American Diabetes Association to Charles T. Roberts Jr.

## REFERENCES

Adamo, M., Werner, H., Farnsworth, W., Roberts, C. T., Jr., Raizada, M., and LeRoith, D. (1988). Dexamethasone reduces steady state insulin-like growth factor I messenger ribonucleic acid levels in rat neuronal and glial cells in primary culture. *Endocrinology* **123**, 2565–2570.

Adamo, M., Raizada, M. K., and LeRoith, D. (1989). Insulin and insulin-like growth factor receptors in the nervous system. *Mol. Neurobiol.* **3**, 71–100.

Aizenman, Y., and de Vellis, J. (1987). Brain neurons develop in a serum and glial free environment: Effects of transferrin, insulin, insulin-like growth factor-1 and thyroid hormone on neuronal survival, growth and differentiation. *Brain Res.* **406**, 32–42.

Andersson, I. K., Edwall, D., Norstedt, G., Rozell, B., Skottner, A., and Hansson, H.-A. (1988). Differing expression of insulin-like growth factor I in the developing and in the adult rat cerebellum. *Acta Physiol. Scand.* **132**, 167–173.

Bach, M. A., Shen-Orr, Z., Lowe, W. L., Jr., Roberts, C. T., Jr., and LeRoith, D. (1991). Insulin-like growth factor I mRNA levels are developmentally regulated in specific regions of the rat brain. *Mol. Brain Res.* **10**, 43–48.

Baker, S. P., Sumners, C., Pitha, J., and Raizada, M. K. (1986). Characteristics of β-adrenoreceptors from neuronal and glial cells in primary cultures of rat brain. *J. Neurochem.* **47**, 1318–1326.

Baskin, D. G., Woods, S. C., West, D. B., Van Houten, M., Posner, B. I., Dorsa, D. M., and Porte, D., Jr. (1983). Immunocytochemical detection of insulin in rat hypothalamus and its possible uptake from cerebrospinal fluid. *Endocrinology* **113**, 1818–1825.

Baskin, D. G., Brewitt, B., Davidson, D. A., Corp, E., Paquette, T., Figlewicz, D. P., Lewellen, T. K., Graham, M. K., Woods, S. G., and Dorsa, D. M. (1986). Quantitative autoradiographic evidence for insulin receptors in the choroid plexus of the rat brain. *Diabetes* **35**, 246–249.

Baskin, D. G., Bohannon, N. J., King, M. G., Wimpy, T. H., and Rosenfeld, R. G. (1990). Characterization of insulin-like growth factor-II binding sites in the rat brain by quantitative autoradiography: Evidence for an IGF-II binding protein. *Neurosci. Res. Commun.* **7**, 53–60.

Beck, F., Samani, N. J., Senior, P., Byrne, S., Morgan, K., Gebhard, R., and Brammar, W. J. (1988). Control of IGF-II mRNA levels by glucocorticoids in the neonatal rat. *J. Mol. Endocrinol.* **1**, R5–R8.

Behringer, R. R., Lewin, T. M., Quaife, C. J., Palmiter, R. D., Brinster, R. L., and D'Ercole, A. J. (1990). Expression of insulin-like growth factor I stimulates normal somatic growth in growth hormone-deficient transgenic mice. *Endocrinology* **127**, 1033–1040.

Bell, G. E., Kayano, T., Buse, J. B., Burant, C. F., Takeda, J., Lin, D., Fukumoto, H., and Seino, S. (1990). Molecular biology of mammalian glucose transporters. *Diabetes Care* **13**, 198–208.

Binoux, M., Hardouin, S., Lassarre, C., and Hossenlopp, P. (1982). Evidence for production by the liver of two IGF binding proteins with similar molecular weights but different affinities for IGF-I and IGF-II. Their relations with serum and cerebrospinal fluid IGF binding proteins. *J. Clin. Endocrinol. Metab.* **55**, 600–602.

Birch, N. P., Christie, D. L., and Renwick, A. G. C. (1984a) Immunoreactive insulin from mouse brain cells in culture and whole rat brain. *Biochem. J.* **218**, 19–27.

Birch, N. P., Christie, D. L., and Renwick, A. G. C. (1984b). Proinsulin-like material in mouse fetal brain cell cultures. *FEBS Lett.* **168**, 299–302.

Bohannon, N. J., Figlewicz, D. P., Corp, E. S., Wilcox, B. J., Porte, D., Jr., and Baskin, D. G. (1986). Identification of binding sites for an insulin-like growth factor (IGF-I) in the median eminence of the rat brain by quantitative autoradiography. *Endocrinology* **119**, 943–945.

Bohannon, N. J., Corp, E. S. Wilcox, B. J., Figlewicz, D. P., Dorsa, D. M., and Baskin, D. G. (1988). Localization of binding sites for insulin-like growth factor I (IGF-I) in the rat brain by quantitative autoradiography. *Brain Res.* **444**, 205–213.

Bondy, C. A. (1991). Transient IGF-I gene expression during the maturation of functionally related central projection neurons. *J. Neurosci.* **11**, 3442–3455.

Bondy, C. A., and Chin, E. (1991). IGF-I mRNA localization in sympathetic and trigeminal target zones during rat embryonic development. *In "Molecular Biology and Physiology of Insulin and Insulin-Like Growth Factors."* (M. K. Raizada and D. LeRoith, eds.) pp. 431–437. Plenum, New York.

Bondy, C. A., Werner, H., Roberts, C. T., Jr., and LeRoith, D. (1990). Cellular pattern of insulin-like growth factor I (IGF-I) and type I IGF receptor gene expression in early organogenesis: Comparison with IGF-II gene expression. *Mol. Endocrinol.* **4**, 1386–1398.

Bondy, C., Werner, H., Roberts, C. T., Jr., and LeRoith, D. (1992). Cellular pattern of type-I insulin-like growth factor receptor gene expression during maturation of the rat brain: Comparison with insulin-like growth factors I and II. *Neuroscience* **46**, 909–923.

Brewer, M. T., Stetler, G. L., Squires, C. H., Thompson, R. C., Busby, W. H., and Clemmons, D. R. (1988). Cloning, characterization and expression of a human insulin-like growth factor binding protein. *Biochem. Biophys. Res. Commun.* **152**, 1289–1297.

Budd, G. C., Pansky, B., and Cordell, B. (1986). Detection of insulin synthesis in mammalian anterior pituitary cells by immunohistochemistry and demonstration of insulin-related transcripts by *in situ* RNA–DNA hybridization. *J. Histochem. Cytochem.* **34**, 673–678.

Burgess, S. K., Jacobs, S., Cuatrecasas, P., and Sahyoun, N. (1987). Characterization of a neuronal subtype of insulin-like growth factor I receptor. *J. Biol. Chem.* **262**, 1618–1622.

Carlsson-Skwirut, C., Jornvall, H., Holmgren, A., Andersson, C., Bergman, T., Lundquist, G., Sjogren, B., and Sara, V. R. (1986). Isolation and characterization of variant IGF-I as well as IGF-II from adult human brain. *FEBS Lett.* **201**, 46–50.

Carlsson-Skwirut, C., Lade, M., Hartmanis, M., Hall, K., and Sara, V. R. (1989). A comparison of the biological activity of the recombinant intact and truncated insulin-like growth factor I (IGF-I). *Biochim. Biophys. Acta* **1011**, 192–197.

Caroni, P., and Grandes P. (1990). Nerve sprouting in innervated adult skeletal muscle induced by exposure to elevated levels of insulin-like growth factors. *J. Cell Biol.* **110**, 1307–1317.

Clarke, D. W., Mudd, L., Boyd, F. T., Jr., Fields, M., and Raizada, M. K. (1986). Insulin is released from rat brain neuronal cells in culture. *J. Neurochem.* **47**, 831–836.

Dahmer, M. K., and Perlman, R. L. (1988). Bovine chromaffin cells have insulin-like growth factor-I (IGF-I) receptors: IGF-I enhances catecholamine secretion. *J. Neurochem.* **51**, 321–323.

Danias, J., and Stylianopoulou, F. (1990). Expression of IGF-I and IGF-II genes in the adult rat eye. *Curr. Eye Res.* **9**, 379–386.
Daughaday, W. H., and Rotwein, P. (1989). Insulin-like growth factors I and II. Peptide, messenger ribonucleic acid and gene structures, serum and tissue concentrations. *Endocrine Rev.* **10**, 68–91.
Debbage, P. L. (1986). The generation and regeneration of oligodendroglia. *J. Neurol. Sci.* **72**, 319–336.
Devaskar, S. U., Holekamp, N., Karycki, L., and Devaskar, U. P. (1986). Ontogenesis of the insulin receptor in the rabbit brain. *Hormone Res.* **24**, 319–327.
Devenport, L. D., and Devenport, J. A. (1985). Adrenocortical hormones and brain growth: Reversibility and differential sensitivity during development. *Expt. Neurol.* **90**, 44–52.
DiCicco-Bloom, E., and Black, I. B. (1989). Depolarization and insulin-like growth factor-I (IGF-I) differentially regulate the mitotic cycle in cultured rat sympathetic neuroblasts. *Brain Res.* **491**, 403–406.
Duffy, K. R., and Pardridge, W. M. (1987). Blood brain barrier transcytosis of insulin in developing rabbits. *Brain Res.* **420**, 32–38.
Duffy, K. R., Pardridge, W. M., and Rosenfeld, R. G. (1988). Human blood brain barrier insulin-like growth factor receptor. *Metabolism* **37**, 136–140.
Ebina, Y., Ellis, L., Jarnagin, K., Edery, M., Graf, L., Clauser, E., Ou, H.-H., Masiarz, F., Kan, Y. W., Goldfine, I. D., Roth, R. A., and Rutter, W. J. (1985). The human insulin receptor cDNA: The structural basis for hormone-activated transmembrane signalling. *Cell* **40**, 747–758.
Fagin, J. A., Fernandez-Mejia, C., and Melmed, S. (1989a). Pituitary insulin-like growth factor-I gene expression: Regulation by triiodothyronine and growth hormone. *Endocrinology* **125**, 2385–2391.
Fagin, J. A., Roberts, C. T., Jr., LeRoith, D., and Brown, A. T. (1989b). Coordinate decrease of insulin-like growth factor-I post-transcriptional alternative mRNA transcripts in diabetes mellitus. *Diabetes* **38**, 428–434.
Gammeltoft, S., Staun-Olsen, P., Ottesen, B., and Fahrenkrug, J. (1984). Insulin receptors in rat brain cortex: Kinetic evidence for a receptor subtype in the central nervous system. *Peptides* **5**, 937–944.
Gammeltoft, S., Haselbacher, G. K., Humbel, R. E., Fehlmann, M., and Van Obberghen, E. (1985). Two types of receptor for insulin-like growth factors in mammalian brain. *EMBO J.* **4**, 3407–3412.
Giacobini, M. M. J., Olson, L., Hoffer, B. J., and Sara, V. R. (1990). Truncated IGF-I exerts trophic effects on fetal brain tissue grafts. *Exp. Neurol.* **108**, 33–37.
Grant, M. B., Guay, C., and Marsh, R. (1990). Insulin-like growth factor I stimulates proliferation, migration, and plasminogen activator release by human retinal pigment epithelial cells. *Curr. Eye Res.* **9**, 323–335.
Han, V. K. M., Lauder, J. M., and D'Ercole, A. J. (1988). Rat astroglial somatomedin/insulin-like growth factor binding proteins: Characterization and evidence of biological function. *J. Neurosci.* **8**, 3135–3143.
Hansson, H. A., Holmgren, A., Norstedt, G., and Rozell, B. (1989). Changes in the distribution of insulin-like growth factor-I, thioredoxin, thioredoxin reductase and ribonucleotide reductase during the development of the retina. *Exp. Eye Res.* **48**, 411–420.
Haselbacher, G., and Humbel, R. (1982). Evidence for two species of insulin-like growth factor II (IGF-II and "Big" IGF-II) in human spinal fluid. *Endocrinology* **110**, 1822–1827.
Haselbacher, G. K., Schwab, M. E., Pasi, A., and Humbel, R. E. (1985). Insulin-like growth factor II (IGF II) in human brain: Regional distribution of IGF II and of higher molecular mass forms. *Proc. Natl. Acad. Sci. U.S.A.* **82**, 2153–2157.
Havrankova, J., Schmechel, D., Roth, J., and Brownstein, M. (1978). Identification of insulin in rat brain. *Proc. Natl. Acad. Sci. U.S.A.* **75**, 5737–5741.
Heidenreich, K. A., and Brandenburg, D. (1986). Oligosaccharide heterogeneity of insulin receptors. Comparison of N-linked glycosylation of insulin receptors in adipocytes and brain. *Endocrinology* **118**, 1835–1842.
Heidenreich, K. A., and Toledo, S. P. (1989a). Insulin receptors mediate growth effects in cultured fetal neurons. I. Rapid stimulation of protein synthesis. *Endocrinology* **125**, 1451–1457.
Heidenreich, K. A., and Toledo, S. P. (1989b). Insulin receptors mediate growth effects in cultured fetal

neurons. II. Activation of a protein kinase that phosphorylates ribosomal protein S6. *Endocrinology* **125,** 1458–1463.

Heidenreich, K. A., Zahniser, N. R., Berhanu, P., Brandenburg, D., and Olefsky, J. M. (1983). Structural differences between insulin receptors in the brain and peripheral target tissues. *J. Biol. Chem.* **258,** 8527–8530.

Heidenreich, K. A., Freidenberg, G. R., Figlewicz, D. P., and Gilmore, P. R. (1986). Evidence for a subtype of insulin-like growth factor I receptor in brain. *Reg. Peptides* **15,** 301–310.

Heidenreich, K. A., Toledo, S. P., Brunton, L. L., Watson, M. J., Daniel-Issakani, S., and Strulovici, B. (1990). Insulin stimulates the activity of a novel protein kinase C, PKC-ε, in cultured fetal chick neurons. *J. Biol Chem.* **265,** 15076–15082.

Hendricks, S. A., Agardh, C.-D., Taylor, S. I., and Roth, J. (1984). Unique features of the insulin receptor in rat brain. *J. Neurochem.* **43,** 1302–1309.

Hill, J. M., Lesniak, M. A., Pert, C. B., and Roth, J. (1986). Autoradiographic localization of insulin receptors in rat brain: Prominence in olfactory and limbic areas. *Neurosci.* **17,** 127–138.

Hossenlopp, P., Seurin, D., Segovia-Quinson, B., and Binoux, M. (1986). Identification of an insulin-like growth factor binding protein in human cerebrospinal fluid with a selective affinity for IGF II. *FEBS Lett.* **208,** 439–444.

Hynes, M. A., Van Wyk, J. J., Brooks, P. J., D'Ercole, A. J., Jansen, M., and Lund, P. K. (1987). Growth hormone dependence of somatomedin-C/insulin-like growth factor-I and insulin-like growth factor-II messenger ribonucleic acids. *Mol. Endocrinol.* **1,** 233–242.

Hynes, M. A., Brooks, P. J., Van Wyk, J. J., and Lund, P. K. (1988). Insulin-like growth factor-II messenger ribonucleic acids are synthesized in the choroid plexus of the rat brain. *Mol. Endocrinol.* **2,** 47–54.

Ichimiya, Y., Emson, P. C., Northrop, A. J., and Gilmour, R. S. (1988). Insulin-like growth factor II in the rat choroid plexus. *Mol. Brain. Res.* **4,** 167–170.

Ishii, D. N. (1989). Relationship of insulin-like growth factor II gene expression in muscle to synaptogenesis. *Proc. Natl. Acad. Sci. U.S.A.* **86,** 2898–2902.

Ishii, D. N., and Recio-Pinto, E. (1987). Role of insulin, insulin-like growth factors and nerve growth factor in neurite formation. *In* "Insulin, Insulin-Like Growth Factors and Their Receptors in the CNS" (M. K. Raizada, M. I. Phillips, and D. LeRoith, eds.), pp. 315–348. Plenum Press, New York.

Kanje, M., Skottner, A., Sjoberg, J., and Lundborg, G. (1989). Insulin-like growth factor I (IGF-I) stimulates regeneration of the rat sciatic nerve. *Brain Res.* **486,** 396–398.

Kappy, M., Sellinger, S., and Raizada, M. (1984). Insulin binding in four regions of the developing rat brain. *J. Neurochem.* **42,** 198–203.

Kiess, W., Lee, L., Graham, D. E., Greenstein, L., Tseng, L. Y.-H., Rechler, M. M., and Nissley, S. P. (1989). Rat C6 glial cells synthesize insulin-like growth factor I (IGF-I) and express IGF-I receptors and IGF-II/mannose 6-phosphate receptors. *Endocrinology* **124,** 1727–1736.

King, G. L., and Johnson, S. M. (1985). Receptor-mediated transport of insulin across endothelial cells. *Science* **227,** 1583–1586.

Knusel, B., Michel, P. P., Schwaber, J. S., and Hefti, F. (1990). Selective and nonselective stimulation of central cholinergic and dopaminergic development in vitro by nerve growth factor, basic fibroblast growth factor, epidermal growth factor, insulin and the insulin-like growth factors I and II. *J. Neurosci.* **10,** 558–570.

Lamson, G., Pham, H., Oh, Y., Ocrant, I., Schwander, J., and Rosenfeld, R. G. (1989). Expression of the BRL-3A insulin-like growth factor binding protein (rBP-30) in the rat central nervous system. *Endocrinology* **123,** 1100–1102.

Lauterio, T. J., Aravich, P. F., and Rotwein, P. (1990). Divergent effects of insulin on insulin-like growth factor-II gene expression in the rat hypothalamus. *Endocrinology* **126,** 392–398.

Lee, Y. L., Hintz, R. L., James, P. M., Lee, P. D. K., Shively, J. E., and Powell, D. R. (1988). Insulin-like growth factor (IGF) binding protein complementary deoxyribonucleic acid from human Hep G2 hepatoma cells: Predicted protein sequence suggests an IGF binding domain different from those of the IGF-I and IGF-II receptors. *Mol. Endocrinol.* **2,** 404–411.

Lenoir, D., and Honegger, P. (1983). Insulin-like growth factor I (IGF-I) stimulates DNA synthesis in fetal rat brain cell cultures. *Dev. Brain Res.* **7**, 205–213.

Lesniak, M. A., Hill, J. M., Kiess, W., Rojeski, M., Pert, C. B., and Roth, J. (1988). Receptors for insulin-like growth factors I and II: Autoradiographic localization in rat brain and comparison to receptors for insulin. *Endocrinology* **123**, 2089–2099.

Levinovitz, A., and Norstedt, G. (1989). Developmental and steroid hormonal regulation of insulin-like growth factor-II expression. *Mol. Endocrinol.* **3**, 797–804.

Lowe, W. L., Jr., Boyd, F. T., Clarke, D. W., Raizada, M. K., Hart, C., and LeRoith, D. (1986). Development of brain insulin receptors: Structural and functional studies of insulin receptors from whole brain and primary cell cultures. *Endocrinology* **119**, 25–35.

Lowe, W. L., Jr., Adamo, M., Werner, H., Roberts, C. T., Jr., and LeRoith, D. (1989). Regulation by fasting of rat insulin-like growth factor I and its receptor: Effects on gene expression and binding. *J. Clin. Invest.* **84**, 619–626.

Luo, J., and Murphy, L. J. (1989). Dexamethasone inhibits growth hormone induction of insulin-like growth factor-I (IGF-I) messenger ribonucleic acid (mRNA) in hypophysectomized rats and reduces IGF-I mRNA abundance in the intact rat. *Endocrinology* **125**, 165–171.

McCusker, R. H., Camacho-Hubner, C., Bayne, M. L., Cascieri, M. A., and Clemmons, D. R. (1990). Insulin-like growth factor (IGF) binding to human fibroblast and glioblastoma cells: The modulating effect of cell released IGF binding proteins (IGFBPs). *J. Cell. Physiol.* **144**, 244–253.

McElduff, A., Poronnik, P., and Baxter, R. C. (1987). The insulin-like growth factor-II (IGF II) receptor from rat brain is of lower apparent molecular weight than the IGF II receptor from rat liver. *Endocrinology* **121**, 1306–1311.

McElduff, A., Poronnik, P., and Baxter, R. C. (1988). A comparison of the insulin and insulin-like growth factor-I receptors from rat brain and liver. *Endocrinology,* **122**, 1933–1939.

McMorris, F. A., and Dubois-Dalcq, M. (1988). Insulin-like growth factor I promotes cell proliferation and oligodendroglial commitment in rat glial progenitor cells developing in vitro. *J. Neurosci. Res.* **21**, 199–209.

McMorris, F. A., Smith, T. M., DeSalvo, S., and Furlanetto, R. W. (1986). Insulin-like growth factor I/somatomedin C: A potent inducer of oligodendrocyte development. *Proc. Natl. Acad. Sci. U.S.A.* **83**, 822–826.

Martin, J. L., and Baxter, R. C. (1986). Insulin-like growth factor-binding protein from human plasma. Purification and characterization. *J. Biol. Chem.* **261**, 8754–8760.

Mathews, L. S., Norstedt, G., and Palmiter, R. D. (1986). Regulation of insulin-like growth factor I gene expression by growth hormone. *Proc. Natl. Acad. Sci. U.S.A.* **83**, 9343–9347.

Merrill, M. J., and Edwards, N. A. (1990). Insulin-like growth factor-I receptors in human glial tumors. *J. Clin. Endocrinol. Metab.* **71**, 199–209.

Moss, A. M., Unger, J. W., Moxley, R. T., and Livingston, J. N. (1990). Location of phosphotyrosine-containing proteins by immunocytochemistry in the rat forebrain corresponds to the distribution of the insulin receptor. *Proc. Natl. Acad. Sci. U.S.A.* **87**, 4453–4457.

Nilsson, L., Sara, V. R., and Nordberg, A. (1988). Insulin-like growth factor I stimulates the release of acetylcholine from rat cortical slices. *Neurosci. Lett.* **88**, 221–226.

Noguchi, T., Kurata, L. M., and Sugisaki, T. (1987). Presence of somatomedin-C-immunoreactive substance in the central nervous system: Immunohistochemical mapping studies. *Neuroendocrinol.* **46**, 277–282.

Ocrant, I., Pham, H., Oh, Y., and Rosenfeld, R. G. (1989). Characterization of insulin-like growth factor binding proteins of cultured rat astroglial and neuronal cells. *Biochem. Biophys. Res. Commun.* **159**, 1316–1322.

Ocrant, I., Fay, C. T., and Parmelee, J. T. (1990). Characterization of insulin-like growth factor binding proteins produced in the rat central nervous system. *Endocrinology* **127**, 1260–1267.

Ocrant, I., Fay, C. T., and Parmelee, J. T. (1991). Expression of insulin and insulin-like growth factor receptors and binding proteins by retinal pigment epithelium. *Exp. Eye Res.* **52**, 581–589.

Okamoto, T., Katada, T., Murayama, Y., Ui, M., Ogata, E., and Nishimoto, I. (1990). A simple structure encodes G protein-activating function of the IGF-II/mannose-6-phosphate receptor. *Cell* **62**, 709–717.

Okamoto, T., Asano, T., Harada, S. I., Ogata, E., and Nishimoto, I. (1991). Regulation of transmembrane signal transduction of insulin-like growth factor II by competence type growth factors or viral *ras* p21. *J. Biol. Chem.* **266**, 1085–1091.

Orlowski, C. C., Brown, A. L., Ooi, G. T., Yang, Y. W.-H., Tseng, L. Y.-H., and Rechler, M. M. (1990). Tissue, developmental, and metabolic regulation of messenger ribonucleic acid encoding a rat insulin-like growth factor-binding protein. *Endocrinology* **126**, 644–652.

Ota, A., Shemer, J., Pruss, R. M., Lowe, W. L., Jr., and LeRoith, D. (1988a). Characterization of the altered oligosaccharide composition of the insulin receptors on neural derived cells. *Brain Res.* **443**, 1–11.

Ota, A., Wilson, G. L., Spilberg, O., Pruss, R., and LeRoith, D. (1988b). Functional insulin-like growth factors I receptors are expressed by neural-derived continuous cell lines. *Endocrinology* **122**, 145–152.

Pardridge, W. M., Boado, R. J., and Farrell, C. R. (1990). Brain-type glucose transporter (GLUT-1) is selectively localized to the blood–brain barrier. *J. Biol. Chem.* **265**, 18035–18040.

Rechler, M. M., and Nissley, S. P. (1985). The nature and regulation of the receptors for insulin-like growth factors. *Ann. Rev. Physiol.* **47**, 425–442.

Recio-Pinto, E., Rechler, M. M., and Ishii, D. N. (1986). Effects of insulin, insulin-like growth factor-II and nerve growth factor on neurite formation and survival in cultured sympathetic and sensory neurons. *J. Neurosci.* **6**, 1211–1219.

Roberts, C. T., Jr., and LeRoith, D. (1988). Molecular aspects of the insulin-like growth factors, their binding proteins and receptors. *Balliere's Clin. Endocrinol. Metab.* **2**, 1069–1085.

Roghani, M., Hossenlopp, P., Lepage, P., Balland, A., and Binoux, M. (1989). Isolation from human cerebrospinal fluid of a new insulin-like growth factor-binding protein with a selective affinity for IGF-II. *FEBS Lett.* **255**, 253–258.

Romanus, J. A., Tseng, L. Y.-H., Yang, Y. W.-H., and Rechler, M. M. (1989). The 34 kilodalton insulin-like growth factor binding proteins in human cerebrospinal fluid and the A673 rhabdomyosarcoma cell line are human homologues of the rat BRL-3A binding protein. *Biochem. Biophys. Res. Commun.* **163**, 875–881.

Rosen, O. M. (1987). After insulin binds. *Science* **237**, 1452–1458.

Rosenzweig, S. A., Zetterstrom, C., and Benjamin, A. (1990). Identification of retinal insulin receptors using site-specific antibodies to a carboxyl-terminal peptide of the human insulin receptor α subunit. Up-regulation of neuronal insulin receptors in diabetes. *J. Biol. Chem.* **265**, 18030–18034.

Rotwein, P., Burgess, S. K., Milbrandt, J. D., and Krause, J. E. (1988). Differential expression of insulin-like growth factor genes in rat central nervous system. *Proc. Natl. Acad. Sci. U.S.A.* **85**, 265–269.

Sadiq, F., Holtzclaw, L., Chundu, K., Muzzafar, A., and Devaskar, S. (1990). The ontogeny of the rabbit brain glucose transporter. *Endocrinology* **126**, 2417–2424.

Saneto, R. P., and deVellis, J. (1987). Neuronal and glial cells: Cell cultures of the central nervous system. *In* "Neurochemistry, A Practical Approach" (A. J. Turner and H. S. Bacheland, eds.), pp. 27–63. IRL Press, Washington, D.C.

Saneto, R. P., Low, K. G., Melner, M. H., and deVellis, J. (1988). Insulin/insulin-like growth factor I and other epigenetic modulators of myelin basic protein expression in isolated oligodendrocyte progenitor cells. *J. Neurosci. Res.* **21**, 210–219.

Sara, V., and Hall, K. (1990). Insulin-like growth factors and their binding proteins. *Physiol. Rev.* **70**, 591–614.

Sara, V. R., Uvnas-Moberg, K., Uvnas, B., Hall, K., Wetterberg, L., Posslancec, B., and Goiny, M. (1982). The distribution of somatomedins in the nervous system of the cat and their release following neural stimulation. *Acta Physiol. Scand.* **115**, 467–470.

Sara, V. R., Hall, K., Misaki, M., Fryklund, L., Christensen, N., and Wetterberg, L. (1983). Ontogenesis of somatomedin and insulin receptors in the human fetus. *J. Clin. Invest.* **71**, 1084–1094.

Sara, V. R., Carlsson-Skwirut, C., Andersson, C., Hall, K., Sjogren, B., Holmgren, A., and Jornvall, H. (1986). Characterization of somatomedins from human fetal brain: Identification of a variant form of insulin-like growth factor I. *Proc. Natl. Acad. Sci. U.S.A.* **83,** 4904–4907.

Sara, V. R., Carlsson-Skwirut, C., Bergman, T., Jornvall, H., Roberts, P. J., Crawford, M., Hakansson, L. N., Civalero, I., and Nordberg, A. (1989). Identification of gly-pro-glu (GPE), the aminoterminal tripeptide of insulin-like growth factor I which is truncated in brain, as a novel neuroactive peptide. *Biochem. Biophys. Res. Commun.* **165,** 766–771.

Schechter, R., Holtzclaw, L., Sadiq, F., Kahn, A., and Devaskar, S. (1988). Insulin synthesis by isolated rabbit neurons. *Endocrinology* **123,** 505–513.

Seino, S., Seino, M., Nishi, S., and Bell, G. I. (1989). Structure of the human insulin receptor gene and characterization of its promoter. *Proc. Natl. Acad. Sci. U.S.A.* **86,** 114–118.

Shemer, J., Adamo, M., Wilson, G. L., Heffez, D., Zick, Y., and LeRoith, D. (1987a). Insulin and insulin-like growth factor-I stimulate a common endogenous phosphoprotein substrate (pp185) in intact neuroblastoma cells. *J. Biol. Chem.* **262,** 15476–15482.

Shemer, J., Raizada, M. K., Masters, B. A., Ota, A., and LeRoith, D. (1987b). Insulin-like growth factor I receptors in neuronal and glial cells. Characterization and biological effects in primary culture. *J. Biol. Chem.* **262,** 6793–6799.

Shemer, J., Adamo, M., Raizada, M. K., Heffez, D., Zick, Y., and LeRoith, D. (1989). Insulin and IGF-I stimulated phosphorylation of their respective receptors in intact neuronal and glial cells in primary culture. *J. Mol. Neurosci.* **1,** 3–8.

Shimasaki, S., Uchiyama, F., Shimanaka, M., and Ling, N. (1991). Molecular cloning of the cDNAs encoding a novel insulin-like growth factor binding protein from rat and human. *Mol. Endocrinol.* **4,** 1451–1458.

Simpson, I. A., and Cushman, S. W. (1986). Hormonal regulation of mammalian glucose transport. *Ann. Rev. Biochem.* **55,** 1059–1089.

Sjoberg, J., and Kanje, M. (1989). Insulin-like growth factor (IGF-I) as a stimulator of regeneration in the freeze-injured rat sciatic nerve. *Brain Res.* **485,** 102–108.

Strauss, D. S., and Takemoto, C. (1990). Effect of dietary protein deprivation on insulin-like growth factor (IGF)-I and II, IGF binding protein-2 and serum albumin gene expression in rat. *Endocrinology* **127,** 1849–1860.

Stylianopoulou, F., Herbert, J., Soares, M. B., and Efstratiadis, A. (1988). Expression of the insulin-like growth factor II gene in the choroid plexus and the leptomeninges of the adult rat central nervous system. *Proc. Natl. Acad. Sci. U.S.A.* **85,** 141–145.

Svrzic, D., and Schubert, D. (1990). Insulin-like growth factor I supports embryonic nerve cell survival. *Biochem. Biophys. Res. Commun.* **172,** 54–60.

Torres-Aleman, I., Naftolin, F., and Robbins, R. J. (1990). Trophic effects of basic fibroblast growth factor on fetal rat hypothalamic cells: Interactions with insulin-like growth factor I. *Dev. Brain. Res.* **52,** 253–257.

Treadway, J. L., Morrison, B. D., Goldfine, I. D., and Pessin, J. E. (1989). Assembly of insulin/insulin-like growth factor-I receptors *in vitro*. *J. Biol. Chem.* **264,** 21450–21453.

Tseng, L. Y.-H., Brown, A. L., Yang, Y. W.-H., Romanus, J. A., Orlowski. C. C., Taylor, T., and Rechler, M. M. (1989). The fetal rat binding protein for insulin-like growth factors is expressed in the choroid plexus and cerebrospinal fluid of adult rats. *Mol. Endocrinol.* **3,** 1559–1568.

Ullrich, A., Bell, J. R., Chen, E. Y., Herrera, R., Petruzzelli, L. M., Dull, T. J., Gray, A., Coussens, L., Liao, Y.-C., Toubokawa, M., Mason, A., Seeburg, P. H., Grundfeld, C., Rosen, O. M. and Ramachandran, J. (1985). Human insulin receptor and its relationship to the tyrosine kinase family of oncogenes. *Nature (London)* **313,** 756–761.

Ullrich, A., Gray, A., Tam, A. W., Yang-Feng, T., Tsubokawa, M., Collins, M., Henzel, W., KeBon, T., Kathuria, S., Chem, E., Jacobs, S., Francke, U., Ramachandran, J., and Fujita-Yamaguchi, Y. (1986). Insulin-like growth factor I receptor primary structure: Comparison with insulin receptor suggests structural determinants that define functional specificity. *EMBO J.* **5,** 2503–2512.

Valentino, K. L., Ocrant, I., and Rosenfeld, R. G. (1990). Developmental expression of insulin-like growth factor-II receptor immunoreactivity in the rat central nervous system. *Endocrinology* **126,** 914–920.

Van der Pal, R. H. M., Koper, J. W., VanGolde, L. M. G., and Lopes-Cardozo, M. (1988). Effects of insulin and insulin-like growth factor-I (IGF-I) on oligodendrocyte-enriched glial cultures. *J. Neurosci. Res.* **19,** 483–490.

Waldbillig, R. J., Fletcher, R. T., Chader, G. J., Rajagopalan, S., Rodrigues, M., and LeRoith, D. (1987a). Retinal insulin receptors. 1. Structural heterogeneity and functional characterization. *Exp. Eye Res.* **45,** 823–835.

Waldbillig, R. J., Fletcher, R. T., Chader, G. J., Rajagopalan, S., Rodrigues, M., and LeRoith, D. (1987b). Retinal insulin receptors. 2. Characterization and insulin-induced tyrosine kinase activity in bovine retinal rod outer segments. *Exp. Eye Res.* **45,** 837–844.

Waldbillig, R. J., Schoen, T. J., Chader, G. J., and Pfeffer, B. A. (1992). Monkey retinal pigment epithelial cells *in vitro* synthesize, secrete, and degrade insulin-like growth factor binding proteins. *J. Cell. Physiol.* **150,** 76–83.

Walicke, P. A., Feige, J.-J., and Baird, A. (1989). Characterization of the neuronal receptor for basic fibroblast growth factor and comparison to receptors on mesenchymal cells. *J. Biol. Chem.* **264,** 4120–4126.

Walker, P. S., Donovan, J. A., Van Ness, B. G., Fellows, R. E., and Pessin, J. E. (1988). Glucose-dependent regulation of glucose transport activity, protein and mRNA in primary cultures of rat brain glial cells. *J. Biol. Chem.* **263,** 15594–15601.

Werner, H., Adamo, M., Lowe, W. L., Jr., Roberts, C. T., Jr., and LeRoith, D. (1989a). Developmental regulation of rat brain/HepG2 glucose transporter gene expression. *Mol. Endocrinol.* **3,** 273–279.

Werner, H., Raizada, M. K., Mudd, L. M., Foyt, H. L., Simpson, I. A., Roberts, C. T., Jr., and LeRoith, D. (1989b). Regulation of rat brain/HepG2 glucose transporter gene expression by insulin and insulin-like growth factor-I in primary cultures of neuronal and glial cells. *Endocrinology* **125,** 314–320.

Werner, H., Woloschak, M., Adamo, M., Shen-Orr, Z., Roberts, C. T., Jr., and LeRoith, D. (1989c). Developmental regulation of the rat insulin-like growth factor I receptor gene. *Proc. Natl. Acad. Sci. U.S.A.* **86,** 7451–7455.

Werther, G. A., Abate, M., Hogg, A., Cheesman, H., Oldfield, B., Hards, D., Hudson, P., Power, B., Freed, K., and Herington, A. C. (1990). Localization of insulin-like growth factor-I mRNA in rat brain by in situ hybridization—relationship to IGF-I receptors. *Mol. Endocrinol.* **4,** 773–778.

Wood, T. L., Brown, A. L., Rechler, M. M., and Pintar, J. E. (1990). The expression pattern of an insulin-like growth factor (IGF)-binding protein gene is distinct from IGF-II in the midgestational rat embryo. *Mol. Endocrinol.* **4,** 1257–1263.

Wood, W. I., Cachianes, G., Henzel, W. J., Winslow, G. A., Spencer, S. A., Hellmiss, R., Martin, J. L., and Baxter, R. C. (1988). Cloning and expression of the growth hormone-dependent insulin-like growth factor-binding protein. *Mol. Endocrinol.* **2,** 1176–1185.

Young, W. S., III (1986). Periventricular hypothalamic cells in the rat brain contain insulin mRNA. *Neuropeptides* **8,** 93–97.

# 14 Neurobiology of Insulin and Insulin-Like Growth Factors

Douglas N. Ishii

## I. INTRODUCTION

The extraordinary success of higher life is in no small measure due to the neural circuitry, which permits organisms to adapt their behavior in response to changes in their internal and external environment. The repertoire of permissible behavior is dependent on the underlying structure of the neural network. It is evident that epigenetic factors determine this structure, because the size of the mammalian genome ($10^5$ to $10^6$ genes) is many orders of magnitude too small to encode individual synapses directly (estimated as $> 10^{12}$ to $10^{14}$).

This chapter will have as its focus the observations that support the hypothesis that the insulin-like growth factors (IGFs) are contributors to the development of the neural circuitry. IGF genes are proposed to be prototypical mammalian genes that play a significant role in both the formation and the regeneration of neuromuscular synapses. A model for the regulation of IGF gene expression by neuromuscular activity is advanced, and this model might explain how activity modulates these synapses. Other aspects of IGF neurobiology, and relationship to pathophysiology, will be considered more briefly.

Recent findings in the nerve growth factor (NGF) field have brought a significant challenge to the classic neurotrophic theory. The neurobiology of insulin/IGFs will be discussed in the context of this theory. Finally, the dim outline of the long hidden biochemical pathway for neurite formation is now better perceived. This pathway is shared among NGF, insulin, and IGFs, and two dominant mechanisms appear to be used to up-regulate the expression of genes encoding major proteins found in axons and dendrites.

## II. NEURAL CIRCUITRY AND INSULIN-LIKE GROWTH FACTORS

The mechanisms concerned with the development and regeneration of vertebrate synapses long have remained shrouded in mystery. Although their neurophysiologic role is still a matter for further exploration, one hypothesis is that insulin homologs play a role in the development, maintenance, and regulation of the neural circuitry (Recio-Pinto and Ishii, 1984; Ishii et al., 1985). This hypothesis was based initially on in vitro observations, but has been strengthened significantly through in vivo studies.

### A. Neurite Outgrowth in Vitro

Insulin and IGFs can increase neurite outgrowth in cultured peripheral and central neurons. Supraphysiologic concentrations of insulin can enhance neurite growth in motor (Caroni and Grandes, 1990), parasympathetic (Collins and Dawson, 1983), and sensory (Bothwell, 1982; Recio-Pinto et al., 1986) neurons. High concentrations may do so as a result of cross-occupancy of type I IGF receptors. Insulin does not, however, cross-occupy type II IGF receptors.

On the other hand, physiologic concentrations of insulin can increase neurite outgrowth in sympathetic neurons (Recio-Pinto et al., 1986) and human neuroblastoma SH-SY5Y cells (Recio-Pinto and Ishii, 1984). Cloned SH-SY5Y neuroblastoma cells (Biedler et al., 1978) are most likely of sympathetic origin (Imashuku et al., 1975) and, as a consequence, may share the same sensitivity to insulin as sympathetic neurons.

Physiologic concentrations of IGF-I and IGF-II can induce neurite outgrowth in cultured motor (Caroni and Grandes, 1990), sensory (Bothwell, 1982; Recio-Pinto et al., 1986), sympathetic (Recio-Pinto et al., 1986), and SH-SY5Y (Recio-Pinto and Ishii, 1984; Ishii and Recio-Pinto, 1987) cells. The proportion of cells with neurites and the average length of neurites are increased (Recio-Pinto and Ishii, 1984).

Insulin and IGFs clearly act directly on neurons. Neurite outgrowth can be elicited even in cultures essentially devoid of nonneurons (Recio-Pinto et al., 1986), as well as in cloned neuroblastoma cells (Recio-Pinto and Ishii, 1984). Neurites can be induced reversibly by insulin homologs under serum-free culture conditions, which shows that other factors are not required. It will be seen that insulin homologs support survival, but the effects on neurite outgrowth are not secondary to survival (discussed in Recio-Pinto et al., 1986). In addition, although insulin and IGFs are mitogens, neurite outgrowth is induced in postmitotic neurons and in neuroblastoma cells, irrespective of whether these cells are in the logarithmic or the stationary phase of growth. Neurite outgrowth is a plastic property of neurons and subject to modulation in mature vertebrates.

## B. Development of Neuromuscular Synapses

In vitro studies alone cannot prove that insulin and IGFs play a significant neurophysiologic role. For that reason, attention was focused next on the neuromuscular junction (NMJ), the best characterized synapse. IGF-II transcripts are high in fetal, but low in adult, muscle (Soares et al., 1985, 1986). Might these transcripts be involved in the developmental innervation of skeletal muscle?

### 1. Prenatal development of polyneuronal innervation

Discrete pools of motor neurons connect to specific muscles, a topographic arrangement that seems to have arisen at about the time of evolution of amniotes from earlier vertebrates (Fetcho, 1987). The precise mechanism by which specific motor neurons reach their appropriate targets is still unknown. The relocation of spinal cord segments (with a particular pool of motor neurons) to ectopic spinal cord regions does not prevent motor axons from reaching their appropriate targets in chicks (Lance-Jones and Landmesser, 1980, 1981). Similarly, motor axons find their targets after surgical deflection of nerves in axolotls (Cass and Mark, 1975). A unique pathway does not appear to specify neuronal connections. Neurotrophic factors may help guide axons to their appropriate targets, but the data are by no means conclusive. Excellent review articles on the structure (Carry and Morita, 1984; Salpeter, 1987), development (Landmesser, 1980; Lomo and Jansen, 1980; Dennis, 1981; Betz, 1987), and physiology (McArdle, 1984) of the NMJ are available.

The IGF-II gene is expressed in muscle in a manner closely correlated with the development of the neuromuscular synapse (Ishii, 1989). Unspecialized motor endplates are observed first on embryonic day (ED) 16 in rats (Kelly and Zacks, 1969). In order for the IGF-II gene to be instructive for synaptogenesis, it would have to be expressed prior to the formation of the first NMJ. In fulfillment of this requirement, IGF-II transcripts are synthesized already in ED 14 limb buds. Polyneuronal innervation is a condition in which a single muscle fiber is innervated by more than one motor neuron (Redfern, 1970). The number of motor neurons forming synapses at a single endplate increases from ED 16 until approximately the time of birth (Brown et al., 1976; Dennis et al., 1981). During this interval, there is a 5-fold increase in the IGF-II mRNA content of developing muscle (Ishii, 1989). IGF-I mRNA also increases in muscle between ED 14 and birth, suggesting that IGF-I and IGF-II may have cooperative effects (Glazner and Ishii, 1989).

The timing of synaptogenesis is set potentially by the fusion of myoblasts and the differentiation of myotubes, events that trigger IGF gene expression. The high level of IGF gene expression might help attract growing axons and predispose toward relatively indiscriminate formation of synapses on uninnervated as well as on previously innervated myotubes, resulting in polyneuronal innervation.

Histochemical data show that IGF-I immunoreactivity is low or absent in mononucleated cells, high in myotubes, and low in mature muscle fibers in neonatal rats (Jennische and Olivecrona, 1987). This immunoreactivity is likely to arise from the IGF-I transcripts produced in muscle cells. Levels of these mRNAs are increased during differentiation of myoblasts in culture (Tollefsen et al., 1989). In addition, IGF-II mRNAs are localized by in situ hybridization to regions of myoblasts, myotubes, and developing skeletal muscle fibers in rats (Beck et al., 1987).

As axons attracted or induced to sprout by IGFs advance over the surface of myotubes, agrin, which potentially may be released from motor axons, might promote acetylcholine receptor clustering and other aspects of postsynaptic cell specialization at endplates (McMahan and Wallace, 1989).

### 2. Postnatal elimination of polyneuronal innervation

Polyneuronal innervation may originate as a means of insuring that all available target cells receive at least one synapse during early prenatal development. However, greater precision in control of movement in higher vertebrates is attained, in part, by the postnatal elimination of superfluous synaptic contacts. Multiple synapses begin to be eliminated at about the time of birth, and the mature state, in which a single motor neuron innervates a single muscle fiber, is attained by postnatal day (PD) 15 in rats (Brown et al., 1976; O'Brien et al., 1978). Elimination of multiple synapses and down-regulation of IGF-II mRNA levels follow the same developmental time course (Ishii, 1989). Postnatal down-regulation of IGF-I mRNA levels is observed also (Glazner and Ishii, 1989).

The hypothesis that increased neuromuscular activity associated with synapse maturation causes feedback inhibition of IGF-II transcript levels in muscle is being tested. Multiple synapses may be eliminated because they can no longer be supported by the decreased levels of IGFs. The lone remaining axon may no longer require high levels of IGFs once a strong synaptic contact is established, because of the close proximity between its terminal and the site of IGF release.

This model is consistent with the observation that innervation of tissue, although distinctly not random, is still somewhat imprecise. There is preferential elimination of inappropriate superfluous synapses, improving specificity (Betz et al., 1990; Van Essen et al., 1990). It is unclear how preferential elimination is brought about, but one possibility is that correct axons release small amounts of factors that are weakly inhibitory to incorrect axons or to their sites of adhesion. Alternatively, the muscle surface contains adhesion sites that bind strongly to correct, but weakly to incorrect, axons. As the high IGF content in muscle begins to wane during postnatal development, these weakly inhibitory axonal factors, adhesive muscle factors, or their combination may come into play to eliminate unwanted synapses. Elements of competition would be observed between incorrect and correct axons but, in the absence of competition, vacant endplate sites would tend to accept or retain even incorrect terminals.

### 3. Neuromuscular activity modulates both polyneuronal innervation and IGF gene expression

The postnatal elimination of polyneuronal innervation is influenced by neuromuscular activity. For example, electrical stimulation of nerve (O'Brien et al.,

1978) or muscle (Thompson, 1983) results in accelerated loss of polyneuronal innervation in rats. On the other hand, presynaptic blockade with botulinum toxin (Thompson et al., 1979; Brown et al., 1981b) or postsynaptic blockage with curare (Ding et al., 1983) or alpha-bungarotoxin (Sohal et al., 1979; Holland and Brown, 1980) retards loss of polyneuronal innervation in avian muscle.

Although the first synapses are formed at ED 15–16, spontaneous transmitter release remains low until about the first postnatal week of life. Then there is approximately a 100-fold increase in miniature endplate potentials (Diamond and Miledi, 1962; Nakajima et al., 1980), and the amount of acetylcholine released per stimulus increases 5- to 10-fold (Betz et al., 1989). This substantial increase in neuromuscular activity might be the postnatal maturation signal responsible for down-regulation of IGF-II gene expression, leading to loss of polyneuronal innervation (Ishii, 1989).

To study whether neuromuscular activity inhibits IGF-II gene expression, left but not right sciatic nerves were transected. IGF-II mRNA content was observed to increase in ipsilateral denervated, but not contralateral intact, muscle (Ishii, 1989). This study showed that neuromuscular activity indeed is a feedback inhibitory signal.

The normal postnatal decline of IGF-II transcripts did not occur. These transcripts remained elevated in rat muscle following presynaptic blockade with botulinum toxin (Ishii, 1990). Botulinum toxin inhibits release of acetylcholine, but has no detectable effect on the ultrastructure of nerve terminals (Zacks et al., 1962; Jirmanova et al., 1964). These results show that the physical contact between terminals and endplate per se is unlikely to regulate IGF-II gene expression, but interruption of the release of acetylcholine or other substances from nerve terminals may be important. Botulinum toxin not only prevents the postnatal decline in IGF-II mRNAs in muscle, but also retards the elimination of polyneuronal innervation (Thompson et al., 1979; Brown et al., 1981b).

## C. Regeneration of Motor Synapses

Mature muscle will not accept additional innervation from a foreign nerve, but will do so following denervation. Rerouted nerves may form synapses at ectopic sites far from the original endplates on denervated muscle. Immature and denervated muscles both can accept innervation and share a biochemical state quite different from that of mature muscles. Is a high content of IGF-II mRNA a common denominator?

Transection of the sciatic nerve results in selective up-regulation of IGF-II mRNA in denervated muscles of 20-day-old rats (Ishii, 1989). Up-regulation also occurs in denervated muscles of mature rats (D. N. Ishii, unpublished observations). The high content of IGF-II correlates with the capacity of denervated muscle to stimulate regeneration of synapses. If high IGF-II mRNA content is the cause of polyneuronal innervation during development, one might expect to observe transient polyneuronal innervation while IGF-II mRNAs are elevated during regeneration; this indeed is observed (McArdle, 1975; Benoit and Changeux, 1978; Taxt, 1983).

### D. Sprouting of Motor Nerve Terminals

One prediction of the Ishii (1989) hypothesis is that administration of IGFs to mammals would promote nerve sprouting. After all, IGFs increase neurite outgrowth in cultured cells and IGF gene expression is correlated with synaptogenesis. It is, moreover, highly pertinent that botulinum toxin increases both IGF-II mRNA content (Ishii, 1990) and sprouting (Duchen and Strich, 1968). This prediction was tested by Caroni and Grandes (1990), who injected IGF-I or IGF-II into rats. Motor nerve terminal sprouting was observed, with increased GAP43 immunoreactivity, validating the prediction.

The Ishii (1989) hypothesis is of heuristic value in understanding other phenomena concerned with sprouting. For example, denervation of some muscle fibers promotes sprouting from intact axons terminating on adjacent muscle fibers (Edds, 1950; Hoffman, 1950). Sprouting resulting in polyneuronal innervation is also stimulated when nerve activity is blocked with a cuff of tetrodotoxin (Brown and Ironton, 1977; Taxt, 1983). These treatments block neuromuscular activity and are predicted to increase IGF-II mRNA content and stimulate sprouting. The local increase in IGFs also would explain the observation that sprouting is generally of two types: terminal sprouting from axon terminals and collateral sprouting from intramuscular nodes of Ranvier. Both forms of sprouting are consistent with responses to a diffusible substance (Slack et al., 1979). The interesting literature on motor nerve sprouting has been reviewed (Brown et al., 1981a).

### E. Relationship between Synapse Development, Regeneration, and Sprouting

Heretofore, the development of neuromuscular synapses, regeneration of synapses, and sprouting of motor nerve terminals have, by and large, been treated as separate topics in the literature. This seemed reasonable because the underlying mechanisms were not known and might have been different for each case. However, a role for IGFs in each of these events has been uncovered, and the expression of IGF genes may be the recurrent theme that ties together these seemingly disparate phenomena. In this light, then, it is not surprising that transient polyneuronal innervation is observed during synapse regeneration (McArdle, 1975; Benoit and Changeux, 1978; Brown and Ironton, 1979), as during development, because IGF transcripts are elevated transiently in muscle, during regeneration (G. W. Glazner and D. N. Ishii, unpublished observations), as well as during development.

## F. Regeneration of Sensory Axons

IGF-I immunoreactivity is increased on Schwann cells near the site of crush lesion, suggesting that IGF-I might play a role in sciatic nerve regeneration (Hansson et al., 1986). Growth hormone is known to regulate IGF-I levels, and infusion of growth hormone increases the regeneration distance of sensory axons in crushed sciatic nerves of normal as well as hypophysectomized rats (Kanje et al., 1988). Infusion of IGF-I increases the distance of regeneration of sensory axons, whereas anti-IGF-I antiserum inhibits spontaneous regeneration (Kanje et al., 1989). It is not presently known whether IGF-I decreases the time to onset of regeneration or increases the rate of regeneration.

## III. IGFs AND INSULIN VIEWED WITHIN THE CONTEXT OF NEUROTROPHIC THEORY

### A. NGF

Surprisingly, recent studies have shown that NGF mRNA and protein production in whisker pads, as well as expression of NGF receptors on the neurons, is initiated only *after* sensory efferents from the trigeminal ganglion reach these targets (Davies et al., 1987). This supports a role for NGF in survival, but not in chemotactic guidance, at least for these sensory neurons. These findings do not rule out the possibility that other neurotrophic factors are involved.

It is curious that NGF mRNA and protein are elevated in distal sciatic nerve segments below a crush site (Heumann et al., 1987), yet infusion of NGF does not increase the distance of sensory axon regeneration in sciatic nerve (Kanje et al., 1989), probably because NGF receptors are down-regulated and net retrograde transport actually is decreased (Raivich et al., 1990).

These observations challenge the long cherished belief that NGF plays a chemotactic role in guidance of axons to targets during development. On the other hand, it must be kept in mind that much less is understood about the role of NGF in guiding developing sympathetic axons. It is well known that administration of NGF to adult vertebrates can cause sprouting and growth of long sympathetic fibers (reviewed in Levi-Montalcini and Angeletti, 1968; Mobley et al., 1977) and can attract sympathetic axons chemotactically in vivo (Menesini-Chen et al., 1978).

### B. IGFs and Insulin

Despite the uncertainties just mentioned, the neurotrophic theory developed around NGF still is used as a basis for evaluating the role of more recently discovered neuroactive factors. Like NGF, IGFs are produced and released from target tissues. There is a close correlation between IGF-II gene expression in muscle and the developmental formation of neuromuscular synapses.

Whether IGFs, like NGF, are chemotactic for advancing axons is not yet known. However, there is a gradient of IGFs close to the surface of muscle. Circulating IGFs are inactivated as a result of sequestration by IGF binding proteins, which have differential affinity for IGF-I and IGF-II (Hardouin et al., 1987). Cerebrospinal fluid is enriched particularly in a binding protein that selectively binds to IGF-II (Hossenlopp et al., 1986). Binding proteins would sequester IGFs quickly within a short distance of their muscle release sites and produce a gradient of free IGFs, which is precisely what is needed to attract, support, or maintain nerve terminals, particularly during development and nerve regeneration.

IGF receptors are present on the nerve terminals of motor and other neurons, indicating that IGFs may be internalized and transported in a retrograde fashion, similar to NGF. However, retrograde transport of ligand is not obligatory to the neurotrophic theory, as discussed in subsequent text. Peripheral and central neurons respond by neurite outgrowth to IGFs, both in vitro and in vivo. Moreover, anti-IGF-I antibodies can inhibit spontaneous regeneration of sensory axons in vivo.

It is curious that insulin and IGFs are much poorer survival factors than NGF for embryonic sensory and sympathetic neurons (Recio-Pinto et al., 1986). Moreover, IGFs increase neurite outgrowth but do not seem to support survival of embryonic chick motor neurons (Caroni and Grandes, 1990). Cortical neuron survival is increased (Aizenman and de Vellis, 1987). These data support the suggestion that the same neurotrophic molecule may not necessarily regulate neurite outgrowth and survival to the same extent in a given population of neurons. Survival and neurite outgrowth are unlinked responses to NGF and IGFs (discussed in Recio-Pinto et al., 1986). For example, NGF is needed for survival of embryonic rat sensory neurons; adult sensory neurons, which do not require NGF for survival, still retain their neurite outgrowth response (Lindsay, 1988). This might explain the role of multiple receptors for these factors on the same cell. In a given cell type, insulin, IGFs, and NGF may cooperate, but each may stimulate survival, neurite outgrowth, and other responses to a different extent by acting through an independent receptor.

Many organs contain high amounts of IGF transcripts in the fetus, but low amounts in the adult. For example, IGF-II transcripts are down-regulated developmentally in liver, heart, kidney, and lung (Soares et al., 1985, 1986; Brown et al., 1986). This implies that IGF-II may contribute to development of the innervation in these and other organs, as well as in muscle. The higher content of IGF transcripts present in adult brain and spinal cord, relative to most peripheral tissues (Soares et al., 1986; Rotwein et al., 1988; Stylianopoulou et al., 1988), may be due to the need to support the turnover and formation of the large number of synapses in the CNS, particularly for learning and memory. Consistent with this interpretation, regions of rat brain in which synapse turnover is relatively high, for example, the olfactory bulb, are enriched particularly in IGF transcripts and receptors. On the other hand, the paracrine and autocrine actions of insulin may be more restricted than these actions of IGFs, since insulin mRNA is detected in only a small percentage of cultured brain neurons and not in glia (Schechter et al., 1988).

## IV. SHARED BIOCHEMICAL PATHWAY FOR NEURITE OUTGROWTH

Since the discovery of NGF, an intense effort has been made to determine the cellular and molecular mechanisms of neurite formation that are under the direction of neurotrophic factors. Each of these factors triggers many actions in addition to neurite outgrowth, and one is faced with the daunting task of assigning a particular second messenger or other biochemical event to a specific action, namely, neurite outgrowth. A useful strategy has emerged based on the observation that insulin, IGF-I, IGF-II, and NGF all can induce neurite outgrowth in the cloned human neuroblastoma SH-SY5Y cell (Sonnenfeld and Ishii, 1982; Recio-Pinto and Ishii, 1984; Ishii and Recio-Pinto, 1987). The strategy is described in detail elsewhere (Ishii et al., 1985; Ishii and Mill, 1987) and is based on the hypothesis that these factors are likely to share a common biochemical pathway that leads to neurite outgrowth.

### A. Receptors

#### 1. Insulin and IGF receptors

One approach to the study of mechanism is to identify and characterize the receptors that are the sites of initial contact between ligand and cell. The properties of insulin, type I IGF, and type II IGF receptors are described in Chapter 13. Insulin and type I IGF receptors are related structurally to one another, but are dissimilar to type II IGF receptors (Massague et al., 1980). Both insulin and type I IGF receptors are present on SH-SY5Y cells (Recio-Pinto and Ishii, 1988b). Although insulin at supraphysiologic concentrations can cross-occupy type I IGF receptors, at low physiologic concentrations neurite outgrowth is enhanced through insulin receptors. Likewise, IGF-I and IGF-II can cross-occupy insulin receptors at supraphysiologic concentrations, but at low physiologic concentrations neurite outgrowth is enhanced through type I IGF receptors. Cross-occupancy of insulin and type I IGF sites probably explains the broad dose–response curves for neurite outgrowth of these factors (Recio-Pinto and Ishii, 1984).

Type II IGF receptors are likely to be present on SH-SY5Y cells as well, but insulin does not cross-occupy these sites. With preparations of impure IGF-I it was initially thought by many workers in the field, including ourselves, that IGF-I could cross-occupy type II sites with moderately high affinity. However, cross-occupancy is either not observed or has low affinity with highly purified IGF-I from recombinant sources (Rosenfeld et al., 1987). Because highly purified recombinant IGF-I does induce neurites from SH-SY5Y cells (D. N. Ishii, unpublished observations), it seems probable that physiologic concentrations of IGF-I act through type I sites. IGF-II also acts through the type I site, albeit with slightly lower affinity than IGF-I, and its actions on neurite outgrowth can be explained on this basis. The role of type II IGF receptors, which are simultaneously mannose 6-phosphate receptors and lysosomal transport proteins (Morgan et al., 1987; MacDonald et al., 1988), in neurite outgrowth, if any, is unknown.

## 2. NGF receptors

High (type I, slow) and low (type II, fast) affinity NGF receptors are present on human neuroblastoma cells (Sonnenfeld and Ishii, 1982). The high affinity receptors mediate neurite outgrowth. SH-SY5Y cells do not display the low affinity sites; nevertheless, neurite outgrowth is increased by NGF (Sonnenfeld and Ishii, 1985). Moreover, under serum-free conditions these cells reversibly lose capacity to bind and respond to NGF, but retain responsiveness to other neuritogenic factors (Recio-Pinto et al., 1984). NGF binding capacity is regulated by insulin and IGFs; the number, but not affinity, of receptors is increased.

High affinity NGF receptors are confined to neurons, whereas low affinity NGF receptors are found also on nonneurons. Fast NGF receptors are speculated to play a role in a combination of contact and chemotactic guidance (Ishii and Mill, 1987). The SH-SY5Y cells provide an interesting model in which to study neurite growth under multiple regulation by insulin, IGF, and NGF receptors.

## B. Phosphorylation

The precise nature of the transmembrane event leading to neurite outgrowth is still not known, but is suspected to involve phosphorylation of common substrates (reviewed in Ishii and Mill, 1987). The $\beta$ subunits of insulin and type I IGF receptors are tyrosine kinases, and an overlapping pattern of phosphorylation is produced by insulin and IGF-I in neuroblastoma (Shemer et al., 1987a) and other cells (Kadowaki et al., 1987). Although there is no evidence of a kinase encoded by the nucleotide sequences in the cloned low affinity NGF receptor, NGF and insulin nevertheless promote phosphorylation of an overlapping population of proteins in PC12 cells (Halegoua and Patrick, 1980). The *trk* proto-oncogene encodes a tyrosine kinase that binds NGF (Kaplan et al., 1991). Thus, occupancy of insulin, IGF, and NGF receptors might trigger tyrosine phosphorylation of the same intracellular substrates. Several phosphorylated substrates have been identified, but determining their functions is a difficult challenge.

## C. Is Retrograde Axonal Transport of Ligand Essential?

NGF released by target tissues is internalized at nerve terminals and transported in a retrograde fashion to the soma of neurons, mainly to perinuclear and lysosomal sites. Retrograde transport often is considered an essential component of the neurotrophic theory (reviewed in Mobley et al., 1977). IGFs released from targets also may act on nerve terminals. Receptors for insulin and IGFs are present on the shaft and terminals of axons (van Houten et al., 1980; Boyd et al., 1985; Caroni and Grandes, 1990); subcutaneously administered IGFs induce nerve terminal sprouting (Caroni and Grandes, 1990), clearly demonstrating that IGFs can act on nerve terminals. Both anterograde and retrograde transport of IGF-I is detected following ligation of sciatic nerves (Hansson et al., 1987). Thus, on the surface, it would appear that retrograde axonal transport may be essential to the actions of IGFs as well.

However, retrograde transport of NGF and IGFs is not necessarily related to the biologic actions of these ligands. This conclusion follows from the observations that the insulin and type I IGF receptors are kinases, and that NGF can activate phosphorylation as well. Binding of insulin and IGFs to the α subunits of insulin and type I IGF receptors occurs on the extracellular surface of cells. This binding activates the kinase on the β subunits situated in the intracellular domain. If subsequent phosphorylation is the transmembrane event that is important for neurite outgrowth, internalization of the ligands themselves may not be needed for activity. It is not yet proven that phosphorylation is the critical transmembrane event, but the observations that both anti-insulin-receptor antibodies (Kahn et al., 1977) and wheat germ agglutinin (Wilden et al., 1989) can mimic the actions of insulin (in the absence of insulin) show that internalization of insulin is not needed for the response, irrespective of whether phosphorylation is the transmembrane event.

Tight binding is needed to promote association of polypeptide ligands with their receptors, particularly when both are present at exceedingly low concentrations. Diffusion is often rate-limiting for association rate constants; thus, tight binding is achieved at the expense of a low dissociation rate constant, which does not favor a rapid response reversal. Ligands might, therefore, be internalized to promote dissociation or degradation of tightly bound ligand, permitting receptors to recycle to the cell surface and become freshly available for response to an ever-changing environment.

At any rate, internalization and retrograde transport of NGF and IGFs molecules may not be essential either to the actions of these factors or to the neurotrophic theory. The retrograde signal may reside entirely in relatively stable second messengers.

## D. Shared Pathway Regulating Neurite Outgrowth

A complementary approach to the study of mechanism is to identify the biochemical events in the pathway of neurite growth and work backward to the receptor events. Our main emphasis has been to determine whether there is a shared biochemical pathway for neurite outgrowth commonly activated by insulin, IGFs, and NGF. The long-term growth of neurites, as occurs during development or regeneration, is likely to require expression of genes encoding structural proteins of axons and dendrites.

### 1. Tubulin gene expression

Microtubules, composed of α- and β-tubulin heterodimers, constitute one of the three major classes of fibrillar cytoskeletal elements found in axons and dendrites. In addition to stabilization of neurites, they also play an important role in anterograde and retrograde transport. The literature on microtubules, and their role in neurite growth, is substantial and summarized elsewhere (Ishii and Mill, 1987).

The levels of α- and β-tubulin mRNAs during neurite outgrowth are increased by insulin, IGF-I, and IGF-II in SH-SY5Y cells (Mill et al., 1985; Fernyhough et al., 1989) and by NGF in PC12 (Fernyhough and Ishii, 1987) and sensory (Ishii and Mill, 1987) cells. Moreover, NGF increases tubulin mRNA content in sympathetic cells, both in culture and in vivo (P. Fernyhough and D. N. Ishii, unpublished observations). There is a close correlation between insulin and NGF concentrations that occupy their respective receptors, induce neurites, and elevate tubulin transcript levels. Tubulin transcript levels become elevated relative to total RNA, poly(A)+ RNA, histone mRNA, and actin mRNA. With respect to the latter, there actually may be a small increase, which would not be surprising because actins polymerize to form a second important class of filamentous proteins (microfilaments) in axons. Interestingly, insulin by itself can increase neither tubulin mRNA levels nor neurite outgrowth in PC12 cells; however, with NGF it can potentiate both responses synergistically (Recio-Pinto et al., 1984; Fernyhough and Ishii, 1987).

NGF increases tubulin protein levels in PC12 cells (Drubin et al., 1985) and microtubule content in neurons (Angeletti et al., 1971). It also stabilizes microtubules against depolymerization (Black and Greene, 1982). The expression of tubulin genes is observed in rat brain during the period of rapid neurite extension (Bond and Farmer, 1983) and in goldfish optic nerves during regeneration (Neumann et al., 1983). Moreover, the peak in tubulin content in developing rat brain (Burgoyne et al., 1981) is reached at the time that insulin and IGF receptor levels are elevated (Kappy et al., 1984; Bassas et al., 1985).

These data show that tubulin mRNA levels are increased by NGF, insulin, IGF-I, and IGF-II in a variety of cell types. Is the mechanism for elevation also common to all factors? NGF stabilizes tubulin transcripts in PC12 cells (Fernyhough and Ishii, 1987). Insulin and IGF-I stabilize α- and β-tubulin, but not histone or actin, transcripts in SH-SY5Y cells (Fernyhough et al., 1989). Polymerase II activity is unaltered, and no change in nuclear run-off rate is observed. Thus, the predominant mechanism shared by insulin, IGFs, and NGF appears to be stabilization of tubulin mRNAs.

### 2. Neurofilament gene expression

Would insulin, IGFs, and NGF, as a group, increase gene expression of other structural proteins as well? Neurofilaments (NFs) are present primarily in neurons, and make up the third major class of fibrillar elements of axons and dendrites. The central core of NFs is formed by polymerized 68-kDa and 170-kDa NF proteins, whereas cross-bridges are formed by the 200-kDa NF protein (Metuzals et al., 1981; Hirokawa et al., 1984; Trojanowski et al., 1985). It is suggested that the caliber of axons is regulated by NF gene expression (Hoffman et al., 1987). NF protein content increases during neurite outgrowth (Angeletti et al., 1971; Black et al., 1986; Lindenbaum et al., 1987).

The neuritogenic polypeptides under consideration here all increase NF gene expression. mRNAs for the 68-kDa and 170-kDa NF are increased by NGF in PC12 (Lindenbaum et al., 1988) and by insulin and IGFs in SH-SY5Y (Wang et al., 1992; Ishii et al., 1989) cells. These data provide further support for the concept of a shared biochemical pathway regulating neurite outgrowth.

### 3. Two shared mechanisms for regulating gene expression

It would seem logical that the mechanism for elevation of NF mRNA and tubulin mRNA levels would be the same, but that is not the case (C. Wang, B. Wible, K. Angelides, and D. N. Ishii, unpublished observations). There is a substantial increase in nuclear run-off rates of transcripts for the 68-kDa and 170-kDa NFs in response to NGF (Lindenbaum et al., 1988) and insulin (D. N. Ishii, unpublished observations). In contrast, these transcripts are stabilized only weakly, if at all.

These emerging data support the hypothesis that there are at least two shared mechanisms for regulating gene expression. Tubulin mRNA levels are elevated predominantly through transcript stabilization by insulin, IGFs, and NGF; an increase in transcription rate has not been observed. On the other hand, NF mRNA levels are elevated predominantly through increased transcription rates by these neuritogenic polypeptides; any effect on stability of these transcripts seems small.

The elucidation of a common biochemical pathway is likely to contribute significantly to the search for second messengers mediating neurite outgrowth. It would be vastly easier to use an assay involving elevation of specific genes as an endpoint, rather than neurite outgrowth, which is an amorphous and complex event.

### 4. Protein kinase C

Some time ago, it was discovered that tumor-promoting phorbol esters could modulate differentiation in cells of neural crest origin (Ishii, 1978). It was shown subsequently that the receptor for tumor-promoting phorbol esters, such as 12-O-tetradecanoyl phorbol-13-acetate (TPA), is the $Ca^{2+}$-activated phospholipid-dependent protein kinase C (PKC) (Castagna et al., 1982; Niedel et al., 1983). Tumor-promoting phorbol esters bind to and persistently activate this kinase. The brain, among various tissues, has the highest content of PKC (Blumberg et al., 1981) and its activity increases during development (Nagle et al., 1981; Murphy et al., 1983).

*a. Activators of PKC enhance neurite outgrowth in cell, but not ganglia culture* A close correlation is found between the capacity of various tumor-promoting phorbol esters to enhance neurite outgrowth (Pahlman et al., 1981; Spinelli et al., 1982) and to occupy PKC sites (Spinelli and Ishii, 1983; Ishii et al., 1985) in SH-SY5Y cells. Other compounds, such as mezerein and teleocidin, that bind and activate PKC also enhance neurite outgrowth. It was proposed that PKC modulates the response to neurotrophic (Ishii et al., 1985) and other factors, possibly by interaction with the receptor-activated pathways that regulate specialized cell functions (Ishii, 1982).

Neurite outgrowth is increased by tumor-promoting phorbol esters in cultured rat embryonic neurons as well (Burgess et al., 1986). TPA by itself cannot induce neurites in PC12 cells, possibly because PKC activity is low (Hama et al., 1986). NGF induces PKC activity (Contreras and Guroff, 1987; Chan et al., 1989), and TPA subsequently becomes able to potentiate the neurite outgrowth response to NGF (Hall et al., 1988). The need to induce PKC may help explain the lag in the neurite outgrowth response to NGF in PC12 cells relative to other cell types.

In contrast to enhancing neurite outgrowth in cultures containing single cells, PKC agonists inhibit NGF-directed neurite outgrowth in cultured dorsal root ganglia (Ishii, 1978). Hsu et al. (1984) reported that PKC agonists could increase neurite outgrowth in ganglia cultured under slightly different conditions. We have attempted several times to replicate the observations of Hsu et al. (1984) exactly, but without success. It is not understood presently why PKC agonists inhibit neurites in ganglion cultures but enhance neurites in single cell cultures. One possibility is that PKC agonists may act on the ganglionic nonneurons, which are in intimate contact with neurons. The nonneurons may in some way inhibit the response of the neurons.

*b. PKC antagonists inhibit NGF-directed neurite outgrowth in PC12 cells* It was not known whether neurite outgrowth as a result of PKC activation occurred through a pathway different from or the same as that of insulin, IGFs, and NGF. This line of investigation did not advance significantly until the discovery of antagonists such as sphingosine. Sphingosine binds to the regulatory subunit of PKC and competitively inhibits both the binding of tumor-promoting phorbol esters and their persistent activation of the enzyme (Hannun et al., 1986; Merrill et al., 1986; Wilson et al., 1986).

Sphingosine reversibly inhibits NGF-directed neurite outgrowth in PC12 cells, whereas several structural analogs are inactive (Hall et al., 1988). Concentrations of sphingosine that are inhibitory to neurite outgrowth additionally block binding of [$^3$H]phorbol dibutyrate (a tumor-promoting phorbol ester) and the capacity of TPA to activate PKC-dependent phosphorylation of proteins, including tyrosine hydroxylase. Other PKC antagonists, such as K-252a (Koizumi et al., 1988) and staurosporine (Hashimoto and Hagino, 1989), inhibit neurite outgrowth as well.

*c. PKC may mediate neurite outgrowth directed by NGF, insulin, and IGFs* Ongoing work in cultured SH-SY5Y cells (P. Fernyhough, F. L. Hall, P. R. Vulliet, and D. N. Ishii, unpublished observations) and embryonic chick sympathetic neurons (P. Fernyhough, F. L. Hall, and D. N. Ishii, unpublished observations) indicates that PKC is part of the common biochemical pathway for neurite outgrowth directed by NGF, insulin, and IGFs. The inhibition of neurite outgrowth, directed by these factors in sympathetic neurons, displays the same sensitivity to inhibition by various doses of sphingosine in each case, indicating a common mechanism (Ishii et al., 1989).

Sphingosine might block neurite outgrowth by inhibiting at sites other than PKC, but this is not the case. TPA can reverse sphingosine inhibition of NGF-dependent neurite outgrowth in PC12 cells (Hall et al., 1988); various phorbol esters can competitively reverse sphingosine inhibition of insulin-dependent neurite outgrowth in SH-SY5Y cells (D. N. Ishii, unpublished observations). The reversal of inhibition by these phorbol ester congeners demonstrates specificity for PKC, and competitive reversal would not occur if sphingosine were blocking at sites other than PKC.

Certain actions are not inhibited by sphingosine, for example, cell flattening and capacity of NGF and insulin to increase tubulin mRNAs in PC12 and SH-SY5Y cells. The rate of neurite outgrowth in PC12 cells is faster than normal after washout of sphingosine, an observation that is consistent with the accumulation of essential metabolites during the block (Hall et al., 1988). These results show that PKC is not in the segment of the common biochemical pathway that leads to elevation of tubulin mRNAs. The locus of PKC is likely to be in a more distal, or separate, segment of the pathway. These results are consistent with the particularly high concentration of PKC in nerve terminals (Wood et al., 1986; Girard et al., 1988) and support the suggestion that PKC acts at or near the growth cone (Ishii, 1978).

*d. Down-regulation of PKC* NGF is able to induce neurite outgrowth despite down-regulation of PKC in PC12 cells (Reinhold and Neet, 1989; Damon et al., 1990). This observation led Reinhold and Neet (1989) to suggest that PKC activity is unnecessary for the neurite outgrowth response to NGF. However, down-regulation is not a particularly reliable test of total PKC activity.

Seven isozymes of PKC have separate regional distributions in brain and are regulated developmentally (Nishizuka, 1988). The multiple PKC isozymes (Jaken and Kiley, 1987; Kikkawa et al., 1987; Nishizuka, 1988) have differential susceptibilities to down-regulation by TPA (Huang et al., 1989a). Reinhold and Neet (1989) selected histone to measure down-regulation in PC12 cells, but this selection may not be relevant because histone is not shown to be in the biochemical pathway of neurite formation. Moreover, in other cell types, TPA can cause down-regulation of PKC-dependent histone, but not vinculin, phosphorylation (Cochet et al., 1986; Cooper et al., 1987).

The majority of PKC is membrane bound in neural tissues (Neary et al., 1988); the activity of this detergent insoluble species of PKC (Huang et al., 1989b; Moss et al., 1990) evidently was not measured by Reinhold and Neet (1989). Thus, difference in methods, as well as the presence of PKC isozymes with differential susceptibilities to down-regulation, might explain the apparent contradiction between the observations of Reinhold and Neet (1989) and those of others (Hall et al., 1988; Koizumi et al., 1988; Hashimoto and Hagino, 1989). Clearly, a PKC isotype is down-regulated by TPA in PC12 cell, (Reinhold and Neet, 1989), but it remains to be determined whether neurite growth is regulated by PKC isotypes that are not susceptible to down-regulation. These interesting data support the possibility that PKC isozymes are compartmentalized and serve separate functions in neurons.

## V. OTHER NEUROBIOLOGICAL ROLES

### A. Mitogenic

Early studies indicate that IGFs can increase DNA synthesis in fetal brain tissues (Sara et al., 1979; Raizada et al., 1980; Roger and Fellows, 1980; Lenoir and Honnegar, 1983; Enberg et al., 1985). Ornithine decarboxylase activity, uridine uptake, and thymidine uptake are increased by insulin, IGF-I, and IGF-II. IGF-I increases proliferation (McMorris et al., 1986) and development (van der Pal et al., 1988) of oligodendrocytes and DNA synthesis in astroglial cells (Han et al., 1987). Mitosis in cultured sympathetic neuroblasts is, in addition, stimulated by IGFs (DiCicco-Bloom and Black, 1988).

SH-SY5Y cells may prove useful to study the mitogenic mechanism. Ornithine decarboxylase activity (Mattsson et al., 1986), thymidine, uridine, and leucine uptake (Recio-Pinto and Ishii, 1984), and cell proliferation (Ishii and Recio-Pinto, 1987) are increased by insulin and IGFs. These results suggest that neuroblastoma cells are arrested in an early developmental state in which the mitogenic response to insulin or IGFs has not been lost, as would occur in normal progression of sympathetic neuroblasts to postmitotic neurons. Thymidine incorporation is also increased in rat pheochromocytoma PC12 cells (Dahmer and Perlman, 1988).

The in vitro data, as a whole, suggest that insulin or IGFs may play an important role in the early proliferation and development of brain cells. Infusion of IGF-I increases brain and body weights of rats (Philipps et al., 1988); similar effects are observed in transgenic IGF-I mice (Mathews et al., 1988).

It is interesting that insulin and IGFs are mitogens in peripheral as well as CNS tissues. Long ago it was shown that the sizes of the lateral motor columns and sensory ganglia are influenced by the size of the peripheral target field being innervated (Hamburger, 1934; Harrison, 1935). It is hypothesized that insulin and IGFs released from targets may act simultaneously on target tissues and their innervating neurons to coordinate growth (Recio-Pinto and Ishii, 1988a).

### B. Survival

Reports showing that insulin and IGFs can increase survival of cultured central and peripheral neurons already have been discussed, in parts of this chapter and elsewhere (Recio-Pinto and Ishii, 1988a). It is, however, still not known whether support of survival in the face of preprogrammed neuronal cell death is an important mechanism. In vivo studies are lacking.

## C. Neurotransmitters

Previously it was shown that IGF-II released from postsynaptic target cells may regulate polyneuronal innervation (Ishii, 1989) and induce sprouting by acting at presynaptic nerve terminals (Caroni and Grandes, 1990). The neurotransmitter phenotype is another plastic property influenced by the environment surrounding a neuron; it would not be surprising if other aspects of the specialized functions of neurons were regulated by insulin and IGFs acting on nerve terminals.

Insulin is suggested to be a neuromodulator (Young et al., 1980). Insulin increases serotonin synthesis in brain tissue (Kwok and Juorio, 1987), uptake of amino acid neurotransmitters in synaptosomes (Rhoads et al., 1984), and activity of choline acetyltransferase in PC12 cells (Shubert et al., 1980). Moreover, release of catecholamines is stimulated in hypothalamic slices (Sauter et al., 1983). Uptake of norepinephrine is inhibited by insulin in brain cell cultures (Boyd et al., 1985) but not by IGF-I (Shemer et al., 1987b). These data may be related to the observations that severe hypoglycemia induced with insulin is associated with increased norepinephrine turnover (Agardh et al., 1979) and plasma catecholamine levels (Rowe et al., 1981). The firing rates of dopaminergic neurons in nigrostriatum (Saller and Chiodo, 1980), and of neurons in the hippocampus (Palovcik et al., 1984), are altered by insulin.

IGF-I, but not IGF-II or insulin, can stimulate the potassium-dependent (but not basal) release of acetylcholine from cortical slices (Nilsson et al., 1988). These data indicate differential effects through selective insulin and IGF receptors. A very interesting observation is that insulin and IGF-I can cause latent precursor cells to differentiate to catecholaminergic neurons in cultured embryonic quail dorsal root ganglia (Xue et al., 1988).

## D. Other Actions

The effects of insulin on feeding behavior and on other aspects of electrical activity of neurons have been reviewed, along with effects on glucose, RNA, and protein metabolism. (Recio-Pinto and Ishii, 1988a).

## VI. PATHOPHYSIOLOGY

### A. Disturbances during Early Development

Alterations in the content of IGFs during early life might have adverse effects on brain development. The IGF-I content in humans normally is low at birth and rises in serum to a peak at puberty, then declines to adult levels in the late teens (Hall and Sara, 1984). In contrast, IGF-II content remains relatively constant throughout life, beyond 1 year of age, with a slight peak at puberty (Zapf et al., 1981).

The serum content of IGFs in children with Down's syndrome, however, appears abnormal (Sara et al. 1983). In particular, IGF-I does not undergo the usual increase. IGF-II activity, in the only case examined to date, is abnormally low in a fetus with Down's syndrome. Growth hormone levels appear normal in this disorder, yet exogenously administered growth hormone still can stimulate somatic growth, suggesting that sensitivity to growth hormone is blunted (Anneren et al., 1986). A correlation seems to emerge between abnormal IGF-I levels during development and brain dysfunction, because IGF-I content is evidently low as well in other disorders associated with mental retardation (Sara et al., 1981; Rasmussen et al., 1983).

The decreased size of neurons, poorly developed axons and dendrites, and retarded maturation of brain and spinal cord that are observed in dw/dw mice (Noguchi et al., 1983) are associated with low IGF-I levels (Noguchi et al., 1987). On the other hand, elevated IGF-II content is observed in CSF, but not in serum in a case of megalencephaly (Schoenle et al., 1986).

When growth hormone is deficient, myelination is reduced significantly in neonatal rats (Pelton et al., 1977; Noguchi et al., 1982). Because growth hormone regulates IGF-I levels, and IGF-I stimulates proliferation and specialized activities of oligodendrocytes, it is suggested (McMorris et al., 1986) that conditions of low serum IGF-I levels, such as those associated with undernutrition (Phillips and Young, 1976; Maes et al., 1984), may explain attendant decreased myelination (Wiggins, 1982) as a result of delayed maturation of oligodendrocytes.

## B. Disturbances in Later Life

Diabetic peripheral neuropathy is a well-known syndrome associated with decreased insulin activity. The distribution of insulin receptors in the central nervous system and its direct action on cells of neural origin lead to the prediction that neuropathy might extend to the central nervous system as well (Carsten et al., 1989). Conduction velocity was diminished in spinal cord and in peripheral nerves in streptozotocin diabetic rats, validating the prediction. Infusion of insulin reversed this decline. Auditory evoked potentials are reduced (Pozzessere et al., 1988), and degeneration of nerve fibers in the CNS is evident on autopsy (Reske-Nielsen and Lundbaek, 1968) in clinical diabetes. Although it is argued that diabetic neuropathy may be the consequence of diminished direct action of insulin on neurons (Carsten et al., 1989), the possibility that these actions are secondary to ischemia, hyperglycemia, or another metabolic disturbance cannot be discounted. The preliminary suggestion that there are elevated IGF-II levels in serum and CSF in Alzheimer's disease (Sara et al., 1982) was not confirmed in a follow-up study on a larger number of patients (Tham et al., 1988).

## ACKNOWLEDGMENTS

Many thanks to Diane M. Guertin for technical assistance in the preparation of this manuscript. This work was supported by National Institute of Neurological Disorders and Stroke grants NS24327 and NS24787.

# REFERENCES

Agardh, C. D., Carlsson, A., Lindqvist, M., and Siesjo, B. K. (1979). The effect of pronounced hypoglycemia on monoamine metabolism in rat brain. *Diabetes* **28,** 804–809.

Aizenman, Y., and de Vellis, J. (1987). Brain neurons develop in a serum and glial free environment: Effects of transferrin, insulin, insulin-like growth factor-I and thyroid hormone on neuronal survival, growth and differentiation. *Brain Res.* **406,** 32–42.

Angeletti, P. U., Levi-Montalcini, R., and Caramia, F. (1971). Analysis of the effects of the antiserum to the nerve growth factor in adult mice. *Brain Res.* **27,** 343–355.

Anneren, G., Sara, V. R., Hall, K., and Tuvemo, T. (1986). Growth and somatomedin responses to growth hormone in children with Down's syndrome. *Arch. Dis. Child.* **61,** 48–52.

Bassas, L., DePablo, F., Lesniak, M. A., and Roth, J. (1985). Ontogeny of receptors for insulin-like peptides in chick embryo tissues: Early dominance of insulin-like growth factor over insulin receptor in brain. *Endocrinology* **117,** 2321–2329.

Beck, F., Samani, N. J., Penschow, J. D., Thorley, B., Tregear, G. W., and Coghlan, J. P. (1987). Histochemical localization of IGF-I and -II mRNA in the developing rat embryo. *Development* **101,** 175–184.

Benoit, P., and Changeux, J. P. (1978). Consequences of blocking the nerve with a local anesthetic on the evolution of multiinnervation at the regenerating neuromuscular junction of the rat. *Brain Res.* **149,** 89–96.

Betz, W. J. (1987). Motoneuron death and synapse elimination in vertebrates. *Neurol. Neurobiol.* **23,** 117–162.

Betz, W. J., Chua, M., and Ridge, R. M. A. P. (1989). Inhibitory interactions between motoneurone terminals in neonatal rat lumbrical muscle. *J. Physiol.* **418,** 25–51.

Betz, W. J., Ribchester, R. R., and Ridge, R. M. A. P. (1990). Competitive mechanisms underlying synapse elimination in the lumbrical muscle of the rat. *J. Neurobiol.* **21,** 1–17.

Biedler, J. L., Roffler-Tarlov, S., Schachner, M., and Freedman, L. S. (1978). Multiple neurotransmitter synthesis by human neuroblastoma cell lines and clones. *Cancer Res.* **38,** 3751–3757.

Black, M. M., and Greene, L. A. (1982). Changes in the colchicine susceptibility of microtubules associated with neurite outgrowth: Studies with nerve growth factor-responsive PC12 pheochromocytoma cells. *J. Cell Biol.* **95,** 379–386.

Black, M. M., Keyser, P., and Sobel, E. (1986). Interval between the synthesis and assembly of cytoskeletal proteins in cultured neurons. *J. Neurosci.* **6,** 1004–1012.

Blumberg, P. M., Declos, K. B., and Jaken, S. (1981). Tissue and species specificity for phorbol ester receptors. *In* "Organ and Species Specificity in Chemical Carcinogenesis" (R. Langenbach, S. Nesnow, and J. M. Rice, eds.), pp. 201–227. Plenum Press, New York.

Bond, J. F., and Farmer, S. R. (1983). Regulation of tubulin and actin mRNA production in rat brain: Expression of a new β-tubulin mRNA with development. *Mol. Cell. Biol.* **3,** 1333–1342.

Bothwell, M. (1982). Insulin and somatomedin MSA promote nerve growth factor-independent neurite formation by cultured chick dorsal root ganglionic sensory neurons. *J. Neurosci. Res.* **8,** 225–231.

Boyd, F. T., Jr., Clarke, D. W., Muther, T. F., and Raizada, M. K. (1985). Insulin receptors and insulin modulation of norepinephrine uptake in neuronal cultures from rat brain. *J. Biol. Chem.* **260,** 15880–15884.

Brown, A. L., Graham, D. E., Nissley, S. P., Hill, D. J., Strain, A. J., and Rechler, M. M. (1986). Developmental regulation of insulin-like growth factor II mRNA in different rat tissues. *J. Biol. Chem.* **261,** 13144–13150.

Brown, M. C., and Ironton, R. (1977). Motor neurone sprouting induced by prolonged tetrodotoxin block of nerve action potentials. *Nature (London)* **265,** 459–461.

Brown, M. C., and Ironton, R. (1978). Sprouting and regression of neuromuscular synapses in partially denervated mammalian muscles. *J. Physiol.* **278,** 325–348.

Brown, M. C., Jansen, J. K. S., and Van Essen, D. (1976). Polyneuronal innervation of skeletal muscle in new-born rats and its elimination during maturation. *J. Physiol.* **261,** 387–422.

Brown, M. C., Holland, R. L., and Hopkins, W. G. (1981a). Motor nerve sprouting. *Ann. Rev. Neurosci.* **4,** 17–42.

Brown, M. C., Holland, R. L., and Hopkins, W. G. (1981b). Restoration of focal multiple innervation in rat muscles by transmission block during a critical stage of development. *J. Physiol.* **318,** 355–364.

Burgess, S. K., Sahyoun, N., Blanchard, S. G., LeVine, H., III, Chang, K.-J., and Cuatrecasas, P. (1986). Phorbol ester receptors and protein kinase C in primary neuronal cultures: Development and stimulation of endogenous phosphorylation. *J. Cell Biol.* **102,** 312–319.

Burgoyne, R. D., Rudge, J. S., and Murphy, S. (1981). Developmental changes in polypeptide composition of, and precursor incorporation into, cellular and subcellular fractions of rat cerebral cortex. *J. Neurochem.* **36,** 661–669.

Caroni, P., and Grandes, P. (1990). Nerve sprouting in innervated adult skeletal muscle induced by exposure to elevated levels of insulin-like growth factors. *J. Cell Biol.* **110,** 1307–1317.

Carry, M. R., and Morita, M. (1984). Structure and morphogenesis of the neuromuscular junction. *In* "The Neuromuscular Junction" (R. A. Brumback and J. W. Gerst, eds.), pp. 25–64. Futura, New York.

Carsten, R. E., Whalen, L. R., and Ishii, D. N. (1989). Impairment of spinal cord conduction velocity in diabetic rats. *Diabetes* **38,** 730–736.

Cass, D. T., and Mark, R. F. (1975). Re-innervation of axolotl limbs. I. Motor nerves. *Proc. R. Soc. London (Ser. B)* **190,** 45–58.

Castagna, M., Takai, Y., Kaibuchi, K., Sano, K., Kikkawa, U., and Nishizuka, Y. (1982). Direct activation of calcium-activated, phospholipid-dependent protein kinase by tumor-promoting phorbol esters. *J. Biol. Chem.* **257,** 7847–7851.

Chan, B. L., Chao, M. V., and Saltiel, A. R. (1989). Nerve growth factor stimulates the hydrolysis of glycosylphosphatidylinositol in PC-12 cells: A mechanism of protein kinase C regulation. *Proc. Natl. Acad. Sci. U.S.A.* **86,** 1756–1760.

Cochet, C., Souvignet, C., Keramidas, M., and Chambaz, E. M. (1986). Altered catalytic properties of protein kinase C in phorbol ester treated cells. *Biochem. Biophys. Res. Commun.* **134,** 1031–1037.

Collins, F., and Dawson, A. (1983). An effect of nerve growth factor on parasympathetic neurite outgrowth. *Proc. Natl. Acad. Sci. U.S.A.* **80,** 2091–2094.

Contreras, M. L., and Guroff, G. (1987). Calcium-dependent nerve growth factor-stimulated hydrolysis of phosphoinositides in PC12 cells. *J. Neurochem.* **48,** 1466–1472.

Cooper, D. R., de Ruiz Galaretta, C. M., Fanjul, L. F., Mojsilovic, L., Standaert, M. L., Pollet, R. J., and Farese, R. V. (1987). Insulin but not phorbol ester treatment increases phosphorylation of vinculin by protein kinase C in BC3H-1 myocytes. *FEBS Lett.* **214,** 122–126.

Dahmer, M. K., and Perlman, R. L. (1988). Insulin and insulin-like growth factors stimulate deoxyribonucleic acid synthesis in PC12 pheochromocytoma cells. *Endocrinology* **122,** 2109–2113.

Damon, D. H., D'Amore, P. A., and Wagner, J. A. (1990). Nerve growth factor and fibroblast growth factor regulate neurite outgrowth and gene expression in PC12 cells via both protein kinase C- and cAMP-independent mechanisms. *J. Cell Biol.* **110,** 1333–1339.

Davies, A. M., Brandtlow, C., Heumann, R., Korsching, S., Rohrer, H., and Thoenen, H. (1987). Timing and site of nerve growth factor systhesis in developing skin in relation to innervation and expression of the receptor. *Nature (London)* **326,** 353–358.

Dennis, M. J. (1981). Development of the neuromuscular junction: Inductive interactions between cells. *Ann. Rev. Neurosci.* **4,** 43–68.

Dennis, M. J., Ziskind-Conhaim, L., and Harris, A. J. (1981). Development of neuromuscular junctions in rat embryos. *Dev. Biol.* **81,** 266–279.

Diamond, J., and Miledi, R. (1962). A study of foetal and new-born rat muscle fibres. *J. Physiol.* **162,** 393–408.

DiCicco-Bloom, E., and Black, I. B. (1988). Insulin growth factors regulate the mitotic cycle in cultured rat sympathetic neuroblasts. *Proc. Natl. Acad. Sci. U.S.A.* **85,** 4066–4070.

Ding, R., Jansen, J. K. S., Laing, N. G., and Tonnesen, H. (1983). The innervation of skeletal muscles in chickens curarized during early development. *J. Neurocytol.* **12,** 887–919.

Drubin, D. G., Feinstein, S. C., Shooter, E. M., and Kirschner, M. W. (1985). Nerve growth factor-induced neurite outgrowth in PC12 cells involves the coordinate induction of microtubule assembly and assembly-promoting factors. *J. Cell Biol.* **101**, 1799–1807.

Duchen, L. W., and Strich, S. J. (1968). The effects of botulinum toxin on the pattern of innervation of skeletal muscle in the mouse. *Q. J. Exp. Physiol.* **53**, 84–89.

Edds, M. V. (1950). Collateral regeneration of residual motor axons in partially denervated muscles. *J. Exp. Zool.* **113**, 517–537.

Enberg, G., Tham, A., and Sara, V. R. (1985). The influence of purified somatomedins and insulin on fetal brain DNA synthesis in vitro. *Acta Physiol. Scand.* **125**, 305–308.

Fernyhough, P., and Ishii, D. N. (1987). Nerve growth factor modulates tubulin transcript levels in pheochromocytoma PC12 cells. *Neurochem. Res.* **12**, 891–899.

Fernyhough, P., Mill, J. F., Roberts, J. L., and Ishii, D. N. (1989). Stabilization of tubulin mRNAs by insulin and insulin-like growth factor I during neurite formation. *Mol. Brain Res.* **6**, 109–120.

Fetcho, J. R. (1987). A review of the organization and evolution of motoneurons innervating the axial musculature of vertebrates. *Brain Res. Rev.* **12**, 243–280.

Girard, P., Wood, J., Frenschi, J., and Kuo, J. (1988). Immunocytochemical localization of protein kinase C in developing brain tissue and in primary neuronal cultures. *Dev. Biol.* **126**, 98–107.

Glazner, G. W., and Ishii, D. N. (1989). Relationship of insulin-like growth factor I mRNA content to synaptogenesis in rat muscle. *Soc. Neurosci. Abstr.* **15**, 1353.

Halegoua, S., and Patrick, J. (1980). Nerve growth factor mediates phosphorylation of specific proteins. *Cell* **22**, 571–581.

Hall, F. L., Fernyhough, P., Ishii, D. N., and Vulliet, P. R. (1988). Suppression of nerve growth factor-directed neurite outgrowth in PC12 cells by sphingosine, an inhibitor of protein kinase C. *J. Biol. Chem.* **263**, 4460–4466.

Hall, K., and Sara, V. R. (1984). Somatomedin levels in childhood, adolescence and adult life. *Clin. Endocrinol. Metab.* **13**, 91–112.

Hama, T., Huang, K.-P., and Guroff, G. (1986). Protein kinase C as a component of a nerve growth factor-sensitive phosphorylation system in PC12 cells. *Proc. Natl. Acad. Sci. U.S.A.* **83**, 2353–2357.

Hamburger, V. (1934). The effects of wing bud extirpation on the development of the central nervous system in chick embryos. *J. Exp. Zool.* **68**, 449–494.

Han, V. K. M., Lauder, J. M., and D'Ercole, A. J. (1987). Characterization of somatomedin/insulin-like growth factor receptors and correlation with biologic action in cultured neonatal rat astroglial cells. *J. Neurosci.* **7**, 501–511.

Hannun, Y. A., Loomis, C. R., Merrill, A. H., Jr., and Bell, R. M. (1986). Sphingosine inhibition of protein kinase C activity and of phorbol dibutyrate binding in vitro and in human platelets. *J. Biol. Chem.* **261**, 12604–12609.

Hansson, H.-A., Dahlin, L. B., Danielsen, N., Fryklund, L., Nachemson, A. K., Polleryd, P., Rozell, B., Skottner, A., Stemme, S., and Lundborg, G. (1986). Evidence indicating trophic importance of IGF-I in regenerating peripheral nerves. *Acta Physiol. Scand.* **126**, 609–614.

Hansson, H.-A., Rozell, B., and Skottner, A. (1987). Rapid axoplasmic transport of insulin-like growth factor I in the sciatic nerve of adult rats. *Cell Tissue Res.* **247**, 241–247.

Hardouin, S., Hossenlopp, P., Segovia, B., Seurin, D. Portolan, G., Lassarre, C., and Binoux, M. (1987). Heterogeneity of insulin-like growth factor binding proteins and relationships between structure and affinity. 1. Circulating forms in man. *Eur. J. Biochem.* **170**, 121–132.

Harrison, R. G. (1935). On the origin and development of the nervous system studied by the methods of experimental embryology. *Proc. R. Soc. London (Biol.)* **118**, 155–196.

Hashimoto, S., and Hagino, A. (1989). Blockage of nerve growth factor action in PC12h cells by staurosporine, a potent protein kinase inhibitor. *J. Neurochem.* **53**, 1675–1685.

Heumann, R., Korsching, S., Bandtlow, C., and Thoenen, H. (1987). Changes of nerve growth factor synthesis in non-neuronal cells in response to sciatic nerve transection. *J. Cell Biol.* **104**, 1623–1631.

Hirokawa, N. M., Glicksman, M. A., and Willard, M. B. (1984). Organization of mammalian neurofilament polypeptides within neuronal cytoskeleton. *J. Cell Biol.* **98,** 1523–1536.
Hoffman, H. (1950). Local re-innervation in partially denervated muscle: A histo-physiological study. *Aust. J. Exp. Biol. Med. Sci.* **28,** 383–397.
Hoffman, P. N., Cleveland, D. W., Griffin, J. W., Landes, P. W., Cowan, N. J., and Price, D. L. (1987). Neurofilament gene expression: A major determinant of axonal caliber. *Proc. Natl. Acad. Sci. U.S.A.* **84,** 3472–3476.
Holland, R. L., and Brown, M. C. (1980). Postsynaptic transmission block can cause terminal sprouting of a motor nerve. *Science* **207,** 649–651.
Hossenlopp, P., Seurin, D., Segovia-Quinson, B., and Binoux, M. (1986). Identification of an insulin-like growth factor-binding protein in human cerebrospinal fluid with a selective affinity for IGF-II. *FEBS Lett.* **208,** 439–444.
Hsu, L., Natyzak, D., and Laskin, J. D. (1984). Effect of the tumor promotor 12-O-tetradecanoyl-phorbol-13-acetate on neurite outgrowth from chick embryo sensory ganglia. *Cancer Res* **44,** 4607–4614.
Huang, F. L., Yoshida, Y., Cunha-Melo, J. R., Beaven, M. A., and Huang, K.-P. (1989a). Differential down-regulation of protein kinase C isozymes. *J. Biol. Chem.* **264,** 4238–4243.
Huang, F. L., Chuang, D.-M., and Huang, K.-P. (1989b). Protein kinase C isozymes in primary cultures of cerebellar granule cells: Glutamate-stimulated membranous association of type II PKC. *Soc. Neurosci. Abstr.* **15,** 832.
Imashuku, S., Takada, H., Sawada, T., Nakamura, T., and La Brosse, E. H. (1975). Studies on tyrosine hydroxylase in neuroblastoma, in relation to urinary levels of catecholamine metabolites. *Cancer* **36,** 450–457.
Ishii, D. N. (1978). Effect of tumor promoters on the response of cultured embryonic chick ganglia to nerve growth factor. *Cancer Res.* **38,** 3886–3893.
Ishii, D. N. (1982). Inhibition of iodinated nerve growth factor binding by the suspected tumor promoters saccharin and cyclamate. *J. Natl. Cancer Instit.* **68,** 299–303.
Ishii, D. N. (1989). Relationship of insulin-like growth factor II gene expression in muscle to synaptogenesis. *Proc. Natl. Acad. Sci. U.S.A.* **86,** 2898–2902.
Ishii, D. N. (1990). Trophic factors in the CNS: Role of insulin-like growth factor II in neurite formation and synaptogenesis. In "Brain Aging: Molecular Biology, The Aging Process and Neurodegenerative Disease" (H. C. Hendrie, L. G. Mendelsohn, and C. Readhead, eds.), pp. 211–220. Hogrefe and Huber, New York.
Ishii, D. N., and Mill, J. F. (1987). Molecular mechanisms of neurite formation stimulated by insulin-like factors and nerve growth factor. *Curr. Top. Membrane Transport* **31,** 31–78.
Ishii, D. N., and Recio-Pinto, E. (1987). Role of insulin, insulin-like growth factors, and nerve growth factor in neurite formation. In "Insulin, IGFs, and Their Receptors in the Central Nervous System" (M. K. Raizada, M. I. Phillips, and D. LeRoith, eds.), pp. 315–348. Plenum, New York.
Ishii, D. N., Recio-Pinto, E., Spinelli, W., Mill, J. F., and Sonnenfeld, K. H. (1985). Neurite formation modulated by nerve growth factor, insulin, and tumor promoter receptors. *Int. J. Neurosci.* **26,** 109–127.
Ishii, D. N., Glazner, G. W., Wang, C., and Fernyhough, P. (1989). Neurotrophic effects and mechanism of insulin, insulin-like growth factors, and nerve growth factor in spinal cord and peripheral neurons. In "Molecular and Cellular Biology of Insulin-Like Growth Factors and Their Receptors" (D. LeRoith and M. K. Raizada, eds.), pp. 403–425. Plenum, New York.
Jaken, S., and Kiley, S. C. (1987). Purification and characterization of three types of protein kinase C from rabbit brain cytosol. *Proc. Natl. Acad. Sci. U.S.A.* **84,** 4418–4422.
Jennische, E., and Olivecrona, H. (1987). Transient expression of insulin-like growth factor I immunoreactivity in skeletal muscle cells during postnatal development in the rat. *Acta Physiol. Scand.* **131,** 619–622.
Jirmanova, I., Sobotkova, M., Thesleff, S., and Zelena, J. (1964). Atrophy in skeletal muscles poisoned with botulinum toxin. *Physiol. Bohemoslov.* **13,** 467–472.
Kadowaki, T., Koyasu, S., Nishida, E., Tobe, K., Izumi, T., Takaku, F., Sakai, H., Yahara, I., and Kasuga,

M. (1987). Tyrosine phosphorylation of common and specific sets of cellular proteins rapidly induced by insulin, insulin-like growth factor 1, and epidermal growth factor in intact cells. *J. Biol. Chem.* **262,** 7342–7350.

Kahn, C. R., Baird, K., Flier, J. S., and Jarrett, D. B. (1977). Effects of autoantibodies to the insulin receptor on isolated adipocytes: Studies of insulin binding and insulin action. *J. Clin. Invest.* **60,** 1094–1106.

Kanje, M., Skottner, A., and Lundborg, G. (1988). Effects of growth hormone treatment on the regeneration of rat sciatic nerve. *Brain Res.* **475,** 254–258.

Kanje, M., Skottner, A., Sjoberg, J., and Lundborg, G. (1989). Insulin-like growth factor I (IGF-I) stimulates regeneration of the rat sciatic nerve. *Brain Res.* **486,** 396–398.

Kaplan, D. R., Hempstead, B. L., Martin-Zanca, D., Chao, M. V., and Parada, L. F. (1991). The *trk* proto-oncogene product: A signal transducing receptor for nerve growth factor. *Science* **252,** 554–557.

Kappy, M., Sellinger, S., and Raizada, M. (1984). Insulin binding in four regions of the developing rat brain. *J. Neurochem.* **42,** 198–203.

Kelly, A. M., and Zacks, S. I. (1969). The fine structure of motor endplate morphogenesis. *J. Cell Biol.* **42,** 154–169.

Kikkawa, U., Ono, Y., Ogita, K., Fujii, T., Asaoka, Y., Sekiguchi, K., Kosaka, Y., Igarashi, K., and Nishizuka, Y. (1987). Identification of the structures of multiple subspecies of protein kinase C expressed in rat brain. *FEBS Lett.* **217,** 227–231.

Koizumi, S., Contreras, M. L., Matsuda, Y., Hama, T., Lazarovici, P., and Guroff, G. (1988). K-252a: A specific inhibitor of the action of nerve growth factor on PC12 cells. *J. Neurosci.* **8,** 715–721.

Kwok, R. P. S., and Juorio, A. V. (1987). Facilitating effect of insulin on brain 5-hydroxytryptamine metabolism. *Endocrinology* **45,** 267–273.

Lance-Jones, C., and Landmesser, L. (1980). Motoneurons projection patterns in the chick hindlimb following early partial reversals of the spinal cord. *J. Physiol.* **302,** 581–602.

Lance-Jones, C., and Landmesser, L. (1981). Pathway selection by embryonic chick motoneurons in an experimentally altered environment. *Proc. R. Soc. London (Biol.)* **214,** 19–52.

Landmesser, L. T. (1980). The generation of neuromuscular specificity. *Ann. Rev. Neurosci.* **3,** 279–302.

Lenoir, D., and Honegger, P. (1983). Insulin-like growth factor I (IGF I) stimulates DNA synthesis in fetal rat brain cell cultures. *Dev. Brain Res.* **7,** 205–213.

Levi-Montalcini, R., and Angeletti, P. U. (1968). Nerve growth factor. *Physiol. Rev.* **48,** 534–569.

Lindenbaum, M. H., Carbonetto, S., and Mushynski, W. E. (1987). Nerve growth factor enhances the synthesis, phosphorylation, and metabolic stability of neurofilament proteins in PC12 cells. *J. Biol. Chem.* **262,** 605–610.

Lindenbaum, M. H., Carbonetto, S., Grosveld, F., Flavell, D., and Mushynski, W. E. (1988). Transcriptional and post-transcriptional effects of nerve growth factor on expression of the three neurofilament subunits in PC-12 cells. *J. Biol. Chem.* **263,** 5662–5667.

Lindsay, R. M. (1988). Nerve growth factors (NGF, BDNF) enhance axonal regeneration but are not required for survival of adult sensory neurons. *J. Neurosci.* **8,** 2394–2405.

Lomo, T., and Jansen, J. K. S. (1980). Requirements for the formation and maintenance of neuromuscular connections. *Curr. Top. Dev. Biol.* **16,** 253–281.

McArdle, J. J. (1975). Complex end-plate potentials at the regenerating neuromuscular junction of the rat. *Exp. Neurol.* **49,** 629–638.

McArdle, J. J. (1984). Overview of the physiology of the neuromuscular junction. *In* "The Neuromuscular Junction" (R. A. Brumback and J. W. Gerst, eds.), pp. 65–119. Futura, New York.

MacDonald, R. G., Pfeffer, S. R., Coussens, L., Tepper, M. A., Brocklebank, C. M., Mole, J. E., Anderson, J. K., Chen, E., Czech, M. P., and Ullrich, A. (1988). A single receptor binds both insulin-like growth factor II and mannose-6-phosphate. *Science* **239,** 1134–1137.

McMahan, U. J., and Wallace, B. G. (1989). Molecules in basal lamina that direct formation of synaptic specializations at neuromuscular junctions. *Dev. Neurosci.* **11,** 227–247.

McMorris, F. A., Smith, T. M., DeSalvo, S., and Furlanetto, R. W. (1986). Insulin-like growth factor

I/somatomedin C: A potent inducer of oligodendrocyte development. *Proc. Natl. Acad. Sci. U.S.A.* **83,** 822–826.
Maes, M., Underwood, L. E., and Ketelslegers, J.-M. (1984). Low serum somatomedin C in protein deficiency: Relationship with changes in liver somatogenic and lactogenic binding sites. *Mol. Cell. Endocrinol.* **37,** 301–309.
Massague, J., Pilch, P. F., and Czech, M. P. (1980). Electrophoretic resolution of three major insulin receptor structures with unique subunit stoichiometries. *Proc. Natl. Acad. Sci. U.S.A.* **77,** 7137–7141.
Mathews, L. S., Hammer, R. E., Behringer, R. R., D'Ercole, A. J., Bell, G. I., Brinster, R. L., and Palmiter, R. D. (1988). Growth enhancement of transgenic mice expressing human insulin-like growth factor I. *Endocrinology* **123,** 2827–2833.
Mattsson, M. E. K., Enberg, G., Ruusala, A.-I., Hall, K., and Pahlman, S. (1986). Mitogenic response of human SH-SY5Y neuroblastoma cells to insulin-like growth factor I and II is dependent on the stage of differentiation. *J. Cell Biol.* **102,** 1949–1954.
Menesini-Chen, M. G. M., Chen, J. S., and Levi-Montalcini, R. (1978). Sympathetic nerve fibers ingrowth in the central nervous system of neonatal rodent upon intracerebral NGF injections. *Arch. Ital. Biol.* **116,** 53–84.
Merrill, A. H., Jr., Sereni, A. M., Stevens, V. L., Hannun, Y. A., Bell, R. M., and Kinkade, J. M., Jr. (1986). Inhibition of phorbol ester-dependent differentiation of human promyelocytic leukemic (HL-60) cells by sphinganine and other long-chain bases. *J. Biol. Chem.* **261,** 12610–12615.
Metuzals, J., Montpetit, V., and Clapin, D. F. (1981). Organization of the neurofilamentous network. *Cell Tissue Res.* **214,** 455–482.
Mill, J. F., Chao, M. V., and Ishii, D. N. (1985). Insulin, insulin-like growth factor II, and nerve growth factor effects on tubulin mRNA levels and neurite formation. *Proc. Natl. Acad. Sci. U.S.A.* **82,** 7126–7130.
Mobley, W. C., Server, A. C., Ishii, D. N., Riopelle, R. J., and Shooter, E. M. (1977). Nerve growth factor. *N. Engl. J. Med.* **297,** 1096–1104; 1149–1158; 1211–1218.
Morgan, D. O., Edman, J. C., Standring, D. N., Fried, V. A., Smith, M. C., Roth, R. A., and Rutter, W. J. (1987). Insulin-like growth factor II receptor as a multifunctional binding protein. *Nature (London)* **329,** 301–307.
Moss, D. J., Fernyhough, P., Chapman, K., Baizer, L., Bray, D., and Allsopp, T. (1990). Chicken growth-associated protein GAP-43 is tightly bound to the actin-rich neuronal membrane skeleton. *J. Neurochem.* **54,** 729–736.
Murphy, K. M. M., Gould, R. J., Oster-Granite, M. L., Gearhart, J. D., and Snyder, S. H. (1983). Phorbol ester receptors: Autoradiographic identification in the developing rat. *Science* **222,** 1036–1038.
Nagle, D. S., Jaken, S., Castagna, M., and Blumberg, P. M. (1981). Variation with embryonic development and regional localization of specific ($^3$H)phorbol-12,13-dibutyrate binding to brain. *Cancer Res.* **41,** 89–93.
Nakajima, Y., Kodokoro, Y., and Klier, G. (1980). The development of functional neuromuscular junctions in vitro: An ultrastructural and physiological study. *Dev. Biol.* **77,** 52–72.
Neary, J., Norenberg, L., and Norenberg, M. (1988). Protein kinase C in primary astrocyte cultures: Cytoplasmic localization and translocation by a phorbol ester. *J. Neurochem.* **50,** 1179–1184.
Neumann, D., Scherson, T., Ginzburg, I., Littauer, U. Z., and Schwartz, M. (1983). Regulation of mRNA levels for microtubule proteins during nerve regeneration. *FEBS Lett.* **162,** 270–276.
Niedel, J. E., Kuhn, L. J., and Vanderbark, G. R. (1983). Phorbol diester receptor copurifies with protein kinase C. *Proc. Natl. Acad. Sci. U.S.A.* **80,** 36–40.
Nilsson, L., Sara, V. R., and Nòrdberg, A. (1988). Insulin-like growth factor 1 stimulates the release of acetylcholine from rat cortical slices. *Neurosci. Lett.* **88,** 221–226.
Nishizuka, Y. (1988). The molecular heterogeneity of protein kinase C and its implications for cellular recognition. *Nature (London)* **334,** 661–665.
Noguchi, T., Sugisaki, T., Takamatsu, K., and Tsukada, Y. (1982). Factors contributing to the poor myelination in the brain of the Snell dwarf mouse. *J. Neurochem.* **39,** 1693–1699.

Noguchi, T., Sekiguchi, M., Sugisaki, T., Tsukada, Y., and Shimai, K. (1983). Faulty development of cortical neurons in the Snell dwarf cerebrum. *Dev. Brain Res.* **10**, 125–138.

Noguchi, T., Kurata, L. M., and Sugisaki, T. (1987). Presence of a somatomedin C-immunoreactive substance in the central nervous system: Immunohistochemical mapping studies. *Neuroendocrinology* **46**, 277–282.

O'Brien, R. A. D., Ostberg, A. J. C., and Vrbova, G. (1978). Observations on the elimination of polyneuronal innervation in developing mammalian skeletal muscle. *J. Physiol.* **282**, 571–582.

Pahlman, S., Odelstad, L., Larsson, E., Grotte, G., and Nilsson, K. (1981). Phenotypic changes of human neuroblastoma cells in culture induced by 12-O-tetradecanoyl-phorbol-13-acetate. *Int. J. Cancer* **28**, 583–589.

Palovcik, R. A., Phillips, M. I., Kappy, M. S., and Raizada, M. K. (1984). Insulin inhibits pyramidal neurons in hippocampal slices. *Brain Res.* **309**, 187–191.

Pelton, E. W., Grindeland, R. E., Young, E., and Bass, N. H. (1977). Effects of immunologically induced growth hormone deficiency on myelinogenesis in developing rat cerebrum. *Neurology (Minneapolis)* **27**, 282–288.

Phillips, A. F., Persson, B., Hall, K., Lake, M., Skottner, A., Sanengen, T., and Sara, V. R. (1988). The effects of biosynthetic insulin-like growth factor-I supplementation on somatic growth, maturation and erythropoiesis on the neonatal rat. *Pediatrics Res.* **23**, 298–305.

Phillips, L. S., and Young, H. S. (1976). Nutrition and somatomedin. II. Serum somatomedin activity and cartilage growth activity in streptozotocin-diabetic rats. *Diabetes* **25**, 516–527.

Pozzessere, G., Rizzo, P. A., Valle, E., Mollica, M. A., Meccia, A., Morano, S., DiMario, U., Andreani, D., and Morocutti, C. (1988). Early detection of neurological involvement in IDDM and NIDDM: Multimodal evoked potentials versus metabolic control. *Diabetes Care* **11**, 473–480.

Raivich, G., Hellweg, R., Graeber, M. B., and Kreutzberg, G. W. (1990). The expression of growth factor receptors during nerve regeneration. *Restor. Neurol. Neurosci.* **1**, 217–223.

Raizada, M. K., Yang, J. W., and Fellows, R. E. (1980). Binding of [$^{125}$I]insulin to specific receptors and stimulation of nucleotide incorporation in cells cultured from rat brain. *Brain Res.* **200**, 389–400.

Rasmussen, F., Sara, V. R., and Gustavson, K.-H. (1983). Serum levels of radioreceptor-assayable somatomedins in children with minor neurodevelopmental disorder. *Uppsala J. Med. Sci.* **88**, 121–126.

Recio-Pinto, E., and Ishii, D. N. (1984). Effects of insulin, insulin-like growth factor-II and nerve growth factor on neurite outgrowth in cultured human neuroblastoma cells. *Brain Res.* **302**, 323–334.

Recio-Pinto, E., and Ishii, D. N. (1988a). Insulin and related growth factors: Effects on the nervous system and mechanism for neurite growth and regeneration. *Neurochem. Int.* **12**, 397–414.

Recio-Pinto, E., and Ishii, D. N. (1988b). Insulin and insulinlike growth factor receptors regulating neurite formation in cultured human neuroblastoma cells. *J. Neurosci. Res.* **19**, 312–320.

Recio-Pinto, E., Lang, F. F., and Ishii, D. N. (1984). Insulin and insulinlike growth factor-II permit nerve growth factor binding and the neurite formation response in cultured human neuroblastoma cells. *Proc. Natl. Acad. Sci. U.S.A.* **81**, 2562–2566.

Recio-Pinto, E., Rechler, M. M., and Ishii, D. N. (1986). Effects of insulin, insulin-like growth factor-II, and nerve growth factor on neurite formation and survival in cultured sympathetic and sensory neurons. *J. Neurosci.* **6**, 1211–1219.

Redfern, P. A. (1970). Neuromuscular transmission in newborn rats. *J. Physiol.* **209**, 701–709.

Reinhold, D. S., and Neet, K. E. (1989). The lack of a role for protein kinase C in neurite extension and in the induction of ornithine decarboxylase by nerve growth factor in PC12 cells. *J. Biol. Chem.* **264**, 3538–3544.

Reske-Nielsen, E., and Lundbaek, K. (1968). Pathological changes in the central and peripheral nervous system of young long-term diabetics. II. The spinal cord and peripheral nerves. *Diabetologia* **4**, 34–43.

Rhoads, D. E., DiRocco, R. J., Osburn, L. D., Peterson, N. A., and Raghupathy, E. (1984). Stimulation

of synaptosomal uptake of neurotransmitter amino acids by insulin: Possible role of insulin as a neuromodulator. *Biochem. Biophys. Res. Commun.* **119,** 1198–1204.
Roger, L. J., and Fellows, R. E. (1980). Stimulation of ornithine decarboxylase activity by insulin in developing rat brain. *Endocrinology* **106,** 619–625.
Rosenfeld, R. G., Conover, C. A., Hodges, D., Lee, P. K. K., Misra, P., Hintz, R. L., and Li, C. H. (1987). Heterogeneity of IGF-I affinity for the IGF-II receptor: Comparison of natural, synthetic and recombinant DNA-derived IGF-I. *Biochem. Biophys. Res. Commun.* **143,** 199–205.
Rotwein, P., Burgess, S. K., Milbrandt, J. D., and Krause, J. E. (1988). Differential expression of insulin-like growth factor genes in rat central nervous system. *Proc. Natl. Acad. Sci. U.S.A.* **85,** 265–269.
Rowe, J. W., Young, J. B., Minaker, K. L., Stevens, A. L., Pallotta, J., and Landsberg, L. (1981). Effect of insulin and glucose infusions on sympathetic nervous system activity in normal man. *Diabetes* **30,** 219–225.
Saller, C. F., and Chiodo, L. A. (1980). Glucose suppresses basal firing and haloperidol-induced increase in the firing rate of central dopaminergic neurons. *Science* **210,** 1269–1271.
Salpeter, M. M. (1987). Development and neural control of the neuromuscular junction and of the junctional acetycholine receptor. *Neurol. Neurobiol.* **23,** 55–115.
Sara, V. R., Hall, K., Wetterberg, L., Fryklund, L., Sjogren, B., and Skottner, A. (1979). Fetal growth: The role of the somatomedins and other growth-promoting peptides. *In* "Somatomedins and Growth" (G. Giordano, J. J. Van Wyk, and F. Minuto, eds.), pp. 225–230. Academic Press, London.
Sara, V. R., Hall, K., and Wetterberg, L. (1981). Fetal brain growth: Proposed model for regulation by embryonic somatomedin. *In* "The Biology of Normal Human Growth" (M. Ritzen, A. Aperia, K. Hall, A. Larsson, and R. Zetterstrom, eds.), pp. 241–252. Raven Press, London.
Sara, V. R., Hall, K., and Enzell, K. (1982). Somatomedins in aging and dementia disorders of Alzheimer type. *Neurobiol. Aging* **3,** 117–120.
Sara, V. R., Gustavson, K.-H., Anneren, G., Hall, K., and Wetterberg, L. (1983). Somatomedins in Down's syndrome. *Biol. Psychiatry* **18,** 803–811.
Sauter, A., Goldstein, M., Engel, J., and Ueta, K. (1983). Effect of insulin on central catecholamines. *Brian Res.* **260,** 330–333.
Schechter, R., Holtzclaw, L., Sadiz, F., Kahn, A., and Devaskar, S. (1988). Insulin synthesis by isolated rabbit neurons. *Endocrinology* **123,** 505–513.
Schoenle, E. J., Haselbacher, G. K., Briner, J., Janzer, R. C., Gammeltoft, S., Humbel, R. E., and Prader, A. (1986). Elevated concentration of IGF II in brain tissue from an infant with megaencephaly. *J. Pediatrics* **108,** 737–740.
Shemer, J., Adamo, M., Wilson, G. L., Heffez, D., Zick, Y., and LeRoith, D. (1987a). Insulin and insulin-like growth factor 1 stimulate a common endogenous phosphoprotein substrate (pp185) in intact neuroblastoma cells. *J. Biol. Chem.* **262,** 15476–15482.
Shemer, J., Raizada, M. K., Masters, B. A., Ota, A., and LeRoith, D. (1987b). Insulin-like growth factor I receptors in neuronal and glial cells. Characterization and biological effects in primary culture. *J. Biol. Chem.* **262,** 7693–7699.
Shubert, D., LaCorbiere, M., Klier, F. G., and Steinbach, J. H. (1980). The modulation of neurotransmitter synthesis by steroid hormones and insulin. *Brain Res.* **190,** 67–79.
Slack, J. R., Hopkins, W. G., and Williams, M. N. (1979). Nerve sheaths and motoneurone collateral sprouting. *Nature (London)* **282,** 506–507.
Soares, M. B., Ishii, D. N., and Efstratiadis, A. (1985). Developmental and tissue-specific expression of a family of transcripts related to rat insulin-like growth factor II mRNA. *Nucleic Acids Res.* **13,** 1119–1134.
Soares, M. B., Turken, A., Ishii, D. N., Mills, L., Episkopou, V., Cotter, S., Zeitlin, S., and Efstratiadis, A. (1986). Rat insulin-like growth factor II gene: A single gene with two promoters expressing a multitranscript family. *J. Mol. Biol.* **192,** 737–752.
Sohal, G. S., Creazzo, T. L., and Oblak, T. G. (1979). Effects of chronic paralysis with α-bungarotoxin on development of innervation. *Exp. Neurol.* **66,** 619–628.

Sonnenfeld, K. H., and Ishii, D. N. (1982). Nerve growth factor effects and receptors in cultured human neuroblastoma cell lines. *J. Neurosci. Res.* **8**, 375–391.
Sonnenfeld, K. H., and Ishii, D. N. (1985). Fast and slow nerve growth factor binding sites in human neuroblastoma and rat pheochromocytoma cell lines: Relationship of sites to each other and to neurite formation. *J. Neurosci.* **5**, 1717–1728.
Spinelli, W., and Ishii, D. N. (1983). Tumor promoter receptors regulating neurite formation in cultured human neuroblastoma cells. *Cancer Res.* **43**, 4119–4125.
Spinelli, W., Sonnenfeld, K. H., and Ishii, D. N. (1982). Effects of phorbol ester tumor promoters and nerve growth factor on neurite outgrowth in cultured human neuroblastoma cells. *Cancer Res.* **42**, 5067–5073.
Stylianopoulou, F., Herbert, J., Soares, M. B., and Efstratiadis, A. (1988). Expression of the insulin-like growth factor II gene in the choroid plexus and the leptomeninges of the adult rat central nervous system. *Proc. Natl. Acad. Sci. U.S.A.* **85**, 141–145.
Taxt, T. (1983). Local and systemic effects of tetrodotoxin on the formation and elimination of synapses in reinnervated adult rat muscle. *J. Physiol.* **340**, 175–194.
Tham, A., Sparring, K., Bowen, D., Wetterberg, L., and Sara, V. R. (1988). Insulin-like growth factors and somatomedin B in the cerebrospinal fluid of patients with dementia of the Alzheimer type. *Acta Psychiatr. Scand.* **77**, 719–723.
Thompson, W. (1983). Synapse elimination in neonatal rat muscle is sensitive to pattern of muscle use. *Nature (London)* **302**, 614–616.
Thompson, W., Kuffler, D. P., and Jansen, J. K. S. (1979). The effect of prolonged, reversible block of nerve impulses on the elimination of polyneuronal innervation of new-born rat skeletal muscle fibres. *Neuroscience* **4**, 271–281.
Tollefsen, S. E., Sadow, J. L., and Rotwein, P. (1989). Coordinate expression of insulin-like growth factor II and its receptor during muscle differentiation. *Proc. Natl. Acad. Sci. U.S.A.* **86**, 1543–1547.
Trojanowski, J. Q., Obracka, M. A., and Lee, V. M. (1985). Distribution of neurofilament subunits in neurons and neuronal processes. *J. Histochem. Cytochem.* **33**, 557–563.
van der Pal, R. H. M., Koper, J. W., van Golde, L. M. G., and Lopes-Cardozo, M. (1988). Effects of insulin and insulin-like growth factor (IGF-I) on oligodendrocyte-enriched glial cultures. *J. Neurosci. Res.* **19**, 483–490.
Van Essen, D. C., Gordon, H., Soha, J. M., and Fraser, S. E. (1990). Synaptic dynamics at the neuromuscular junction: Mechanisms and models. *J. Neurobiol.* **21**, 223–249.
van Houten, M., Kopriwa, B. M., and Brawer, J. R. (1980). Insulin binding sites localized to nerve terminals in rat median eminence and arcuate nucleus. *Science* **207**, 1081–1083.
Wang, C., Li, Y., Wible, B., Angelides, K. J., and Ishii, D. N. (1992). Effects of insulin and insulinlike growth factors on neurofilament mRNA and tubulin mRNA content in human neuroblastoma SH-SY5Y cells. *Molec. Brain Res.* **13**, 289–300.
Wiggins, R. C. (1982). Myelin development and nutritional insufficiency. *Brain Res.* **257**, 151–175.
Wilden, P. A., Morrison, B. D., and Pessin, J. E. (1989). Wheat germ agglutinin stimulation of $\alpha\beta$ heterodimeric insulin receptor $\beta$-subunit autophosphorylation by noncovalent association into an $\alpha_2\beta_2$ heterotetrameric state. *Endocrinology* **124**, 971–979.
Wilson, E., Olcott, M. C., Bell, R. M., Merrill, A. H., Jr., and Lambeth, J. D. (1986). Inhibition of the oxidative burst in human neutrophils by sphingoid long-chain bases: Role of protein kinase C in activation of the burst. *J. Biol. Chem.* **261**, 12616–12623.
Wood, J., Girard, P., Mazzei, G., and Kuo, J. (1986). Immunocytochemical localization of protein kinase C in identified neuronal compartments of rat brain. *J. Neurosci.* **6**, 2571–2577.
Xue, Z. G., Le Douarin, N. M., and Smith, J. (1988). Insulin and insulin-like growth factor-I can trigger the differentiation of catecholaminergic precursors in cultures of dorsal root ganglia. *Cell Diff. Dev.* **25**, 1–10.
Young, W. S., Kuhar, M. J., Roth, J., and Brownstein, J. (1980). Radiohistochemical localization of insulin receptors in the adult and developing rat brain. *Neuropeptides* **1**, 15–22.

Zacks, S. I., Metzger, J. F., Smith, C. W., and Blumberg, J. M. (1962). Localization of ferritin labelled botulinum toxin in the neuromuscular junction of the mouse. *J. Neuropath. Exp. Neurol.* **21**, 610–633.

Zapf, J., Walter, H., and Froesch, E. R. (1981). Radioimmunological determination of insulin-like growth factors I and II in normal subjects and in patients with growth disorders and extrapancreatic tumor hypoglycemia. *J. Clin. Invest.* **68**, 1321–1330.

# 15 Ciliary Neuronotrophic Factor

Marston Manthorpe, Jean-Claude Louis, Theo Hagg, and Silvio Varon

Now is an opportune time for a review that *exclusively* addresses the ciliary neuronotrophic factor (CNTF), since nearly 15 years have passed since the first evidence revealed the existence of the CNTF protein. In past review articles (Manthorpe and Varon, 1985; Manthorpe et al., 1985, 1989), earlier information on CNTF had suggested that CNTF functionally may resemble nerve growth factor (NGF), which governs selected neuronal cell behaviors. Interested readers are encouraged to consult these previous publications, since they may contain information and perspectives not emphasized in this chapter. Information on CNTF also can be found in general reviews on growth factors, neuronotrophic factors (Varon and Adler, 1980, 1981; Varon et al., 1984, 1988a; Thoenen et al., 1987; Hefti et al., 1989; Snider and Johnson, 1989; Walicke, 1989), and NGF (Greene and Shooter, 1980; Bradshaw et al., 1985; Thoenen et al., 1985; Perez-Polo and Werrbach-Perez, 1988; Hefti et al., 1989), as well as in more specific reviews on methodologies for detection and analysis of the biological activities of CNTFs and other growth factors (Manthorpe et al., 1991). Although we have tried to cover all publications on CNTF through 1990, we apologize for any we might have omitted inadvertently.

## I. HISTORICAL BACKGROUND: 1940–1975

### A. Target-Derived Protein Factor Concept

During the 1940s and early 1950s, a series of in vitro and in vivo experiments on developing chick embryos led to the conceptualization of target-derived protein factors (later named neurotrophic or neuronotrophic factors; NTFs) that are essential for the development, maintenance, growth, and functional performances of selected populations of neurons (Hamburger, 1989). During early development, death of a large proportion of neurons occurs at the time that their axons reach the innervation territory. Escape from this developmental neuronal death appeared to depend on acquiring NTFs from the innervation territory, that is, the extent of neuronal death could be increased by target removal or by treatment with antibodies against the protein factor and could be reduced by implantation of additional target tissue or by addition of target tissue extracts or proteins purified from them (Cowan, 1973; Hamburger, 1975; Cunnigham, 1982; Hamburger and Oppenheim, 1982; Cowan et al., 1984). The best-characterized among the very few currently identified NTFs is NGF, a basic 26-kDa dimer protein addressing peripheral sympathetic and dorsal root sensory ganglion neurons (Levi-Montalcini, 1987) as well as cholinergic neurons of the basal forebrain (Hefti et al., 1989; Whittemore and Seiger, 1987). The concept of protein factors involved in the development and function of cells was a major advancement in biology, duly acknowledged by presentation of the 1986 Nobel prize for Medicine to Rita Levi-Montalcini and Stanley Cohen for their discovery and further investigation of NGF and epidermal growth factor, respectively.

## II. FROM CILIARY NEURON FACTOR CONCEPT TO FIRST IDENTIFICATION: 1976–1980

### A. Need for Purified Cholinergic Neuronal Cultures

Recognizing from past work (Hess, 1965; Marwitt et al., 1971; Landmesser and Pilar, 1972) that the avian ciliary ganglion contains a purely cholinergic parasympathetic motor neuron population innervating intraocular muscle cells, two groups of investigators developed cultures of chick ciliary ganglia (Hooisma et al., 1975; Betz, 1976). Explants of 7-day-old embryonic (ED 7) chick ciliary ganglia (CG) were placed in co-culture with dissociated ED 11 chick hindlimb skeletal muscle cells (normally the target for cholinergic spinal motor neurons). At least some CG neurons could survive, since neuritic outgrowth from the ganglia expanded over several days and since some of these neurites were axons capable of forming electrophysiologically and pharmacologically functional cholinergic synapses with some of the skeletal muscle cells. These studies were the first to demonstrate that a population of unambiguously cholinergic motor neurons in culture would retain the ability to form proper functional contacts with muscle cells. These studies appeared to encourage other investigators to use the ciliary ganglion as a source of purified motor neuron cultures for the investigation of putative NTFs (for a general strategy, see Varon et al., 1983).

## B. Heart Muscle-Derived "Parasympathetic Factor" for Ciliary Cholinergic Neurons

Helfand et al. (1976) were the first to report biologic materials with an ability to maintain cultures of dissociated CG neurons. Dissociated ED 8 chick CG were plated at a low density ($< 5000$ cells/cm$^2$) on a polylysine substratum (Letourneau, 1975) and used to test the effects on neuronal survival of a serum-containing medium "conditioned" over dissociated ED 8 chick heart cell cultures. In control serum-containing medium, all CG neurons died within 48 hr of plating, a phenomenon similar to the death of sensory or sympathetic neurons when cultured in the absence of NGF or NGF-producing cells. However, in heart-conditioned medium, up to 15% of the plated neurons survived and extended processes over a 24-hr period; a few even survived, with neurites, for 12 days. The survival-promoting activity of the heart-conditioned medium did not adsorb to the culture surface and was not contributed by the serum. It was nondialyzable and could be concentrated 10-fold by ultrafiltration, indicating an active agent of macromolecular nature. CG neuron survival could not be supported by NGF, which at the time of these studies was the only available purified neuronotrophic factor. These investigators suggested that CG neurons might require a distinct parasympathetic growth factor for their survival and that such heart muscle- derived factors may be "equivalent to factors normally derived from end organs."

Other investigators (Collins, 1978a; Varon et al., 1978, 1979; Adler and Varon, 1980) determined that heart muscle cell-conditioned medium contains at least two separable factors acting on ciliary ganglion neurons. One factor remains in the medium and supports long-term neuronal survival but not neurite outgrowth. The other factor promotes neurite outgrowth only after binding to the polycationic culture substratum, but is incapable of supporting neuronal survival by itself. These early studies promoted the concept that in vivo regulation of neuronal survival and neurite outgrowth in the ciliary ganglion also might be controlled by different agents. This concept was important at that time (and still is), since the only characterized factor known to act on neurons was NGF, which appeared to support both survival *and* neurite outgrowth (note the name "nerve growth" factor) in responsive sensory and sympathetic neurons (Levi-Montalcini, 1987). Later work has determined that the neurite-promoting activity in medium conditioned by muscle or glial (Manthorpe et al., 1981b) cultures resides in laminin, a component of the extracellular matrix, and that laminin is a very potent inducer of neurite outgrowth for a variety of peripheral and central nervous system neurons (Manthorpe et al., 1983; for reviews, see also Davis et al., 1985; Manthorpe and Varon, 1985; Manthorpe et al., 1990). Further identification of the heart-derived survival-promoting factor for parasympathetic ciliary motor neurons has been followed up only partially (Watters and Hendry, 1987).

## C. "Skeletal Muscle-Derived Factor" for Cholinergic Ciliary Neurons

Nishi and Berg (1977) subsequently determined that a similar parasympathetic neuronal survival-promoting factor was produced in cultures of ED 8–14 chick *skeletal* muscle cells. By visually counting neuronal cells in dissociated muscle–neuron co-cultures, these investigators showed that essentially all the neurons known to exist in ovo in the ED 8 ciliary ganglia by histology (Landmesser and Pilar, 1974) could be collected by dissociation. They also found that virtually all these neurons, when cultured with skeletal muscle cells (Nishi and Berg, 1977) or in the presence of skeletal muscle cell-conditioned medium (Nishi and Berg, 1979), would survive for more than 3 weeks and that most of them would extend neuritic processes and form functional synapses on available muscle cells. They suggested that *all* the ganglionic neurons, including those destined to die in ovo, could be rescued and would retain normal functional properties in response to factors produced by skeletal muscle cells. There is recent evidence suggesting that at least part of the CG neuron-supporting activity extractable from skeletal muscle resides in basic fibroblast growth factor (bFGF). Muscle extract contains a mitogenic activity for fibroblasts that is immunologically related to bFGF (McManaman et al., 1989). Moreover, bFGF stimulates cholinergic development in rat spinal cord cultures (McManaman et al., 1989) and chick CG neurons in vitro (Vaca et al., 1989) and, finally, bFGF supports CG neuron survival (Unsicker et al., 1987a; Eckenstein et al., 1990).

## D. Ciliary Neuron Target Tissue (Eye) Is a Better Source for Purification of the Survival Promoting Substance

The number of chick embryo CG motor neurons is known to decrease by 50% in vivo during the critical period of neuron–target cell connection (Landmesser and Pilar, 1978), a loss analogous to that described for other neuronal populations that undergo developmental neuronal death. Target eye removal (Landmesser and Pilar, 1978) or the addition of an eye primordium (Narayanan and Narayanan, 1978) partially exacerbated or alleviated, respectively, this developmental ciliary ganglion neuronal death, again suggesting that neuronal survival required the availability of an adequate supply of some target-derived substance.

Chick embryo extract, a commonly used neural cell culture supplement, was found to possess CG neuronal survival-promoting activity (Varon et al., 1979). It was reasonable to speculate that an even better tissue source would be the target organ for ciliary innervation. A quantitative assay for this survival-promoting activity was established and used for such investigations (Adler et al., 1979). The assay consisted of counting the number of ciliary ganglionic neurons surviving 24 hr under defined culture conditions in the presence of serially diluted test material, determining the dilution required for half-maximal survival, and defining 1 trophic unit (TU) as the activity present in 1 ml medium for such half-maximal effect. Such an assay, with further refinements, continues to be in general use for CNTF activity determinations (Manthorpe and Varon, 1989). In the first study (Adler et al., 1979), extracts from selected ED 12 chick embryo tissues were examined for total (TU/embryo equivalent) or specific (TU/mg protein) trophic activity for CG neurons. Whole chick embryo extract contained about 8000 TU/embryo, one-third of which resided in the eye. Further, the specific activity of the eye extract was about 7-fold greater than that of the embryo extract. The various layers of the chick embryo eye were dissected, extracted, and examined for trophic activity (Adler et al., 1979; Landa et al., 1980). Those components of the eye that contain the muscle cells innervated by the CG neuron in vivo (Manthorpe et al., 1985) had the highest levels of trophic activity. Extracts of these eye subcomponents (including choroid, iris, and ciliary body) contained a combined activity of about 80% of the total activity present in the eye and had a specific activity 3-fold that in whole eye and almost 20-fold that in whole chick embryo extract (2400 TU/mg vs. 800 or 125 TU/mg, respectively). These intraocular tissues were the richest sources of trophic activity found thus far and appeared to provide a practical source for the isolation and characterization of the survival-promoting activity. Moreover, the eye-derived activity increased during embryonic development over the same period (Hamburger–Hamilton stages 37 to 39) in which the developmental death of ciliary ganglionic neurons was tapering off, in strong support of a physiologic role of the putative trophic factor (Landa et al., 1980). Further evidence in that direction came from in vitro studies with the intact chick ciliary ganglion, showing that a neuronal requirement for the eye-derived factor appeared after the 8th day of embryonic development, at which time neuronal death begins to occur in vivo (Adler and Varon, 1982).

### E. Identification of Chick CNTF, a Trophic Factor Protein for Ciliary Ganglionic Neurons

In an initial biochemical study (Manthorpe et al., 1980), the substance supporting ciliary ganglion neuronal survival was purified partially from concentrated extracts of the intraocular tissues listed earlier (termed CIPE, for choroid, iris, ciliary body, and attached pigment epithelium) and defined as ciliary neuronotrophic factor or CNTF. By gel filtration and ion exchange chromatographies, as well as by isoelectric focusing (Manthorpe et al., 1982), the chick eye CNTF exhibited a size of about 40,000 daltons and an isoelectric point of 5.0–6.0. The partially purified material exhibited a specific activity of 20,000–60,000 TU/mg, a potency still 10-fold lower than that of βNGF (i.e., 500,000 TU/mg).

A more complete purification of the chick eye CNTF protein was reported in 1984 (Barbin et al., 1984). ED 15 chick eye CIPE extracts were submitted to sequential steps of DE52 ion exchange chromatography ultrafiltration concentration, sucrose density centrifugation, and preparative SDS–PAGE. An interesting characteristic of the chick CNTF was its ability to withstand heating in SDS (with or without reduction) and subsequent SDS gel electrophoresis. About 13 μg purified CNTF was recovered from 300 eye equivalents, with a specific activity of $7.7 \times 10^6$ TU/mg, an apparent molecular mass of 20,400 daltons, and an isoelectric point of 5.0. By specific activity, this CNTF was purified about 400-fold from CIPE. 1200-fold from whole eye, and 60,000-fold from whole chick embryo extract. Electrophoretic analysis of this CNTF preparation revealed one band by SDS–PAGE but multiple bands by isoelectric focusing gels. Unfortunately, the amino acid sequence from this chick CIPE-derived CNTF has not been determined to date. However, others have isolated a CNTF-like protein named growth promoting activity (GPA) from adult chicken sciatic nerves and shown it to be identical to that obtainable from embryonic chick sciatic nerve (Eckenstein et al., 1990; Nishi et al., 1990).

## III. MOLECULAR STUDIES LEADING TO MAMMALIAN CNTF PURIFICATION: 1981–1990, AND GENE CLONING: 1989–1991

### A. Mammalian CNTF Purification

During a series of studies examining the distribution of CNTF activity within rat brain and spinal cord tissue extracts, it was noted that grossly dissected distal areas of the cord (including cauda equina) contained very high CNTF activity. After further investigation it was determined that hypoglossal (pure motor), sural (pure sensory), and sciatic (mixed motor, sensory, and sympathetic) nerves and dorsal and ventral rootlets all contained high CNTF activity (Williams et al., 1984). The specific CNTF activity in the extract of adult rat sciatic nerve was about 20,000 TU/mg, or nearly the same as that obtainable from chick embryo eye tissue after partial purification. Thus, sciatic nerve represented a readily available source of mammalian CNTF for purification. The purification scheme for rat sciatic nerve CNTF (Manthorpe et al., 1986a) required some modifications of the one used for chick eye extracts (Barbin et al., 1984), including the use of a lower salt concentration for the elution from the DE52 column and the collection of CNTF activity from a higher molecular weight region of the SDS gel, to accommodate the slightly different properties of the rat CNTF. This procedure allowed the isolation of about

2 µg CNTF protein from 1.5 gm nerves (about 20 nerves; 29 mg protein), exhibiting one 24,000 dalton band by SDS–PAGE and a specific activity of about $2 \times 10^7$ TU/mg, representing a 1000-fold purification from the nerve extract.

We have developed a more convenient, rapid, and high yield procedure for purifying rat sciatic nerve CNTF that involves sequential chromatographic steps through MonoQ anion exchange (in 5% ethylene glycol), phenyl sepharose, and C2 reverse-phase chromatographies (M. Manthorpe, unpublished observations). This new 2-day procedure can be performed on a large scale (i.e., 8000 adult rat sciatic nerves; 600 g wet weight) and results in a 7000-fold purification of CNTF, with a recovery of about 30% and a specific activity of about $2 \times 10^7$ TU/mg.

Another group (Lin et al., 1989, 1990) has purified rabbit CNTF from adult sciatic nerve by sequential steps of ammonium sulfate precipitation followed by CL-4B phenyl sepharose. MonoP ion exchange, alkyl-sepharose HR10/10, and preparative SDS–PAGE chromatographies and, finally, C8 reverse-phase HPLC. Approximately 1.2 µg CNTF was obtained from 200 gm wet weight) nerves (600 nerves; 2.8 gm protein). This preparation exhibited two major bands at 22,000 and 24,000 daltons by SDS–PAGE, had a specific activity of about $1.2 \times 10^7$ TU/mg, and was purified about 25,000-fold.

Several other groups have reported attempts to isolate proteins possessing CNTF-like activity. Partially purified protein preparations have been derived from chicken gizzard (Miki et al., 1981), pig lung (Wallace and Johnson, 1987), and bovine heart (Watters and Hendry, 1987). These partially purified proteins have not been characterized sufficiently to compare with the sequenced CNTF to be described next.

## B. CNTF Sequencing, Cloning, and Expression

### 1. Rat CNTF

Different research groups now have reported the purification, sequencing, cloning, and recombinant expression of CNTF. In one study (Stöckli et al., 1989), the investigators used the same CNTF purification procedure used previously by others for rat sciatic nerve extract (Manthorpe et al., 1986a), except that the final SDS gel eluate was submitted to an additional fractionation through C4 reverse-phase HPLC. The purified product exhibited one spot on two-dimensional gels but a specific activity was not reported. CNTF fragments were generated by cleavage with cyanogen bromide or trypsin, the fragments separated by C4 HPLC, and selected peaks representing about 50% of the CNTF protein were sequenced using a gas-phase sequenator (Applied Biosystems). Complementary oligonucleotides corresponding to selected amino acid sequences were synthesized with reverse transcriptase and used as primers with a rat brain astrocyte RNA library for amplification of specific CNTF cDNA segments. The CNTF cDNAs were subcloned and several inserts were sequenced; a unique CNTF amino acid sequence was deduced. The cDNA predicts a protein of 200 amino acids ($M_r$, 22,800 daltons) that contains only one cysteine ($Cys_{17}$) and lacks a signal sequence and consensus sequences for glycosylation. These investigators were able to express a full-length 847-bp CNTF transcript containing the 600-bp coding region using a riboprobe system in HeLa cells. As expected from the deduced lack of a signal sequence for CNTF, transfected HeLa cells expressed the active CNTF protein but did not release it into the culture medium.

## 2. Rabbit CNTF

Almost simultaneously, Lin and colleagues (Lin et al., 1989, 1990) purified (see previous section) and sequenced adult rabbit CNTF. Endoprotease and chymotryptic fragments were generated and separated by C18 HPLC and amino acid sequences were determined for several of the isolated peptides. Degenerate oligonucleotides corresponding to selected peptides were prepared and used as primers for PCR-expansion of a rabbit genomic library. The reaction products were subcloned and screened with an intervening oligonucleotide. One subclone was radiolabeled and used to screen a rabbit nerve cDNA library from which a 1.5-kb *Eco*RI fragment was obtained and shown to contain the entire CNTF coding sequence. The CNTF cDNA was deduced to encode a protein of $M_r$ 22,700 daltons, with an isoelectric point of 5.78, both of which agreed with the properties displayed by the purified rabbit CNTF protein. The 1.5-kb *Eco*RI fragment was subcloned into a transient expression vector and transfected into COS-7 cells. Active CNTF could be extracted from the transfected cells but no CNTF activity was released into the culture medium.

## 3. Human recombinant CNTF

More recently, Lam and colleagues (Lam et al., 1991) generated labeled oligonucleotide probes corresponding to the published rabbit CNTF gene sequence and used them to screen a human lung fibroblast Lambda F1x genomic library. The human CNTF gene was prepared from phage DNA using PCR and amplified by use of primers to the 5′ and 3′-translated regions of rabbit CNTF cDNA, which contain two specific *Eco*RI recognition sites. The amplified fragment was subcloned into pUC and M13 plasmids. Restriction enzyme fragments were generated using *Eco*RI–*Hin*dIII (>>5.5 kb), *Sau*3A, and *Xba*I (>>4.5 kb), and subcloned into pUC or M13. DNA sequences were determined using the dideoxynucleotide chain-termination method. The human CNTF gene encodes a 200-residue protein and appears to be a single-copy gene with simple genetic organization. The human CNTF coding domain, like that of the rat (Stöckli et al., 1989), is interrupted by only a single intron; the gene is located on chromosome 11. Figure 1 shows that the amino acid sequence of human CNTF has been well conserved in evolution, since

```
MAFAEQTPLTLHRRDLCSRSIWLARKIRSD          30    rat
MAFMEHSALTPHRRELCSRTIWLARKIRSD          30    rabbit
MAFTEHSPLTPHRRDLCSRSIWLARKIRSD          30    human
----------------------------------------------------
LTALMESYVKHQGLNKNINLDSVDGVPVAS          60
LTALTESYVKHQGLNKNINLDSVDGVPMAS          60
LTALMESYVKHQGLNKNINLDSADGMPVAS          60
----------------------------------------------------
TDRNSEMTEAERLQENLQAYRTFQGMLTKL          90
TDQWSELTEAERLQENLQAYRTFHIMLARL          90
TDQWSELTEAERLQENLQAYRTFHVLLARL          90
----------------------------------------------------
LEDQRVHFTPTEGDFHQAIHTLMLQVSAFA         120
LEDQQVHFTPAEGDHFQAIHTLLLQVAAFA         120
LEDQRVHFTPTEGDFHQAIHTLLLQVAAFA         120
----------------------------------------------------
YQLEELMVLLEQKIPENEADGMPATVGDGG         150
YQIEELMVLLECNIPPKDADGTPVIGGDG?         150
YQIEELMILLEYKIPRNEADGMPINVGDGG         150
----------------------------------------------------
LFEKKLNGLKVLQELSQWTVRSIHDLRVIS         180
LFEKKLWGLKVLQELSHWTVRSIHDLRVIS         180
LFEKKLNGLKVLQELSQWTVRSIHDLRFIS         180
----------------------------------------------------
SHQMGISALESHYGAKDKQM                   200
CHQTGIPAHGSHYIANDKEM                   200
SHQTGIPARGSHYIANNKM                    200
```

FIGURE 1

Rat, rabbit, and human CNTF sequence comparison. Rat sequences are from Stöckli et al. (1989), rabbit sequences are from Lin et al. (1989), and human sequences are from Lam et al. (1991). Underlined residues represent differences. The genes have an overall 80-85% sequence homology.

rat and rabbit sciatic nerve CNTFs display ~85% homology to the inferred amino acid sequence for human CNTF. Three other groups have reported the preparation of recombinant human CNTF (Masiakowski et al., 1991; McDonald et al., 1991; Negro et al., 1991). Recombinant CNTFs now are being produced by four pharmaceutical companies: California Biotechnology (Mountain View), Regeneron (Tarrytown, New York), Synergen (Boulder, Colorado) and Fidia (Italy).

### 4. Antibodies against CNTF

Several groups have reported on the generation or use of antibodies to the CNTF protein (see also Hendry et al., 1988; Watters et al., 1989). Four groups have generated antibodies to synthetic peptide fragments corresponding to regions in the CNTF protein. One group (Lin et al., 1989, 1990) prepared affinity purified rabbit polyclonal antibodies to synthetic peptides corresponding to rat CNTF residue 45–60 (peptide A) and 181–200 (C-terminal peptide B). Both peptide antibodies recognized purified rat CNTF by ELISA as one 24–28,000-dalton band in purified rat CNTF immunoblots and in rat sciatic nerve extracts. These antibodies were used to immunostain adult sciatic nerve sections, demonstrating the localization of the CNTF antigen exclusively in Schwann cells (Rende et al., 1992). A third group (Stöckli et al., 1991) produced antisera to amino acid sequences 127–153 and 186–199 of rat CNTF. These antisera stained a subpopulation of type 1 astrocytes in optic nerve and olfactory bulb of adult rats, as well as in vitro. A fourth group reported the production of monoclonal antibodies generated against the C terminus of human CNTF (Furth et al., 1990). These antibodies are specific to human recombinant CNTF (they do not react with rat recombinant CNTF) and recognize CNTF bound to its receptor.

### 5. CNTF receptors

The lack of reagents and molecular probes for the CNTF receptor has hindered the characterization of the molecular mechanism by which CNTF exerts its actions on nerve cells. Virtually nothing is known thus far about properties and localization of CNTF receptors. In one study (Squinto et al., 1990), a CNTF molecule containing a 10-amino-acid epitope tag at its C terminus was engineered genetically and expressed in *E. coli*. Several human neuronal tumor cell lines, as well as chick embryo and adult dorsal root ganglionic (DRG) neurons (see subsequent text), were identified as targets for tagged CNTF binding by a rosetting assay. Rosette-positivity correlated with the rapid induction of c-*fos* and c-*jun* by the modified CNTF, indicating the presence of functional receptors. This tagging strategy was used by that group for the purification and molecular cloning of a low-affinity CNTF receptor (Davis et al., 1991). The CNTF receptor shows homology to the interleukin-6 receptor, a finding which, together with the predicted structural homology of CNTF with interleukin-6 and CDF/LIF, suggest that CNTF belongs to a superfamily of neuroactive cytokines (Bazan, 1991; Hall and Rao, 1992).

## IV. BIOLOGIC PROPERTIES

Early studies recognized and defined CNTF by its survival-promoting activity on developing ciliary ganglionic neurons. This relationship has been the one most studied until now. The CNTF-like activities present in extracts and partially purified fractions, however, need not be true CNTF; thus, careful scrutiny will be required to identify which cells respond to CNTF during development in vivo and in vitro and in the adult in vivo. Many in vivo studies since have reported the effects of purified CNTF on developing (pre- or perinatal) neurons from the peripheral nervous system (PNS) as well as the central nervous system (CNS). More recently, CNTF effects in vivo on lesioned neonatal and adult CNS neurons have been described. CNTF displays a variety of activities addressing survival as well as differentiation. Moreover, CNTF affects neurons of more than one transmitter phenotype, indicating that CNTF is not a transmitter-specific neuronotrophic factor. CNTF addresses neurons that also respond to other factors, such as NGF and bFGF, raising the concept of multiple modulatory interactions controlling the development and maintenance of neurons. The trophic factor prototype, NGF, which was discovered about 40 years ago, also was recognized initially and studied on developing peripheral neurons, but has been implicated subsequently as a trophic factor for neurons of the adult CNS (Varon et al., 1984; Whittemore and Seiger, 1987; Hefti et al., 1989). Moreover, a growing number of reports indicates an NGF role in the immune and other nonnervous systems (Ayer et al., 1988b; Otten et al., 1989). By analogy, it is to be expected that the list of CNTF-responsive cells and CNTF-induced responses will expand in the future. In this section, we shall describe (1) the cells that currently are known to respond to CNTF, (2) cellular and tissue sources of CNTF, and (3) the types of responses induced by CNTF in various nerve cell populations.

## A. CNTF-Responsive Cells

### 1. CNTF-responsive neurons

*a. Developing ciliary ganglionic neurons*   As already mentioned, initial studies used the survival of cultured cholinergic parasympathetic motor neurons from the chick CG to recognize and measure CNTF-like trophic activities from different tissues. When cultured in vitro, ED 8 CG motor neurons will die within 1–2 days with no addition of the factor, but will survive in medium supplemented with CNTF from chick eye or rat sciatic nerve and are equally responsive to human and rat recombinant CNTF. In addition to providing trophic support for CG neurons, CNTF also stimulates their choline acetyltransferase (ChAT) activity (Nishi and Berg, 1979; Unsicker et al., 1987a). CNTF-supported CG neurons, however, usually do not extend neurites unless they are grown on laminin (Adler and Varon, 1980).

*b. Developing sympathetic neurons*   Sympathetic ganglionic neurons are dependent on the availability of NGF for survival during their in vivo and in vitro development. Chick eye-derived CNTF also supports sympathetic ganglion neurons from ED 12 chick and neonatal rat as well as NGF will (Barbin et al., 1984), thereby overlapping the neuronal target spectrum of NGF. CNTF affects the differentiation of cultured ED 7 chick sympathetic neurons by inhibiting the proliferation of the sympathoblasts and by inducing the expression of vasoactive intestinal peptide (VIP) in a large population (60%) of neurons that otherwise express tyrosine hydroxylase (TH) (Ernsberger et al., 1989). In cultures of neonatal rat sympathetic neurons, CNTF also induces cholinergic differentiation by increasing their ChAT content and activity and neuropeptide expression and decreasing TH (Saadat et al., 1989; Rao et al., 1990, 1992).

*c. Developing peripheral sensory neurons*  Sensory neurons from the DRG are also dependent on the availability of NGF during defined periods of their ontogenesis (reviewed in Levi-Montalcini, 1987). CNTF does not support chick DRG neurons in vitro (whereas NGF does) in their earlier stages (ED 8), but does so at ED 10–15 (Manthorpe et al., 1981a; Barbin et al., 1984; Eckenstein et al., 1990). CNTF does not seem to address a specific subset of DRG neurons, since all the DRG neurons that are supported maximally by NGF and brain-derived neurotrophic factor (BDNF, Lindsay et al., 1985) also are supported maximally by CNTF (Skaper et al., 1986). Moreover, with the epitope-tagged CNTF as a ligand, binding sites could be revealed on the surface of all embryonic chick DRG neurons supported by NGF (Squinto et al., 1990). The CNTF binding sites were distributed all over the cell bodies as well as along the neurites, in embryonic as well as in adult DRG neurons. Functional CNTF receptors, able to cause rapid activation of early gene expression, also have been shown in adult rat neurons (Squinto et al., 1990). Adult DRG neurons, although not requiring NGF for their survival, respond to NGF in vitro by extending neurites (Lindsay, 1988).

*d. Spinal cord neurons*  Recent studies, both in vivo and in vitro, indicate that somatic spinal cord motor neurons are targets of CNTF. The survival of a significant fraction (60%) of ED 6 chick motor neurons in enriched cultures is supported in a dose-dependent manner by human recombinant CNTF (Arakawa et al., 1990). Moreover, administration of CNTF to chick embryos at the time of naturally occurring spinal cord motor neuron death (ED 6–10) rescues 40% of the motor neurons that normally die (Wewetzer et al., 1990; Oppenheim et al., 1991). The somatic motor neurons present in cultures of ED 14 rat spinal cord also depend on CNTF for survival and for ChAT expression (Wong et al., 1990; Magal et al., 1991a). However, the action of CNTF on spinal cord cultures does not appear to be restricted to somatic and visceral motor neurons, since a much larger population (25–35%) of the neurons dissociated from ED 14 rat cord survives in culture in the presence of CNTF (Magal et al., 1991a). This population represents noncholinergic smaller neurons that have not been characterized. In addition, there is evidence that survival of neonatal motor neurons in vivo also is promoted by CNTF (Sendtner et al., 1990). Application of CNTF to the proximal stump of a transected facial nerve rescues most of the motor neurons of the facial nucleus of the brainstem, which would otherwise degenerate.

Studies in the adult rat suggest that preganglionic viscero-motor neurons of the intermediate lateral column that innervate the adrenal medulla also require a target-derived factor for their maintenance in vivo (Blottner et al., 1989). Destruction of the adrenal medulla causes a loss of 25% of these preganglionic neurons (located at Th7–10) after 4 weeks. CNTF (as well as bFGF) can rescue these neurons; interruption of the axonal pathway from the spinal cord to the adrenal gland abolishes the effect of CNTF, supporting its proposed role as a retrograde trophic factor for preganglionic neurons.

*e. CNS neurons* Only little evidence exists for well-defined CNS targets of CNTF. CNTF increases GABA uptake and neurofilament protein expression in cultures of ED 18 rat hippocampus and promotes the survival of hippocampal GABAergic, cholinergic, and calbindin-positive neurons (Ip et al., 1991). Under certain culture conditions embryonic rat noradrenergic neurons from the locus coeruleus as well as dopaminergic ones from the substantia nigra respond to CNTF with an increased tyrosine hydroxylase expression (Louis; Magal, unpublished observations). This suggests that the spectrum of responsive CNS neurons is broader. In the adult rat intraventricular infusion of CNTF after a unilateral fimbria–fornix transection can prevent the loss and atrophy of all medial septum neurons (cholinergic as well as noncholinergic) and, in the cholinergic neurons, can prevent the axotomy-induced loss of low affinity NGF receptor (LNGFR) but not ChAT staining (Hagg et al., 1992). In addition, intraventricular CNTF induced LNGFR (but not ChAT) staining in the cholinergic neurons of the striatum, both markers being known to respond to NGF (Hagg et al., 1989b; Williams et al., 1989).

CNTF may play a role in the development and differentiation of the neural retina. CNTF is able to support about 40% of ED 10 chick retinal ganglionic neurons in culture (Lehwalder et al., 1989). This effect is shared by NGF and a crude preparation of BDNF, but not by bFGF. CNTF may address the same subpopulation of retinal ganglion cells as NGF, since their effects are not additive. In addition another class of retinal neurons, the amacrine cells, are induced by CNTF to increase their cholinergic characteristics in culture (Hofmann, 1988).

*f. Neuronal cell lines* The epitope tagging strategy was used to demonstrate that several neuronal cell lines expressed CNTF receptors (Squinto et al., 1990). Whereas only one (of four) neuroblastoma line (SH-SY5Y) was found to be positive, all three neuroepithelioma lines tested (CHP-100, SK-N-LO, SK-N-MC) and two of three Ewing's sarcoma lines tested (5838, IARC-EW-1) expressed CNTF binding sites. None of the nonneuronal cell lines (astrocytoma, melanoma, fibroblast) was found to be receptor positive. A rapid and significant increase of c-*fos* mRNA in response to CNTF was detected in all cell lines that displayed CNTF binding sites, suggesting the presence of transducing receptors.

## 2. CNTF-responsive nonneuronal cells

*a. Adrenal chromaffin cells* CNTF promotes in vitro survival and neurite outgrowth of cultured chromaffin cells of 8-day-old rats (Unsicker et al., 1985a) in a fashion similar to that of NGF. CNTF and NGF synergistically affect the formation of neurites in chromaffin cells from older (16–100-day-old) rats that no longer require these factors for their survival (Unsicker et al., 1985b). In contrast to NGF, CNTF does not induce tyrosine hydroxylase activity, but partially protects the catecholamine storage capacity of chromaffin cells (Seidl et al., 1987).

*b. Oligodendrocyte-type 2-astrocyte progenitor cells* The neuroepithelial cells of the neural tube give rise to two major classes of glial cells, astrocytes and oligodendrocytes, which structurally and functionally support the neurons and their axons in the CNS (Raff, 1989). Cell culture studies of the optic nerve or cerebral cortex of the newborn rat have identified a bipotential precursor cell, the oligodendrocyte–type-2 astrocyte (O–2A) progenitor, which can develop into an oligodendrocyte or a type-2 astrocyte under the appropriate inducing conditions (Raff, 1989; Noble et al., 1990). The differentiation into oligodendrocytes appears not to require environmental inducing signals and seems to be driven by an intrinsic mechanism that limits O–2A cell proliferation (Temple and Raff, 1985). Oligodendrocyte differentiation is thought to follow automatically as a consequence of O–2A cell withdrawal from the cell cycle.

Conversion of O–2A cells to type-2 astrocytes, on the other hand, appears to involve CNTF action (Hughes and Raff, 1987; Hughes et al., 1988; Lillien et al., 1988, 1990; Lillien and Raff, 1990). Optic nerve extract is capable of inducing O–2A cells to express the astrocytic marker GFAP (Hughes and Raff, 1987). It has been shown that type-1 astrocytes of the optic nerve are a source of CNTF (Lillien et al., 1988). The differentiation-inducing activity is very low in optic nerve of 1-week-old rats and reaches its adult high level at 3 weeks. Optic nerve-derived CNTF activity and the GFAP-inducing activity co-purify, and their effects are mimicked by purified CNTF, suggesting that CNTF is the physiologic signal that induces type-2 astrocyte differentiation (Hughes et al., 1988). The differentiation induced by CNTF is, however, transient and other signals are required to drive the process to completion. Recent studies have shown that molecules produced by nonglial cells of the optic nerve in vitro and associated with the extracellular matrix cooperate with CNTF to induce stable type-2 astrocyte differentiation (Lillien et al., 1990).

## B. CNTF Sources

### 1. Target Cells

The dependence for survival of developing chick ciliary ganglionic and spinal neurons on a target-derived substance was postulated in the early studies (e.g., Helfand et al., 1976). Indeed, subsequent studies (Adler et al., 1979; Landa et al.,

1980) showed that 80% of the CNTF-like activity of the eye resided in the structures that contain the muscles innervated by the ciliary ganglionic neurons (choroid, iris, and ciliary body). CNTF or GPA (Eckenstein et al., 1990) may be available in substantial amounts during development in the innervation target tissue of chick CG neurons. The dependence of CG neurons on CNTF is maximal during a specific time window during development and declines thereafter (Manthorpe et al., 1981a).

No information is currently available on the presence of CNTF or its mRNA in the eye or in the target innervation territories of sympathetic and sensory neurons, on which CNTF exerts a trophic action. However, no CNTF mRNA was found in the skin (Stöckli et al., 1989).

Skeletal muscle-derived factors had been shown to support chick CG motor neuron survival, that is, they had a CNTF-like activity. With the target-derived factor hypothesis in mind, skeletal muscle was examined for CNTF mRNA and activity. Using Northern blot analysis, no detectable levels of CNTF mRNA were found in adult rat skeletal muscle tissue (Stöckli et al., 1989). In fact, as discussed earlier, the survival-promoting activity for CG neurons that was detected previously in muscle extracts may be associated mainly with FGF.

### 2. Peripheral nerve

Northern blot analysis of the distribution of CNTF mRNA revealed that sciatic nerve contained by far the largest amount of it in tissues of adult rat (Stöckli et al., 1989). CNTF mRNA was undetectable in sciatic nerves of neonatal rats, only becoming apparent by day 4 and reaching a maximum at day 13. The undetectable levels of CNTF mRNA in adult skin and muscle indicate that the large amounts of CNTF present in the adult sciatic nerve do not represent CNTF transported retrogradely from the periphery (as is the case for NGF), but rather CNTF synthesized locally. Immunohistochemical studies have revealed that, in the adult nerve, almost all CNTF is associated with the Schwann cells and none can be found in skeletal muscle (Rende et al., 1992). These results, coupled with the finding that Schwann cells can produce CNTF-like activity in vitro (Muir et al., 1989), strongly suggest that, in the adult peripheral nervous system, CNTF is produced by the Schwann cells.

### 3. CNS

In the CNS, two presumptive sources of CNTF are glial cells (astrocytes or oligodendrocytes) and the neurons that are postsynaptic targets for CNTF-using neurons. The CNTF protein/activity and its mRNA in the adult CNS has recently been shown to reside predominantly in the optic nerve and olfactory bulb with detectable, but much lower, levels in a few other regions (e.g., hippocampus) (Stöckli et al., 1991). There is increasing evidence that CNTF can be produced by type-1 astrocytes (Stöckli et al., 1991). In vitro, neonatal rat cortex type-1 astrocytes produce CNTF-like activity (Rudge et al., 1985, 1987; Manthorpe et al., 1986b, 1988), neonatal rat hippocampal type-1 astrocytes express mRNA for CNTF and express CNTF immunoreactivity on their cell surface (Rudge et al., 1990), and adult optic nerve and olfactory bulb astrocytes have been found immunopositive for CNTF. Astrocytes also are known to produce NGF in vitro (Rudge et al., 1985; Furukawa et al., 1986; Assouline et al., 1987) but they do not contain detectable levels of NGF mRNA in the normal adult brain (Ayer et al., 1988a). By analogy, it is possible that most adult astrocytes in vivo would not express detectable amounts of CNTF mRNA.

### 4. Nonneural tissues

Low levels of a CNTF activity (400-fold lower than sciatic nerve) with an electrophoretic mobility similar to that of purified CNTF have been found in kidney (Rudge et al., 1987). In addition, very low levels of CNTF activity were found in liver, gizzard, skin muscle, lung, and heart (Rudge et al., 1987). In the adrenal gland, only the innervated medulla contains CNTF-like trophic activity (Unsicker et al., 1988).

### 5. CNTF availability

The cDNA of CNTF predicts a protein that lacks the consensus hydrophobic signal sequence necessary for processing of proteins through the endoplasmic reticulum and subsequent vesicular secretion (Stöckli et al., 1989), suggesting that CNTF is a cell-associated cytosolic rather than a secreted protein. This property has been recognized in an early study that reported that CNTF-like activity produced by cultured astrocytes after several passages was extractable from the cells but not from their conditioned media (Rudge et al., 1985; Manthorpe et al., 1988). Additional evidence for the cell-associated nature of CNTF is the finding that cultured HeLa cells (Stöckli et al., 1989) or COS-7 cells (Lin et al., 1989) transfected with CNTF cDNA produce CNTF without releasing it into the medium.

After an injury of the adult rat peripheral nerve, levels of CNTF-like activity measured in a silicone chamber fitted to the nerve stump increased dramatically during the first few hours to days (Longo et al., 1983a,b; Lin et al., 1989; Manthorpe et al., 1989). The acute and transient nature of the increase suggests that CNTF is released immediately or in the first days after the lesion, for example, from damaged Schwann cells in the nerve stumps. In the adult brain, mechanical or chemical lesions to the entorhinal cortex resulted in an increased CNTF-like activity (much lower than that seen in the silicone chamber) in the tissue surrounding the wound, in the wound fluid, and in the deafferented brain regions (Nieto-Sampedro et al., 1982, 1983; Longo et al., 1983a,b; Lin et al., 1989; Manthorpe et al., 1989). The maximum levels were reached progressively after about 10 days, suggesting an active production and release mechanism. This time course and the location of the increased CNTF activity correlate with and may reflect the appearance of reactive astrocytes after similar traumatic and chemical injuries. The difference between the PNS and the CNS with regard to the time course and maximal levels of the injury-induced CNTF activity is likely to reflect the difference in amounts of stored CNTF in Schwann and astroglial cells, respectively. Thus, CNTF in the adult may play the role of a lesion faction, to be found in soluble form (or in the extracellular milieu) only under pathologic conditions.

Other protein factors lacking a signal sequence are known to be released from their producing cells, for example, FGF (of which the mechanism of processing and secretion is unknown) and interleukin-1 alpha (IL-1). Phosphorylation of the IL-1 precursor is thought to facilitate its intracellular processing, transport, and insertion into the membrane; IL-1 may be released into the extracellular fluid by proteolytic cleavage of the membrane-associated precursor (Beuscher et al., 1988; Kobayashi et al., 1988; Streck et al., 1988). Phosphorylation also may be a mechanism by which bFGF is exported from the cell (Feige et al., 1989). Although nothing yet is known about CNTF, it is of interest that, analogous to the membrane-bound form of IL-1, cultured newborn rat astrocytes express CNTF immunoreactivity on or associated with their cell structure (Rudge et al., 1990).

After its release, under normal or pathologic conditions, CNTF could act as a soluble factor and interact with its specific receptors or, like FGF (Baird et al., 1987; Flaumenhaft et al., 1989), become bound to the extracellular matrix. The bound CNTF could act locally from its bound position or be collected actively and transported retrogradely by the responsive cells. No evidence is yet available for binding of CNTF to the extracellular matrix. Interestingly, the presence of matrix-associated molecules is required for stable CNTF-induced differentiation of O–2A progenitor cells into type-2 astrocytes in serum-free culture conditions (Lillien et al., 1990). Also consistent with this concept is the finding that CNTF bound to nitrocellulose paper retains its trophic activity for ciliary, dorsal root, and sympathetic ganglionic neurons (Carnow et al., 1985; Rudge et al., 1987; Pettmann et al., 1988). On the other hand, other investigators have found no survival-promoting activity for CG neurons by chick CNTF (CIPE) bound to laminin- or polyornithine-coated substrates (Unsicker et al., 1987a).

## C. Physiologic Functions of CNTF

### 1. Neuronal survival

Most of the knowledge about CNTF has been obtained on prenatal or neonatal tissue. There is no doubt that CNTF—or a homologous factor—plays an important role during neuronal development. CNTF affects survival of peripheral CG neurons during a defined period of their ontogenesis, during which the process of naturally occurring cell death takes place. This effect suggests that eye-derived CNTF is the target-derived neurotrophic factor for CG motor neurons. CNTF could play a similar role during the ontogenesis of sympathetic and sensory neurons. The view that CNTF acts as a natural target-derived neurotrophic factor for these peripheral neurons should, however, be tempered by the recent and rather unexpected observation that the naturally occurring cell death in the chick ciliary ganglion (ED 9–14), sympathetic ganglion (ED 6–10), and dorsal root ganglion (ED 6–10) cannot be prevented by the administration of exogenous recombinant human CNTF to the chorioallantoic membrane (Oppenheim et al., 1991). It is possible that the survival of these neuronal populations requires the endogenous native CNTF molecule of the chick and that recombinant CNTF does not perform all the functions of the chick CNTF. Indeed, GPA that has been purified from chick eye and sciatic nerve, and may be the true neurotrophic factor for chicken, has been shown to display slightly different molecular characteristics from but similar biologic properties in vitro to mammalian CNTFs (Eckenstein et al., 1990; Parent et al., 1990). On the other hand, multiple influences may be required for the full development of CG neurons. Depolarization induced by exposure to high $K^+$ concentrations (25–40 m$M$) improves survival of ED 8 CG neurons (Nishi and Berg, 1981) but fails to induce neurite extension. Basic FGF also is capable of supporting CG neurons in vitro (Schubert et al., 1987; Unsicker et al., 1987a). Moreover, it should be noted that the reports identifying CNTF as the trophic activity contained in extracts of various tissues or in conditioned media must be reevaluated carefully, since bFGF may be present in such extracts also.

A CNTF-like factor in skeletal muscle may play a similar role as target-derived neurotrophic factor for spinal motor neurons during development. In vitro, ED 6 chick (Arakawa et al., 1990) and ED 14 rat (Wong et al., 1990; Magal et al., 1991) spinal motor neurons are dependent on CNTF for their survival and for ChAT and neurofilament protein expression. In addition, extracts of muscle or media conditioned by cultured muscle cells are able to support survival and enhance cholinergic development of rat and mouse spinal cord motor neurons (Giller et al., 1977; Kaufman et al., 1985; McManaman et al., 1988, 1990; Martinou et al., 1989). The factors responsible for these activities have not been purified fully (with the exception of the ChAT development factor, CDF, note: not the same as CDF/LIF) and their relationship to CNTF still must be clarified. CDF is a 22,000-dalton polypeptide, isolated from rat muscle, that increases the levels of ChAT activity in cultures of rat spinal cord neurons (McManaman et al., 1988, 1990). The application of this factor to developing chick embryo during the period of naturally occurring spinal motor neuron death (ED 6–10) significantly increased the survival of motor neurons (McManaman et al., 1990). Similarly, administration of CNTF to chick embryos rescues 40% of the motor neurons that normally die during this period, suggesting that CNTF may act as, or substitute for, a target-derived neurotrophic factor for spinal motor neurons (Wewetzer et al., 1990; Oppenheim et al., 1991). Although CNTF and CDF display some different physical and chemical features (McManaman et al., 1988), the relationship between these two factors remains to be established. The failure of CNTF to prevent the death of the remaining motor neurons is again consistent with the idea that more than one factor regulate the development of motor neurons (Oppenheim et al., 1991). bFGF, for example, is a good candidate for such a role (Arakawa et al., 1990). However, the very low levels of CNTF protein and mRNA found during perinatal development and the lack of a signal sequence necessary for secretion are not in favor of a role in naturally occurring death of motor neurons.

In adult rat, very little CNTF mRNA is found in the muscle (Stöckli et al., 1989). In contrast, CNTF protein and mRNA are present in large amounts in peripheral nerve. This nerve (Schwann cell)-derived CNTF is likely to play a crucial role in the protection of motor neurons against injury-induced degeneration as well as in their normal functional maintenance. Evidence for such a role is provided by already mentioned experiments that show that axotomy of the facial nerve of 1-week-old rats results in the nearly complete loss of motor neurons in the facial nucleus of the brain stem, but that most of them are rescued by the application of CNTF to the proximal stump (Sendtner et al., 1990). This axotomy-induced death of facial motor neurons in the young rat coincides with a very low level of CNTF protein and mRNA in the peripheral nerves (Stöckli et al., 1989). In the adult rat, on the other hand, axotomy is not followed by motor neuron death. At this stage, CNTF is present in large quantities in the Schwann cells of the peripheral nerves and can be released after lesion in amounts sufficient to prevent the degeneration of the motor neurons. Schwann cell-associated CNTF also could be made available to normal adult spinal motor neurons and play a role in their maintenance and physiologic function. A similar emergency function may be attributed to CNTF in the maintenance and protection, after lesion, of adult CNS neurons of the medial septum (Hagg et al., 1992), indicating that the trophic action of CNTF may have a broader spectrum of neuronal targets that is still to be recognized.

### 2. Neurite outgrowth

For NGF, a specific effect on neurite extension distinct from that on survival has been described in developing sympathetic and sensory neurons (Lindsay, 1988). Unlike NGF, CNTF does not seem to have a clear-cut action on neurite outgrowth. Adrenal chromaffin cells (Unsicker et al., 1985b, 1987b) extend neurites when cultured in the presence of CNTF. CNTF enhances neurofilament protein synthesis in cultured rat hippocampal neurons (Ip et al., 1991) and spinal cord neurons (Wong et al., 1990). However, it is unclear whether the outgrowth of neurites seen in these cultures is due to a general trophic effect of CNTF on these cells, thus providing optimal conditions for them to extend fibers, or to a specific CNTF neurite-promoting activity. Arguing against such a role for CNTF are the analyses of CNTF activity by cell blot (Pettmann et al., 1988). CNTF and NGF bands can be transferred from an electrophoretic gel to nitrocellulose strips and probed with dissociated neural cell cultures for survival- and neurite-promoting activities. Neurons from chick ED 10 DRG and ED 12 sympathetic ganglia survive equally well on the CNTF band and on the 26- and 13-kDa bands (corresponding to the dimer and monomer forms) of NGF. However, extensive neurite outgrowth from the neurons, evident on the NGF bands, was not observed on the CNTF band. In addition, embryonic chick CG neurons, supported by CNTF for their survival, do not extend neurites unless they are grown on laminin (Adler and Varon, 1980). Further, unlike NGF, CNTF does not include sprouting of lesioned medial septum cholinergic neurons when infused intraventricularly in the adult rat (Hagg et al., 1992; see also Schonfeld et al., 1985). In cultures of older CG neurons (ED 15), which do not depend on CNTF for their survival, CNTF does induce neurite outgrowth (Manthorpe et al., 1981a).

### 3. Neurotransmitter regulation

CNTF acts on several types of PNS and CNS neurons that share a cholinergic neurotransmitter function: CG motor neurons, spinal motor neurons, medial septal neurons, amacrine cells, and preganglionic sympathetic neurons. For some of these neurons (CG and spinal motor neurons), the increase in the expression of ChAT by CNTF cannot be dissociated from the CNTF effect on survival. For the amacrine cells, however, the effect of CNTF is clearly on ChAT activity (Hofmann, 1988). Likewise, CNTF has only a very modest effect on the survival of neonatal sympathetic neurons in culture, but promotes cholinergic differentiation, that is, an increase of ChAT content and activity and a decrease of tyrosine hydroxylase (Saadat et al., 1989; Rao et al., 1990). This transmitter-phenotype switching effect of CNTF on the sympathetic neurons is shared by at least two other molecules: the cholinergic neuronal differentiation factor, similar to the leukemia-inhibitory factor (CDF/LIF; Yamamori et al., 1989; Rao et al., 1990, 1992), and a 29-kDa membrane-associated neurotransmitter-stimulating factor (MANS), isolated from rat spinal cord (Wong and Kessler, 1987; Adler et al., 1989). Comparison of their deduced amino acid sequences and of their biologic (CDF/LIF does not support CG neurons) and immunologic properties indicates that CDF/LIF and CNTF are distinct factors (Lin et al., 1989; Stöckli et al., 1989; Yamamori et al., 1989; Rao et al., 1990, 1992). The relationship between MANS and CNTF is, as yet, unclear. Like CNTF, MANS is competent to induce cholinergic properties in sympathetic neurons and supports CG neuron survival.

CNTF, however, is not to be considered a cholinergic neuron-specific, or ChAT-inducing, neurotrophic factor. In cultures of embryonic rat septum, CNTF, unlike NGF, fails to support the survival of the cholinergic septal neurons (or, alternatively, fails to enhance the ChAT activity by which they are recognized) (Knüsel and Hefti, 1988). Likewise, cholinergic pedunculopontine neurons in culture do not appear to respond to CNTF (Knüsel and Hefti, 1988). Further, CNTF prevents the degeneration of axotomized adult rat cholinergic as well as noncholinergic medial septal neurons and, in the cholinergic ones, does not prevent axotomy-induced losses of their ChAT staining (Hagg et al., 1992). Also, CNTF supports the survival in vitro of GABAergic hippocampal neurons (Ip et al., 1991) and of noncholinergic spinal cord neurons (Magal et al., 1991).

## 4. NGF receptor regulation

Use of monoclonal antibody 192-IgG has led to recent demonstrations, in vivo and in vitro, of an effect of CNTF on immunostainable NGF receptors. This antibody recognizes the LNGFR, but not the high affinity binding sites (Meakin and Shooter, 1991) with which biologic actions of NGF have been associated (Sonnenfeld and Ishii, 1985; Green et al., 1986). Interestingly, NGF (Cavicchioli et al., 1989; Hagg et al., 1989a, 1992) and bFGF (Taiji et al., 1990) also have been shown to be involved in the up-regulation of LNGFR. In cultures of rat spinal cord, LNGFR is expressed by somatic motor neurons and by a population of more numerous but smaller cholinergic neurons. Whereas the somatic motor neurons depend on CNTF for their survival, the small cholinergic neurons survive without CNTF, but do not express LNGFR or ChAT. Presentation of CNTF to these neurons induces the expression of LNGFR and ChAT in 24 hr (Magal et al., 1991a). The induction of LNGFR by CNTF also has been shown in cultured neurons from other CNS regions, such as the cortex, the septum, the striatum, and the brainstem (Magal et al., 1991b).

As already mentioned, intraventricular infusion with CNTF in fimbria–fornix transected adult rats (Hagg and Varon, 1991) can prevent the degeneration of nearly all lesioned medial septum neurons, noncholinergic as well as cholinergic. In the axotomized cholinergic medial septum neurons, CNTF almost completely prevented the loss of LNGFR but not ChAT immunoreactivity, while also preventing their axotomy-induced somal atrophy. In these CNTF-protected LNGFR-positive neurons, addition of a low dose of NGF was not accompanied by any greater increase of ChAT-positive neurons (a well-known response to NGF) than with NGF alone, indicating that the cholinergic neurons had not become more responsive to NGF. Thus, CNTF may be involved in the regulation of only the low affinity nontransducing LNGFRs in these axotomized adult cholinergic neurons. In the intermediate lateral septum, CNTF also increased LNGFR staining dramatically in a cluster of normally faintly LNGFR-positive neurons.

Although the cholinergic neurons of the normal adult rat neostriatum express high affinity binding sites for NGF (indicating transducing receptors) (Richardson et al., 1986; Altar et al., 1991), they normally do not (or, at best, very faintly) express immunostainable LNGFR, in contrast with the cholinergic neurons of the basal forebrain. In response to CNTF infusion, however, LNGFR was found to be expressed in the parenchyma and in several cholinergic neuronal cell bodies of the neostriatum. It is possible that a differential availability of CNTF to neostriatal and basal forebrain neurons from their surroundings or their target territories constitutes a major determinant of the difference in their LNGFR expression. One could speculate further that LNGFR expression in other cell types similarly reflects CNTF-dependent regulations. The corpus callosum of CNTF-infused rats, but only the portion neighboring a penetrating lesion, also displays numerous LNGFR-positive small cells with such a morphology are the oligodendrocytes, type-2 astrocytes, and microglial cells. LNGFR expression by microglial cells, thus far, has been reported only in the neural lobe of the pituitary gland (Yan et al., 1990). The apparent need for both a lesion and CNTF to induce LNGFR in the callosal nonheuronal cells may be the CNS counterpart to the increased LNGFR expression that is displayed by Schwann cells in response to injury or axonal deprivation (Raivich and Kreutzberg, 1987; Taniuchi et al., 1988). That a release of CNTF after the lesion may play an LNGFR-regulating role in the Schwann cells is congruent with the already reviewed findings that peripheral nerve is rich in CNTF, that adult Schwann cells in vivo contain CNTF, that cultured Schwann cells can produce CNTF, and that CNTF activity is released into chambers fitted between the stumps of a transected nerve. The significance of the effect of CNTF on LNGFR expression by various neuronal (and possibly glial) populations is not clear at present.

## V. SUMMARY AND PERSPECTIVES

The available (and still limited) literature on CNTF reviewed in this chapter challenges the validity of the concept of a unique relationship between a target-derived neurotrophic factor and a limited number of neurons. The spectrum of CNTF-responsive cells is rather broad. In addition to CG neurons, other peripheral neurons and neuron-like cells (sensory, sympathetic, adrenal chromaffin) and central neurons (septal, hippocampal, spinal cord), as well as nonneuronal cells (O–2A progenitors), that respond to CNTF have been identified. More CNTF responsive targets are likely to be discovered. In fact, with the recognition that CNTF may belong to the cytokine superfamily, one should expect to find its actions on hematopoietic cell function. Like NGF, CNTF addresses neurons of more than one transmitter phenotype, thus arguing against responses restricted to neurons because of their transmitter phenotype. Although CNTF can support these several neuronal populations in vitro, it may not be the natural survival factor in vivo for all of them (e.g., Oppenheim et al., 1991). Eye-derived CNTF fulfills the requirements for the true target-derived neurotrophic factor for ciliary ganglionic motor neurons, that is, it is produced and supplied by their innervation territory and it supports their survival and growth. Likewise, a muscle-derived CNTF might regulate the survival of spinal motor neurons during development. The nerve CNTF, which appears later during postnatal development and remains present throughout adulthood, may serve the function of maintaining adult spinal cord neurons may be relevant to such a role. The CNTF present in CNS sources (astrocytes), which have yet to be characterized fully, could play a similar role for a broader spectrum of neurons.

Some of the neurons responding to CNTF are affected also by other neurotrophic factors, such as NGF and bFGF. Even CG neurons, by which CNTF activity and purification were followed, also respond to bFGF, and may even be addressed by NGF (Collins, 1978b). These observations suggest that a single neuron can respond to multiple trophic influences and that survival, functional maintenance, growth, and repair of selected populations of neurons may result from multiple modulatory interactions.

Now that purified CNTF has become available in relatively large quantities through recombinant technology, future studies are expected to be pursued with as much vigor (and with as many problems) as those addressing the other available factors such as NGF and FGF. Many areas in the CNTF field investigated currently and many others remain to be explored. For example, further details on the sites of production of CNTF and its mode of storage need to be acquired. By which mechanism is CNTF presented to its targets? Is it secreted or is it only released by disintegrating or leaking cells? If it is secreted, what is the mechanism of release and delivery? Does CNTF bind to the extracellular matrix? Is an interaction between CNTF and extracellular matrix molecules necessary for its action? What is the nature of the CNTF receptors? Which signal transduction pathway(s) is CNTF using? What is the physiologic meaning of the induction by CNTF of LNGFRs? Are there multiple members of a CNTF gene family? Are there oncogenes encoding CNTF-related factors? The struggle to address these and many other questions will prove to be as rewarding as the ongoing studies of the other protein factors and will contribute to an increased understanding of the agents and mechanisms by which survival, maintenance, growth, function, and repair of adult nerve cells can be controlled and secured.

## ACKNOWLEDGMENTS

Much of the work reviewed here has been supported by grants from the National Institute for Neurological and Communicative Disease and Stroke (NS-16349 and NS-27047).

## REFERENCES

Adler, J. E., Schleifer, L. S., and Black, I. B. (1989). Partial purification and characterization of a membrane-derived factor regulating neurotransmitter phenotypic expression. *Proc. Natl. Acad. Sci. U.S.A.* **86,** 1080–1083.

Adler, R., and Varon, S. (1980). Cholinergic neuronotrophic factors. V. Segregation of survival and neurite-promoting activities in heart conditioned media. *Brain Res.* **188,** 437–448.

Adler, R., and Varon, S. (1982). Neuronal survival in intact ciliary ganglia *in vivo* and *in vitro:* CNTF as a target surrogate. *Dev. Biol.* **92,** 470–475.

Adler, R., Landa, K. B., Manthorpe, M., and Varon, S. (1979). Cholinergic neuronotrophic factors: Intraocular distribution of trophic activity for ciliary neurons. *Science* **204,** 1434–1436.

Altar, C. A., Dugich-Djordjevic, M., Armanini, M., and Bakhit, C. (1991). Medial-to-lateral gradient of neostriatal NGF-receptors: Relationship to cholinergic neurons and NGF-like immunoreactivity. *J. Neurosci.* **11,** 828–836.

Arakawa, Y., Sendtner, M., and Thoenen, H. (1990). Survival effect of ciliary neurotrophic factor (CNTF) on chick embryonic motoneurons in culture: Comparison with other neurotrophic factors and cytokines. *J. Neurosci.* **10,** 3507–3515.

Assouline, J. G., Bosch, P., Lim, R., Kim, I. S., Jensen, R., and Pantazis, N. J. (1987). Rat astrocytes and Schwann cells in culture synthesize nerve growth factor-like neurite-promoting factors. *Brain Res.* **428**, 103–118.

Ayer, L. C., Olson, L., Ebendal, T., Seiger, A., and Persson, H. (1988a). Expression of the beta-nerve growth factor gene in hippocampal neurons. *Science* **240**, 1339–1341.

Ayer, L. C., Olson, L., Ebendal, T., Hallbook, F., and Persson, H. (1988b). Nerve growth factor mRNA and protein in the testis and epididymis of mouse and rat. *Proc. Natl. Acad. Sci. U.S.A.* **85**, 2628–2632.

Baird, A., Ueno, N., Esch, F., and Ling, N. (1987). Distribution of fibroblast growth factors (FGFs) in tissues and structure–function studies with synthetic fragments of basic FGF. *J. Cell Physiol. (Suppl.)* **5**, 101–106.

Barbin, G., Manthorpe, M., and Varon, S. (1984). Purification of the chick eye ciliary neuronotrophic factor (CNTF). *J. Neurochem.* **43**, 1468–1478.

Bazan, J. F. (1991). Neuropoietic cytokines in the hematopoietic fold, *Neuron* **7**, 197–208.

Betz, W. (1976). The promotion of synapses between chick embryo skeletal muscle and ciliary ganglion grown in vitro. *J. Physiol. (London)* **254**, 63–73.

Beuscher, H. U., Nickells, M. W., and Colten, H. R. (1988). The precursor of interleukin-1 alpha is phosphorylated at residue serine 90. *J. Biol. Chem.* **263**, 4023–4028.

Blottner, D., Bruggemann, W., and Unsicker, K. (1989). Ciliary neurotrophic factor supports target-deprived preganglionic sympathetic spinal cord neurons. *Neurosci. Lett.* **105**, 316–320.

Bradshaw, R. A., Dunbar, J. C., Isaakson, P. J., Kouchalakos, R. N., and Morgan, C. J. (1985). Nerve growth factor: Mechanism of action. In "Mediators of Cell Growth and Differentiation" (R. J. Ford and A. L. Maizd, eds.), pp. 87–101. Raven Press, New York.

Carnow, T. B., Manthorpe, M., Davis, G. E., and Varon, S. (1985). Localized survival of ciliary ganglionic neurons identifies neuronotrophic factor bands on nitrocellulose blots. *J. Neurosci.* **5**, 1965–1971.

Cavicchioli, L., Flanigan, T., Vantini, G., Fusco, M., Polato, P., Toffano, G., Walsh, F., and Leon, A. (1989). NGF amplifies the expression of NGF receptor messenger RNA in mammalian forebrain cholinergic neurons. *Eur. J. Neurosci.* **1**, 258–262.

Collins, F. (1978a). Induction of neurite outgrowth by a conditioned-medium factor bound to the culture substratum. *Proc. Natl. Acad. Sci. U.S.A.* **75**, 5210–5213.

Collins, F. (1978b). Axon initiation by ciliary neurons in culture. *Dev. Biol.* **65**, 50–57.

Cowan, W. M. (1973). Neuronal death as a regulative mechanism in the control of cell number in the nervous system. In "Development and Aging in the Nervous System" (M. Rockstein and M. C. Sussman, eds.), pp. 19–44. Academic Press, New York.

Cowan, W. M., Fawcett, J. W., O'Leary, D. D. M., and Stanfield, B. B. (1984). Regressive events in neurogenesis. *Science* **225**, 1258–1265.

Cunnigham, T. J. (1982). Naturally occurring neuron death and its regulation by developing neural pathways. *Int. Rev. Cytol.* **74**, 163–186.

Davis, G. E., Varon, S., Engvall, E., and Manthorpe, M. (1985). Substratum-binding neurite promoting factors: Relationships to laminin. *Trends Neurosci.* **8**, 528–532.

Davis, S., Aldrich, T. H., Valenzuela, D. M., Wong, V., Furth, M. E., Squinto, S. P., and Yoncopoulos, G. D. (1991). Receptor for ciliary neurotrophic factor. *Science* **253**, 60–63.

Eckenstein, F. P., Esch, F., Holbert, T., Blacher, R. W., and Nishi, R. (1990). Purification and characterization of a trophic factor for embryonic peripheral neurons: Comparison with fibroblast growth factors. *Neuron* **4**, 623–631.

Ernsberger, U., Sendtner, M., and Rohrer, H. (1989). Proliferation and differentiation of embryonic chick sympathetic neurons: Effects of ciliary neurotrophic factor. *Neuron* **2**, 1275–1284.

Feige, J. J., and Baird, A. (1989). Basic fibroblast growth factor is a substrate for protein phosphorylation and is phosphorylated by capillary endothelial cells in culture. *Proc. Natl. Acad. Sci. U.S.A.* **86**, 3174–3178.

Flaumenhaft, R., Moscatelli, D., Saksela, O., and Rifkin, D. B. (1989). Role of extracellular matrix in the action of basic fibroblast growth factor: Matrix as a source of growth factor for long-term stimulation of plasminogen activator production and DNA synthesis. *J. Cell Physiol.* **140,** 75–81.

Furth, M. E., Morrissey, D. M., Aldrich, T. A., Uffenheimer, L., Mickle, A. P., and Panayatos, N. (1990). Species-specific monoclonal antibodies to the carboxyl-terminus of human ciliary neurotrophic factor (CNTF). *Soc. Neurosci. Abstr.* **16,** 990.

Furukawa, S., Furukawa, Y., Satoyoshi, E., and Hayashi, K. (1986). Synthesis and secretion of nerve growth factor by mouse astroglial cells in culture. *Biochem. Biophys. Res. Commun.* **136,** 57–63.

Giller, E. L., Neale, J. H., Bullock, P. N., Schrier, B. K., and Nelson, P. G. (1977). Choline acetyltransferase activity of spinal cord cell cultures increased by co-culture with muscle and by muscle-conditioned medium. *J. Cell. Biol.* **74,** 16–29.

Green, S. H., Rydel, R. E., Connolly, J. L., and Greene, L. A. (1986). PC12 cell mutants that possess low- but not high-affinity nerve growth factor receptors neither respond to nor internalize nerve growth factor. *J. Cell Biol.* **102,** 830–843.

Greene, L. A., and Shooter, E. M. (1980). The nerve growth factor: Biochemistry, synthesis, and mechanism of action. *Ann. Rev. Neurosci.* **3,** 353–402.

Hagg, T., Fass-Holmes, B., Vahlsing, H. L., Manthorpe, M., Conner, J. M., and Varon, S. (1989a). Nerve growth factor (NGF) reverses axotomy-induced decreases in choline acetyltransferase, NGF receptor and size of medial septum cholinergic neurons. *Brain Res.* **505,** 29–38.

Hagg, T., Hagg, F., Vahlsing, H. L., Manthorpe, M., and Varon, S. (1989b). Nerve growth factor effects on cholinergic neurons of neostriatum and nucleus accumbens in the adult rat. *Neurosci.* **30,** 95–103.

Hagg, T., Quon, D., Higaki, J., and Varon, S. (1992). Ciliary neurotrophic factor prevents neuronal degeneration and promotes low affinity NGF receptor expression in the adult rat CNS. *Neuron* **8,** 145–158.

Hall, A. K., and Rao, M. S. (1992). Cytokines and neurokines: Related ligands and related receptors. *Trends Neurosci.* **15,** 35–37.

Hamburger, V. (1975). Cell death in the development of the lateral motor column of the chick embryo. *J. Comp. Neurol.* **160,** 535–546.

Hamburger, V. (1989). The journey of a neuroembryologist. *Ann. Rev. Neurosci.* **12,** 1–12.

Hamburger, V., and Oppenheim, R. W. (1982). Naturally occurring neuronal death in vertebrates. *Neurosci. Commun.* **1,** 39–55.

Hefti, F., Hartikka, J., and Knusel, B. (1989). Function of neurotrophic factors in the adult and aging brain and their possible use in the treatment of neurodegenerative diseases. *Neurobiol. Aging* **10,** 515–533.

Helfand, S. L., Smith, G. A., and Wessels, N. K. (1976). Survival and development in culture of dissociated parasympathetic neurons from ciliary ganglia. *Dev. Biol.* **50,** 541–547.

Hendry, I. A., Hill, C. E., Belford, D., and Watters, D. J. (1988). A monoclonal antibody to a parasympathetic neurotrophic factor causes immunoparasympathectomy in mice. *Brain Res.* **475,** 160–163.

Hess, A. (1965). Developmental changes in the structure of the synapse on the myelinated cell bodies of the chicken ciliary ganglion. *J. Cell. Biol.* **25,** 1–19.

Hofmann, H. D. (1988). Ciliary neuronotrophic factor stimulates choline acetyltransferase activity in cultured chicken retina neurons. *J. Neurochem.* **51,** 109–113.

Hooisma, J., Slaaf, D. W., Meeter, E., and Stevens, W. F. (1975). The innervation of chick striated muscle fibers by the chick ciliary ganglion in tissue culture. *Brain Res.* **85,** 79–85.

Hughes, S. M., and Raff, M. C. (1987). An inducer protein may control the timing of fate switching in a bipotential glial progenitor cell in rat optic nerve. *Development* **101,** 157–167.

Hughes, S. M., Lillien, L. E., Raff, M. C., Rohrer, H., and Sendtner, M. (1988). Ciliary neurotrophic factor induces type-2 astrocyte differentiation in culture. *Nature (London)* **335,** 70–73.

Ip, N. Y., Li, Y., van de Stadt, I., Panayotatos, N., Alderson, R. F., and Lindsay, R. M. (1991). Ciliary neurotrophic factor enhances neuronal survival in embryonic rat hippocampal cultures. *J. Neurosci.* **11,** 3124–3134.

Kaufman, L. M., Barry. S. R., and Barrett, J. N. (1985). Characterization of tissue-derived macromolecules affecting transmitter synthesis in rat spinal cord neurons. *J. Neurosci.* **5,** 160–166.

Knusel, B., and Hefti, F. (1988). Development of cholinergic pedunculopontine neurons in vitro: Comparison with cholinergic septal cells and response to nerve growth factor, ciliary neuronotrophic factor, and retinoic acid. *J. Neurosci. Res.* **21,** 365–375.

Kobayashi, Y., Appella, E., Yamada, M., Copeland, T. D., Oppenheim, J. J., and Matsushima, K. (1988). Phosphorylation of intracellular precursors of human IL-1. *J. Immunol.* **140,** 2279–2287.

Lam, A., Fuller, F., Miller, J., Kloss, J., Manthorpe, M., Varon, S., and Cordell, B. (1991). Sequence and structural organization of the human gene encoding ciliary neurotrophic factor. *Gene* **102,** 271–276.

Landa, K. B., Adler, R., Manthorpe, M., and Varon, S. (1980). Cholinergic neuronotrophic factors. III. Developmental increase of trophic activity for chick embryo ciliary ganglion neurons in their intraocular target tissues. *Dev. Biol.* **74,** 401–408.

Landmesser, L., and Pilar, G. (1972). The onset and development of transmission in the chick ciliary ganglion. *J. Physiol. (London)* **222,** 691–713.

Landmesser, L., and Pilar, G. (1974). Synapse formation during embryogenesis on ganglion cells lacking a periphery. *J. Physiol.* **241,** 715–736.

Landmesser, L., and Pilar, G. (1978). Interactions between neurons and their targets during in vivo synaptogenesis. *Fed. Proc.* **37,** 2016–2022.

Lehwalder, D., Jeffrey, P. L., and Unsicker, K. (1989). Survival of purified embryonic chick retinal ganglion cells in the presence of neurotrophic factors. *J. Neurosci. Res.* **24,** 329–337.

Letourneau, P. (1975). Possible roles for cell-to-substratum adhesion in neuronal morphogenesis. *Dev. Biol.* **44,** 77–91.

Levi-Montalcini, R. (1987). The nerve growth factor 35 years later. *Science* **237,** 1154–1162.

Lillien, L. E., and Raff, M. C. (1990). Differentiation signals in the CNS: Type-2 astrocyte development in vitro as a model system. *Neuron* **5,** 111–119.

Lillien, L. E., Sendtner, M., Rohrer, H., Hughes, S. M., and Raff, M. C. (1988). Type-2 astrocyte development in rat brain cultures is initiated by a CNTF-like protein produced by type-1 astrocytes. *Neuron* **1,** 485–494.

Lillien, L. E., Sendtner, M., and Raff, M. C. (1990). Extracellular matrix-associated molecules collaborate with ciliary neurotrophic factor to induce type-2 astrocyte development. *J. Cell Biol.* **111,** 635–644.

Lin, L. F., Mismer, D., Lile, J. D., Armes, L. G., Butler, E. T. III, Vannice, J. L., and Collins, F. (1989). Purification, cloning, and expression of ciliary neurotrophic factor (CNTF). *Science* **246,** 1023–1025.

Lin, L. F., Armes, L. G., Sommer, A., Smith, D. J., and Collins, F. (1990). Isolation and characterization of ciliary neurotrophic factor from rabbit sciatic nerves. *J. Biol. Chem.* **265,** 8942–8947.

Lindsay, R. M. (1988). Nerve growth factors (NGF, BDNF) enhance axonal regeneration but are not required for survival of adult sensory neurons. *J. Neurosci.* **8,** 2394–2405.

Lindsay, R. M., Thoenen, H., and Barde, Y. A. (1985). Placode and neural crest-derived sensory neurons are responsive at early developmental stages to brain-derived neurotrophic factor. *Dev. Biol.* **112,** 319–328.

Longo, F. M., Manthorpe, M., Skaper, S. D., Lundborg, G., and Varon, S. (1983a). Neuronotrophic activities accumulate in vivo within silicone nerve regeneration chambers. *Brain Res.* **261,** 109–117.

Longo. F. M., Skaper, S. D., Manthorpe, M., Williams, L. R., Lundborg, G., and Varon, S. (1983b). Temporal changes in neuronotrophic activities accumulating in vivo within nerve regeneration chambers. *Exp. Neurol.* **81,** 756–769.

McDonald, J. R., Ko, C., Mismer, D., Smith, D. J., and Collins, F. (1991). Expression and characterization of recombinant human ciliary neurotrophic factor from Escherichia coli. *Biochim. Biophys. Acta* **1090,** 70–80.

McManaman, J., Crawford, F., Stewart, S. S., and Appel, S. (1988). Purification of a skeletal muscle polypeptide which stimulates choline acetyltransferase activity in cultured spinal cord neurons. *J. Biol. Chem.* **263**, 5890–5897.

McManaman, J., Crawford, F., Clark, R., Richker, J., and Fuller, F. (1989). Multiple neurotrophic factors from skeletal muscle: Demonstration of effects of basic fibroblast growth factor and comparisons with the 22-kD choline acetyltransferase development factor. *J. Neurochem.* **53**, 1763–1771.

McManaman, J., Oppenheim, R. W., Prevette, D., and Marchetti, D. (1990). Rescue of motoneurons from cell death by a purified skeletal muscle polypeptide: Effects of the ChAT development factor, CDF. *Neuron* **4**, 891–898.

Magal, E., Burnham, P., and Varon, S. (1991a). Effect of ciliary neuronotrophic factor on rat spinal cord neurons in vitro: Survival and expression of choline acetyltransferase and low-affinity nerve growth factor receptors. *Dev. Brain Res.* **63**, 141–150.

Magal, E., Burnham, P., and Varon, S. (1991b). Effect of CNTF on low-affinity NGF receptor expression by cultured neurons from different rat brain regions. *J. Neurosci. Res.* **30**, 560–566.

Manthorpe, M., and Varon, S. (1985). Regulation of neuronal survival and neuritic growth in the avian ciliary ganglion by trophic factors. *In* "Growth and Maturation Factors" (G. Gurroff, eds.), Vol. 3, pp. 77–117. John Wiley & Sons, New York.

Manthorpe, M., and Varon, S. (1989). Use of ciliary ganglion neurons for the in vitro assay of neuronotrophic factors. *In* "A Dissection and Tissue Culture Manual for the Nervous System" (J. DeVellis, A. Shahar, A. Vernadakis, and B. Haber, eds.), pp. 317–321. Liss, New York.

Manthorpe, M., Skaper, S. D., Adler, R., Landa, K. B., and Varon, S. (1980). Cholinergic neuronotrophic factors: Fractionation properties of an extract from selected chick embryonic eye tissues. *J. Neurochem.* **34**, 69–75.

Manthorpe, M., Adler, R., and Varon, S. (1981a). Cholinergic neuronotrophic factors. VI. Age-dependent requirements for chick embryo ciliary ganglionic neurons. *Dev. Biol.* **85**, 156–163.

Manthorpe, M., Varon, S., and Adler, R. (1981b). Neurite-promoting factor (NPF) in conditioned medium from RN22 schwannoma cultures: Bioassay, fractionation and other properties. *J. Neurochem.* **37**, 759–767.

Manthorpe, M., Barbin, G., and Varon, S. (1982). Isoelectric focusing of the chick eye ciliary neuronotrophic factor. *J. Neurosci. Res.* **8**, 233–239.

Manthorpe, M., Engvall, E., Ruoslahti, E., Longo, F. M., Davis, G. E., and Varon, S. (1983). Laminin promotes neuritic regeneration form cultured peripheral and central neurons. *J. Cell Biol.* **97**, 1882–1890.

Manthorpe, M., Davis, G. E., and Varon, S. (1985). Purified proteins acting on cultured ciliary ganglion neurons. *Fed. Proc.* **44**, 2753–2759.

Manthorpe, M., Skaper, S. D., Williams, L. R., and Varon, S. (1986a). Purification of adult rat sciatic nerve ciliary neuronotrophic factor. *Brain Res.* **367**, 282–286.

Manthorpe, M., Rudge, J., and Varon, S. (1986b). Astroglial cell contributions to neuronal survival and neuritic growth. *In* "Astrocytes" (S. Fedoroff and A. Vernadakis, eds.), Vol. 2, pp. 315–376. Academic Press, San Diego.

Manthorpe, M., Pettmann, B., and Varon, S. (1988). Modulation of astroglial cell output of neuronotrophic and neurite promoting factors. *In* "The Biochemical Pathology of Astrocytes" (M. Norenberg, L. Hertz, and A. Schousboe, eds.), Vol. 39, pp. 41–57. Liss, New York.

Manthorpe, M., Ray, J., Pettmann, B., and Varon, S. (1989). Ciliary neuronotrophic factors. *In* "Nerve Growth Factors" (R. Rush, eds.), pp. 31–56. John Wiley & Sons, New York.

Manthorpe, M., Muir, D., Hagg, T., Engvall, E., and Varon, S. (1990). Glial cell laminin and neurite outgrowth. *In* "Differentiation and Functions of Glial Cells" (G. Levi, eds.), pp. 135–146. Liss, New York.

Manthorpe, M., Muir, D., Pettmann, B., and Varon, S. (1991). Detection and analyses of growth factors affecting neural cells. *In* "Neuromethods: Neural Cultures" (A. A. Boulton, G. B. Baker, and W. Walz, eds.), Vol. 21, pp. in press. Humana Press, Clifton, New Jersey.

Martinou, J. C., Le Van Thai, A., Cassar, G., Roubinet, F., and Weber, M. J. (1989). Characterization of two factors enhancing choline acetyltransferase activity in cultures of purified rat motoneurons. *J. Neurosci.* **9**, 3645–3656.

Marwitt, G. R., Pilar, G., and Weakly, J. N. (1971). Characterization of two cell populations in the avian ciliary ganglion. *Brain Res.* **25**, 317–334.

Masiakowski, P., Liu, H. X., Radziejewski, C., Lottspeich, F., Oberthuer, W., Wong, V., Lindsay, R. M., Furth, M. E., and Panayotatos, N. (1991). Recombinant human and rat ciliary neurotrophic factors. *J. Neurochem.* **57**, 1003–1012.

Meakin, S. O., and Shooter, E. M. (1991). Molecular investigations on the high-affinity nerve growth factor receptor. *Neuron* **6**, 153–163.

Miki, N., Hayashi, Y., and Higashida, H. (1981). Characterization of chick gizzard extract that promotes neurite outgrowth in cultured ciliary neurons. *J. Neurochem.* **37**, 627–633.

Muir, D., Gennrich, C., Varon, S., and Manthorpe, M. (1989). Rat sciatic nerve Schwann cell microcultures: Responses to mitogens and production of trophic and neurite-promoting factors. *Neurochem. Res.* **14**, 1003–1012.

Narayanan, C. H., and Narayanan, Y. (1978). Neuronal adjustments in developing nuclear centers of the chick embryo following transplantation of an optic primordium. *J. Embryol. Exp. Morphol.* **44**, 53–70.

Negro, A., Corona, G., Bigon, E., Martini, I., Grandi, C., Skaper, S. D., and Callegaro, L. (1991). Synthesis, purification, and characterization of human ciliary neuronotrophic factor from E. coli. *J. Neurosci. Res.* **29**: 251–260.

Nieto-Sampedro, M., Lewis, E. R., Cotman, C. W., Manthorpe, M., Skaper, S. D., Barbin, G., Longo, F. M., and Varon, S. (1982). Brain injury causes a time-dependent increase in neuronotrophic activity at the lesion site. *Science* **217**, 860–886.

Nieto-Sampedro, M., Manthorpe, M., Barbin, G., Varon, S., and Cotman, C. W. (1983). Injury-induced neuronotrophic activity in the adult rat brain. Correlation with survival of delayed implants in a wound cavity. *J. Neurosci.* **3**, 2219–2229.

Nishi, R., and Berg, D. K. (1977). Dissociated ciliary ganglion neurons in vitro: Survival and synapse formation. *Proc. Natl. Acad. Sci. U.S.A.* **74**, 5171–5175.

Nishi, R., and Berg, D. K. (1979). Survival and development of ciliary ganglion neurons growth alone in cell culture. *Nature (London)* **227**, 232–234.

Nishi, R., and Berg, D. K. (1981). Effects of high $K^+$ concentrations on the growth and development of ciliary ganglion neurons in cell culture. *Dev. Biol.* **87**, 301–307.

Nishi, R., Holbert, T., and Eckenstein, F. P. (1990). Ciliary neurotrophic factor in adult sciatic nerves is biochemically identical to trophic activity in embryonic chick eyes. *Soc. Neurosci. Abstr.* **16**, 1135.

Noble, M., Fok-Seang, J., Wolswijk, G., and Wren, D. (1990). Development and regeneration in the central nervous system. *Phil. Trans. R. Soc. London (Series B)* **327**, 127–143.

Oppenheim, R. W., Prevette, D., Qin-Wei, Y., Collins, F., and MacDonald, J. (1991). Control of embryonic motoneurons survival in vivo by ciliary neurotrophic factor (CNTF). *Science* **251**, 1616–1618.

Otten, U., Ehrhard, P., and Peck, R. (1989). Nerve growth factor induces growth and differentiation of human B lymphocytes. *Proc. Natl. Acad. Sci. U.S.A.* **86**, 10059–10063.

Parent, A., Eckenstein, F. P., and Nishi, R. (1990). Gene expression of a trophic factor during development of embryonic chick ciliary ganglion neurons. *Soc. Neurosci. Abstr.* **16**, 1135.

Perez-Polo, J. R., and Werrbach-Perez, K. (1988). Role of nerve growth factor in neuronal injury and survival. *In* "Neural Development and Regeneration Cellular and Molecular Aspects" (A. Gorio, J. R., Perez-Polo, J. de Vellis, and B. Haber, eds.), pp. 339–410. Springer-Verlag, Heidelberg.

Pettmann, B., Manthorpe, M., Powell, J. A., and Varon, S. (1988). Biological activities of nerve growth factor bound to nitrocellulose paper by Western blotting. *J. Neurosci.* **8**, 3624–3632.

Raff, M. C. (1989). Glial cell diversification in the rat optic nerve. *Science* **243**, 1450–1455.

Raivich, G., and Kreutzberg, G. W. (1987). Expression of growth factor receptors in injured nervous tissue. II. Induction of specific platelet-derived growth factor binding in the injured PNS is associated with a breakdown in the blood-nerve barrier and endoneurial interstitial oedema. *J. Neurocytol.* **16**, 701-711.

Rao, M. S., Landis, S. C., and Patterson, P. H. (1990). The cholinergic neuronal differentiation factor from heart cell conditioned medium is different from the cholinergic factors in sciatic nerve and spinal cord. *Dev. Biol.* **139**, 65-74.

Rao, M. S., Tyrrell, S., Landis, S. C., and Patterson, P. H. (1992). Effects of ciliary neurotrophic factor (CNTF) and depolarization on neuropeptide expression in cultured sympathetic neurons. *Dev. Biol.* **150**, 281-293.

Rende, M., Muir, D., Ruoslahti, E., Hagg, T., Varon, S., and Manthorpe, M. (1992). Immunolocalization of ciliary neurotrophic factor in adult rat sciatic nerve. *Glia* **5**, 25-32.

Richardson, P. M., Issa, V. M., and Riopelle, R. J. (1986). Distribution of neuronal receptors for nerve growth factor in the rat. *J. Neurosci.* **6**, 2312-2321.

Rudge, J. S., Manthorpe, M., and Varon, S. (1985). The output of neuronotrophic and neurite-promoting agents from rat brain astroglial cells: A microculture method for screening potential regulatory molecules. *Dev. Brain Res.* **19**, 161-172.

Rudge, J. S., Davis, G. E., Manthorpe, M., and Varon, S. (1987). An examination of ciliary neuronotrophic factors from avian and rodent tissue extracts using a blot and culture technique. *Brain Res.* **429**, 103-110.

Rudge, J. S., Alderson, R. F., Ip, N., and Lindsay, R. M. (1990). Characterization of ciliary neurotrophic factor in rat astrocytes. *Soc. Neurosci. Abstr.* **16**, 484.

Saadat, S., Sendtner, M., and Rohrer, H. (1989). Ciliary neurotrophic factor induces cholinergic differentiation of rat sympathetic neurons in culture. *J. Cell Biol.* **108**, 1807-1816.

Schonfeld, A. R., Heacock, A. M., and Katzman, R. (1985). Neurotrophic factors: Effects on central cholinergic regeneration in vivo. *Brain Res.* **336**, 297-301.

Schubert, D., Ling, N., and Baird, A. (1987). Multiple influences of a heparin-binding growth factor on neuronal development. *J. Cell Biol.* **104**, 635-643.

Seidl, K., Manthorpe, M., Varon, S., and Unsicker, K. (1987). Differential effects of nerve growth factor and ciliary neuronotrophic factor on catecholamine storage and catecholamine synthesizing enzymes of cultured rat chromaffin cells. *J. Neurochem.* **49**, 169-174.

Sendtner, M., Kreutzberg, G. W., and Thoenen, H. (1990). Ciliary neurotrophic factor prevents the degeneration of motor neurons after axotomy. *Nature (London)* **345**, 440-441.

Skaper, S. D., Montz, H. P., and Varon, S. (1986). Control of $Na^+$, $K^+$-pump activity in dorsal root ganglionic neurons by different neuronotrophic agents. *Brain Res.* **386**, 130-135.

Snider, W. D., and Johnson, E. J. (1989). Neurotrophic molecules. *Ann. Neurol.* **26**, 489-506.

Sonnenfeld, K. H., and Ishii, D. N. (1985). Fast and slow nerve growth factor binding sites in human neuroblastoma and rat pheochromocytoma cell lines: Relationship of sites to each other and to neurite formation. *J. Neurosci.* **5**, 1717-1728.

Squinto, S., Aldrich, T. H., Lindsay, R. M., Morrissey, D. M., Panayatatos, N., Bianco, S. M., Furth, M. E., and Yancopoulos, G. (1990). Identification of functional receptors for ciliary neurotrophic factor on neuronal cell lines and primary neurons. *Neuron* **5**, 757-766.

Stöckli, K. A., Lottspeich, F., Sendtner, M., Masiakowski, P., Carroll, P., Götz, R., Lindholm, D., and Thoenen, H. (1989). Molecular cloning, expression and regional distribution of rat ciliary neurotrophic factor. *Nature (London)* **342**, 920-923.

Stöckli, K. A., Lillien, L. E., Näher-Noé, M., Britfeld, G., Hughes, R. A., Raff, M. C., Thoenen, H., and Sendtner, M. (1991). Regional distribution, developmental changes, and cellular localization of CNTF-mRNA and protein in the rat brain. *J. Cell Biol.* **115**, 447-459.

Streck, H., Gunther, C., Beuscher, H. U., and Rollinghoff, M. (1988). Studies on the release of cell-associated interleukin 1 by paraformaldehyde-treated murine macrophages. *Eur. J. Immunol.* **18**, 1609-1613.

Taiji, M., Taiji, K., Deyerle, K. L., and Bothwell, M. (1990). Basic fibroblast growth factor enhances the

NGF receptor promoter activity in human neuroblastoma cells; CPH100. *Soc. Neurosci. Abstr.* **16,** 992.

Taniuchi, M., Clark, H. B., Schweitzer, J. B., and Johnson, E. M. (1988). Expression of nerve growth factor receptors by Schwann cells of axotomized peripheral nerves: Ultrastructural location, suppression by axonal contact, and binding properties. *J. Neurosci.* **8,** 664–681.

Temple, S., and Raff, M. C. (1985). Differentiation of a bipotential glial progenitor cell in single cell microculture. *Nature (London)* **313,** 323–325.

Thoenen, H., Korsching, S., Heumann, R., and Acheson, A. (1985). Nerve growth factor. *In* "Growth Factors in Biology and Medicine," (D. Evered, J. Nugent, and J. Whelan, eds.), pp. 113–128. Pitman, London.

Thoenen, H., Barde, Y. A., Davies, A. M., and Johnson, J. E. (1987). Neurotrophic factors and neuronal death. *CIBA Found. Symp.* **126,** 82–95.

Unsicker, K., Skaper, S. D., and Varon, S. (1985a). Neuronotrophic and neurite-promoting factors: Effects on early postnatal chromaffin cells from rat adrenal medulla. *Dev. Brain Res.* **17,** 117–129.

Unsicker, K., Skaper, S. D., and Varon, S. (1985b). Developmental changes in the responses of rat chromaffin cells to neurotrophic and neurite-promoting factors. *Dev. Biol.* **111,** 425–433.

Unsicker, K., Reichert-Preibsch, H., Schmidt, R., Pettmann, B., Labourdette, G., and Sensenbrenner, M. (1987a). Astroglial and fibroblast growth factors have neurotrophic functions for cultured peripheral and central nervous system neurons. *Proc. Natl. Acad. Sci. U.S.A.* **84,** 5459–5463.

Unsicker, K., Stahnke, G., and Muller, T. H. (1987b). Survival, morphology, and catecholamine storage of chromaffin cells in serum-free culture: Evidence for a survival and differentiation promoting activity in medium conditioned by purified chromaffin cells. *Neurochem. Res.* **12,** 995–1003.

Unsicker, K., Blottner, D., Gehrke, D., Grothe, C., Heymann, D., Stögbauer, F., and Westermann, R. (1988). Neurectodermal cells: Storage and release of growth factors. *In* "Neural Development and Regeneration" (A. Gorio, ed.), NATO Series, Vol. H22, pp. 43–52. Springer-Verlag, Heidelberg.

Vaca, K., Stewart, S. S., and Appel, S. H. (1989). Identification of basic fibroblast growth factor as a cholinergic growth factor from human muscle. *J. Neurosci.* **23,** 55–63.

Varon, S., and Adler, R. (1980). Nerve growth factor and control of nerve growth. *Curr. Top. Dev. Biol.* **16,** 207–252.

Varon, S., and Adler, R. (1981). Trophic and specifying factors directed to neuronal cells. *Adv. Cell. Neurobiol.* **2,** 115–163.

Varon, S., Adler, R., and Manthorpe, M. (1978). Trophic agents directed to ciliary ganglionic cells in monolayer cultures. *Soc. Neurosci. Abstr.* **4,** 589.

Varon, S., Manthorpe, M., and Adler, R. (1979). Cholinergic neuronotrophic factors. I. Survival, neurite outgrowth and choline acetyltransferase activity in monolayer cultures from chick embryo ciliary ganglia. *Brain Res.* **173,** 29–45.

Varon, S., Adler, R., Manthorpe, M., and Skaper, S. D. (1983). Culture strategies for trophic and other factors directed to neurons. *In* "Neuroscience Approached through Cell Culture" (S. E. Pfeiffer, ed.), Vol. 2, pp. 53–77. CRC Press, Boca Raton, Florida.

Varon, S., Manthorpe, M., and Williams, L. R. (1984). Neuronotrophic and neurite-promoting factors and their clinical potentials. *Dev. Neurosci.* **6,** 73–100.

Varon, S., Manthorpe, M., Davis, G. E., Williams, L. R., and Skaper, S. D. (1988a). Growth factors. *In* "Advances in Neurology: Functional Recovery in Neurological Disease" (S. G. Waxman, ed.), Vol. 47, pp. 493–521. Raven Press, New York.

Varon, S., Manthorpe, M., Williams, L. R., and Gage, F. H. (1988b). Neuronotrophic factors and their involvement in the adult central nervous system. *In* "Aging and the Brain" (R. D. Terry, ed.), pp. 493–421. Raven Press, New York.

Walicke, P. A. (1989). Novel neurotrophic factors, receptors and oncogenes. *Ann. Rev. Neurosci.* **12,** 103–126.

Wallace, T. L., and Johnson, E. J. (1987). Partial purification of a parasympathetic neurotrophic factor in pig lung. *Brain Res.* **411,** 351–363.

Watters, D. J., and Hendry. I. A. (1987). Purification of a ciliary neurotrophic factor from bovine heart. *J. Neurochem.* **49,** 705–713.

Watters, D., Belford, D., Hill, C., and Hendry, I. (1989). Monoclonal antibody that inhibits biological activity of a mammalian ciliary neurotrophic factor. *J. Neurosci. Res.* **22,** 60–64.

Wewetzer, K., MacDonald, J. R., Collins, F., and Unsicker, K. (1990). CNTF rescues motoneurons from ontogenetic cell death in-vivo, but not in-vitro. *Neuroreport,* **1,** 203–206.

Whittemore, S. R., and Seiger, A. (1987). The expression, localization and functional significance of beta-nerve growth factor in the central nervous system. *Brain Res.* **434,** 439–464.

Williams, L. R., Manthorpe, M., Nieto-Sampedro, M., Cotman, C. W., and Varon, S. (1984). High ciliary neuronotrophic specific activity in rat peripheral nerve. *Int. J. Dev. Sci.* **2,** 177–180.

Williams, L. R., Jodelis, K. S., and Donald, M. R. (1989). Axotomy-dependent stimulation of choline acetyltransferase activity by exogenous mouse nerve growth factor in adult rat basal forebrain. *Brain Res.* **498,** 243–256.

Wong, V., and Kessler, J. A. (1987). Solubilization of a membrane factor that stimulates levels of substance P and choline acetyltransferase in sympathetic neurons. *Proc. Natl. Acad. Sci. U.S.A.* **84,** 8726–8729.

Wong, V., Arriaga, R., and Lindsay, R. M. (1990). Effects of ciliary neurotrophic factor (CNTF) on ventral spinal cord neurons in culture. *Soc. Neurosci. Abstr.* **16,** 384.

Yamamori, T., Fukada, K., Aebersold, R., Korsching, S., Fann, M. J., and Patterson, P. H. (1989). The cholinergic neuronal differentiation factor from heart cells is identical to leukemia inhibitory factor. *Science* **246,** 1412–1416.

Yan, Q., Clark, H. B., and Johnson, E. J. (1990). Nerve growth factor receptor in neural lobe of rat pituitary gland: Immunohistochemical localization, biochemical characterization and regulation. *J. Neurocytol.* **19,** 302–312.

# 16 Skeletal Muscle-Derived Neurotrophic Factors and Motoneuron Development

James L. McManaman and Ronald W. Oppenheim

The survival of motoneurons, like that of other neurons, is dependent on signals from their target organ—skeletal muscle. Approximately half the motoneurons in the spinal cords of developing chick or rat embryos degenerate and die during embryonic development by a process of naturally occurring cell death (Hamburger, 1975; Harris and McCaig, 1984; Oppenheim, 1991). Before the onset of cell death, the cells that die are indistinguishable morphologically, biochemically, and functionally from those that survive (cf. Oppenheim, and Chu-Wang, 1983). All cells, including the ones destined to die, differentiate normally, send axons into the developing limb buds, and are capable of retrograde transport (Oppenheim and Heaton, 1975; Chu-Wang and Oppenheim, 1978). The extent of motoneuron death is regulated at the level of the target organ by both the size of the target field and synaptic activity. Removal of the limb bud before innervation results in the death of all motoneurons (Hamburger, 1975; Oppenheim et al., 1978). Similarly, there is an almost complete loss of motoneurons in limbless mutant chick embryos (Lanser and Fallon, 1987). Conversely, the addition of a supernumerary limb before the time of cell death decreases subsequent motoneuron loss (Hollyday and Hamburger, 1979). Pharmacologic paralysis through neuromuscular blockage also reduces cell death, whereas electrical stimulation hastens the loss of cells (Pittman and Oppenheim, 1979; Oppenheim and Nunez, 1982). These findings have been interpreted as evidence that (1) skeletal muscle tissue contains substances, neurotrophic factors, that are necessary for the survival of embryonic motoneurons; (2) the levels, or availabilities, of these neurotrophic factors in the developing limb are limited; and (3) the number of surviving motoneurons results from competition for neurotrophic factors. Nerve growth factor (NGF) is the most thoroughly studied target-derived neurotrophic factor and, thus far, the only such factor with a firmly established role in neuronal development (Levi-Montalcini, 1987). Embryonic motoneurons express NGF receptor mRNA (Ernfors et al., 1989; Large et al., 1989), exhibit both high and low affinity binding sites for NGF (Raivich et al., 1985; Marchetti and McManaman, 1990), and transport NGF retrogradely (Wayne and Heaton, 1990). However, NGF does not appear to regulate motoneuron survival or affect motoneuron properties (Oppenheim et al., 1982; Yan et al., 1988). Several laboratories have documented the presence of other substances, in extracts of skeletal muscle and in culture media obtained from skeletal muscle myotubes that affect motoneurons. Several of these substances have been isolated or identified. The goal of this chapter is to review the data identifying skeletal muscle-derived factors affecting the in vivo and in vitro properties of motoneurons. Although this chapter focuses primarily on skeletal muscle-derived factors, other cell types encountered by motoneurons (e.g., astrocytes, Schwann cells, fibroblasts, and interneurons) may exert trophic influences on the development and maintenance of motoneurons through the actions of trophic molecules. Since the neurotrophic effects of many of these molecules are covered in other chapters, we will discuss their effects only as they apply to motoneurons.

## I. IN VITRO EVIDENCE FOR ACTIONS OF SKELETAL MUSCLE NEUROTROPHIC FACTORS

Extracts of skeletal muscle or myotube-conditioned media promote survival, enhance neurite elongation, and stimulate cholinergic development of cultured spinal cord neurons. The survival of motoneurons in vitro has been investigated in dissociated cultures of both embryonic rat and chick spinal cord cells. Cultures were prepared from each of these organisms at the beginning of their period of naturally occurring motoneuron cell death [embryonic day (ED) 6 for chick embryos and ED 14 for rat embryos]. However, by this stage of development, motoneurons already have sent axons into their targets (Oppenheim and Heaton, 1975). Consequently, dissociated spinal cord cultures contain cells that have suffered varying degrees of damage. Thus, some in vitro trophic effects may reflect amelioration of injury-induced changes in addition to influences on normal development.

Both conditioned media from cultured striated myotubes and extracts of skeletal muscle significantly enhance the survival of retrogradely labeled motoneurons in whole spinal cord cultures prepared from ED 14 rats (Smith et al., 1986; Martinou et al., 1989) and ED 6 or ED 9 chick spinal cords (Bennett et al., 1980; Tanaka and Obata, 1983; O'Brien and Fischbach, 1986; Tanaka, 1987). Factors from skeletal muscle extracts and myotube-conditioned media also enhance the survival of isolated motoneurons, prepared by either density gradient sedimentation or fluorescence activated cell sorting (Calof and Reichardt, 1984; Dohrmann et al., 1986; O'Brien and Fischbach, 1986; Schaffner et al., 1987; Martinou et al., 1989; Arakawa et al., 1990). However, the survival of isolated motoneurons, even in the presence of muscle extract, is less than that in heterogenous cell cultures (Dohrmann et al., 1987; Martinou et al., 1989). These findings raise the possibility that non-target-derived factors also may contribute to the survival of motoneurons in vitro. Dohrmann et al. (1986) have shown that extracts of spinal cord also enhance the survival of isolated motoneurons and that these effects are synergistic with those of skeletal muscle extract. Similarly, extracts from brain or spinal cord also promote motoneuron survival in vivo (Oppenheim et al., 1989b); deafferentation of motoneurons in vivo increases the naturally occurring loss of motoneurons (Okado and Oppenheim, 1984). The survival of isolated motoneurons also is enhanced by conditioned medium from fibroblast and spinal cord cultures, not only by conditioned medium from myotube cultures (Calof and Reichardt, 1984; Dohrmann et al., 1987). Synaptic contact with skeletal muscle fibers also may influence the survival of motoneurons (Oppenheim, 1989; Oppenheim et al., 1989c). O'Brien and Fischbach (1986) showed that isolated motoneurons co-cultured with skeletal muscle myotubes survived for at least 1 week in culture, whereas the number of motoneurons began to decline after 3 days when cultured on polylysine in the presence of myotube-conditioned medium.

The survival of other spinal cord neurons does not appear to be sensitive to the effects of skeletal muscle factors. Smith et al. (1986) found that treatment with muscle extract enhanced the survival of retrogradely labeled embryonic rat motoneurons in whole spinal cord cultures, but not the total number of tetanus toxin-labeled neurons. Dohrmann et al. (1986) found that isolated chick embryo motoneurons did not survive in culture without muscle extract, whereas nonmotoneuron cultures survived for at least 12 days. Similarly, O'Brien and Fischbach (1986) found that co-culturing dissociated embryonic chick spinal cords with skeletal muscle myotubes supported the survival of labeled motoneurons, but did not alter total neuron survival or the number of neurons that transported GABA. Although these findings suggest that skeletal muscle factors have a limited range of cellular reactivity, their actions are not limited solely to motoneurons. Extracts of skeletal muscle also enhance neuronal survival in cultures of sympathetic (Dohrmann et al., 1987) and ciliary ganglion neurons (Vaca et al., 1985).

Several laboratories have demonstrated that myotube-conditioned media or extracts of skeletal muscle enhance process outgrowth in cultures of embryonic rat and chick spinal cords (Dribin and Barrett, 1982; Smith and Appel, 1983; Calof and Reichardt, 1984; Smith et al., 1985,1986; Dohrmann et al., 1986). This activity is enriched in skeletal muscle compared with other tissues and is influenced by the developmental stage and state of innervation. Smith et al. (1985) showed that extracts of skeletal muscle and spinal cord, but not liver, lung, heart, skin, or brain, increase neurite outgrowth from both motoneurons and nonmotoneurons in embryonic rat spinal cord cultures. Henderson et al. (1983) and Smith et al. (1985) found that the potency of the neurite outgrowth activity in extracts prepared from adult muscle was less than that of those from newborn muscle; this activity was increased following denervation.

Conditioned media or extracts of skeletal muscle tissue also contain substances that affect the levels of acetylcholine synthesis, choline acetyltransferase (ChAT, the enzyme responsible for the production of acetylcholine), and sodium-dependent high affinity choline uptake in embryonic spinal cord cultures (Giller et al., 1977; Smith and Appel, 1983; Giess and Weber, 1984; Kaufman et al., 1985; Smith et al., 1986; McManaman et al., 1988a,b; Martinou et al., 1989). The level of cholinergic-enhancing activity in extracts of skeletal muscle tissue, like the neurite-promoting activity, is regulated developmentally and influenced by the state of innervation. Smith et al. (1985) showed an age-dependent increase in the ability of muscle extract to stimulate acetylcholine synthesis in embryonic rat spinal cord cultures. Denervation, on the other hand, led to a decrease in the level of cholinergic-enhancing activity in muscle extracts (Smith et al., 1985). Substances enhancing acetylcholine synthesis in spinal cord cultures are found also in extracts of other tissues that receive cholinergic innervation (Smith et al., 1985). However, extracts of tissues that are not innervated by cholinergic nerves do not appear to affect acetylcholine synthesis (Smith et al., 1985).

## II. IN VIVO EVIDENCE FOR ACTIONS OF SKELETAL MUSCLE NEUROTROPHIC FACTORS

Neurotrophic effects of skeletal muscle extracts also have been documented in vivo. The administration of rat or chick skeletal-muscle extracts to developing chick embryos during the period of motoneuron cell death (ED 6–ED 9) rescues up to 45% of the motoneurons that normally would die by ED 10 (Oppenheim et al., 1988). However, the administration of identically prepared extracts of control tissues (liver, kidney, lung, heart, gizzard) or heat-inactivated muscle extract does not influence motoneuron survival. The effects of muscle extracts appear to be selective for motoneurons, since the survival of sensory neurons in the dorsal root ganglion, sympathetic neurons, nodose ganglion neurons, or preganglionic cholinergic neurons, all of which undergo cell death during this period of development, was not affected. Initial characterization of the active factor shows that it is precipitated from solution by 25–75% ammonium sulfate, is recovered in the < 30,000-dalton fraction after sizing chromatography, and is destroyed by trypsin treatment (Oppenheim et al., 1988). Conditioned medium from skeletal myotubes also promotes motoneuron survival in vivo, suggesting that the active factor is secreted from muscle cells (R. Oppenheim, unpublished observations).

Skeletal muscle extracts also ameliorate the severe loss of motoneurons that results from limb-bud removal. After unilateral limb-bud removal on ED 2, the loss of motoneurons as a result of target removal does not begin until the normal period of cell death 3 days later (Oppenheim and Chu-Wang, 1983). Effects of extract treatment, begun on ED 4, are not apparent on the unoperated side of the embryo by ED 6–7. However, on the operated side, extract treatment reduces the 50% loss of motoneurons (resulting, at this stage, from the early limb-bud removal) by half (Oppenheim et al., 1988). By ED 8–9, at which time approximately 90% of the original population would have been lost, extract treatment doubles the number of surviving motoneurons. The activity that prevents chick embryonic motoneuron cell death in vivo is not restricted to the muscle of that species. Extracts of limb muscle of fetal and neonatal mouse, postnatal rat, and fetal human are equally effective in ameliorating naturally occurring cell death (McManaman et al., 1991).

## III. ISOLATION AND IDENTIFICATION OF SKELETAL MUSCLE-DERIVED TROPHIC FACTORS

The initial attempts at purification of the trophic activity in skeletal muscle extracts demonstrated chromatographically separate species with distinctive effects on cultured embryonic spinal cord cells. Giller et al. (1977) demonstrated that factors in skeletal myotube-conditioned medium that affect ChAT activity in mouse spinal cord cells exhibit an apparent molecular mass greater than 50 kDa on size-selective membranes. Calof and Reichardt (1984) separated a laminin-like substance that enhances neurite outgrowth from other factors that support motoneuron survival. Smith et al. (1985) identified a neurite-promoting factor in extracts of rat skeletal muscle that is distinct from factors affecting cholinergic development. Skeletal muscle extracts contain more than one substance capable of enhancing cholinergic development in spinal cord cultures. Smith et al. (1985,1986) found several peaks of activity by size-exclusion chromatography and isoelectric focusing (Smith et al., 1986). Although incompletely characterized, these factors appeared to have different effects on cholinergic properties and to act in an additive fashion (Smith et al., 1986). Since factors enhancing motoneuron survival also might enhance the level of cholinergic activity indirectly, some of these cholinergic factors may, in reality, be survival factors. In support of this possibility, Martinou et al. (1989) separated factors in myotube-conditioned media that support the survival of isolated motoneurons from those that affect cholinergic development.

These results clearly show that muscle contains several biochemically distinct molecules that influence motoneuron properties in vitro. However, skeletal muscle-derived trophic factors have been isolated and identified only recently. The first skeletal muscle-derived factor to be purified to homogeneity is CDF (ChAT development factor; see also Chapter 19). This factor was purified based on its ability to enhance the development of ChAT in ED 14 rat spinal cord cultures; isolation was done using combinations of acid precipitation, conventional chromatography, and either preparative SDS gel electrophoresis (McManaman et al,. 1988a,b) or reverse-phase HPLC (McManaman and Crawford, 1991). CDF is an acidic protein with a molecular mass of 22–24 kDa and a pI of 4.8–5.0. The purified protein migrates as a single species on one-dimensional SDS gel electrophoresis and isoelectric focusing electrophoresis, using either silver staining or autoradiography of $^{125}$I-labeled CDF as the detection method. Reduction does not alter either the migration of the purified protein on analytical SDS gel electrophoresis or the elution position of biologically active CDF from preparative SDS gel electrophoresis (McManaman et al., 1988a,b). In addition, immunoblot analyses of incompletely purified fractions of CDF, using a monoclonal antibody to CDF, have detected a single 22–24 kDa species (McManaman and Crawford, 1991). Thus, the active molecule appears to be composed of a single homogeneous polypeptide chain and does not appear to exhibit significant size heterogeneity. The molecular weight, isoelectric point, subunit composition, and immunologic properties of CDF differ from those of most of the other neurotrophic factors that have been isolated to date [i.e., NGF, brain-derived neurotrophic factor (BDNF), neurotrophin-3 (NT-3), acidic and basic fibroblast growth factor (FGF), insulin-like growth factors (IGF), transforming growth factor (TGF) β, and epidermal growth factor (EGF)] (McManaman et al., 1988a,b,1989). The biochemical properties of CDF are similar, although not identical, to those of ciliary neurotrophic factor (CNTF). Both factors have a molecular mass of ~22kDa (Lin et al., 1989). However, the isoelectric point of CDF (pI 4.8–5.0) is more acidic than that of CNTF (pI 5.8–6.0) (Lin et al., 1989). In spite of similar biochemical properties, these two factors do not appear to be related; CNTF mRNA was not found in rat skeletal muscle by Northern blotting (Stockli et al., 1989) and preliminary structural analysis of CDF indicates that its amino acid sequence differs from that of CNTF or of any of the other previously identified growth factors (J. McManaman, unpublished observations).

Although the biologic activities of CDF are characterized incompletely, it selectively enhances the survival of embryonic motoneurons (see subsequent text) and stimulates the development of ChAT in motoneurons and cholinergic human neuroblastoma cells (McManaman et al., 1989,1990; McManaman and Crawford, 1991). CDF does not, however, affect ChAT development in all cholinergic cell types. The levels of ChAT activity in cultures of septal neurons or pheochromocytoma (PC12) cells, for instance, are not affected by CDF treatment (McManaman and Crawford, 1991). In addition, CDF does not affect the growth or cell division of 3T3 fibroblasts, embryonic spinal cord cells, or human neuroblastoma cells (McManaman et al., 1989; McManaman and Crawford, 1991). Thus, in this regard, the range of action of this factor is more restricted than that of mitogenic neurotrophic factors (e.g., FGFs, IGFs, EGF).

Extracts of skeletal muscle tissue also contain bFGF (McManaman et al., 1989; Vaca et al., 1989; see also Chapters 9 and 10). Although best known as a mitogen (Gospodarowicz et al., 1987), bFGF also affects the in vitro properties of cortical, hippocampal, ciliary ganglion, and spinal cord neurons (Morrison et al., 1986; Walicke et al., 1986; Unsicker et al., 1987; Vaca et al., 1989) and the in vivo survival of both lesioned septal cholinergic neurons and dorsal root ganglion neurons (Dreyer et al., 1989; Otto et al., 1989). bFGF also enhances the level of ChAT activity in cultures embryonic spinal cord cells (McManaman et al., 1989). The magnitude the effect of bFGF on ChAT is comparable to, and additive with, that of CDF (McManaman et al., 1989). At maximal concentrations, the combined effect of CDF and bFGF on ChAT activity of rat spinal cord cultures is equal to that of crude extracts of skeletal muscle. Although the molecular mass of bFGF is only 17 kDa, it behaves as a much larger species in aqueous extracts of skeletal muscle. The substances in rat muscle extract that stimulate ChAT in spinal cord cultures separate into high (> 70 kDa) and low (< 40 kDa) molecular mass species by filtration chromatography (McManaman et al., 1989). bFGF-like immunoreactivity in muscle extracts and $^{125}$I-labeled bFGF, added to muscle extract as a tracer, coelute with the high molecular weight substances that enhance ChAT activity. Further, the bFGF in the high molecular weight fraction is biologically active and is inhibited by anti-bFGF antibodies (McManaman et al., 1989). Thus, bFGF appears to account for at least a portion of the high molecular weight neurotrophic activity in extracts of skeletal muscle.

Skeletal muscle also expresses IGFs (Beck et al., 1988; Ishii, 1989); IGF immunoreactivity is found in adult motoneurons (Hansson et al., 1988). Ishii (1989) found that IGF-2 mRNA levels are correlated inversely with the loss of multiply innervated neuromuscular junctions and can be elevated by denervation. Experiments by Caroni (Caroni and Grandes, 1990) documented that injections of IGF-1 into the gluteus muscle of adult rats stimulated neuronal sprouting and increased the levels of the 43-kDa growth-associated protein (GAP43). IGF-1 also stimulates sciatic nerve regeneration (Sjoberg and Kanje, 1989) and enhances the development of ChAT in embryonic rat spinal cord cultures. Although the effect of IGF-1 on ChAT is similar to, and additive with, that of CDF (Wood and McManaman, 1990), the levels of IGF-1 in muscle extracts are unknown. Thus, it is unclear whether it accounts for any of the observed neurotrophic activities of skeletal muscle extract.

Two other factors that affect the in-vitro survival of motoneurons are CNTF and TGFβ. CNTF is a 22.8-kDa monomeric protein found in brain, optic tissue, and Schwann cells (Adler et al., 1979; Barbin et al., 1984; Lin et al., 1989; see also Chapter 15). Purified CNTF affects the properties of both neuronal and glial cells in culture (Barbin et al., 1984; Hughes et al., 1988; Ernsberger et al., 1989; Saadat et al., 1989). Arakawa et al., (1990) found that CNTF enhances the in vitro survival of isolated chick embryo motoneurons. Thus, although muscle cells do not appear to express CNTF mRNA (Stockli et al., 1989), the expression of this molecule by Schwann cells and its affect on motoneuron survival suggest that it may play a role in the development and maintenance of motoneurons.

TGFβ is a homodimeric 25-kDa protein, first characterized by its mitogenic action on fibroblasts and subsequently shown to influence the proliferation and differentiation of many cell types (Massague, 1990; see also Chapter 12). Recently Martinou et al. (1990) demonstrated that TGFβ enhances the in vitro survival of isolated fetal rat motoneurons. However, similar studies with purified populations of chick motoneurons indicate that TGFβ has no effect on survival (Arakawa et al., 1990). Although extracts of fetal skeletal muscle contain a TGFβ-like activity (Hill et al., 1986), it is unclear if this factor is genuine TGFβ, since it was not purified. Whether muscle cells express TGFβ mRNA is also uncertain (Massague, 1990). Thus, additional studies will be required to determine what contribution, if any, TGFβ makes to the neurotrophic activity of skeletal muscle.

## IV. EFFECTS OF PURIFIED FACTORS ON MOTONEURON SURVIVAL IN VIVO

A necessary and critical step in validating the biologic importance of putative neurotrophic factors is the demonstration of their ability to influence the properties of their respective target neurons in vivo. Although the list of neurotrophic factors with documented effects in vivo is still small, the number is growing (Barde, 1989). Recent studies have begun to test the in vivo effects of many skeletal muscle-derived factors identified by in vitro experiments. The daily administration of CDF to chick embryos from ED 6 to ED 9 rescued an average of 2762 motoneurons that otherwise would have died during this period (McManaman et al., 1990). Because CDF also reduced the number of degenerating (pyknotic) motoneurons (McManaman et al., 1990), it appears to act by preventing the death of motoneurons, not by increasing their proliferation or migration or by altering neuronal phenotypes. CDF does not rescue certain other spinal cord neurons that undergo naturally occurring cell death during this period (i.e., sensory neurons in the dorsal root ganglia and cholinergic preganglionic sympathetic neurons). CDF treatment also does not alter the morphology of motoneurons or affect motoneuron growth, as reflected by nuclear diameter. Although motoneuron survival is increased also by neuromuscular blockage (Pittman and Oppenheim, 1979; Oppenheim, 1987), CDF does not affect embryonic movements; thus, it is unlikely that its effects are an indirect result of actions on neuromuscular activity.

The concentration of CDF used in the in vivo studies is comparable to the concentration of NGF required to enhance the in vivo survival of dorsal root ganglion neurons in both chick and quail embryos (Hamburger et al., 1981; Hofer and Barde, 1988) and to the amount of BDNF required to enhance the survival of dorsal root and nodose ganglion neurons (Hofer and Barde, 1988). Thus, the in vivo potency of CDF is within the range of that exhibited by other neurotrophic molecules.

Although the in vitro effects of bFGF on spinal cord ChAT activity are comparable to those of CDF, the administration of purified recombinant human bFGF (rh-bFGF) to developing chick embryos does not enhance motoneuron survival or prevent motoneuron degeneration (McManaman et al., 1990). This lack of effect in vivo is not a result of the use of human bFGF, since there was a marked increase in the proliferation of glial cells in the spinal cords treated with rh-bFGF over saline-treated controls (Oppenheim et al., unpublished observation). Further, human bFGF enhances the in vitro survival and cholinergic development of embryonic chick ciliary ganglion neurons (Vaca et al., 1989). Whether bFGF affects other aspects of motoneuron function is presently unknown. Although the possibility remains that bFGF is biologically important to the development of motoneurons, it does not appear to be involved in regulating their survival during the period of naturally occurring cell death.

IGFs and CNTF also may be involved in regulating the survival of embryonic motoneurons. In preliminary experiments, the application of IGF-1 to chick embryos during the period of naturally occurring cell death produced a small increase in the number of surviving motoneurons on ED 10 (R. Oppenheim, unpublished observations). The administration of CNTF to developing chick embryos, on the other hand, produced a much greater increase in motoneuron survival, rescuing approximately 3800 motoneurons from cell death (Oppenheim et al., 1991). Further, Sendtner et al. (1990) found that CNTF prevents the death of lesioned neonatal motoneurons. Thus, motoneurons may be a major target for CNTF.

In summary, motoneurons are sensitive to a variety of neurotrophic molecules, many of which exhibit activity in vivo. However, to what extent these factors are responsible, either alone or in combination, for the observed effects of skeletal muscle on motoneuron development remains to be determined. Antibodies to NGF have helped clarify its biologic role by producing functional deficits in both peripheral (Nja and Purves, 1978; Yip et al., 1984) and central nervous system neurons (Vantini et al., 1989). However, the findings that several factors have in vivo effects on motoneurons raises the possibility that loss of any single factor may not have a dramatic effect on motoneuron function. Motoneuron survival and differentiation may be mediated by influences derived from their targets and afferents as well as from nonneuronal cells. Accordingly, a full understanding of how motoneuron survival is regulated may require testing of these putative influences in different combinations. Investigations of the cellular and molecular control of motoneuron survival and differentiation by targets and other extrinsic factors address questions about normal development that arose at the turn of the present century (Oppenheim, 1981). The resolution of these questions, however, may provide a deeper understanding of the role such factors in the pathology and treatment of motoneuron disease (Appel, 1981; Oppenheim et al., 1989a; Haverkamp and Oppenheim, 1990; McManaman et al., 1991).

## ACKNOWLEDGMENTS

This work was supported by grants from the National Institutes of Health to J. L. M. and R. W. O. and the Muscular Dystrophy Association to J. L. M. The authors would like to thank Lanny Haverkamp and Frances Crawford for critically reviewing the manuscript and for their helpful comments.

## REFERENCES

Adler, R., Landa, K. B., Manthorpe, M. and Varon, S. (1979). Cholinergic neuronotrophic factors: Intraocular distribution of trophic activity for ciliary neurons. *Science* **204**, 1434–1436.

Appel, S. H. (1981). A unifying hypothesis for the cause of amyotrophic lateral sclerosis, parkinsonism, and Alzheimer disease. *Ann. Neurol.* **10**, 499–505.

Arakawa, Y., Sendtner, M., and Thoenen, H. (1990). Survival effect of ciliary neurotrophic factor (CNTF) on chick embryonic motoneurons in culture: Comparisons with other neurotrophic factors and cytokines. *J. Neurosci.* **10**: 3507–3515.

Barbin, G., Manthorpe, M., and Varon, S. (1984). Purification of the chick eye ciliary neuronotrophic factor. *J. Neurochem.* **43**, 1468–1478.

Barde, Y.-A. (1989). Trophic factors and neuronal survival. *Neuron* **2**, 1525–1534.
Beck, F., Samani, N. J., Byrne, S., Morgan, K., Gebhard, R., and Brammar, W. J. (1988). Histochemical localization of IGF-II and IGF-II mRNA in the rat between birth and adulthood. *Development* **104**, 29–39.
Bennett, M. R., Lai, K., and Nurcombe, V. (1980). Identification of embryonic motoneurons in vitro: Their survival is dependent on skeletal muscle. *Brain Res.* **190**, 537–542.
Calof, A. L., and Reichardt, L. F. (1984). Motoneurons purified by cell sorting respond to two distinct activities in myotube-conditioned medium. *Dev. Biol.* **106**, 194–210.
Caroni, P., and Grandes, P. (1990). Nerve sprouting in innervated adult skeletal muscle induced by exposure to elevated levels of insulin-like growth factors. *J. Cell Biol.* **110**, 1307–1317.
Chu-Wang, I.-W., and Oppenheim, R. W. (1978). Cell death of motoneurons in the chick embryo spinal cord. I. A light and electron microscopic study of naturally occurring and induced cell loss during development. *J. Comp. Neurol.* **177**, 33–58.
Dohrmann, U., Edgar, D., Sendtner, M., and Thoenen, H. (1986). Muscle-derived factors that support survival and promote fiber outgrowth from embryonic chick spinal motor neurons in culture. *Dev. Biol.* **118**, 209–221.
Dohrmann, U., Edgar, D., and Thoenen, H. (1987). Distinct neurotrophic factors from skeletal muscle and the central nervous system interact synergistically to support the survival of cultured embryonic spinal motor neurons. *Dev. Biol.* **124**, 145–152.
Dreyer, D., Lagrange, A., Grothe, C., and Unsicker, K. (1989). Basic fibroblast growth factor prevents ontogenetic neuron death in vivo. *Neurosci. Lett.* **99**, 35–38.
Dribin, L. B., and Barrett, J. N. (1982). Characterization of neuritic outgrowth-promoting activity of conditioned medium on spinal cord explants. *Dev. Brain Res.* **4**, 435–441.
Ernfors, P., Henschen, A., Olson, L., and Persson, H. (1989). Expression of nerve growth factor receptor mRNA is developmentally regulated and increased after axotomy in rat spinal cord motoneurons. *Neuron* **2**, 1605–1613.
Ernsberger, U., Sendtner, M., and Rohrer, H. (1989). Proliferation and differentiation of embryonic chick sympathetic neurons: Effects of ciliary neurotrophic factor. *Neuron* **2**, 1275–1284.
Giess, M.-C., and Weber, M. J. (1984). Acetylcholine metabolism in rat spinal cord cultures: Regulation by a factor involved in the determination of the neurotransmitter phenotype of sympathetic neurons. *J. Neurosci.* **4**, 1442–1452.
Giller, E. L., Neale, J. H., Bullock, P. N., Schrier, B. K., and Nelson, P. G. (1977). Choline acetyltransferase activity of spinal cord cell cultures increased by co-culture with muscle and by muscle-conditioned medium. *J. Cell Biol.* **74**, 16–29.
Gospodarowicz, D., Ferrara, N., Schweigerer, L., and Neufeld, G. (1987). Structural characterization and biochemical functions of fibroblast growth factor. *Endocrinol. Rev.* **8**, 95–114.
Hamburger, V. (1975). Cell death in the development of the lateral motor column of the chick embryo. *J. Comp. Neurol.* **160**, 535–546.
Hamburger, V. Brunso-Bechtold, J. K., and Yip, J. W. (1981). Neuronal death in the spinal ganglia of the chick embryo and its reduction by nerve growth factor. *J. Neurosci.* **1**, 60–71.
Hansson, H. A. Nilsson, A., Isgaard, J., Billig, H., Isaksson, O., Skottner, A., Anderson, I. K., and Rozell, B. (1988). Immunohistochemical localization of insulin-like growth factor 1 in the adult rat. *Histochemistry* **89**, 403–410.
Harris, A. J., and McCaig, C. D. (1984). Motoneuron death and motor unit size during embryonic development of the rat. *J. Neurosci.* **4**, 13–24.
Haverkamp, L. J., and Oppenheim, R. W. (1990). Death of motoneurons by accident and design: Neurotrophic interaction of anterior horn cells with their targets. *In* "Amyotrophic Lateral Sclerosis: Concepts in Pathogenesis and Etiology" (A. J. Hudson, ed.), pp. 24–38. University of Toronto Press, Toronto.
Henderson, C. E., Huchet, M., and Changeux, J.-P. (1983). Denervation increases a neurite-promoting activity in extracts of skeletal muscle. *Nature (London)* **303**, 609–611.
Hill, D. J., Strain, A. J., and Milner, R. D. G. (1986). Presence of transforming growth factor-β-like activity in multiple fetal rat tissues. *Cell Biol. Int. Rep.* **10**, 915–922.

Hofer, M. M., and Barde, Y.-A. (1988). Brain-derived neurotrophic factor prevents neuronal death in vivo. *Nature (London)* **331**, 261–262.

Hollyday, M., and Hamburger, V. (1979). Reduction of the naturally occurring motor neuron loss by enlargement of the periphery. *J. Comp. Neurol.* **170**, 311–320.

Hughes, S. M., Lillien, L. E., Raff, M. C., Rohrer, H., and Sendtner, M. (1988). Ciliary neurotrophic factor induces type-2 astrocyte differentiation. *Nature (London)* **335**, 70–73.

Ishii, D. N. (1989). Relationship of insulin-like growth factor 2 gene expression in muscle to synaptogenesis. *Proc. Natl. Acad. Sci. U.S.A.* **86**, 2898–2902.

Kaufman, L. M., Barry, S. R., and Barrett, J. N. (1985). Characterization of tissue-derived macromolecules affecting transmitter synthesis in rat spinal cord neurons *J. Neurosci* **5**, 160–166.

Lanser, M. E., and Fallon, J. F. (1987). Development of the brachial lateral motor column in the wingless mutant chick embryo: Motoneuron survival under varying degrees of peripheral load. *J. Comp. Neurol.* **261**, 423–434.

Large, T. H., Weskamp, G., Helder, J. C., Radeke, M. J., Misko, T., Shooter, E. M., and Reichardt, L. F. (1989). Structure and developmental expression of the nerve growth factor receptor in the chicken central nervous system. *Neuron* **2**, 1123–1134.

Levi-Montalcini, R. (1987). The nerve growth factor 35 years later. *Science* **237**, 1154–1162.

Lin, L.-F. H., Mismer, D., Lile, J., Armes, L. G., Butler, E. T., Vannice, J. L., and Collins, F. (1989). Purification, cloning, and expression of ciliary neurotrophic factor (CNTF). *Science* **246**, 1023–1025.

McManaman, J. L., and Crawford, F. G. (1991). Skeletal muscle proteins stimulate cholinergic differentiation of human neuroblastoma cells. *J. Neurochem.* (in press).

McManaman, J. L., Crawford, F. G., Stewart, S. S., and Appel, S. H. (1988a). Purification of a skeletal muscle polypeptide which stimulates choline acetyltransferase activity in cultured spinal cord neurons. *J. Biol. Chem.* **263**, 5890–5897.

McManaman, J. L., Haverkamp, L. J., and Appel, S. (1988b). Developmental discord among markers for cholinergic differentiation: *In vitro* time courses for early expression and responses to skeletal muscle extract. *Dev. Biol.* **125**, 311–320.

McManaman, J. L., Crawford, F. G., Clark, R., Richker, J., and Fuller, F. (1989). Multiple neurotrophic factors form skeletal muscle: Demonstration of effects of bFGF and comparisons with the 22-kdalton ChAT development factor. *J. Neurochem.* **53**, 1763–1771.

McManaman, J. L., Oppenheim, R. W., Prevette, D., and Marchetti, D. (1990). Rescue of motoneurons from naturally occurring cell by a purified skeletal muscle polypeptide: In vivo and in vitro effects of the choline acetyltransferase development factor. *Neuron* **4**, 891–898.

McManaman, J. L., Haverkamp, L. J., and Oppenheim, R. W. (1991). Skeletal muscle proteins rescue motoneurons from cell death in vivo. *In* "Amyotrophic Lateral Sclerosis and Other Motor Neuron Diseases" (L. P. Rowland, ed.), pp. 81–88. Raven Press, New York.

Marchetti, D., and McManaman, J. L. (1990). Characterization of nerve growth factor binding to embryonic rat spinal cord neurons. *J. Neurosci. Res.* **27**, 211–218.

Martinou, J. C., Le Van Thai, A., Cassar, G., Roubinet, F., and Weber, M. J. (1989). Characterization of two factors enhancing choline acetyltransferase activity in cultures of purified rat motoneurons. *J. Neuroscience* **9**, 3645–3656.

Martinou, J. C., Le Van Thai, A., Valette, A., and Weber, M. J. (1990). Transforming growth factor beta 1 is a potent survival factor for rat embryo motorneurons in culture. *Dev. Brain Res.* **52**, 175–181.

Massague, J. (1990). The transforming growth factor-β family. *Ann. Rev. Cell Biol.* **6**, 597–641.

Morrison, R. S., Sharma, A., de Vellis, J., and Bradshaw, R. A. (1986). Basic fibroblast growth factor supports the survival of cerebral cortical neurons in primary culture. *Proc. Natl. Acad. Sci. U.S.A.* **83**, 7537–7541.

Nja, A., and Purves, D. (1978). The effects of nerve growth factor and its antiserum on synapses in the superior cervical ganglion of the guinea-pig. *J. Physiol.* **277**, 53–75.

O'Brien, R. J., and Fischbach, G. D. (1986). Isolation of embryonic chick motoneurons and their survival in vitro. *J. Neurosci.* **6**, 3265–3274.

Okado, N., and Oppenheim, R. W. (1984). Cell death of motoneurons in the chick embryo spinal cord. IX. The loss of motoneurons following removal of afferent inputs. *J. Neurosci.* **4**, 1639–1652.

Oppenheim, R. W. (1981). Neuronal cell death and some related regressive phenomena during neurogenesis: A selective historical review and progress report. In "Studies in Developmental Neurobiology: Essays in Honor of Viktor Hamburger" (W. M. Cowan, ed.), pp. 74–133. Oxford University Press, New York.

Oppenheim, R. W., (1987). Muscle activity and motor neuron death in the spinal cord of the chick embryo. In "Selective Neuronal Death" (G. Bock and M. O'Connor, eds.), pp. 96–108. Wiley, New York.

Oppenheim, R. W., (1989). The neurotrophic theory and naturally occurring motoneuron cell death. *Trends Neurosci.* **12**, 252–255.

Oppenheim, R. W. (1991) Cell death during development of the nervous system. *Ann. Rev. Neurosci.* **14**, 453–501.

Oppenheim, R. W., and Chu-Wang, I.-W. (1983). Aspects of naturally-occurring motoneuron death in the chick spinal cord during embryonic development In "Somatic and Autonomic Nerve-Muscle Interactions" (G. Burnstock, ed.), 57–107. Elsevier Science, New York.

Oppenheim, R. W., and Heaton, M. B. (1975). The retrograde transport of horseradish peroxidase from the developing limb of the chick embryo. *Brain Res.* **98**, 291–302.

Oppenheim, R. W., and Nunez, R. (1982). Electrical stimulation of hindlimb increases neuronal cell death in the chick embryo. *Nature (London)* **295**, 57–59.

Oppenheim, R. W., Chu-Wang, I-W., and Maderdrut, J. L. (1978). Cell death of motoneurons in the chick embryo spinal cord. III. The differentiation of motoneurons prior to their induced degeneration following limb-bud removal. *J. Comp. Neurol.* **177**, 87–112.

Oppenheim, R. W., Maderdrut, J. L., and Wells, D. J. (1982). Cell death of motoneurons in the chick embryo spinal cord. VI. Reduction of naturally occurring cell death in the thoracolumbar column of Terni by nerve growth factor. *J. Comp. Neurol.* **210**, 174–189.

Oppenheim, R. W., Haverkamp, L. J., Prevette, D., McManaman, J. L., and Appel, S. (1988). Reduction of naturally occurring motoneuron death *in vivo* by a target-derived neurotrophic factor. *Science* **240**, 919–922.

Oppenheim, R. W., Haverkamp, L. J., Prevette, D., and McManaman, J. L. (1989a). Neurotrophic regulation of motoneuron survival during normal embryonic development: A possible model for infantile spinal muscular atrophy. In "Current Concepts in Childhood Spinal Muscular Atrophy" (L. Merlini, C. Granata, and V. Dubowitz, eds.), pp. 43–55. Springer-Verlag, New York.

Oppenheim, R. W., Dohrmann, U., Prevette, D., and Qin-Wei, Y. (1989b). Prevention of motoneuron death in vivo by a putative CNS-derived neurotrophic factor. *Soc. Neurosci. Abstr.* **15**, 18.

Oppenheim, R. W., Burszatajn, S., and Prevette, D. (1989c). Cell death of motoneurons in the chick embryo spinal cord. XI. Acetylcholine receptors and synaptogenesis in skeletal muscle following the reduction of motoneuron death by neuromuscular blockage. *Development* **107**, 331–341.

Oppenheim, R. W., Prevette, D., Qin-Wei, Y., Collins, F., and MacDonald, J. (1991). Control of embryonic motoneuron survival in vivo by ciliary neurotrophic factor. *Science* **251**, 1616–1618.

Otto, D., Frotscher, M., and Unsicker, K. (1989). Basic fibroblast growth factor and nerve growth factor administered in gel foam rescue medial septal neurons after fimbria fornix transection. *J. Neurosci. Res.* **22**, 83–91.

Pittman, R., and Oppenheim, R. W. (1979). Cell death of motoneurons in the chick embryo spinal cord. IV. Evidence that a functional neuromuscular interaction is involved in the regulation of naturally occurring cell death and the stabilization of synapses. *J. Comp. Neurol.* **187**, 425–446.

Raivich, G., Zimmermann, A., and Sutter, A. (1985). The spatial and temporal pattern of NGF receptor expression in the developing chick embryo. *EMBO J.* **4**, 637–644.

Saadat, S., Sendtner, M., and Rohrer, H. (1989). Ciliary neurotrophic factor induces cholinergic differentiation of rat sympathetic neurons in culture. *J. Cell Biol.* **108**, 1807–1816.

Schaffner, A. E., St. John, P. A., and Barker, J. L. (1987). Fluorescence-activated cell sorting of embryonic mouse and rat motoneurons and their long-term survival in vitro. *J. Neurosci.* **7**, 3088–3104.

Sendtner, M., Kreutzberg, G. W., and Thoenen, H. (1990). Ciliary neurotrophic factor prevents the degeneration of motor neurons after axotomy. *Nature (London)* **345**, 440–441.

Sjoberg, J., and Kanje, M. (1989). Insulin-like growth factor (IGF-1) as a stimulator of regeneration in the freeze-injured rat sciatic nerve. *Brain Res.* **485**, 102–108.

Smith, R. G., and Appel, S. H. (1983). Extracts of skeletal muscle increase neurite outgrowth and cholinergic activity of fetal rat spinal motor neurons. *Science* **219**, 1079–1081.

Smith, R. G., McManaman, J., and Appel, S. H. (1985). Trophic effects of skeletal muscle extracts on ventral spinal cord neurons in vitro: Separation of a protein with morphologic activity from proteins with cholinergic activity. *J. Cell Biol.* **101**, 1608–1621.

Smith, R. G., Vaca, K., McManaman, J., and Appel, S. H. (1986). Selective effects of skeletal muscle extract fractions on motoneuron development in vitro. *J. Neurosci.* **6**, 439–447.

Stockli, K. A., Lottspeich, F., Sendtner, M., Masiakowski, P., Carroll, P., Gotz, R., Lindholm, D., and Thoenen, H. (1989). Molecular cloning, expression and regional distribution of rat ciliary neurotrophic factor. *Nature (London)* **342**, 920–923.

Tanaka, H. (1987). Chronic application of curare does not increase the level of motoneuron survival-promoting activity in limb muscle extracts during the naturally occurring motoneuron cell death period. *Dev. Biol.* **124**, 347–357.

Tanaka, H., and Obata, K. (1983). Survival of HRP-labeled spinal motoneurons of chick embryo in tissue and cell cultures. *Dev. Brain Res.* **9**, 390–395.

Unsicker, K., Reichert-Preibsch, H., Schmidt, R., Pettmann, B., Labourdette, G., and Sensenbrenner, M. (1987). Astroglial and fibroblast growth factors have neurotrophic functions for cultured peripheral and central nervous system neurons. *Proc. Natl. Acad. Sci. U.S.A.* **84**, 5459–5463.

Vaca, K., McManaman, J. L., Bursztajn, S., and Appel, S. H. (1985). Differential morphological effects of two fractions from fetal calf muscle on cultured chick ciliary ganglion cells. *Dev. Brain Res.* **19**, 37–46.

Vaca, K., Stewart, S. S., and Appel, S. H. (1989). Identification of basic fibroblast growth factor as a cholinergic growth factor from human muscle. *J. Neurosci. Res.* **23**, 55–63.

Vantini, G., Schiavo, N., DiMartino, A., Polato, P., Triban, C., Callegro, L., Toffano, G., and Leon, A. (1989). Evidence for a physiological role of nerve growth factor in the central nervous system of neonatal rats. *Neuron* **3**, 267–273.

Walicke, P., Cowan, W. M., Ueno, N., Baird, A., and Guillemin, R. (1986). Fibroblast growth factor promotes survival of dissociated hippocampal neurons and enhances neurite extension. *Proc. Natl. Acad. Sci. U.S.A.* **83**, 3012–3016.

Wayne, D. B., and Heaton, M. B. (1990). The ontogeny of specific retrograde transport of nerve growth factor by motoneurons of the brainstem and spinal cord. *Dev. Biol.* **138**, 484–498.

Wood, J. B., and McManaman, J. L. (1990). A possible neurotrophic role for IGF-1 and IGF-2 on motoneurons. *J. Cell Biochem.* **14**, 96.

Yan, Q., Snider, W. D., Pinzone, J. J., and Johnson, E. M. (1988). Retrograde transport of nerve growth factor (NGF) in motoneurons of developing rats: Assessment of potential neurotrophic effects. *Neuron* **1**, 335–343.

Yip, H. K., Rich, K., Lampe, P. A., and Johnson, E. M., (1984). The effects of nerve growth factor and its antiserum on the postnatal development and survival after injury of sensory neurons in rat dorsal root ganglia. *J. Neurosci.* **4**, 2986–2992.

# 17 Growth Factors for Myelinating Glial Cells in the Central and Peripheral Nervous Systems

Ellen J. Collarini and William D. Richardson

## I. INTRODUCTION

Glial cells play essential roles during development and in the mature central (CNS) and peripheral (PNS) nervous systems. For example, astrocytes probably provide structural guidance for migrating neurons and signals for axon extension during development of the CNS. In the adult, they presumably perform other functions, perhaps regulating extracellular ion concentrations at sites of axonal activity (see Bevan, 1990; Hatten et al., 1990, for reviews). Schwann cells in the PNS and oligodendrocytes in the CNS synthesize the myelin sheaths around axons. Much effort has gone into identifying the factors that control glial cell proliferation and differentiation and induce glial cells to assume their proper relationships with neurons. Understanding these issues has important implications for future treatment of degenerative diseases such as multiple sclerosis (in which oligodendrocytes die and are not replaced), for treatment of glial tumors, and for assisting regeneration of damaged nerves. This review focuses on the growth factors involved in controlling the proliferation, differentiation, and survival of the myelinating cells of the PNS and the CNS.

Although Schwann cells and oligodendrocytes both myelinate axons, there are differences in the way they interact with neurons and differences in their requirements for neuron-derived factors. Individual oligodendrocytes contribute to the myelin sheaths around several axons, whereas each myelinating Schwann cell is associated with a single axon. The major protein constituents of CNS myelin are proteolipid protein (PLP) and myelin basic protein (MBP). The major myelin protein in the PNS is $P_0$, which is not present in the CNS. PLP and MBP are present in the PNS, but do not appear to be necessary for normal myelin formation, since mice with mutations affecting PLP (*jimpy*) and MBP (*shiverer*) have normal PNS myelin, despite severe myelin defects in the CNS (see Hudson, 1990 for a review). In addition, the regulation of these myelin genes differs between the PNS and CNS. Schwann cells seem to require continual axonal contact to maintain myelin gene expression whereas oligodendrocytes, at least in culture, can express myelin proteins in the absence of neurons (Mirsky et al., 1980). Following peripheral nerve injury, Schwann cells revert to an undifferentiated phenotype, proliferate, and remyelinate the regenerating axons (reviewed in Fawcett and Keynes, 1990; Jessen and Mirsky, 1991). Axon regeneration is much less efficient in the CNS, partly because oligodendrocyte membranes contain proteins that are inhibitory to axonal outgrowth (Schwab, 1990).

## II. CENTRAL NERVOUS SYSTEM GLIA

Glial cells in the CNS are subdivided into microglia and macroglia. Microglial cells are specialized resident macrophages that are derived from blood-borne monocytes (reviewed by Perry and Gordon, 1991). Macroglia, like CNS neurons, are derived from the neuroectodermal cells of the neural tube and are a diverse group of cells that includes ependymal cells (the ciliated cells that line the ventricles), oligodendrocytes, and astrocytes. Many different categories of astrocytes can be defined by morphology in Golgi stained tissue preparations. Cell type is defined more reliably by cellular function and cell lineage during development, but astrocyte function and lineage are poorly understood. More is known about oligodendrocytes and their precursors since there are specific antibodies available that can be used to identify and manipulate these cells in vitro.

### A. Oligodendrocyte Type-2 Astrocyte Lineage

In several regions of the developing rat CNS, including the optic nerve, cerebellum, cerebral cortex, and forebrain, are glial progenitor cells that can give rise to oligodendrocytes in vitro when cultured in chemically defined medium containing low concentrations ($< 0.5\%$) of fetal calf serum (Raff et al., 1983; Levi et al., 1987; Levine and Stallcup, 1987; Behar et al., 1988; Gard and Pfeiffer, 1989; Hardy and Reynolds, 1991). When cultured in the presence of 10% fetal calf serum, these same progenitor cells differentiate instead into type-2 astrocytes; the progenitor cells are therefore known as oligodendrocyte–type-2 astrocyte (O–2A) progenitors. Type-2 astrocytes have a distinctive process-bearing morphology in vitro and are so named to distinguish them from type-1 astrocytes, which are fibroblast-like in culture and arise from their own dedicated precursor cells (Raff et al., 1984) (see Fig. 1). It is not clear, however, whether type-2 astrocytes exist in vivo (reviewed in Richardson et al., 1990).

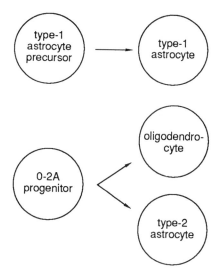

FIGURE 1

Glial cell lineages in the rat optic nerve. Type-1 astrocytes develop from one type of precursor cell, whereas oligodendrocytes develop from a different precursor cell that, at least in culture, can also give rise to type-2 astocytes.

## B. Role of PDGF in Oligodendrocyte Development

Small numbers of proliferating O–2A progenitor cells are present in the developing rat optic nerve at embryonic day 15 (ED 15) (Small et al., 1987). They first start to give rise to oligodendrocytes on the day of birth (ED 21/P0), and continue to proliferate and differentiate into oligodendrocytes for several weeks after birth (Miller et al., 1985). In contrast, when embryonic rat optic nerve cells are dissociated and placed in culture in chemically defined low-serum medium, most O–2A progenitors stop dividing prematurely and differentiate rapidly into oligodendrocytes (Raff et al., 1983). It seems that the defined medium lacks a mitogen that acts in vivo to maintain O–2A progenitor division and to prevent premature differentiation into oligodendrocytes. Because of the key role played by this mitogen in regulating the final number of oligodendrocytes, as well as the time and rate of their appearance, there has been considerable effort expended in trying to identify the growth factors that influence development of the O–2A lineage in vitro and in vivo.

O–2A progenitor cells freshly isolated from newborn rat optic nerves possess cell-surface receptors for platelet-derived growth factor (PDGF) (Hart et al., 1989b; McKinnon et al., 1990). PDGF is a potent mitogen for O–2A progenitors in vitro (Noble et al., 1988; Richardson et al., 1988), and can reconstitute the normal timing of oligodendrocyte development in vitro (Raff et al., 1988). For example, if optic nerve cells from ED 17 rats are cultured in chemically defined medium containing PDGF, the O–2A progenitors continue to divide for some time; oligodendrocytes do not appear until the fourth day in culture, corresponding to the day of birth in vivo ED 21). If ED 18 optic nerve cells are cultured in the same way, the first oligodendrocytes appear on the third day in vitro, and so on. Since PDGF mRNA (Richardson et al., 1988; Pringle et al., 1989) and protein (H. Mudhar and W. D. Richardson, unpublished observations) are present in the perinatal brain and optic nerve at the time that O–2A progenitors are proliferating rapidly in vivo, and PDGF-like mitogenic activity can be detected in extracts of optic nerves (Raff et al., 1988), it seems likely that PDGF is important for normal development of the O–2A lineage in vivo.

PDGF is a disulfide-linked dimer of A and B chains with the structure AA, AB, or BB, depending on its source (for a review of PDGF, see Heldin and Westermark, 1989). For example, PDGF from human platelets is a mixture of all three dimeric isoforms, of which PDGF-AB is the major species, whereas some human tumor cell lines synthesize mainly PDGF-AA. PDGF elicits its biologic effects by binding to transmembrane receptors with extracellular ligand-binding domains and intracellular tyrosine kinase domains. The unoccupied receptors are monomeric and inactive, but PDGF binding induces dimerization and activates their tyrosine kinase activity. There are two types of PDGF receptor subunits with different ligand specificities: the $\alpha$ receptor subunit (PDGF$\alpha$R) binds both A and B chains of PDGF whereas the $\beta$ receptor subunit (PDGF$\beta$R) binds only B chains. Thus, the response of a cell to PDGF depends on the relative numbers of $\alpha$ and $\beta$ receptors that it expresses as well as on the precise mixture of PDGF isoforms that it encounters (see Fig. 2). O–2A progenitor cells express predominantly PDGF$\alpha$R (Hart et al., 1989b; McKinnon et al., 1990) and, consequently, respond to all three dimeric isoforms of PDGF, although PDGF-AA is effective at lower concentrations than PDGF-BB because PDGF$\alpha$R has a higher affinity for PDGF A chains than B chains (Heldin et al., 1988). Consistent with this, we could detect mRNA encoding the PDGF A chain, but not the B chain, in the rat optic nerve (Pringle et al., 1989), suggesting that PDGF-AA may be the predominant PDGF isoform in the nerve.

If PDGF is responsible for stimulating O–2A progenitor proliferation during development, what causes the O–2A progenitors to stop dividing eventually and differentiate into oligodendrocytes, and what dictates the timing of this decision? An O–2A progenitor differentiates into an oligodendrocyte when it is cultured on its own in a microwell in defined medium (Temple and Raff, 1985), so oligodendrocyte differentiation seems to proceed by default when an O–2A progenitor is deprived of PDGF and other exogenous signals. It is unlikely, however, that oligodendrocyte differentiation in vivo is triggered by PDGF withdrawal. First, PDGF A chain mRNA is present in the brain and optic nerve from before birth into adulthood (Richardson et al., 1988; Pringle et al., 1989), suggesting that PDGF may be continuously available throughout life. Second, O–2A progenitors in optic nerve cell cultures do not proliferate indefinitely in vitro; they eventually stop dividing and differentiate into oligodendrocytes, no matter how much or how often PDGF

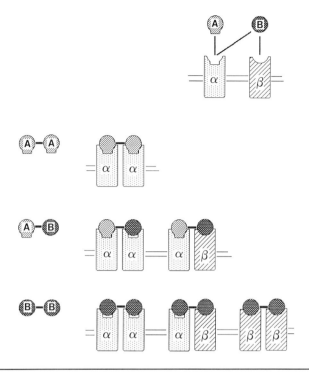

**FIGURE 2**

PDGF and its receptors. PDGF is a dimer of A and B chains, with the structure AA, BB, or AB. PDGF α-receptor subunits (PDGFαR) can bind both PDGF A and B chains, whereas PDGF β-receptor subunits (PDGFβR) can bind only PDGF B chains. Consequently, PDGF-AA can cross-link and activate only PDGFαR, whereas PDGF-BB can activate both PDGFαR and PDGFβR. PDGF-AB can cross-link two A chains or one A and one B chain. In addition, PDGF-AB may be able to act as a receptor antagonist for cells with only PDGFβR, since it may be able to bind to, but not dimerize, PDGFβR (Heldin and Westermark, 1989).

is added to the culture medium (Raff et al., 1988). Moreover, in these cultures, dividing and differentiating O–2A progenitors coexist in the same dish. This suggests that the decision to differentiate is cell autonomous and does not depend on timed signals from other cells. Thus, in order to understand how oligodendrocyte differentiation is timed, we need to know why proliferating O–2A progenitor cells eventually lose their ability to divide in response to PDGF. In principle, this loss of responsiveness to PDGF could be caused by a loss of PDGF receptors at the cell surface, a block in one of the intracellular second messenger systems that transduces the mitogenic signal, or a deficiency in part of the cell replication machinery. PDGF receptor loss does not seem to be the key, because newly formed post mitotic oligodendrocytes in vitro are still able to bind $^{125}$I-labled PDGF (Hart et al., 1989b). PDGF receptors eventually are lost from maturing oligodendrocytes, but this is a consequence of differentiation, not the primary cause. The PDGF receptors on newly formed oligodendrocytes are functional and linked into at least part of the signal transduction apparatus because, in young oligodendrocytes in vitro, PDGF stimulation leads to elevation of cytosolic $Ca^{2+}$ (Hart et al., 1989a) and activation of the proto-oncogene products c-JUN and c-FOS in the nucleus (I. K. Hart, W. D. Richardson, and M. C. Raff, unpublished observations). Thus, it appears that the mitotic block that develops in an O–2A progenitor cell, causing it to drop out of division and differentiate into an oligodendrocyte, lies downstream of nuclear proto-oncogene activation or, alternatively, in an unidentified parallel signaling pathway.

### C. Regulation of SCIP Transcription Factor by Growth Factors

Progress in understanding the molecular events that initiate oligodendrocyte differentiation has been hampered by the difficulty in obtaining O–2A progenitors in sufficient numbers or purity for biochemical analysis. However, recent technical developments now allow us to purify O–2A progenitors by immunoselection and keep them proliferating in vitro with a combination of basic fibroblast growth factor (bFGF) and PDGF (Bögler et al., 1990; see Section II,D). This has enabled us to examine the expression of transcription factors that might be involved in controlling the switch between proliferation and differentiation of O–2A progenitors in vitro. For example, we have found that the transcription factor SCIP/Tst-1 (He et al., 1989; Monuki et al., 1989) is highly expressed in O–2A progenitors that are kept proliferating in vitro with a combination of bFGF and PDGF, but that SCIP mRNA declines to background levels within 6 hr after oligodendrocyte differentiation is initiated by growth factor withdrawal (E. J. Collarini et al., 1992). This down-regulation of SCIP mRNA is followed by a decline in SCIP protein over the next 24 hr, and subsequent appearance of myelin-specific products. SCIP previously was shown to be expressed in proliferating Schwann cells and to be down-regulated when they start to express myelin-specific genes (Monuki et al., 1989) (see Section III,B). There is evidence that SCIP is a repressor of myelin gene expression in proliferating Schwann cells (Monuki et al., 1990); the same may be true in proliferating O–2A progenitors. It is not yet known whether a high level of SCIP is necessary or sufficient to keep O–2A progenitors dividing and inhibit oligodendrocyte differentiation. However, it is known that the related POU transcription factor Oct-3 is required for mouse eggs to complete their first embryonic division (Rosner et al., 1991). Thus, it is still too early to say whether SCIP is involved causally in the decision of an O–2A progenitor cell to stop dividing and differentiate into an oligodendrocyte, but this possibility is under investigation.

### D. Sources of PDGF in the CNS: Neurons versus Glia

It has been known to some time that there is mitogenic activity for Schwann cells on the surfaces of PNS axons, although it is not known whether this activity is synthesized by the neurons themselves or deposited by other cells (see Sections

III,A and III,D). Nevertheless, it is an attractive idea that neurons in both the PNS and CNS might produce mitogens, and possibly chemoattractants, for the glial cells with which they physically interact. The possibility that axons might influence the development of the oligodendrocyte lineage has been tested in vivo (Privat et al., 1981; David et al., 1984). Newborn rat optic nerves, which carry axons from ganglion neurons in the retina to the brain, were transected just behind the eye, causing the axons in the nerve to degenerate rapidly. The number of oligodendrocytes and their progenitors present in the nerve stump 1 week later was reduced more than 8-fold compared with untransected nerves, whereas the number of type-1 astrocytes (which belong to a different cell lineage) was reduced less than 2-fold (David et al., 1984). Surprisingly, despite the large decrease in the population of oligodendrocyte lineage cells, the mitotic index of these cells did not differ significantly between transected and untransected nerves. This result was interpreted to mean that O–2A progenitors depend on axons not for proliferation per se but for survival. These experiments suggested that intact axons are not obligatory for mitogenic stimulation of O–2A progenitors in the optic nerve; instead, glial cells in the nerve might be the source of the mitogen(s). Support for this latter view came from the finding that cultured astrocytes from rat cerebral cortex, which resemble type-1 astrocytes in optic nerve cell cultures, secrete a mitogen for O–2A progenitors (Noble and Murray, 1984). This astrocyte-derived mitogen subsequently was shown to be a form of PDGF, possibly PDGF-AA (Richardson et al., 1988; Pringle et al., 1989). Type-1 astrocytes are the most abundant glial cells in the newborn rat optic nerve (Miller et al., 1985), so these cells would be expected to influence significantly the composition of the interstitial fluid in the nerve. Axons are not thought to contain mRNA, so the fact that PDGF A chain mRNA can be detected in the newborn rat optic nerve in situ (Pringle et al., 1989) suggests that glial cells, possibly type-1 astrocytes, synthesize PDGF in vivo. In support of this concept, we have been able to demonstrate immunostaining of astrocyte processes in postnatal mouse and rat optic nerves with antisera raised against the mouse PDGF A chain (H. Mudhar and W. D. Richardson, unpublished observations).

It was reported that many neurons in the CNS and PNS also synthesize PDGF A or B chains (Yeh et al., 1991; Sasahara et al., 1991), so we should not neglect the possibility that neurons also might be an important source of PDGF (or other growth factors) for O–2A progenitors in the developing CNS. There is some experimental evidence in support of this idea; O–2A progenitor cells from developing rat cerebellum or optic nerve are stimulated to divide in vitro by medium conditioned by cultures of young cerebellar interneurons. At least half of this mitogenic activity can be neutralized by antibodies against PDGF (Levine, 1989). Retinal ganglion neurons contain mRNA encoding the PDGF A chain and immunostain with anti-PDGF sera (H. Mudhar and W. D. Richardson, unpublished observations), so these neurons conceivably could supply PDGF to the optic nerve, although this probably would require that PDGF be transported anterogradely along, and secreted from, the axons of retinal ganglion neurons. It remains to be seen whether this process can occur. It may transpire that neuron-derived PDGF is not involved in neuron–glial cell interactions, but instead has some autocrine function or mediates interactions between neurons and their targets. For example, FGF is known to promote the survival of hippocampal neurons in vitro (Walicke et al., 1986) and can prevent ontogenetic neuron death in vivo (Dreyer et al., 1989). There is also a report that PDGF has neurotrophic activity in vitro (Smits et al., 1991).

### E. Role of FGF in O–2A Lineage Development

FGF is reported to be mitogenic for O–2A lineage cells in culture (Eccleston and Silberberg, 1985; Saneto and DeVellis, 1985; McKinnon et al., 1990), although at least one report contradicts this (Hunter and Bottenstein, 1990). This disagreement may result from differences in the purity and source of the O–2A progenitor cells or the particular preparation of FGF. It was found that the combination of bFGF and PDGF has a striking cooperative effect, stimulating prolonged proliferation of O–2A progenitors in the apparent absence of oligodendrocyte differentiation (Bögler et al., 1990). bFGF is present in the developing and mature CNS (Gospodarowicz, 1984; Gonzalez et al., 1990); by immunohistochemistry it appears to be present in the cell bodies of neurons (Janet et al., 1988). However, it is not known whether bFGF can be released from neurons, or any living cell, since its polypeptide precursor lacks a recognizable signal for entry into the constitutive secretory pathway. In any event, it does not seem likely that bFGF and PDGF act together on O–2A progenitors in the postnatal optic nerve, or in other developing white matter tracts. Otherwise, oligodendrocyte differentiation presumably would be inhibited. Perhaps bFGF is released only from dying cells as a response to CNS injury. Alternatively, FGF may be released from living cells by an unconventional mechanism, but its biologic effects are restricted to a compartment(s) of the CNS in which oligodendrocytes are not required. O–2A progenitors are thought to be migratory cells in vivo, moving into the optic nerve and other developing white matter tracts from germinal centers elsewhere in the brain (Small et al., 1987; LeVine and Goldman, 1988; Reynolds and Wilkin, 1988). These germinal centers probably lie in parts of the subventricular zone (Altman, 1966; Paterson et al., 1973). As O–2A lineage cells migrate, they may encounter a succession of microenvironments in which they could be exposed to different mixtures of mitogens or be induced to express different sets of growth factor receptors. There is some evidence that O–2A progenitors may alter their responsiveness to certain growth stimuli as they progress along the oligodendrocyte differentiation pathway (Gard and Pfeiffer, 1990). It is possible that, in vivo, receptors for bFGF (bFGFR) are expressed on O–2A lineage cells only at an early stage in their development, perhaps on a population of self-renewing cells near the ventricles whose function is to generate a steady stream of migrating O–2A progenitors or preprogenitors. The migrating cells normally might lose bFGFR as they move away from the subventricular layer, perhaps gaining PDGFαR instead. This hypothesis would coincide with the observations that mRNA encoding bFGFR is found preferentially in the subventricular zones of developing rat and chicken brain (Heuer et al., 1990; Wanaka et al., 1991), whereas mRNA encoding PDGFαR is located in glial cells almost entirely outside the subventricular zone (N. Pringle et al., 1992). This hypothesis also requires that O–2A progenitor cells inappropriately re-express bFGFR when they are dissociated and placed in culture. More needs to be learned about the FGF receptors on O–2A lineage cells in vitro and in vivo, and about the availability of FGF in the CNS, before this and other possibilities can be explored fully.

### F. Insulin, Insulin-Like Growth Factors, and the O–2A Lineage

Insulin and insulin-like growth factor-I (IGF-I) have been reported to be mitogenic for oligodendrocytes in culture (McMorris and Dubois-Dalcq, 1988). However, alternative explanations are compatible with the reported data. For example, insulin and IGF-I could be mitogenic for O–2A progenitor cells; the increase in oligodendrocyte numbers could result from differentiation of a greater number of progenitors. Alternatively, insulin and IGF-I could act as survival factors for O–2A progenitors or oligodendrocytes. Recent experiments with cultures of O–2A progenitors purified from rat optic nerves by immunoselection (B. Barres et al., 1992) show that insulin, IGF-I, and IGF-II (all at concentrations that saturate IGF-I receptors) are survival factors, but not mitogens, for purified O–2A progenitor cells and young oligodendrocytes derived from them. In contrast, PDGF is a survival factor *and* a mitogen for O–2A progenitor cells, and a survival factor for young postmitotic oligodendrocytes (B. Barres et al., 1992). IGF-I mRNA is present in the brain (Lund et al., 1986; Ayer-LeLievre et al., 1991) and is synthesized and secreted by rat cortical (type-1-like) astrocytes in culture (Ballotti et al., 1987). O–2A progenitor cells have been shown to express both insulin and IGF-I receptors in vivo (Baron-Van Evercooren et al., 1991), so insulin and IGF-I are likely to act directly on these cells. These survival-promoting effects mediated through the IGF-I receptor probably explain the increased brain size and hypermyelination observed in transgenic mice that constitutively overexpress IGF-I (Matthews et al., 1988; McMorris et al., 1990).

### III. PERIPHERAL NERVOUS SYSTEM GLIA

Schwann cells, which are derived from the neural crest, proliferate and migrate along developing peripheral nerves in contact with the growing axons. Eventually the cells stop dividing and differentiate; each Schwann cell then myelinates a single large (> 1 μm) diameter axon, or simply ensheaths several smaller axons. These latter non-myelin-forming Schwann cells express some characteristic markers, such as glial fibrillary acidic protein and nerve growth factor (NGF) receptors, that are not present in myelin-forming cells (Jessen et al., 1990). In the developing rat sciatic nerve, Schwann cell proliferation continues for several weeks after birth; the first myelin-forming Schwann cells appear on the day of birth, and the first nonmyelinating Schwann cells during the third postnatal week (see Jessen and Mirsky, 1991, for a review of Schwann cell development). In the adult PNS, Schwann cells do not proliferate unless the nerve is injured or a Schwann cell tumor develops.

## A. Interactions between Schwann Cells and Neurons

Axonal contact seems to be necessary for maintaining the differentiated phenotype of Schwann cells. When rat Schwann cells lose contact with axons, either after axon degeneration resulting from nerve injury or when nerve cells are dissociated and placed in culture, they down-regulate myelin gene expression (Lemke and Chao, 1988; Trapp et al., 1988) and up-regulate a set of proteins, including glial fibrillary acidic protein and NGF receptors, that are characteristic of nonmyelinating Schwann cells (Jessen et al., 1990). Following nerve transection, Schwann cells in the distal nerve stump proliferate, possibly stimulated by myelin debris (Salzer and Bunge, 1980) or by factors provided by invading macrophages that have ingested myelin fragments (Baichwal et al., 1988). In contrast, Schwann cells placed in culture divide very slowly, even in medium containing 10% fetal calf serum. However, Schwann cells can be stimulated to divide in vitro by neurons or neuronal membrane fragments (Salzer and Bunge, 1980; Salzer et al., 1980b). When Schwann cells are co-cultured with dorsal root ganglion neurons, they adhere to neurites and proliferate until the neurites are covered completely by Schwann cells (Salzer and Bunge, 1980). Cell–cell contact appears to be necessary for the mitogenic stimulus, because Schwann cells will not divide in neuron-conditioned medium or when cultured with neurons on opposite sides of a semi-permeable membrane (Wood and Bunge, 1975; Salzer et al., 1980a). Membrane-enriched fractions of dorsal root ganglion homogenates also possess mitogenic activity (Salzer et al., 1980a). The mitogenic factor has been solubilized from dorsal root ganglion neuronal membrane fractions and from brain preparations; it is a basic protein that seems to be associated with cell-surface heparan sulfate proteoglycans, since it can be solubilized with the enzyme heparitinase and has a high affinity for immobilized heparin (Ratner et al., 1988; De Coster and De Vries, 1989). The identity of the neuron-associated factor is not known, but it is not thought to be bFGF (which has high affinity for heparin) because antibodies to bFGF do not neutralize the mitogenic activity in dorsal root ganglion membrane fractions (Ratner et al., 1988).

## B. cAMP-Dependent Regulation of Schwann Cell Development

Proliferation of Schwann cells also can be stimulated in vitro by raising the level of intracellular cAMP with cAMP analogs or drugs such as forskolin, which activates adenylate cyclase (Raff et al., 1978). Elevation of intracellular cAMP is not mitogenic in the absence of serum, but synergizes with serum or certain polypeptide growth factors to stimulate Schwann cell division (see Section III,C). Elevation of intracellular cAMP also stimulates the expression of SCIP/Tst-1 (Monuki et al., 1989), a POU-domain transcription factor that has been shown to repress $P_0$ and

MBP expression in cultured Schwann cells (Monuki et al., 1990; He et al., 1991). In apparent contradiction with these results, there is other in vitro evidence that cAMP elevation can up-regulate expression of myelin markers (including $P_0$ and MBP) (Lemke and Chao, 1988) that normally are associated with postmitotic Schwann cells.

Recent work (Morgan et al., 1991) has resolved this paradox to some degree. When Schwann cells are cultured in conditions permissive for cell division (i.e., in the presence of serum or certain growth factors), elevating intracellular cAMP stimulates proliferation. In conditions that inhibit cell division, that is, in the absence of serum, elevating cAMP stimulates expression of $P_0$ and represses expression of markers of non-myelin-forming Schwann cells. These observations, coupled with the apparent requirement for neuronal contact, has led to the suggestion that, in vivo, contact with axonal factors might increase Schwann cell intracellular cAMP, priming them to respond to other signals on neuronal membranes or in the extracellular environment (Morgan et al., 1991). In concert with certain growth factors, cAMP elevation would permit cell division. SCIP might play a role in this response, repressing expression of the myelin genes. If cell division were arrested, then continued cAMP elevation might trigger myelin gene expression. In support of this model, there is a report that exposure to neuronal membranes results in an elevation of cAMP in cultured Schwann cells (Ratner et al., 1984), although another report conflicts with this conclusion (Meador-Woodruff et al., 1984). It also is not clear how Schwann cells might be signaled to stop proliferating. There are several possibilities, for example, contact inhibition, the disappearance of environmental mitogens or their receptors on Schwann cells, or the appearance of inhibitory signals or receptors. There is some evidence suggesting that inhibitory factors secreted by the Schwann cells themselves may be involved (Muir et al., 1990; Eccleston et al., 1991; see Section III,D).

## C. Synergy between Polypeptide Growth Factors and cAMP

Many purified polypeptide growth factors have been tested for mitogenic activity on cultured Schwann cells. In 10% serum-containing medium, glial growth factor (GGF), a basic 31-kDa protein that has been purified from pituitary extracts (Lemke and Brockes, 1984), and transforming growth factors β1 and β2 (TGFβ1 and TGFβ2) are potent mitogens (Lemke and Brockes, 1984; Eccleston et al., 1989; Ridley et al., 1989). PDGF-BB and bFGF have been reported to be weakly mitogenic (Ratner et al., 1988; Eccleston et al., 1990) or not mitogenic (Davis and Stroobant, 1990; Weinmaster and Lemke, 1990); this uncertainty might be accounted for by differences in cell preparation or variation among batches of serum. In 10% serum, aFGF is not mitogenic (Ratner et al., 1988; Davis Stroobant, 1990). The effects of all these growth factors are increased greatly by simultaneously elevating intracellular cAMP using forskolin (Ridley et al., 1989; Davis and Stroobant, 1990). It has been reported that, in the complete absence of serum, the mitogenic effects of GGF, PDGF, and bFGF are all dependent on concomitant cAMP elevation (Stewart et al., 1991). Perhaps factors present in fetal calf serum can elevate cAMP in Schwann cells to a limited extent that is sufficient to potentiate the effects of GGF but not PDGF or FGF. TGFβ1, which is a Schwann cell mitogen in 10% serum-containing medium, is not mitogenic in the absence of serum, even with the addition of forskolin (Stewart et al., 1991).

Cultured Schwann cells have been shown to possess PDGF β receptors (PDGFβR) (Eccleston et al., 1990; Weinmaster and Lemke, 1990), which can be activated by PDGF-BB but not PDGF-AA (see Section II,B and Fig. 2). One response of Schwann cells to increased intracellular cAMP is to up-regulate PDGFβR mRNA and protein (Weinmaster and Lemke, 1990). This increase in PDGFβR leads to increased receptor autophosphorylation and increased c-*fos* expression in response to PDGF-BB (Weinmaster and Lemke, 1990). The synergy between cAMP and PDGF therefore might be explained by the induction of PDGFβR. Perhaps the synergy of cAMP with other growth factors can be explained in a similar way. An alternative explanation is that simultaneous activation of two signal transduction pathways, one of them cAMP-dependent, may be required to stimulate Schwann cell division (Rozengurt, 1986).

The roles played in vivo by cAMP and the growth factors just discussed are still unclear. As mentioned, the neuronal mitogen is still unidentified. Whether it is a combination of a previously identified Schwann cell mitogen with an unidentified agent that can elevate cAMP or some novel factor that combines both activities is not known. PDGF-like activity can be recovered in extracts of rat sciatic nerves (Eccleston et al., 1990), suggesting that PDGF may play a role in the control of Schwann cell growth in vivo. PDGF can remain bound to the surfaces of cells that produce it (LaRochelle et al., 1991), possibly in association with heparan sulfate proteoglycans (Shing et al., 1984), and therefore could act as a surface-bound mitogen. Although PDGF elevates cAMP in other cell types such as fibroblasts (Rozengurt, 1986), it apparently does not do so in Schwann cells. Otherwise, it might be expected that PDGF would be mitogenic on its own. Calcitonin gene-related peptide (CGRP) elevates Schwann cell cAMP in vitro (M. Khan and A. Mudge, unpublished results), is present in PNS and CNS neurons, and is up-regulated following axotomy (Haas et al., 1990), so CGRP could be involved in controlling Schwann cell cAMP levels in vivo.

### D. Autocrine Regulation of Schwann Cell Growth

The cellular source of Schwann cell growth factors in peripheral nerves needs to be addressed. It is not known whether the axon-bound Schwann cell mitogen originates in the neurons themselves or if it binds to the axon surface after secretion by neighboring nonneuronal cells. Neurons are known to synthesize polypeptide growth factors, however. For example, PDGF A chain mRNA and protein have been detected in mouse PNS and CNS neurons (Yeh et al., 1991); PDGF B chain also has been detected in CNS neurons (Sasahara et al., 1991), although it is not known whether it is present also in developing PNS neurons. However, whether growth factors synthesized in the neuronal cell body can be transferred to the axonal surface is not clear. Alternative sources of growth factors in peripheral nerves could include fibroblasts, macrophages, and Schwann cell themselves. Schwann cells are known to synthesize autocrine growth factors in vitro (Porter et al., 1987; Kimura et al., 1990), including a PDGF-like activity (Eccleston et al., 1990).

Why do Schwann cells eventually stop dividing, although they remain in contact with axons? It is possible that proliferation is inhibited in an autocrine fashion when the Schwann cells reach maximum density along their axons. It was reported that quiescent cultured Schwann cells secrete a heat-stable trypsin-sensitive factor that, when added to proliferating Schwann cells in culture, inhibits their proliferation (Eccleston et al., 1991). Spontaneously immortalized Schwann cells that have high endogenous cAMP levels (Stewart et al., 1991), enabling them to respond to growth factors without added forskolin, did not secrete this inhibitory activity, although they could respond to it (Eccleston et al., 1991). The inhibitory activity was not identified, although it was shown not to be γ-interferon or collagen, two molecules that are known to inhibit Schwann cell proliferation in vitro (Eccleston et al., 1991). Previously, it had been reported that cultured Schwann cells secrete a 55-kDa heat-sensitive factor that inhibits Schwann cell proliferation (Muir et al., 1990), although this factor had no effect on a cultured schwannoma line or Schwann cells immortalized in vitro. The inhibitory factors described by Eccleston et al. (1991) and Muir et al. (1990) may be different, or the immortalized cells used in these two studies may have different properties. In any event, immortalized Schwann cells appear to have lost either the ability to produce an inhibitory factor or the ability to respond to it, suggesting that inhibitory factors may play a role in controlling normal Schwann cell growth in vivo.

## IV. CONCLUSIONS AND OUTSTANDING QUESTIONS

In vitro studies have provided important clues to the identity of factors that may influence glial cell proliferation and differentiation in vivo. It seems likely that PDGF is an important mitogen for O–2A progenitors in the CNS. Other growth factors, such as FGF and IGF-I, probably play additional and complementary roles during development; we must ascertain what these functions might be. Also, we must discover how the availability of all these factors and the cellular responses to them are regulated in time and space, and to what extent this regulation dictates the progress of O–2A lineage development.

It remains to be determined whether neurons or other glial cells (e.g., astrocytes) are the major source of growth factors for O–2A progenitor cells in the developing CNS. In the PNS, it seems clear that the Schwann cell mitogen is associated with the surfaces of axons, although it is not known what the mitogen is or whether it originates in the axons or is deposited on the axonal surface by other nonneuronal cells. PNS axons also can be mitogenic for O–2A lineage cells, as can CNS axons for Schwann cells (Mason et al., 1990), although there is no evidence

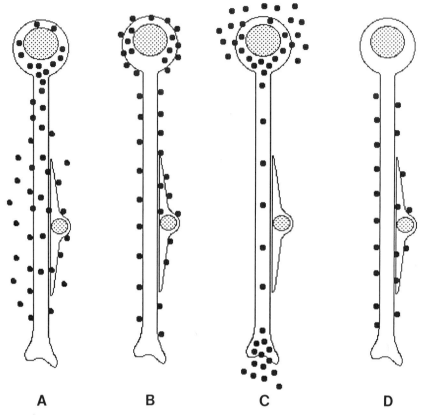

**FIGURE 3**

Some possible scenarios regarding the synthesis and release of glial cell mitogens from neurons. (A) Growth factors, synthesized in the neuronal cell body, might be transported into and released from the axon. The growth factor could either be secreted into the interstitial fluid or remain attached to the axonal membrane; in either case it would be available to a Schwann cell or oligodendrocyte in contact with the axon. (B) Growth factors might be secreted from the neuronal cell body, but remain anchored to the cell surface. The membrane-bound factors might then be transferred to the axon surface by membrane redistribution. (C) Growth factors synthesized in the neuronal cell body might be released locally, to act on other neurons or glia in the vicinity of the cell body. Alternatively, growth factors might be transported along (but not secreted from) the axon, and be released from the nerve terminal. In neither case would the neuron-derived growth factors be available to interact with glial cells surrounding the axon. (D) In addition to the preceding models, growth factors that are synthesized by nonneuronal cells around the axon might adhere to the axonal membrane where they could subsequently act on glial cells by cell–cell contact.

as yet that intact CNS axons are mitogenic for O–2A progenitor cells. Nevertheless, it is an attractive idea that axons in both CNS and PNS might provide growth factors for the myelinating glial cells with which they interact. An important question, therefore, is whether PDGF or other glial mitogens that are synthesized in neurons can be transferred to the axonal surface and, if so, how (see Fig. 3).

The molecular details of the "clocks" that time the onset of glial cell differentiation remain to be elucidated. This is an important general issue because the orderly development of an organism requires that cells of all lineages stop dividing and begin differentiating according to a predetermined schedule. Oligodendrocyte differentiation seems to follow automatically when an O–2A progenitor cell loses the ability to divide in response to PDGF and, possibly, other mitogens in its environment. The key question here is what initially causes the loss of mitogenic responsiveness? Schwann cell differentiation seems to require cAMP-dependent and -independent signals from neurons and other Schwann cells; the identity of the signaling molecules must be determined. The expression of myelin genes in both oligodendrocytes and Schwann cells seems to involve the SCIP/Tst-1 transcription factor; it will be interesting to discover how expression of this regulatory protein is regulated by growth factors (in oligodendrocytes) and by intracellular cAMP (in Schwann cells).

## ACKNOWLEDGMENTS

We would like to thank our colleagues in the Medawar Building for helpful comments, and in particular Barbara Barres, Mike Khan, Anne Mudge, Hardeep Mudhar, and Martin Raff for critical reading of the manuscript. Our own contribution to the work reviewed here was supported by the Medical Research Council, the Multiple Sclerosis Society of Great Britain and Northern Ireland, and the Wellcome Trust.

## REFERENCES

Altman, J. (1966). Proliferation and migration of undifferentiated precursor cells in the rat during postnatal gliogenesis. *Exp. Neurol.* **16,** 263–278.

Ayer-LeLievre, C., Ståhlbom, P.-A., and Sara, V. R. (1991). Expression of IGF-I and -II mRNA in the brain and craniofacial region of the rat fetus. *Development* **111,** 105–115.

Baichwal, R. R., Bigbee, J. W., and DeVries, G. H. (1988). Macrophage-mediated myelin-related mitogenic factor for cultured Schwann cells. *Proc. Natl. Acad. Sci. U.S.A.* **85,** 1701–1705.

Ballotti, R., Nielsen, F. C., Pringle, N., Kowalski, A., Richardson, W. D., Van Obberghen, E., and Gammeltoft, S. (1987). Insulin-like growth factor I in cultured rat astrocytes: expression of the gene and receptor tyrosine kinase. *EMBO J.* **6,** 3633–3639.

Baron-Van Evercooren, A., Olichon-Berthe, C., Kowalski, A., Visciano, G., and Van Obberghen, E. (1991). Expression of IGF-I and insulin receptor genes in the rat central nervous system: A developmental, regional, and cellular analysis. *J. Neurosci. Res.* **28,** 244–253.

Barres, B. A., Hart, I. K., Coles, H. S. R., Burne, J. F., Voyvodic, J. T., Richardson, W. D., and Raff, M. C. (1992). Cell death and control of cell survival in the oligodendrocyte lineage. *Cell,* in press.

Behar, T., McMorris, F. A., Novotny, E. A., Barker, J. L., and Dubois-Dalcq, M. (1988). Growth and differentiation properties of O–2A progenitors purified from rat cerebral hemispheres. *J. Neurosci. Res.* **21**, 168–180.

Bevan, S. (1990). Ion channels and neurotransmitter receptors in glia. *Sem. Neurosci.* **2**, 467–481.

Bögler, O., Wren, D., Barnett, S. C., Land, H., and Noble, M. (1990). Cooperation between two growth factors promotes extended self-renewal and inhibits differentiation of O–2A progenitor cells. *Proc. Natl. Acad. Sci. U.S.A.* **87**, 6368–6372.

Collarini, E. J., Kuhn, R., Marshall, C. J., Monuki, E. S., Lemke, G., and Richardson, W. D. (1992). Down-regulation of the POU transcription factor SCIP is an early event in oligodendrocyte differentiation *in vitro*. *Development,* in press.

David, S., Miller, R. H., Patel, R., and Raff, M. (1984). Effects of neonatal transection on glial cell development in the rat optic nerve: Evidence that the oligodendrocyte-type 2 astrocyte cell lineage depends on axons for its survival. *J. Neurocytol.* **13**, 961–974.

Davis, J. B., and Stroobant, P. (1990). Platelet-derived growth factors and fibroblast growth factors are mitogens for rat Schwann cells. *J. Cell Biol.* **110**, 1353–1360.

De Coster, M. A., and DeVries, G. H. (1989). Evidence that the axolemmal mitogen for cultured Schwann cells is a positively-charged, heparin sulfate proteoglycan-bound, heparin-displaceable molecule. *J. Neurosci. Res.* **22**, 283–288.

Dreyer, D., Lagrange, A., Grothe, C., and Unsicker, K. (1989). Basic fibroblast growth factor prevents ontogenetic neuron death *in vivo*. *Neurosci. Lett.* **99**, 35–38.

Eccleston, P. A., and Silberberg, D. H. (1985). Fibroblast growth factor is a mitogen for oligodendrocytes *in vitro*. *Brain Res.* **210**, 315–318.

Eccleston, P. A., Jessen, K. R., and Mirsky, R. (1989). Transforming growth factor-β and gamma-interferon have dual effects on growth of peripheral glia. *J. Neurosci. Res.* **24**, 524–530.

Eccleston, P. A., Collarini, E. J., Jessen, K. R., Mirsky, R., and Richardson, W. D. (1990). Schwann cells secrete a PDGF-like factor: Evidence for an autocrine growth mechanism involving PDGF. *Eur. J. Neurosci.* **2**, 985–992.

Eccleston, P. A., Mirsky, R., and Jessen, K. R. (1991). Spontaneous immortalisation of Schwann cells in culture: Short-term cultured Schwann cells secrete growth inhibitory activity. *Development* **112**, 33–42.

Fawcett, J. W., and Keynes, R. J. (1990). Peripheral nerve regeneration. *Ann. Rev. Neurosci.* **13**, 43–60.

Gard, A. L., and Pfeiffer, S. E. (1989). Oligodendrocyte progenitors isolated directly from developing telencephalon at a specific phenotypic stage: Myelinogenic potential in a defined environment. *Development* **106**, 119–132.

Gard, A. L., and Pfeiffer, S. E. (1990). Two proliferative stages of the oligodendrocyte lineage (A2B5$^+$O4$^-$ and O4$^+$GalC$^-$) under different mitogenic control. *Neuron* **5**, 615–625.

Gonzalez, A.-M., Buscaglia, M., Ong, M., and Baird, A. (1990). Distribution of basic fibroblast growth factor in the 18-day rat fetus: Localization in the basement membranes of diverse tissues. *J. Cell. Biol.* **110**, 753–765.

Gospodarowicz, D. (1984). Brain and pituitary growth factors. *In* "Hormonal Proteins and Peptides" (C. H. Li, ed.), Vol. XII, pp. 205–230. Academic Press, New York.

Haas, C. A., Streit, W. J., and Kreuzberg, G. W. (1990). Rat facial motoneurons express increased levels of calcitonin gene-related peptide mRNA in response to axotomy. *J. Neurosci. Res.* **27**, 270–275.

Hardy, R., and Reynolds, R. (1991). Proliferation and differentiation potential of rat forebrain oligodendroglial progenitors both *in vitro* and *in vivo*. *Development* **111**, 1061–1080.

Hart, I. K., Richardson, W. D., Bolsover, S. R., and Raff, M. C. (1989a). PDGF and intracellular signaling in the timing of oligodendrocyte differentiation. *J. Cell Biol.* **109**, 3411–3417.

Hart, I. K., Richardson, W. D., Heldin, C.-H., Westermark, B., and Raff, M. C. (1989b). PDGF receptors on cells of the oligodendrocyte–type-2 astrocyte (O–2A) cell lineage. *Development* **105**, 595–603.

Hatten, M. E., Fishell, G., Stitt, T. N., and Mason, C. A. (1990). Astroglia as a scaffold for development of the CNS. *Sem. Neurosci.* **2**, 455–465.

He, X., Treacy, M. N., Simmons, D. M., Ingraham, H. A., Swanson, L. W., and Rosenfeld, M. G. (1989). Expression of a large family of POU-domain regulatory genes in mammalian brain development. *Nature (London)* **340**, 35–42.

He, X., Gerrero, R., Simmons, D. M., Park, R. E., Lin, C. R., Swanson, L. W., and Rosenfeld, M. G. (1991). Tst-1, a member of the POU domain gene family, binds the promoter of the gene encoding the cell surface adhesion molecule $P_0$. *Mol. Cell Biol.* **11**, 1739–1744.

Heldin, C.-H., and Westermark, B. (1989). Platelet-derived growth factor: Three isoforms and two receptor types. *Trends Genet.* **5**, 108–111.

Heldin, C.-H., Bäckström, G., Östman, A., Hammacher, A., Rönnstrand, L., Rubin, K., Nistér, M., and Westermark. B. (1988). Binding of different dimeric forms of PDGF to human fibroblasts: evidence for two separate receptor types. *EMBO J.* **7**, 1387–1393.

Heuer, F. G., von Bartheld, C. S., Kinoshita, Y., Evers, P. C., and Bothwell, M. (1990). Alternating phases of FGF receptor and NGF receptor expression in the developing chicken nervous system. *Neuron* **5**, 283–296.

Hudson, L. D. (1990). Molecular biology of myelin proteins in the central and peripheral nervous system. *Sem. Neurosci.* **2**, 483–496.

Hunter, S. F., and Bottenstein, J. E. (1990). Growth factor responses of enriched bipotential glial progenitors. *Dev. Brain Res.* **54**, 235–248.

Janet, T., Grothe, C., Pettmann, B., Unsicker, K., and Sensenbrenner, M. (1988). Immunocytochemical demonstration of fibroblast growth factor in cultured chick and rat neurons. *J. Neurosci. Res.* **19**, 195–201.

Jessen, K. R., and Mirsky, R. (1991). Schwann cell precursors and their development. *Glia* **4**, 185–194.

Jessen, K. R., Morgan, L., Stewart, H. J. S., and Mirsky. R. (1990). Three markers of adult non-myelin-forming Schwann cells, 217c(Ran-1), A5E3, and GFAP: Development and regulation by neuron-Schwann cell interactions. *Development* **109**, 91–103.

Kimura, H., Fischer, W. H., and Schubert, D. (1990). Structure, expression and function of a schwannoma-derived growth factor. *Nature (London)* **348**, 257–260.

LaRochelle, W. J., May-Siroff, M., Robbins, K. C., and Aaronson, S. A. (1991). A novel mechanism regulating growth factor association with the cell surface: Identification of a PDGF retention domain. *Genes Dev.* **5**, 1191–1199.

Lemke, G. E., and Brockes, J. P. (1984). Identification and purification of glial growth factor. *J. Neurosci.* **4**, 75–83.

Lemke, G., and Chao, M. (1988). Axons regulate Schwann cell expression of the major myelin and NGF receptor genes. *Development* **102**, 499–504.

Levi, G., Aloisi, F., and Wilkin, G. P. (1897). Differentiation of cerebellar bipotential glial precursors into oligodendrocytes in primary culture: Developmental profile of surface antigens and mitotic activity. *J. Neurosci. Res.* **18**, 407–417.

Levine, J. M. (1989). Neuronal influences on glial progenitor cell development. *Neuron* **3**, 103–113.

Levine, J. M., and Stallcup, W. B. (1987). Plasticity of developing cerebellar cells *in vitro* studied with antibodies against the NG2 antigen. *J. Neurosci.* **7**, 2721–2731.

Le Vine, S. M., and Goldman, J. E. (1988). Spatial and temporal patterns of oligodendrocyte differentiation in rat cerebrum and cerebellum. *J. Comp. Neurol.* **277**, 441–455.

Lund, P. K., Moats-Staats, B. M., Hynes, M. A., Simmons, J. G., Jansen, M., D'Ercole, A. J., and Van Wyk, J. J. (1986). Somatomedin-C/insulin-like growth factor-I and insulin-like growth factor-II mRNAs in rat fetal and adult tissues. *J. Biol. Chem.* **261**, 14539–14544.

McKinnon, R. D., Matsui, T., Dubois-Dalcq, M., and Aaronson, S. A. (1990). FGF modulates the PDGF-driven pathway of oligodendrocyte development. *Neuron* **5**, 603–614.

McMorris, F. A., and Dubois-Dalcq, M. (1988). Insulin-like growth factor I promotes cell proliferation and oligodendroglial commitment in rat glial progenitor cells developing in vitro. *J. Neurosci. Res.* **21**, 199–209.

McMorris, F. A., Furlanetto, R. W., Mozell, R. L., Carson, M. J., and Raible, D. W. (1990). Regulation of oligodendrocyte development by insulin-like growth factors and cyclic AMP. *In* "Cellular and

Molecular Biology of Myelination" (G. Jeserich, H. H. Althaus, and T. V. Waehneldt, eds.), pp. 281–292. Springer-Verlag, Berlin.

Mason, P. W., Chen, S. J., and DeVries, G. H. (1990). Evidence for the colocalization of the axonal mitogen for Schwann cells and oligodendrocytes. *J. Neurosci. Res.* **26**, 296–300.

Mathews, L. S., Hammer, R. E., Behringer, R. E., D'Ercole, A. J., Bell, G. I., Brinster, R. L., and Palmiter, R. D. (1988). Growth enhancement of transgenic mice expressing insulin-like growth factor-I. *Endocrinology* **123**, 2827–2833.

Meador-Woodruff, J. H., Lewis, B. L., and DeVries, G. H. (1984). Cyclic AMP and calcium as potential mediators of stimulation of cultured Schwann cell proliferation by axolemma-enriched and myelin-enriched membrane fractions. *Biochem. Biophys. Res. Commun.* **122**, 373–380.

Miller, R. H., David, S., Patel, R., Abney, E. R., and Raff, M. (1985). A quantitative immunohistochemical study of macroglial cell development in the rat optic nerve: in vivo evidence for two distinct astrocyte lineages. *Dev. Biol.* **111**, 35–41.

Mirsky, R., Winter, J., Abney, E. R., Pruss, R. M., Gavrilovic, J., and Raff, M. C. (1980). Myelin-specific proteins and glycolipids in rat Schwann cells and oligodendrocytes in culture. *J. Cell Biol.* **84**, 483–494.

Monuki, E. S., Weinmaster, G., Kuhn, R., and Lemke, G. (1989). SCIP: A glial POU domain gene regulated by cyclic AMP. *Neuron* **3**, 783–793.

Monuki, E. S., Kuhn, R., Weinmaster, G., Trapp, B. D., and Lemke, G. (1990). Expression and activity of the POU transcription factor SCIP. *Science* **249**, 1300–1303.

Morgan, L., Jessen, K. R., and Mirsky, R. (1991). The effects of cAMP on differentiation of cultured Schwann cells: Progression from an early phenotype ($O4^+$) to a myelin phenotype ($P_0$, $GFAP^-$, $N-CAM^-$, $NGF-receptor^-$) depends on growth inhibition. *J. Cell Biol.* **112**, 457–467.

Muir, D., Varon, S., and Manthorpe, M. (1990). Schwann cell proliferation in vitro is under negative autocrine control. *J. Cell Biol.* **111**, 2663–2671.

Noble, M., and Murray, K. (1984). Purified astrocytes promote the in vitro division of a bipotential glial progenitor cell. *EMBO J.* **3**, 2243–2247.

Noble, M., Murray, K., Stroobant, P., Waterfield, M. D., and Riddle, P. (1988). Platelet-derived growth factor promotes division and motility and inhibits premature differentiation of the oligodendrocyte/type-2 astrocyte progenitor cell. *Nature (London)* **333**, 560–562.

Paterson, J. A., Privat, A., Ling, E. A., and Leblond, C. P. (1973). Investigation of glial cells in semithin sections. III. Transformation of subependymal cells into glial cells, as shown by autoradiography after [$^3$H]thymidine injection into the lateral ventricle of the brain of young rats. *J. Comp. Neurol.* **149**, 83–102.

Perry, V. H., and Gordon, S. (1991). Macrophages and the nervous system. *Int. Rev. Cytol.* **125**, 203–244.

Porter, S., Glaser, L., and Bunge, R. P. (1987). Release of autocrine growth factors by primary and immortalized Schwann cells. *Proc. Natl. Acad. Sci. U.S.A.* **84**, 7768–7772.

Pringle, N., Collarini, E. J., Mosley, M. J., Heldin, C.-H., Westermark, B., and Richardson, W. D. (1989). PDGF A chain homodimers drive proliferation of bipotential (O–2A) glial progenitor cells in the developing rat optic nerve. *EMBO J.* **8**, 1049–1056.

Pringle, N. P., Mudhar, H. S., Collarini, E. J., and Richardson, W. D. (1992). PDGF receptors in the rat CNS: during late neurogenesis, PDGF alpha-receptor expression appears to be restricted to glial cells of the oligodendrocyte lineage. *Development,* in press.

Privat, A., Valat, J., and Fulcrand, J. (1981). Proliferation of neuroglial cell lines in the degenerating optic nerve of young rats. *J. Neuropathol. Exp. Neurol.* **40**, 46–60.

Raff, M. C., Hornby-Smith, A., and Brockes, J. P. (1978). Cyclic AMP as a mitogenic signal for cultured rat Schwann cells. *Nature (London)* **273**, 672–673.

Raff, M. C., Miller, R. H., and Noble, M. (1983). A glial progenitor cell that develops in vitro into an astrocyte or an oligodendrocyte depending on the culture medium. *Nature (London)* **303**, 390–396.

Raff, M. C., Abney, E. R., and Miller, R. H. (1984). Two glial cell lineages diverge prenatally in rat optic nerve. *Dev. Biol.* **106**, 53–60.

Raff, M. C., Lillien, L. E., Richardson, W. D., Burne, J. F., and Noble, M. (1988). Platelet-derived growth factor from astrocytes drives the clock that times oligodendrocyte development in culture. *Nature (London)* **333,** 562–565.

Ratner, N., Glaser, L., and Bunge, R. P. (1984). PC12 cells as a source of neurite-derived cell surface mitogen, which stimulates Schwann cell division. *J. Cell Biol.* **98,** 1150–1155.

Ratner, N., Hong, D., Lieberman, M. A., Bunge, R. P., and Glaser, L. (1988). The neuronal cell-surface molecule mitogenic for Schwann cells is a heparin-binding protein. *Proc. Natl. Acad. Sci. U.S.A.* **85,** 6992–6996.

Reynolds, R., and Wilkin, G. P. (1988). Development of macroglial cells in rat cerebellum. II. An in situ immunohistochemical study of oligodendroglial lineage from precursor to mature myelinating cell. *Development* **102,** 409–425.

Richardson, W. D., Pringle, N., Mosley, M. J., Westermark, B., and Dubois-Dalcq, M. (1988). A role for platelet-derived growth factor in normal gliogenesis in the central nervous system. *Cell* **53,** 309–319.

Richardson, W. D., Raff, M., and Noble, M. (1990). The oligodendrocyte–type-2 astrocyte lineage. *Sem. Neurosci.* **2,** 445–454.

Ridley, A. J., Davis, J. B., Stroobant, P., and Land, H. (1989). Transforming growth factors-β1 and β2 are mitogens for rat Schwann cells. *J. Cell Biol.* **109,** 3419–3424.

Rosner, M. H., De Santo, R. J., Arnheiter, H., and Staudt, L. M. (1991). Oct-3 is a maternal factor required for the first mouse embryonic division. *Cell* **64,** 1103–1110.

Rozengurt, E. (1986). Early signals in the mitogenic response. *Science* **234,** 161–166.

Salzer, J. L., and Bunge, R. P. (1980). Studies of Schwann cell proliferation. I. An analysis in tissue culture of proliferation during development, Wallerian degeneration, and direct injury. *J. Cell Biol.* **84,** 739–752.

Salzer, J. L., Bunge, R. P., and Glaser, L. (1980a). Studies of Schwann cell proliferation. III. Evidence for the surface localization of the neurite mitogen. *J. Cell Biol.* **84,** 767–778.

Salzer, J. L., Williams, A. K., Glaser, L., and Bunge, R. P. (1980b). Studies of Schwann cell proliferation. II. Characterization of the stimulation and specificity of the response to a neurite membrane fraction. *J. Cell Biol.* **84,** 753–766.

Saneto, R., and DeVellis, J. (1985). Characterization of cultured rat oligodendrocytes proliferating in a serum free chemically defined medium. *Proc. Natl. Acad. Sci. U.S.A.* **82,** 3509–3515.

Sasahara, M., Fries, J. W. U., Raines, E. W., Gown, A. M., Westrum, L. E., Frosch, M. P., Bonthron, D. T., Ross, R., and Collins, T. (1991). PDGF B-chain in neurons of the central nervous system, posterior pituitary, and in a transgenic model. *Cell* **64,** 217–227.

Schwab, M. E. (1990). Myelin-associated inhibitors of neurite growth and regeneration in the CNS. *Trends Neurosci.* **13,** 452–456.

Shing, Y., Folkman, J., Sullivan, R., Butterfield, C., Murray, J., and Klagsbrun, M. (1984). Heparin affinity: Purification of a tumor-derived capillary endothelial cell growth factor. *Science* **223,** 1296–1299.

Small, R. K., Riddle, P., and Noble, M. (1987). Evidence for migration of oligodendrocyte–type-2 astrocyte progenitor cells into the developing rat optic nerve. *Nature (London)* **328,** 155–157.

Smits, A., Kato, M., Westermark, B., Nistér, M., Heldin, C.-H., and Funa, K. (1991). Neurotrophic activity of PDGF: Rat neuronal cells possess functional PDGF β-receptors and respond to PDGF. *Proc. Natl. Acad. Sci. U.S.A.* **88,** 8159–8163.

Stewart, H. J. S., Eccleston, P. A., Jessen, K. R., and Mirsky, R. (1991). The interaction between cAMP elevation, identified growth factors and serum components in regulating Schwann cell growth (1991). *J. Neurosci. Res.* **30,** 346–352.

Temple, S., and Raff, M. C. (1985). Differentiation of a bipotential glial progenitor cell in single cell microculture. *Nature (London)* **313,** 223–225.

Trapp, B. D., Hauer, P., and Lemke, G. (1988). Axonal regulation of myelin mRNA levels in actively myelinating Schwann cells. *J. Neurosci.* **8,** 3515–3521.

Walicke, P., Cowan, W. M., Ueno, N., Baird, A., and Guillemin, R. (1986). Fibroblast growth factor promotes survival of dissociated hippocampal neurons and enhances neurite extension. *Proc. Natl. Acad. Sci. U.S.A.* **83,** 3012–3016.

Wanaka, A., Johnson, E. M., Jr., and Milbrandt, J. (1990). Localization of FGF receptor mRNA in the adult rat central nervous system by *in situ* hybridization. *Neuron* **5**, 267–281.

Weinmaster, G., and Lemke, G. (1990). Cell-specific cyclic AMP-mediated induction of the PDGF receptor. *EMBO J.* **9**, 915–920.

Wood, P. M., and Bunge, R. P. (1975). Evidence that sensory axons are mitogenic for Schwann cells. *Nature (London)* **256**, 662–664.

Yeh, H.-J., Ruit, K. G., Wang, Y.-X., Parks, W. C., Snider, W. D., and Deuel, T. F. (1991). PDGF A-chain gene is expressed by mammalian neurons during development and in maturity. *Cell* **64**, 209–216.

# 18 Adhesion Factors

Hans W. Müller

## I. INTRODUCTION

The survival of neurons is enhanced not only by soluble molecules such as peptide growth factors, hormones, or transmitters, but also by adhesion factors expressed in the extracellular milieu or on cell membranes (Jessel, 1988). The observation that adhesion molecules, particularly those that mediate cell–matrix interactions, express novel survival supporting functions for neurons has not been appreciated widely in neurotrophic factor research in the past. The role of cell–matrix adhesion factors in the developing or adult nervous system often was considered to be confined to functions in extracellular matrix assembly, cell attachment and spreading, motility, and stimulation of neurite outgrowth or axonal guidance (Carbonetto, 1984; Kleinmann et al., 1985; Liesi, 1985; Hammarback et al., 1988; Sanes, 1989).

In this chapter, proteins and other molecules with cell adhesive properties will be discussed that support, either alone or in combination with various peptide growth factors, the survival and stability of peripheral or central neurons. Among those are the extracellular matrix (ECM) glycoproteins laminin and fibronectin, heparan sulfate and chondroitin/dermatan sulfate proteoglycans, heparin-binding lipid carrier proteins, and gangliosides.

None of these adhesion factors with survival-enhancing activity fits into the classical hypothesis of neurotrophic factors originally put forward by Levi-Montalcini and Hamburger (see Barde, 1989; Oppenheim, 1989) since, in many cases, the adhesion molecules are not target derived and may not be present in limiting amounts. Thus, to consider survival-enhancing adhesion molecules neurotrophic factors, a broader definition of the classical neurotrophic concept is necessary (see Walicke, 1989; see also Chapter 2).

## II. LAMININ AND FIBRONECTIN

Laminin is a large multidomain glycoprotein (900 kDa) found in most basement membranes. The laminin molecule initially isolated from mouse tumor tissue (Timpl et al., 1979) has a cross-shaped structure composed of three distinct polypeptide chains, one A chain (400 kDa) and two B chains (B1, 215 kDa; B2, 205 kDa), that are held together by disulfate bonds (reviewed in Martin and Timpl, 1987; Beck et al., 1990).

Laminin has attracted much interest because of its diverse biologic functions, for example, in cell adhesion and growth, migration, and differentiation, as well as in the assembly of the ECM. These biologic activities are important for the development and maintenance of cellular organization in different tissues, including the nervous system (for review, see Kleinman et al., 1985; Beck et al., 1990).

The various biologic functions of laminin reside in the multidomain structure of this molecule. Distinct sites have been mapped for cell adhesion, promotion of neurite outgrowth, and heparin binding using proteolytic fragmentation techniques, synthetic peptides, or domain-specific antibodies (Edgar et al., 1984; Engvall et al., 1986; Aumailley et al., 1987; Graf et al., 1987).

Laminin interacts with cells through a variety of cell-surface receptors (reviewed in Buck and Horwitz, 1987; Edgar, 1989), including integrins (heterodimeric receptors with relatively low binding affinity for laminin and fibronectin), a high affinity 67-kDa laminin receptor that recognizes the pentapeptide YIGSR in the cell-attachment region of laminin, and various less characterized laminin-binding proteins with molecular mass of approximately 100 kDa (cranin; Smalheiser and Schwartz, 1987) and 180 kDa (Kleinman et al., 1988).

Several groups have investigated the involvement of carbohydrates in neuronal cell adhesion to laminin using carbohydrate-specific probes. Antibodies recognizing the carbohydrate L2 (or HNK-1), which is associated with a variety of adhesion molecules (Kruse et al., 1984), have been shown to react with integrin (Pesheva et al., 1989). In addition, specific gangliosides and their antibodies have been found to interfere with cell adhesion to laminin (see Section VIII). There is ample evidence that neurons express more than one receptor to interact with laminin and that carbohydrates may modulate this interaction.

In addition to laminin, the multifunctional ECM (and plasma) glycoprotein fibronectin plays a central role in cell adhesion (for review, see Hynes, 1987a; Ruoslahti, 1988a). Fibronectin contains multiple domains, allowing diverse interactions with a number of proteins among which are collagen, glycosaminoglycans, proteoglycans, plasminogen, plasminogen activator, fibrin, and several other components of cell surfaces.

The fibronectin molecule contains two polypeptide chains that associate through two disulfate bonds near the carboxy terminus to form a dimer with a molecular mass of approximately 550 kDa. Slight variations in the structure of fibronectin depend on the cellular source of the protein. Some variant forms are derived from alternative splicing of the fibronectin mRNA.

The general structure of fibronectin polypeptides is a series of tightly folded globular units (repeats) specialized for binding specific molecular partners. Distinct functional domains have been mapped, for example, for collagen binding and cell attachment, heparin binding, and axonal growth. By means of controlled proteolysis and affinity binding, peptide fragments of fibronectin have been generated that support neuronal attachment and axonal growth (Rogers et al., 1985).

A fragment from the central region contains the RGDS tetrapeptide sequence that interacts with cell-surface integrin receptors (Ruoslahti and Pierschbacher, 1986; Hynes, 1987b). Another cell-binding domain of fibronectin is located on a 33 kDa fragment from the carboxy terminus that contains one of the heparin binding sites, but lacks the RGDS sequence. This region of fibronectin is suggested to interact with cell-surface proteoglycans (Letourneau et al., 1988; Ruoslahti, 1988b). A cell-surface proteoglycan with these binding properties has been isolated (Rapraeger et al., 1985).

Cell adhesion of fibronectin that is mediated through membrane-bound proteoglycans is subject to modulation by extracellular soluble proteoglycans, for example, the small chondroitin/dermatan sulfate proteoglycan decorin (PG40 or PG20), which compete with membrane-bound proteoglycans for the glycosaminoglycan binding site on the carboxy-terminal 33 kDa fragment of fibronectin (see subsequent text).

## III. POTENTIATION OF NEUROTROPHIC ACTIVITY OF PEPTIDE GROWTH FACTORS BY LAMININ AND FIBRONECTIN

In contrast to fibronectin, laminin alone is able to support short-term survival of some neurons in culture (Baron-Van Evercooren et al., 1982; Pixley and Cotman, 1986), but cell survival can be potentiated when laminin or fibronectin acts in combination with neurotrophic peptide growth factors or other ECM constituents such as proteoglycans.

Edgar et al. (1984) reported that, in the presence of βNGF (nerve growth factor), the survival of cultured sympathetic neurons from embryonic chick [embryonic day (ED) 11–13] could be stimulated markedly (approximately 4-fold) by laminin coated onto polyornithine culture substrates. The survival-enhancing effect of laminin appeared to be confined near the heparin-binding globular domain at the end of the long arm of the laminin molecule, as revealed by perturbation studies using affinity-purified antibodies to the respective laminin fragment. In the absence of NGF, virtually all sympathetic neurons died, indicating the requirement of the peptide growth factor.

Interesting data from Ernsberger et al. (1989) demonstrate that neuronal survival of sympathetic ganglia from young chicken embryos (ED 7) may not depend on NGF but, instead, depend on substratum-bound laminin. With respect to enhancement of cell survival, fibronectin and heart cell-conditioned medium were much less effective than laminin in these cultures.

Cooperativity between NGF and substratum-bound laminin and fibronectin in promoting neuronal cell survival and neurite outgrowth has been reported for sensory neurons from embryonic chick (ED 10) dorsal root ganglia as well (Millaruelo et al., 1988). Further, laminin enhanced survival of dissociated neurons from early chick embryo neural tube containing the trigeminal motor nucleus (Heaton, 1989). In addition, laminin potentiated NGF-induced neurite growth.

Laminin potentiated the neurotrophic survival effect of brain-derived neurotrophic factor (BDNF), another member of the neurotrophin family of growth factors (reviewed in Chapter 8) for dorsal root ganglion neurons from embryonic chick (Lindsay et al., 1985).

Fibronectin also has been shown to potentiate survival of sensory neurons from newborn and adult mouse dorsal root ganglia (Horie and Kim, 1984; Smith and Orr, 1987).

These studies demonstrate the importance of ECM adhesion molecules such as laminin and fibronectin in supporting neuronal cell survival through potentiation of peptide growth factor activity for various types of neurons at different developmental stages.

Binding of laminin to its cell-surface receptor(s) has been shown to cause activation of protein kinase C as an important step, at least in stimulating neurite outgrowth of ciliary ganglion neurons on a laminin substrate (Bixby, 1989). Further, it is interesting to note that activation of protein kinase C also appears to support survival of neurons (Bhave et al., 1990). It has been shown also that basic fibroblast growth factor (bFGF) can be phosphorylated by protein kinases A and C, as regulated through interaction with adhesion factors such as laminin, fibronectin, and heparin (Feige et al., 1989), suggesting a putative indirect mechanism of action of ECM glycoproteins in supporting cell survival.

## IV. PROTEOGLYCANS

Proteoglycans belong to a major class of macromolecules that constitutes the extracellular matrix. These matrix molecules are composed of glycosaminoglycans, long unbranched polysaccharide chains containing repeating disaccharide units that are covalently linked to a core protein. This class of molecules has the capacity for multiple noncovalent interactions with other ECM molecules, cell adhesion molecules, and growth factors (for review, see Ruoslahti, 1988b, 1989; Margolis and Margolis, 1989). Most, but not all, proteoglycan interactions are mediated via the glycosaminoglycan side chain(s). In some instances, however, the core protein interacts with other macromolecules. One example is the binding of the decorin core protein to fibronectin and collagen (see subsequent text).

Proteoglycans have been localized either on the cell surface, in intracellular granules, or as constituents of the ECM. Cell-surface (heparan sulfate and chondroitin sulfate) proteoglycans may be expressed as integral membrane proteins (Rapraeger et al., 1986) or anchored to the cell membrane through a phosphoinositol linkage (Ishihara et al., 1987). Proteoglycans of this category could function possibly as cell-surface receptors for proteins carrying binding domains for glycosaminoglycans. Heparan sulfate proteoglycans are considered to be the predominant type of cell-surface proteoglycans that reinforces cell attachment (Lark and Culp, 1984; Cole et al., 1985; Izzard et al., 1986; Saunders and Bernfield, 1987), because treatment of cells with glycosaminoglycan lyases that specifically degrade heparan sulfate prevents cell attachment to glycosaminoglycan-binding proteins.

Proteoglycans in intracellular granules occur in many cell types (Oldberg et al., 1981; Schmidt et al., 1985) and can be released from cells on stimulation. A synaptic vesicle proteoglycan has been shown to be incorporated into the ECM after secretion (Stadler and Kiene, 1987), suggesting a link between intracellular and ECM proteoglycans.

ECM proteoglycans are probably the most heterogeneous group of heparan sulfate and chondroitin/dermatan sulfate proteoglycans that are expressed by many cell types, including nerve cells. Chondroitin/dermatan sulfate proteoglycans, either soluble in the interstitial space or bound to ECM constituents, appear to have an inhibitory effect on cell adhesion (Schmidt et al., 1987; Ruoslahti, 1988b).

Proteoglycans and their component glycosaminoglycan chains generally have not been considered to support the survival of individual cells, since cells failing to express glycosaminoglycans grow quite well in culture (Esko et al., 1987); the addition of proteoglycans even may inhibit neurite growth on other substrates (Carbonetto et al., 1983; Akeson and Warren, 1986). On the other hand, evidence suggests that proteoglycans may regulate cell functions related to differentiation and proliferation. With respect to neural systems, it has been shown that, for example, Schwann cells fail to produce myelin sheath in dorsal root explant cultures when treated with a xyloside derivative that competes with proteoglycan core proteins for glycosaminoglycan synthesizing enzyme (Carey et al., 1987).

Proteoglycans further appear to affect growth and differentiation of nerve cells by providing attachment sites for heparin-binding neurotrophic growth factors such as bFGF, transforming growth factor $\beta$ (TGF$\beta$), and the lipid carrier protein purpurin. ECM proteoglycans may bind growth factors and, thus, sequester biologically active neurotrophic molecules to provide a reservoir for a localized and persistent stimulation effect on nerve cells.

Endothelial cells have been shown to express a molecular complex containing bFGF and heparan sulfate proteoglycan that is released from these cells by plasminogen activator-mediated proteolysis (Saksela and Rifkin, 1990). It remains to be seen whether similar FGF–proteoglycan complexes are synthesized and released by neural cells and whether they express functions according to the well-known neurotrophic activity of bFGF (see Chapters 9 and 10; Morrison et al., 1986; Walicke et al., 1986; Unsicker et al., 1987). In fact, immunocytochemical studies have revealed bFGF associated with basement membrane constituents of neuroectoderm-derived cells of fetal rat (Gonzales et al., 1990).

It has been shown that binding of heparin to bFGF can induce structural changes that alter the substrate specificity for protein kinases of this growth factor (Feige et al., 1989).

Another link between proteoglycans and specific growth factors has been described for the action of TGFβ. This growth factor is known to induce ECM chondroitin/dermatan sulfate proteoglycans in various cell types (Bassols and Massagué, 1988), presumably including neural cells (see subsequent text). Further, a negative feedback regulation of growth-stimulating TGFβ activity through binding of decorin to TGFβ has been described (Yamaguchi et al., 1990).

Using a differential hybridization screening of a cDNA library from regenerating crushed rat sciatic nerve (Spreyer et al., 1990), it has been possible to demonstrate the expression of transcripts of the decorin core protein in Schwann cells of peripheral nerve (Hanemann et al., 1990). The decorin core protein contains repeats of a 24-amino-acid unit characterized by an arrangement of conserved leucine residues. Because similar repeats exist in other proteins that act as morphogenic factors or as membrane receptors, it has been suggested that the 24-amino-acid repeat may be a primordial protein binding domain (Ruoslahti, 1988b).

Since TGFβ is expressed by glial cells in the peripheral nervous system (Wramm et al., 1987; Underwood et al., 1990) and, in turn, stimulates DNA synthesis and proliferation of Schwann cells (Eccleston et al., 1989; Ridley et al., 1989), an autocrine mechanism of action of TGFβ in peripheral nerve that may regulate decorin expression and Schwann cell proliferation is suggested.

Decorin further binds to collagen (Ts'ao and Eisenstein, 1981), leading to a delay in collagen fibril formation and reducing the thickness of fibrils, thus modulating the structure of the ECM (Ruoslahti, 1988b).

Moreover, decorin is able to interact with ECM adhesion factors such as fibronectin, leading to inhibition of cell attachment to insoluble fibronectin (Schmidt et al., 1987). Chernoff (1988) and Rogers et al. (1989) showed that, in addition to the known interaction of fibronectin with integrin receptors via the RGDS region, nerve cells bind to a 33 kDa fragment of this adhesion protein. Ruoslahti (1988b) and Margolis and Margolis (1989) have proposed a model for regulating cell adhesion to fibronectin through competitive binding of soluble chondroitin/dermatan sulfate proteoglycans, such as decorin, and membrane-associated proteoglycans to glycosaminoglycan binding sites located on matrix proteins such as fibronectin. The inhibition of cell adhesion may be caused either by blocking the binding of membrane proteoglycans to the glycosaminoglycan binding site or by masking the nearby RGD region recognized by the membrane integrin receptors.

In addition to the more indirect mechanisms of proteoglycan action in regulating cellular growth and differentiation through complexes with other ECM constituents, evidence suggests a direct role of soluble proteoglycans in neurotrophic support of nerve cells. Schulz et al. (1990) reported on a retinal ganglion cell neurotrophic factor purified from the superior colliculus that could be identified as a chondroitin sulfate proteoglycan with a molecular mass > 400 kDa. This proteoglycan, which is not considered to require the binding of other growth factors or adhesion molecules to express neurotrophic activity, supports survival of cultured retinal ganglion cells in short-term bioassays. However, this proteoglycan did not support sensory neurons from dorsal root ganglia, indicating cell-type specificity of the neurotrophic activity (Schulz et al., 1990).

Astrocytes and meningeal cells from newborn rat brain release into their conditioned medium very potent neurite-promoting and neurotrophic activities for central neurons (Banker, 1980; Müller and Seifert, 1982; Matthiessen et al., 1989, 1991). Further, the concomitant release and accumulation of heparan sulfate and chondroitin sulfate proteoglycans in these conditioned media has been reported (Müller et al., 1990; Matthiessen et al., 1991). The proteoglycans could be shown to be associated with the major astroglial and meningeal cell-derived neurite-promoting activities. Recent FPLC fractionation studies using astroglial and meningeal cell-conditioned media suggest that a soluble chondroitin sulfate proteoglycan may contribute to the neurotrophic activity that supports long-term survival of cultured neurons from different regions of the brain (unpublished observations).

## V. ACTIVE COMPLEXES OF LAMININ AND FIBRONECTIN WITH PROTEOGLYCANS

Heparan sulfate and chondroitin sulfate proteoglycans have been detected on the cell surface of many different cell types (see Ruoslahti, 1988b), including neuronal and glial cells (Gallo et al., 1987; Stallcup and Beasley, 1987; Vallen et al., 1988; Herndon and Lander, 1990). Both laminin and fibronectin contain heparin binding domains. Additional binding sites for chondroitin/dermatan sulfate have been identified on fibronectin (Schmidt et al., 1987) and, more recently, have been suggested to occur on astroglial and meningeal cell laminin (Müller et al., 1990; Matthiessen et al., 1991).

Membrane-bound heparan sulfate proteoglycans have been shown to modulate the affinity binding of fibronectin to neuroblastoma cells (Vallen et al., 1988). Binding studies using domain-specific peptide probes revealed that fibronectin strongly interacts with spinal cord neurons, probably through the heparin binding domain, whereas binding of these cells to the RGDS-containing region of fibronectin that is recognized by the integrin receptor is less prominent (Letourneau et al., 1988). In contrast, cell attachment and neurite outgrowth of sensory neurons on fibronectin (and laminin) appear to be mediated predominantly by integrin receptor-mediated interactions rather than by the heparin binding domains, suggesting that distinct classes of neurons express different levels of interactions with cell adhesion molecules (Letourneau et al., 1988).

In addition to interactions of laminin and fibronectin with cell surface-bound proteoglycans, biologically active complexes containing proteoglycans and laminin or fibronectin have been released into conditioned media of various cell types, including those derived from peripheral and central nervous system (Lander et al.,

1985; Davis et al., 1987; Matthiessen et al., 1989,1991). These tight but noncovalent laminin– or fibronectin–proteoglycan complexes express neurite growth-promoting activities for cultured peripheral and central neurons. Antibody perturbation studies have revealed that a laminin–heparan sulfate proteoglycan complex not only promotes neurite growth in vitro (Matthew and Patterson, 1983) but functions in axonal regeneration in vivo (Sandrock and Matthew, 1987).

The isolation of two major neurite-promoting adhesion factors released by cerebral astrocytes and meningeal cells in culture has been reported (Matthiessen et al., 1989,1991; Müller et al., 1990). The most active factor in both conditioned media could be identified as a high molecular weight complex containing laminin associated with heparan sulfate proteoglycan and, to a lesser degree, with chondroitin sulfate proteoglycan.

Interestingly, laminin molecules derived from peripheral and central glia or meningeal cells from rat brain revealed structural differences from Eugelbreth-Holm-Swarm tumor laminin (Edgar et al., 1988; Liesi and Risteli, 1989; Matthiessen et al., 1989,1991). The laminin molecules from neural sources contained B chains of normal size (220 kDa) but little, if any, A chain (440 kDa). Since laminin in glial and meningeal cell-conditioned media binds to proteoglycans despite this structural modification, it is suggested that binding domains for glycosaminoglycans should not reside exclusively on the A chain (Edgar et al., 1984) but should be found also on the B1 or B2 chain. In fact, a heparin binding site has been localized on the B1 chain of laminin (Charonis et al., 1988), but the putative binding domain for chondroitin sulfate on laminin (Müller et al., 1990; Matthiessen et al., 1991) has not been mapped. On the other hand, recognition sites for laminin-binding integrins and other receptors, as well as cell adhesion domains (Charonis et al., 1988) and neurite growth-promoting sites (Liesei et al., 1989), have been identified on the B1 and B2 chains of laminin, respectively, allowing the truncated laminin molecule to express its growth promoting functions.

Heparan sulfate or chondroitin sulfate proteoglycans not only enhance the growth-promoting activity of laminin but also express inhibitory activity for laminin-induced neurite outgrowth in schwannoma cell-conditioned medium, as seen (Muir et al., 1989). This is interesting, because the same cell type was shown previously to secrete a heparan sulfate-chondroitin sulfate proteoglycan–laminin complex with strong neurite-promoting activity (Davis et al., 1987). This observation points to the fact that laminin-mediated growth-promoting activities may be subject to regulation, possibly through binding of laminin to different proteoglycans, leading to either stimulatory or inhibitory responses.

## VI. CELL–CELL CONTACT-MEDIATED NEURONAL SURVIVAL

Investigation of the molecular and cellular requirements for long-term survival of central neurons studied in defined serum-free primary cell cultures from embryonic rat brain revealed that nerve cells not only require appropriate substrate-bound neurite growth-promoting factors and diffusible neurotrophic molecules but also critical cell-contact density (Müller, 1990; Müller et al., 1990). Neocortical or hippocampal neurons cultured at subcritical cell density (10,000 or fewer cells/cm$^2$) did not survive for more than 1 week, even when plated on a laminin substrate in the presence of neurotrophic conditioned medium factors. However, increasing the cell density to 20,000 (or more) cells/cm$^2$ significantly enhanced neuronal survival for more than 30 days; control experiments ruled out that the increased neuronal survival was caused solely by the concomitant rise in endogenous glial cells. Under these conditions, a synergistic effect of soluble astroglia-derived neurotrophic factors and contact-mediated neuron–neuron interactions on cell survival could be observed (Müller et al., 1990).

The identified neuron–neuron supportive interactions could, however, be replaced by neuron–astroglia cell contact, given that a sufficiently high number of astrocytes was added to a low density neuronal population (Müller et al., 1991). On the other hand, fibroblasts could not replace astrocytes in this respect, suggesting that the cell contact-mediated survival-enhancing activity is not neuron-specific but cell-type restricted.

The molecular nature of the surface-bound molecules on central neurons and astrocytes that support long-term survival of central neurons in culture is not yet known. It is possible that cell-surface glycosaminoglycans, proteoglycans, or cell adhesion molecules may participate in these cell–cell interactions (Schachner, 1989). In this respect, it is interesting to note that cell-surface glycosaminoglycans can act as binding partners for neural cell adhesion molecules such as L1 and NCAM (Werz and Schachner, 1988). In fact, NCAM was shown to copurify with a heparan sulfate proteoglycan (Cole and Burg, 1989). The binding of this proteoglycan by NCAM appears to be required for NCAM-mediated cell adhesion.

## VII. PURPURIN AND APOLIPOPROTEINS

Purpurin is a 20 kDa heparin-binding protein that is released by neural retina cells (Schubert and LaCorbiere, 1985). This protein is able to associate with ECM constituents and to promote cell-substratum adhesion. Purpurin prolongs the survival of cultured neural retina cells from embryonic chick in nanomolar concentrations (Schubert et al., 1986). This protein is highly concentrated in and restricted to the retina, where the transcript is synthesized by photoreceptor cells (Berman et al., 1987).

Purpurin belongs to the $\alpha_2$-microglobulin superfamily of small hydrophobic molecule carriers. This superfamily of proteins is characterized by structural features such as the beta-sheet barrel binding domain and two short but highly conserved amino acid consensus sequences (Godovac-Zimmermann, 1988).

The mechanism of the neurotrophic action of purpurin is not known. Whether retinol, the hydrophobic molecule transported across the interphotoreceptor cell matrix by purpurin, is required for the neuronal survival activity remains to be investigated. The isolation and biochemical characterization of cell-surface purpurin receptors may help shed light on the mechanism of cell adhesion and survival activity of this protein.

The expression of another member of this superfamily, apolipoprotein D (apo D), has been discovered in the peripheral nervous system (Spreyer et al., 1990). The amino acid sequence of human and rat apo D (33 kDa) revealed that this protein is not related to the apolipoprotein gene family (Drayna et al., 1986; Spreyer et al., 1990).

Whether the biologic activity of apo D in the nervous system is similar to that of purpurin is not known; however, a supporting function in neural repair has been suggested due to the stimulation of high level expression of apo D mRNA in regenerating peripheral nerve (Spreyer et al., 1990). In contrast, expression of apo D transcripts in normal peripheral nerve or transected nerves that are prevented from regenerating is much lower. Apo D transcripts could be localized in endoneurial fibroblasts but not in Schwann cells or macrophages. Apo D protein is secreted into the endoneurial interstitial space, where it was shown to accumulate in endoneurial lipoprotein complexes (Spreyer et al., 1990). These endoneurial lipoproteins, which, in addition to apo D, contain the heparin-binding apo E protein, were shown to interact with cultured Schwann cells and sensory neurons from rat dorsal root ganglia (Rothe and Müller, 1991).

The elevation of synthesis of apo E protein (37 kDa) has been described as one of the most prominent molecular reactions following injury of peripheral nerve (Skene and Shooter, 1983; Ignatius et al., 1986; Müller et al., 1986; Snipes et al., 1986). Apo E protein is expressed by hematogenous macrophages infiltrating the damaged peripheral nerve (Stoll and Müller, 1986); this protein has been localized in astrocytes (Boyles et al., 1985; Stoll and Müller, 1986) and oligodendrocytes (Stoll et al., 1989) in the normal and injured central nervous system, respectively.

Because of several heparin binding sites located on apo E (Cardin et al., 1986; Weisgraber et al., 1986), this protein has the capacity to adhere to glycosaminoglycans and heparan sulfate proteoglycans. Indeed, it was shown that the extracellular accumulation of apo E can be regulated by heparin (Majack et al., 1988). Further, apo E immunoreactivity could be localized in regions of basal lamina deposits around axon–Schwann cell units in sciatic nerve (Stoll and Müller, 1986). The interaction of apo E with glycosaminoglycans or proteoglycans present in large amounts in the basal lamina of peripheral nerve may explain the observation that much more apo E accumulates in damaged peripheral nerves that are known to regenerate than in injured fiber tracts of the adult and poorly regenerating central nervous system (Müller et al., 1985). Nerve fibers in the central nervous system lack basal lamina and ECM constituents required to bind apo E. Apo D/E-containing endoneurial lipoproteins deposited in injured peripheral nerve have been shown in vitro to interact with sensory neurons and Schwann cells through cell surface low-density lipoprotein receptors (Rothe and Müller, 1991).

Although the precise function(s) of apolipoproteins in the intact and regenerating nervous system has yet to be defined, current knowledge suggests an expanding role of this group of proteins in modulating cell growth and differentiation (Mahley, 1988).

## VIII. GANGLIOSIDES

Gangliosides are a family of sialic acid-containing glycosphingolipids that are embedded in plasma membranes and exposed on the cell surface (Ledeen, 1985). Although the physiologic role of gangliosides in cell membranes is not clear, they have been implicated as receptors for toxins and hormones (Van Heyningen, 1974), viruses (Holmgren et al., 1980), and growth factors (Morgan and Seifert, 1979). Gangliosides are particularly abundant on nerve cells and are enriched at the nerve endings (Ledeen, 1985).

Evidence presented by several laboratories suggests that gangliosides, in particular those with two or more sialic acid moieties, may play a role in cellular adhesion mechanisms. Addition of exogenous gangliosides or specific antibodies to cultured cells inhibit cell attachment to extracellular matrix proteins such as fibronectin, laminin, and collagen (Kleinman et al., 1979; Cheresh et al., 1986; Laitinen et al., 1987), suggesting a receptor function for gangliosides for ECM molecules (Perkins et al., 1982). Gangliosides have been shown to stimulate neurite outgrowth in primary neuronal cultures and established nerve cell lines and have been reported further to enhance axonal sprouting and minimize behavioral deficits after brain injury (reviewed in Baker, 1988).

In addition to the neuritogenic properties of gangliosides, neuroprotective and neurotrophic survival activities of certain gangliosides have been reported from in vivo and in vitro studies (reviewed in Baker, 1988). Unsicker and Wieland (1988) describe a survival enhancing effect of the di- and trisialogangliosides $G_{D1b}$ and $G_{T1b}$ for cultured peripheral neurons from embryonic chick dorsal root ganglia and ciliary ganglia. These gangliosides were able to compensate almost fully for NGF or CNTF (ciliary neuronotrophic factor) neurotrophic activity after withdrawal of these peptide growth factors.

The mechanism by which gangliosides influence neuronal survival is unknown, but it may be linked to complex events associated with metabolism of the neuron, modulation of cell-surface receptors, or activation of second messenger pathways related to neuronal survival and regeneration. It has been shown that specific gangliosides may be complexed with an integrin-like receptor (Cheresh et al., 1987).

Cell-surface gangliosides could play a further role in the electrostatic requirements for cell–substratum interactions, because periodate-induced ganglioside oxidation inhibited attachment of neuroblastoma and melanoma cells to laminin and fibronectin (Kleinman et al., 1979; Cheresh et al., 1986). Work of Cheresh et al., (1986) suggests that an intact unsubstituted terminal sialic acid residue, which may bind calcium ions, is critical for ganglioside involvement in cell attachment. Thus, it is possible that the electrostatic environment created by a ganglioside–calcium ion complex may modulate the interaction of cell–surface receptors, such as integrins and heparan sulfate proteoglycans, with ECM adhesion glycoproteins such as fibronectin and laminin (Mugnai et al., 1988).

## IX. CONCLUSIONS

An increasing body of evidence points to a new and significant role for adhesion factors in the nervous system to support survival and differentiation of central and peripheral neurons. The neurotrophic survival activity associated with adhesion factors, however, is not confined to a specific class of adhesion proteins but is found among a diverse group of molecules, including ECM glycoproteins such as laminin and fibronectin, proteoglycans, lipid carrier proteins, and gangliosides, as well as, presumably, adhesion molecules mediating cell–cell surface interactions. In addition, the interaction of adhesion molecules with established peptide growth factors appears to potentiate neurotrophic survival activity very effectively.

Because of the diverse types of adhesion molecules that express survival-enhancing activities and because of the heterogeneity of their cell-surface receptors, a common mechanism of neurotrophic action is unlikely. In addition to preliminary studies on the intracellular responses following the interaction of neurons with adhesion molecules (Bixby, 1989; Mueller et al., 1989; Schuch et al., 1989) further work is required (1) to link cell attachment of adhesion factors to survival-relevant receptor-mediated second messenger pathways that become activated; (2) to identify the possible role of the cytoskeleton anchored to surface receptors in cell adhesion, signal transduction, and cell survival; and (3) to determine the role of transmembrane ion fluxes leading to changes in intracellular pH or $Ca^{2+}$ concentration that could influence the expression of gene regulatory proteins.

Since the vast majority of observations related to neurotrophic survival support by adhesion factors was obtained from in vitro studies, future experiments need to demonstrate and confirm the relevance of distinct adhesion factors to neuronal survival in vivo.

## ACKNOWLEDGMENTS

The preparation of this review and some of the work from our laboratory presented in it were supported by the Deutsche Forschungsgemeinschaft and the Bundesminister für Forschung und Technologie.

# REFERENCES

Akeson, R., and Warren, S. L. (1986). PC12 adhesion and neurite formation on selected substrates are inhibited by some glycosaminoglycans and a fibronectin-derived tetrapeptide. *Exp. Cell Res.* **162,** 347–362.

Aumailley, M., Nurcombe, V., Edgar, D., Paulsson, M., and Timpl, R. (1987). The cellular interactions of laminin fragments. Cell adhesion correlates with two fragment-specific high affinity binding sites. *J. Biol. Chem.* **262,** 11532–11538.

Baker, R. E. (1988). Gangliosides as cell adhesion factors in the formation of selective connections within the nervous system. *Prog. Brain Res.* **73,** 491–508.

Banker, G. (1980). Trophic interactions between astroglial cells and hippocampal neurons in culture. *Science* **209,** 809–810.

Barde, Y.-A. (1989). Trophic factors and neuronal survival. *Neuron* **2,** 1525–1534.

Baron-Van Evercooren, A., Kleinman, H. K., Ohno, S., Marangos, P., Schwartz, J. P., and Dubois-Dalcq, M. (1982). Nerve growth factor, laminin, and fibronectin promote neurite growth in human fetal sensory ganglia cultures. *J. Neurosci. Res.* **8,** 179–193.

Bassols, A., and Massagué, J. (1988). Transforming growth factor-β regulates the expression and structure of extracellular matrix chondroitin/dermatan sulfate proteoglycans. *J. Biol. Chem.* **263,** 3039–3045.

Beck, K., Hunter, I., and Engel, J. (1990). Structure and function of laminin: Anatomy of a multidomain glycoprotein. *FASEB J.* **4,** 148–160.

Berman, P., Gray, P., Chen, E., Keyser, K., Ehrlich, D., Karten, H., LaCorbiere, M., Esch, F., and Schubert, D. (1987). Sequence analysis, cellular localization, and expression of a neuroretina adhesion and cell survival molecule. *Cell* **51,** 135–142.

Bhave, S. V., Malhotra, R. K., Wakade, T. D., and Wakade, A. R. (1990). Survival of chick embryonic sensory neurons in culture is supported by phorbol esters. *J. Neurochem.* **54,** 627–632.

Bixby, J. L. (1989). Protein kinase C is involved in laminin stimulation of neurite outgrowth. *Neuron* **3,** 287–297.

Boyles, J. K., Pitas, R. E., Wilson, E., Mahley, R. W., and Taylor, J. M. (1985). Apolipoprotein E associated with astrocytic glia of the central nervous system and with nonmyelinating glia of the peripheral nervous system. *J. Clin. Invest.* **76,** 1501–1513.

Buck, C. A., and Horwitz, A. F. (1987). Cell surface receptors for extracellular matrix molecules. *Ann. Rev. Cell Biol.* **3,** 179–205.

Carbonetto, S. (1984). The extracellular matrix of the nervous system. *Trends Neurosci.* **7,** 382–387.

Carbonetto, S., Gruver, M. M., and Turner, D. C. (1983). Nerve fiber growth in culture on fibronectin, collagen and glycosaminoglycan substrates. *J. Neurosci.* **3,** 2324–2335.

Cardin, A. D., Hirose, N., Blankenship, D. T., Jackson, R. L., and Harmony, A. K. (1986). Binding of a high reactive heparin to human apolipoprotein E: Identification of two heparin binding domains. *Biochem. Biophys. Res. Commun.* **134,** 783–789.

Carey, D. J., Rafferty, C. M., and Todd, M. S. (1987). Effects of inhibition of proteoglycan synthesis on the differentiation of cultured rat Schwann cells. *J. Cell Biol.* **105,** 1013–1021.

Charonis, A. S., Skubitz, A. P. N., Koliakos, G. G., Reger, L. A., Dege, J., Vogel, A. M., Wohlhueter, R., and Furcht, L. T. (1988). A novel synthetic peptide from the B1 chain of laminin with heparin-binding and cell adhesion-promoting activities. *J. Cell Biol.* **107,** 1253–1260.

Cheresh, D. A., Pierschbacher, M. D., Herzig, M. A., and Mujoo, K. (1986). Disialogangliosides GD2 and GD3 are involved in the attachment of human melanoma and neuroblastoma cells to extracellular matrix proteins. *J. Cell Biol.* **102,** 688–696.

Cheresh, D. A., Pytela, R., Pierschbacher, M. D., Klier, F. G., Ruoslahti, E., and Reisfeld, R. A. (1987). An Arg–Gly–Asp-directed receptor on the surface of human melanoma cells exists in a divalent cation-dependent functional complex with the disialoganglioside GD2. *J. Cell Biol.* **105,** 1163–1173.

Chernoff, E. A. (1988). The role of endogenous heparan sulfate proteoglycan in adhesion and neurite outgrowth from dorsal root ganglia. *Tissue Cell* **20**, 165–178.
Cole, G. J., and Burg, M. (1989). Characterization of a heparan sulfate proteoglycan that copurifies with the neural cell adhesion molecule. *Exp. Cell Res.* **182**, 44–60.
Cole, G. J., Schubert, D., and Glaser, L. (1985). Cell-substratum adhesion in chick neural retina depends upon protein-heparan sulfate interactions. *J. Cell Biol.* **110**, 1192–1199.
Davis, G. E., Klier, F. G., Engvall, E., Cornbrooks, C., Varon, S., and Manthorpe, M. (1987). Association of laminin with heparan and chondroitin sulfate-bearing proteoglycans in neurite-promoting factor complexes from rat Schwannoma cells. *Neurochem. Res.* **12**, 909–921.
Drayna, D., Fielding, C., McLean, J., Baer, B., Castro, G., Chen, E., Comstock, L., Henzel, W., Kohr, W., Rhee, L., Wion, K., and Lawn, R. (1986). Cloning and expression of human apolipoprotein D cDNA. *J. Biol. Chem.* **35**, 16535–16539.
Eccleston, P. A., Jessen, K. R., and Mirsky, R. (1989). Transforming growth factor-beta and gamma-interferon have dual effects on growth of peripheral glia. *J. Neurosci. Res.* **24**, 524–530.
Edgar, D. (1989). Neuronal laminin receptors. *Trends Neurosci.* **12**, 248–251.
Edgar, D., Timpl, R., and Thoenen, H. (1984). The heparin-binding domain of laminin is responsible for its effects on neurite outgrowth and neuronal survival. *EMBO J.* **3**, 1463–1468.
Edgar, D., Timpl, R., and Thoenen, H. (1988). Structural requirements for the stimulation of neurite outgrowth by two variants of laminin and their inhibition by antibodies. *J. Cell Biol.* **106**, 1299–1306.
Engvall, E., Davis, G. E., Dickerson, K., Ruoslahti, E., Varon, S., and Manthorpe, M. (1986). Mapping of domains in human laminin using monoclonal antibodies: Localization of the neurite-promoting site. *J. Cell Biol.* **103**, 2457–2465.
Ernsberger, U., Edgar, D., and Rohrer, H. (1989). The survival of early chick sympathetic neurons in vitro is dependent on a suitable substrate but independent of NGF. *Dev. Biol.* **135**, 250–262.
Esko, J. D., Weinke, J. L., Taylor, W. H., Ekborg, G., Rodén, L., Anantharamaiah, G., and Gawish, A. (1987). Inhibition of chondroitin and heparan sulfate biosynthesis in Chinese hamster ovary cell mutants defective in galactosyltransferase I. *J. Biol. Chem.* **262**, 12189–12195.
Feige, J.-J., Bradley, J. D., Fryburg, K., Farris, J., Cousens, L. C., Barr, P. J., and Baird, A. (1989). Differential effects of heparin, fibronectin, and laminin on the phosphorylation of basic fibroblast growth factor by protein kinase C and the catalytic subunit of protein kinase A. *J. Cell Biol.* **109**, 3105–3114.
Gallo, V., Bertolotto, A., and Levi, G. (1987). The proteoglycan chondroitin sulfate is present in a subpopulation of cultured astrocytes and in their precursors. *Dev. Biol.* **123**, 282–285.
Godovac-Zimmermann, J. (1988). The structural motif of β-lactoglobulin and retinol-binding protein: A basic framework for binding and transport of small hydrophobic molecules? *Trends Biochem. Sci.* **13**, 64–66.
Gonzales, A.-M., Buscaglia, M., Ong, M., and Baird, A. (1990). Distribution of basic fibroblast growth factor in the 18-day rat fetus: Localization in the basement membranes of diverse tissues. *J. Cell Biol.* **110**, 753–765.
Graf, J., Ogle, R. C., Robey, F. A., Sasaki, M., Martin, G. R., Yamada, Y., and Kleinman, H. K. (1987). A pentapeptide from the laminin B1 chain mediates cell adhesion and binds the 67000 laminin receptor. *Biochemistry* **26**, 6896–6900.
Hammarback, J. A., McCarthy, J. B., Palm, S. L., Furcht, L. T., and Letourneau, P. C. (1988). Growth cone guidance by substratum-bound laminin pathways is correlated with neuron-to-pathway adhesivity. *Dev. Biol.* **126**, 29–39.
Hanemann, C. O., Spreyer, P., Gillen, C., Kuhn, G., and Müller, H. W. (1990). The proteoglycan decorin is differentially expressed during peripheral nerve regeneration. Abstracts of the 35th Harden Conference on "Cell–cell Interactions in the Nervous System," London.
Heaton, M. B. (1989). Influence of laminin on the responsiveness of early chick embryo neural tube neurons to nerve growth factor. *J. Neurosci. Res.* **22**, 390–396.
Herndon, M. E., and Lander, A. D. (1990). A diverse set of developmentally regulated proteoglycans is expressed in the rat central nervous system. *Neuron* **4**, 949–961.

Holmgren, J., Svennerholm, L., Elwing, H., Fredman, P., and Strannegard, O. (1980). Sendai virus receptor: Proposed recognition structure based on binding to plastic-absorbed gangliosides. *Proc. Natl. Acad. Sci. U.S.A.* **77**, 1947–1950.
Horie, H., and Kim, S. U. (1984). Improved survival and differentiation of newborn and adult mouse neurons in F12 defined medium by fibronectin. *Brain Res.* **294**, 178–181.
Hynes, R. O. (1987a). Fibronectins: A family of complex and versatile adhesive glycoproteins derived from a single gene. *Harvey Lect. Ser.* **81**, 133–152.
Hynes, R. O. (1987b). Integrins: A family of cell surface receptors. *Cell* **48**, 549–554.
Ignatius, M. J., Gebicke-Haerter, P. J., Skene, J. H. P., Schilling, J. W., Weisgraber, K. H., Mahley, R. W., and Shooter, E. M. (1986). Expression of apolipoprotein E during nerve degeneration and regeneration. *Proc. Natl. Acad. Sci. U.S.A.* **83**, 1125–1129.
Ishihara, M., Fedarko, N. S., and Conrad, H. E. (1987). Involvement of phosphatidylinositol and insulin in the coordinate regulation of proteoheparan sulfate metabolism and hepatocyte growth. *J. Biol. Chem.* **262**, 4708–4716.
Izzard, C. S., Radinsky, R., and Culp, L. A. (1986). Substratum contacts and cytoskeletal reorganization of BALB/c 3T3 cells on a cell-binding fragment and heparin-binding fragments of plasma fibronectin. *Exp. Cell Res.* **165**, 320–336.
Jessel, T. M. (1988). Adhesion molecules and the hierarchy of neural development. *Neuron* **1**, 3–13.
Kleinman, H. K., Martin, G. R., and Fishman, P. H. (1979). Ganglioside inhibition of fibronectin-mediated cell adhesion to collagen. *Proc. Natl. Acad. Sci. U.S.A.* **76**, 3367–3371.
Kleinman, H. K., Cannon, F. B., Laurie, G. W., Hassell, J. R., Aumailley, M., Terranova, U. P., Martin, G. R., and Dubois-Dalc, M. (1985). Biological activities of laminin. *J. Cell Biochem.* **27**, 317–325.
Kleinman, H. K., Ogle, R. C., Cannon, F. B., Little, C. D., Sweeney, T. M., and Luckenbill-Edds, L. (1988). Laminin receptors for neurite formation. *Proc. Natl. Acad. Sci. U.S.A.* **85**, 1282–1286.
Kruse, J., Mailhammer, R., Wernecke, H., Faissner, A., Sommer, I., Goridis, C., and Schachner, M. (1984). Neural cell adhesion molecules and myelin-associated glycoprotein share a common carbohydrate moiety recognized by monoclonal antibodies L2 and HNK-1. *Nature (London)* **311**, 153–155.
Laitinen, J., Lopponen, R., Merenmies, J., and Rauvala, H. (1987). Binding of laminin to brain gangliosides and inhibition of laminin-neuron interaction by the gangliosides. *FEBS Lett.* **217**, 94–100.
Lander, A. D., Fujii, D. K., and Reichardt, L. F. (1985). Laminin is associated with the "neurite outgrowth-promoting factors" found in conditioned media. *J. Cell Biol.* **82**, 2183–2187.
Lark, M. W., and Culp, L. A. (1984). Turnover of heparan sulfate proteoglycans from substratum adhesion sites of murine fibroblasts. *J. Biol. Chem.* **259**, 212–217.
Ledeen, R. W. (1985). Gangliosides of the neuron. *Trends Neurosci.* **8**, 169–174.
Letourneau, P., Rogers, S., Pech, I., Palm, S., McCarthy, J., and Furcht, L. (1988). Cellular biology of neuronal interactions with fibronectin and laminin. In "Current Issues in Neural Regeneration Research" (P. J. Reier, R. P. Bunge, and F. J. Seil, eds.) Vol. 48, pp. 137–146. Liss, New York.
Liesi, P. (1985). Do neurons in the vertebrate CNS migrate on laminin? *EMBO J.* **4**, 1163–1170.
Liesi, P., and Risteli, L. (1989). Glial cells of mammalian brain produce a variant form of laminin. *Exp. Neurol.* **105**, 86–92.
Liesi, P., Närvänen, A., Soos, J., Sariola, H., and Snounou, G. (1989). Identification of a neurite outgrowth-promoting domain of laminin using synthetic peptides. *FEBS Lett.* **244**, 141–148.
Lindsay, R. M., Thoenen, H., and Barde, Y.-A. (1985). Placode and neural crest-derived sensory neurons are responsive at early developmental stages to brain-derived neurotrophic factor. *Dev. Biol.* **112**, 319–328.
Mahley, R. W. (1988). Apolipoprotein E: Cholesterol transport protein with expanding role in cell biology. *Science* **240**, 622–630.
Majack, R. A., Castle, C. K., Goodman, L. V., Weisgraber, K. H., Mahley, R. W., Shooter, E. M., and Gebicke-Härter, P. J. (1988). Expression of apolipoprotein E by cultured vascular smooth muscle cells is controlled by growth state. *J. Cell Biol.* **107**, 1207–1213.
Margolis, R. U., and Margolis, R. K. (1989). Nervous tissue proteoglycans. *Dev. Neurosci.* **11**, 276–288.

Martin, G. R., and Timpl, R. (1987). Laminin and other basement membrane components. *Ann. Rev. Cell Biol.* **3,** 57–85.

Matthew, W. D., and Patterson, P. H. (1983). The production of a monoclonal antibody that blocks the action of a neurite outgrowth-promoting factor. *Cold Spring Harbor Symp. Quant. Biol.* **48,** 625–631.

Matthiessen, H. P., Schmalenbach, C., and Müller, H. W. (1989). Astroglia-released neurite growth-inducing activity for embryonic hippocampal neurons is associated with laminin bound in a sulfated complex and free fibronectin. *Glia* **2,** 177–188.

Matthiessen, H. P., Schmalenbach, C., and Müller, H. W. (1991). Identification of meningeal cell released neurite promoting activities for embryonic hippocampal neurons. *J. Neurochem.* **56,** 759–768.

Millaruelo, A. I., Nieto-Sampedro, M., and Cotman, C. W. (1988). Cooperation between nerve growth factor and laminin or fibronectin in promoting sensory neuron survival and neurite outgrowth. *Dev. Brain Res.* **38,** 219–228.

Morgan, J. I., and Seifert, W. (1979). Growth factors and gangliosides: A possible new perspective in neuronal growth control. *J. Supramol. Struct.* **10,** 111–124.

Morrison, R. S., Sharma, A., deVellis, J., and Bradshaw, R. A. (1986). Basic fibroblast growth factor supports the survival of cerebral cortical neurons in primary culture. *Proc. Natl. Acad. Sci. U.S.A.* **83,** 7537–7541.

Müller, H. W. (1990). Development and long-term survival of brain neurons in culture: Role of astroglia-derived neurotrophic and neurite-promoting factors, neuronal cell/contact density and bioelectric activity. *In* "Brain Repair" (A. Björklund, A. J. Aguayo, and D. Ottoson, eds.), Vol. 56, pp. 167–174. Macmillan, London.

Müller, H. W., and Seifert, W. (1982). A neurotrophic factor (NTF) released from primary glial cultures supports survival and fiber outgrowth of cultured hippocampal neurons. *J. Neurosci. Res.* **8,** 195–204.

Müller, H. W., Gebicke-Härter, P. J., Hangen, D. H., and Shooter, E. M. (1985). A specific 37,000-Dalton protein that accumulates in regenerating but not in nonregenerating mammalian nerves. *Science* **228,** 499–501.

Müller, H. W., Ignatius, M. J., Hangen, D. H., and Shooter, E. M. (1986). Expression of specific sheath cell proteins during peripheral nerve growth and regeneration in mammals. *J. Cell Biol.* **102,** 393–402.

Müller, H. W., Matthiessen, H. P., and Schmalenbach, C. (1990). Astroglial factors supporting neurite growth and long-term neuronal survival. *In* "Advances in Neural Regeneration Research" (F. J. Seil, ed.) Vol. 60, pp. 147–159. Wiley-Liss, New York.

Müller, H. W., Matthiessen, H. P., Schmalenbach, C., and Schroeder, W. O. (1991). Glial support of CNS neuronal survival, neurite growth and regeneration. *Restor. Neurol. Neurosci.* **2,** 229–232.

Mueller, S. C., Kelly, T., Dai, M., Dai, H., and Chen, W.-T. (1989). Dynamic cytoskeleton–integrin associations induced by cell binding to immobilized fibronectin. *J. Cell Biol.* **109,** 3455–3464.

Mugnai, G., Lewandowska, K., Choi, H. U., Rosenberg, L. C., and Culp, L. A. (1988). Ganglioside-dependent adhesion events of human neuroblastoma cells regulated by the RGDS-dependent fibronectin receptor and proteoglycans. *Exp. Cell Res.* **175,** 229–247.

Muir, D., Engvall, E., Varon, S., and Manthorpe, M. (1989). Schwannoma cell-derived inhibitor of the neurite-promoting activity of laminin. *J. Cell Biol.* **109,** 2353–2362.

Oldberg, A., Hayman, E. G., and Ruoslahti, E. (1981). Isolation of a chondroitin sulfate proteoglycan from a rat yolk sac tumor and immunochemical demonstration of its cell surface localization. *J. Biol. Chem.* **256,** 10847–10852.

Oppenheim, R. W. (1989). The neurotrophic theory and naturally occurring motoneuron death. *Trends Neurosci.* **12,** 252–255.

Perkins, R. M., Kellie, S., Patel, B., and Critchley, D. R. (1982). Gangliosides as receptors for fibronectin. *Exp. Cell Res.* **141,** 231–243.

Pesheva, P., Horwitz, A. F., and Schachner, M. (1989). Integrin, the cell surface receptor for fibronectin and laminin, is a member of the L2/HNK-1 family of adhesion molecules. *Neurosci. Lett.* **83**, 303–306.

Pixley, S. K. R., and Cotman, C. W. (1986). Laminin supports short-term survival of rat septal neurons in low-density, serum-free cultures. *J. Neurosci. Res.* **15**, 1–17.

Rapraeger, A., Jalkanen, M., Endo, E., Koda, J., and Bernfield, M. (1985). The cell surface proteoglycan from mouse mammary epithelial cells bears chondroitin sulfate and heparan sulfate glycosaminoglycans. *J. Biol. Chem.* **260**, 11046–11052.

Rapraeger, A., Jalkanen, M., and Bernfield, M. (1986). Cell surface proteoglycan associates with the cytoskeleton at the basolateral cell surface of mouse mammary epithelial cells. *J. Cell Biol.* **103**, 2683–2696.

Ridley, A. J., Davis, J. B., Stroobant, P., and Land, H. (1989). Transforming growth factor-$\beta$1 and $\beta$2 are mitogens for rat Schwann cells. *J. Cell Biol.* **109**, 3419–3424.

Rogers, S. L., McCarthy, J. B., Palm, S. L., Furcht, L. T., and Letourneau, P. C. (1985). Neuron-specific interactions with two neurite-promoting fragments of fibronectin. *J. Neurosci.* **5**, 369–378.

Rogers, S. L., Letourneau, P. C., and Pech, I. V. (1989). The role of fibronectin in neural development. *Dev. Neurosci.* **11**, 248–265.

Rothe, T., and Müller, H. W. (1991). Uptake of endoneurial lipoprotein into Schwann cells and sensory neurons is mediated by LDL-receptors and stimulated after axonal injury. *J. Neurochem.* **57**, 2016–2025.

Ruoslahti, E. (1988a). Fibronectin and its receptors. *Ann. Rev. Biochem.* **57**, 575–413.

Ruoslahti, E. (1988b). Structure and biology of proteoglycans. *Ann. Rev. Cell Biol.* **4**, 229–255.

Ruoslahti, E. (1989). Minireview: Proteoglycans in cell regulation. *J. Biol. Chem.* **264**, 13369–13372.

Ruoslahti, E., and Pierschbacher, M. D. (1986). Arg–Gly–Asp: A versatile cell recognition signal. *Cell* **44**, 517–518.

Saksela, O., and Rifkin, D. B. (1990). Release of basic fibroblast growth factor–heparan sulfate complexes from endothelial cells by plasminogen activator-mediated proteolytic activity. *J. Cell Biol.* **110**, 767–775.

Sandrock, A. W., and Matthew, W. D. (1987). An in vitro neurite-promoting antigen functions in axonal regeneration in vivo. *Science* **237**, 1605–1608.

Sanes, J. R. (1989). Extracellular matrix molecules that influence neural development. *Ann. Rev. Neurosci.* **12**, 491–516.

Saunders, S., and Bernfield, M. (1987). Cell surface proteoglycan binds mouse mammary epithelial cells to fibronectin and behaves as an interstitial matrix receptor. *J. Cell Biol.* **106**, 423–430.

Schachner, M. (1989). Families of neural adhesion molecules. *CIBA-Found. Symp.* **145**, 156–169.

Schmidt, G., Robenek, H., Harrach, B., Glössl, J., Nolte, V., Hormann, H., Richter, H., and Kresse, H. (1987). Interaction of small dermatan sulfate proteoglycan from fibroblasts with fibronectin. *J. Cell Biol.* **104**, 1683–1691.

Schmidt, R. E., MacDermott, R. P., Bartley, G., Bertovich, M., Amato, D. A., Austen, K. F., Schlossman, S. F., Stevens, R. L., and Ritz, J. (1985). Specific release of proteoglycans from human natural killer cells during target lysis. *Nature (London)* **318**, 289–291.

Schubert, D., and LaCorbiere, M. (1985). Isolation of an adhesion-mediating protein from chick neural retina adherons. *J. Cell Biol.* **101**, 1071–1077.

Schubert, D., LaCorbiere, M., and Esch, F. (1986). A chick neural retina adhesion and survival molecule is a retinol-binding protein. *J. Cell Biol.* **102**, 2295–2301.

Schuch, U., Lohse, M. J., and Schachner, M. (1989). Neural cell adhesion molecules influence second messenger systems. *Neuron* **3**, 13–20.

Schulz, M., Raju, T., Ralston, G., and Bennett, M. R. (1990). A retinal ganglion cell neurotrophic factor purified from the superior colliculus. *J. Neurochem.* **55**, 832–841.

Skene, J. H. P., and Shooter, E. M. (1983). Denervated sheath cells secrete a new protein after nerve injury. *Proc. Natl. Acad. Sci. U.S.A.* **80**, 4169–4173.

Smalheiser, N. R., and Schwartz, N. B. (1987). Cranin: A laminin-binding protein of cell membranes. *Proc. Natl. Acad. Sci. U.S.A.* **84**, 6457–6461.

Smith, R. A., and Orr, D. J. (1987). The survival of adult mouse sensory neurons in vitro is enhanced by natural and synthetic substrata, particularly fibronectin. *J. Neurosci. Res.* **17**, 265–270.

Snipes, G. J., McGuire, C. B., Norden, J. J., and Freeman, J. A. (1986). Nerve injury stimulates the secretion of apolipoprotein E by nonneuronal cells. *Proc. Natl. Acad. Sci. U.S.A.* **83**, 1130–1134.

Spreyer, P., Schaal, H., Kuhn, G., Rothe, T., Unterbeck, A., Olek, K., and Müller, H. W. (1990). Regeneration-associated high level expression of apolipoprotein D mRNA in endoneurial fibroblasts of peripheral nerve. *EMBO J.* **9**, 2479–2484.

Stadler, H., and Kiene, M.-L. (1987). Synaptic vesicles in electromotoneurones. II. Heterogeneity of populations is expressed in uptake properties; exocytosis and insertion of a core proteoglycan into the extracellular matrix. *EMBO J.* **6**, 2217–2221.

Stallcup, W. B., and Beasley, L. (1987). Bipotential glial precursor cells of the optic nerve express the NG2 proteoglycan. *J. Neurosci.* **7**, 2737–2744.

Stoll, G., and Müller, H. W. (1986). Macrophages in the peripheral nervous system and astroglia in the central nervous system of rat commonly express apolipoprotein E during development but differ in their response to injury. *Neurosci. Lett.* **72**, 233–238.

Stoll, G., Müller, H. W., Trapp, B. D., and Griffin, J. W. (1989). Oligodendrocytes but not astrocytes express apolipoprotein E after injury of rat optic nerve. *Glia* **2**, 170–176.

Timpl, R., Rohde, H., Gehron Robey, P., Rennard, S. I., Foidart, J.-M., and Martin, G. R. (1979). Laminin—A glycoprotein from basement membranes. *J. Biol. Chem.* **254**, 9933–9937.

Ts'ao, C.-H., and Eisenstein, R. (1981). Attachment of proteoglycans to collagen fibrils. *Lab Invest.* **45**, 450–455.

Underwood, J. L., Rappolee, D. A., Flannery, M. L., and Werb, Z. (1990). A role for the tissue inhibition of metalloproteinases (TIMP) in regeneration of peripheral nerve. *J. Cell Biol.* **111**, 15.

Unsicker, K., and Wieland, H. (1988). Promotion of survival and neurite outgrowth of cultured peripheral neurons by exogenous lipids an detergents. *Exp. Cell Res.* **178**, 377–389.

Unsicker, K., Reichert-Preibsch, H., Schmidt, R., Pettmann, B., Labourdette, G., and Sensenbrenner, M. (1987). Astroglial and fibroblast growth factors have neurotrophic functions for cultured peripheral and central nervous system neurons. *Proc. Natl. Acad. Sci. U.S.A.* **84**, 5459–5463.

Vallen, E. A., Eldridge, K. A., and Culp, L. A. (1988). Heparan sulfate proteoglycans in the substratum adhesion sites of human neuroblastoma cells: Modulation of affinity binding to fibronectin. *J. Cell. Physiol.* **135**, 200–212.

Van Heyningen, W. E. (1974). Gangliosides as membrane receptors for tetanus toxin, cholera toxin and serotonin. *Nature (London)* **249**, 415–417.

Walicke, P. A. (1989). Novel neurotrophic factors, receptors and oncogenes. *Ann. Rev. Neurosci.* **12**, 103–126.

Walicke, P. A., Cowan, W. M., Ueno, N., Baird, A. and Guillemin, R. (1986). Fibroblast growth factor promotes survival of dissociated hippocampal neurons and enhances neurite extension. *Proc. Natl. Acad. Sci. U.S.A.* **83**, 3012–3016.

Weisgraber, K. H., Rall, S. C., Mahley, R. W., Milne, R. W., Marcel, Y. L., and Sparrow, J. T. (1986). Human apolipoprotein E: Determination of the heparin binding sites of apolipoprotein E. *J. Biol. Chem.* **261**, 2068–2076.

Werz, W., and Schachner, M. (1988). Adhesion of neural cells to extracellular matrix constituents. Involvement of glycosaminoglycans and cell adhesion molecules. *Dev. Brain Res.* **43**, 225–234.

Wramm, M., Bodmer, S., De Martin, R., Siepl, C., Hofer-Warbinek, R., Freii, K., Hofer, E., and Fontana, A. (1987). T cell suppressor factor from human glioblastoma cells is a 12.5-kd protein closely related to transforming growth factor β. *EMBO J.* **6**, 1633–1636.

Yamaguchi, Y., Mann, D. M., and Ruoslahti, E. (1990). Negative regulation of transforming growth factor-beta by the proteoglycan decorin. *Nature (London)* **346**, 281–284.

# 19 Instructive Neuronal Differentiation Factors

Paul H. Patterson

## I. PHENOTYPIC PLASTICITY

In addition to the numerous trophic factors that influence neuronal survival and growth, another category of factors has the ability to instruct neurons as to their choice of phenotype during development. These phenotype-specifying factors can control many aspects of neuronal identity (Patterson and Landis, 1992). Several types of neurons have been found to respond to cues in their environment by alteration in morphology, via changes in the number and size of their processes. Differentiation signals also can influence the number and spatial distribution of the synaptic connections a neuron makes and receives. For example, the particular target encountered by a neuron's axons can influence the type of synaptic inputs that neuron receives on its dendrites. Moreover, the neurotransmitters and neuropeptides used at synapses can be controlled by the identity of the target that is being innervated. That is, not only do different neurons make distinctive sets of connections, but the participants in the synaptic contact can influence the chemical identities of each other. Manipulations of neuron–target relationships in vivo have defined some of the critical interactions that control decisions of which transmitters and peptides to produce (Landis, 1990). For instance, target cells can induce, as well as down-regulate, the neurotransmitter enzymes and neuropeptides expressed by the neurons innervating them (see also Chapter 20). Further, this remarkable phenotypic plasticity can be maintained into maturity. Cross-innervation, grafting, and tissue culture experiments have shown that neurons in adult animals also can alter their phenotypes in response to a changing environment.

The molecular nature of these trans-synaptic (as well as glial–neuronal) instructive signals has been studied using cell culture assays (Patterson, 1978, 1990). Such assays enabled the identification of a number of proteins with phenotype-specifying activities. The genes for several of these have been cloned, and the proteins and their mRNAs are beginning to be localized in vivo. Although this chapter focuses primarily on the purified differentiation factors, other activities and relevant phenomena also will be considered briefly. In addition, the use of differentiation factors to generate neuronal diversity is reminiscent of the control of phenotypic decisions in the hematopoietic system. Analogies and homologies between these two systems will be discussed.

## II. CHOLINERGIC DIFFERENTIATION FACTOR/ LEUKEMIA INHIBITORY FACTOR

Intensive study of gene expression in cultured sympathetic neurons has revealed that they are capable of synthesizing, storing, and releasing numerous neurotransmitters and neuropeptides at functional synapses (Bunge et al., 1978; Patterson, 1978; Potter et al., 1983; Black et al., 1987). It is now clear that this phenotypic plasticity observed in culture reflects not only the extremely broad range of phenotypes expressed by these neurons in vivo (Furness et al., 1989), but the fact that these neurons change their transmitter and peptide identities during normal development in situ as they encounter different environments (Patterson and Landis, 1992). For instance, under appropriate culture conditions, dissociated rat sympathetic neurons can synthesize, take up, and release serotonin (Sah and Matsumoto, 1987). Such neurons also produce serotonin in the embryo, but down-regulate their expression of this transmitter postnatally (Soinila et al., 1989). The transient expression of certain transmitters and neuropeptides during normal development is not, however, a particularity of sympathetic neurons. This phenomenon has been observed in mammalian sensory, parasympathetic, and enteric ganglia and in the central nervous system, as well as in a variety of identified neurons in invertebrates (Patterson and Landis, 1992). A key aspect of this phenomenon is that changes in transmitter and peptide expression often are correlated with contact by the neuron with a particular target cell (Morris and Gibbons, 1989). These findings raise the possibility, confirmed in the case of the sweat gland (described subsequently), that targets can control the phenotype of the neurons that innervate them.

The most detailed study of transient expression of a transmitter is that of the noradrenergic-to-cholinergic switch that occurs when the axons of sympathetic neurons innervate sweat glands in the foot pad. Although this study will be discussed more extensively in a subsequent section, the essential point for this context is that the target tissue can induce initially noradrenergic sympathetic neurons to down-regulate tyrosine hydroxylase (TH) and catecholamine (CA) levels while inducing the expression of choline acetyltransferase (ChAT), vasoactive intestinal peptide (VIP), and acetylcholine (ACh) synthesis (Landis, 1990). This switch in transmitters was observed originally in cultured sympathetic neurons, in which it could be evoked by co-culture with a number of different non-neuronal cells (or their conditioned media), sera, or tissue extracts (Patterson and Chun, 1974; Patterson et al., 1975; Bunge et al., 1978). The first such cholinergic inducing activity to be purified was from conditioned medium (CM) of cultured heart cells (Fukada, 1985). This cholinergic differentiation factor, or CDF, does not act as a growth or trophic factor for postmitotic sympathetic neurons; it induces ChAT activity and ACh synthesis and suppresses CA synthesis without affecting neuronal survival or growth, as measured by neuronal protein and lipid (Patterson and Chun, 1977a). In this case, then, CDF is an instructive factor, distinguishable from nerve growth factor (NGF) which, in this context, is permissive, allowing the neurons to survive and grow but not influencing their choice of phenotype (Chun and Patterson, 1977). Under other circumstances, NGF also can act as an instructive factor, as described in a subsequent section. CDF can, however, influence the growth rate of other types of cells, such as myeloid tumor cells (Gough and Williams, 1989). In fact, this protein first was cloned on the basis of its ability to inhibit the proliferation of leukemic myeloid cells and to induce them to acquire a macrophage phenotype (Gearing et al., 1987). Hence it was named, by Metcalf and colleagues, leukemia inhibitory factor or LIF (Gearing et al., 1987). LIF can alter proliferation and gene expression in cells from many different tissues, in adults as well as embryos. Experiments with recombinant LIF and an affinity-purified antiserum against the 11 N-terminal amino acids of CDF established that CDF and LIF are the same protein (Yamamori et al., 1989; Nawa et al., 1991a,b).

As determined by its deduced amino acid sequence, the molecular mass of CDF/LIF is 19.9 kDa. Because it is glycosylated by cultured heart cells at at least six different sites, CDF/LIF displays an apparent size of 43–45 kDa in column chromatography as well as on SDS-PAGE (Fukada, 1985; Weber et al., 1985). The large difference in apparent size between the unglycosylated and glycosylated proteins is probably due, however, to a change in the shape of the factor (as deduced from its frictional ratio, $f/f_0$) caused by glycosylation (Weber et al., 1985). The oligosaccharide residues are not required for cholinergic activity of CDF/LIF in cultured sympathetic neurons (Fukada, 1985; Yamamori et al., 1989). The biophysical data were obtained on a partially purified protein secreted by skeletal muscle cells that is very likely to be identical to the protein secreted by cultured rat heart cells (Giess and Weber, 1984; Fukada, 1985; Weber et al., 1985).

The deduced amino acid sequence of CDF/LIF of rat is 92% identical to that of mouse and 82% identical to that of human (Yamamori et al., 1989). Although only one gene has been detected (Gearing et al., 1987; Stahl et al., 1990), the 3′-untranslated portion of the mouse gene displays a complex hybridization pattern on Southern blots (Stahl et al., 1990). Both human and mouse genes are composed of three exons, two introns, and an unusually long 3′-untranslated region, specifying a mRNA of 4.1 kb (Stahl et al., 1990). Two start sites of CDF/LIF transcription have been determined, both adjacent to TATA-like elements. Alternative transcripts are detected in embryonic fibroblasts; these apparently generate both a secreted protein and a form bound to the extracellular matrix (Rathjen et al., 1990). The considerable sequence conservation of parts of the untranslated and flanking regions suggests that these regions are possible sites of cis-acting controls (Stahl et al., 1990). Corticosteroid would be one candidate for such a control, since this hormone is known to inhibit the production or secretion of CDF/LIF in cultured heart cells (Fukada, 1980) and probably glial cells (McLennan et al., 1980).

CDF/LIF up- and down-regulates many genes in cultured sympathetic neurons. In addition to increasing ChAT activity and mRNA ≥ 100-fold, the factor can decrease TH activity and mRNA 10- to 50-fold. DOPA decarboxylase and dopamine β-hydroxylase activities are depressed as well (Swerts et al., 1983; Raynaud et al., 1987; Brice et al., 1989; K. Fukada, J. Rushbrook, and M. Towle, unpublished observations). These effects are consistent with those observed in situ during the noradrenergic–cholinergic conversion in sympathetic neurons innervating sweat glands, as discussed subsequently. The converted neurons express an apparently undiminished level of the vesicular CA uptake transporter (Scherman and Weber, 1987), as well as CA uptake across the cell-surface membrane (Reichardt and Patterson, 1977; Wakshull et al., 1978; Landis, 1980). The cholinergic sympathetic neurons also express a different form of monoamine oxidase (Pintar et al., 1987).

CDF/LIF also alters neuropeptide expression in cultured sympathetic neurons. The recombinant or purified protein induces VIP, substance P (SP), cholecystokinin (CCK), and somatostatin (SOM) mRNA and peptide, and lowers neuropeptide Y (NPY) peptide levels (Nawa et al., 1991a,b; Freidin and Kessler, 1991; M-J. Fann and P. H. Patterson, 1992; but see Marek and Mains, 1989). Heart cell (CM) induces met-enkephalin (metENK) mRNA, but calcitonin gene-related peptide (CGRP) and dynorphin mRNA remains undetectable (Nawa and Sah, 1990). The induction of metENK has not yet been shown to be caused by CDF/LIF, however. Treatment of sympathetic neurons with CDF/LIF also results in a change in the size of the mRNAs for ChAT, SOM, and NPY; in each case a larger message is found by Northern analysis (Nawa et al., 1991b). A 2.7-kb ChAT mRNA is found in embryonic brain and a 4-kb form is observed in adult spinal cord (Ishii et al., 1990; Lorenzi et al., 1990). Although both sizes can be seen in cultured sympathetic neurons, CDF/LIF induces the 4-kb adult form. Thus, there are qualitative as well as quantitative aspects of mRNA regulation by CDF/LIF.

In addition, heart cell CM changes surface membrane glycoprotein and glycolipid expression by cultured sympathetic neurons, as well as the pattern of secreted glycoproteins (Braun et al., 1981; Sweadner, 1981; Zurn, 1982). These findings require confirmation with recombinant CDF/LIF. Surface or secreted markers specific for the noradrenergic or cholinergic phenotype would be very useful.

How does CDF/LIF effect these many changes in gene expression? The binding of the factor to myeloid cells displays both high and low affinity components (Hilton et al., 1988); the latter receptor is homologous to the gp130 subunit of the IL-6 receptor (Gearing et al., 1991). Work with cultured sympathetic neurons has identified some of the initial transcriptional effects of CDF/LIF. It induces the immediate early genes c-*fos* and *jun-B* without affecting c-*myc, fra-1,* v-*jun,* or actin mRNA levels (during the 30 min incubation; Yamamori, 1991). CDF/LIF induces some of the same early response genes, as well as tyrosine phosphorylation of a 160 kDa protein, in myeloid cells (Lord et al., 1991). Calcium also could play a role in the actions of CDF/LIF. Electrical or chemical depolarization of cultured sympathetic neurons significantly inhibits responsiveness to the cholinergic factor (Walicke et al., 1977; Raynaud et al., 1987). This effect requires extracellular calcium, is enhanced by barium and calcium ionophores, and is blocked by magnesium (Walicke and Patterson, 1981b). Although various manipulations of cyclic AMP and GMP do not affect cholinergic induction in rat neurons (Walicke and Patterson, 1981a), forskolin evokes a rapid increase in ACh production in chick sympathetic neurons (Zurn, 1990). Despite the attractiveness of neuronal activity as a means of controlling phenotypic decisions at the level of individual neurons (Patterson, 1978), there is no evidence for such an action of depolarization in vivo as yet. Denervation of sympathetic neurons in situ does not elevate ChAT levels, although denervating the ganglia and placing them in culture does (Hill and Hendry, 1979). It should be pointed out, however, that denervation could alter spontaneous activity or calcium fluxes in the neurons. An informative experiment would be to stimulate the sympathetic nerves electrically soon after they innervate the sweat glands in the foot pad.

Perhaps relevant to the mechanism of CDF/LIF action is the finding that at least some of the effects of CDF/LIF are reversible. Removal of the factor from cultured sympathetic neurons after maximal induction of SP and SOM results in a major decrease in these neuropeptides (Nawa et al., 1991a). This result is consistent with that obtained by cross-innervation of sensory neuron targets, where neuropeptide phenotypes can be reversed in postmitotic neurons in vivo (McMahon and Gibson, 1987). Similar plasticity was observed for adult sensory neurons in culture (Schoenen et al., 1989). Moreover, the initial noradrenergic phenotype of the sympathetic neurons that innervate the sweat glands is clearly reversible, as mentioned previously. In addition, sympathetic neurons from adult rats can develop a cholinergic phenotype in vitro (Wakshull et al., 1979; Johnson et al., 1980; Potter et al., 1986; see also Lindsay et al., 1989). Thus, CDF/LIF not only *induces* changes in neuronal phenotype, but may be required to *maintain* those phenotypes. Such a requirement in the adult animal raises the interesting possibility that these factors could be used in the repair of neurological damage or in the therapy of mental disease. It should be noted, however, that ChAT induction in cultured sympathetic neurons by CM is not reversed readily by CM withdrawal (Vidal et al., 1987). Indeed, the effects of CM and its withdrawal are lessened as neurons grow older in culture (Patterson and Chun, 1977b). Although these results require further investigation using recombinant CDF/LIF, it is possible that reversibility of gene expression could differ between neuropeptides and neurotransmitter enzymes.

A critical area of CDF/LIF research now concerns the sites of action of this pleiotropic cytokine in vivo (Gough and Williams, 1989). It is clear that this factor can act on several different types of neurons. In addition to its effects on postmitotic sympathetic neurons, CDF/LIF enhances cholinergic function in embryonic adrenal chromaffin cells and the sympathoadrenal progenitor cell line, MAH (Vandenbergh et al., 1991). The partially purified protein also induces ChAT activity and ACh synthesis in cultured nodose ganglion sensory neurons (Mathieu et al., 1984), and recombinant CDF/LIF induces ACh synthesis and VIP expression in cultured dorsal root ganglion neurons (as it does in sympathetic neurons) (Nawa et al., 1991a). In contrast to its effect on sympathetic neurons, however, the factor does not induce SOM or SP in dorsal root ganglion neurons (Nawa et al., 1991a). Thus, CDF/LIF can induce cholinergic properties in neurons that are not known to express them in vivo, but the profile of the its actions depends on the lineage history of the target neurons. Moreover, CDF/LIF acts on central nervous system (CNS) neurons; the partially purified protein enhances ChAT activity in purified rat motor neurons in culture without affecting their survival (Martinou et al., 1989), and the recombinant factor increases ChAT levels in these neurons (Martinou et al., 1991, 1992).

In addition to its effects on directing differentiation choices, CDF/LIF has been shown to have neurotrophic activity for sensory neurons (Murphy et al., 1991). Evidence for a direct survival effect on cultured dorsal root ganglion neurons was obtained, particularly for late embryonic and postnatal mouse ganglia. In addition, the recombinant protein was found to enhance neuronal numbers in cultures of mouse neural crest (Murphy et al., 1991).

It would appear, then, that many types of neurons have CDF/LIF receptors and the second messenger/early gene response capability to respond to the factor. Further questions include whether these and other neurons respond to the factor in situ and whether CDF/LIF is found in the environment of responsive neurons in developing and adult animals. Addressing the second question, a recent study measured the levels of CDF/LIF mRNA in various postnatal tissues using a semiquantitative, reverse transcription polymerase chain reaction (RT-PCR) method (Yamamori, 1991). Two independent sets of CDF/LIF primers were employed, and actin and tubulin mRNA were used as internal controls to normalize mRNA levels among tissues. A significant signal was detected in rat foot pads containing the targets of cholinergic sympathetic neurons, the sweat glands. No signal was detected in targets of noradrenergic sympathetic neurons such as heart, liver, and lachrymal, parotid, and submaxillary glands. Thus, there is a striking correlation between the presence of CDF/LIF mRNA and the cholinergic phenotype of the sympathetic innervation in these targets, at least in postnatal rats. Interestingly, CDF/LIF mRNA also was detected specifically in the adult visual cortex and superior colliculus, but not in olfactory bulb, hippocampus, cerebellum, basal ganglia, frontal cortex, or ventral brain stem (Yamamori, 1991). An independent RT-PCR study has confirmed the presence of CDF/LIF in brain (Minami et al., 1991), and RNase protection experiments have extended these findings further (Bhatt, et al., 1991; Patterson and Fann, 1992). The presence of this factor in the CNS is consistent with its capacity to act on at least some central neurons. Whereas the presence of the CDF/LIF mRNA does not necessarily connote the presence of the protein, these early results from tissue extracts suggest that this factor could play a selective role in differentiation and/or function in vivo.

## III. CILIARY NEUROTROPHIC FACTOR

Ciliary neurotrophic factor (CNTF) has both trophic (survival) and instructive (phenotype-specifying) activities. The protein was originally identified and purified because of its ability to enhance the survival of chick ciliary (parasympathetic) ganglion neurons (Adler et al., 1979; Varon et al., 1979; Barbin et al., 1984; Manthorpe et al., 1986; see also Chapter 15). It is now known that many types of neurons can respond to CNTF with enhanced survival or neurite outgrowth in culture, a finding that has raised questions as to whether this factor belongs to the selective neurotrophic factor class exemplified by NGF (Barde, 1988; Davies, 1988). The deduced molecular mass of CNTF is 22.7 kDa, and it does not have high levels of sequence homology with known proteins (Lin et al., 1989; Stockli et al., 1989) (but see also section XVI). Northern blots of adult rat and rabbit tissues reveal a band of CNTF mRNA (Lin et al., 1989; Stockli et al., 1989). The amino acid sequence of human CNTF is 85% identical to that of rat and rabbit (Cordell et al., 1990). A protein with CNTF activity was purified from chick sciatic nerve, and a 21-amino-acid peptide of the chick protein was found to be 57% identical to the mammalian protein (Eckenstein et al., 1990). Although this may be the avian form of CNTF, it is also possible that distinct CNTF-like proteins exist, as discussed in Section V.

Several other findings with CNTF are inconsistent with the now classical NGF paradigm. First, as described below, CNTF apparently is not produced in the target tissues of the neurons on which it acts as a trophic factor. Second, rather than being a trophic factor in rate-limiting, low concentrations, CNTF appears to be a major cellular product, at least in adult mammalian sciatic nerve (Stockli et al., 1989). Third, the CNTF sequence does not contain an $N$-terminal signal sequence typical for secreted proteins, and CNTF is not detected in the culture medium on expression in COS and HeLa cells (Lin et al., 1989; Stockli et al., 1989). Moreover, type-1 astrocytes release CNTF activity into the medium only when injured (Lillien and Raff, 1990a,b). It has been suggested that CNTF may be released from Schwann cells (see below) in response to damage to the nerve and, thus, act as part of a repair mechanism during nerve regeneration. Exogenously applied CNTF can enhance the survival of damaged motor neurons in vivo (Sendtner et al., 1990) and in culture (Arakawa et al., 1990). In considering growth factors without signal sequences, another unconventional possibility should be considered: intracellular, autocrine action. In fact, acidic fibroblast growth factor (aFGF) contains a consensus sequence for targeting the protein to the nucleus; and anti-aFGF antibodies stain the nuclei of mesenchymal cells in vivo (Sano et al., 1990). Constructs of aFGF lacking this nuclear translocation sequence are unable to induce mitosis in vitro (Imamura et al., 1990).

In addition to its neurotrophic activities, CNTF has cholinergic activity on cultured chick and rat sympathetic neurons (Ernsberger et al., 1989; Saadat et al., 1989). The factor increases ChAT activity and ACh synthesis > 100-fold and lowers TH levels and CA synthesis 3- to 4-fold (Saadat et al., 1989; Rao et al., 1990).

CNTF also induces VIP expression in chick sympathetic neurons (Ernsberger et al., 1989), and VIP, SP, and SOM in rat sympathetic neurons, as well as lowering NPY in the latter (Rao et al., 1992). Thus, the effects of CNTF on sympathetic neurons closely mimic those of CDF/LIF. CNTF and CDF/LIF do not share a high level or sequence homology, however, and CDF/LIF has no effect on the survival of ciliary neurons, nor does an affinity-purified anti-CDF/LIF antiserum precipitate CNTF (Rao et al., 1990). Moreover, each protein binds to distinct receptors (Davis, et al., 1991, Gearing et al., 1991), leading to at least one very similar cascade of second messengers, and numerous but selective changes in neuronal gene expression. One difference in signal transduction is supported by the striking finding that, whereas depolarization blocks the cholinergic effects of CDF/LIF, it does not inhibit the effects of CNTF on sympathetic neurons (Patterson, 1992; Rao et al., 1992). CDF/LIF alters neuropeptide expression in cultured dorsal root ganglion sensory neurons (Nawa et al., 1991a,b), whereas CNTF does not (Rao et al., 1992). CNTF also enhances cholinergic development of cultured retinal neurons (Hofman, 1988), but CDF/LIF has not yet been tested on these cells.

The similarity in the actions of CNTF and CDF/LIF on cultured sympathetic neurons raises the key question of the site of CNTF production in vivo. Northern analysis indicates high CNTF mRNA levels in adult rat sciatic and optic nerve, very low levels in adult spinal cord, and no detectable signal in liver, muscle, or skin (Lin et al., 1989; Stockli et al., 1989; Stockli et al., 1991). The lack of CNTF mRNA in skeletal muscle and in the embryonic eye (the original source for its purification) raises further doubts about its role as a target-related neurotrophic factor. In situ hybridization results suggest a selective concentration of CNTF mRNA in neurons of the hippocampus (Dobrea et al., 1990), but this is not supported by northern analysis (Stockli et al., 1991). In the sciatic nerve, affinity-purified antibodies against a C-terminal peptide of CNTF stain only the Schwann cells in a perinuclear pattern (Rende et al., 1990). In marked contrast to the apparently selective localization of the mRNA, anti-CNTF antibodies stain nearly all neurons and glia in brain and spinal cord (Varon et al., 1990). Astrocytes produce high levels of CNTF (Stockli et al., 1991), and express CNTF immunoreactivity on their surfaces in culture (Rudge et al., 1990). The latter finding is of interest with respect to the lack of CNTF secretion from astrocytes and COS and HeLa cells discussed earlier. In summary, it is difficult to draw firm conclusions regarding the distribution of CNTF at this point. As mentioned previously, some of the apparently contradictory results on CNTF localization could be due to multiple forms of CNTF-like proteins. There also may be some confusion in the literature due to the similarity between CNTF and FGF (Eckenstein et al., 1990).

The high concentration of CNTF in the sciatic nerve apparently is not able to induce the noradrenergic sympathetic axons in the nerve to become cholinergic. Experiments are underway to investigate the role of CNTF and other neurotrophic factors in vivo by injection of the factors or antibodies against them into embryos and adult animals. As mentioned previously, application of CNTF to axotomized motor neurons in adult rats prevents neuronal degeneration (Sendtner et al., 1990), a finding consistent with culture studies (Arakawa et al., 1990). Also supporting the notion that motor neurons are able to respond to this factor in situ is the finding that injection of recombinant CNTF into chick embryos promotes the survival of spinal motor neurons without affecting naturally occurring neuronal death in ciliary, sensory, or sympathetic ganglia (Oppenheim et al., 1991). Further critical experiments concerning the role of the factors in normal development and in the maintenance of adult neurons will be antibody perturbation experiments and gene knockout experiments (Rossant, 1990; Thomas and Capecchi, 1990). In one of the first of these experiments, injection of a blocking monoclonal antibody against aFGF into pre- or neonatal mice was found to reduce ChAT activity in the iris by 50% without affecting TH activity (Hendry et al., 1988; Belford, 1990). Consistent with this result is the enhanced survival of ciliary ganglion neurons elicited by FGF administration in vivo (Dreyer et al., 1989), a finding that subsequently has been disputed (Oppenheim et al., 1990).

## IV. MEMBRANE-ASSOCIATED NEUROTRANSMITTER-STIMULATING FACTOR

In addition to the soluble proteins CDF/LIF and CNTF, membrane-bound factors also can induce sympathetic neurons to become cholinergic (Hawrot, 1980; Adler and Black, 1985; Acheson and Rutishauser, 1988). Cell contacts also can increase TH levels (Acheson and Thoenen, 1983). One such activity, the membrane-associated neurotransmitter-stimulating (MANS) factor, was solubilized and partially purified from rat spinal cord membranes (Wong and Kessler, 1987), and has an apparent size of 29 kDa on SDS-PAGE. Another factor with somewhat different properties, but of a similar size and a similar biologic activity, also has been partially characterized (Adler et al., 1989). Although MANS decreases CA synthesis and induces SP and SOM in cultured sympathetic neurons (Wong and Kessler, 1987; Rao et al., 1990), this factor is clearly distinguishable from CDF/LIF. In addition to differences in size, charge, and subcellular localization, a MANS fraction enhances the survival of ciliary neurons whereas CDF/LIF does not, and anti-CDF/LIF antibodies do not immunoprecipitate the MANS protein or activity (Rao et al., 1990). In fact, the ciliary neuron survival activity of MANS raises the possibility that it may be a membrane-bound form of CNTF. Both CNTF and MANS are found in sciatic nerve. Anti-CNTF antibodies display immunorcactivity for the surfaces of astrocytes (Rudge et al., 1990) and bind to MANS preparations in immunoblots (Rao, et al., 1992).

## V. SWEAT GLAND FACTORS

As mentioned previously, noradrenergic sympathetic neurons switch to the cholinergic phenotype upon innervating the sweat glands during normal development (Landis, 1990). Transplantation experiments demonstrated that sweat glands also induce other noradrenergic sympathetic neurons to become cholinergic when the glands are placed in ectopic sites (Schotzinger and Landis, 1988). Moreover, when target tissues that normally are innervated by noradrenergic sympathetic neurons, such as the parotid glands, are grafted in place of the sweat glands in the foot pads, the sympathetic neurons that innervate the parotids do not switch phenotypes, and remain noradrenergic (Schotzinger and Landis, 1990). These results clearly point to the sweat glands as a site of cholinergic- and peptidergic-conversion activity. What is the nature of this factor and is it identical to any of the known cholinergic factors? Soluble extracts of foot pads do, in fact, contain a cholinergic activity when tested on cultured sympathetic neurons. This factor not only induces ChAT activity, but it also induces VIP expression and reduces CA synthesis and NPY levels (Rao and Landis, 1990). These effects mimic those of CDF/LIF and CNTF, as well as those of the foot pad grafts in vivo (Rao et al., 1992; Rohrer, 1992). The activity in the sweat gland extract appears as early as P5 and increases to adult levels by P21, a time course that parallels that of the transmitter conversion observed in the sympathetic axons innervating the gland (Rao and Landis, 1990; Rohrer, 1992). This type of cholinergic activity is not found in soluble extracts of liver, hairy skin, and parotid gland, consistent with the inability of the latter tissue to induce the cholinergic conversion in vivo.

The size and charge of the foot pad activity differ from those of CDF/LIF. Moreover, an antiserum against the N terminus of CDF/LIF does not precipitate the major cholinergic activity from foot pad extracts; labeled CDF/LIF is precipitated when added to the same extracts (Rao and Landis, 1990). In apparent conflict with the results obtained with sweat gland factor, CDF/LIF mRNA can be detected specifically in foot pad extracts by the RT-PCR method, as described in Section II (Yamamori, 1991). It could be that no detectable CDF/LIF protein is produced from this message, or that the protein is modified in the sweat gland so its physical properties and immunoreactivity differ from those of the protein characterized in heart and skeletal muscle cells. It is also possible that the bulk of the soluble activity in the foot pad extracts is normally intracellular (as CNTF may be), is therefore not accessible to the sympathetic axons, and is not responsible for the transmitter switch. Thus, the remaining cholinergic activity in the extract that is not precipitated by the anti-CNTF antiserum (see subsequent text) could be the activity relevant to the transmitter switch in situ. Conceivably, this activity corresponds to the CDF/LIF mRNA detected by the RT-PCR method.

CNTF is presently the other major candidate for sweat gland factor. The two factors are very similar in physical properties, and anti-CNTF antibodies precipitate 80% of the ChAT-inducing activity (but only 50% of the VIP-inducing activity and 20% of the NPY-inducing activity) (Rao et al., 1992; Rohrer, 1992). Nonetheless, no signal is detected in immunoblots of foot pads using two different anti-CNTF antisera (Rao et al., 1992). These antisera yield strong signals in immunoblots of sciatic nerve and recombinant rat CNTF. Moreover, analyses of CNTF mRNA by Northern blots and in situ hybridization are negative for sweat glands and positive for sciatic nerve (Stockli et al., 1989; Rao et al., 1992; Rohrer, 1992). The immunoblots and Northern assays should have detected a CNTF signal if it were present because the sciatic nerve and foot pad extracts contain equivalent levels of cholinergic inducing activity (Rao and Landis, 1990; Rao et al., 1992). One possibility raised by these results is that there are multiple forms of CNTF (Eckenstein et al., 1990). Such forms have not yet been described in the cloning work on the mammalian factor.

One characteristic of the foot pad extract, as well as of CNTF and CDF/LIF, that is not consistent with the phenotypic conversion that occurs in situ is that each factor induces the expression of SP in cultured sympathetic neurons; this neuropeptide is not detected in the sympathetic axons that innervate the glands in the rat (Landis et al., 1988; Rao et al., 1991). SP is, however, found in the sympathetic innervation of sweat glands in the cat (Lindh et al., 1989). The lack of SP in the rat could mean that the soluble activity in the foot pad extract is not the physiologically relevant factor. It is also possible, however, that the sympathetic neurons are prevented specifically from up-regulating SP in vivo by an inhibitory influence. Specific suppression of genes for particular neurotransmitter enzymes and neuropeptides has, in fact, been observed. CDF/LIF and CNTF, for instance, suppress CA synthesis and TH and NPY expression in cultured sympathetic neurons, as discussed earlier. Depolarization of sympathetic neurons curtails their responsiveness to CDF/LIF, as mentioned previously. Finally, although CM from ganglionic nonneuronal cells can induce preproSOM mRNA and peptide in cultures of pure sympathetic neurons, the CM is ineffective when added to the neurons cultured on living ganglionic nonneuronal cells (Spiegel et al., 1990). If the CM is given to the neurons cultured on a substrate of killed nonneuronal cells, however, the mRNA is induced, but SOM peptide is not enhanced. Therefore, contact with the ganglionic nonneuronal cells apparently inhibits the induction of the peptide, but not of the preproSOM mRNA, by CM. The data also suggest that SOM mRNA and peptide can be controlled independently.

A role for glia in the regulation of neuronal phenotype is also supported by findings that purified cultures of Schwann cells can exhibit both stimulatory and inhibitory influences on SOM expression (Spiegel et al., 1990). Glia have been implicated in the regulation of a number of phenotypic characteristics (Patterson and Chun, 1974; Mudge, 1984; Kessler, 1984; Cooper and Lau, 1986; Kessler et al., 1986; Mizuno et al., 1989), including neuronal morphology, which is discussed in a later section.

Suggestive evidence for suppressive or inhibitory influences on the development of particular phenotypes also comes from in vivo perturbations of the retina. Chemical ablation of selected populations, such as CA-containing amacrine cells or glutamate-sensitive cells, results in an increased production of the missing phenotypes (Negishi et al., 1982; Reh and Tully, 1986; Reh, 1987).

## VI. FACTORS ACTING ON MOTOR NEURONS

Considerable effort has been devoted to isolating potential motor neuron trophic factors, and some of the results have raised the possibility that instructive factors may also influence these neurons. CNTF supports the survival of 50–60% of partially purified embryonic chick motor neurons in culture, as does bFGF; aFGF supports about 35% of these neurons. The effects of bFGF and CNTF are additive, together supporting 100% of the neurons (Arakawa et al., 1990). Many other cytokines and growth factors were tested and found to support few or no motor neurons. Since ChAT induction was not assessed in this study, it is not clear whether CNTF or bFGF may up-regulate phenotypic characteristics of these cells as well as survival. ChAT activity has, however, been monitored in rat spinal cord cultures; both CDF/LIF and CNTF significantly enhance enzyme activity (Martinou et al., 1989, 1992). In addition to the culture results, CNTF promotes motor neuron survival in vivo, as discussed previously (Sendtner et al., 1990; Oppenheim et al., 1991). It should be noted, however, that a more highly purified preparation of cultured chick motor neurons failed to respond to CNTF (Bloch-Gallego et al., 1991), raising the possibility that this factor supports survival via an indirect action on other cell types. On the other hand, this latter study did not employ recombinant CNTF. Although CNTF is able to rescue motor neurons from naturally occurring cell death during normal development (Oppenheim et al., 1991), the mRNA for this factor has not been detected in skeletal muscle (Lin et al., 1989; Stockli et al., 1989).

A 22-kDa protein, purified to apparent homogeneity from rat skeletal muscle (McManaman et al., 1988), rescues about one-third of the motor neurons that normally die when it is applied to chick embryos in vivo. This treatment has no affect on the survival of sensory or preganglionic sympathetic neurons (McManaman et al., 1990). This protein, also called CDF, increases ChAT levels in cultures of enriched rat motor neurons 3-fold while enhancing metabolic activity of the cultures less than 50% (McManaman et al., 1990). Thus, it is expected that the factor increases ChAT specific activity, although this was not measured directly. This factor is likely to be distinct from several other proteins known to increase ChAT levels in motor neuron cultures (CDF/LIF, aFGF, and bFGF; the latter also stimulates ChAT in ciliary neuron cultures; Vaca et al., 1989). It is the same size as nonglycosylated CDF/LIF but has a different charge (McManaman et al., 1988). Moreover, the motor neuron factor does not display cholinergic inducing activity on rat sympathetic neurons in culture (Rao et al., 1992), and adult skeletal muscle does not contain detectable levels of CDF/LIF mRNA (Yamamori, 1991). The motor neuron factor also differs from both aFGF and bFGF in its chromatographic and immunological properties, as well as in its effects on fibroblasts and motor neurons (McManaman et al., 1989). Finally, the motor neuron factor differs from CNTF in that it does not have cholinergic activity on cultured sympathetic neurons (Rao et al., 1992) and CNTF mRNA has not been detected in skeletal muscle (Stockli et al., 1989).

The possibility should also be considered that this motor neuron factor could be acting indirectly, via contaminating nonneuronal cells in the rat spinal cord cultures. Evidence for this type of effect was obtained for γ-interferon; this protein increases ChAT activity in cultures of human spinal cord, but appears to act via the nonneuronal cells (Erkman et al., 1989). Similarly, CM from activated splenocytes, as well as recombinant IL-1, induces SP mRNA and peptide expression while decreasing TH mRNA in cultured sympathetic ganglia; the effect is blocked by dexamethasone (Jonakait and Hart, 1990; Barbany et al., 1991; Hart et al., 1991; Jonakait and Schotland, 1990). The induction by IL-1 is not observed with pure neuronal cultures, however, suggesting that the cytokine acts indirectly, via the ganglionic nonneuronal cells (Freidin and Kessler, 1991). The corticosteroid inhibition also could be indirect, since dexamethasone was shown to inhibit the production or release of CDF/LIF by heart cells and probably by ganglionic nonneuronal cells as well (Fukada, 1980; McLennan et al., 1980).

A selective increase in ChAT activity in cultured rat spinal cord neurons is observed after growth in muscle cell CM (Kaufman et al., 1985). Although this CM also increases protein synthesis, the synthesis of gamma amino butyric acid (GABA) is decreased. The same study found that extracts of neonatal rat muscle increased ChAT levels without changing GABA, whereas extracts of adult muscle had little effect on either parameter in the cord cultures. The activities in the CM and neonatal muscle extracts had apparent molecular sizes of 40 kDa (Kaufman et al., 1985). It will be important to determine if ChAT induction is due to CDF/LIF, since CDF/LIF (as discussed earlier) is found in muscle CM, increases ChAT in motor neurons, and has a similar size when glycosylated. MANS also increases ChAT activity in cultures of dissociated rat ventral spinal cord, without increasing the number of neurons that stain with anti-ChAT antibodies, suggesting an increase in ChAT specific activity (Lombard-Golly et al., 1990).

## VII. NORADRENERGIC FACTORS

Either expression of TH is extremely labile, or a great many investigators are focusing on this marker in development, because there are numerous reports of the induction of TH by environmental cues and its transient appearance during development. For example, many neurons of the embryonic rat cerebral cortex that do not normally express TH can be made to do so when transplanted into the adult mouse cortex or when placed in the appropriate culture environment (Park et al., 1986; Iacovitti et al., 1987; Hermann et al., 1988). The induction of TH in cortical neurons is also stimulated by co-culture with primary smooth, skeletal, or cardiac muscle cells or cell lines, but not by co-culture with fibroblasts or glial cells (Iacovitti et al., 1989; Iacovitti, 1991). Increases in TH immunoreactivity are paralleled by increased TH mRNA, but not by enhanced neuronal survival. The physical properties of the factor causing TH induction, termed muscle-derived differentiation factor (MDF), remain to be determined.

A number of manipulations in vivo have demonstrated a key role for olfactory afferent innervation in the induction and maintenance of the dopamine phenotype by their target neurons in the olfactory bulb. Activity in the afferents is likely to be an important feature of this control. The marker for this phenotype, TH, is reduced greatly in bulb neurons on unilateral neonatal olfactory deprivation (Baker, 1990).

This effect is quite specific, since GABA, aromatic L-amino acid decarboxylase (both substances found in the same neurons as the TH), and olfactory marker protein (localized in the afferent fibers) are not reduced by this treatment. This effect has been reproduced in culture; TH and dopamine uptake by olfactory bulb neurons is enhanced by co-culture with olfactory epithelial neurons (Denis-Donini, 1989). In fact, the effect of the afferents can be mimicked by addition of CGRP, a peptide produced and released by the primary olfactory neurons (Denis-Donini, 1989). Thus, the effect on activity may be mediated by the neuropeptide, and the phenomenon may resemble the well-known, activity-dependent trans-synaptic induction of specific enzymes such as TH (Black et al., 1985; Thoenen and Acheson, 1987; Zigmond et al., 1989). However, this case is particularly interesting because in vivo manipulations have shown that the olfactory afferents also can induce TH expression in cortical neurons, which would not express TH normally (Guthrie and Leon, 1989).

TH is expressed transiently in several regions of the brain. In the mouse mutants tottering and leaner, however, TH persists throughout adulthood (Hess and Wilson, 1991). Purkinje cells, for instance, do not down-regulate TH expression in these mutants. These findings suggest that the product of the *tg* locus may be involved in suppressing the noradrenergic phenotype in central neurons. Liver CM can selectively enhance CA production in cultured chick sympathetic neurons; the active factor has been identified as inosine (Zurn and Do, 1988). Addition of this purine increases protein synthesis but does not increase ACh production. Since the inosine effect is blocked by an inhibitor of nucleoside transport, the agent appears to be acting intracellularly rather than by stimulating a surface receptor. Since adenosine has the same effect as inosine, it has been suggested that the nucleosides may be incorporated into ATP and thereby stimulate TH activity or transcription (Zurn and Do, 1988; Zurn, 1991).

TH and CA levels can be regulated in cultured neural crest cells by fibronectin (Sieber-Glum et al., 1981), extracellular matrix (Loring et al., 1982; Maxwell and Forbes, 1987, 1990; Morrison-Graham et al., 1990), and chick embryo and tissue extracts (Howard and Bronner-Fraser, 1985, 1986; Xue et al., 1985; Ziller et al., 1987), as well as by various nonneuronal cells and CMs (Mackey et al., 1988). Chick serum and various CMs also promote neuronal differentiation and higher ChAT levels in a subpopulation of mesencephalic crest cells (Barald, 1989). Insulin and insulin-like growth factor I also can induce TH in cultures of embryonic quail dorsal root ganglia (Xue et al., 1988). These factors also influence mitosis and differentiation of sympathoadrenal progenitors (DiCicco-Bloom et al., 1990; Carnahan and Patterson, 1991b). The role of various growth factors in the control of neurogenesis has been reviewed by Rohrer (1990).

Neuronal activity and transmitters may also play a role in TH regulation during early development. CA uptake blockers can inhibit the expression of TH in neural crest cultures (Sieber-Blum, 1989), suggesting that CA may autostimulate its own differentiation pathway. A feedback mechanism has been suggested for the development of serotonin neurons. Serotonin can increase the number of neurons expressing its biosynthetic enzyme, tryptophan hydroxylase, in cultures of embryonic mouse hypothalamus (De Vitry et al., 1986b). This transmitter also amplifies its own synthesis in a hypothalamic neuronal cell line (De Vitry et al., 1986a).

Other cases of possible activity-dependent changes in phenotype include the increase in CA production in primary cultures of rat hypothalamus and brainstem neurons by angiotensin II (MacLean et al., 1990). VIP stimulates proEnk A mRNA levels in cultured chromaffin cells (Wan and Livett, 1989), choline acetyltransferase in cultured chick sympathetic neurons (Beretta and Zurn, 1991), and several neuronal markers in spinal cord cultures (Foster et al., 1989). It is quite likely that some of these effects are due to stimulation of neuronal survival and/or growth, however.

## VIII. PEPTIDERGIC FACTORS

The selective induction and suppression of particular neuropeptide genes by CDF/LIF, CNTF, MANS, and the sweat gland factor have been discussed previously. In addition, a number of partially characterized factors have been found to regulate neuropeptide expression selectively in several types of neurons. Skin CM specifically induces metENK mRNA in cultured rat sympathetic neurons (Nawa and Sah, 1990). Factors biochemically distinct from CDF/LIF can induce VIP or SOM selectively in these neurons as well (Nawa and Patterson, 1990). None of these factors alters the survival of the sympathetic neurons. CM from one of the target tissues for ciliary neurons, the choroid muscle, contains a factor termed the SOM stimulating activity (SSA) (Coulombe and Nishi, 1991). This factor can induce > 90% of cultured ciliary neurons to express SOM, although only about half these neurons normally would innervate the choroid and express SOM. Since SSA has little effect on neuronal survival and has an apparent size of 30–40 kDa, it is likely to be a phenotype-specifying protein.

## IX. CORTICOSTEROID

Corticosteroid plays a central role in phenotypic decisions in the sympathoadrenal lineage. This hormone both enhances the expression of chromaffin-specific traits in progenitor cells and inhibits the expression of neuronal genes (Doupe and Patterson, 1982; Anderson, 1989; Patterson, 1990; Vogel and Weston, 1990). The high concentration of corticosteroid in the adrenal medulla is therefore likely to promote the differentiation of chromaffin cells in that environment, whereas the multipotential progenitors differentiate into neurons in the environment of developing sympathetic ganglia (Doupe et al., 1985a). At intermediate corticosteroid concentrations, a third sympathoadrenal derivative can differentiate: the small intensely fluorescent (SIF) cells (Doupe et al., 1985b). Corticosteroid can also prolong the transient appearance of TH in early neurons populating the embryonic gut (Jonakait et al., 1981). This effect may be related to the ability of the hormone to delay neuronal differentiation in sympathoadrenal progenitors. Sympathoadrenal progenitor cells and chromaffin cells express a set of antigens (termed SA antigens), and corticosteroid treatment prolongs the expression of the SA antigens in culture (Carnahan and Patterson, 1991a,b). Strikingly, the neurons that transiently express TH in the gut also express the SA antigens, providing further evidence of the link between the sympathoadrenal lineage and the enteric neurons (Carnahan, et al., 1991).

As discussed previously, corticosteroid can influence transmitter choice in sympathetic neurons indirectly by regulating the production of CDF/LIF (Fukada, 1980; McLennan et al., 1980). The hormone also can inhibit the effects of IL-1 on ganglionic nonneuronal cells, as discussed in Section VI. Moreover, it is likely that this hormone influences the developmental appearance of particular neuropeptides (Henion and Landis, 1990; Garcia-Arraras, 1991).

Corticosterone can alter the ratios of neuropeptide mRNAs within particular neurons that coexpress the peptides. In parvicellular neurons of the paraventricular nucleus, for example, this steroid decreases the mRNAs for corticotropin-releasing hormone (CRH), vasopressin, and preproenkephalin without affecting the mRNAs for CCK, β-preprotachykinin, angiotensin, and TH (Swanson and Simmons, 1989). In other neurons of this nucleus, in contrast, corticosterone increases CRH mRNA; the stimulatory effects of corticosterone have different thresholds in the various neuronal groups. These results have led to the concept of "biochemical switching," wherein the transmitters produced in anatomically fixed, adult circuits are modified (Swanson and Simmons, 1989).

## X. GONADAL STEROIDS

Gonadal steroids are well-known trophic factors for sexually dimorphic neurons, both in the CNS and in the periphery (Arnold and Breedlove, 1985; Konishi, 1989; Hodgkin, 1991; Reisert and Pilgrim, 1991; Goldstein et al., 1990; Beyer et al., 1991). The sex hormones can control neuronal survival and growth, and can regulate neuropeptide expression (DeVries, 1990; Brown et al., 1991). For example, estrogen can regulate the expression of CCK and SP differentially at the transcriptional level in a sexually dimorphic olfactory pathway in the amygdala (Simerly et al., 1989, 1990). In this case, the peptides are co-expressed in the same neurons, so the effect is gene-specific and not likely to be caused by alterations in neuronal growth and/or survival. Since these hormones control a variety of sexually dimorphic muscles and glands, there are also widespread indirect steroid effects on the neurons that innervate these tissues (e.g., Breedlove, 1986; Kelley, 1986; Thorn and Truman, 1989). Androgens and estrogen affect ChAT and TH development during a critical period in the development of the pelvic ganglion, a mixed sympathetic and parasympathetic structure. Androgens can restore ChAT and TH to normal levels in castrated male rats, whereas estrogen therapy reverses the deficit in ChAT but does not affect TH (Melvin and Hamill, 1989; Hamill and Schroeder, 1990). This effect could be due to differential neuronal survival, since the ganglion, like the superior cervical ganglion, is sexually dimorphic (Greenwood et al., 1985; Wright and Smolen, 1983). It is also possible, however, that estrogen may be able to alter the transmitter phenotype of neurons in this ganglion. Estrogen also enhances ChAT levels in the rat peoptic area (Luine et al., 1980) and in specific basal forebrain nuclei and projection areas (Luine, 1985).

## XI. MORPHOLOGICAL FACTORS

Although there is presumptive evidence that at least part of the program for generating neuronal shape may be genetic and not subject to environmental cues (Banker and Waxman, 1988), there is also emerging support for a role for both glial and target cells in regulating neuronal morphology. For instance, Schwann cells can stimulate cultured dorsal root ganglion neurons to undergo the normal transition from bipolar to the pseudo-unipolar shape characteristic of these neurons in the adult (Mudge, 1984). Glial cells also promote the formation of dendrites by rat sympathetic neurons in culture. Without nonneuronal cells, most of these neurons produce only a single axon. In the presence of Schwann cells or astrocytes, but not heart cells or fibroblasts, the neurons become multipolar and extend many dendrites (Tropea et al., 1988; Johnson et al., 1989). The development of dendrites does not, however, alter the number or growth of axons, demonstrating that the growth of axons and dendrites can be controlled independently. The glial effect can be mimicked by the matrix molecule(s) in a basement membrane extract, but not by laminin or the known growth factors in the extract (Lein and Higgins, 1989).

In CNS cultures, mesencephalic dopaminergic neurons display distinct branching patterns when grown on astrocytes from the area in which their somas are located (mesencephalon) or on astrocytes from their target area (striatum). On mesencephalic glial cells, the neurons form highly branched dendrites, whereas, on striatal glia, the neurons often display a single unbranched axon (Denis-Donini et al., 1984; Chamak et al., 1987; Autillo-Touati et al., 1988). In addition to suggesting a novel developmental role for astrocytes, these results add to a growing literature on the regional heterogeneity of astrocytes, Schwann cells, and enteric glial cells (c.f., Suzue et al., 1990; Jessen and Mirsky, 1991; Miller and Szigeti, 1991). As in the case of the sympathetic and sensory neurons, the identity of the molecules that control the morphology of the dopaminergic neurons is unclear at present. The differential effects of mesencephalic and striatal glia cannot be fully mimicked by CMs from these cells (Rousselet et al., 1988, 1990). Experiments with surfaces coated with matrix molecules suggest a role for differential adhesion in the formation of axons versus dendrites (Chamak and Prochiantz, 1989). Nonneuronal cells from spinal cord can regulate not only dendritic morphology but also responses to neurotransmitter in cultures of chick sympathetic preganglionic neurons (Clendening and Hume, 1990).

It is also clear that target tissues can control the morphology of the neurons that innervate them. In addition to effects on neuronal size (Purves, 1988), particular targets can evoke distinctive dendritic branching patterns in their innervating neurons (Jellies et al., 1987; Loer et al., 1987; Blaser et al., 1990). These dendritic differences are associated with distinctive patterns of synaptic inputs that the neurons receive (Loer and Kristan, 1989). The nature of the retrograde signal is not yet known, but it will be interesting to see if such a morphologic cue also regulates the neurotransmitter/neuropeptide identity of the neurons. Could a phenotype-specifying, target factor that controls transmitter/peptide choice (as in the sweat gland), control the dendritic pattern of the neurons that innervate the glands?

## XII. MELANIZATION FACTORS

Pigment production by melanocytes is another phenotype available to neural crest cells. Since there is evidence for a reciprocal relationship between the differentiation along a melanocyte path and that along a neuronal/glial path (Perris and Lofberg, 1986; Vogel and Weston, 1988; Ciment, 1990), consideration of melanization is potentially relevant for neuronal differentiation. Both positive and negative factors for melanogenesis have been identified (Bagnara et al., 1985; Campbell and Bard, 1985; Jerden et al., 1985; Saunders et al., 1985; Fukuzawa and Ide, 1986; Fukuzawa and Bagnara, 1989). For instance, bFGF, but not EGF or NGF, enhances pigmentation in ganglionic and peripheral nerve cultures, whereas TGF-β1 inhibits these effects (Stocker, et al., 1991). An innovative approach for analysis of extracellular matrix influences involves inserting a filter into the melanocyte migration pathway in the subepidermal space of the axolotl embryo. The matrix deposited on such filters can enhance pigment cell differentiation by axolotl neural crest cells (Perris and Lofberg, 1986; Perris et al., 1988). Filters inserted into a different location, the presumptive dorsal root ganglion, do not promote melanocyte differentiation from such crest cell populations, but enhance neuronal development.

In a study of the matrix molecules laminin and fibronectin, cultured chick crest cell dispersal was found to be correlated inversely with pigment formation (Rogers et al., 1990). Fibronectin provided the best substrate for migration, and the fewest pigment cells were observed on this surface. Uncoated glass dishes did not allow dispersal, and the highest level of pigmentation was found. Laminin was intermediate in both dispersal and melanization. These results differ from those seen in the axolotl, where the subepidermal matrix promoted both dispersal and melanization. The *Steel (Sl)* mutation in mouse is also relevant in this context. This mutation is non-cell autonomous, and results in alterations in pigmentation, hematopoiesis, and germ cell development. When normal mouse crest cells are plated on matrix secreted by normal and *Steel-Dickie* fetal skin cells, the latter matrix fails to promote melanogenesis to the same extent that the normal matrix does (Morrison-Graham et al., 1990). The matrices from both sources promote the appearance of TH-positive cells equally.

Recent biochemical and molecular cloning results have demonstrated that the product of the *Sl* locus is a novel, multipotent cytokine. This factor is variously termed stem cell factor (SCF), mast cell growth factor (MGF), *kit* ligand (KL), and steel factor (SLF) (Witte, 1990). The protein is heavily glycosylated, and alternative splicing of its mRNA yields secreted and cell-bound forms. Although SCF affects early myeloid cells and lymphocytes to some degree, it synergizes extremely well with erythropoietin, GM-CSF, and IL-7 (Martin et al., 1990; Zsebo et al., 1990b). In vivo injections can reverse anemia and mast cell deficiencies (Zsebo et al., 1990a). The receptor for SCF is a previously defined tyrosine kinase termed the c-*kit* receptor, now known to be the product of the white spotting locus *(W)* (Witte, 1990). Mutations in the *W* locus also cause pigmentation, hematopoietic, and germ cell deficits. In the hematopoietic system, the defect in *W* is localized in the stem cells, whereas the defect in *Sl* is in the microenvironment of the bone marrow. In situ analysis of SCF expression localizes its mRNA to the migratory pathways and homing sites of melanoblasts, hematopoietic stem cells, and germ cells (Matsui et al., 1990). Label is observed over the dorsal regions of the somites but not over the migrating neural crest cells themselves, nor in dorsal root ganglia. In contrast, c-*kit* mRNA is found in presumptive melanoblasts (Orr-Urtreger et al., 1990). Both SCF and c-*kit* are expressed in a variety of other tissues, including the embryonic brain and spinal cord (Nocka et al., 1989; Matsui et al., 1990; Orr-Urtreger et al., 1990). Curiously, homozygous *Sl/Sl* embryos have not yet been reported to display gross CNS abnormalities. Analysis of potential defects in the peripheral nervous system also has not been reported.

## XIII. NEURONAL ACTIVITY

Activity plays a critical role in the development of connectivity patterns in the nervous system (c.f., Constantine-Paton et al., 1990; Jansen and Fladby, 1990; Shatz, 1990). In addition, membrane depolarization and associated ion fluxes and second messenger changes can influence both the acute activity and the transcription of neuropeptides and neurotransmitter enzymes (c.f., Zigmond et al., 1989; Goodman, 1990). Activity also can influence phenotypic choice. In fact, activity would be a particularly effective method for controlling phenotypes, since (1) the firing pattern of a neuron depends in part on the connections it makes and receives and (2) that pattern is regulated at the level of individual cells. Thus phenotypic specification could make use of the specificity of the wiring pattern, and could generate mixed phenotypes within a layer or nucleus of otherwise similar cells. This model also would provide a method for the establishment of retrograde and anterograde functional streams, with matched transmitters and transmitter receptors, within otherwise homogeneous cell masses.

The inhibitory role for activity in the noradrenergic–cholinergic switch by sympathetic neurons was discussed in Section II. Depolarization of sensory neurons likewise blocks the induction of SP in these cells (Adler et al., 1984). In a particularly dramatic example of plasticity, norepinephrine addition to cultured rat pineal cells suppresses their expression of rhodopsin and serotonin-containing processes (Araki and Tokunaga, 1990). This result suggests that the normal innervation of mammalian pinealocytes inhibits their capacity for neuronal or photoreceptor differentiation. This model is particularly interesting because the pineal gland of lower vertebrates contains neurons and is part of the photoendocrine system.

Previously discussed evidence shows that certain neuropeptides and transmitters can influence phenotypic choice in a number of systems, for example, the induction of TH in the olfactory bulb by CGRP and in the cortex by olfactory afferents, serotonin induction of tryptophan hydroxylase in the hypothalamus, angiotensin effects on the hypothalamus, and VIP effects on chromaffin cells and spinal cord neurons. Further, depolarization and the presynaptic agonists SP and substance K induce TH in cultured dopaminergic neurons (Friedman et al., 1988). Activity also regulates mENK expression in spinal cord cultures (Foster et al., 1990; Agoston et al., 1991). Further investigation in this area will involve assessing phenotypes after perturbations in activity levels of specific neuronal populations in vivo, as well as examination of the second messenger pathways involved.

## XIV. OTHER FACTORS

Numerous factors, both known and uncharacterized, can increase the activity or mRNA for particular transmitter synthetic enzymes or neuropeptides, but it is not yet clear if these effects are due to a traditional neurotrophic effect on the survival and growth of a distinct neuronal population, or to the control of phenotypic decisions made by the neurons. For instance, although NGF can increase the specific activities of TH and ChAT in sympathetic neurons, it does not push the neurons selectively along either the noradrenergic or the cholinergic pathway (Chun and Patterson, 1977). Similarly, NGF can enhance several neuropeptides simultaneously (e.g., Lindsay et al., 1989). NGF application can, however, alter the relative levels of peptides in ganglia in vivo (Hayashi et al., 1985). In addition, there are other cases in which an effector molecule can enhance one phenotypic characteristic selectively, but it may not yet be clear whether the effector is simply activating an enzyme or whether it is turning on an entire set of phenotypic characteristics (e.g., Zurn and Do, 1988).

A number of other CMs, substrata, and tissue extracts that exert interesting effects on various phenotypic characteristics of neurons have been described. Some of these phenotypic markers include transmitter biosynthetic and degradative enzymes, uptake sites, receptors, and neuropeptides, as well as other enzymes and antigens (Mudge, 1981; Crean et al., 1982; Hendry, 1985; Cooper and Lau, 1986; Phillips and Freschi, 1986; Shaw and Letourneau, 1986; Gray and Tuttle, 1987; Salvaterra et al., 1987; Nishi and Willard, 1988; Wakade and Bhave, 1988; Barakat and Droz, 1989; MacLean et al., 1989; Whitaker-Azmitia and Azmitia, 1989; Clendening and Hume, 1990; DeMello et al., 1990; Mangoura et al., 1990; Smith and Kessler, 1990; Trifaro et al., 1990; Halvorsen, et al., 1991; Gardette et al., 1991; Katz,

1991). It will be of great interest to determine the molecular nature of these various factors. Are they related to the known factors? Will they form families of homologous genes? How large will this group of factors be?

Similar questions arise concerning the presumed receptors for the instructive factors. For instance, how many of these receptors will a particular neuron express? Another area of considerable interest concerns the second messenger systems that presumably mediate the effects of the factors. Can a receptor for a particular factor be linked with various G proteins, for instance, to produce distinct messages in different neurons?

## XV. MUTANTS IN INVERTEBRATES

Intensive study of mutants affecting differentiation choices in the nervous systems of *Drosophila* and *Coenarhablitis elegans* is beginning to yield potential intercellular signalling factors and receptors (Greenwald and Rubin, 1992). For instance, differentiation of the eight distinct photoreceptor cells (R1–8) in the *Drosophila* ommatidium can be altered by mutation so that one or more of the cells fails to develop (Banerjee and Zipursky, 1990). These cells can be distinguished on the basis of their position in the ommatidium and by the opsins expressed. Although normal development proceeds in an invariant sequence, complex cellular interactions are required for appropriate differentiation choices (Ready, 1989). In the *sevenless (sev)* mutant, the precursor cell that normally becomes R7 fails to migrate and differentiates into a cone cell. Mosaic analysis indicates that the *sev* product exerts its function in the R7 precursor cell. In fact, the *sev* protein is a membrane protein with a large extracellular domain and a cytoplasmic domain that contains a tyrosine kinase motif. This result implies that the precursor cell may be receiving instructional signals, since such kinases have been found to be growth factor receptors in several systems, including those of epidermal growth factor, insulin, and the previously discussed stem cell factor involved in melanogenesis. The leading candidate for the ligand for the *sev* tyrosine kinase receptor is the product of the *bride-of-sevenless (boss)* gene (Banerjee and Zipursky, 1990). *boss*-negative mutants also lack the R7 cell, but only if the mutation is expressed in photoreceptor R8. Significantly, R8 is the only cell to contact the R7 precursor cell directly. The deduced structure of *boss* is a membrane protein, and thus is in a position to interact with the *sev* receptor.

Interestingly, as in the case of the instructive factors acting on sympathetic neurons, the R7 differentiation pathway also appears to involve inhibitory or suppressive activities. In the *seven-up (svp)* mutant, photoreceptors R1, R3, R4, and R6 also acquire the R7 phenotype. This suggests that the *svp* product is required to suppress the differentiation of these cells into R7 (Banerjee and Zipursky, 1990). The deduced product of *svp* encodes a DNA-binding protein of the steroid receptor superfamily. Here, also, there are echos of the sympathoadrenal lineage in which, as discussed previously, corticosteroid can suppress the action of protein growth factors (Patterson, 1990), and there is likely to be involvement of tyrosine kinase growth factor receptors. Both FGF and NGF appear to turn on the neuronal pathway in that system; infection of embryonic quail sympathetic ganglion cells with v-*src*-containing retrovirus results in the cessation of mitosis and neuronal differentiation (Haltmeier and Rohrer, 1990). v-*src* is a constitutively active tyrosine kinase that also promotes neuronal differentiation of PC12 cells (Alema et al., 1985), cells originally derived from the sympathoadrenal lineage. Another inhibitory instructive factor is encoded by the *scabrous (sca)* gene. Mutation of the *sca* locus appears to disrupt a signal produced by R8 cells that normally inhibits other cells from assuming the R8 phenotype. *sca* appears to encode a secreted protein that is related to fibrinogen (Baker et al., 1990).

Many other genes important in neural development are found to encode transcription factors (e.g., Jan and Jan, 1990), some of which may play similar roles in mammalian systems (e.g., Johnson et al., 1990). It will be of considerable interest to identify the vertebrate homologs of the *Drosophila* and *C. elegans* genes encoding the surface membrane receptors and ligands as well.

## XVI. HEMATOPOIETIC ANALOGY

The fact that several different proteins (CDF/LIF, CNTF, MANS, and possibly a distinct sweat gland factor) can induce sympathetic neurons to become cholinergic raises questions about the physiologic role of this apparent redundancy. Moreover, CDF/LIF and CNTF also induce a number of neuropeptides in these neurons (SP, SOM, VIP), peptides that also can be induced by several other distinct peptidergic factors. Thus, although the analysis of these instructive factors is still at an early stage, an extensive, if partial, overlap or redundancy in the biologic effects of these factors is apparent. These observations are reminiscent of the control of phenotypic decisions in the hematopoietic system. In that case, too, multipotential cells have a number of possible fates, and a family of proteins, called hematopoietic regulators or lymphokines, can influence choices. Moreover, these regulators have partially overlapping effects, so several of them stimulate, for instance, macrophage development as well as other phenotypes (c.f., Metcalf 1987, 1989). Experiments in vivo as well as in culture have demonstrated combinatorial effects, so that several factors together can produce a different set of phenotypes than when the factors are added separately.

The prospect of combinatorial factor interactions could be critical for the nervous system, where the number of possible phenotypes is enormous. For example, if a neuron were to produce two classical transmitters of the known repertoire of about 12, and simultaneously produce three neuropeptides of the known repertoire of > 30, the number of possible combinations, or phenotypes, available to the cell would be > 250,000. One mechanism would be to have a single instructive factor for each transmitter and peptide. This hypothesis has the drawback that it would involve a great many factors, and most of the candidate factors described so far affect multiple transmitters. Another model would use factors that each induce a commonly found combination of transmitter/peptide genes. CDF/LIF and CNTF emulate this model almost perfectly, inducing expression of the various genes found in cholinergic sympathetic neurons, while suppressing those genes found in noradrenergic neurons. (The exception to this simple picture is that these factors both induce SP, a peptide found in cat but not in rat cholinergic sympathetic neurons, as discussed in Section V.) Modifications to produce distinct neuronal subpopulations could then be made by selectively placing additional stimulatory and inhibitory factors to yield unique combinations of phenotypic characteristics. In addition, because of its characteristic lineage history, each class of neuron is likely to respond somewhat differently to the same factors, as discussed in Section II. In summary, the fact that certain factors have partially overlapping effects allows for more economical and refined combinatorial possibilities.

The analogy between the hematopoietic and nervous systems has recently extended to the level of homology of at least some of the phenotype-specifying factors involved in both systems. In addition to the observations that CDF/LIF and IL-6 can act on both neurons and myeloid cells, certain of the factors can be grouped as putative neuropoietic or hematopoietic factors based on predicted structural themes and homologies of their receptors (Patterson and Fann, 1992; Patterson, 1992). Comparing the deduced amino acid sequences of CDF/LIF, CNTF, oncostatin M (ONC), IL-6, myelomonocytic growth factor (MGF), and granulocyte colony stimulating factor (GM-CSF), Bazan (1991) predicts that each of these proteins will have the 4 α-helical bundle structure of growth hormone. Moreover, the genes for each of these proteins display some similarity in the pattern of exon junctions, suggesting a common evolutionary origin. The neuropoietic group of CDF/LIF, CNTF and ONC contains a characteristic sequence in the D helix that distinguishes these three proteins from the other group of three.

These predicted structural themes suggested a relationship at the receptor level as well. This has been confirmed in a very striking manner with the cloning and sequencing of several of the corresponding receptors. The CDF/LIF receptor is homologous to the transducing subunit of the IL-6 receptor (Gearing et al., 1991), and the CNTF receptor is homologous to the other subunit of the IL-6 receptor (Davis et al., 1991). Moreover, ONC can displace CDF/LIF from its high affinity receptor on myeloid cell lines (Gearing and Bruce, 1991). These multiple and surprising relationships suggest that all 6 of these proteins and their receptors should be intensively investigated in the nervous system.

It should be noted that most of the experiments on instructive factors in the vertebrate nervous system have been carried out on postmitotic neurons, whereas the hematopoietic regulators act, at least initially, on dividing multipotential progenitor cells. Thus, the hematopoietic factors expand selected populations of cells as well as up-regulate the sets of genes particular for that phenotype (Metcalf 1987, 1989). It is not yet known how closely this feature of the hematopoietic model applies to the nervous system. It has been found, however, that CNTF specifically inhibits the proliferation of cells in embryonic chick sympathetic ganglia that express neuronal markers (Ernsberger et al., 1989). CNTF also acts on dividing progenitors in the O–2A glial lineage, promoting type-2 astrocyte differentiation (Lillien and Raff, 1990a,b). CDF/LIF is known to act on dividing populations of cells in several tissues (Gough and Williams, 1989), and can regulate gene expression in cultured rat sympathoadrenal progenitor cells (Vandenberg, et al., 1991). Clearly much remains to be learned about the role of instructive factors in decision making by neuroblasts and stem cells in the nervous system. Perhaps more could also be learned about possible effects of the hematopoietic regulators on post-mitoitic cells in that lineage.

## REFERENCES

Acheson, A. L., and Rutishauser, U. (1988). Neural cell adhesion molecule regulates cell contact-mediated changes in choline acetyltransferase activity of embryonic chick sympathetic neurons. *J. Cell Biol.* **106**, 479–486.

Acheson, A. L., and Thoenen, H. (1983). Cell contact-mediated regulation of tyrosine hydroxylase synthesis in cultured bovine adrenal chromaffin cells. *J. Cell Biol.* **97**, 925–928.

Adler, J. E., and Black, I. B. (1985). Sympathetic neuron density differentially regulates transmitter phenotypic expression. *Proc. Natl. Acad. Sci. U.S.A.* **82**, 4296–4300.

Adler, J. E., Kessler, J. A., and Black, I. B. (1984). Development and regulation of substance P in sensory neurons *in vitro*. *Dev. Biol.* **102**, 417–425.

Adler, J. E., Schleifer, L. S., and Black, I. B. (1989). Partial purification and characterization of a membrane-derived factor regulating neurotransmitter phenotypic expression. *Proc. Natl. Acad. Sci. U.S.A.* **86**, 1080–1083.

Adler, R., Landa, K. B., Manthorpe, M., and Varon, S. (1979). Cholinergic neuronotrophic factors: Intraocular distribution of trophic activity for ciliary neurons. *Science* **204**, 1434–1436.

Agoston, D. V., Eiden, L. E., and Brenneman, D. E. (1991). Calcium-dependent regulation of the enkephalin phenotype by neuronal activity during early ontogeny. *J. Neurosci. Res.* **28**, 140–148.

Alema, S., Casalbore, P., Agostini, E., and Tato, F. (1985). Differentiation of PC12 phaeochromocytoma cells induced by v-*src* oncogene. *Nature (London)* **316**, 557–559.

Anderson, D. J. (1989). Cellular "neotony". A possible developmental basis for chromaffin cell plasticity. *Trends Genet.* **5**, 174–178.

Arakawa, Y., Sendtner, M., and Thoenen, H. (1990). Survival effect of ciliary neurotrophic factor (CNTF) on chick embryonic motoneurons in culture: Comparison with other neurotrophic factors and cytokines. *J. Neurosci.* **10**, 3507–3515.

Araki, M., and Tokunaga, F. (1990). Norepinephrine suppresses both photoreceptor and neuron-like properties expressed by cultured rat pineal glands. *Cell Diff. Dev.* **31**, 129–135.

Arnold, A. P., and Breedlove, S. M. (1985). Organizational and activational effects of sex steroid hormones on vertebrate behavior: A re-analysis. *Hormones Behav.* **19**, 469–498.

Autillo-Touati, A., Chamak, B., Araud, D., Vuillet, J., Seite, R., and Prochiantz, A. (1988). Region-specific neuro-astroglial interactions: Ultrastructural study of the *in vitro* expression of neuronal polarity. *J. Neurosci. Res.* **19**, 326–342.

Bagnara, J., Klaus, S. N., Paul, E., and Schartl, M., eds. (1985). "Pigment Cell 1985: Biological, Molecular, and Clinical Aspects of Pigmentation." University of Tokyo Press, Tokyo.

Baker, H. (1990). Unilateral, neonatal olfactory deprivation alters tyrosine hydroxylase expression but not aromatic amino acid decarbaoxylase or GABA immunoreactivity. *Neuroscience* **36**, 761–771.

Baker, N. E., Mlodzik, M., and Rubin, G. M. (1990). Spacing differentiation in the developing *Drosophila* eye: A fibrinogen-related lateral inhibitor encoded by *scabrous*. *Science* **250**, 1370–1377.

Banerjee, U., and Zipursky, S. L. (1990). The role of cell–cell interaction in the development of the *Drosophila* visual system. *Neuron* **4**, 177–187.
Banker, G. A., and Waxman, A. B. (1988). Hippocampal neurons generate natural shapes in cell culture. *In* "Intrinsic Determinants of Neuronal Form and Function" (R. J. Lasek and M. M. Black, eds.), pp. 61–82. Liss, New York.
Barakat, I., and Droz, B. (1989). Calbindin-immunoreactive sensory neurons in dissociated dorsal root ganglion cell cultures of chick embryo: Role of culture conditions. *Dev. Brain Res.* **50**, 205–216.
Barald, K. F. (1989). Culture conditions affect the cholinergic development of an isolated subpopulation of chick mesencephalic neural crest cells. *Dev. Biol.* **135**, 349–366.
Barbany, G., Friedman, W. J., and Persson, H. (1991). Lymphocyte-mediated regulation of neurotransmitter gene expression in rat sympathetic ganglia. *J. Neuroimmunol.* **32**, 97–104.
Barbin, G., Manthorpe, M., and Varon, S. (1984). Purification of the chick eye ciliary neuronotrophic factor. *J. Neurochem.* **43**, 1468–1478.
Barde, Y.-A. (1988). What, if anything, is a neurotrophic factor? *Trends Neurosci.* **11**, 343–346.
Bazan, J. F. (1991). Neuropoietic cytokines in the hematooietic fold. *Neuron* **7**, 197–208.
Belford, D. A. (1990). Identification and characterization of a parasympathetic neurotrophic factor. Ph.D. Thesis. Australian National University. Canberra, A.C.T.
Beretta, C., and Zurn, A. D. (1991). The neuropeptide VIP modulates the neurotransmitter phenotype of cultured sympathetic neurons. *Devel. Biol.* **148**, 87–94.
Beyer, C., Pilgrim, C., and Reisert, I. (1991). Dopamine content and metabolism in mesencephalic and diencephalic cell cultures: sex differences and effects of sex steroids. *J. Neurosci.* **11**, 1325–1333.
Bhatt, H., Brunet, L. J., and Stewart, C. L. (1991). Uterine expression of leukemia inhibitory factor coincides with the onset of blastocyst implantation. *Proc. Natl. Acad. Sci. U.S.A.* **88**, 11408–11412.
Black, I. B., Chikaraishi, D. M., and Lewis, E. J. (1985). Transsynaptic regulation of growth and development of adrenergic neurons in rat sympathetic ganglion. *Brain Res.* **339**, 151–153.
Black, I. B., Adler, J. E., Dreyfus, E. F., Friedman, W. F., LaGamma, E. F., and Roach, A. H. (1987). Biochemistry of information storage in the nervous system. *Science* **236**, 1263–1268.
Blaser, P. F., Catsicas, S., and Clarke, P. G. H. (1990). Retrograde modulation of dendritic geometry in the vertebrate brain during development. *Developmental Brain Res.* **57**, 139–142.
Bloch-Gallego, E., Huchet, M., El M'Hamdi, H., Xie, F.-K., Tanaka, H., and Henderson, C. E. (1991). Survival *in vitro* of motoneurons identified or purified by novel antibody-based methods is selectively enhanced by muscle-derived factors. *Development* **111**, 221–232.
Braun, S. J., Swedner, K. J., and Patterson, P. H. (1981). Neuronal cell surfaces: Distinctive glycoproteins of cultured adrenergic and cholinergic sympathetic neurons. *J. Neurosci.* **1**, 1397–1406.
Breedlove, S. M. (1986). Cellular analyses of hormone influence on motoneuronal development and function. *J. Neurobiol.* **17**, 157–176.
Brice, A., Berrard, S., Raynaud, B., Ansieau, S., Coppola, T., Weber, M. J., and Mallet, J. (1989). Complete sequence of a cDNA encoding an active rat choline acetyltransferase: A tool to investigate the plasticity of cholinergic phenotype expression. *J. Neurosci. Res.* **23**, 266–273.
Brown, E. R., Roth, K. A., and Krause, J. E. (1991). Sexually dimorphic distribution of substance P in specific anterior pituitary cell populations. *Proc. Natl. Acad. Sci. U.S.A.* **88**, 1222–1226.
Bunge, R. P., Johnson, M., and Ross, C. D. (1978). Nature and nurture in development of the autonomic neuron. *Science* **199**, 1409–1416.
Campbell, S., and Bard, J. B. L. (1985). The acellular stroma of the chick cornea inhibits melanogenesis of the neural crest cells that colonize it. *J. Embryol. Exp. Morphol.* **86**, 143–154.
Carnahan, J., and Patterson, P. H. (1991a). The generation of monoclonal antibodies that bind preferentially to adrenal chromaffin cells and the cells of embryonic sympathetic ganglia. *J. Neurosci.* **11**, 3493–3506.
Carnahan, J., Anderson, D. J., and Patterson, P. H. (1991). Evidence that enteric neurons may derive from the sympathoadrenal lineage. *Devel. Biol.* **148**, 552–561.
Carnahan, J., and Patterson, P. H. (1991b). Isolation of progenitor cells of the sympathoadrenal lineage from embryonic sympathetic ganglia with the SA monoclonal antibodies. *J. Neurosci.* **11**, 3520–3530.

Chamak, B., and Prochiantz, A. (1989). Influence of extracellular matrix proteins on the expression of neuronal polarity. *Development* **106**, 483–491.

Chamak, B., Fellous, A., Glowinski, J., and Prochiantz, A. (1987). MAP2 expression and neuritic outgrowth and branching are coregulated through region-specific neuro-glial interactions. *J. Neurosci.* **7**, 3163–3170.

Chun, L. L. Y., and Patterson, P. H. (1977). The role of nerve growth factor in the development of rat sympathetic neurons *in vitro*. III. Effect on acetylcholine induction. *J. Cell Biol.* **75**, 712–718.

Ciment, G. (1990). The melanocyte Schwann cell progenitor: a bipotent intermediate in the neural crest lineage. *Comments Devel. Neurobiol.* **1**, 207–223.

Ciment, G., and Weston, J. A. (1985). Segregation of developmental abilities in neural crest-derived cells: Identification of partially restricted intermediate cell types in the branchial arches of avian embryos. *Dev. Biol.* **111**, 73–83.

Clendening, B., and Hume, R. I. (1990). Cell interactions regulate dendritic morphology and responses to neurotransmitters in embryonic chick sympathetic preganglionic neurons *in vitro*. *J. Neurosci.* **10**, 3992–4005.

Constantine-Paton, M., Cline, H. T., and Debski, E. (1990). Patterned activity, synaptic convergence, and the NMDA receptor in developing visual pathways. *Ann. Rev. Neurosci.* **13**, 129–154.

Cooper, E., and Lau, M. (1986). Factors affecting the expression of acetylcholine receptors on rat sensory neurones in culture. *J. Physiol.* **377**, 409–420.

Cordell, B., Lam, A., Fuller, F., Miller, J., Varon, S., and Manthorpe, M. (1990). Structural organization and sequence of the human ciliary neurotrophic factor gene. *Soc. Neurosci. Abstr.* **16**, 278.7.

Coulombe, J. N., and Nishi, R. (1991). Stimulation of somatostatin expression in developing ciliary ganglion neurons by cells of the choroid layer. *J. Neurosci.* **11**, 553–562.

Crean, G., Pilar, G., Tuttle, J. B., and Vaca, K. (1982). Enhanced chemosensitivity of chick parasympathetic neurones in co-culture with myotubes. *J. Physiol.* **331**, 87–104.

Davies, A. M. (1988). The emerging generality of the neurotrophic hypothesis. *Trends Neurosci.* **11**, 243–244.

Davis, S., Aldrich, T. H., Valenzuela, D. M., Wong, V., Furth, M. E., Squinto, S. P., and Yancopoulos, G. D. (1991). The receptor for ciliary neurotrophic factor. *Science* **253**, 59–63.

DeMello, F. G., DeMello, M. C. F., Hudson, R., and Klein, W. L. (1990). Selective expression of factors preventing cholinergic dedifferentiation. *J. Neurochem.* **54**, 886–892.

Denis-Donini, S. (1989). Expression of dopaminergic phenotypes in the mouse olfactory bulb induced by the calcitonin gene-related peptide. *Nature (London)* **339**, 701–703.

Denis-Donini, S., Glowinski, J., and Prochiantz, A. (1984). Glial heterogeneity may define the three-dimensional shape of mouse mesencephalic dopaminergic neurones. *Nature (London)* **307**, 641–643.

De Vitry, F., Catelon, J., Dubois, M., Thibault, J., Barritault, D., Courty, J., Bourgoin, S., and Hamon, M. (1986a). Partial expression of monoaminergic (serotonergic) properties by the multipotent hypothalmic cell line F7—An example of learning at the cellular level. *Neurochem. Int.* **9**, 43–53.

De Vitry, F., Hamon, M., Catelon, J., Dubois, M., and Thibault, J. (1986b). Serotonin initiates and autoamplifies its own synthesis during mouse central nervous system development. *Proc. Natl. Acad. Sci. U.S.A.* **83**, 8629–8633.

DeVries, G. J. (1990). Sex differences in neurotransmitter systems. *J. Neuroendocrinol.* **2**, 1–13.

DiCicco-Bloom, E., Townes-Anderson, E., and Black, I. B. (1990). Neuroblast mitosis in dissociated culture: Regulation and relationship to differentiation. *J. Cell Biol.* **110**, 2073–2086.

Dobrea, G. M., Wilcox, B. J., and Unnerstall, J. R. (1990). Ciliary neurotrophic factor (CNTF) mRNA is localized to neurons in rat hippocampus. *Soc. Neurosci. Abstr.* **16**, 209.4

Doupe, A. J., and Patterson, P. H. (1982). Glucocorticoids and the developing nervous system. *Curr. Top. Neuroendocrinol.* **2**, 23–43.

Doupe, A. J., Landis, S. C., and Patterson, P. H. (1985a). Environmental influences in the development of neural crest derivatives: Glucocorticoids, growth factors, and chromaffin cell plasticity. *J. Neurosci.* **5**, 2119–2142.

Doupe, A. J., Patterson, P. H., and Landis, S. C. (1985b). Small intensely fluorescent cells in culture: Role of glucocorticoids and growth factors in their development and interconversions with other neural crest derivatives. *J. Neurosci.* **5,** 2143–2160.

Dreyer, D., Lagrange, A., Grothe, C., and Unsicker, K. (1989). Basic fibroblast growth factor prevents ontogenetic neuron death *in vivo*. *Neurosci. Lett.* **99,** 35–38.

Eckenstein, F. P., Esch, F., Holbert, T., Blacher, R. W., and Nishi, R. (1990). Purification and partial characterization of a trophic factor for embryonic peripheral neurons: Comparison with fibroblast growth factors. *Neuron* **4,** 623–631.

Erkman, L., Wuarin, L., Cadelli, D., and Kato, A. C. (1989). Interferon induces astrocyte maturation causing an increase in cholinergic properties of cultured human spinal cord cells. *Dev. Biol.* **132,** 375–388.

Ernsberger, U., Sendtner, M., and Rohrer, H. (1989). Proliferation and differentiation of embryonic chick sympathetic neurons: Effects of ciliary neurotrophic factor. *Neuron* **2,** 1275–1284.

Fann, M-H. and Patterson, P. H. (1992). Results from a rapid screening technique for assay of neuronal differentiation factors. *Soc. Neurosci.,* in press.

Fletcher, T. L., and Banker, G. A. (1989). The establishment of polarity by hippocampal neurons: The relationship between the stage of a cell's development *in situ* and its subsequent development in culture. *Dev. Biol.* **136,** 446–454.

Foster, G. A., Eiden, L. E., and Brenneman, D. E. (1989). Regulation of discrete subpopulations of transmitter-identified neurones after inhibition of electrical activity in cultures of mouse spinal cord. *Cell Tiss. Res.* **256,** 543–552.

Foster, G. A., Eiden, L. E., and Brenneman, D. E. (1990). Enkephalin expression in spinal cord neurons is modulated by drugs related to classical and peptidergic transmitters. *Europ. J. Neurosci.* **3,** 32–39.

Freidin, M., and Kessler, J. A. (1991). Cytokine regulation of substance P expression in sympathetic neurons. *Proc. Natl. Acad. Sci. U.S.A.* **88,** 3200–3203.

Friedman, W. J., Dreyfus, C. F., McEwen, B., and Black, I. B. (1988). Presynaptic transmitters and depolarizing influences regulate development of the substantia nigra in culture. *J. Neurosci.* **8,** 3616–3623.

Fukada, K. (1980). Hormonal control of neurotransmitter choice in sympathetic neurone cultures. *Nature (London)* **287,** 553–555.

Fukada, K. (1985). Purification and partial characterization of a cholinergic neuronal differentiation factor. *Proc. Natl. Acad. Sci. U.S.A.* **82,** 8795–8799.

Fukuzawa, T., and Bagnara, J. T. (1989). Control of melanoblast differentiation in amphibia by α-melanocyte stimulating hormone, a serum melanization factor, and a melanization inhibiting factor. *Pig. Cell Res.* **2,** 171–181.

Fukuzawa, T., and Ide, H. (1988). A ventrally localized inhibitor of melanization in *Xenopus laevis* skin. *Dev. Biol.* **129,** 25–36.

Furness, J. B., Morris, J. L., Gibbons, I. L., and Costa, M. (1989). Chemical coding of neurons and plurichemical transmission. *Ann. Rev. Pharmac. Toxicol.* **29,** 289–306.

Garcia-Arraras, J. E. (1991). Modulation of neuropeptide expression in avian embryonic sympathetic cultures. *Dev. Brain Res.* **60,** 19–27.

Gardette, R., Listerud, M. D., Brussaard, A. B., and Role, L. W. (1991). Developmental changes in transmitter sensitivity and synaptic transmission in embryonic chicken sympathetic neurons innervated in vitro. *Devel. Biol.* **147,** 83–95.

Gearing, D. P., Gough, N. M., King, J. A., Hilton, D. J., Nicola, N. A., Simpson, R. J., Nice, E. C., Kelso, A., and Metcalf, D. (1987). Molecular cloning and expression of cDNA encoding a murine myeloid leukaemia inhibitory factor (LIF). *EMBO J.* **6,** 3995–4002.

Gearing, D. P., and Bruce, A. G. (1991). Oncostatin M binds the high-affinity leukemia inhibitory factor receptor. *New Biolgist* **3,** 1–5.

Gearing, D. P., Thut, C. J., VandenBos, T., Gimpel, S. D., Delaney, P. B., King, J., Price, V., Cosman, D., and Beckman, M. P. (1991). Leukemia inhibitory factor receptor is structurally related to the IL-6 signal transducer, gp130. *EMBO J,* **10,** 2839–2848.

Giess, M.-C., and Weber, M. J. (1984). Acetylcholine metabolism in rat spinal cord cultures: Regulation by a factor involved in the determination of the neurotransmitter phenotype of sympathetic neurons. *J. Neurosci.* **4**, 1442–1452.

Goldstein, L. A., Kurz, E. M., and Sengelaub, D. R. (1990). Androgen regulation of dendritic growth and retraction in the development of a sexually dimorphic spinal nucleus. *J. Neurosci.* **10**, 935–946.

Goodman, R. H. (1990). Regulation of neuropeptide gene expression. *Ann. Rev. Neurosci.* **13**, 111–128.

Gough, N. M., and Williams, R. L. (1989). The pleiotropic actions of leukemia inhibitory factor. *Cancer Cells* **1**, 77–80.

Gray, D. B., and Tuttle, J. B. (1987). [$^3$H]Acetylcholine synthesis in cultured ciliary ganglion neurons: Effects of myotube membranes. *Dev. Biol.* **119**, 290–298.

Greenwald, I., and Rubin, G. M. (1992). Making a difference: the role of cell-cell interactions in establishing separate identifies for equivalent cells. *Cell* **68**, 271–281.

Greenwood, D., Coggeshall, R. E., and Hulsebosch, C. E. (1985). Sexual dimorphism in numbers of neurons in the pelvic ganglion of adult rats. *Brain Res.* **340**, 160–162.

Guthrie, K. M., and Leon, M. (1989). Induction of tyrosine hydroxylase expression in rat forebrain neurons. *Brain Res.* **497**, 117–131.

Haltmeier, H., and Rohrer, H. (1990). Distinct and different effects of the oncogenes v-*myc* and v-*src* on avian sympathetic neurons: Retroviral transfer of v-*myc* stimulates neuronal proliferation whereas v-*src* transfer enhances neuronal differentiation. *J. Cell Biol.* **10**, 2087–2098.

Halvorsen, S. W., Schmid, H. A., McEachern, A. E., and Berg, D. K. (1991). Regulation of acetylcholine receptors on chick ciliary ganglion neurons by components from the synaptic target tissue. *J. Neurosci.* **11**, 2177–2186.

Hamill, R. W. and Schroeder, B. (1990). Hormonal regulation fo adult sympathetic neurons: the effects of castration on neuropeptide Y, norepinephrine, and tyrosine hydroxylase activity. *J. Neurobiol.* **21**, 731–742.

Hart, R. P., Shadiack, A. M., and Jonakait, G. M. (1991). Substance P gene expression is regulated by interleukin-1 in cultured sympathetic ganglia. *J. Neurosci. Res.* **29**, 282–291.

Hawrot, E. (1980). Cultured sympathetic neurons: Effects of cell-derived and synthetic substrata on survival and development. *Dev. Biol.* **74**, 136–151.

Hayashi, M., Edgar, D., and Thoenen, H. (1985). Nerve growth factor changes the relative levels of neuropeptides in developing sensory and sympathetic ganglia of the chick embryo. *Dev. Biol.* **108**, 49–55.

Hendry, I. A. (1985). Phenotypic plasticity of sympathetic adrenergic and cholinergic neurones. *Trends Pharmacol. Sci.* **6**, 126–129.

Hendry, I. A., Hill, C. E., Belford, D., and Watters, D. J. (1988). A monoclonal antibody to a para-sympathetic neurotrophic factor causes immunoparasympathectomy in mice. *Brain Res.* **475**, 160–163.

Henion, P. D., and Landis, S. C. (1990). Asynchronous appearance and topographic segregation of neuropeptide-containing cells in the developing rat adrenal medulla. *J. Neurosci.* **10**, 2886–2896.

Herman, J. P., Abrous, N., Vigny, A., Dulluc, J., and LeMoal, M. (1988). Distorted development of intracerebral grafts: Long-term maintenance of tyrosine hydroxylase-containing neurons in grafts of cortical tissue. *Dev. Brain Res.* **40**, 81–88.

Hess, E. J., and Wilson, M. C. (1991). Tottering and leaner mutations perturb transient developmental expression of tyrosine hydroxylase in embryologically distinct Purkinje cells. *Neuron* **6**, 123–132.

Hill, C. E., and Hendry, I. A. (1979). The influence of preganglionic nerves on the superior cervical ganglion of the rat. *Neurosci. Lett.* **13**, 133–139.

Hilton, D. J., Nicola, N. A., and Metcalf, D. (1988). Specific binding of murine leukemia inhibitory factor to normal and leukemic monocytic cells. *Proc. Natl. Acad. Sci. U.S.A.* **85**, 5971–5975.

Hodgkin, J. (1991). Sex determination and the generation of sexually dimorphic nervous systems. *Neuron* **6**, 177–185.

Hofman, H. D. (1988). Ciliary neuronotrophic factor stimulates choline acetyltransferase activity in cultured chicken retina neurons. *J. Neurochem.* **51**, 109–113.

Howard, M. J., and Bronner-Fraser, M. (1985). The influence of neural tube-derived factors on

differentiation of neural crest cells in vitro. I. Histochemical study of the appearance of adrenergic cells. *J. Neurosci.* **5**, 3302–3309.
Howard, M. J., and Bronner-Fraser, M. (1986). Neural tube-derived factors influence differentiation of neural crest cells *in vitro:* Effects on activity of neurotransmitter biosynthetic enzymes. *Dev. Biol.* **117**, 45–54.
Iacovitti, L. (1991). Effects of a novel differentiation factor on the development of catecholamine traits in noncatecholamine neurons from various regions of the rat brain: studies in tissue culture. *J. Neurosci.* **11**, 2403–2409.
Iacovitti, L., Lee, J., Joh, T. H., and Reis, D. J. (1987). Expression of tyrosine hydroxylase in neurons of cultured cerebral cortex: Evidence for phenotypic plasticity in neurons of the CNS. *J. Neurosci.* **7**, 1264–1270.
Iacovitti, L, Evinger, M. J., Joh, T. H., and Reis, D. J. (1989). A muscle-derived factor(s) induces expression of a catecholamine phenotype in neurons of cultured rat cerebral cortex. *J. Neurosci.* **9**, 3529–3537.
Imamura, T., Engleka, K., Zhan, X., Tokita, Y., Forough, R., Roeder, D., Jackson, A., Maier, J. A. M., Hla, T., and Maciag, T. (1990). Recovery of mitogenic activity of a growth factor mutant with a nuclear translocation sequence. *Science* **249**, 1567–1570.
Ishii, K., Oda, Y., Ichikawa, T., and Deguchi, T. (1990). Complementary DNAs for choline acetyltransferase from spinal cords of rat and mouse: Nucleotide sequences, expression in mammalian cells and *in situ* hybridization. *Mol. Brain Res.* **7**, 151–159.
Jan, Y. N., and Jan, L. Y. (1990). Genes required for specifying cell fates in *Drosophila* embryonic sensory nervous system. *Trends Neurosci.* **13**, 493–498.
Jansen, J. K. S., and Fladby, T. (1990). The perinatal reorganization of the innervation of skeletal muscle in mammals. *Prog. Neurobiol.* **34**, 39–90.
Jellies, J., Loer, C. M., and Kristan, W. B., Jr. (1987). Morphological changes in leech Retzius neurons after target contact during embryogenesis. *J. Neurosci.* **7**, 2618–2629.
Jerden, J. A., Varner, H. H., Greenberg, J. H., Horn, V. J., and Martin, G. R. (1985). Isolation and characterization of a factor from calf serum that promotes the pigmentation of embryonic and transformed melanocytes. *J. Cell Biol.* **100**, 1493–1498.
Jessen, K. R., and Mirsky, R. (1991). Schwann cell precursors and their development. *Glia* **4**, 185–194.
Johnson, J. E., Birren, S. J., and Anderson, D. J. (1990). Two rat homologues of *Drosophila achaete-scute* specifically expressed in neuronal precursors. *Nature (London)* **346**, 858–861.
Johnson, M. I., Ross, D., Myers, M., Rees, R., Bunge, R. P., Wakshull, E., and Burton, H. (1976). Synaptic vesicle cytochemistry changes when cultured sympathetic neurons develop cholinergic interactions. *Nature (London)* **262**, 308–310.
Johnson, M., Ross, D., and Bunge, R. P. (1980). Morphological and biochemical studies on the development of cholinergic properties in cultured sympathetic neurons. II. Dependence on postnatal age. *J. Cell Biol.* **84**, 692–704.
Johnson, M. I., Higgins, D., and Ard, M. D. (1989). Astrocytes induce dendritic development in cultured sympathetic neurons. *Dev. Brain Res.* **47**, 289–292.
Jonakait, G. M., and Hart, R. P. (1990). Interleukin-1 (IL-1) stimulates preprotachykinin gene expression in cultured sympathetic ganglia. *Soc. Neurosci. Abstr.* **16**, 499.10.
Jonakait, G. M., and Schotland, S. (1990). Conditioned medium from activated splenocytes increases substance P in sympathetic ganglia. *J. Neurosci. Res.* **26**, 24–30.
Jonakait, G. M., Bohn, M. C., Markey, K., Goldstein, M., and Black, I. B. (1981). Elevation of maternal glucocorticoid hormones alters neurotransmitter phenotypic expression in embryos. *Dev. Biol.* **88**, 288–296.
Kaufman, L. M., Barry, S. R., and Barrett, J. N. (1985). Characterization of tissue-derived macromolecules affecting transmitter synthesis in rat spinal cord neurons. *J. Neurosci.* **5**, 160–166.
Kelley, D. B. (1986). Neuroeffectors for vocalization in *Xenopus laevis:* Hormonal regulation of sexual dimorphism. *J. Neurobiol.* **17**, 231–248.
Kessler, J. A. (1984). Non-neuronal cell conditioned medium stimulates peptidergic expression in sympathetic and sensory neurons *in vitro*. *Dev. Biol.* **106**, 61–69.

Kessler, J. A., Conn, G., and Hatcher, V. B. (1986). Isolated plasma membranes regulate neurotransmitter expression and facilitate effects of a soluble brain cholinergic factor. *Proc. Natl. Acad. Sci. U.S.A.* **83,** 8726–8729.

Konishi, M. (1989). Birdsong for neurobiologists. *Neuron* **3,** 541–549.

Landis, S. C. (1980). Developmental changes in the neurotransmitter properties of dissociated sympathetic neurons: A cytochemical study of the effects of medium. *Dev. Biol.* **77,** 348–361.

Landis, S. C. (1990). Target regulation of neurotransmitter phenotype. *Trends Neurosci.* **13,** 344–350.

Landis, S. C., Siegel, R. E., and Schwab, M. (1988). Evidence for neurotransmitter plasticity *in vivo.* II. Immunocytochemical studies of rat sweat gland innervation during development. *Dev. Biol.* **126,** 129–140.

Lein, P. J., and Higgins, D. (1989). Laminin and a basement membrane extract have different effects on axonal and dendritic outgrowth from embryonic rat sympathetic neurons *in vitro. Dev. Biol.* **136,** 330–345.

Lillien, L. E., and Raff, M. C. (1990a). Analysis of the cell–cell interactions that control type-2 astrocyte development in vitro. *Neuron* **4,** 525–534.

Lillien, L. E., and Raff, M. C. (1990b). Differentiation signals in the CNS: Type-2 astrocyte development in vitro as a model system. *Neuron* **5,** 111–119.

Lin, L.-F. H., Mismer, D., Lile, J. D., Armes, L. G., Butler, E. T., Vannice, J. L., and Collins, F. (1989). Purification, cloning, and expression of ciliary neurotrophic factor (CNTF). *Science* **246,** 1023–1025.

Lindh, B., Lundberg, J., and Hokfelt, T. (1989). NPY-, galanin-, VIP/PHI-, CGRP- and substance P-immunoreactive neuronal subpopulations in cat autonomic and sensory ganglia and their projections. *Cell Tiss. Res.* **256,** 259–273.

Lindsay, R., Lockett, C., Sternberg, J., and Winter, J. (1989). Neuropeptide expression in cultures of adult sensory neurons—Modulation of substance-P and calcitonin gene-related peptide levels by nerve growth-factor. *Neuroscience* **33,** 53–65.

Loer, C. M., and Kristan, W. B., Jr. (1989). Central synaptic inputs to identified leech neurons determined by peripheral targets. *Science* **244,** 64–66.

Loer, C. M., Jellies, J., and Kristan, W. B., Jr. (1987). Segment-specific morphogenesis of leech Retzius neurons requires particular peripheral targets. *J. Neurosci.* **7,** 2630–2638.

Lombard-Golly, D., Wong, V., and Kessler, J. A. (1990). Regulation of cholinergic expression in cultured spinal cord neurons. *Dev. Biol.* **139,** 396–406.

Lord, K. A., Abdollah, A., Thomas, S. M., Demarco, M., Brugge, J. S., Hoffman-Liebermann, B., and Liebermann, D. A. (1991). *Mol. Cell Biol.* **11,** 4371–4379.

Lorenzi, M. V., Hefti, F., Knusel, B., and Strauss, W. L. (1990). A developmental change in choline acetyltransferase gene transcription: a potential role for nerve growth factor? *Soc. Neurosci. Abstr.* **16,** 70.12.

Loring, J., Glimelius, B., and Weston, J. A. (1982). Extracellular matrix materials influence quail neural crest cell differentiation *in vitro. Dev. Biol.* **90,** 165–174.

Luine, V. N. (1985). Estradiol increases choline acetyltransferase activity in specific basal forebrain nuclei and projection areas of female rats. *Exper. Neurol.* **89,** 484–490.

Luine, V. N., Park, D., Joh, T., Reis, D., and McEwen, B. (1980). Immunochemical demonstration of increased choline acetyltransferase concentration in rat preoptic area after estradiol administration. *Brain Res.* **191,** 273–279.

Mackey, H. M., Payette, R. F., and Gershon, M. D. (1988). Tissue effects on the expression of serotonin, tyrosine hydroxylase and GABA in cultures of neurogenic cells from the neuraxis and branchial arches. *Development* **104,** 205–217.

MacLean, D. B., Bennett, B., Morris, M., and Wheeler, F. B. (1989). Differential regulation of calcitonin gene-related peptide and substance P in cultured neonatal rat vagal sensory neurons. *Brain Res.* **478,** 349–355.

MacLean, M. R., Raizada, M. K., and Sumners, C. (1990). The influence of angiotensin II on catecholamine synthesis in neuronal cultures from rat brain. *Biochem. Biophys. Res. Commun.* **167,** 492–497.

McLennan, I. S., Hill, C. E., and Hendry, I. A. (1980). Glucocorticoids modulate transmitter choice in developing superior cervical ganglion. *Nature (London)* **283**, 206–207.

McMahon, S. B., and Gibson, S. (1987). Peptide expression is altered when afferent nerves reinnervate inappropriate tissue. *Neurosci. Lett.* **73**, 9–15.

McManaman, J., Crawford, F., Stewart, S. S., and Appel, S. H. (1988). Purification of a skeletal muscle polypeptide which stimulates choline acetyltransferase activity in cultured spinal cord neurons. *J. Biol. chem.* **263**, 5890–5897.

McManaman, J., Crawford, F., Clark, R., Richker, J., and Fuller, F. (1989). Multiple neurotrophic factors from skeletal muscle: demonstration of effects of basic fibroblast growth factor and comparisons with the 22-kilodalton choline acetyltransferase development factor. *J. Neurochem.* **53**, 1763–1771.

McManaman, J., Oppenheim, R. W., Prevette, D., and Marchetti, D. (1990). Rescue of motoneurons from cell death by a pruified skeletal muscle polypeptide: Effects of the ChAT development factor, CDF. *Neuron* **4**, 891–898.

Mangoura, D., Sakellaridis, N., and Vernadakis, A. (1990). Evidence for plasticity in neurotransmitter expression in neuronal cultures derived from 3-day-old chick embryo. *Dev. Brain Res.* **51**, 93–101.

Manthorpe, M., Skaper, S., Williams, L. R., and Varon, S. (1986). Purification of adult rat sciatic nerve ciliary neuronotrophic factor. *Brain Res.* **367**, 282–286.

Marek, K. L., and Mains, R. E. (1989). Biosynthesis, development, and regulation of neuropeptide Y in superior cervical ganglion culture. *J. Neurochem.* **52**, 1807–1816.

Martin, F. H., Suggs, S. V., Langley, K. E., Lu, H. S., Ting, J., Okino, K. H., Morris, C. F., McNiece, I. K., Jacobsen, F. W., Mendiaz, E. A., Birkett, N. C., Smith, K. A., Johnson, M. J., Parker, V. P., Flores, J. C., Patel, A. C., Fisher, E. F., Erjavec, H. O., Herrera, C. J., Wypych, J., Sachdev. R. K., Pope, J. A., Leslie, J., Wen, D., Lin, C.-H., Cupples, R. L., and Zsebo, K. M. (1990). Primary structure and functional expression of rat and human stem cell factor DNAs. *Cell* **63**, 203–211.

Martinou, J. C., Le Van Thai, A., Cassar, G., Roubinet, F., and Weber, M. J. (1989). Characterization of two factors enhancing choline acetyltransferase activity in cultures of purified rat motor-neurons. *J. Neurosci.* **9**, 3645–3656.

Martinou, J. C., Martinou, I., Patterson, P. H., and Kato, A. C. (1991). Cholinergic differentiation factor (CDF/LIF) promotes choline acetyltransferase activity in rat spinal motoneurons. *3rd IBRO World Congress of Neuroscience* p. 4, 45.

Martinou, J. C., Martinou, I., and Kato, A. C. (1992). Cholinergic differentiation factor (CDF/LIF) promotes survival of isolated rat embryonic motoneurons in vitro. *Neuron* **8**, 737–744.

Mathieu, C., Moisand, A., and Weber, M. J. (1984). Acetylcholine metabolism by cultured neurons from rat nodose ganglia: Regulation by a macromolecule from muscle-conditioned medium. *Neuroscience* **13**, 1373–1386.

Matsui, Y., Zsebo, K. M., and Hogan, B. L. M. (1990). Embryonic expression of a haematopoietic growth factor encoded by the *Sl* locus and the ligand for c-*kit*. *Nature (London)* **347**, 667–669.

Maxwell, G. D., and Forbes, M. E. (1987). Exogenous basement-membrane-like matrix stimulates adrenergic development in avian neural crest cultures. *Development* **101**, 767–776.

Maxwell, G. D., and Forbes, M. E. (1990). The phenotypic response of cultured quail trunk neural crest cells to a reconstituted basement membrane-like matrix is specific. *Dev. Biol.* **141**, 233–237.

Melvin, J. E., and Hamill, R. W. (1989). Hypogastric ganglion perinatal development: evidence for androgen specificity via androgen receptors. *Brain Res.* **485**, 11–19.

Metcalf, D. (1987). The molecular control of normal and leukemic granulocytes and macrophages. *Proc. R. Soc. London B* **230**, 389–423.

Metcalf, D. (1989). The molecular control of cell division, differentiation commitment and maturation in haemopoietic cells. *Nature (London)* **339**, 27–30.

Miller, R. H., and Szigeti, V. (1991). Clonal analysis of astrocyte diversity in neonatal rat spinal cord cultures. *Development* **113**, 353–362.

Minami, M., Kurashi, Y., and Satoh, M. (1991). Effects of kainic acid on messenger RNA levels of IL-1b, IL-6, TNFa and LIF in the rat brain. *Biochem. Biophys. Res. Comm.* **176,** 593–598.

Mizuno, N., Matsuoka, I., and Kurihara, K. (1989). Possible involvements of intracellular $Ca^{2+}$ and $Ca^{2+}$ dependent protein phosphorylation in cholinergic differentiation of clonal rat pheochromocytoma cells (PC12) induced by glioma-conditioned medium and retinoic acid. *Dev. Brain Res.* **50,** 1–10.

Mlodzik, M., Hiromi, Y., Weber, U., Goodman, C. S., and Rubin, G. M. (1990). The *Drosophila seven-up* gene, a member of the steroid receptor gene superfamily, controls photoreceptor cell fates. *Cell* **60,** 211–224.

Morris, J. L., and Gibbons, I. L. (1989). Co-localization and plasticity of transmitters in peripheral autonomic and sensory neurons. *Int. J. Dev. Neurosci.* **7,** 521–531.

Morrison-Graham, K., West-Johnsrud, L., and Weston, J. A. (1990). Extracellular matrix from normal but not *Steel* mutant mice enhances melanogensis in cultured mouse neural crest cells. *Dev. Biol.* **139,** 299–307.

Mudge, A. W. (1981). Effect of the chemical environmenton levels of substance P and somatostatin in cultured neurones. *Nature (London)* **292,** 764–767.

Mudge, A. W. (1984). Schwann cells induce morphological transformation of sensory neurones in vitro. *Nature (London)* **309,** 367–369.

Murphy, M., Reid, K., Hilton, D. J., and Bartlett, P. F. (1991). Generation of sensory neurons is stimulated by leukemia inhibitory factor. *Proc. Natl. Acad. Sci. U.S.A.* **88,** 3498–3501.

Nawa, H., and Patterson, P. H. (1990). Separation and partial characterization of neuropeptide-inducing factors in heart cell conditioned medium. *Neuron* **4,** 269–277.

Nawa, H., and Sah, D. (1990). Different biological activities in conditioned media control the expression of a variety of neuropeptides in cultured sympathetic neurons. *Neuron* **4,** 279–287.

Nawa, H., Yamamori, T., Le, T., and Patterson, P. H. (1991a). The generation of neuronal diversity: Analogies and homologies with hematopoiesis. *Cold Spring Harbor Symp. Quant. Biol.* **55,** 247–253.

Nawa, H., Nakanishi, S., and Patterson, P. H. (1991b). Recombinant cholinergic differentiation factor (LIF) regulates sympathetic neuron phenotype by alterations in the size and amounts of neuropeptide mRNAs. *J. Neurochem.* **56,** 2147–2150.

Negishi, K., Teranishi, T., and Kato, S. (1982). New dopaminergic and indolamine-accumulating cells in the growth zone of goldfish retinas after neurotoxic destruction. *Science* **216,** 747–749.

Nishi, R., and Willard, A. L. (1988). Conditioned medium alters electrophysiological and transmitter-related properties expressed by rat enteric neurons in cell culture. *Neuroscience* **25,** 759–769.

Nocka, K., Majumder, S., Chabot, B., Ray, P., Cervone, M., Bernstein, A. and Besmer, P. (1989). Expression of c-*kit* gene products in known cellular targets of W mutations in normal and W mutant mice—evidence for an impaired c-*kit* kinase in mutant mice. *Genes Dev.* **3,** 816–826.

O'Lague, P. H., Obata, K., Claude, P., Furshpan, E. J., and Potter, D. D. (1974). Evidence for cholinergic synapses between dissociated rat sympathetic neurons in cell culture. *Proc. Natl. Acad. Sci. U.S.A.* **71,** 3602–3606.

Oppenheim, R. W., Prevette, D., and Fuller, F. H. (1990). In vivo treatment with basic fibroblast growth factor does not alter naturally occurring neuronal death. *Soc. Neurosci. Abstr.* **16,** 467.4.

Oppenheim, R. W., Prevette, D., and Qin-Wei, Y., Collins, F., and MacDonald, J. (1991). Control of embryonic motoneuron survival in vivo by ciliary neurotrophic factor *Science* **251,** 1616–1618.

Orr-Urtreger, A., Avivi, A., Zimmer, Y., Givol, D., Yarden, Y., and Lonai, P. (1990). Developmental expression of c-*kit*, a proto-oncogene encoded by the *W* locus. *Development* **109,** 911–923.

Park, J. K., Joh, T. H., and Ebner, F. F. (1986). Tyrosine hydroxylase is expressed by neocortical neurons after transplantation. *Proc. Natl. Acad. Sci. U.S.A.* **83,** 7495–7498.

Patterson, P. H. (1978). Environmental determination of autonomic neurotransmitter functions. *Ann. Rev. Neurosci.* **1,** 1–17.

Patterson, P. H. (1990). Control of cell fate in a vertebrate neurogenic lineage. *Cell* **62,** 1035–1038.

Patterson, P. H. (1992). The emerging neuropoietic cytokine family: first CDF/LIF, CNTF and IL-6, next ONC, MGF, GCSF? *Curr. Opinion Neurobiol.* **2,** 94–97.

Patterson, P. H., and Fann, M.-J. (1992). Further studies of the distribution of CDF/LIF mRNA. *Ciba Foundation Symposium* **167**, 125–140.

Patterson, P. H., and Chun, L. L. (1974). The influence of non-neuronal cells on catecholamine and acetylcholine synthesis and accumulation in cultures of dissociated sympathetic neurons. *Proc. Natl. Acad. Sci. U.S.A.* **71**, 3607–3610.

Patterson, P. H., and Chun, L. L. Y. (1977a). The induction of acetylcholine synthesis in primary cultures of dissociated rat sympathetic neurons. I. Effects of conditioned medium. *Dev. Biol.* **56**, 263–280.

Patterson, P. H., and Chun, L. L. Y. (1977b). The induction of acetylcholine synthesis in primary cultures of dissociated rat sympathetic neurons. II. Developmental aspects. *Dev. Biol.* **60**, 473–481.

Patterson, P. H., and Landis, S. C. (1992). Phenotype specifying factors and the control of neuronal differentiation decision. *In* "Development, Regeneration and Plasticity of the Autonomic Nervous System" (I. A. Hendry and C. E. Hill, eds.) Harwood Acad. Publ. Chur. Science, in press.

Patterson, P. H., Reichardt, L. F., and Chun, L. L. Y. (1975). Biochemical studies on the development of primary sympathetic neurons in cell culture. *Cold Spring Harbor Symp. Quant. Biol.* **40**, 389–397.

Perris, R., and Lofberg, J. (1986). Promotion of chromatophore differentiation in isolated premigratory neural crest cells by extracellular material explanted on microcarriers. *Dev. Biol.* **113**, 327–341.

Perris, R., von Boxberg, Y., and Lofberg, J. (1988). Local embryonic matrices determine region-specific phenotypes in neural crest cells. *Science* **241**, 86–89.

Phillips, C. E., and Freschi, J. E. (1986). Cytochemical study of developing neurotransmitter properties of dissociated sympathetic neurons grown in co-culture with dissociated pineal cells. *Neuroscience* **17**, 1139–1146.

Pintar, J. E., Breakefield, X. O., and Patterson, P. H. (1987). Differences in monoamine oxidase activity between cultured noradrenergic and cholinergic sympathetic neurons. *Dev. Biol.* **120**, 305–308.

Potter, D. D., Furshpan, E. J., and Landis, S. C. (1983). Transmitter status in cultured rat sympathetic neurons: plasticity and multiple function. *Fed. Proc.* **42**, 1626–1632.

Potter, D. D., Landis, S. C., Matsumoto, S. G., and Furshpan, E. J. (1986). Synaptic functions in rat sympathetic neurons in microcultures. II. Adrenergic/cholinergic dual status and plasticity. *J. Neurosci.* **6**, 1080–1090.

Prochiantz, A., Dagnet, M.-C., Herbert, A., and Glowinski, J. (1981). Specific stimulation of in vitro maturation of mesencephalic dopaminergic neurones by striatal membranes. *Nature (London)* **293**, 570–572.

Purves, D. (1988). "Body and Brain. A Trophic Theory of Neural Connections." Harvard University Press, Cambridge, Massachusetts.

Rao, M., and Landis, S. C. (1990). Characterization of a cholinergic neuronal differentiation factor in sweat gland extracts. *Neuron* **5**, 899–910.

Rao, M., Landis, S. C., and Patterson, P. H. (1990). The cholinergic neuronal differentiation factor from heart cell conditioned medium is different from the cholinergic factors in sciatic nerve and spinal cord. *Dev. Biol.* **139**, 65–74.

Rao, M., Landis, S. C., and Patterson, P. H. (1992). Comparison of the sweat gland neuronal differentiation factor to known cholinergic factors. *Development* (in press).

Rao, M., Patterson, P. H., and Landis, S. C. (1992). Membrane-associated neurotransmitter stimulating factor (MANS) is very similar to ciliary neurotrophic factor (CNTF). *Dev. Biol,* in press.

Rao, M. S., Tyrrell, S., Landis, S. C., and Patterson, P. H. (1992). Effects of ciliary neurotrophic factor (CNTF) and depolarization on neuropeptide expression in cultured sympathetic neurons. *Devel. Biol.,* **150**, 281–293.

Rathjen, P. D., Toth, S., Willis, A., Heath, J. K., and Smith, A. G. (1990). Differentiation inhibiting activity is produced in matrix-associated and diffusible forms that are generated by alternate promoter usage. *Cell* **62**, 1105–1114.

Raynaud, B., Clarous, D., Vidal, S., Ferrand, C., and Weber, M. J. (1987). Comparison of the effects of elevated K$^+$ ions and muscle-conditioned medium on the neurotransmitter phenotype of cultured sympathetic neurons. *Dev. Biol.* **121,** 548–558.

Ready, D. F. (1989). A multifaceted approach to neural development. *Trends Neurosci.* **12,** 102–110.

Reh, T. A. (1987). Cell-specific regulation of neuronal production in the larval frog retina. *J. Neurosci.* **7,** 3317–3324.

Reh, T. A., and Tully, T. (1986). Regulation of tyrosine hydroxylase-containing amacrine cell number in larval frog retina. *Dev. Biol.* **114,** 463–469.

Reichardt, L. F., and Patterson, P. H. (1977). Neurotransmitter synthesis and uptake by isolated rat sympathetic neurons developing in microcultures. *Nature (London)* **270,** 147–151.

Reisert, I., and Pilgrim, C. (1991). Sexual differentiation of monoamine neurons—genetic or epigenetic? *Trends Neurosci.* **14,** 468–473.

Rende, M., Hagg, T., Magal, E., Varon, S., and Manthorpe, M. (1990). Ciliary neuronotrophic factor (CNTF) immunoreactivity in the rat sciatic nerve. *Soc. Neurosci. Abstr.* **16,** 209.8.

Rogers, S. L., Bernhard, L., and Weston, J. A. (1990). Substratum effects on cell dispersal, morphology, and differentiation in cultures of avian neural crest cells. *Dev. Biol.* **141,** 173–182.

Rohrer, H. (1990). The role of growth factors in the control of neurogenesis. *Eur. J. Neurosci.* **2,** 1005–1015.

Rohrer, H. (1992). Cholinergic neuronal differentiation factors: evidence for the presence of both CNTF-like and non-CNTF-like factors in developing rat footpad. *Development* **114,** 689–698.

Rossant, J. (1990). Manipulating the mouse genome: Implications for neurobiology. *Neuron* **4,** 323–334.

Rousselet, A., Fetler, L., Chamak, B., and Prochiantz, A. (1988). Rat mesencephalic neurons in culture exhibit different morphological traits in the presence of media conditioned on mesencephalic or striatal astroglia. *Dev. Biol.* **129,** 495–504.

Rousselet, A., Autillo-Touati, A., Araud D., and Prochiantz, A. (1990). In vitro regulation of neuronal morphogenesis and polarity by astrocyte-derived factors. *Dev. Biol.* **137,** 33–45.

Rudge, J. S., Alderson, R. F., Ip, N., and Lindsay, R. M. (1990). Characterization of ciliary neurotrophic factor in rat astrocytes. *Soc. Neurosci. Abstr.* **16,** 209.7.

Saadat, S., and Thoenen, H. (1986). Selective induction of tyrosine hydroxylase by cell–cell contact in bovine adrenal chromaffin cells is mimicked by plasma membranes. *J. Cell Biol.* **103,** 1991–1997.

Saadat, S., Stechle, A., Lumouroux, A., Mallet, J., and Thoenen, H. (1987). Influence of cell–cell contact on levels of tyrosine hydroxylase in cultured bovine adrenal chromaffin cells. *J. Biol. Chem.* **262,** 13007–13014.

Saadat, S., Sendtner, M., and Rohrer, H. (1989). Ciliary neurotrophic factor induces cholinergic differentiation of rat sympathetic neurons in culture. *J. Cell Biol.* **108,** 1807–1816.

Sah, D. W. Y., and Matsumoto, S. G. (1987). Evidence for serotonin synthesis, uptake, and release in dissociated rat sympathetic neurons in culture *J. Neurosci.* **7,** 391–399.

Sano, H., Forough, R., Maier, J. A. M., Case, J. P., Jackson, A., Engleka, K., Maciag, T., and Wilder, R. L. (1990). Detection of high levels of heparin binding growth factor-1 (acidic fibroblast growth factor) in inflammatory arthritic joints. *J. Cell Biol.* **110,** 1417–1426.

Salvaterra, P. M., Bournias-Vardiabasis, N., Nair, T., Hou, G., and Lieu, C. (1987). In vitro neuronal differentiation of *Drosophila* embryo cells. *J. Neurosci.* **7,** 10–22.

Saunders, J. W., Jr., Quevedo, W. C., Pierro, L., and Morbeck, F. E. (1955). The effects of tyrosine and phenylalanine on the synthesis of pigment in melanocytes of embryonic chick skin cultured *in vitro*. *J. Natl. Cancer Inst.* **16,** 475–487.

Scherman, D., and Weber, M. J. (1987). Characterization of the vesicular monoamine transporter in cultured rat sympathetic neurons: Persistence upon induction of cholinergic phenotypic traits. *Dev. Biol.* **119,** 68–74.

Schoenen, J., Delree, P., Leprince, P., and Moonen, G. (1989). Neurotransmitter phenotype plasticity in cultured dissociated adult rat dorsal root ganglia: An immunocytochemical study. *J. Neurosci. Res.* **22,** 473–487.

Schotzinger, R. J., and Landis, S. C. (1988). Cholinergic phenotype developed by noradrenergic sympathetic neurons after innervation of a novel target in vivo. *Nature (London)* **335,** 637–639.

Schotzinger, R. J., and Landis, S. C. (1990). Acquisition of cholinergic and peptidergic properties by

sympathetic innervation of rat sweat glands requires interaction with normal target. *Neuron* **5,** 91–100.
Sendtner, M. (1989). Elucidation of the primary structure of ciliary neurotrophic factor. *Soc. Neurosci. Abstr.* **15,** 287.14.
Sendtner, M., Kreutzberg, G. W., and Thoenen, H. (1990). Ciliary neurotrophic factor prevents the degeneration of motor neurons after axotomy. *Nature (London)* **345,** 440–441.
Shatz, C. J. (1990). Impulse activity and the patterning of connections during CNS development. *Neuron* **5,** 745–756.
Shaw, T. J., and Letourneau, P. C. (1986). Chromaffin cell heterogeneity of process formation and neuropeptide content under control and nerve growth factor-altered conditions in cultures of chick embryonic adrenal gland. *J. Neurosci. Res.* **6,** 337–355.
Sieber-Blum, M. (1989). Inhibition of the adrenergic phenotype in cultured neural crest cells by norepinephrine uptake inhibitors. *Dev. Biol.* **136,** 372–380.
Sieber-Blum, M., Sieber, F., and Yamada, K. M. (1981). Cellular fibronectin promotes adrenergic differentiation of quail neural crest cells in vitro. *Exp. Cell Res.* **133,** 285–295.
Simerly, R. B. (1990). Hormonal control of neuropeptide gene expression in sexually dimorphic olfactory pathways. *Trends Neurosci.* **13,** 104–110.
Simerly, R. B., Young, B. J., Capozza, M. A., and Swanson, L. W. (1989). Estrogen differentially regulates neuropeptide gene expression in a sexually dimorphic olfactory pathway. *Proc. Natl. Acad. Sci. U.S.A.* **86,** 4766–4770.
Smith, K. E., and Kessler, J. A. (1990). Non-neuronal cell-conditioned medium regulates muscurinic receptor expression in cultured sympathetic neurons. *J. Neurosci.* **8,** 2406–2413.
Soinila, S., Ahonen, M., Lahtinen, T., and Happola, O. (1989). Developmental changes in 5-hydroxytryptamine immunoreactivity of sympathetic cells. *Int. J. Dev. Neurosci.* **7,** 553–563.
Spiegel, K., Wong, V., and Kessler, J. A. (1990). Translational regulation of somatostatin in cultured sympathetic neurons. *Neuron* **4,** 303–311.
Squinto, S. P., Aldrich, T. H., Lindsay, R. M., Morrissey, D. M., Panayotatos, N., Bianco, S. M., Furth, M. E., and Yancopoulos, G. D. (1990). Identification of functional receptors for ciliary neurotrophic factor on neuronal cell lines and primary neurons. *Neuron* **5,** 757–766.
Stahl, J., Gearing, D. P., Willson, T. A., Brown, M. A., King, J. A., and Gough, N. M. (1990). Structural organization of the genes for murine and human leukemia inhibitory factor. *J. Biol. Chem.* **265,** 8833–8841.
Stocker, K. M., Sherman, L., Rees, S., and Ciment, G. (1991). Basic FGF adn TGF-β1 influence commitment to melanogenesis in neural crest-derived cells of avian embryos. *Development* **111,** 635–645.
Stockli, K. A., Lottspeich, F., Sendtner, M., Masiakowski, P., Carroll, P., Lindholm, D., and Thoenen, H. (1989). Molecular cloning, expression and regional distribution of rat ciliary neurotrophic factor. *Nature (London)* **342,** 920–923.
Stockli, K. A., Lillien, L. E., Naher-Noe, M., Breifeld, G., Hughes, R. A., Raff, M. C., Thoenen, H., and Sendtner, M. (1991). *J. Cell Biol.* **115,** 447–459.
Stockli, K. A., Lottspeich, F., Sendtner, M., Masiakowski, P., Carroll, P., Lindholm, D., and Thoenen, H. (1989). Molecular cloning, expression and regional distribution of rat ciliary neurotrophic factor. *Nature (London)* **342,** 920–923.
Suzue, T., Kaprielian, Z., and Patterson, P. H. (1990). A monoclonal antibody that defines rostrocaudal gradients in the mammalian nervous system. *Neuron* **5,** 421–431.
Swanson, L. W., and Simmons, D. M. (1989). Differential steroid hormone and neural influences on peptide mRNA levels in CRH cells of the paraventricular nucleus: A hybridization histochemical study in the rat.
Sweadner, K. J. (1981). Environmentally regulated expression of soluble extracellular proteins of sympathetic neurons. *J. Biol. Chem.* **256,** 4063–4070.
Swerts, J. P., Le Van Thai, A., Vigny, A., and Weber, M. J. (1983). Regulation of enzymes responsible for neurotransmitter synthesis and degradation in cultured rat sympathetic neurons. *Dev. Biol.* **100,** 1–11.
Teitelman, G., Joh, T. H., Grayson, L., Park, D. H., Reis, D. J., and Iacovitti, L. (1985). Cholinergic

neurons of the chick ciliary ganglia express adrenergic traits *in vivo* and *in vitro*. *J. Neurosci.* **5,** 29–39.
Thoenen, H., and Acheson, A. (1987). Activity-dependent regulation fo gene expression. *In* "The Neural and Molecular Bases of Learning" (J.-P. Changeux and M. Konishi, eds.), pp. 85–98. John Wiley & Sons, New York.
Thomas, K. R., and Capecchi, M. R. (1990). Targeted disruption of the murine *int*-1 proto-oncogene resulting in severe abnormalities in midbrain and cerebellar development. *Nature (London)* **346,** 847–850.
Thorn, R. S., and Truman, J. W. (1989). Sex-specific neuronal respecification during the metamorphosis of the genital segments of the tobacco hornworm moth *Manduca sexta*. *J. Comp. Neurol.* **284,** 489–503.
Trifaro, J.-M., Tang, R., and Novas, M. L. (1990). Monolayer co-culture of rat heart cells and bovine adrenal chromaffin paraneurons. *In Vitro Cell. Dev. Biol.* **26,** 335–347.
Tropea, M., Johnson, M. I., and Higgins, D. (1988). Glial cells promote dendritic development in rat sympathetic neurons in vitro. *Glia* **1,** 380–392.
Tucker, R. P., and Erickson, C. A. (1986). Pigment cell pattern formation in *Taricha torosa:* The role of the extracellular matrix in controlling pigment cell migration and differentiation. *Dev. Biol.* **118,** 268–285.
Vaca, K., Stewart, S. S., and Appel, S. H. (1989). Identification of basic fibroblast growth factor as a cholinergic growth factor from human muscle. *J. Neurosci. Res.* **23,** 55–63.
Vandenbergh, D. J., Mori, N., and Anderson, D. J. (1991). Co-expression of multiple neurotransmitter enzyme genes in normal and immortialized sympathoadrenal progenitor cells. *Devel. Biol.* **148,** 10–22.
Varon, S., Manthorpe, M., and Adler, R. (1979). Cholinergic neuronotrophic factors. I. Survival, neurite outgrowth and choline acetyltransferase activity in monolayer cultures from chick embryo ciliary ganglia. *Brain Res.* **173,** 29–45.
Varon, S., Hagg, T., Prospero-Garcia, O., Conner, J. M., and Manthorpe, M. (1990). Ciliary neuronotrophic factor (CNTF) immunoreactivity in rat CNS. *Soc. Neurosci. Abstr.* **16,** 342.5.
Vidal, S., Raynaud, B., Clarous, D., and Weber, M. J. (1987). Neurotransmitter plasticity of cultured sympathetic neurons. Are the effects of muscle-conditioned medium reversible? *Development* **101,** 617–625.
Vogel, K. S., and Weston, J. A. (1988). A subpopulation of cultured avian neural crest cells has transient neurogenic potential. *Neuron* **1,** 569–577.
Vogel, K. S., and Weston, J. A. (1990). The sympathoadrenal lineage in avian embryos. II. Effects of glucocorticoids on cultured neural crest cells. *Dev. Biol.* **139,** 1–12.
Wakade, A. R., and Bhave, S. V. (1988). Facilitation of noradrenergic character of sympathetic neurons by co-culturing with heart cells. *Brain Res.* **458,** 115–122.
Wakshull, E., Johnson, M. I., and Burton, H. (1978). Persistence of an amine uptake system in cultured sympathetic neurons which use acetylcholine as their transmitter. *J. Cell Biol.* **79,** 121–131.
Wakshull, E., Johnson, M. I., and Burton, H. (1979). Postnatal rat sympathetic neurons in culture. II. Synaptic transmission by postnatal neurons. *J. Neurophysiol.* **42,** 1426–1436.
Walicke, P. A., and Patterson, P. H. (1981a). On the role of cyclic nucleotides in the transmitter choice made by cultured sympathetic neurons. *J. Neurosci.* **1,** 333–342.
Walicke, P. A., and Patterson, P. H. (1981b). On the role of $Ca^{2+}$ in the transmitter choice made by cultured sympathetic neurons. *J. Neurosci.* **1,** 343–350.
Walicke, P. A., Campenot, R. B., and Patterson, P. H. (1977). Determination of transmitter function by neuronal activity. *Proc. Natl. Acad. Sci. U.S.A.* **74,** 5767–5771.
Wan, D. C.-C., and Livett, B. G. (1989). Vasoactive intestinal peptide stimulates proenkephalin A mRNA expression in bovine adrenal chromaffin cells. *Neurosci. Lett.* **101,** 218–222.
Weber, M. J., Raynaud, B., and Delteil, C. (19850. Molecular properties of a cholinergic differentiation factor from muscle conditioned medium. *J. Neurochem.* **45,** 1541–1547.
Whitaker-Azmitia, P. M., and Azmitia, E. C. (1989). Stimulation of astroglial serotonin receptors produces culture media which regulates growth of serotonergic neurons. *Brain Res.* **497,** 80–85.

Witte, O. N. (1990). Steel locus defines new multipotent growth factor. *Cell* **63**, 5–6.
Wong, V., and Kessler, J. A. (1987). Solubilization of a membrane factor that stimulates levels of substance P and choline acetyltransferase in sympathetic neurons. *Proc. Natl. Acad. Sci. U.S.A.* **84**, 8726–8729.
Wright, L. L., and Smolen, A. J. (1983). Neonatal testosterone treatment increases neuron and synapse numbers in male rat superior cervical ganglion. *Dev. Brain Res.* **8**, 145–153.
Xue, Z. G., Smith, J., and LeDouarin, N. M. (1985). Differentiation of catecholaminergic cells in cultures of embryonic avian sensory ganglia. *Proc. Natl. Acad. Sci. U.S.A.* **82**, 8800–8804.
Xue, Z. G., LeDouarin, N. M., and Smith, J. (1988). Insulin and insulin-like growth factor-I can trigger the differentiation of catecholaminergic precursors in cultures of dorsal root ganglia. *Cell Diff. Dev.* **25**, 1–10.
Yamamori, T. (1991). CDF/LIF selectively increases the levels of c-*fos* and *jun-B* transcripts in cultured sympathetic neurons. *NeuroReport* **2**, 173–176.
Yamamori, T. (1991). Localization of CDF/LIF mRNA in the rat brain and peripheral tissues. *Proc. Natl. Acad. Sci. U.S.A.* **88**, 7298–7302.
Yamamori, T., Fukada, K., Aebersold, R., Korsching, S., Fann, M.-J., and Patterson, P. H. (1989). The cholinergic neuronal differentiation factor from heart cells is identical to leukemia inhibitory factor. *Science* **246**, 1412–1414.
Zigmond, R. E., Schwarzschild, M. A., and Rittenhouse, A. R. (1989). Acute regulation of tyrosine hydroxylase by nerve activity and by neurotransmitters via phosphorylation. *Ann. Rev. Neurosci.* **12**, 415–462.
Ziller, C., Fauquet, M., Kalcheim, C., Smith, J., and LeDouarin, N. M. (1987). Cell lineages in peripheral nervous system ontogeny: medium-induced modulation of neuronal phenotypic expression in neural crest cell cultures. *Dev. Biol.* **120**, 101–111.
Zurn, A. D. (1982). Identification of glycolipid binding sites for soybean agglutinin and differences in the surface glycolipids of cultured adrenergic and cholinergic sympathetic neurons. *Dev. Biol.* **94**, 483–498.
Zurn, A. D. (1990). Differential increase in the cholinergic properties of cultured chick sympathetic neurons by forskolin. *Dev. Biol.* **140**, 53–56.
Zurn, A. D. (1991). Neurotransmitter plasticity in the sympathetic nervous system: influence of external factors and possible physiological implications. *Life Sciences* **48**, 1799–1808.
Zurn, A. D., and Do, K. Q. (1988). Purine metabolite inosine is an adrenergic neurotrophic substance for cultured chicken sympathetic neurons. *Proc. Natl. Acad. Sci. U.S.A.* **85**, 8301–8305.
Zsebo, K. M., Williams, D. A., Geissler, E. N., Broudy, V. C., Martin, F. H., Atkins, H. L., Hsu, R.-Y., Birkett, N. C., Okino, K. H., Murdock, D. C., Jacobsen, F. W., Langley, K. E., Smith, K. A., Takeishi, T., Cattanach, B. M., Galli, S. J., and Suggs, S. V. (1990a). Stem cell factor is encoded at the *Sl* locus of the mouse and is a ligand for the c-*kit* tyrosine kinase receptor. *Cell* **63**, 213–224.
Zsebo, K. M., Wypych, J., McNiece, I. K., Lu, H. S., Smith, K. A., Karkare, S. B., Sachdev, R. K., Yuschenkoff, V. N., Birkett, N. C., Williams, L. R., Satyagal, V. N., Tung, W., Bosselman, R. A., Mendiaz, E. A., and Langley, K. E. (1990b). Identification, purification, and biological characterization of hematopoietic stem cell factor from Buffalo rat liver-conditioned medium. *Cell* **63**, 195–201.

# 20 Neurotransmitters as Neurotrophic Factors

Frances M. Leslie

## I. INTRODUCTION

Abundant evidence suggests that cellular development and plasticity in brain and other tissues is influenced greatly by afferent neuronal activity (Black et al., 1984; Rakic, 1988; O'Leary, 1989; Constantine-Paton et al., 1990; Sur et al., 1990). Although characterized growth factors (GFs) undoubtedly are involved in these processes, a growing body of data has implicated neurotransmitters (NTs) as important trophic agents.

A number of observations suggest that NTs may play a broader role than merely rapid intercellular communication. First, the developmental appearance of many NTs in the central nervous system (CNS) precedes the onset of synaptogenesis (Lauder, 1987). Receptors for many NTs also are expressed at early developmental stages, prior to the onset of neurotransmission, with numerous reports of transient expression during critical periods of development (Cynader et al., 1990; Bodenant et al., 1991; Leslie and Loughlin, 1992). Many types of NT receptors have been found to be localized on glia, as well as on neurons, further indicating a more long-term trophic role for NTs in the CNS (Kimelberg, 1988; Stone and Ariano, 1989). Finally, pharmacologic manipulations of NT receptor populations, both in vivo and in vitro, have been found to influence developmental plasticity significantly (see subsequent text).

As will be discussed in this chapter, substantial evidence supports the idea that NTs modulate structural, as well as physiologic, plasticity. Morphogenic influences of NTs have been examined most extensively in vitro using cell culture techniques. In contrast, evidence for NT-induced functional plasticity has been obtained primarily in vivo. A particularly useful model for analysis of functional changes has been the developing visual system, which exhibits considerable experience-dependent plasticity during a "critical" postnatal period (Constantine-Paton et al., 1990).

Whereas NT-induced plasticity is particularly obvious in developing tissue, it also occurs to a more limited extent in the adult. Since the existing literature on trophic actions of NTs is too extensive for comprehensive review, it is necessary to confine this analysis to a discussion of selected examples. For further information, the reader is referred to other excellent articles on this topic (Lauder, 1987; Mattson, 1988,1989b; Lipton and Kater, 1989).

## II. AMINO ACID TRANSMITTERS

### A. Glutamate

L-Glutamate is widely distributed throughout mammalian brain, in both local and projection neurons, and is believed to be the principal excitatory NT of the CNS (Fonnum, 1984). It mediates its widespread effects via activation of both ionotropic and metabotropic receptors (Watkins et al., 1990). Ionotropic glutamate receptors are ligand-gated cation channels, consisting of a homologous family of kainate /α-amino-3-hydroxy-5-methyl-4-isoxazole propionic acid (AMPA) receptors that mediate fast synaptic transmission (Barnard and Henley, 1990; Boulter et al., 1990; Keinanen et al., 1990; Sommer et al., 1990) and a separate class of $N$-methyl-D-aspartate (NMDA) receptors that mediate a slower long-duration response (Collingridge et al., 1988; Forsythe and Westbrook, 1988). In contrast, the metabotropic receptor is G protein-coupled to stimulation of the inositol phosphate/intracellular $Ca^{2+}$ signaling pathway (Schoepp et al., 1990; Masu et al., 1991).

In the adult CNS, glutamate receptors have been implicated in numerous processes in addition to synaptic transmission. These include use-dependent long-term modifications in synaptic efficacy (Collingridge and Singer, 1990) and neurotoxicity (Choi, 1988). During development, excitatory amino acid (EAA) receptors are expressed transiently in a number of forebrain regions (Tremblay et al., 1989); in some instances, there is increased sensitivity to the agonist actions of glutamate (Hamon and Heinemann, 1988; Fox et al., 1991). As will be reviewed in subsequent text, there is now increasing evidence that glutamate may subserve a specific trophic role in developing CNS.

#### 1. Morphogenesis

*a. Growth stimulation* Unlike mature neurons, most immature neurons in culture are insensitive to receptor-mediated glutamate toxicity (Balazs et al., 1988; Brenneman et al., 1990a; Murphy et al., 1990). Instead, glutamate appears to exert trophic effects on various neuronal populations. The trophic actions of glutamate on cells in vitro are extremely complex and are dependent on a number of factors, including cell type, age, and the presence or absence of GFs or ongoing electrical activity. EAAs induce increases in both survival (Balazs and Hack, 1990; Brenneman et al., 1990a; Cohen-Cory et al., 1991) and dendritic outgrowth (Brewer and Cotman, 1989; Cambray-Deakin and Burgoyne, 1990; Cohen-Cory et al., 1991) of cultured neurons derived from hippocampus, cerebellum, and spinal cord. These effects are observed following treatment with low doses of glutamate agonists or following stimulus-evoked release of endogenous EAAs from cultured cells, and are mediated by receptors of both NMDA and non-NMDA types (Balazs and Hack, 1990). EAA-induced increases in neuronal survival appear to be calcium mediated, since they are inhibited by calcium channel blockers (Balazs and Hack, 1990) and mimicked by calcium ionophores (Brenneman et al., 1990a). NMDA-induced process outgrowth in cerebellar granule cells results from activation of protein kinase C (Cambray-Deakin and Burgoyne, 1990).

With increased time in culture, these neuronal populations become increasingly sensitive to the toxic effects of glutamate (Balazs et al., 1988; Brenneman et al., 1990a; Schramm et al., 1990). The reason for the switch from neurotrophism to neurotoxicity is presently unclear. It has been suggested that this change may result from an increase in glutamate receptor density (Peterson et al., 1989). Alternatively, it may reflect developmental changes in the binding or signal transduction properties of glutamate receptors (Cambray-Deakin et al., 1990). It is also possible that maturational changes in intracellular calcium buffering activity or in cellular sensitivity to intracellular calcium changes may occur in these cells.

*b. Growth inhibition* In some cell populations, negative growth effects of EAAs also may be of trophic significance. Hippocampal pyramidal cells in vitro do not exhibit EAA-induced growth stimulation at any phase of development (Mattson et al., 1988a). In contrast, high doses of glutamate induce pyramidal cell death, whereas lower doses produce a selective retraction of dendrites (Mattson et al., 1988a; Mattson and Kater, 1989b). These effects are mediated by both NMDA and non-NMDA receptors, and result from stimulation of cellular calcium influx. This EAA-induced dendritic pruning may be of physiologic relevance, since it has been observed to occur in co-cultures in which glutamate is released from entorhinal cortex afferents (Mattson et al., 1988b).

Glutamate also may serve a physiologic role in modulating morphologic changes that occur in horizontal cells of the vertebrate retina in response to changes in light. In teleost retina, axon-bearing horizontal cells form reciprocal connections with cones that contain glutamate as the NT (Dowling, 1987). During light adaptation, the terminal dendrites of the horizontal cells elaborate numerous new neurites whereas, during dark adaptation, the opposite process occurs (Weiler and Wagner, 1984). Glutamate and kainate stimulate neurite retraction in these cells, both in vivo and in vitro (Dos Santos Rodrigues and Dowling, 1990).

*c. Interactions with GFs* Interactions between glutamate and GFs are complex. In hippocampal pyramidal cell cultures, the outgrowth-inhibiting actions of glutamate are opposed by basic fibroblast growth factor (bFGF) (Mattson et al., 1989). This finding has led to speculation that equal and opposite GF–glutamate inter-

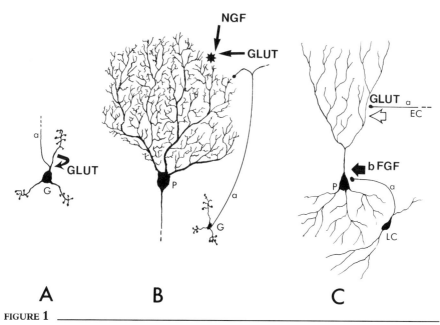

FIGURE 1

Trophic actions of glutamate on three types of neuronal cell in culture. (A) Autocrine effects on a cerebellar granule neuron (G). In this case, glutamate exerts a positive trophic influence in the absence of GFs. (B) Paracrine effects of glutamate released from a granule cell (G) on a cerebellar Purkinje neuron (P). In this case, synergy of glutamate and NGF actions promote neuronal survival and differentiation. (C) Paracrine effects of glutamate released from entorhinal cortex (EC) terminals on hippocampal pyramidal neurons (P). In this case, dendritic pruning effects of glutamate are opposed by bFGF, which is released from a local circuit (LC) neuron.

actions are required for normal maintenance of cell function (Fig. 1B), and that imbalance may result in neurodegenerative disease (Mattson, 1989b). Although this may be true for some cell types, the hypothesis is not generally applicable to all cells at all stages of development (Fig. 1). For example, in cerebellar Purkinje cells, in which growth-promoting actions of EAAs have been observed, there are synergistic interactions between glutamate and nerve growth factor (NGF) (Cohen-Cory et al., 1991; Fig. 1C).

### 2. Functional plasticity

Glutamate receptor activation has been implicated as a mediator of functional plasticity, both in the adult and in the developing CNS. These functional changes are characterized by alterations in synaptic efficacy, and may or may not involve structural changes (see Constantine-Paton et al., 1990, for discussion). In most instances, these phenomena have been described primarily at a systems level; the underlying cellular mechanisms remain controversial.

A theoretical framework for use-dependent plasticity has been provided by Hebb (1949), who proposed that concurrent presynaptic and postsynaptic activation is required for synaptic modification to occur. In this regard, considerable attention has focused on the NMDA receptor, which has properties ideally suited for signaling presynaptic activity levels to a postsynaptic cell (reviewed by Cotman and Monaghan, 1988). This receptor is a component of a combined ligand/voltage-gated $Ca^{2+}$ channel. While in the resting state, the channel ion pore is blocked by extracellular $Mg^{2+}$. Depolarization forces $Mg^{2+}$ out and permits intracellular entry of $Ca^{2+}$. Thus, both presynaptic release of transmitter and an adequate degree of postsynaptic depolarization are required for NMDA receptor-induced responses to occur. This mechanism could underlie the strengthening of active synapses in both the developing and the adult CNS.

Long-term potentiation (LTP) is an example of synaptic plasticity that is dependent on NMDA receptor function (Cotman and Monaghan, 1988; Collingridge and Singer, 1990). This phenomenon, which involves long-lasting enhancement of a postsynaptic response following high frequency presynaptic stimulation, is characteristic of many of the synapses of adult hippocampus and may be the mechanism that underlies higher cognitive function (Cotman and Monaghan, 1988). It has been observed also in other forebrain regions during critical phases of development (Tsumoto et al., 1990). NMDA receptor activation has been shown to be a necessary prerequisite for induction of LTP, although it is not the only factor involved (Collingridge and Singer, 1990).

The developing visual system has been used widely as a model for examining afferent-induced changes in synaptic plasticity (Constantine-Paton et al., 1990). One example of experience-dependent plasticity in this system is the change that occurs in the responsivity of kitten visual cortex neurons following monocular deprivation. Whereas in normal animals the majority of neurons is driven by binocular inputs, following monocular deprivation the majority of cells becomes responsive only to the open eye (Cynader et al., 1990; see Section II,C,2,a). This functional plasticity is blocked by cortical infusion of the NMDA receptor antagonist, 2-amino-5-phosphonovaleric acid (APV) (Bear et al., 1990). Since this drug treatment has been shown to alter the response properties of visual cortical neurons, there has been considerable controversy over the mechanism of APV-induced blockade of ocular dominance plasticity and over the physiologic significance of these findings (reviewed by Constantine-Paton et al., 1990). However, some findings do suggest that NMDA receptors may serve to modulate thalamocortical connectivity in developing brain. In normal visual cortex, a developmental loss of NMDA receptor function in layer IV, the thalamocortical recipient layer, parallels the loss of ocular dominance plasticity (Fox et al., 1989). Dark-rearing, which delays the loss of cortical plasticity, also delays this developmental loss of functional NMDA receptors (Fox et al., 1991).

NMDA receptor activation has been implicated in control of selective synapse stabilization in the tadpole optic tectum, another visual system model of synaptic plasticity (see Constantine-Paton et al., 1990; Cline, 1991, for review). Exposure of the optic tectum to NMDA receptor antagonists has been found to disrupt retinal topography in normal animals and to cause stripe desegregation in those animals implanted with a supernumerary eye. Conversely, NMDA treatment sharpens the stripe borders in three-eyed animals by reducing the growth of retinal ganglion cell terminals into those regions that are innervated primarily by the normal eye (Cline and Constantine-Paton, 1989).

## B. γ-Amino Butyric Acid

γ-Amino butyric acid (GABA) is present in high concentrations in mammalian brain (Seiler and Lajtha, 1987). It is localized both in fiber tracts and in interneurons, and acts as the principal inhibitory NT of the CNS. The physiologic actions of GABA are mediated through two classes of receptor, $GABA_A$ and $GABA_B$. $GABA_A$ receptors are integral channel proteins gating $Cl^-$ ion conductance, that consist of multimeric arrays of α and β subunits (Barnard, 1988). A third γ subunit, which confers benzodiazepine sensitivity, also has been characterized (Pritchett et al., 1989). The molecular identification of a large number of subunit variants implies that the pharmacology of the $GABA_A$ receptor is complex (Verdoorn et al., 1990), a conclusion that is supported by the findings of receptor binding studies (Sieghart and Karobeth, 1980; see below). $GABA_B$, in contrast, is a G protein-linked receptor that is negatively coupled to adenyl cyclase; its activation reduces $Ca^{2+}$-spike duration and increases $K^+$ conductance (Bowery, 1989). This receptor type is localized both pre- and postsynaptically.

In the embryonic rat CNS, the GABA system has been found to differentiate before that of most other NTs (Lauder et al., 1986). By embryonic day 13 (ED 13), a well-established fiber system is present in brainstem, mesencephalon, and diencephalon. Benzodiazepine binding sites exhibit a similar developmental timecourse in these regions (Schlumpf et al., 1983). GABA uptake and release mechanisms have been demonstrated in growth cones isolated from neonatal rat brain (Taylor et al., 1990). Prior to the onset of synaptogenesis and exocytotic release, GABA is released from these growth cones by a $Ca^{2+}$ independent mechanism. Such findings, coupled with the observation that GABA stimulates protein synthesis in adult rat brain (Campbell et al., 1966), suggest that GABA may serve a role as a diffuse regulator in the CNS, in addition to its NT function. As will be reviewed in subsequent text, considerable evidence supports the concept that GABA modulates both structural and functional plasticity of adult and developing nervous systems.

### 1. Morphogenesis

*a. In vivo studies* Trophic effects of GABA first were demonstrated by Wolff et al. (1978) in a study of the superior cervical ganglion (SCG) of adult rat. Chronic in vivo application of GABA to the SCG produces numerous morphologic and biochemical changes in ganglionic neurons, including increased density of cholinergic receptors and decreased density of the degradative enzyme acetylcholinesterase, as well as increased appearance of postsynaptic thickenings and of presynaptic-like vesicle aggregations in dendrites (Wolff et al., 1987). These changes appear to increase the receptivity of ganglionic cells to cholinergic innervation, since double innervation of the SCG occurs after hypoglossal nerve implantation in GABA-treated animals but not in untreated controls (Dames et al., 1985).

***b. In vitro studies*** In primary neuronal cultures, GABA induces significant effects on cell growth and differentiation via activation of $GABA_A$ receptors. Cultured cerebellar granule cells exhibit increased densities of protein synthetic apparatus and an increased rate of general maturation (Hansen et al., 1984; Meier et al., 1985; Meier and Jorgensen, 1986). In dissociated neuronal cultures derived from chick retina, cerebral cortex, and optic tectum, significant increases in neuronal number and degree of process outgrowth are observed, as is increased synaptogenesis (Michler-Stuke and Wolff, 1987; Spoerri, 1988). These trophic effects of GABA are dependent on culture conditions, however. In optic tectal cultures, GABA stimulates neuronal growth in serum-containing medium but has an opposite effect in the absence of serum (Michler-Stuke and Wolff, 1987). In cultured cerebellar granule cells, the stimulatory effect of GABA is restricted to early developmental time-points, and is not detectable at 14 days in vitro (Hansen et al., 1988).

In rat hippocampal cultures, GABA does not exert a direct stimulatory effect on neuronal growth and development (Mattson and Kater, 1989a). Indeed, a high concentration of GABA, in combination with diazepam, has been shown to inhibit hippocampal neurite outgrowth. In lower doses, however, GABA/diazepam combinations reduce EAA-induced hippocampal cell loss and dendritic retraction. Although this finding indicates significant NT interactions in developing cells in culture, GABA inhibition of EAA neurotoxicity is not a universal phenomenon. In rat cortical cultures, activation of $GABA_A$ receptors has been found to accelerate excitotoxic cell death (Erdo et al., 1991). Thus, GABA–EAA interactions are antagonistic in one cell type but synergistic in another.

**2. Functional plasticity**

***a. Receptor expression*** Cortical cells in culture, some of which produce endogenous GABA, exhibit both high and low affinity [$^3$H]GABA binding (Snodgrass et al., 1980). In contrast, cerebellar granule cells, which do not synthesize and release GABA, do not exhibit low affinity ($^3$H]GABA binding (reviewed in Meier et al., 1987). Exposure of these cells to GABA, or the specific $GABA_A$ agonists muscimol or 4,5,6,7-tetrahydroisoxazolo[5,4-C]pyridin-3-ol (THIP), however, rapidly induces the appearance of low affinity binding. This receptor expression appears to involve the synthesis of novel proteins, since it can be blocked by pretreatment with either actinomycin D or cycloheximide. As with GABA-induced changes in cerebellar neuron morphology, this increased expression of low affinity GABA binding sites is restricted to a critical developmental period, and is not observed after 7 days in vitro. Although the exact nature of these low affinity sites is not presently clear, they appear to modulate glutamate release. Only those cultures grown in the presence of GABA exhibit inhibition of $K^+$-evoked EAA release by $GABA_A$ receptor agonists.

In rabbit retina, GABA exposure during a critical period of development increases expression of a high affinity $GABA_A$ site (reviewed by Madtes, 1987). Both in vivo administration of the GABA-uptake blocker nipecotic acid and in vitro exposure to $GABA_A$ receptor agonists produce a rapid up-regulation of [$^3$H]muscimol binding sites, an effect that is blocked by the antagonist bicuculline. Unlike the induction of low affinity sites in cerebellum, this process does not involve de novo protein synthesis but appears to result from the "unmasking" of previously inaccessible sites. This effect is observed only in light-reared animals and is most apparent at postnatal days 9–12, a period that coincides with the time of eye opening. In older animals, classical receptor desensitization is observed.

Whereas cerebellar and retinal cells respond to GABA with increases in receptor binding, this phenomenon has not been observed in other regions of the brain. In the majority of studies of cultured cells, GABA-induced receptor down-regulation occurs (Maloteaux et al., 1987; Tehrani and Barnes, 1988; Hablitz et al., 1989; Roca et al., 1990), possibly as a result of decreased receptor mRNA expression (Montpied et al., 1991). Further, regionally specific decreases in low affinity $GABA_A$ receptor binding have been shown to result from in vivo perinatal exposure to diazepam (Gruen et al., 1990).

***b. Physiological plasticity*** In adult rat hippocampal CA1, the role of GABA in the induction of LTP has been examined (Davies et al., 1991). During low frequency transmission, significant activation of the NMDA receptor system is blocked by GABA-mediated synaptic inhibition. At high stimulus frequencies, however, GABA release is reduced by activation of presynaptic $GABA_B$ autoreceptors, permitting sufficient depolarization to remove the $Mg^{2+}$ blockade of NMDA-activated channels. Thus, $GABA_B$ autoreceptor activation is critical for this form of synaptic plasticity.

The role of GABA receptors in developmental functional plasticity has not yet been delineated as clearly. However, studies have suggested a role for this NT in ocular dominance plasticity (Ramoa et al., 1988; Reiter and Stryker, 1988). Infusion of the $GABA_A$-selective antagonist bicuculline into kitten visual cortex has been found to reduce the orientation selectivity of cortical neurons, and to reduce the ocular dominance shift following monocular deprivation (Ramoa et al., 1988). Suppression of evoked cortical activity by infusion of the $GABA_A$-selective agonist muscimol produces an unexpected effect: inputs from the less active eye become dominant (Reiter and Stryker, 1988).

## III. MONOAMINES

### A. Serotonin

Many serotonin cell bodies are localized in the brainstem raphe nuclei, sending extensive projections throughout the CNS via both ascending and descending fiber pathways (Steinbusch and Nieuwenhuys, 1983). This transmitter exhibits a complex pharmacology, interacting with heterogeneous G protein-coupled $S_1$ and $S_2$ receptors and with an $S_3$ receptor that is a member of the ligand-gated ion channel family (Frazer et al., 1990). Although serotonin is one of the principal NTs of the CNS (Bobker and Williams, 1990), increasing evidence suggests that it may serve an important function as a regulator of cellular growth and development.

Significant effects of serotonin on the growth of nonneuronal cells have been demonstrated. Acting synergistically with growth factors, serotonin stimulates the proliferation of bovine aortic smooth muscle cells and CCL39 hamster lung fibroblast cells (Nemecek et al., 1986; Seuwen et al., 1988). Although this effect appears to be mediated by $S_{1b/1d}$ receptors, which modulate adenyl cyclase activity (Seuwen and Pouyssegur, 1990), serotonin also may influence growth by activation of other second messenger pathways. In NIH 3T3 fibroblasts transfected with the $S_{1c}$ receptor gene, serotonin stimulation of phospholipase C results in malignant transformation of these cells (Julius et al., 1989).

During ontogeny, serotonin is detectable at early stages of development in several vertebrate and invertebrate species (Shuey et al., 1990). Its presence in sea urchin embryos at early cleavage division stages has led to its designation as a prenervous transmitter (Buznikov, 1984). A functional role of this chemical system in control of embryogenesis has been indicated by pharmacologic studies, in which manipulations with receptor agonists and antagonists have been shown to influence embryonic cleavage and cellular migration (Palen et al., 1979; Renaud et al., 1983; Buznikov, 1984; Zimmerman and Wee, 1984).

In rat brain, cells expressing serotonin immunoreactivity are detectable at ED 12 (Aitken and Tork, 1988; Konig et al., 1988), shortly after their birth (Lauder and Bloom, 1974). Ascending fiber bundles are apparent in the mesencephalon at ED 13, whereas descending fiber bundles are detectable at ED 14 (reviewed by Lauder, 1990). By ED 19, all major pathways are established. In rodent neocortex, a transient expression of serotonergic innervation has been detected at a critical period of development (D'Amato et al., 1987; Hohmann et al., 1988b; Rhoades et al., 1990). During the first two postnatal weeks, this innervation overlaps the representational map of the body surface in somatosensory cortex, then disappears. The appearance of serotonin pathways prior to the appearance of the cells that they innervate, in combination with transient expression of the type displayed in sensory cortex, has led to the hypothesis that this neurochemical system may play a significant role in the growth and differentiation of the CNS (Lauder, 1990). As will be reviewed in the following sections, an increasing number of studies provides support for this view.

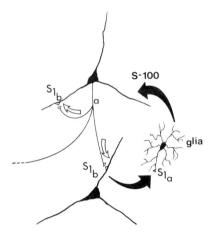

FIGURE 2

Homologous regulation of serotonergic neuron growth. Serotonin released from raphe neurons activates $S_{1b}$ autoreceptors to inhibit neurite outgrowth (open arrow). In higher concentrations, serotonin activates glial $S_{1a}$ receptors, causing release of protein S-100, which stimulates dendritic outgrowth (closed arrow).

### 1. Morphogenesis

*a. Homologous growth regulation*   There is evidence, from both cell culture and in vivo studies, for autoregulatory control of the development of serotonergic raphe cells (Whitaker-Azmitia and Azmitia, 1986a; Jonakait et al., 1988; Shemer et al., 1991). The effects of serotonin are complex, and are dependent on the concentration of agonist used. However, present data suggest that this complexity reflects two mechanisms of action. In low concentrations, serotonin selectively activates $S_{1b}$ autoreceptors to inhibit neuronal differentiation and neurite outgrowth (Whitaker-Azmitia and Azmitia, 1986a; Fig. 2). At higher serotonin concentrations, stimulation of neuronal growth occurs via activation of glial $S_{1a}$ receptors (Whitaker-Azmitia and Azmitia, 1989; Fig. 2). Experiments with glial-conditioned medium indicate that this indirect stimulatory effect results from release of the growth factor S-100 (Whitaker-Azmitia, et al., 1990).

In certain hypothalamic cells, serotonin also regulates its own synthesis (De Vitry et al., 1986b). These cells express some markers of the serotonergic phenotype, such as reuptake sites and the synthetic enzyme aromatic acid decarboxylase, but do not contain the primary synthetic enzyme tryptophan hydroxylase. Following in vitro exposure of cells to the selective $S_{1a}$ receptor agonist 8-hydroxy-(2-N-dipropylamine)-tetralin-(8-OH-DPAT) induction of aromatic acid decarboxylase activity results in cell expression of 5-hydroxytryptamine after pretreatment with the precursor 5-hydroxytryptophan. In the hypothalamic F7 neural cell line, treatment with a serotonergic agonist and eye-derived growth factor produces a similar initiation of serotonin biosynthesis (De Vitry et al., 1986a).

***b. Heterologous growth regulation*** Early studies in this field have indicated that inhibition of serotonin synthesis during prenatal brain ontogeny delays the onset of neuronal differentiation in serotonergic target regions (Lauder and Krebs, 1976,1978). Such findings have led to speculation that serotonin may serve a generalized role in controlling neuronal maturation.

Recent findings have provided some support for this view. In identified neurons of the buccal ganglion of the snail *Helisoma,* serotonin selectively inhibits neurite outgrowth, growth cone motility, and electrical synaptogenesis (Haydon et al., 1984, 1987). In cultured fetal rat cortical neurons, stimulation of serotonin $S_{1a}$ receptors decreases neurite branching and total neuritic length by more than 50% (Sikich et al., 1990). Although these effects appear to result from direct interaction with neuronal receptors, $S_{1a}$ receptors are detectable also in primary glial cultures derived from newborn rat brain, and are present in highest density in immature cells (Whitaker-Azmitia and Azmitia, 1986b). Stimulation of these glial serotonergic receptors decreases expression of glial fibrillary acidic protein, a marker for astroglial differentiation (Le Prince et al., 1990).

Although these studies in dissociated cells indicate that serotonin inhibits cellular maturation, an opposite effect has been observed in organotypic cultures of rat cerebral cortex and hippocampus (Gromova et al., 1983; Chubakov et al., 1986). Systematic addition of serotonin to these cultures has been shown to stimulate parameters of morphologic development, such as glial proliferation, neuron differentiation, axon myelination, and synaptogenesis, as well as physiologic maturation. The mechanism of action of serotonin in this more complex tissue culture system has yet to be determined.

### 2. Functional plasticity

In vivo studies have implicated serotonin in environmental regulation of neural development and plasticity (Mitchell et al., 1990; Peters, 1990). Early environmental stimulation both increases serotonin turnover in the brain and induces a permanent increase in the binding capacity of type II corticosteroid receptors in hippocampus. A relationship between these two phenomena is indicated by the finding that neonatal treatment with the serotonergic neurotoxin 5,7-dihydroxytryptamine produces a permanent reduction in the expression of hippocampal type II corticosteroid sites. Further, neonatal stress-induced elevation of hippocampal corticosteroid binding is blocked by concurrent administration of the $S_2$ receptor antagonist ketanserin.

## B. Catecholamines

The two principal catecholamine NTs of the CNS are norepinephrine (NE) and dopamine (DA). A high proportion of central noradrenergic innervation arises from the nucleus locus coeruleus (LC), which consists of a compact group of cells located in the pontine tegmentum (Loughlin and Fallon, 1985). This nucleus gives rise to a diffuse and massive innervation of diverse brain structures via both ascending and descending pathways. More discrete dopaminergic projections to limbic and motor regions arise from cell groups in the ventral tegmental area (VTA) and substantia nigra (SN), respectively (Loughlin and Fallon, 1985). The physiological effects of NE and DA in the brain are mediated via interaction with two separate familes of G protein-coupled receptors. Although there appears to be great catecholamine receptor diversity (Minneman, 1988; Emorine et al., 1989; Harrison et al., 1991; Sunahara et al., 1991; Van Tol et al., 1991), there are no reports of either NE or DA receptors which are members of the ligand-gated ion channel family.

In developing rat brain, catecholamine neurons differentiate at, or before, E12 and give rise to projections shortly thereafter (Specht et al., 1981a; Konig et al., 1988; Voorn et al., 1988). As with the serotoninergic system, many of these pathways reach their destinations prior to the differentiation of target neurons (Schlumpf et al, 1980). In view of their early appearance during the ontogeny of the brain, and their diffuse and widespread distributions, the potential trophic role of CNS catecholamines has been closely examined. While this is an issue which has been surrounded with controversy, there is now substantial evidence to suggest that catecholamine systems do influence the plasticity of developing and adult neural tissue.

### 1. Morphogenesis

*a. In vitro studies* In cultured cells, DA has been shown to inhibit neurite outgrowth significantly. This effect was demonstrated first in cells of the central ganglia of the snail *Helisoma* (McCobb et al., 1988), in which growth effects on individual neurons ranged from a sustained arrest of neurite outgrowth to a similar, but transient, response. Two subsequent studies have demonstrated a similar inhibitory effect of DA on growth cone motility in cultured horizontal neurons derived from chick and fish retina (Lankford et al., 1988; Dos Santos Rodrigues and Dowling, 1990). Although both studies implicate $D_1$ receptor involvement, there is disagreement on the underlying mechanism. In chick retinal cells, the effect of DA was mimicked by forskolin, suggesting that cAMP serves as a second messenger (Lankford et al., 1988). In contrast, in teleost horizontal cells, DA-induced neurite retraction was mimicked by phorbol esters but not by forskolin, suggesting involvement of protein kinase C, activated by phosphatidyl inositol hydrolysis (Dos Santos Rodrigues and Dowling, 1990). Interpretation of these data is complicated further by the observation that, in vivo, protein kinase C activation *stimulates* neurite outgrowth in teleost horizontal cells (Weler et al., 1991) whereas DA receptor blockade has the opposite effect (Weiler et al., 1988).

Adrenergic receptors, particularly of the β type, exhibit a widespread distribution on glial cells in the CNS (Bockaert and Ebersol, 1988; McCarthy et al., 1988; Stone and Ariano, 1989). These receptors, which may receive innervation from the locus coeruleus (reviewed by Stone and Ariano, 1989), control a number of aspects of glial function, including glycogenolysis (Stone and Ariano, 1989), membrane potential (Bowman and Kimelberg, 1988), and possibly growth factor synthesis and release (Schwartz and Costa, 1977). Prominent effects of β-adrenergic receptor stimulation on cell morphology have been demonstrated in both primary glial cultures and clonal cell lines (Shain et al., 1987; McCarthy et al., 1988). NE induced cAMP accumulation markedly stimulates morphologic differentiation, possibly via phosphorylation of the intermediate filament proteins, glial fibrillary acidic protein, and vimentin. Although cultured neurons also express adrenergic receptors (Atkinson and Minneman, 1991), there have been few reports of NE influences on neuronal morphology. However, one early study of explant cultures has indicated a significant NE-induced decrement in cerebellar growth (Vernadakis and Gibson, 1974). Subsequent in vivo studies have suggested a similar inhibitory influence of central NE systems on cerebellar development (see subsequent text).

*b. In vivo studies*  Both drug administration and lesion studies have provided some evidence that catecholamines may modulate CNS development. Intracisternal administration of adrenergic agonists to neonatal rats induces a rapid increase in the activity of ornithine decarboxylase (Morris and Slotkin, 1984), an important mediator of cellular development (Janne et al., 1978; Russell and Durie, 1978). This effect exhibits regional specificity, being most pronounced in the cerebellum, and is mediated via activation of $β_2$-adrenergic receptors (Morris and Slotkin, 1985). Conversely, intracisternal administration of the catecholaminergic neurotoxin 6-hydroxydopamine (6-OHDA), in a dose that produces a permanent depletion of NE in neocortex and cerebellum, transiently inhibits ornithine decarboxylase activity in these brain regions (Lau et al., 1990). This finding suggests that the noradrenergic system may exert some tonic control over the development of neocortex and cerebellum.

Morphologic studies indicate that neonatal 6-OHDA treatments significantly inhibit the development of rodent cerebellum, but have variable effects on neocortex. Lesions of the adrenergic afferents to cerebellum during the early postnatal period produce a decrease in tissue weight, a delayed disappearance of the external granular cell layer, and a decrease in the area of the inner granular cell layer (Lovell, 1982; Lau et al., 1990). Although such effects on cerebellar development are observed with incomplete catecholaminergic lesions, effects on neocortical development are detectable only when noradrenergic depletion is virtually complete (Felten ct al., 1982). In such cases, a range of subtle defects has been reported to occur, including abnormal dendritic outgrowth and organization (Maeda et al., 1974; Wendlandt et al., 1977; Felten et al., 1982; Loeb et al., 1987) and a transient increase in synaptogenesis (Parnavelas and Blue, 1982).

The observed differences in the effects of 6-OHDA administration on cerebellar and neocortical development may reflect differences in the maturational profiles of these areas. Although ornithine decarboxylase activity is reduced in both brain regions by postnatal 6-OHDA treatment, cerebellum is more sensitive than neocortex to disruption of polyamine synthesis during this period (Bell et al., 1986). Since neocortical development is more sensitive to inhibition of ornithine decarboxylase activity during the prenatal period (Bell et al., 1986), earlier blockade of catecholamine transmission may be necessary to observe major effects on cortical development. To date, however, no gross morphologic changes have been reported to result from specific prenatal NE depletion (Lidov and Molliver, 1982).

### 2. Functional plasticity

*a. Receptor expression*  Early studies have suggested that the developmental expression of catecholamine receptors in the CNS is dependent on the maturation of presynaptic terminals (Rosengarten and Friedhoff, 1979; Deskin et al., 1981). Although later studies have provided little evidence that noradrenergic receptor expression in target regions requires the presence of catecholaminergic terminals (Jones et al., 1985; Lorton et al., 1988), there is some indication that DA receptor expression may be dependent on afferents. In rats, pharmacologic blockade of $D_1$ receptors during the postnatal period has been shown to result in a substantial decrease in forebrain receptor density (Kostrzewa and Saleh, 1989). A similar loss of $D_1$ receptors has been observed following substantial depletion of endogenous DA with 6-OHDA (Gelbard et al., 1990). Although other 6-OHDA studies have failed to observe a similar effect (Broaddus and Bennett, 1990; Caboche et al., 1991), this could be because of the incomplete lesions in these studies. In such cases, near normal levels of dopaminergic transmission may be maintained through up-regulation of the residual terminals (Broaddus and Bennett, 1990).

The role of endogenous DA systems in regulating $D_2$ receptor development has been an area of considerable controversy. In their early study, Rosengarten and Friedhoff (1979) found that prenatal haloperidol treatment inhibited $D_2$ receptor development, whereas postnatal treatment was without effect. This observation prompted these authors to propose a critical period for endogenous DA influences on $D_2$ receptor development. Although some have questioned the validity of this finding (Madsen et al., 1981), others have confirmed a small regionally specific decrement in $D_2$ receptor expression following prenatal haloperidol exposure (Scalzo et al., 1989). However, there does not seem to be a prenatal critical period for this effect, since postnatal DA depletion or $D_2$ receptor blockade also significantly inhibits $D_2$ receptor development (Kostrzewa and Saleh, 1989; Broaddus and Bennett, 1990).

*b. Physiological plasticity*  A role of NE in experience-dependent plasticity has been proposed in both adult and developing animals (see Harley, 1987, for review). In adult rat hippocampus, NE modulates LTP via activation of β-adrenergic receptors (Hopkins and Johnston, 1984). In developing visual cortex, NE activity at β-adrenergic receptors has been implicated in control of ocular dominance plasticity (Kasamatsu and Pettigrew, 1976; Kasamatsu and Shirokawa, 1985). Although this developmental role has been the subject of much controversy, there is now substantial evidence that noradrenergic and cholinergic systems work in conjunction to modulate the plasticity of kitten visual cortex (see Section II,C,2,a for more complete discussion).

## C. Acetylcholine

The central cholinergic system consists of major projections from cell bodies localized in the basal forebrain and pontine tegmentum, and of local circuit neurons in striatum and elsewhere (Butcher and Woolf, 1986). Acetylcholine (ACh) interacts with two classes of receptor, G protein-coupled muscarinic receptors (Bonner et al., 1987) and nicotinic receptors that are multimeric proteins of the ligand-gated ion channel family (Deneris et al., 1991). These receptors are distributed widely throughout the brain (Bonner et al., 1987; Deneris et al., 1991).

In the mature CNS, ACh has been implicated strongly as a controlling influence in higher cognitive function (Gage et al., 1984). In the developing brain, in contrast, the role of this NT is less clear. The relatively late developmental maturation of the cholinergic system (Hohmann and Ebner, 1985; Hohmann et al., 1985) has led to the suggestion that ACh may act as a termination signal for neural development (Hohmann and Ebner, 1988). However, recent observations of the transient appearance of cortical cholinergic interneurons (Dori and Parnavelas, 1989), acetylcholinesterase (Robertson, 1987), and nicotinic receptor binding sites (Fuchs, 1989) indicate a potentially more complex role. As outlined in the subsequent sections, a growing body of evidence indicates that ACh is a significant regulator of neural plasticity, both in the developing CNS and in the adult.

### 1. Morphogenesis

*a. In vitro studies* A potential role for ACh in control of neurite outgrowth has been suggested by the studies of Van Hooff et al. (1989), who have demonstrated that muscarinic receptor activation of phosphatidyl inostol metabolism stimulates GAP43 phosphorylation in isolated growth cones derived from neonatal rat forebrain. Such an effect is observed following both exogenous administration of muscarinic agonists and $K^+$-evoked release of endogenous stores of ACh. Although the physiologic significance of this observation is not understood fully, GAP43 phosphorylation has been implicated in signal transduction at the growth cone membrane (Van Hooff et al., 1988).

In cell culture studies, functional consequences of both muscarinic and nicotinic cholinergic receptor activation have been observed. In embryonic hippocampal pyramidal neurons, ACh acts via muscarinic receptors to potentiate EAA-induced neurodegeneration and to lower the threshold for neurotoxicity (Mattson, 1989a). In cultures of retinal ganglion cells, nicotinic receptor antagonists do not influence neuronal survival but markedly enhance neurite outgrowth (Lipton et al., 1988), an effect that appears to result from blockade of the actions of endogeneously released ACh. In contrast to these inhibitory effects on neurons, ACh stimulates the growth of glial cells in vitro. In both primary glial cultures and clonal astrocytoma cell lines, cholinergic agonists stimulate DNA synthesis and mitosis via muscarinic receptor activation of phosphatidyl inositol hydrolysis (Ashkenazi et al., 1989). This receptor signaling mechanism also induces the expression of mRNA for a number of primary response genes, including c-*fos* (Arenader et al., 1989).

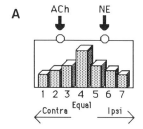
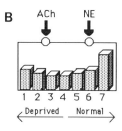
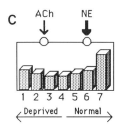
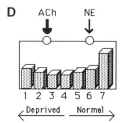
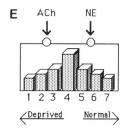
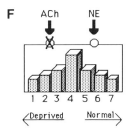
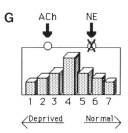

FIGURE 3

Effects of monocular deprivation on the responsiveness to visual stimulation of neurons in kitten visual cortex. Bars represent ocular dominance histograms compiled from units recorded in visual cortex. Neurons in category 1 are driven only by stimulation of the contralateral eye, those in category 7 are driven only by stimulation of the ipsilateral eye, whereas those in category 4 are driven equally by either eye. (A) Normal animal in which visual deprivation has not occurred. Most neurons are binocularly driven. (B) Animal in which monocular deprivation has occurred. Most neurons are driven by stimulation of the normal nondeprived eye. (C) Animal in which monocular deprivation has occurred in combination with a lesion of neocortical cholinergic afferents (small arrow). Most neurons are driven by stimulation of the normal nondeprived eye. (D) Animal in which monocular deprivation has occurred in combination with a lesion of neocortical noradrenergic afferents (small arrow). Most neurons are driven by stimulation of the normal nondeprived eye. (E) Animal in which monocular deprivation has occurred in combination with lesions of both cholinergic and noradrenergic afferents (small arrows). Most neurons are binocularly driven. (F) Animal in which monocular deprivation has occurred in combination with pharmacological blockade of neocortical muscarinic receptors (X). Most neurons are binocularly driven. (G) Animal in which monocular deprivation has occurred in combination with pharmacological blockade of neocortical β-adrenergic receptors (X). Most neurons are binocularly driven.

***b. In vivo studies*** Lesions of cholinergic basal forebrain afferents to neocortex during the early postnatal period produce significant and long-lasting impairment of cortical differentiation (Hohmann et al., 1988a). In tissue transplantation studies, a similar lesion of the adult host increases the degree of thalamocortical innervation of transplanted fetal neocortex (Hohmann and Ebner, 1988). Although such lesion studies tend to indicate a significant role of cholinergic afferents in neocortical development, additional pharmacologic analysis is necessary to identify the receptor type that is involved.

In other studies, pharmacologic administration of nicotinic agonists has been shown to influence CNS ontogeny. Whereas some of the effects of prenatal nicotine exposure may result from generalized growth impairment and fetal hypoxia (Slotkin et al., 1987), significant retardation of cellular maturation is observed in the absence of these phenomena (Navarro et al., 1989).

### 2. Functional plasticity

***a. Ocular dominance plasticity*** Since 6-OHDA infusion into the lateral ventricle reduces ocular dominance following monocular deprivation in the kitten, it has been proposed that NE is an important modulator of plasticity in developing cortex (Kasamatsu and Pettigrew, 1976). Since other NE-depleting lesions do not produce this effect (Bear et al., 1983; Daw et al., 1984; Adrian et al., 1985; Trombley et al., 1986) and 6-OHDA has significant antimuscarinic activity (Bear and Singer, 1986), there has been controversy over the relative contributions of cholinergic and adrenergic systems to ocular dominance changes (Gu and Singer, 1989). Whereas selective lesions of noradrenergic or cholinergic afferents alone have been found to be ineffective, combined lesions abolish the response to monocular deprivation (Bear and Singer, 1986; Fig. 3), suggesting that there is synergism between the two NT systems. In contrast, pharmacologic blockade of either β-adrenergic (Kasamatsu and Shi-okawa, 1985) or muscarinic receptors (Gu and Singer, 1989) alone is sufficient to block plasticity (Fig. 3). Thus, neither the adrenergic nor the cholinergic system is sufficient to maintain plasticity in developing visual cortex when either muscarinic or β-adrenergic transmission is blocked reliably (Fig. 3). It has been suggested that the lack of influence of single lesions on ocular dominance plasticity may reflect incomplete NT depletion combined with hypersensitivity of the corresponding receptors (Gu and Singer, 1989).

***b. Adult Neocortex*** In adult sensory cortices, topographic maps are modified significantly in response to alterations in the periphery (Kaas, 1991). Evidence suggests that cortical ACh may be an important modulator of this plasticity. Ionophoretic application of exogenous ACh, or stimulation of forebrain cholinergic afferents, enhances the response of sensory cortex to appropriate peripheral stimulation via activation of muscarinic receptors (Metherate et al., 1988; Rasmusson and Dykes, 1988). Conversely, muscarinic receptor blockade, or depletion of cortical

ACh by lesion of cholinergic afferents, reduces cortical responsiveness to sensory stimulation (Sato et al., 1987; Juliano et al., 1990). A functional correlate of this reduced neuronal responsivity has been described (Juliano et al., 1991). Unilateral removal of a digit in cats, followed by stimulation of an adjacent digit, normally results in a pattern of metabolic activity that is expanded greatly in comparison with the control contralateral hemisphere. Following depletion of cortical ACh, however, this manipulation produces a reduced, rather than expanded, area of activity.

## IV. NEUROPEPTIDES

The CNS contains a rich variety of neuropeptides that are localized primarily in neuronal pathways (see Herkenham, 1987, for review), but also may be present in glia (Shinoda et al., 1989; Hauser et al., 1990; Masters et al., 1990; Spruce et al., 1990). These peptides and their receptors appear early in development, often exhibiting transient elevated peaks of expression, then diminishing to adult levels (Chun et al., 1987; Berry and Haynes, 1989; Bodenant et al., 1991; Leslie and Loughlin, 1992). This early and transient developmental appearance, combined with the GF-like effects of many peptides in nonneural cells (see Rozengurt, 1991, for review), suggests a possible physiologic role as modulators of CNS growth and differentiation. Whereas most studies have focused on endogenous opioids and proopiomelanocortin (POMC)-derived peptides, there is evidence that other peptides may also significantly influence CNS development.

### A. Opioid Peptides

Endogenous opioid peptides are present from the earliest stages of development and may influence numerous ontogenetic processes. The POMC-derived peptide β-endorphin has been identified in ovarian follicles (Petraglia et al., 1985) and has been shown to influence oocyte maturation (O, 1990). Proenkephalin-derived products are synthesized in spermatozoa, and have been implicated in the fertilization process (Kew et al., 1990). In the developing chick, opioid receptors are detectable in both brain and body tissue at early stages of embryogenesis (Gibson and Vernadakis, 1982). In the developing rat, proenkephalin-derived peptides exhibit a widespread organ distribution during a restricted period of postnatal development (Kew and Kilpatrick, 1990), a finding consistent with a role for endogenous opioids as modulators of organogenesis (Zagon and McLaughlin, 1989).

In developing mouse brain, μ opioid receptor binding sites are first detectable at ED 12.5 (Rius et al., 1991), 2 days after the initial expression of POMC in hypothalamic neurons (Elkabes et al., 1989). A similar early development of CNS opioid receptors is seen in other species, with transient receptor expression apparent in many regions (see Leslie and Loughlin, 1992, for review). Much current work is directed toward characterizing possible developmental roles for these receptors and their associated peptides. The conclusion of the majority of studies is that endogenous opioid systems may act as inhibitory modulators of mitosis and differentiation.

***a. In vivo studies*** Administration of opioid agonists during restricted periods of early postnatal development has been shown to result in decreases in brain ornithine decarboxylase activity (Bartolome et al., 1986) and DNA synthesis (Vertes et al., 1982; Kornblum et al., 1987; Lorber et al., 1990). Although some studies have failed to demonstrate effects of opioid receptor blockade (Bartolome et al., 1987; Kornblum et al., 1987; Lorber et al., 1990), others have shown significant increases in brain DNA synthesis following receptor antagonist treatment (Vertes et al., 1982). This finding is consistent with a tonic role for endogenous opioids in modulating postnatal cell division, a conclusion that is supported by the results of anatomical studies.

Low doses of endogenous opioid peptides, administered during the first postnatal week, significantly inhibit cellular proliferation in rat cerebellar cortex and retina (Isayama et al., 1991; Zagon and McLaughlin, 1991). Conversely, chronic administration of high doses of opioid antagonists produces increased cell numbers in cortex, hippocampus, and cerebellum (Zagon and McLaughlin, 1983; Schmahl et al., 1989; Shepanek et al., 1989). Interpretation of these antagonist studies is somewhat complicated by the paradoxical effects of lower doses of drug, which significantly reduce cellular development. One possible explanation of such findings is that a rebound supersensitivity of endogenous opioid systems occurs between periods of receptor blockade (Zagon and McLaughlin, 1983).

Opioid receptor manipulations also have been shown to influence cellular differentiation significantly. Perinatal treatment with the opioid agonist morphine produces delayed growth of cortical dendrites (Ricalde and Hammer, 1990). Conversely, chronic administration of high doses of the opioid antagonist naltrexone results in regionally specific increases in neuronal dendrite spine elaboration (Hauser et al., 1989). With intermittent receptor blockade, subnormal dendritic growth is observed.

***b. In vitro studies*** In cell culture studies, both neurons and glia have been shown to express opioid peptides and their receptors (see Leslie and Loughlin, 1992, for review). Whereas one study has reported opioid peptide-induced stimulation of neural development (Ilyinsky et al., 1987), the majority of in vitro studies provides evidence consistent with in vivo data that opioid agonists inhibit growth and differentiation (Sakellaridis et al., 1986; Steine-Martin and Hauser, 1990; Vernadakis and Kentroti, 1990; Zagon et al., 1990). The in vitro effects of opioids may be dependent on both dose and time of exposure, however. In dissociated embryonic rat mesencephalic cultures, leu-enkephalin has been shown to modulate process outgrowth of serotonergic neurons with a complex dose–response relationship. A single application of leu-enkephalin at the time of plating produces significant stimulation of serotonergic growth, whereas chronic daily administration inhibits serotonergic development (Davila-Garcia and Azmitia, 1989).

## B. POMC-Derived Peptides

The majority of studies of developmental effects of endogenous opioid peptides implicates proenkephalin-derived peptides as the active agents (Davila-Garcia and Azmitia, 1989; Steine-Martin and Hauser, 1990; Vernadakis and Kentroti, 1990; Zagon et al., 1990; Isayama et al., 1991; Zagon and McLaughlin, 1991). However, POMC-derived products, both opioid and nonopioid, also have been shown to influence CNS development (Berry and Haynes, 1989; Davila-Garcia and Azmitia, 1990).

When administered in vivo to rats during early postnatal development, β-endorphin significantly *inhibits* both ornithine decarboxylase activity and DNA synthesis in the brain (Bartolome et al., 1986; Lorber et al., 1990). In contrast, N-acetyl-β-endorphin, another POMC product, *stimulates* central ornithine decarboxylase activity when administered in a similar paradigm (Bartolome et al., 1987). Although this developmental effect of N-acetyl-β-endorphin is blocked by the opioid antagonist naloxone, this peptide has been reported to have no affinity for opioid receptors (Smyth et al., 1979). Its mechanism of action in influencing neural development is therefore presently unclear.

Other POMC-derived peptides, adrenal corticotropic hormone (ACTH) and α-melanocyte stimulating hormone (α-MSH) [N-acetyl-ACTH(1–13)], also influence developmental processes. These peptides long have been known to have effects on the development and regeneration of peripheral neurons and their targets (see Gispen et al., 1987, for review). More recent studies have reported stimulatory effects of these peptides on cultured CNS neurons. Increased neurite outgrowth results from addition of α-MSH or ACTH(4–10) to cultures of rat serotonergic raphe neurons (Davila-Garcia and Azmitia, 1990), rat cerebral cortical neurons (Richter-Landsberg et al., 1987), and rat spinal cord explants (Van der Neut et al., 1988). Similar trophic actions of ACTH(1–24) on chick cerebral neurons have been observed (Daval et al., 1983).

Although ACTH and α-MSH are both neurotrophic POMC-derived peptides, studies of their effects on serotonergic neurons in vitro suggest that they may have different sites of action (Azmitia and de Kloet, 1987). ACTH and its synthetic analogs stimulate initial [$^3$H]5-hydroxytryptamine uptake by serotonergic neurons cultured alone, but not in co-culture with target hippocampal neurons. Conversely, α-MSH has no effect on the maturation of serotonergic raphe neurons cultured alone, but stimulates [$^3$H]-5-hydroxytryptamine uptake by serotonergic raphe neurons co-cultured with target neurons. These data suggest that α-MSH may influence serotonergic differentiation *indirectly* by stimulating release of a target-derived neurotrophic factor. In contrast, ACTH may influence these developing neurons *directly,* acting as a weak trophic agent in the absence of target-derived factors.

## C. Other Peptides

Neurotrophic actions of a number of other peptides have been reported. Vasoactive intestinal peptide (VIP), which is a preganglionic transmitter in the sympathetic nervous system of the adult rat (Ip et al., 1982), has been shown to exert multiple actions on developing sympathetic neuroblasts in culture, including stimulation of neuronal mitosis, survival, and process outgrowth (Pincus et al., 1990). In the CNS, VIP also increases the survival of cultured spinal cord neurons during a critical period of development (Brenneman et al., 1985; Brenneman and Eiden, 1986). This neurotrophic action of VIP is mediated indirectly via nonneuronal cells (Brenneman et al., 1990b). Two contributing mechanisms have been observed: VIP is a potent glial mitogen and a secretagogue for a glial-derived trophic substance.

Other neuropeptides exhibit trophic actions on cultured spinal cord neurons (Iwasaki et al., 1989). Substance P (SP), thyrotropin releasing hormone (TRH), and bombesin all have been found to exhibit neurite-promoting activities in spinal cord explants, whereas neurotensin does not. Of the peptides tested, SP was the most potent, with significant effects at a concentration of $10^{-12} M$, whereas bombesin exhibited the greatest maximal effect. Whether these peptides act directly on the neuron or via interaction with nonneuronal cells is not clear. However, all three peptides may activate the same physiologic mechanism, since their effects are nonadditive.

SP also has been reported to influence the development of peripheral and central catecholamine systems. In cultured chick embryonic dorsal root ganglion cells, this peptide induces neurite outgrowth similar to that produced by NGF (Narumi and Fujita, 1978). In explant cultures derived from embryonic mouse mesencephalon, SP and another tachykinin peptide, substance K, significantly increase expression of the catecholamine synthetic enzyme tyrosine hydroxylase. This stimulation of DA synthesis appears to result from membrane depolarization, since it can be blocked by the $Na^+$ channel blocker tetrodotoxin. In intact neonatal rats, intracisternal administration of SP induces permanent alterations in the plasticity of central noradrenergic neurons following chemical or surgical lesions of locus coeruleus (Jonsson and Hallman, 1982,1983). The mechanism of this effect has not been examined, but has been proposed to result from direct stimulatory actions of SP on damaged immature NE neurons.

Another neuropeptide, calcitonin gene-related peptide (CGRP), has been shown to influence the phenotypic expression of dopaminergic properties in mouse olfactory bulb (Denis-Donini, 1989). In olfactory bulb, tyrosine hydroxylase expression is regulated trans-synaptically, exhibiting a marked decrease after deafferentation and a return to normal levels after regeneration of the primary afferents (Nadi et al., 1981). During ontogeny, olfactory bulb tyrosine hydroxylase is expressed after birth, following synapse formation by the olfactory epithelial afferents (Specht et al., 1981b). In vitro, olfactory bulb cells do not express the DA phenotype unless co-cultured with olfactory epithelial neurons or with media conditioned by these cells (Denis-Donini, 1989). CGRP, which is a soluble factor in the epithelial neurons, can induce the expression of DA phenotypic markers in cultured olfactory bulb cells. Further, antisera directed at CGRP blocks the DA phenotypic expression induced by epithelial neuron co-culture. These data suggest that CGRP is the physiologic signal for afferent-induced modulation of dopaminergic properties in the olfactory bulb.

## V. CONCLUSIONS

A lack of correspondence between the anatomical distributions of NTs and their receptors in brain has prompted the suggestion that NTs may play a broader physiologic role than merely promoting fast synaptic transmission (Herkenham, 1987). The evidence reviewed in this chapter provides strong support for this view, and indicates that NTs have an important trophic function in the CNS. Under appropriate conditions, NTs of the amino acid, monamine, and peptide families all can produce long-lasting changes in the structure and function of the brain.

It is increasingly apparent that there may be little distinction among many of the neurotrophic actions of GFs and NTs. Like GFs, NTs can influence mitosis, survival, process outgrowth, and phenotypic expression of specific neuronal and glial populations. Many of these effects are manifested during a critical developmental period and disappear in mature cells. In the adult, however, NTs also exhibit numerous GF-like effects, and can influence process outgrowth, maintenance of connections, and reactive plasticity. The receptors that mediate the neurotrophic actions of NTs are, in most cases, similar to those that mediate transfer of synaptic information. In many instances, the underlying signaling mechanisms are similar to those activated by characterized GFs.

In the review of the literature that is presented here, it is clear that there is no single or systematic neurotrophic action of NTs. Rather, like those of GFs, effects of NTs vary widely depending on cell type and experimental conditions. The relationship between the neurotrophic actions of NTs and those of characterized GFs, as determined in tissue culture studies, is also complex and heterogeneous. In some cells, there is no apparent interaction between NTs and GFs. In others, the trophic actions of NTs are mediated via release of glial-derived GFs. In some instances, there is synergy between the actions of a NT and a characterized GF, whereas in others there are opposing effects. Even interactions between the neurotrophic effects of two NTs on a single cell type are sometimes apparent.

Several articles have addressed the conceptual difficulties in defining neurotrophic actions of GFs (Walicke, 1989; Chapters 1 and 2). Similar theoretical issues pertain to the definition of GF-like actions of NTs. Detailed characterization of functionally relevant neurotrophic actions of NTs are hampered also by a lack of appropriate tools for analysis. Although in vitro tissue culture enables detailed examination of NT actions under controlled experimental conditions, the nonphysiologic nature of such models must be considered, since physiologic responses in an artificial environment may differ from those observed in situ. In vivo studies, however, similarly are fraught with interpretational problems, as is evident from this review of the literature. One problem is that total blockade of neurochemical transmission by neurotoxic lesion is difficult to achieve, particularly in developing animals. Compensatory mechanisms may occur that result in maintenance of normal transmission or even hyperinnervation (Kostrezwa, 1988; Broaddus and Bennett, 1990). In pharmacologic studies, indirect effects mediated by alterations of feeding, respiration, and hormonal status must be considered and controlled (Lichtblau and Sparber, 1984). Pharmacokinetic influences on drug levels in plasma also may produce paradoxical responses that are difficult to interpret (Slotkin et al., 1987; Zagon and McLaughlin, 1989). Given such considerations, it is clear that no single in vitro or in vivo experimental approach is optimal for analysis of the neurotrophic actions of NTs. Much additional work, using a multidisciplinary approach, will be needed to characterize the mechanisms by which NTs influence synaptic plasticity in both developing and adult brain.

## ACKNOWLEDGMENTS

This work was supported by PHS grants NS19319 and DC00450. I gratefully acknowledge Jim Fallon for his contribution of schematic diagrams.

## REFERENCES

Adrian, J., Blanc, G., Buisseret, P., Fregnac, Y., Gary-Bobo, E., Imbert, M., Tassin, J. P., and Trotter, Y. (1985). Noradrenaline and functional plasticity in kitten visual cortex: A re-examination. *J. Physiol.* **367,** 73–98.
Aitken, A. R., and Tork, I. (1988). Early development of serotonin-containing neurons and pathways as seen in wholemount preparations of the fetal rat brain. *J. Comp. Neurol.* **274,** 32–47.
Arenander, A. T., de Vellis, J., and Herschman, H. R. (1989). Induction of c-*fos* and TIS genes in cultured rat astrocytes by neurotransmitter. *J. Neurosci. Res.* **24,** 107–114.
Ashkenazi, A., Ramachandran, J., and Capon, D. J. (1989). Acetylcholine analogue stimulates DNA synthesis in brain-derived cells via specific muscarinic receptor subtypes. *Nature (London)* **340,** 146–150.
Atkinson, B. N., and Minneman, K. P. (1991). Multiple adrenergic receptor subtypes controlling cyclic AMP formation: Comparison of brain slices and primary neuronal and glial cultures. *J. Neurochem.* **56,** 587–595.
Azmitia, E. C., and de Kloet, R. (1987). ACTH neuropeptide stimulation of serotonergic neuronal maturation in tissue culture: Modulation of by hippocampal cells. *Prog. Brain Res.* **72,** 311–318.
Balazs, R., and Hack, N. (1990). Trophic effects of excitatory amino acids in the developing nervous system. *In* "Excitatory Amino Acids and Neuronal Plasticity" (Y. Ben-Ari, ed.), pp. 221–228. Plenum Press, New York.
Balazs, R., Jorgensen, O. S., and Hack, N. (1988). N-Methyl-D-aspartate promotes the survival of cerebellar granule cells in culture. *Neurosci.* **27,** 437–452.
Barnard, E. A. (1988). Structural basis of the GABA-activated chloride channel: Molecular biology and molecular electrophysiology. *In* "Chloride Channels and their Modulation by Neurotransmitters and Drugs." (G. Biggio, and E. Costa, eds.), pp. 1–18. Raven Press, New York.
Barnard, E. A., and Henley, J. M. (1990). The non-NMDA receptors: Types, protein structure and molecular biology. *Trends Pharmacol. Sci.* **11,** 500–507.
Bartolome, J. V., Bartolome, M. B., Daltner, L. A., Evans, C. J., Barchas, J. D., Kuhn, C. M., and Schanberg, S. M. (1986). Effects of β-endorphin on ornithine decarboxylase in tissues of developing rats: A potential role for this endogenous neuropeptide in the modulation of tissue growth. *Life Sci.* **38,** 2355–2362.
Bartolome, J. V., Bartolome, M. B., Harris, E. B., and Schanberg, S. M. (1987). Nα-Acetyl-β-endorphin stimulates ornithine decarboxylase activity in preweanling rat pups: Opioid- and non-opioid-mediated mechanisms. *J. Pharm. Exp. Ther.* **240,** 895–899.
Bear, M. F., and Singer, W. (1986). Modulation of visual cortical plasticity by acetylcholine and noradrenaline. *Nature (London)* **320,** 172–176.
Bear, M. F., Paradiso, M. A., Schwartz, M., Nelson, S. B., Carnes, K. M., and Daniels, J. D. (1983). Two

methods of catecholamine depletion in kitten visual cortex yield different effects on plasticity. *Nature (London)* **302**, 245–247.

Bear, M. F., Kleinschmidt, A., Gu, Q., and Singer, W. (1990). Disruption of experience-dependent synaptic modifications in striate cortex by infusion of an NMDA receptor antagonist. *J. Neurosci.* **10(3)**, 909–925.

Bell, J. M., Whitmore, W. L., and Slotkin, T. A. (1986). Effects of α-difluoromethylornithine, a specific irreversible inhibitor of ornithine decarboxylase, on nucleic acids and proteins in developing rat brain: Critical perinatal periods for regional selectivity. *Neuroscience* **17**, 399–407.

Berry, S., and Haynes, L. W. (1989). The opiomelanocortin peptide family: Neuronal expression and modulation of neural cellular development and regeneration in the central nervous system. *Comp. Biochem. Physiol.* **93A**, 267–272.

Black, I. B., Adler, J. E., Dreyfus, C. F., Jonakait, G. M., Katz, D. M., LaGamma, E. F., and Markey, K. M. (1984). Neurotransmitter plasticity at the molecular level. *Science* **225**, 1266–1270.

Bobker, D. H., and Williams, J. T. (1990). Ion conductances affected by 5-HT receptor subtypes in mammalian neurons. *Trends Neurosci.* **13**, 169–173.

Bockaert, J., and Ebersol, C. (1988). α-Adrenergic receptors on glial cells. *In* "Glial Cell Receptors" (H. K. Kimelberg, ed.), pp. 35–52. Raven Press, New York.

Bodenant, C., Leroux, P., Gonzalez, B. J., and Vaudry, H. (1991). Transient expression of somatostatin receptors in the rat visual system during development. *Neuroscience* **41**, 595–606.

Bonner, T. I., Buckley, N. J., Young, A. C., and Brann, M. R. (1987). Identification of a family of muscarinic acetylcholine receptor genes. *Science* **237**, 527–532.

Boulter, J., Hollmann, M., O'Shea-Greenfield, A., Hartley, M., Deneris, E., Maron, C., and Heinemann, S. (1990). Molecular cloning and functional expression of glutamate receptor subunit genes. *Science* **249**, 1033–1037.

Bowery, N. (1989). GABA$_B$ receptors and their significance in mammalian pharmacology. *Trends Pharmacol. Sci.* **10**, 401–406.

Bowman, C. L., and Kimelberg, H. K. (1988). Adrenergic-receptor-mediated depolarization of astrocytes. *In* "Glial Cell Receptors" (H. K. Kimelberg, ed.), pp. 53–76. Raven Press, New York.

Brenneman, D. E., and Eiden, L. E. (1986). Vasoactive intestinal peptide and electrical activity influence neuronal survival. *Proc. Natl. Acad. Sci. U.S.A.* **83**, 1159–1162.

Brenneman, D. E., Eiden, L. E., and Siegel, R. E. (1985). Neurotrophic action of VIP on spinal cord cultures. *Peptides* **6**, 35–39.

Brenneman, D. E., Forsythe, I. D., Nicol, T., and Nelson, P. G. (1990a). N-Methyl-D-aspartate receptors influence neuronal survival in developing spinal cord cultures. *Dev. Brain Res.* **51**, 63–68.

Brenneman, D. E., Nicol, T., Warren, D., and Bowers, L. M. (1990b). Vasoactive intestinal peptide: A neurotrophic releasing agent and an astroglial mitogen. *J. Neurosci. Res.* **25**, 386–394.

Brewer, G. J., and Cotman, C. W. (1989). NMDA receptor regulation of neuronal morphology in cultured hippocampal neurons. *Neurosci. Lett.* **99**, 268–273.

Broaddus, W. C., and Bennet, J. P., Jr. (1990). Postnatal development of striatal dopamine function. II. Effects of neonatal 6-hydroxydopamine treatments on $D_1$ and $D_2$ receptors, adenylate cyclase activity and presynaptic dopamine function. *Dev. Brain Res.* **52**, 273–277.

Butcher, L. L., and Woolf, N. J. (1986). Central cholinergic systems: Synopsis of anatomy and overview of physiology and pathology. *In* "The Biological Substrates of Alzheimer's Disease" (A. B. Scheibel and A. F. Wechsler, eds.), pp. 73–86. Academic Press, New York.

Buznikov, G. A. (1984). The action of neurotransmitters and related substances on early embryogenesis. *Pharmac. Ther.* **25**, 23–59.

Caboche, J., Rogard, M., and Besson, M.-J. (1991). Comparative development of $D_1$-dopamine and μ opiate receptors in normal and in 6-hydroxydopamine-lesioned neonatal rat striatum: Dopaminergic fibers regulate μ but not $D_1$ receptor distribution. *Dev. Brain Res.* **58**, 111–122.

Cambray-Deakin, M. A., and Burgoyne, R. D. (1990). Regulation of neurite outgrowth from cerebellar granule cells in culture: NMDA receptors and protein kinase C. *In* "Excitatory Amino Acids and Neuronal Plasticity" (Y. Ben-Ari, ed.), pp. 245–253. Plenum Press, New York.

Cambray-Deakin, M. A., Foster, A. C., and Burgoyne, R. D. (1990). The expression of excitatory amino

acid binding sites during neuritogenesis in the developing rat cerebellum. *Dev. Brain Res.* **54,** 265–271.
Campbell, M. K., Mahler, H. R., Moore, W. J., and Tewari, S. (1966). Protein synthesis systems from rat brain. *Biochem.* **5,** 1174–1184.
Choi, D. W. (1988). Calcium-mediated neurotoxicity: Relationship to specific channel types and role in ischemic damage. *Trends Neurosci.* **11,** 465–469.
Chubakov, A. R., Gromova, E. A., Konovalov, G. V., Sarkisova, E. F., and Chumasov, E. I. (1986). The effects of serotonin on the morpho-functional development of rat cerebral neocortex in tissue culture. *Brain Res.* **369,** 285–297.
Chun, J. J. M., Nakamura, M. J., and Shatz, C. J. (1987). Transient cells of the developing mammalian telencephalon are peptide-immunoreactive neurons. *Nature (London)* **325,** 617–620.
Cline, H. (1991). Activity-dependent plasticity in the visual systems of frogs and fish. *Trends Neurosci.* **14,** 104–111.
Cline, H. T., and Constantine-Paton, M. (1989). NMDA receptor antagonists disrupt the retinotectal topographic map. *Neuron* **3,** 413–426.
Cohen-Cory, S., Dreyfus, C. F., and Black, I. B. (1991). NGF and excitatory neurotransmitters regulate survival and morphogenesis of cultured cerebellar Purkinje cells. *J. Neurosci.* **11,** 462–471.
Collingridge, G. L., and Singer, W. (1990). Excitatory amino acid receptors and synaptic plasticity. *Trends Pharmacol. Sci.* **11,** 290–296.
Collingridge, G. L., Herron, C. E., and Lester, R. A. J. (1988). Frequency-dependent $N$-methyl-D-aspartate receptor-mediated synaptic transmission in rat hippocampus. *J. Physiol.* **399,** 301–312.
Constantine-Paton, M., Cline, H. T., and Debski, E. (1990). Patterned activity, synaptic convergence, and the NMDA receptor in developing visual pathways. *Ann. Rev. Neurosci.* **131,** 129–154.
Cotman, C. W., and Monaghan, D. T. (1988). Excitatory amino acid neurotransmission: NMDA receptors and Hebb-type synaptic plasticity. *Ann. Rev. Neurosci.* **11,** 61–80.
Cynader, M., Shaw, C., Prusky, G., and Van Huizen, F. (1990). Neural mechanisms underlying modifiability of response properties in developing cat visual cortex. *In* "Vision and the Brain" (B. Cohen and I. Bodis-Wollner, eds.), pp. 85–108. Raven Press, New York.
D'Amato, R. J., Blue, M. E., Largent, B. L., Lynch, D. R., Ledbetter, D. J., Molliver, M. E., and Snyder, S. H. (1987). Ontogeny of the serotonergic projection to rat neocortex: Transient expression of a dense innervation to primary sensory areas. *Proc. Natl. Acad. Sci. U.S.A.* **84,** 4322–4326.
Dames, W., Joo, F., Feher, O., Toldi, J., and Wolff, J. R. (1985). γ-Aminobutyric acid enables synaptogenesis in the intact superior cervical ganglion of the adult rat. *Neurosci. Lett.* **54,** 159–164.
Daval, J.-L., Louis, J.-L., Gerard, M.-J., and Vincendon, G. (1983). Influence of adrenocorticotropic hormone on the growth of isolated neurons in culture. *Neurosci. Lett.* **36,** 299–304.
Davies, C. H., Starkey, S. J., Pozza, M. F., and Collingridge, G. L. (1991). $GABA_B$ autoreceptors regulate the induction of LTP. *Nature (London)* **349,** 609–611.
Davila-Garcia, M. I., and Azmitia, E. C. (1989). Effects of acute and chronic administration of leu-enkephalin on cultured serotonergic neurons: Evidence for opioids as inhibitory neuronal growth factors. *Dev. Brain Res.* **49,** 97–103.
Davila-Garcia, M. I., and Azmitia, E. C. (1990). Neuropeptides as positive or negative neuronal growth regulatory factors: Effects of ACTH and leu-enkephalin on cultured serotonergic neurons. *In* "Moleuclar Aspects of Development an Aging of the Nervous System" (J. M. Lauder, ed.), pp. 75–92. Plenum Press, New York.
Daw, N. W., Robertson, T. W., Rader, R. K., Videen, T. O., and Coscia, C. J. (1984). Substantial reduction of cortical noradrenaline by lesions of adrenergic pathway does not prevent effects of monocular deprivation. *J. Neurosci.* **4,** 1354–1360.
Deneris, E. S., Connolly, J., Rogers, S. W., and Duvoisin, R. (1991). Pharmacological and functional diversity of neuronal nicotinic acetylcholine receptors. *Trends Pharmacol. Sci.* **12,** 34–40.
Denis-Donini, S. (1989). Expression of dopaminergic phenotypes in the mouse olfactory bulb induced by the calcitonin gene-related peptide. *Nature (London)* **339,** 701–703.
De Vitry, F., Catelon, J., Dubois, M., Thibault, J., Barritault, D., Courty, J., Bourgoin, S., and Hamon, M. (1986a). Partial expression of monoaminergic (serotoninergic) properties by the multipotent hypothalamic cell line $F_7$. An example of learning at the cellular level. *Neurochem. Int.* **9,** 43–53.

De Vitry, F., Hamon, M., Catelon, J., Dubois, M., and Thibault, J. (1986b). Serotonin initiates and autoamplifies its own synthesis during mouse central nervous system development. *Proc. Natl. Acad. Sci. U.S.A.* **83,** 8629–8633.

Deskin, R., Seidler, F. J., Whitmore, W. L., and Slotkin, T. A. (1981). Development of $\alpha$-noradrenergic and dopaminergic receptor systems depends on maturation of their presynaptic nerve terminals in the rat brain. *J. Neurochem.* **36,** 1683–1690.

Dori, I., and Parnavelas, J. G. (1989). The cholinergic innervation of the rat cerebral cortex shows two distinct phases in development. *Exp. Brain Res.* **76,** 417–423.

Dos Santos Rodrigues, P. D. S., and Dowling, J. E. (1990). Dopamine induces neurite retraction in retinal horizontal cells via diacylglycerol and protein kinase C. *Proc. Natl. Acad. Sci. U.S.A.* **87,** 9693–9697.

Dowling, J. E. (1987). "The Retina." Harvard University Press, Cambridge, Massachusetts.

Elkabes, S., Loh, Y. P., Niebergs, A., and Wray, S. (1989). Prenatal ontogenesis of proopiomelanocortin in the mouse central nervous system and pituitary gland: an in situ hybridization and immunohistochemical study. *Dev. Brain Res.,* **46,** 85–95.

Emorine, L. J., Marullo, S., Briend-Sutren, M.-M., Patey, G., Tate, K., Belavier-Klutchko, G., and Strosberg, D. A. (1989). Molecular characterization of the human $\beta_3$-adrenergic receptor. *Science* **245,** 1118–1121.

Erdo, S. L., Michler, A., and Wolff, J. R. (1991). GABA accelerates excitotoxic cell death in cortical cultures: Protection by blockers of GABA-gated chloride channels. *Brain Res.* **542,** 254–258.

Fallon, J. H., and Loughlin, S. E. (1985). Substantia nigra. *In* "The Rat Nervous System" (G. Paxinos, ed.), Vol. 1, pp. 353–374. Academic Press, Australia.

Felten, D. L., Hallman, H., and Jonsson, G. (1982). Evidence for a neurotrophic role of noradrenaline neurons in the postnatal development of rat cerebral cortex. *J. Neurocytol.* **11,** 119–135.

Fonnum, F. (1984). Glutamate: A neurotransmitter in mammalian brain. *J. Neurochem.* **42,** 1–11.

Forsythe, I. D., and Westbrook, G. L. (1988). Slow excitatory postsynaptic currents mediated by $N$-methyl-D-aspartate receptors on cultured mouse central neuronses. *J. Physiol.* **396,** 515–533.

Fox, K., Sato, H., and Daw, N. (1989). The location and function of NMDA receptors in cat and kitten visual cortex. *J. Neurosci.* **9,** 2443–2454.

Fox, K., Daw, N., Sato, H., and Czepita, D. (1991). Dark-rearing delays the loss of NMDA-receptor function in kitten visual cortex. *Nature (London)* **350,** 342–344.

Frazer, A., Maayani, S., and Wolfe, B. B. (1990). Subtypes of receptors for serotonin. *Ann. Rev. Pharmacol. Toxicol.* **30,** 307–348.

Fuchs, J. L. (1989). [$^{125}$I]$\alpha$-Bungarotoxin binding marks primary sensory areas of developing rat neocortex. *Brain Res.* **501,** 223–234.

Gage, F. H., Bjorklund, A., Stenevi, U., Dunnett, S. B., and Kelly, P. A. (1984). Intrahippocampal septal grafts ameliorate learning impairments in aged rats. *Science* **225,** 533–536.

Gelbard, H. A., Teicher, M. H., Baldessarini, R. J., Gallitano, A., Marsh, E. R., Zorc, J., and Faedda, G. (1990). Dopamine $D_1$ receptor development depends on endogenous dopamine. *Dev. Brain Res.* **56,** 137–140.

Gibson, D. A., and Vernadakis, A. (1982). [$^3$H]Etorphine binding activity in early chick embryos: Brain and body tissue. *Dev. Brain Res.* **4,** 23–29.

Gispen, W. H., De Koning, P., Kuiters, R. R. F., Van de Zee, C. E. E. M., and Verhaagen, J. (1987). On the neurotrophic actions of melanocortins. *In* "Progress in Brain Research" (E. R. de Kloet, V. M. Wiegart, and D. de Wied, eds.), Vol. 72, pp. 319–331. Elsevier, Amsterdam.

Gromova, H. A., Chubakov, A. R., Chumasov, E. I., and Knonvalov, H. V. (1983). Serotonin as a stimulator of hippocampal cell differentiation in tissue culture. *Int. J. Dev. Neurosci.* **1,** 339–349.

Gruen, R. J., Elsworth, J. D., and Roth, R. H. (1990). Regionally specific alterations in the low-affinity GABA$_A$ receptor following perinatal exposure to diazepam. *Brain Res.* **514,** 151–154.

Gu, Q., and Singer, W. (1989). The role of muscarinic acetylcholine receptors in ocular dominance plasticity. *Experientia* **57** *(Suppl.),* 305–314.

Hablitz, J. J., Tehrani, M. H. J., and Barnes, E. M., Jr. (1989). Chronic exposure of developing cortical neurons to GABA down-regulates GABA/benzodiazepine receptors and GABA-gated chloride currents. *Brain Res.* **501,** 332–338.

Hamon, R., and Heinemann, U. (1988). Developmental changes in neuronal sensitivity to excitatory amino acids in area CA1 of the rat hippocampus. *Dev. Brain Res.* **38**, 286.

Hansen, G. H., Meier, E., and Schousboe, A. (1984). GABA influences the ultrastructure composition of cerebellar granule cells during development in culture. *Int. J. Dev. Neurosci.* **2**, 247–257.

Hansen, G. H., Belhage, B., Schousboe, A., and Meier, E. (1988). γ-Aminobutyric acid agonist-induced alterations in the ultrastructure of cultured cerebellar granule cells is restricted to early development. *J. Neurochem.* **51**, 243–245.

Harley, C. W. (1987). A role for norepinephrine in arousal, emotion and learning? Limbic modulation by norepinephrine and the Kety hypothesis. *Prog. Neuro-Psychopharmacol. Biol. Psychiat.* **11**, 419–458.

Harrison, J. K., Pearson, W. R., and Lynch, K. R. (1991). Molecular characterization of $\alpha_1$-and $\alpha_2$-adrenoceptors. *Trends Pharmacol. Sci.* **12**, 62–67.

Hauser, K. F., McLaughlin, P. J., and Zagon, I. S. (1989). Endogenous opioid systems and the regulation of dendritic growth and spine formation. *J. Comp. Neurol.* **281**, 13–22.

Hauser, K. F., Osborne, J. G., Stiene-Martin, A., and Melner, M. H. (1990). Cellular localization of proenkephalin mRNA and enkephalin peptide products in cultured astrocytes. *Brain Res.* **522**, 347–353.

Haydon, P. G., McCobb, D. P., and Kater, S. B. (1984). Serotonin selectively inhibits growth cone motility and synaptogenesis of specific identified neurons. *Science* **226**, 561–564.

Haydon, P. G., McCobb, D. P., and Kater, S. B. (1987). The regulation of neurite outgrowth, growth cone motility, and electrical synaptogenesis by serotonin. *J. Neurobiol.* **18**, 197–215.

Hebb, D. O. (1949). "The Organization of Behavior." Wiley, New York.

Herkenham, M. (1987). Mismatches between neurotransmitter and receptor localizations in brain: Observations and implications. *Neuroscience* **23**, 1–38.

Hohmann, C. F., and Ebner, F. F. (1985). Development of cholinergic markers in mouse forebrain. I. Acetylcholinesterase straining pattern and choline acetyltransferase activity. *Dev. Brain Res.* **23**, 225–241.

Hohmann, C. F., and Ebner, F. F. (1988). Basal forebrain lesions facilitate adult host fiber ingrowth into neocortical transplants. *Brain Res.* **448**, 53–66.

Hohmann, C. F., Pert, C. B., and Ebner, F. F. (1985). Development of cholinergic markers in mouse forebrain. II. Muscarinic receptor binding in normal and transplanted cortex. *Dev. Brain Res.* **23**, 243–253.

Hohmann, C. F., Brooks, A. R., and Coyle, J. T. (1988a). Neonatal lesions of the basal forebrain cholinergic neurons result in abnormal cortical development. *Dev. Brain Res.* **42**, 253–264.

Hohmann, C. F., Hamon, R., Batshaw, M. L., and Coyle, J. T. (1988b). Transient postnatal elevation of serotonin levels in mouse neocortex. *Dev. Brain Res.* **43**, 163–166.

Hopkins, W. F., and Johnston, D. (1984). Frequency-dependent noradrenergic modulation of long-term potentiation in the hippocampus. *Science* **226**, 350–351.

Ilyinsky, O. B., Kozlova, E. S., Konrikova, E. S., Kalentchuk, V. U., Titov, M. I., and Bespalova, Z. D. (1987). Effects of opioid peptides and naloxone on nervous tissue in culture. *Neuroscience* **22**, 719–735.

Ip, N. Y., Ho, C. K., and Zigmond, R. E. (1982). Secretin and vasoactive intestinal peptide acutely increase tyrosine 3-monoxygenase in the rat superior cervical ganglion. *Proc. Natl. Acad. Sci. U.S.A.* **79**, 7566–7569.

Isayama, T., McLaughlin, P. J., and Zagon, I. S. (1991). Endogenous opioids regulate cell proliferation in the retina of developing rat. *Brain Res.* **544**, 79–85.

Iwasaki, Y., Kinoshita, M., Ikeda, K., Takamiya, K., and Shiojima, T. (1989). Trophic effect of various neuropeptides on the cultured ventral spinal cord of rat embryo. *Neurosci. Lett.* **101**, 316–320.

Janne, J., Poso, H., and Raina, A. (1978). Polyamines in rapid growth and cancer. *Biochem. Biophys. Acta* **473**, 241–293.

Jonakait, G. M., Schotland, S., and Ni, L. (1988). Development of serotonin, substance P and thyrotrophin-releasing hormone in mouse medullary raphe grown in organotypic tissue culture: Developmental regulation by serotonin. *Brain Res.* **473**, 336–343.

Jones, L. S., Gauger, L. L., Davies, J. N., Slotkin, T. A., and Bartolome, J. V. (1985). Postnatal development of brain α-adrenergic receptors: In vitro autoradiography with [$^{125}$I]HEAT in normal rats and rats treated with α-difluoromethyl-ornithine, a specific, irreversible inhibitor of ornithine decarboxylase. *Neuroscience* **15**, 1195–1202.

Jonsson, G., and Hallman, H. (1982). Substance P counteracts neurotoxin damage on norepinephrine neurons in rat brain during ontogeny. *Science* **215**, 75–77.

Jonsson, G., and Hallman, H. (1983). Effect of substance P on neonatally axotomized noradrenaline neurons in rat brain. *Med. Biol.* **61**, 179–185.

Juliano, S. L., and Whitsel, B. L. (1990). Cholinergic manipulation alters stimulus-evoked metabolic activity in cat somatosenory cortex. *J. Comp. Neurol.* **297**, 106–120.

Juliano, S. L., Ma, W., and Eslin, D. (1991). Cholinergic depletion prevents expansion of topographic maps in somatosensory cortex. *Proc. Natl. Acad. Sci. U.S.A.* **88**, 780–784.

Julius, D., Livelli, T. J., Jessell, T. M., and Axel, R. (1989). Ectopic expression of the serotonin 1c receptor and the triggering of malignant transformation. *Science* **244**, 1057–1062.

Kaas, J. H. (1991). Plasticity of sensory and motor maps in adult mammals. *Ann. Rev. Neurosci.* **14**, 137–167.

Kasamatsu, T., and Pettigrew, J. D. (1976). Depletion of brain catecholamines: Failure of ocular dominance shift after monocular occlusion in kittens. *Science* **194**, 206–209.

Kasamatsu, T., and Shirokawa, T. (1985). Involvement of β-adrenoreceptors in the shift of ocular dominance after monocular deprivation. *Exp. Brain Res.* **59**, 507–514.

Keinanen, K., Wisen, W., Sommer, B., Werner, P., Herb, A., Verdoorn, T. A., Sakmann, B., and Seeburg, P. H. (1990). A family of AMPA-selective glutamate receptors. *Science* **249**, 556–560.

Kew, D., and Kilpatrick, D. L. (1990). Widespread organ expression of the rat proenkephalin gene during early postnatal development. *Mol. Endocrinol.* **4**, 337–340.

Kew, D., Muffly, K. E., and Kilpatrick, D. L. (1990). Proenkephalin products are stored in the sperm acrosome and may function in fertilization. *Proc. Natl. Acad. Sci. U.S.A.* **87**, 9143–9147.

Kimelberg, H. K. (1988). "Glial Cell Receptors." Raven Press, New York.

Konig, N., Wilkie, M. B., and Lauder, J. M. (1988). Tyrosine hydroxylase and serotonin containing cells in embryonic rat rhombencephalon: A whole-mount immunocytochemical study. *J. Neurosci. Res.* **20**, 212–223.

Kornblum, H. I., Loughlin, S. E., and Leslie, F. M. (1987). Effects of morphine on DNA synthesis in neonatal rat brain. *Dev. Brain Res.* **31**, 45–52.

Kostrzewa, R. M. (1988). Reorganization of noradrenergic neuronal systems following neonatal chemical and surgical injury. *Prog. Brain Res.* **73**, 405–423.

Kostrzewa, R. M., and Saleh, M. I. (1989). Impaired ontogeny of striatal dopamine $D_1$ and $D_2$ binding sites after postnatal treatment of rats with SCH-23390 and spiroperidol. *Dev. Brain Res.* **45**, 95–101.

Lankford, K. L., DeMello, F. G., and Klein, W. L. (1988). $D_1$-type dopamine receptors inhibit growth cone motility in cultured retina neurons: Evidence that neurotransmitters act as morphogenic growth regulators in the developing central nervous system. *Proc. Natl. Acad. Sci. U.S.A.* **85**, 4567–4571.

Lau, C., Cameron, A., Antolick, L., and Slotkin, T. A. (1990). Trophic control of the ornithine decarboxylase/polyamine system in neonatal rat brain regions: Lesions caused by 6-hydroxydopamine produce effects selective for cerebellum. *Dev. Brain Res.* **52**, 167–173.

Lauder, J. M. (1987). Neurotransmitters as morphogenetic signals and trophic factors. *In* "Model Systems of Development and Aging of the Nervous System" (A. Vernadakis, ed.), pp. 219–237. Nijhoff Publishing, Boston.

Lauder, J. M. (1990). Ontogeny of the serotonergic system in the rat: Serotonin as a developmental signal. *Ann. N.Y. Acad. Sci.* **600**, 297–314.

Lauder, J. M., and Bloom, F. E. (1974). Ontogeny of monoamine neurons in the locus coeruleus, raphe nuclei, and substantia nigra of the rat. I. Cell differentiation. *J. Comp. Neurol.* **155**, 469–481.

Lauder, J. M., and Krebs, H. (1976). Effects of *p*-chlorophenylalanine on time of neuronal origin during embryogenesis in the rat. *Brain Res.* **107**, 638–644.

Lauder, J. M., and Krebs, H. (1978). Serotonin as a differentiation signal in early neurogenesis. *Dev. Neurosci.* **1,** 15–30.

Lauder, J. M., Han, V. K. M., Henderson, P., Verdoorn, T., and Towle, A. C. (1986). Prenatal ontogeny of the GABAergic system in the rat brain: An immunocytochemical study. *Neuroscience* **19,** 465–493.

Le Prince, G., Copin, M.-C., Hardin, H., Belin, M.-F., Bouilloux, J.-P., and Tardy, M. (1990). Neuron-glia interactions: Effect of serotonin on the astroglial expression of GFAP and of its encoding message. *Dev. Brain Res.* **51,** 295–298.

Leslie, F. M., and Loughlin, S. E. (1992). Development of multiple opioid receptor types. In "Development of the Central Nervous System: Effects of Alcohol and Opiates" (M. W. Miller, ed.), pp. 255–283, Liss, New York.

Lichtblau, L., and Sparber, S. B. (1984). Opioids and development: A perspective on experimental models and methods. *Neurobehav. Toxicol. Teratol.* **6,** 3–8.

Lidov, H. G., and Molliver, M. E. (1982). The structure of cerebral cortex in the rat following prenatal administration of 6-hydroxydopamine. *Brain Res.* **255,** 81–108.

Lipton, S. A., and Kater, S. B. (1989). Neurotransmitter regulation of neuronal outgrowth, plasticity and survival. *Trends Neurosci.* **12,** 265–270.

Lipton, S. A., Frosch, M. P., Phillips, M. D., Tauck, D. L., and Aizenman, E. (1988). Nicotinic antagonists enhance process outgrowth by rat retinal ganglion cells in culture. *Science* **239,** 1293–1296.

Loeb, E. P., Chang, F.-L. F., and Greenough, W. T. (1987). Effects of neonatal 6-hydroxydopamine treatment upon morphological organization of the posteromedial barrel subfield in mouse somatosensory cortex. *Brain Res.* **403,** 113–120.

Loh, Y. P. (1991). Prenatal expression of pro-opiomelanocortin (POMC) mRNA, POMC-derived peptides and μ opiate receptors in the mouse embryo. In "Molecular Approaches to Drug Abuse Research" (Bethesda, ed.). National Institute of Drug Abuse, Bethesda, Maryland.

Lorber, B. A., Freitag, S. K., and Bartolome, J. V. (1990). Effects of beta-endorphin on DNA synthesis in brain regions of preweanling rats. *Brain Res.* **531,** 329–332.

Lorton, D., Bartolome, J., Slotkin, T. A., and Davies, J. N. (1988). Development of brain β-adrenergic receptors after neonatal 6-hydroxydopamine treatment. *Brain Res. Bull.* **21,** 591–600.

Loughlin, S. E., and Fallon, J. H. (1985). Locus coeruleus. In "The Rat Nervous System" (G. Paxinos, ed.), Vol. 2, pp. 79–94. Academic Press, Australia.

Lovell, K. L. (1982). Effects of 6-hydroxydopamine-induced norepinephrine depletion on cerebellar development. *Dev. Neurosci.* **5,** 359–368.

McCarthy, K. D., Salm, A., and Lerea, L. S. (1988). Astroglial receptors and their regulation of intermediate filament protein phosphorylation. In "Glial Cell Receptors" (H. K. Kimelberg, ed.), pp. 1–22. Raven Press, New York.

McCobb, D. P., Haydon, P. G., and Kater, S. B. (1988). Dopamine and serotonin inhibition of neurite elongation of different identified neurons. *J. Neurosci. Res.* **19,** 19–26.

Madsen, J. R., Campbell, A., and Baldessarini, R. J. (1981). Effects of prenatal treatment of rats with haloperidol due to altered drug distribution in neonatal brain. *Neuropharmacol.* **20,** 931–939.

Madtes, P., Jr. (1987). Ontogeny of the GABA receptor complex. In "Neurology and Neurobiology" (V. Chan-Palay, and S. L. Palay, eds.), Vol. 32, pp. 161–188. Liss, New York.

Maeda, T., Tohyama, M., and Shimizu, N. (1974). Modification of postnatal development of neocortex in rat brain with experimental deprivation of locus coeruleus. *Brain Res.* **70,** 515–520.

Maloteaux, J.-M., Octave, J.-N., Gossuin, A., Laterre, C., and Trouet, A. (1987). GABA induces down-regulation of the benzodiazepine-GABA receptor complex in the rat cultured neurons. *Eur. J. Pharmacol.* **144,** 173–183.

Masters, B. A., Pruysers, C. R., Millard, W. J., Meyer, E. M., and Poulakos, J. J. (1990). Preproneuropeptide Y mRNA expression in glial cell cultures of the neonate but not 21-day-old rat brain. *Ann. N.Y. Acad. Sci.* **611,** 522–524.

Masu, M., Tanabe, Y., Tsuchida, K., Shigemoto, R., and Nakanishi, S. (1991). Sequence and expression of the metabotropic glutamate receptor. *Nature (London)* **349,** 760–765.

Mattson, M. P. (1988). Neurotransmitters in the regulation of neuronal cytoarchitecture. *Brain Res. Rev.* **13**, 179–212.

Mattson, M. P. (1989a). Acetylcholine potentiates glutamate-induced neurodegeneration in cultured hippocampal neurons. *Brain Res.* **497**, 402–406.

Mattson, M. P. (1989b). Cellular signaling mechanisms common to the development and degeneration of neuroarchitecture. A review. *Mech. Aging Dev.* **50**, 103–157.

Mattson, M. P., and Kater, S. B. (1989a). Excitatory and inhibitory neurotransmitters in the generation and degeneration of hippocampal neuroarchitecture. *Brain Res.* **478**, 337–348.

Mattson, M. P., and Kater, S. B. (1989b). Development and selective neurodegeneration in cell cultures from different hippocampal regions. *Brain Res.* **490**, 110–125.

Mattson, M. P., Dou, P., and Kater, S. B. (1988a). Outgrowth-regulating actions of glutamate in isolated hippocampal pyramidal neurons. *J. Neurosci.* **8**, 2087–2100.

Mattson, M. P., Lee, R. E., Adams, M. E., Guthrie, P. B., and Kater, S. B. (1988b). Interactions between entorhinal axons and target hippocampal neurons: a role for glutamate in the development of hippocampal circuitry. *Neuron* **1**, 865–876.

Mattson, M. P., Murrain, M., Guthrie, P. B., and Kater, S. B. (1989). Fibroblast growth factor and glutamate: Opposing roles in the generation and degeneration of hippocampal neuroarchitecture. *J. Neurosci.* **9**, 3728–3740.

Meier, E., and Jorgensen, O. S. (1986). γ-Aminobutyric acid affects the developmental expression of neuron-associated proteins in cerebellar granule cell cultures. *J. Neurochem.* **46**, 1256–1262.

Meier, E., Hansen, G. H., and Schousboe, A. (1985). The trophic effect of GABA on cerebellar granule cells is mediated by GABA-receptors. *Int. J. Dev. Neurosci.* **3**, 401–407.

Meier, E., Belhage, B., Drejer, J., and Schousboe, A. (1987). The expression of GABA receptors on cultured cerebellar granule cells is influenced by GABA. *In* "Neurology and Neurobiology" (V. Chan-Palay, and S. L. Palay, eds.), Vol. 32, pp. 139–160. Liss, New York.

Metherate, R., Tremblay, N., and Dykes, R. W. (1988). The effects of acetylcholine on response properties of cat somatosensory cortical neurons. *J. Neurophysiol.* **59**, 1231–1252.

Michler-Stuke, A., and Wolff, J. R. (1987). Facilitation and inhibition of neurite elongation by GABA in chick tectal neurons. *In* "Neurology and Neurobiology" (V. Chan-Palay, and S. L. Palay, eds.), Vol. 32, pp. 253–266. Liss, New York.

Minneman, K. P. (1988). $\alpha_1$-Adrenergic receptor subtypes, inositol phosphates, and sources of cell $Ca^{2+}$. *Pharmacol. Rev.* **40**, 87–119.

Mitchell, J. B., Iny, L. J., Meaney, M. J. (1990). The role of serotonin in the development and environmental regulation of type II corticosteroid receptor binding in rat hippocampus. *Dev. Brain Res.* **55**, 231–235.

Montpied, P., Ginns, E. I., Martin, B. M., Roca, D., Farb, D. H., and Paul, S. M. (1991). γ-Aminobutyric acid (GABA) induces a receptor-mediated reduction on $GABA_A$ receptor α subunit messenger RNAs in embryonic chick neurons in culture. *J. Biol. Chem.* **266**, 6011–6014.

Morris, G., and Slotkin, T. A. (1985). Beta-2 adrenergic control of ornithine decarboxylase activity in brain regions of the developing rat. *J. Pharmacol. Exp. Ther.* **233**, 141–147.

Murphy, T. H., Schnaar, R. L., and Coyle, J. T. (1990). Immature cortical neurons are uniquely sensitive to glutamate toxicity by inhibition of cystine uptake. *FASEB J.* **4**, 1624–1633.

Nadi, S. S., Head, R., Grillo, M., Hempstead, J., Grannot-Reisfeld, N., and Margolis, F. L. (1981). Chemical deafferentiation of the olfactory bulb: Plasticity of the levels of tyrosine hydroxylase, dopamine and norepinephrine. *Brain Res.* **213**, 365–377.

Narumi, S., and Fujita, T. (1978). Stimulatory effects of substance P and nerve growth factor (NGF) on neurite outgrowth in embryonic chick dorsal root ganglia. *Neuropharmacol.* **17**, 73–76.

Navarro, H. A., Seidler, F. J., Schwartz, R. D., Baker, F. E., Dobbins, S. S., and Slotkin, T. A. (1989). Prenatal exposure to nicotine impairs nervous system development at a dose which does not affect viability or growth. *Brain Res. Bull.* **23**, 187–192.

Nemecek, G. M., Coughlin, S. R., Handley, D. A., and Moskowitz, M. A. (1986). Stimulation of aortic smooth muscle cell mitogenesis by serotonin. *Proc. Natl. Acad. Sci. U.S.A.* **83**, 674–678.

O, W.-S. (1990). The effect of β-endorphin on rat oocyte maturation in vitro. *Mol. Cell. Endocrinol.* **68**, 181–185.

O'Leary, D. D. M. (1989). Do cortical areas emerge from a protocortex. *Trends Neurosci.* **12,** 400–406.
Palen, K., Thorneby, L., and Emanuelsson, H. (1979). Effects of serotonin and serotonin antagonists on chick embryogenesis. *Roux's Arch.* **187,** 89–103.
Parnavelas, J. G., and Blue, M. E. (1982). The role of the noradrenergic system on the formation of synapses in the visual cortex of the rat. *Dev. Brain Res.* **3,** 140–144.
Peters, D. A. V. (1990). Maternal stress increases fetal brain and neonatal cerebral cortex 5-hydroxytryptamine synthesis in rats: A possible mechanism by which stress influences brain development. *Pharmacol. Biochem. Behav.* **35,** 943–947.
Peterson, C., Neal, J. H., and Cotman, C. W. (1989). Development of $N$-methyl-D-aspartate excitotoxicity in cultured hippocampal neurons. *Dev. Brain Res.* **48,** 187–195.
Petraglia, F., Segre, A., Fachinetti, F., Campanini, D., Ruspa, M., and Genazzani, A. R. (1985). β-endorphin and met-enkephalin in peritoneal and ovarian follicular fluids of fertile and postmenopausal women. *Fertil. Steril.* **44,** 615–621.
Pincus, D. W., DiCicco-Bloom, E., and Black, I. B. (1990). Vasoactive intestinal peptide regulates mitosis, differentiation and survival of cultured sympathetic neuroblasts. *Nature (London)* **343,** 564–567.
Pritchett, D. B., Sontheimer, H., Shivers, B. D., Ymer, S., Kettenmann, H., Schofield, P. R., and Seeburg, P. H. (1989). Importance of a novel $GABA_A$ receptor subunit for benzodiazepine pharmacology. *Nature (London)* **338,** 582–585.
Rakic, P. (1988). Specification of cerebral cortical areas. *Science* **241,** 170–176.
Ramoa, S., Paradiso, M. A., and Freeman, R. D. (1988). Blockade of intracortical inhibition in kitten striate cortex: Effects on receptive field properties and associated loss of ocular dominance plasticity. *Exp. Brain Res.* **73,** 285–296.
Rasmusson, D. D., and Dykes, R. W. (1988). Long-term enhancement of evoked potentials in cat somatosensory cortex produced by co-activation of the basal forebrain and cutaneous receptors. *Exp. Brain Res.* **70,** 276–286.
Reiter, H. O., and Stryker, M. P. (1988). Neural plasticity without postsynaptic action potentials: Less-active inputs become dominant when kitten visual cortical cells are pharmacologically inhibited. *Proc. Natl. Acad. Sci. U.S.A.* **85,** 3623–3627.
Renaud, F., Parisi, E., Capasso, A., and Deprisco, P. (1983). On the role of serotonin and 5-methoxytryptamine in the regulation of cell division in sea urchin eggs. *Dev. Biol.* **98,** 37–46.
Rhoades, R. W., Bennett-Clarke, C. A., Chiaia, N. L., White, F. A., MacDonald, G. J., Haring, J. H., and Jacquin, M. F. (1990). Development and lesion-induced reorganization of the cortical representation of the rat's body surface as revealed by immunocytochemistry for serotonin. *J. Comp. Neurol.* **293,** 190–207.
Ricalde, A. A., and Hammer, R. P., Jr. (1990). Perinatal opiate treatment delays growth of cortical dendrites. *Neurosci. Let.* **115,** 137–143.
Richter-Landsberg, C., Bruns, I., and Flohr, H. (1987). ACTH neuropeptides influence development and differentiation. *Neurosci. Res. Comm.* **1,** 153–162.
Rius, R. A., Barg, J., Bem, W. T., Coscia, C. J., and Loh, Y. P. (1991). The prenatal developmental profile of expression of opioid peptides and receptors in mouse brain. *Dev. Brain Res.,* **58,** 237–241.
Robertson, R. T. (1987). A morphogenetic role for transiently expressed acetylcholinesterase in developing thalamocortical systems? *Neurosci. Lett.* **75,** 259–264.
Roca, D. J., Rozenberg, I., Farrant, M., and Farb, D. H. (1990). Chronic agonist exposure induces down-regulation and allosteric uncoupling of the γ-aminobutyric acid/benzodiazepine receptor complex. *Mol. Pharmacol.* **37,** 37–43.
Rosengarten, H., and Friedhoff, A. R. (1979). Enduring changes in dopamine receptor cells of pups from drug administration to pregnant and nursing rats. *Science* **203,** 1133–1135.
Rozengurt, E. (1991). Neuropeptides as cellular growth factors: Role of multiple signaling pathways. *Eur. J. Clin. Invest.* **21,** 123–134.
Russell, D. H., and Durie, B. G. M. (1978). Polyamines as biochemical markers of normal and malignant growth. Raven Press, New York.

Sakellaridis, N., Mangoura, D., and Vernadakis, A. (1986). Effects of opiates on the growth of neuron-enriched cultures from chick embryonic brain. *Int. J. Dev. Neurosci.*, **4**, 293–302.

Sato, H., Hata, Y., Hagihara, K., and Tsumoto, T. (1987). Effects of cholinergic depletion on neuron activities in the cat visual cortex. *J. Neurophysiol.* **58**, 781–794.

Scalzo, F. M., Holson, R. R., Gough, B. J., and Ali, S. F. (1989). Neurochemical effects of prenatal haloperidol exposure. *Pharmacol. Biochem. Behav.* **34**, 721–725.

Schlumpf, M., Shoemaker, W. J., and Bloom, F. E. (1980). Innervation of embryonic rat cerebral cortex by catecholamine-containing fibers. *J. Comp. Neurol.* **192**, 361–377.

Schlumpf, M., Richards, J. G., Lichtensteiger, W., and Mohler, H. (1983). An autoradiographic study of the prenatal development of benzodiazepine-binding sites in rat brain. *J. Neurosci.* **3**, 1478–1487.

Schmahl, W., Funk, R., Miaskowski, U., and Plendl, J. (1989). Long-lasting effects of naltrexone, an opioid receptor antagonist, on cell proliferation in developing rat forebrain. *Brain Res.* **486**, 297–300.

Schoepp, D., Bockaert, J., and Sladeczek, F. (1990). Pharmacological and functional characteristics of metabotropic excitatory amino acid receptors. *Trends Pharmacol. Sci.* **11**, 508–515.

Schramm, M., Eimerl, S., and Costa, E. (1990). Serum and depolarizing agents cause acute neurotoxicity in cultured cerebellar granule cells: Role of the glutamate receptor responsive to $N$-methyl-$D$-aspartate. *Proc. Natl. Acad. Sci. U.S.A.* **87**, 1193–1197.

Schwartz, J. P., and Costa, E. (1977). Regulation of nerve growth factor content in C6 glioma cells by β-adrenergic receptor stimulation. *Naunyn-Schmied. Arch. Pharmacol.* **300**, 123–129.

Seiler, N., and Lajtha, A. (1987). Functions of GABA in the vertebrate organism. In "Neurology and Neurobiology" (V. Chan-Palay, and S. L. Palay, eds.), Vo. 32, pp. 1–56. Liss, New York.

Seuwen, K., and Pouyssegur, J. (1990). Serotonin as a growth factor. *Biochem. Pharm.* **39**, 985–990.

Seuwen, K., Magnaldo, I., and Pouyssegur, J. (1988). Serotonin stimulates DNA synthesis in fibroblasts acting through 5-HT$_{1B}$ receptors coupled to a G$_i$-protein. *Nature (London)* **335**, 254–256.

Shain, W., Forman, D. S., Madelian, V., and Turner, J. N. (1987). Morphology of astroglial cells is controlled by beta-adrenergic receptors. *J. Cell Biol.* **105**, 2307–2314.

Shemer, A. V., Azmitia, E. C., and Whitaker-Azmitia, P. M. (1991). Dose-related effects of prenatal 5-methoxytryptamine (5-MT) on development of serotonin terminal density and behavior. *Dev. Brain Res.* **59**, 59–63.

Shepanek, N. A., Smith, R. F., Tyer, Z., Royall, D., and Allen, K. (1989). Developmental, behavioral and structural effects of prenatal opiate receptor blockade. *Ann. N.Y. Acad Sci.* **562**, 377–379.

Shinoda, H., Marini, A. M., Cost, C., and Schwartz, J. P. (1989). Brain region and gene specificity of neuropeptide gene expression in cultured astrocytes. *Science* **245**, 415–417.

Shuey, D. L., Yavarone, M., Sadler, T. W., and Lauder, J. M. (1990). Serotonin and morphogenesis in the cultured mouse embryo. In "Molecular Aspects of Development and Aging of the Nervous System" (J. M. Laudner, ed.), pp. 205–215. Plenum Press, New York.

Sieghart, W., and Karobeth, M. (1980). Molecular heterogeneity of benzodiazepine receptors. *Nature (London)* **286**, 285–287.

Sikich, L., Hickok, J. M., and Todd, R. D. (1990). 5-HT$_{1A}$ receptors control neurite branching during development. *Dev. Brain Res.* **56**, 269–274.

Slotkin, T. A., Orband-Miller, L., Queen, K. L., Whitmore, W. L., and Seidler, F. J. (1987). Effects of prenatal nicotine exposure on biochemical development of rat brain regions: Maternal drug infusions via osmotic minipumps. *J. Pharm. Exp. Ther.* **240**, 602–611.

Smyth, D. G., Massey, D. E., Zakarian, S., and Finnie, M. D. A. (1979). Endorphins are stored in biologically active and inactive forms: Isolation of α-$N$-acetyl peptides. *Nature (London)* **279**, 252–254.

Snodgrass, S. R., White, W. F., Biales, B., and Dichter, M. (1980). Biochemical correlates of GABA function in rat cortical neurons in culture. *Brain Res.* **190**, 123–138.

Sommer, B., Keinanen, K., Verdoorn, T. A., Wisen, W., Burnashev, N., Herb, A., Kohler, M., Takagi, T., Sakmann, B., and Seeburg, P. H. (1990). Flip and flop: A cell-specific functional switch in glutamate-operated channels of the CNS. *Science* **249**, 1580–1585.

Specht, L. A., Pickel, V. M., Joh, T. H., and Reis, D. J. (1981a). Light-microscopic immunocytochemical localization of tyrosine hydroxylase in prenatal rat brain. I. Early ontogeny. *J. Comp. Neurol.* **199**, 233–253.

Specht, L. A., Pickel, V. M., Joh, T. H., and Reis, D. J. (1981b). Light-microscopic immunocytochemical localization of tyrosine hydroxylase in prenatal rat brain. I. Late ontogeny. *J. Comp. Neurol.* **199**, 255–276.

Spoerri, P. E. (1988). Neurotrophic effects of GABA in cultures of embryonic chick brain and retina. *Synapse* **2**, 11–22.

Spruce, B. A., Curtis, R., Wilkin, G. P., and Glover, D. M. (1990). A neuropeptide precursor in cerebellum: Proenkephalin exists in subpopulations of both neurons and astrocytes. *EMBO J.* **9**, 1787–1795.

Steinbusch, H. W. M., and Nieuwenhuys, R. (1983). The raphe nuclei of the rat brainstem: A cytoarchitectonic and immunohistochemical study. *In* "Chemical Neuroanatomy" (P. C. Emson, ed.), pp. 131–208. Raven Press, New York.

Steine-Martin, A., and Hauser, K. F. (1990). Opioid-dependent growth of glial cultures: Suppression of astrocyte DNA synthesis by met-enkephalin. *Life Sci.* **46**, 91–98.

Stone, E. A., and Ariano, M. A. (1989). Are glial cells targets of the central noradrenergic system? A review of the evidence. *Brain Res. Rev.* **14**, 297–309.

Sunahara, R. K., Guan, H. C., O'Dowd, B. F., Seeman, P., Laurier, L. G., Ng, G., George, S. R., Torchia, J., Van Tol, H. H., and Niznik, H. B. (1991). Cloning of the gene for a human dopamine $D_5$ receptor with higher affinity for dopamine than $D_1$. *Nature (London)* **350**, 614–619.

Sur, M., Pallas, S. L., and Roe, A. W. (1990). Cross-modal plasticity in cortical development: Differentiation and specification of sensory neocortex. *Trends Neurosci.* **11**, 227–233.

Taylor, J., Docherty, M., and Gordon-Weeks, P. R. (1990). GABAergic growth cones: release of endogenous γ-aminobutyric acid precedes the expression of synaptic vesicle antigens. *J. Neurochem.* **54**, 1689–1699.

Tehrani, M. H. J., and Barnes, E. M., Jr. (1988). GABA down-regulates the GABA/benzodiazepine receptor complex in developing cerebral neurons. *Neursoci. Lett.* **87**, 288–292.

Tremblay, E., Roisin, M. P., Represa, A., Charriaut-Marlangue, C., and Ben-Ari, Y. (1989). Transient increased density of NMDA binding sites in the developing rat hippocampus. *Brain Res.* **461**, 393.

Trombley, P., Allen, E. E., Soyke, J., Blaha, C. D., Lane, R. F., and Gordon, B. (1986). Doses of 6-hydroxydopamine sufficient to deplete norepinephrine are not sufficient to decrease plasticity in the visual cortex. *J. Neurosci.* **6**, 266–273.

Tsumoto, T., Kimura, F., and Nishigori, A. (1990). A role of NMDA receptors and $Ca^{2+}$ influx in synaptic plasticity in the developing visual cortex. *In* "Excitatory Amino Acids and Neuronal Plasticity" (Y. Ben-Ari, ed.), pp. 173–180. Plenum Press, New York.

Van der Neut, R., Bar, P. R., Sodaar, P., and Gipsen, W. H. (1988). Trophic influences of alpha-MSH on ACTH 4-10 or neuronal outgrowth in vitro. *Peptides* **9**, 1015–1020.

Van Hooff, C. O. M., De Graan, P. N. E., Oestreicher, A. B., and Gispen, W. H. (1988). B-50 phosphorylation and polyphosphoinositide metabolism in nerve growth cone membranes. *J. Neurosci.* **8**, 1789–1795.

Van Hooff, C. O. M., De Graan, P. N. E., Oestreicher, A. B., and Gispen, W. H. (1989). Muscarinic receptor activation stimulates B-50/GAP43 phosphorylation in isolated nerve growth cones. *J. Neurosci.* **9**, 3753–3759.

Van Tol, H. H. M., Bunzow, J. R., Guan, H.-C., Sunahara, R. K., Seeman, P., Niznik, H. B., and Civelli, O. (1991). Cloning of the gene for a human dopamine $D_4$ receptor with high affinity for the antipsychotic clozapine. *Nature (London)* **350**, 610–614.

Verdoorn, T. A., Draguhn, A., Ymer, S., Seeburg, P. H., and Sakmann, B. (1990). Functional properties of recombinant rat $GABA_A$ receptors depend upon subunit composition. *Neuron* **4**, 919–928.

Vernadakis, A., and Gibson, D. A. (1974). Role of neurotransmitter substances in neural growth. *In* "Perinatal Pharmacology: Problems and Priorities" (J. Dancis, and J. C. Hwang, eds.), pp. 65–76. Raven Press, New York.

Vernadakis, A., and Kentroti, S. (1990). Opioids influence neurotransmitter phenotypic expression in chick embryonic neuronal cultures. *J. Neurosci. Res.* **26,** 342–348.

Vertes, Z., Melgh, G., and Kovacs, S. (1982). Effect of naloxone and D-Met$^2$-D-Pro$^5$-enkephalinamide treatment on the DNA synthesis in the developing rat brain. *Life Sci.* **31,** 119–126.

Voorn, P., Kalsbeek, A., Jorritsma-Byham, B., and Groenewegen, H. J. (1988). The pre- and postnatal development of the dopaminergic cell groups in the ventral mesencephalon and the dopaminergic innervation of the striatum of the rat. *Neuroscience* **25,** 857–887.

Walicke, P. A. (1989). How well do we understand neurotrophic factors and the control of CNS neuronal growth? *Neurobiol. Aging* **10,** 535–537.

Watkins, J. C., Krogsgaard-Larsen, P., and Honore, T. (1990). Structure–activity relationships in the development of excitatory amino acid receptor agonists and competitive antagonists. *Trends Pharmacol. Sci.* **11,** 25–33.

Weiler, R., and Wagner, H.-J. (1984). Light-dependent change of cone-horizontal cell interactions in carp retina. *Brain Res.* **298,** 1–9.

Weiler, R., Kohler, K., Kirsch, M., and Wagner, H.-J. (1988). Glutamate and dopamine modulate synaptic plasticity in horizontal cell dendrites of fish retina. *Neurosci. Lett.* **87,** 205–209.

Weiler, R., Kohler, K., and Janssen, U. (1991). Protein kinase C mediates transient spinule-type neurite outgrowth in the retina during light adaptation. *Proc. Natl. Acad. Sci. U.S.A.* **88,** 3603–3607.

Wendlandt, S., Crow, T. J., and Stirling, R. V. (1977). The involvement of the noradrenergic system arising from the locus coeruleus in the postnatal development of the cortex in rat brain. *Brain Res.* **125,** 1–9.

Whitaker-Azmitia, P. M., and Azmitia, E. C. (1986a). Autoregulation of fetal serotonergic neuronal development: Role of high affinity serotonin receptors. *Neurosci. Lett.* **67,** 307–312.

Whitaker-Azmitia, P. M., and Azmitia, E. C. (1986b). [$^3$H]5-Hydroxytryptamine binding to brain astroglial cells: Differences between intact and homogenized preparations and mature and immature cultures. *J. Neurochem.* **46,** 1186–1191.

Whitaker-Azmitia, P. M., and Azmitia, E. C. (1989). Stimulation of astroglial serotonin receptors produces culture media which regulates growth of serotonergic neurons. *Brain Res.* **497,** 80–85.

Whitaker-Azmitia, P. M., Murphy, R., and Azmitia, E. C. (1990). Stimulation of astroglial 5-HT$_{1A}$ receptors releases the serotonergic growth factor, protein S-100, and alters astroglial morphology. *Brain Res.* **528,** 155–158.

Wolff, J. R., Joo, E., and Dames, W. (1978). Plasticity in dendrites shown by continuous GABA administration in superior cervical ganglion of adult rat. *Nature (London)* **274,** 72–74.

Wolff, J. R., Joo, F., and Kasa, P. (1987). Synaptic, metabolic, and morphogenetic effects of GABA in the superior cervical ganglion of rats. *In* "Neurology and Neurobiology" (V. Chan-Palay, and S. L. Palay, eds.), Vol. 32, pp. 221–252. Liss, New York.

Zagon, I. S., and McLaughlin, P. J. (1983). Increased brain size and cellular content in infant rats treated with an opiate antagonist. *Science* **221,** 1179–1180.

Zagon, I. S., and McLaughlin, P. J. (1989). Naloxone modulates body and organ growth of rats: Dependency on the duration of opioid receptor blockade and stereospecificity. *Pharm. Biochem. Behav.* **33,** 325–328.

Zagon, I. S., and McLaughlin, P. J. (1991). Identification of opioid peptides regulating proliferation of neurons and glia in the developing nervous system. *Brain Res.* **542,** 318–323.

Zagon, I. S., Goodman, S. R., and McLaughlin, P. J. (1990). Demonstration and characterization of zeta($\zeta$), a growth-related opioid receptor, in a neuroblastoma cell line. *Brain Res.* **511,** 181–186.

Zimmerman, E. F., and Wee, E. L. (1984). Role of neurotransmitters in palate development. *In* "Current Topics in Developmental Biology" (E. F. Zimmerman, ed.), Vol. 19, pp. 37–63. Academic Press, New York.

# Index

Acetylcholine, 11, 31, 33, 37
  nerve growth factor, 40
  ocular dominance, 581
  receptor activation, 579
ACH *see* Acetylcholine
Acidic fibroblast growth factor (aFGF), 10, 12, 13, 15, 16, 19, 285
  γ-aminobutyric acid, 20
Adhesion factor, 8
  as neurotrophic factor, 4
  fibronectin, 510, 511
  laminin, 510
aFGF *see* Acidic fibroblast growth factor
Aging, nerve growth factor, 40
Alzheimer's disease, 41
  brain derived neurotrophic factor, 276
  nerve growth factor, 240
Amino acid transmitter, glutamate, morphogenesis, 566
Anterograde transport, 7
AP-1, 91
Astrocytes, 10, 15, 19, 98
Autocrine secretion, 7
Axon elongation
  soluble molecules, 63
  substrate-bound molecules, 63
Axoplasmic transport
  colchicine, 58
  nerve growth factor effects, 58

Basic fibroblast growth factor (bFGF), 12, 13, 32, 38, 285
  γ-aminobutyric acid, 20
  interaction with glutamate, 567
  substantia nigra, 20
  tyrosine hydroxylase, 36
BDNF *see* Brain derived neurotrophic factor
Benzodiazepam, effects on primary response gene expression, 109
Beta adrenergic receptor, 69
Beta nerve growth factor, structure, 137
bFGF *see* Basic fibroblast growth factor
Blood brain barrier, 10
Brain derived neurotrophic factor (BDNF), 12, 13, 28, 33, 38
  cell death, 268
  dopaminergic neuron survival, 267
  functional role, 263
  gene expression, cholinergic neurons, 266
  history, 257
  localization in brain, 270, 271
  molecular cloning, 269
  nerve growth factor, comparison with, 262
  nerve growth factor family, 54
  neuron survival, 264
  peripheral tissues, 34
  purification, 261
  receptors, 272, 274

C6 glioma, 98
CA *see* Catecholamine
Caffeine, effects on primary response gene expression, 109
cAMP *see* Cyclic AMP
Catecholamine (CA)
  dopamine, 575
  effect in central nervous system, 576
  receptor expression, 578
Caudate putamen, 14, 16, 17
CDF *see* Cholinergic differentiation factor
Cell death, 36
  histogenic, 51
  ontogenic, 51
Cell survival
  basic fibroblast growth factor, 62
  fibroblast growth factor
    interaction, 59
  insulin-like growth factor, 62
  nerve growth factor, interaction, 59
  phorbol esters, 62
  transforming growth factor beta, 62
Central nervous system
  fibroblast growth factor, 314
  glial cells, 490
  nerve growth factor
    neuron maintenance, 238
    neuron survival, 238
Central nervous system diseases
  nerve growth factor role, 239
Central nervous system neurons
  brain derived neurotrophic factor, 265
*c-fos*, 90
  cyclic AMP interaction, 93
  effects of calcium on inhibitory control promoter, 93
  growth factor stimulation of, lack of specificity, 92
  induction pathways, 92
*c-jun*, 90, 91
Cholinergic differentiation factor (CDF), 528
  function, 529
Cholinergic differentiation factor/leukemia inhibitory factor
  function, 530, 531
  role in differentiation, 532
Ciliary neurotrophic factor, 28, 53
  acetylcholine, 37
  antibodies, 451
  availability, 458
  central nervous system, 457
  cholinergic neurons, 444, 445
  gene expression, 449

  growth promoting activity, 448
  history, 443
  neurite outgrowth, 461
  neurotransmitter function, 462
  neurotrophic activities, 533, 534
  peripheral nerve, 457
  physiologic function, 459, 460
  purification, 448
  receptor regulation, 463
  receptor, 452
  responsive neuron, 453
  responsive nonneuronal cells, 456
  target cells, 457
  target tissue, 447
Circadian rhythms
  effects on primary response gene expression, 111
CNTF *see* Ciliary neurotrophic factor
Cocaine
  effects on primary response gene, 107
Cortex, 15
Cortico-striato-pallidal system, 14
Corticosteriod, 541
CREB *see* Cyclic AMP responsive element binding protein
Critical periods
  brain derived neurotrophic factor, 38
  nerve growth factor, 38
  neurotrophin-3, 38
  specificity of action, 38
Cross Talk, 94
Cyclic AMP, responsive element binding protein, 93
  role in memory, 106

DA *see* Dopamine
Death genes, 17
Decorin, 514
2DG
  correlation with primary response gene, 104
Differentiation
  basic fibroblast growth factor, 36
  epidermal growth factor, 36
  nerve growth factor, 36
Dopamine (DA), 15, 20
  effects on primary response gene expression, 107

Early response gene *see* Primary response gene
EGF *see* Epidermal growth factor

*egr1*, 95
Endothelial cells, 10
Enkephalin, 17
Entopeduncular nucleus, 16
Ependymal cells, 10
Ependymal secretion, 7
Epidermal growth factor receptor, 346, 349
   retinoic acid, 61
Epidermal growth factor, 15, 16, 18, 28, 30
   characteristics of, 352
   immunoreactivity, 348
   localization, immunoreactive material, 345
   molecular structure, 339–341
   neuronal effects, 350
   neurotrophic effects, 351
   receptor expression, immunocytochemical, 347
   receptor, 346, 349
   sequence homology, 342 (Figure), 343, 344
   tyrosine hydroxylase, 36
Epilepsy, 13
Extracellular matrix, neurotrophic factor availability, 8
Extrapyramidal motor system, 14, 15 (Figure)
   basic fibroblast growth factor, 20
   epidermal growth factor, 20
   γ-aminobutyric acid, 20


FGF *see* Fibroblast growth factor
Fibroblast growth factor
   acidic fibroblast growth factor, 53
   amino acid sequences, 289, 290
   anterograde transport, 57
   basic fibroblast growth factor, 53
   biologic, 299
   cell lineage, 326
   characteristics of, 285, 313
   chemotactic, 303
   family protein structures, 288
   function, 298
   gene expression, 287
   glial cells, 321
   heparin, 293
   immunocytochemical mapping, 316
   immunostaining, 315
   induction, 300
   interaction with cell survival, 59
   interaction with other growth factors, 327
   lesioned nervous system, 322
   mitogenic, 302
   neurotrophic activity, 53
   neurotrophic effect, 319, 320, 323, 324
   neurotrophic mechanisms, 325
   phosphorylation, 291, 297
   polyclonal antibodies, 53
   receptor domain, 294
   receptors, 292, 295, 317, 318
   signal transduction, 296
   structure, 286
   target cells, 301
Fibronectin, 510–512, 514–516
   proteoglycans interaction, 515, 516
FIRE *see fos* intragenic regulatory element
FOS gene family, 91
*fos* intragenic regulatory element, 93
*fos see* c-*fos*
FOS, 90
FOS–JUN convergence, 94


GABA *see* γ-aminobutyric acid
γ-aminobutyric acid, 571
   physiologic action, 570
   receptor expression, 572
   trophic effects, 570
Gangliosides, 519
Glia, 8, 15
   primary response gene expression, 100
Glial cell, 490 (Figure)
   growth factor, 502 (Figure)
   oligodendrocyte, platelet-derived growth factor, 491, 492
Globus pallidus, 16, 17
Glucocorticoids, effects on primary response gene expression, 111
Glutamate
   functional plasticity, 568
   growth inhibition, 567
   interaction with basic fibroblast growth factor, 567
   receptor mechanism, 569
   trophic actions on neuronal cell, 568 (Figure)
Gonadal steroids, effects of, 542


Hamburger, V., 1
Hematopoietic analogy
   compared to nervous system, 548
   phenotypic homologies, 548
Hippocampus, 11, 39
   localization of neurotrophic factors, 12
Holocrine secretion, 7, 10
Huntington's disease, 14, 16
   enkephalin, 17
   globus pallidus, 17

Huntington's disease *(cont.)*
  patch and matrix, 17
  striatum, 17
  substance, P, 17
  substantia nigra, 17

IFN *see* Interferon
IGF-1 *see* Insulin-like growth factor-1
IL-1 *see* Interleukin-1
Instructive neuronal differentiation factor, phenotypic plasticity, 527
Insulin, neuronal survival, 58
Insulin-like growth factor
  binding proteins, 397, 398 (Table)
  biologic action, 403
  characteristics of, 422
  gene expression
    neurofilament, 426
    tubulin, 425
  mitosis, 430
  nerve growth factor, interaction, 421
  neurite outgrowth, 416
  neuronal development, 417
  neurotransmitter, 431
  pathophysiology, 431
  phosphorylation, 424
  physiologic regulation, 396
  protein kinase C, 427–429
  receptor localization, 401, 402 (Figure)
  receptor, 399 (Figure), 400, 423
  retrograde transport, 424
  sprouting, 420
synapse elimination, 418
  synapse regeneration, 419
  synthesis, 392
Insulin-like growth factor-1, 12, 13, 28, 54
  characteristics of, 393
  localization, 394 (Figure)
  metabolic control, 404 (Table)
  physiologic signaling, 405, 406 (Figure)
Insulin-like growth factor-II
  synthesis, 395
Interferon, 52
Interleukin-1, 28
Intermediately early gene *see* Primary response gene
Intracrine secretion, 7, 10

JUN gene family, 91
JUN, 91
Juxtacrine secretion, 7
  interaction, 8

Kindling, primary response gene expression, 103
*krox20*, 95, 100

Lesion, effects on primary response gene expression, 108
Leukemia inhibitory factor, 529
Leuteinizing hormone releasing hormone, inductin of *c-fos*, 111
Levi-Montalcini, R., 1
LHRH *see* Leuteinizing hormone releasing hormone
LIF *see* Leukemia inhibitory factor
Long-term potentiation, primary response gene expression, 103
LTP *see* Long term potentiation

Melanization factor, neuronal differentiation, 544
Membrane-associated neurotransmitter, 535
Memory, correlation with primary response gene, 105
Monoamine, serotonin, effects of, 573
Monocular deprivation effects, 580 (Figure)
Morphine, 583
  effects on primary response gene, 106, 107
Morphological factor, neuronal effects, 543
Motor neuron survival
  acidic fibroblast growth factor, 538
  basic fibroblast growth factor, 538
  cholinergic differentiation factor, 538
  cholinergic differentiation factor/leukemia inhibitory factor, 538
  ciliary neurotrophic factor, 538
  neurotrophic factor effects, 482
Motor neurons, 15
Motor systems, 15
Mutants in invertebrates, effects of differentiation, 547
Myristoylation, phosphorylation, 72

Nerve growth factor expression
  glia, 198
  hormones, 190
  inducers, 190
  injury, 196
  mechanisms, interleukin-1, 198
  regulated in nervous system development, 196
  seizures, 187
Nerve growth factor induction

glycosyl-phosphatidylinositol, 153
macrophages, 197
phosphorylation, 156
primary response genes, 157
prostaglandins, 154
proteins, 160, 161
Nerve growth factor primary subunit, amino
    acid sequence, 131 (Figure)
Nerve growth factor receptor
  binding, 67, 214, 216
  brain derived neurotrophic factor, 273
  characterization of, 138
  molecular cloning, 139
  retinoic acid, 60
  signal transduction by *trk*, 148 (Figure)
  structure models, 140
  tyrosine kinase, 139
Nerve growth factor regulation
  in vivo targets, 186
  sympathetic targets, 183
Nerve growth factor response
  kinetics, 224
Nerve growth factor signal transduction, 226
  patterns, 181
Nerve growth factor signaling pathways, 162
    (Figure)
Nerve growth factor transport
  retrograde transport, 56
Nerve growth factor, 11–13, 16, 25, 28, 30–33,
    38, 54, 546
  acetylcholine, 36
  aging, 40
  alpha subunit identification, 136
  amino acid sequence, 134, 135 (Figure)
  antibodies to, 209
  antibodies, 218
  axonal growth, gene directed, 222
  biology of, 259
  brain derived neurotrophic factor, 137
  cell death theory, 53 (Figure)
  cell death, 219, 223
  central nervous system target, 185
  characteristics of, 258
  chemotaxic, 37
  cyclic AMP levels, 145
  definition, 129
  developmental gene expression, 185, 221
  differentiation, 36
  discovery, 210
  early development, 200
  effects, cyclic AMP in PC12 cells, 146
  effects in central nervous system, 237
  gamma subunit identification, 136

  gene expression, 227, 233
    cholinergic neurons, 235, 236
  gene induction, 69
  history, 129
  immunoregulation, 34
  interaction with cell survival, 62
  interaction with insulin-like growth factor,
    421
  internalization, 143, 228, 229
  mechanism model, PC12 cells, 141
  mitosis, 217
  modulation by primary response gene prod-
    ucts, 113
  neurite outgrowth, 37
  neurotrophin-3, 137
  neurotrophin-4, 137
  pathway determination, 192
  PC12 cell differentiation, 142
  peripheral nervous system, 211
  precursor sequence, 132 (Figure)
  precursors, 191, 193
  presence in central nervous system, 232
  protein kinase A, 225
    activity, 154
  protein kinase C, 225
  protein synthesis, 142
  receptor
    positive feedback, 32
    regulation, 32
  receptor affinity, 195
  receptor binding, 144, 213, 234
  receptor complexes, 215 (Figure)
  receptor interaction, 133
  receptor subtypes, 133
  regulated pathways, 194
  role in peripheral nervous system, 230
  sensory target cell death, 184
  sequence, 189
  subunits, beta nerve growth factor derivation,
    130
  sympathetic and sensory neuron survival, 53
  sympathetic neuron, 39, 220
  target derived, 212
  target tissue, 182
  transcription, 188
  tyrosine phosphorylation, 144
Nerve growth gene induction pathways, 159
    (Figure)
Nerve injury, nerve growth factor, 231
Neurite outgrowth
  molecular interaction, 64
  nerve growth factor, 37
  substrate requirements, 67

Neuroendocrine secretion, 10
Neuroendocrine system, role of primary response gene, 110
Neuronal activity, role of, 545
Neuronal differentiation, PC12 cells, 147
Neuronal receptor interaction
  calcium, 68
  phosphorylation, 68
Neuronal survival
  apolipoprotein D, 518
  cell–cell contact, 516, 517
  gangliosides, 519
  purpurin, 517
Neuropeptide
  as neurotrophic factor, 4
  neuronal development, 581
Neurotransmitter, as neurotrophic factor, 4
Neurotrophic action
  calcitonin gene-related peptide, 585
  substance P, 584, 585
  thyrotropin releasing hormone, 585
  vasoactive intestinal peptide, 584, 585
Neurotrophic activity
  fibronectin, 511, 512
  laminin, 511
  vs. nerve growth factor, 260
Neurotrophic factor
  aging, 40
  anatomical binding, epidermal growth factor, 18
  anatomical localization, 2
    acidic fibroblast growth factor, 16, 19
    epidermal growth factor, 16, 19
    synthesis, 30
    transforming growth factor alpha, 16, 19
  antagonists, 41
  antibodies, 55
  availability, 54
  binding proteins, 9
  cell death, 51
  cellular actions, 32
  choline acetyltransferase, 37
  classical retrophin, 71 (Figure)
  convergence
    on primary response genes, 6
    on second messenger systems, 6
  definition, 1
    Baird, Y., 26
    examples, 27
    general vs. restrictive, 2, 26
    survival, 26, 35
  dendrites, 9
  differential localization, 12, 13
  discovery in brain, 1
  examples, 3 (Table), 27 (Table)
  functional role, 2
  G protein, 70
  genes
    regulatory elements, 29
  hippocampus, 12
  history, 1, 25
  interaction
    with acetylcholine, 33
    with steroids, 41
    with interleukin-1, 29
    with nerve growth factor, 29
    with neuropeptides, 6
    with neurotransmitters, 6
  intracellular communication
    anterograde transport, 7
    autocrine secretion, 7
    ependymal secretion, 7, 10
    holocrine secretion, 7, 10
    intracrine secretion, 7
    juxtacrine secretion, 7, 34
    nonneuronal, 34
    paracrine secretion, 7, 34
    retrograde transport, 7, 34
    sequential secretion, 7, 10, 15 (Figure), 33
    telocrine secretion, 7
    vascular secretion, 7, 10
  low abundance, 34
  mechanism of action, 5 (Figure), 33, 38
  metabolism, 32
  modes of secretion, 6
  multifunctional role, 34
  multiple domain, 9
  neuroanatomical approach
    chemical neuroanatomy, 11
    neural system, 11, 14
    regional analysis, 11
  neurotransmitters
    effect on basic fibroblast growth factor, 32
    effect on nerve growth factor, 32
    effect on release, 32
    effect on synthesis, 32
  nontransported retrophin, 71
  parallel with neurotransmitters, 2, 30
  pharmacologic effects, 52
  pharmacological exploitation, 40, 41
  physiological function, 28
  protection effects in epilepsy, 13
  receptor, 9, 34
    acidic fibroblast growth factor, 20
    basic fibroblast growth factor, 20

        effect on cholinergic cells, 31
        epidermal growth factor, 20
        mechanism of action, 31
        nerve growth factor, 31, 32
        transducing mechanism, 30
        tyrosine kinase, 31
    release, 28
    retrograde transport, 33, 34
    role in adult nervous system, 39
    selective vs. nonselective effects, 38
    specificity, 38
    structural proteins, 32
    structure–function relationships, 41
    subcellular localization, 9
    sympathetic neurons, 65
    synergy, 6
    synthesis, 28
    target derived, 34
Neurotrophic factor availability
    adhesion factor, 8
    extracellular matrix, 8
Neurotrophic factor receptors, mismatch, 8
Neurotrophic molecule, second messenger
    long-lived, 70
    short-lived, 70
Neurotrophin
    history, 26
    release, 28
    synthesis, 28
Neurotrophin-3, 12, 13, 28
    peripheral tissues, 34
    vs. brain derived growth factor, 275
    vs. nerve growth factor, 275
Neurotrophin-4, 28
Neurotrophin-5, 28
NGF *see* Nerve growth factor
Nigrostriatal system, 14
NMDA, primary response gene, 107
Noradrenergic factor
    catecholamine, 540
    tyrosine hydroxylase, 539, 540
NT-3 *see* Neurotrophin-3
*nur77*, 95

Oligodendrocyte
    differentiation, 494
    fibroblast growth factor, 496
    insulin-like growth factor-1, 497
    peripheral nervous system glia, 497
Opioid peptides
    development in central nervous system, 582
    receptor effects, 582–583
O–2A progenitor cells, 99

Pain, primary response gene transduction, 106
Palidum, 15
Paracrine secretion, 7
Parkinson's disease, 14
PDGF *see* Platelet-derived growth factor
Peptidergic factor, 541
Phospholipase activity, receptor mediated hydrolysis, 150 (Figure)
Phospholipid hydrolysis
    mechanism, 149
    nerve growth factor induction, 151
    1,2-diacylglycerol, 152
Phosphorylation
    nerve growth factor, 66
Pineal
    expression of primary response genes, 111
Platelet-derived growth factor
    receptor, 493
    sources, 495
Polypeptide growth factor
    cyclic AMP, 499
POMC *see* Proopiomelanocortin-derived peptides
Postsynaptic factor, interaction, 55
POU-domain gene, 96
Presynaptic factor, interaction, 55
Primary response gene
    2-DG studies, 104
    additive phenomena, 98
    and pain, 106
    calcium channel activation, 97
    definition, 89
    dopamine, 107
    effects of benzodiazepam, 109
    effects of caffeine, 109
    effects of cocaine, 107
    effects of injury, 109
    effects of morphine, 107
    effects of physiological stimulation in memory, 105
    effects of seizures, 101
    effects of stress, 101, 109
    electrophysiological recording, 104
    expression in adrenal medulla, 97
    expression in glial cells, 98
    expression in neuronal cells, 96, 97
    gene induction, 158
    growth factor stimulation, 97
    heterodimers, 91
    homodimers, 91
    in ontogeny, 99
    in vivo expression, 99
    kindling, 103

Primary response gene *(cont.)*
  lesion effects, 108
  ligand specific, 158
  mapping of neuronal networks, 104
  methodology used to study, 100, 101
  NMDA activation, 97
  quantitative parameters, 116
  refractory period, 102
  role in reproductive cycle, 110 16
  specificity of expression, 115, 117
Proenkephalin, modulation by primary response gene products, 113
Proopiomelanocortin-derived peptides, developmental effects, 584
Protein kinase C
  multiple forms, 155
Protein synthesis
  nerve growth factor, 52
Proteoglycans, 512, 513
  decorin, 514
  fibronectin interaction, 515, 516
  laminin interaction, 515, 516
  nerve cells, 513
Purpurin, 517
Ramon y Cajal, S., 1
Reproductive cycle
  role of primary response gene, 110
Retinoic acid
  acidic fibroblast growth factor, neuronal survival, 60
  gradient effects, 61
Retrograde transport, 7
  growth factor effect, 57
Retrophin, 52
RNA synthesis
  nerve growth factor, 52


Schwann cell vs. neuron, 498
Schwann cell, 98
  cyclic AMP, 498
  nerve growth factor, 199
  regulation, 500
*scip,* 96
Second messenger system
  GTP binding proteins, 66
  tyrosine kinase, 66
Secondary response gene, as targets of primary response gene products, 112
Seizure, effects on primary response gene, 101, 102
Sequential secretion, 7, 10, 17
Serotonin

heterologous growth regulation, 575
homologous growth regulation, 574 (Figure)
Serum responsive element (SRE), 92, 93
Skeletal muscle-derived factor
  identification, 479–481
  nerve growth factor interaction, 475
  neurotrophic effects
    in vitro, 476, 477
    in vivo, 478
SP *see* Substance P
SRE *see* Serum responsive element
Stress, effects on primary response gene expression, 109
Striatal derived neurotrophic factor, 17
Striatum, 14, 15, 16, 17
Substance P, 584, 585
Substantia nigra, 14, 15, 16, 19
  astrocytes, 20
  brain derived neurotrophic factor, 20
  epidermal growth factor, 17
  transforming growth factor alpha, 17
Survival, competition for neurotrophic factors, 35
Sweat glands, 535
  cholinergic activity, 536, 537
  ciliary neurotrophic factor, 536
Sympathetic neurons, 32


Telocrine secretion, 7
TGFα *see* Transforming growth factor alpha
TGFβ *see* Transforming growth factor beta
TPA responsive elements, 90
Trans-synaptic influence
  postsynaptic transmission, 54
  presynaptic transmission, 54
Transforming growth factor alpha, 12, 13, 15, 16, 17, 19
  assays, 361
  behavioral efficacy, 17
  discovery, 359
  function, 364
  gene expression, 366
  history, 360
  Huntington's disease, 17
  molecular structure, 365
  Parkinson's disease, 17
  physiologic effects, 367, 368
  precursor structure, 362 (Figure)
  receptor, 365
  structure, 363 (Figure)
  synthesis, 364
  transplantation, 17

Transforming growth factor beta
  assays, 369
  cell behavior, 376
  cell differentiation, 376
  characteristics of, 371, 378, 379
  function, 373
  gene expression, 375
  history, 370
  molecular structure, 374
  physiologic effects, 375, 377
  receptor, 374
  synthesis, 372
Transin, modulation by primary response gene products, 115
Transplantation
  behavioral efficacy, 17
  increases in transforming growth factor alpha, 17
TRE's *see* TPA responsive elements
TRG *see* Thyrotropin releasing hormone
trk *see* Tyrosine kinase
Tyrosine hydroxylase, modulation by primary response gene products, 114
Tyrosine kinase, 31

Vascular secretion, 7
Vasoactive intestinal peptide, 584, 585
Ventral pallidum, 16
VIP *see* Vasoactive intestinal peptide

$225.00